The action plan for AUSTRALIAN BIRDS 2010

ST Garnett, JK Szabo and G Dutson
Charles Darwin University

With maps by Glenn Ehmke

CSIRO PUBLISHING

Birds Australia
CONSERVATION THROUGH KNOWLEDGE

National Library of Australia Cataloguing-in-Publication entry

Garnett, Stephen, 1955–

The action plan for Australian birds 2010/by Stephen Garnett, Judit Szabo and Guy Dutson.

9780643103689 (pbk.)
9780643103696 (epdf)
9780643103702 (epub)

Includes bibliographical references and index.

Birds – Protection – Australia.
Endangered species – Australia.

Szabo, Judit.
Dutson, Guy.
Birds Australia.

598.0994

Published by

CSIRO PUBLIS HING
36 Gardiner Road, Clayton VIC 3168
Private Bag 10, Clayton South VIC 3169
Australia

Telephone: [+613] 9545 8555
Local call: 1300 788 000 (Australia only)
Fax: +61 3 9662 7555
Email: csiropublishing@csiro.au
Web site: www.publishing.csiro.au

Front cover: Princess Parrot (by John Manger)

Set in 10/13 Adobe Minion and ITC Stone Sans
Cover and text design by James Kelly
Typeset by Desktop Concepts Pty Ltd, Melbourne
Printed by Ingram Lightning Source

Feb26_RP_ILS

Contents

Foreword

At one level this book describes a tragedy. That in the 200 short years since Europeans arrived in Australia we have so diminished our natural capital that 234 Australian bird taxa are either Extinct, threatened with extinction or Near Threatened, is a national disgrace. On the other hand, as its title suggests, this book is a call for action. We do not need to lose any more Australian birds. This book outlines how we are to achieve this objective.

And there is much reason to hope. We would have lost far more had there not been enormous effort over the last few decades. After all it is only 20 years ago since all the information available on Australia's threatened birds was compiled in the first *Action Plan for Australian Birds.* Before that there was an awareness that some birds were scarce but there was no comprehensive treatment. For some birds it was already too late – they were either gone or so near to going that nothing could have been done to save them. But since then there has been a consciousness in government and community of the damage we are doing to our biodiversity legacy.

And there have been some wonderful achievements. One bird that does not feature in this book is Muir's Corella *Cacatua pastinator pastinator* from south-western Australia – the southern subspecies of the Western Corella. The reason is that it no longer needs to. In the 1940s only a few birds remained and even they were being persecuted. But slowly enforcement of conservation laws and protection of habitat have led to their recovery. A decade ago, in the *Action Plan for Australian Birds 2000,* the subspecies was listed as Endangered – there were fewer than 2500 birds and the population was much lower than it had been three generations earlier. Today there are over 10 000, numbers are rapidly increasing and the Muir's Corella is considered to have the IUCN Red List status of Least Concern. Other species have also moved down the list of threatened categories. Effective conservation action in Tasmania, for instance, has halted decline of the local subspecies of Wedge-tailed Eagle *Aquila audax fleayi* so that it can be listed as Vulnerable not Endangered. More nests are protected than ever before.

But it should never be expected that the Tasmanian Wedge-tailed Eagle will be removed from the list altogether. There will never be a large population. Threats will always be present. Some of our threatened birds will need active management for the foreseeable future – they will never 'recover' to a state where they can persist without assistance, we have changed their environment too much for that. But that is no different from so much that we as a society value. Our great cultural heritage needs constant care and maintenance if it is not to be lost. A responsibility for threatened species is something with which we are entrusted by our children – it is not a task we should take lightly.

However, while Birds Australia members and many others in the community are willing to devote substantial portions of their lives to helping threatened birds, many birds also need professional help. That takes money. Sadly, over the last decade, threatened birds appear to have gone out of fashion among government policy makers. Successes are ignored and failures emphasised. Instead of species, emphasis is being put on the conservation of landscapes without attention to the precious threatened detail those landscapes contain. We at Birds Australia do not share that view. We believe, like the majority of the public, that a vital role for conservation agencies is prevention of species loss. To do this will require dedicated public and private funding. Yes we need landscapes but we also need to keep all components of those landscape. If we as a nation fail to take the actions this book spells out, Australia will be a poorer place in which to live.

Dr Graeme Hamilton
Chief Executive Officer
Birds Australia

Summary

This Action Plan lists 27 taxa as Extinct, 20 as Critically Endangered, 60 as Endangered, 68 as Vulnerable and 63 as Near Threatened as at 31 December, 2010. The remaining 1028 taxa are deemed to be Least Concern. The Australian list also includes 31 introduced taxa and 151 vagrants. Conservation summaries are presented for all Extinct, threatened and Near Threatened taxa. Of those taxa known to have been present or to have occurred regularly in Australia when Europeans settled in 1788, 2.2% are Extinct and a further 11.8% are considered threatened. Some 5.0% are Near Threatened.

Since the last Action Plan, research and surveys have shown that 61 taxa are less threatened than was thought but a further 26 taxa should have been listed at a higher level. Other differences between the 2000 Action Plan and the present volume can be accounted for by changes to taxonomy (7 taxa) or to more rigorous IUCN criteria that better define the different categories (58 taxa).

Using current knowledge, taxonomy and IUCN criteria, there has been a change in the status of 66 taxa (5.0%) since the last Action Plan. For seven the conservation status can be downlisted as a result of effective conservation management. Abbott's Booby *Papasula abbotti* and the Christmas Island Hawk-Owl *Ninox natalis* shifted from Critically Endangered to Vulnerable as a result action to control invasive ants. The Southern Cassowary *Casuarius casuarius johnsonii*, Tasmanian Wedge-tailed Eagle *Aquilla audax fleayi* and Albert's Lyrebird *Menura alberti* have all benefited greatly from habitat protection; the first two were Endangered but can now be listed as Vulnerable, the last has been shifted from Vulnerable to Near Threatened after effective protection of Whian Whian State Forest in New South Wales. Numbers of Gouldian Finch *Erythrura gouldiae* have increased so much in the last decade that they can be moved from Endangered to Near Threatened. The reason is not entirely clear but they are known to have benefited at some sites from better fire management and conservative stocking rates. The population of the seventh taxon, the southern subspecies of Western Corella *Cacatua pastinator pastinator* has increased so much as a result of effective law enforcement and habitat protection that it has been shifted from Endangered to Least Concern. However the status of 39 taxa has been uplisted. This includes four taxa that are now Critically Endangered – three species, the Grey-headed Albatross *Thalassarche chrysostoma*, the Western Ground Parrot *Pezoporus flaviventris* and the Regent Honeyeater *Anthochaera phrygia*, and one subspecies, the Norfolk Island Tasman Parakeet *Cyanoramphus cookii cookii*. Along with the Orange-bellied Parrot *Neophema chrysogaster*, the ground parrot and the honeyeater have the potential to be Extinct in the Wild by 2000; all have all suffered rapid declines in their already very small populations. While, for each, a captive population has been established, for only the Orange-bellied Parrot has there been a conscious attempt to capture genetic variability before it is too late.

Most of the additions to the list in 2010 are migratory waders, numbers of which have plummeted, largely due to reclamation or degradation of habitat along their migratory pathways. A more detailed understanding of the taxonomy of grasswrens *Amytornis* spp., in which similarities in plumage hide deep genetic rifts, has also led to many additional taxa listed, including some that are already Extinct. Most threatened or Extinct taxa continue to be on islands. Here introduced predators, often aided by habitat destruction, have taken a heavy toll, or they could cause losses if they were introduced to places like Heard Island. However programs to remove or control island predators are having major benefits. At the time of the 2000 Action Plan the problem with an invasion of Christmas Island by Yellow Crazy Ants *Anoplolepis gracilipes* had just become apparent and many endemic taxa there were deemed Critically Endangered. Rapid action has controlled the ants for the present, though, as it happened, few birds were likely to have been affected in the short–medium term. On the mainland, land clearance and the resulting habitat fragmentation may continue to cause species declines for decades as biodiversity pays its extinction debt. Grazing by domestic and feral herbivores and changes in fire management are also major threatening processes. However, the rate of land clearing has dropped dramatically and some major fire management programs offer hope for fire-prone landscapes. At sea, all albatross taxa and several petrels continue to be threatened by high rates of mortality associated with fishing, but there has also been substantial progress in developing and implementing mitigation techniques.

Most threatened marine birds frequent the Southern Ocean, particularly around Tasmania. On the mainland, the majority of threatened and Near Threatened taxa occur in the south-east of the continent, particularly west of the Great Dividing Range, with the highest concentration in the Murray mallee and south-western Australia, particularly in the south east of Albany. Australia's coastlines also support a particularly large number of threatened and Near Threatened taxa because so many shorebird taxa are now listed.

About the authors

Stephen Garnett has been studying Australian birds since 1974. During the 1980s he worked on the *Handbook of Australian, Antarctic and New Zealand Birds* and in 1990, with Joost Brouwer, he drafted the first review of Australia's threatened bird species. He wrote the first *Action Plan for Australian Birds* in 1992 and the second Action Plan, in association with Gabriel Crowley, in 2000. He has studied and written about many threatened Australian bird species, most notably the Golden-shouldered Parrot and the Kangaroo Island Glossy Black-Cockatoo, and helped bring the idea of Important Bird Areas to Australia. He also works on natural-resource based livelihoods in northern Australia and South-East Asia. He is a research professor at Charles Darwin University in Darwin.

Judit Szabo is a research fellow at Charles Darwin University working on Australian threatened birds. She has studied the effects of pesticides used in locust control on Australian birds. During the five years of field work and data analysis she became interested in issues with data collection and study design, which led to her work on optimal monitoring, bias associated with bird survey data, and using volunteer-collected data to detect trends in bird populations.

Guy Dutson is an independent consultant who has been researching threatened Australasian birds since rediscovering the Superb Pitta in PNG in 1990. He developed BirdLife International's Pacific program, managing projects across the Pacific islands. Guy moved to Australia in 2006 to develop the Birds Australia–Rio Tinto *Important Bird Areas* program, and identified all Australian sites of global bird conservation importance. Guy has also worked with bird tourism in the Northern Territory and leads birdwatching tours across the country and globally.

Glenn Ehmke is a senior analyst at Birds Australia where he has worked on conservation and research projects for the last seven years. Glenn's work ranges from analysing national-scale bird distributions, to design and coordination of volunteer-based bird monitoring programs, working with community groups and threatened bird research programs. Glenn is also a regular contributor to the *State of Australia's Birds* series and *Wingspan*.

Acknowledgements

This book, like its predecessors, is dedicated to the biologists who work on threatened birds. Such people have always been driven more by a belief that threatened birds are worth saving than monetary reward. It is now 20 years since the first summary of Australia's threatened birds was written and many of those who were active then are still striving to keep our avian biodiversity intact. To the many who have contributed their knowledge and time to the preparation of this Action Plan we offer our sincere thanks.

We are particularly grateful to the Birds Australia Threatened Species Committee whose members have provided a sounding board for our deliberations over many months. Members of the committee are Barry Baker, Andrew Burbidge, Allan Burbidge, Stuart Butchart, Graham Carpenter, Les Christidis, Guy Dutson, Hugh Ford, Tim Holmes, Sarah Legge, Richard Loyn, Peter Menkhorst, James O'Connor, Penny Olsen, Hugh Possingham, Don Saunders, Andy Symes and David Watson. Many have been working on the committee since its inception in 2003. Birds Australia as an organisation, and its CEO Graeme Hamilton, have been particularly supportive of the whole enterprise and we wish to express our appreciation to the Research and Conservation Committee and the Birds Australia Council for endorsing the process by which we arrived at the IUCN Red List status of each taxon and imbuing this list with their authority.

For the purposes of the current exercise the Threatened Species Committee was joined by Charles Meredith, CEO of Biosis. The donations of Biosis to Birds Australia made this latest review of Australian birds possible, along with a personal donation of Bob Baird of Biosis, and we are most grateful for their belief in the value of this work. Together with Birds Australia, they provided the co-funding that persuaded the Australian Research Council to support the larger project of which this book is a part: Project LP0990395 *Increasing the effectiveness of Australian threatened bird conservation* led by Stephen Garnett, Hugh Possingham and Stuart Butchart. Additional in-kind support for the project has been provided by the Australian Wildlife Conservancy, BirdLife International and Biosis, all of whom have been far kinder with their time than they may have anticipated.

As part of preparation of the plan we undertook workshops around the country or made contact by other means to garner advice from local experts. We are very grateful to the following people who gave up their time to discuss threatened birds: Ian Abbott, David Akers, Rachel Alderman, David Andrew, Mark Antos, Chris Armstrong, Dave Armstrong, Mick Atzeni, John Augusteyn, Barry Baker, Jack Baker, David Baker-Gabb, Mike Bamford, Paul Barden, Bob Bates, Phil Bell, Simon Bennett, Joe Benshemesh, Jez Bird, Andrew Black, Stewart Blackhall, Margaret Blakers, John Blyth, Judy Blyth, Walter Boles, Michael Booker, Lesley Brooker, Alan Briggs, Mike Brooke, Bill Brown, Sarah Brown, Sally Bryant, Peter Buosi, Allan Burbidge, Andrew Burbidge, Mavis Burgess, Nigel Burgess, Peter Cale, Rob Calvert, Margaret Cameron, Matt Cameron, Hamish Campbell, Nicholas Carlile, Graham Carpenter, Ian Carruthers, Andrew Carter, Carla Catterall, Dominic Chaplin, Wayne Cherry, Margaret Christian, Maureen Christie, Peter Christie, Greg Clancy, Mike Clarke, Rohan Clarke, Rob Clemens, Alister Cockburn, Andrew Cockburn, Sarah Comer, Jane Cooper, Peter Copley, Nicola Crockford, Gabriel Crowley, Greg Czechura, Jeff Davies, Rob Davis, James Dawson, Stephen Debus, Anneke de Graaf, Terry Dennis, Richard Donaghey, Bruce Doran, Peter Dostine, Mike Double, Fiona Douglas, Nic Dunlop, David Drynan, Peter Ekert, Glenn Ehmke, Robert Farnes, Gary Fitt, James Fitzsimons, Hugh Ford, Bob Forsyth, Don Franklin, Alastair Freeman, Amanda Freeman, Cliff Frith, Dawn Frith, Rosemary Gales, David Geering, Heather Gibbs, Sharon Gillam, Alan Gillanders, Cheryl Gole, Steve Goosem, Ken Gosbell, Sue Gould, Steve Goosem, Gordon Graham, John Grant, Peter Green, Tony Grice, John Griffiths, Ian Gynther, Don Hadden, Michelle Hall, Neil Hamilton, Sheryl Hamilton, Jeff Hardy, Daniel Harley, Graham Harrington, Ken Harris, Chris Hassell, Claire Hawkins, Chris Hedger, Rob Heinsohn, Janos Hennicke, Kerryn Herman, Faye Hill, Richard Hill, Rod Hobson, Richard Holdaway, Mark Holdsworth, Tim Holmes, Jodie Honan, Philippa Horton, Wayne Houston, Victor Hurley, Dean Ingwersen, Roger Jaensch, David James, Clare Jones, Ron Johnstone, Leo Joseph, Liana Joseph, Rod Kavanagh, Arthur Keates, Sheryl Keates, Simon Kennedy, Richard Kingsford, Marcel Klaassen, Alex Kutt, Peter Kyne, Russ Lamb, Peter Latch, Ivan Lawler, Sarah Legge, Colin Lendon, David Lindenmayer, Vanessa Lipianin, Geoff Lipsett-Moore, Lana Little, Sarah Lloyd, Richard Loyn, Patrick McConnell, Keith McDonald, David McFarland, Grainne McGuire, Andrew McIntyre, Natasha McLean, Max Maddock, Richard Major, Adrian Manning, Martine Maron, Ian Mason, Mike Mathieson, Peter Mawson, Ian May, Peter Menkhorst, Charles Meredith, David Milton, Clive Minton, James Moloney, Nick Mooney, Gethin Morgan, Steve Murphy, Simon Mustoe, Tim Nevard, Brenda Newby,

Mike Newman, Lloyd Nielsen, Richard Noske, Richard Nott, Martin O'Brien, Damon Oliver, Jerry Olsen, Penny Olsen, Paul O'Neill, Carol Palmer, Blair Parsons, David Paton, Chris Pavey, Saravan Peacock, Lynn Pedler, Harvey Perkins, Justin Perry, Robyn Pickering, Marcus Pickett, John Porter, Hugh Possingham, Noel Preece, Ivor Preston, Catherine Price, Steve Priday, David Priddel, Rachel Pritchard, Sarah Pryke, Bruce Quin, Jim Radford, Kate Ravich, Harry Recher, Julian Reid, Tim Reid, Del Richards, Carly Rickerby, Graham Robertson, Doug Robinson, Mick Roderick, Daniel Rogers, Danny Rogers, Graham Rogers, David Rohweder, Eleanor Rowley, Coral Rowston, Paul Sagar, Rosemary Sandford, Debbie Saunders, Denis Saunders, Don Saunders, Michael Saxon, Elinor Scambler, Jonny Schoenjahn, Erica Shedley, Susan Shephard, Tom Shephard, Luke Shoo, Andrew Silcocks, Anja Skroblin, Ian Smales, Geoff Smith, James Smith, Mike Smith, Todd Soderquist, Roy Sonnenburg, Alistair Stewart, David Stewart, Peter Stokie, Ben Sullivan, Scott Sullivan, Elisa Tack, Michael Tarburton, Peter Taylor, Susan Taylor, Janelle Thomas, David Thompson, Mick Todd, Chris Tzaros, Brian Venables, Anne Marie van Doorn, Jason van Weenen, Mark Wapstra, Matthew Ward, Ron Ward, Simon Ward, Brooke Watson, David Watson, James Watson, John Westaway, Matthew Webb, Rick Webster, Dez Wells, David Westcott, Kirsten Wilkes, Steve Williams, Eric Woehler, John Woinarski and Mark Ziembecki. Other individuals who shared their expertise on individual species is acknowledged at the end of each taxon, in some of which we also acknowledge contributions to the previous action plans where that information is still pertinent. Of course the current work builds on that of the previous plans which would not have existed but for Joost Brouwer, who initiated the first synopsis of threatened birds in Australia in the 1980s, and Gabriel Crowley who was a co-author of the Action Plan for Australian Birds 2000.

There are others to whom we are most grateful for their contributions. Preparation of the maps would not have been possible without the mastery of Glenn Ehmke who has the rare combination of skill, artistry and knowledge of birds that was essential for this project. He was greatly supported by Andrew Silcocks, curator of the Birds Australia database. David Drynan, long-term coordinator of the Australian Bird and Bat Banding Scheme, kindly provided the most recent longevity data, thus updating our estimates of generation time. Mark Burgman and Marissa McBride were instrumental in a structured elicitation process needed to determine the status of some contentious taxa, for which we also had advice from Resit Akcakaya. Tracey Rout helped distinguish extinct taxa from those possibly extinct. For preparation of the website on which a summary of this book is now available we thank Tim Schinkel, and for the support of Jen Sutfin. A team at Charles Darwin University has helped with administration of the project including Roanne Ramsay, Heather Robertson and Owen Wrangle. Social researcher Gill Ainsworth encouraged us to be more explicit about values, leading to inclusion of a score for taxonomic uniqueness. And we would like to thank our editor, Tracey Millen, and publisher John Manger for their understanding as we tried to make this book as current as possible. Stephen and Guy would like to thank their incredibly tolerant partners, Kerstin Zander and Wendy Foden, who sometimes understood the need to keep working until four in the morning, and Stephen his children, Jeremy, Leah and Christopher, for whom there has never been a time without bird action plans. Judit would like to thank her friends and family, who for the last year tolerated every conversation with her ending up at threatened birds.

Introduction

AIMS OF THE ACTION PLAN

This is the third iteration of the *Action Plan for Australian Birds*, updating Garnett (1992) and Garnett and Crowley (2000), and follows much the same form and style to provide consistency. The aims of these Action Plans are to:

- provide a national overview of the conservation status of all birds occurring in Australian territory against IUCN Red List categories and criteria
- identify threats and recommend actions to minimise those threats
- identify habitats or areas of particular importance for bird
- identify information gaps, recommend conservation research and management actions.

The IUCN Red List is not a statutory list. As such we often are able to be more precautionary than is possible in legislative schedules, as well as react more quickly to threats, or the diminution of those threats. Some of the recommendations of this Action Plan may be incorporated into the *Environmental Conservation and Biodiversity Protection Act 1999* and other equivalent legislation in the States and Territories. However, such legislative procedures must usually take account of more than the risk of extinction if they are to be legally binding. However, what this Action Plan does is provide a picture of the status of Australia's birds as of 2010 against a single set of criteria using a consistent methodology.

METHODS

Reassessment of the status of Australian species and subspecies of birds was undertaken in four stages.

In early 2010, a complete list of the species and subspecies of Australian birds was assembled. Each taxon was assessed against the IUCN Red List Guidelines Version 8.1 (IUCN Standards and Petitions Subcommittee 2010). The IUCN Red list protocols consist of a set of rules that make up a logic tree; if a species is assessed to be consistent with the specifications of any one of five rule sets, it may be classified at that level. Interpretations of the rules require data on a range of features that reflect the likelihood of extinction including adult population size, geographic range and occupancy, trends in population size and habitat, severe fluctuations and habitat fragmentation. The guidelines used in 2010 have been refined significantly from the version of the IUCN Red List Guidelines used for *The Action Plan for Australian Birds 2000* (IUCN 1994).

The initial assessments were tested in workshops held in all Australian State and Territory capital cities as well as in Townsville, Queensland. Knowledgeable individuals from state and federal governments and the wider ornithological community were invited to these workshops to discuss the status of all taxa in their area, and any other taxa on which they had particular expertise. At the same time the draft list was widely disseminated among the birdwatching community, soliciting comments, as well as being sent to members of the Birds Australia Threatened Species Committee. There was also discussion outside the workshops with experts holding particular knowledge of specific taxa.

Detailed texts were prepared for all taxa that were to be listed as threatened or near threatened in the 2010 Action Plan. In the course of preparing these texts some new information came to light. When this occurred, experts were again consulted to determine whether they concurred with the interpretation of the new material. While the status is current as of 31 December 2010, final comments were incorporated up to 1 June 2011.

For all but nine taxa, agreement was reached during workshops or in subsequent discussions. For these nine,

16 members of the Birds Australia Threatened Species Committee, whose members have wide experience in IUCN Red Listing as well as specialist skills, engaged in a structured elicitation (Burgman 2005) to answer 108 questions needed to estimate the IUCN Red List parameters. The experts were provided with background information derived from literature published on the taxon and submissions by species specialists. The experts then made an initial judgement, viewed the estimates of the same value by other committee members then provided a second, confidential estimate. The medians of the combined final results were then used to construct an IUCN Red List assessment for each taxon. For three of the nine taxa – Southern Cassowary *Casuarius casuarius johnsonii*, Superb Parrot *Polytelis swainsonii* and Gouldian Finch *Erythura gouldiae* – the expert panel's judgements were further analysed using the RAMAS Red List software (Akçakaya 2007). This software takes into account the uncertainty around estimates of the true value. The RAMAS software has explicit settings for dispute tolerance and risk tolerance. In each case this was set at the default level, as it was judged that this would most closely approximate the international standards to which the project aspires. The use of the RAMAS software produced results that differed from simple analysis of the median values provided by the expert panel who contributed to each taxon's results, because it also included the uncertainty around judgements.

CONSERVATION SUMMARIES

Conservation summaries are of two types: species conservation summaries and subspecies conservation summaries. Some taxa have summaries for both a species and all its constituent subspecies. The term 'recovery outline' is not used here. Recovery outlines, like 'recovery plans' imply that taxa are to be returned to an ideal state, after which they are 'recovered' and are self-sustaining. In keeping with the IUCN Red List guidelines, it is acknowledged that all taxa are to some extent threatened, although most have the status of Least Concern and are not at immediate risk of extinction. However, for many of the taxa listed in this book, recovery is an unrealistic expectation. As with many things we value, most threatened species will require care indefinitely.

The components of the conservation summaries are as follows:

Common name

Vernacular names are those in Christidis and Boles (2008), except where there are differences in taxonomic treatment. Subspecies names are geographically-based where possible, though for some a commonly-used epithet is also noted (e.g. Helmeted Honeyeater).

Scientific name

Genera, and the order in which they are listed, follow Christidis and Boles (2008). Species follow BirdLife International (2011) so that assessments in the Action Plan are comparable with international assessments. This affects the albatrosses in particular, but also a small number of other taxa, as noted in the text. At the subspecies level, we followed the Handbook of Australian, New Zealand and Antarctic Birds (Marchant and Higgins 1990, 1993; Higgins and Davies 1996; Higgins 1999; Higgins *et al.* 2001; Higgins and Peter 2002; Higgins *et al.* 2006a, 2006b). Differences from this rule are justified in the text. In recent years a number of studies have failed to detect genetic variation between subspecies. We adopt the conservative precautionary view that, while genetic studies can detect differences between populations, some of which may not readily be apparent in the phenotype, a failure to detect variation may reflect incomplete sampling of the genome. The subspecies are presented in alphabetical order.

Family

Families, and the order in which they are listed, follow Christidis and Boles (2008).

Conservation status

The IUCN Red List categories are listed below, with the criteria used to define them specified in Table 1.

Extinct (EX): A taxon is Extinct when there is no reasonable doubt that the last individual has died. A taxon is presumed Extinct when exhaustive surveys in known and/or expected habitat, at appropriate times (diurnal, seasonal, annual), throughout its historic range have failed to record an individual. Surveys should be over a time frame appropriate to the taxon's life cycles and life form.

Extinct in the Wild (EW): A taxon is Extinct in the Wild when it is known only to survive in cultivation, in captivity or as a naturalised population (or populations) well outside the past range. No Australian bird taxon is Extinct in the Wild.

Critically Endangered (CR): A taxon is Critically Endangered when the best available evidence indicates that it meets any of the criteria A to E for Critically Endangered, and it is therefore considered to be facing an extremely high risk of extinction in the wild.

Endangered (EN): A taxon is Endangered when the best available evidence indicates that it meets any of the criteria A to E for Endangered, and it is therefore considered to be facing a very high risk of extinction in the wild.

Vulnerable (VU): A taxon is Vulnerable when the best available evidence indicates that it meets any of the criteria A to E for Vulnerable, and it is therefore considered to be facing a high risk of extinction in the wild.

Near Threatened (NT): A taxon is Near Threatened when it has been evaluated against the criteria but does not qualify for Critically Endangered, Endangered or Vulnerable now, but is close to qualifying for or is likely to qualify for a threatened category in the near future. To qualify for the Near Threatened category, the taxon should be close to qualifying for the Vulnerable category. The estimates of population size or habitat should be close to the Vulnerable thresholds, especially when there is a high degree of uncertainty, or possibly meet some of the subcriteria. This may be combined with biological susceptibility and threat. The category Near Threatened is not specified by its own criteria, but instead by the proximity of a taxon to the criteria for the category Vul-

Table 1. Summary of the five criteria (A-E) used to evaluate if a taxon belongs in a threatened category (Critically Endangered, Endangered or Vulnerable)

Use any of the criteria A-E	Critically Endangered	Endangered	Vulnerable
A. Population reduction	Declines measured over the longer of 10 years or 3 generations		
A1	>90%	>70%	>50%
A2, A3 & A4	>80%	>50%	>30%
A1. Population reduction observed, estimated, inferred, or suspected in the past where the causes of the reduction are clearly reversible AND understood AND ceased based on and specifying any of the following:			
(a) direct observation			
(b) an index of abundance appropriate to the taxon			
(c) a decline in area of occupancy (AOO), extent of occurrence (EOO) and/or habitat quality			
(d) actual or potential levels of exploitation			
(e) effects of introduced taxa, hybridisation, pathogens, pollutants, competitors or parasites			
A2. Population reduction observed, estimated, inferred, or suspected in the past where the causes of reduction may not have ceased OR may not be understood OR may not be reversible, based on any of (a) to (e) under A1			
A3. Population reduction projected or suspected to be met in the future (up to a maximum of 100 years) based on any of (b) to (e) under A1.			
A4. An observed, estimated, inferred, projected or suspected population reduction (up to a maximum of 100 years) where the time period must include both the past and the future, and where the causes of reduction may not have ceased OR may not be understood OR may not be reversible, based on any of (a) to (e) under A1.			
B. Geographic range in the form of either B1 (extent of occurrence) OR B2 (area of occupancy)			
B1. Either extent of occurrence	<100 km^2	<5000 km^2	<20 000 km^2
B2. or area of occupancy	<10 km^2	<500 km^2	<2000 km^2
and 2 of the following 3:			
(a) severely fragmented or no. of locations	= 1	≤ 5	≤ 10
(b) continuing decline in (i) extent of occurrence (ii) area of occupancy, (iii) area, extent and/or quality of habitat, (iv) number of locations or subpopulations and (v) number of mature individuals.			
(c) extreme fluctuations in any of (i) extent of occurrence, (ii) area of occupancy, (iii) number of locations or subpopulations and (iv) number of mature individuals.			
C. Small population size and decline			
Number of mature individuals	<250	<2500	<10 000
and either **C1** or **C2**:			
C1. An estimated continuing decline of at least	25% in 3 years	20% in 5 years	10% in 10 years
up to a maximum of 100 years	or 1 generation	or 2 generations	or 3 generations
C2. A continuing decline and (a) and/or (b)			
(a i) no. of mature individuals in largest subpopulation	<50	<250	<1000
(a ii) or % mature individuals in 1 subpopulation	90–100%	95–100%	100%
(b) extreme fluctuations in the number of mature individuals			
D. Very small or restricted population			
Either (1) number of mature individuals	<50	<250	<1000
or (2) restricted area of occupancy	na	na	typically: AOO <20 km^2 or no. of locations ≤5
E. Quantitative analysis			
Indicating the probability of extinction in the wild to be at least	50% in 10 years or 3 generations (100 years max)	20% in 20 years or 5 generations (100 years max)	10% in 100 years

nerable. In this book we have extended the use of Near Threatened to assess the probability of threat to a location. To meet the Criteria for Vulnerable D2 there needs to be a plausible threat to one or more of up to 5 locations. We felt that there are plausible threats to all locations, but some are less likely than others. For island endemics in particular, we have used Near Threatened where threats are plausible, but unlikely.

Least Concern (LC): A taxon is Least Concern when it has been evaluated against the criteria and does not qualify for Critically Endangered, Endangered, Vulnerable or Near Threatened. Widespread and abundant taxa are included in this category.

Data Deficient (DD): A taxon is Data Deficient when there is inadequate information to make a direct, or indirect, assessment of its risk of extinction based on its distribution and/or population status. We have eschewed this category, though good information is lacking for many taxa. Action Plans hope to generate action, and analysis of the impact of the 1992 Action Plan for Australian Birds (Garnett 1992) found that action was much less likely for taxa in a category equivalent to Data Deficient (Garnett *et al.* 2003).

Not Evaluated (NE): A taxon is Not Evaluated when it is has not yet been evaluated against the criteria. All Australian bird taxa were evaluated.

Reasons for listing

This is a brief summary of the IUCN Red List criteria that are thought to apply to a particular taxon. Details of how the taxon rates against all five of the criteria are supplied in the first of two tables, which are provided for all extant taxa (*Current eligibility against IUCN Red List Criteria*). A taxon may meet multiple criteria, but some do so at lower levels than the global status. A taxon can meet a criterion, but be downlisted one or two categories depending on the known or presumed level of genetic exchange with other extralimital populations and their status in line with the IUCN guidelines on the regional application of the IUCN Red List criteria (IUCN 2003).

Status 2000 and Status 1990

This section shows the retrospective status in 2000 and 1990 based on current knowledge, taxonomy and the most recent set of IUCN Red List guidelines. This may differ from the status in the previous two Action Plans. Differences are outlined in Appendix 2.

Reasons for change in 2000 and 1990

Reasons why the retrospective status has changed, reflecting what are believed to be genuine changes in status. This is given only where the status category has changed not the criteria.

Taxonomy

For species, this section lists all subspecies and their IUCN Red List status. For subspecies, it lists other threatened Australian taxa and their status, enumerates other Least Concern Australian subspecies and extralimital taxa and provides the global IUCN Red List status of the species. We also list the global taxonomic uniqueness of each taxon using a modified version of the method of Daniels *et al.* (1991). The taxonomic uniqueness value is inversely related to the number of close relatives, having a maximum value of 1.0 for a taxon in a monospecific family and genus, with no subspecies. The taxonomy followed a range of international sources (since there is no international equivalent of HANZAB). High taxonomic value was ascribed to taxa in the top quintile, low taxonomic value for taxa in the lowest quintile.

Range (or Former range for Extinct taxa)

Describes the extent of occurrence and area of occupied by each taxon, justifies estimates and outlines information on trends. Sometimes AOO has been estimated by dividing a population estimate by an estimate of density. Range estimates were established from over 250 000 locational bird records. This data was solicited from a number of sources including the new Atlas of Australian Birds (Barrett *et al.* 2003), the First Atlas of Australian Birds (Blakers *et al.* 1984), government databases, historical datasets, such as museum collections, and in some cases the personal records of individual people. The majority of this information has been collected by highly skilled and dedicated volunteers all over Australia, including areas far from mainland population centres.

These records are shown on maps as:

- Red dots: core records, known to represent extant locations of focal taxa; in the case of intra-continental migrants, only records approximating the taxon's breeding distribution are used
- Orange dots: non-core records, representing vagrant occurrences of the focal taxon obviously outside its usual range, or historical observations of a taxon known to be extinct at that location
- Grey dots: records of other subspecies of the same species as the focal taxon; or in the case of intra-continental migrants, non-breeding records of the focal taxon not included as core records
- Blue polygons: minimum convex polygons underlying the locational records on the maps that depict each extant taxon's Extent of Occurrence (EOO)
- Grey polygons: ranges of other seabird subspecies of the same species as the local taxon in the AFZ
- Light brown polygons: areas formerly occupied by Extinct taxa
- Red arrows: the location of taxa with very small distributions and known seabird breeding locations.

Maps were made using the Geoscience Australia Lambert Conformal Conic (GALCC) projection.

Abundance (or Former abundance for Extinct taxa)

Summarises available information on the number of mature individuals and trends in their abundance. Some populations are estimated from the sum of AOO and a measure of density.

Ecology

Describes use of habitat and, where appropriate, breeding biology. Includes a justification of the estimate of generation length. Data on generation length is derived from the global database of BirdLife International (2011) augmented by longevity data from the Australian Bird and Bat Banding Scheme or recent literature where appropriate. Two methods have been used to calculate generation length

i. $g = b + (l - b)/2$
ii. $g = b + 1/m$

where g = generation length
b = average age at first breeding
l = maximum known longevity in the wild
m = proportion dying annually

Formula (i) was used by preference where good data are available, since it is derived from a population of birds. It was used in combination with (ii) when data quality was lower. Just (ii) was used when no survival data was available. When no information was available

Table 2: IUCN Red List assessment data (supplied for each extant taxon)

Parameter	Estimate	Reliability
Extent of occurrence	the minimum convex polygon around core Australian records in km^2, or for intra-continental migrants, core breeding records only. For landbirds areas of sea are excluded if they fall within the polygon. For seabirds the EOO is the area within the Australian Fishing Zone, over which the species forages. For breeding seabirds this is the area thought to be used while breeding. Data originates from Birds Australia databases with Atlas data augmented by other data as required	high: within 10% of stated figure medium: within 50% of stated figure low: usually within one order of magnitude of stated figure based on knowledge of distribution
trend	decreasing, stable or increasing; trends are justified in the Range section of the text	high: trend in EOO documented medium: trend in EOO considered likely based on documentation low: trend in EOO suspected but evidence equivocal
Area of occupancy	the number of 1 km^2 grid squares, in which the species is thought to occur at the time when its population is at its most constrained (in km^2). For seabirds this is the area of land where they breed. For non-breeding seabirds, or those with both breeding and non-breeding populations, this area is the same as the EOO, even if globally their AOO is constrained to a much smaller area	high: within 10% of stated figure medium: within 50% of stated figure low: usually within one order of magnitude of stated figure based on knowledge of distribution
trend	decreasing, stable or increasing; trends are justified in the Range section of the text	high: trend in AOO documented medium: trend in AOO considered likely based on documentation low: trend in AOO suspected but evidence equivocal
No. of mature individuals:	based on an estimate of the number of the limiting sex multiplied by two. In many cases this will be an overestimate because usually a relatively small proportion of breeding individuals contribute to recruitment, especially in Australia where so many bird taxa breed communally. However, it will underestimate the floating individuals that would breed given the opportunity. Where possible account is taken of these compounding demographic variables	high: within 10% of stated figure medium: within 50% of stated figure low: usually within 100% of stated figure based on such factors as estimated AOO and any available data on density
trend	decreasing, stable or increasing; rate of change, if any, justified in the Abundance section of the text	high: trend documented medium: trend considered likely based on documentation low: trend suspected but evidence equivocal
No. of sub-populations	number of reproductively-isolated groups, between which there is little likelihood of significant genetic exchange, defined by IUCN as one successful exchange within one generation	high: number of subpopulations documented and, based on the biology the taxon, their genetic isolation highly likely medium: number of subpopulations estimated and, based on the biology the taxon, their genetic isolation probable low: uncertainty about the number of subpopulations and/or extent of genetic separation
Largest sub-population	as for number of mature individuals. Not included if only 1 subpopulation	as for number of mature individuals
Generation time	the average age of parents of newborn individuals in the population in years	high: life-history data documented; medium: high-quality life-history data available for related taxa in a similar environment; low: life-history data of poor quality or extrapolated from other genera
Global population share	% taxa which breeds within Australian territory	high: Australian and/or global populations thought to have been estimated accurately to within 10%; includes endemics medium: Australian and/or global populations thought to have been estimated accurately to within 50% low: Australian and/or global populations poorly known
Level of genetic exchange	the probability of genetic exchange with populations outside Australia occurring at least once per generation. Not included for endemics	high: exchange documented or Australian or extralimital population in very close proximity (i.e. northern islands of Torres Strait) where exchange is almost certain to occur medium: exchange considered likely based on documentation low: exchange suspected but evidence equivocal

for a taxon, data from the nearest relatives or ecological equivalents were used.

Threats (or Reasons for extinction)

Summarises processes most likely to threaten the taxon with extinction, as well as those likely to kill individuals, although the relationship between the death of individuals from a cause and the relative contribution of that cause of death to the probability of extinction has rarely been determined.

Conservation objectives

Long-term aims that provide indicators of the taxon's conservation.

Information required

Lists major data gaps needed for effective conservation.

Management actions required

Lists the most important actions needed for effective conservation management, including research tasks.

Comments received from

List people who have read and commented on texts, or provided pertinent information, even if their conclusions on status were not necessarily adopted by the plan.

The Conservation Summaries in this Action Plan omit three sections that were included in the Recovery Outlines of the *Action Plan for Australian Birds 2000*. The *List of actions already taken* is now getting too long for a book like this, although ongoing actions have been included among the *Management action required*. The *Names of the organisations responsible for the conservation* of each bird are too ephemeral for such a list to be useful. *The estimate of the costs of the different actions* also dated rapidly and was seldom useful.

BIBLIOGRAPHY

Barrett G, Silcocks A, Barry S, Cunningham R, Poulter R (2003) *The New Atlas of Australian Birds*. Royal Australian Ornithologists Union, Hawthorn East.

BirdLife International (2011) 'Birdlife data zone'. Retrieved 17 January, 2011 from <http://www.birdlife.org/datazone/speciessearchresults.php?reg=0&cty=13&cri=&fam=0&gen=0&spc=&cmn=&hab=&thr=&bt=&rec=N&vag=N&hdnAction=ADV_SEARCH&SearchTerms=>.

Blakers MS, Davies JJF, Reilly PN (1984) *The Atlas of Australian Birds*. Melbourne University Press, Melbourne.

Christidis L, Boles WE (2008) *Systematics and Taxonomy of Australian Birds*. CSIRO Publishing, Melbourne.

Daniels RRJ, Hedge M, Joshi NV, Gadgil M (1991) Assigning conservation value: a case-study from India. *Conservation Biology* **5**, 464–475.

Garnett S, Crowley G, Balmford A (2003) The costs and effectiveness of funding the conservation of Australian threatened birds. *BioScience* **53**, 658–665.

Garnett ST (1992) *The Action Plan for Australian Birds*. Australian National Parks and Wildlife Service, Canberra.

Higgins PJ (Ed.) (1999) *Handbook of Australian, New Zealand and Antarctic Birds. Volume 4: Parrots to Dollarbird*. Oxford University Press, Melbourne.

Higgins PJ, Davies SJJF (Eds) (1996) *Handbook of Australian, New Zealand and Antarctic Birds. Volume 3: Snipe to Pigeons*. Oxford University Press, Melbourne.

Higgins PJ, Peter JM (Eds) (2002) *Handbook of Australian, New Zealand and Antarctic Birds. Volume 6: Pardalotes to Spangled Drongo*. Oxford University Press, Melbourne.

Higgins PJ, Peter JM, Cowling SJ (Eds) (2006a) *Handbook of Australian, New Zealand and Antarctic Birds. Volume 7: Boatbill to Starlings. Part A. Boatbill to Larks*. Oxford University Press, Melbourne.

Higgins PJ, Peter JM, Cowling SJ (Eds) (2006b) *Handbook of Australian, New Zealand and Antarctic Birds. Volume 7: Boatbill to Starlings. Part B. Dunnock to Starlings*. Oxford University Press, Melbourne.

Higgins PJ, Peter JM, Steele WK (Eds) (2001) *Handbook of Australian, New Zealand and Antarctic Birds. Volume 5: Tyrant-Flycatchers to Chats*. Oxford University Press, Melbourne.

IUCN (1994) 'IUCN Red List categories'. IUCN Species Survival Commission, Gland, Switzerland.

IUCN (2003) 'Guidelines for application of IUCN criteria at regional levels. Version 3.0'. IUCN.

IUCN Standards and Petitions Subcommittee (2010) 'Guidelines for using the IUCN Red List categories and criteria. Version 8.1'. Prepared by the Standards and Petitions Subcommittee in March 2010. Downloadable from http://intranet.iucn.org/webfiles/doc/SSC/RedList/RedListGuidelines.pdf.

Marchant S, Higgins PJ (Eds) (1990) *Handbook of Australian, New Zealand and Antarctic Birds. Volume 1: Ratites to Ducks*. Oxford University Press, Melbourne.

Marchant S, Higgins PJ (Eds) (1993) *Handbook of Australian, New Zealand and Antarctic Birds. Volume 2: Raptors to Lapwings*. Oxford University Press, Melbourne.

The IUCN Red List Status of Extinct, Threatened and Near Threatened Australian bird taxa

Full species are in bold, subspecies are in plain text. For taxa that have populations breeding in Australian territory as well as visiting Australia from breeding grounds elsewhere, the status of the populations have been assessed separately. All species in which one or more ultrataxon or population is not Least Concern have been listed.

Taxon name		IUCN Red List Status of Australian population	Page no.
Southern Cassowary *Casuarius casuarius*		Vulnerable A2c	16
Southern Cassowary (Australian) *C. c. johnsonii*		Vulnerable A2c	16
Emu *Dromaius novaehollandiae*		Least Concern	
Emu (Tasmanian) *D. n. diemenensis*		Extinct	18
King Island Emu *Dromaius ater*		Extinct	19
Kangaroo Island Emu *Dromaius baudinianus*		Extinct	20
Malleefowl *Leipoa ocellata*		Vulnerable A2bce	21
Cape Barren Goose *Cereopsis novaehollandiae*		Least Concern	
Cape Barren Goose (south-western) *C. n. grisea*		Vulnerable D1	23
Blue-billed Duck *Oxyura australis*		Near Threatened C1	24
Red-tailed Tropicbird *Phaethon rubricauda*	breeding	Near Threatened A3be+4be	26
	non-breeding	Least Concern	
White-tailed Tropicbird *Phaethon lepturus*		Endangered B2ab(v)	28
White-tailed Tropicbird (Christmas Island) *P. l. fulvus*		Endangered B2ab(v)	28
White-tailed Tropicbird (Indian Ocean) *P. l. lepturus*	breeding	Endangered D	29
	non-breeding	Least Concern	
White-throated Pigeon *Columba vitiensis*		Extinct	31
White-throated Pigeon (Lord Howe Island) *C. v. godmanae*		Extinct	31
Emerald Dove *Chalcophaps indica*		Least Concern	
Emerald Dove (Christmas Island) *C. i. natalis*		Near Threatened D2	31
Partridge Pigeon *Geophaps smithii*		Vulnerable A2bc+3c+4c	33
Partridge Pigeon (western) *G. s. blaauwi*		Vulnerable C2a(ii)	35
Partridge Pigeon (eastern) *G. s. smithii*		Vulnerable A2bc+3c+4c	37
White-quilled Rock-Pigeon *Petrophassa albipennis*		Least Concern	
White-quilled Rock-Pigeon (Stokes Range) *P. a. boothi*		Near Threatened B1ab(ii,iii,v)+2ab(ii,iii,v), C2a(ii)	39
Norfolk Island Ground-Dove *Gallicolumba norfolciensis*		Extinct	40
Christmas Island Imperial-Pigeon *Ducula whartoni*		Near Threatened D2	41
New Zealand Pigeon *Hemiphaga novaeseelandiae*		Extinct	42
New Zealand Pigeon (Norfolk Island) *H. n. spadicea*		Extinct	42

Taxon name		IUCN Red List Status of Australian population	Page no.
Glossy Swiftlet ***Collocalia esculenta***		Near Threatened D2	43
Glossy Swiftlet (Christmas Island) *C. e. natalis*		Near Threatened D2	43
Matsudaira's Storm-Petrel ***Hydrobates matsudairae***		Vulnerable D2	45
Wilson's Storm-Petrel ***Oceanites oceanicus***		Least Concern	
Wilson's Storm-Petrel (subantarctic) *O. o. oceanicus*	breeding	Vulnerable D2	46
	non-breeding	Least Concern	
Grey-backed Storm-Petrel ***Garrodia nereis***	breeding	Endangered B2ab(iii), C2a(i), D	47
	non-breeding	Least Concern	
White-bellied Storm-Petrel ***Fregetta grallaria***		Vulnerable D2	49
White-bellied Storm-Petrel (Tasman Sea) *F. g. grallaria*	breeding	Vulnerable D2	49
	non-breeding	Vulnerable D2	49
Wandering Albatross ***Diomedea exulans***	breeding	Critically Endangered B2ab(ii,iii), D	50
	non-breeding	Vulnerable A4bd	50
Tristan Albatross ***Diomedea dabbenena***		Critically Endangered A4ade	53
Antipodean Albatross ***Diomedea antipodensis***		Endangered A4ad	54
Antipodean Albatross (Auckland Islands) *D. a. gibsoni*		Endangered A4ad	54
Northern Royal Albatross ***Diomedea sanfordi***		Endangered A4bc, B2ab(iii,v)	56
Southern Royal Albatross ***Diomedea epomophora***		Vulnerable D2	57
Black-browed Albatross ***Thalassarche melanophrys***	breeding	Endangered B2ab(ii,iii)	59
	non-breeding	Endangered A4bd	59
Campbell Albatross ***Thalassarche impavida***		Vulnerable D2	61
Shy Albatross ***Thalassarche cauta***		Vulnerable D2	63
White-capped Albatross ***Thalassarche steadi***		Vulnerable D2	65
Salvin's Albatross ***Thalassarche salvini***		Vulnerable A4bd	67
Grey-headed Albatross ***Thalassarche chrysostoma***	breeding	Critically Endangered B2ab(ii,iii)	68
	non-breeding	Vulnerable A4bd	68
Indian Yellow-nosed Albatross ***Thalassarche carteri***		Endangered A4bde	71
Buller's Albatross ***Thalassarche bulleri***		Near Threatened D2	72
Buller's Albatross (southern) *T. b. bulleri*		Near Threatened D2	74
Buller's Albatross (northern) *T. b. platei*		Near Threatened D2	75
Sooty Albatross ***Phoebetria fusca***		Endangered A4bd	77
Light-mantled Albatross ***Phoebetria palpebrata***	breeding	Endangered B2ab(ii,iii)	78
	non-breeding	Near Threatened A4bd	78
Blue Petrel ***Halobaena caerulea***	breeding	Critically Endangered B2ab(iii,v)	80
	non-breeding	Least Concern	
Antarctic Prion ***Pachyptila desolata***	breeding	Endangered B2ab(iii,v)	82
	non-breeding	Least Concern	
Fairy Prion ***Pachyptila turtur***		Least Concern	
Fairy Prion (southern) *P. t. subantarctica*	breeding	Endangered D	83
	non-breeding	Least Concern	
Fulmar Prion ***Pachyptila crassirostris***		Least Concern	
Fulmar Prion (southern) *P. c. eatoni*	breeding	Vulnerable D2	85
	non-breeding	Least Concern	
White-chinned Petrel ***Procellaria aequinoctialis***		Vulnerable A4bcde	86
Westland Petrel ***Procellaria westlandica***		Vulnerable D2	87
Black Petrel ***Procellaria parkinsoni***		Vulnerable D2	89
Grey Petrel ***Procellaria cinerea***	breeding	Endangered B2ab(iii)	90
	non-breeding	Near Threatened A2cde+3cde+4cde	90
Buller's Shearwater ***Ardenna bulleri***		Vulnerable D2	92
Flesh-footed Shearwater ***Ardenna carneipes***	breeding	Near Threatened A2bc	93
	non-breeding	Near Threatened A2bc	93
Sooty Shearwater ***Ardenna grisea***	breeding	Near Threatened A2de+3de+4de	96
	non-breeding	Near Threatened A2de+3de+4de	96
Hutton's Shearwater ***Puffinus huttoni***		Endangered B2ab(ii,iii)	98
Little Shearwater ***Puffinus assimilis***		Least Concern	
Little Shearwater (Tasman Sea) *P. a. assimilis*		Vulnerable D2	99

Taxon name		IUCN Red List Status of Australian population	Page no.
Buff-banded Rail *Gallirallus philippensis*		Least Concern	
Buff-banded Rail (Cocos Keeling Islands) *G. p. andrewsi*		Vulnerable D1+2	155
Buff-banded Rail (Macquarie Island) *G. p. macquariensis*		Extinct	156
Lord Howe Woodhen *Gallirallus sylvestris*		Endangered D	157
Black-faced Sheathbill *Chionis minor*		Vulnerable D1	159
Black-faced Sheathbill (Heard Island) *C. m. nasicornis*		Vulnerable D1	159
Grey Plover *Pluvialis squatarola*		Near Threatened A2bc+3c+4bc	160
Lesser Sand Plover *Charadrius mongolus*		Endangered A2bc+3c+4bc	162
Lesser Sand Plover (Mongolian) *C. m. mongolus*		Endangered A2bc+3c+4bc	162
Lesser Sand Plover (Kamchatkan) *C. m. stegmanni*		Endangered A2bc+3c+4bc	164
Greater Sand Plover *Charadrius leschenaultii*		Vulnerable A2bc+3c+4bc	166
Greater Sand Plover (Mongolian) *C. l. leschenaultii*		Vulnerable A2bc+3c+4bc	166
Hooded Plover *Thinornis rubricollis*		Vulnerable C1	168
Hooded Plover (eastern) *T. r. rubricollis*		Vulnerable C1+2a(ii)	170
Hooded Plover (western) *T. r. tregellasi*		Vulnerable C2a(ii)	172
Plains-wanderer *Pedionomus torquatus*		Endangered C2a(ii)	174
Australian Painted Snipe *Rostratula australis*		Endangered C2a(ii)	176
Black-tailed Godwit *Limosa limosa*		Near Threatened A2b	178
Black-tailed Godwit (eastern Siberian) *L. l. melanuroides*		Near Threatened A2b	178
Bar-tailed Godwit *Limosa lapponica*		Vulnerable A3c+4bc	180
Bar-tailed Godwit (western Alaskan) *L. l. baueri*		Vulnerable A3c+4bc	180
Bar-tailed Godwit (northern Siberian) *L. l. menzbieri*		Vulnerable A3c+4bc	182
Whimbrel *Numenius phaeopus*		Near Threatened A3c+4bc	184
Whimbrel (eastern Siberian) *N. p. variegatus*		Near Threatened A3c+4bc	184
Eastern Curlew *Numenius madagascariensis*		Vulnerable A2bc+3c+4bc	186
Grey-tailed Tattler *Tringa brevipes*		Near Threatened A2b	188
Ruddy Turnstone *Arenaria interpres*		Near Threatened A2b	190
Ruddy Turnstone (Palaearctic) *A. i. interpres*		Near Threatened A2b	190
Asian Dowitcher *Limnodromus semipalmatus*		Near Threatened A3c+4c, C2a(ii), D1	192
Great Knot *Calidris tenuirostris*		Vulnerable A3c+4bc	194
Red Knot *Calidris canutus*		Vulnerable A3c+4bc	196
Red Knot (New Siberian islands) *C. c. piersmai*		Vulnerable A3c+4bc	196
Red Knot (north-eastern Siberia) *C. c. rogersi*		Vulnerable A3c+4bc	198
Curlew Sandpiper *Calidris ferruginea*		Vulnerable A2bc+3c+4bc	200
Black-breasted Button-quail *Turnix melanogaster*		Near Threatened B2ab(i,ii,iii,iv,v)	202
Buff-breasted Button-quail *Turnix olivii*		Endangered B2ac(ii)	204
Painted Button-quail *Turnix varius*		Least Concern	
Painted Button-quail (Houtman Abrolhos) *T. v. scintillans*		Endangered B1ab(v)+2ab(v), C2a(ii)	206
Lesser Noddy *Anous tenuirostris*		Endangered B2ab(iii,v)	208
Lesser Noddy (Houtman Abrolhos) *A. t. melanops*		Endangered B2ab(iii,v)	208
Fairy Tern *Sternula nereis*		Vulnerable C1	209
Fairy Tern (New Caledonian) *S. n. exsul*		Endangered D	211
Fairy Tern (Australian) *S. n. nereis*		Vulnerable C1	212
White-fronted Tern *Sterna striata*	breeding	Near Threatened D1	214
	non-breeding	Least Concern	
Antarctic Tern *Sterna vittata*		Least Concern	
Antarctic Tern (New Zealand) *S. v. bethunei*	breeding	Endangered D	215
	non-breeding	Least Concern	
Antarctic Tern (Indian Ocean) *S. v. vittata*	breeding	Endangered D	217
	non-breeding	Least Concern	
Norfolk Island Kaka *Nestor productus*		Extinct	218
Palm Cockatoo *Probosciger aterrimus*		Vulnerable C2a(ii), E	219
Palm Cockatoo (Australian) *P. a. macgillivrayi*		Vulnerable C2a(ii), E	219
Red-tailed Black-Cockatoo *Calyptorhynchus banksii*		Least Concern	
Red-tailed Black-Cockatoo (south-eastern) *C. b. graptogyne*		Endangered C2a(ii)	221
Red-tailed Black-Cockatoo (south-western) *C. b. naso*		Vulnerable A2c+3c+4c	223
Glossy Black-Cockatoo *Calyptorhynchus lathami*		Least Concern	
Glossy Black-Cockatoo (Kangaroo Island) *C. l. halmaturinus*		Endangered D	225
Glossy Black-Cockatoo (south-eastern) *C. l. lathami*		Near Threatened C2a(ii)	227
Carnaby's Black-Cockatoo *Calyptorhynchus latirostris*		Endangered A2bcde+3bcde+4bcde	229

Taxon name	IUCN Red List Status of Australian population	Page no.
Baudin's Black-Cockatoo *Calyptorhynchus baudinii*	Endangered A3cde+4cde	231
Major Mitchell's Cockatoo *Lophochroa leadbeateri* Major Mitchell's Cockatoo (eastern) *L. l. leadbeateri*	Least Concern Near Threatened A3c	 234
Double-eyed Fig-Parrot *Cyclopsitta diophthalma* Double-eyed Fig-Parrot (southern, Coxen's) *C. d. coxeni*	Least Concern Endangered D	 236
Eclectus Parrot *Eclectus roratus* Eclectus Parrot (Cape York Peninsula) *E. r. macgillivrayi*	Near Threatened D1 Near Threatened D1	237 237
Regent Parrot *Polytelis anthopeplus* Regent Parrot (eastern) *P. a. monarchoides*	Least Concern Endangered B2ab(iii,v)	 239
Princess Parrot *Polytelis alexandrae*	Near Threatened D1	241
Green Rosella *Platycercus caledonicus* Green Rosella (King Island) *P. c. brownii*	Least Concern Vulnerable D1	 243
Swift Parrot *Lathamus discolor*	Endangered C2a(ii)	244
Golden-shouldered Parrot *Psephotus chrysopterygius*	Endangered B1ab(iii,v)	247
Paradise Parrot *Psephotus pulcherrimus*	Extinct	249
Tasman Parakeet *Cyanoramphus cookii* Tasman Parakeet (Norfolk Island) *C. c. cookii* Tasman Parakeet (Lord Howe Island) *C. c. subflavescens*	Critically Endangered B1ab(iii,v)+2ab(iii,v), C2a(ii) Critically Endangered B1ab(iii,v)+2ab(iii,v), C2a(ii) Extinct	250 251 253
Red-fronted Parakeet *Cyanoramphus novaezelandiae* Red-fronted Parakeet (Macquarie Island) *C. n. erythrotis*	Extinct Extinct	254 254
Orange-bellied Parrot *Neophema chrysogaster*	Critically Endangered A2abc+3bc+4abc, B1ab(i,ii,iii,iv,v), C1+2a(i,ii), D, E	255
Western Ground Parrot *Pezoporus flaviventris*	Critically Endangered A2bc, C1+2a(ii)	257
Eastern Ground Parrot *Pezoporus wallicus* Eastern Ground Parrot (mainland) *P. w. wallicus*	Least Concern Near Threatened C2a(i)	 259
Night Parrot *Pezoporus occidentalis*	Endangered B2ab(iii)c(ii,iii,iv), D	261
Barking Owl *Ninox connivens* Barking Owl (Papuan) *N. c. assimilis* Barking Owl (southern) *N. c. connivens*	Least Concern Near Threatened D1 Near Threatened A2c+3c+4c	 263 264
Southern Boobook *Ninox novaeseelandiae* Southern Boobook (Lord Howe Island) *N. n. albaria* Southern Boobook (Norfolk Island x New Zealand) *N. n. undulata* x *N. n. novaeseelandiae*	Least Concern Extinct Critically Endangered D	 266 267
Christmas Island Hawk-Owl *Ninox natalis*	Vulnerable D1+2	269
Masked Owl *Tyto novaehollandiae* Masked Owl (Tasmanian) *T. n. castanops* Masked Owl (northern) *T. n. kimberli* Masked Owl (Tiwi Islands) *T. n. melvillensis*	Least Concern Endangered C2a(ii) Vulnerable C2a(i) Endangered C2a(ii)	 271 272 274
Azure Kingfisher *Ceyx azureus* Azure Kingfisher (Tasmanian) *C. a. diemenensis*	Least Concern Vulnerable D1	276
Sacred Kingfisher *Todiramphus sanctus* Sacred Kingfisher (Norfolk Island) *T. s. norfolkiensis*	Least Concern Near Threatened D1+2	 278
Albert's Lyrebird *Menura alberti*	Near Threatened B1ab(i,ii,iv)+2ab(i,ii,iv)	279
Rufous Scrub-bird *Atrichornis rufescens* Rufous Scrub-bird (southern) *A. r. ferrieri* Rufous Scrub-bird (northern) *A. r. rufescens*	Endangered B2ab(ii,iv,v) Endangered B2ab(ii,v) Endangered B1ab(ii,iv,v)+2ab(ii,iv,v)	281 282 284
Noisy Scrub-bird *Atrichornis clamosus*	Endangered B1ab(iii,v)+2ab(iii,v)	286
Brown Treecreeper *Climacteris picumnus* Brown Treecreeper (south-eastern) *C. p. victoriae*	Least Concern Near Threatened A2bc+3c+4bc	 288
Fawn-breasted Bowerbird *Ptilonorhynchus cerviniventris*	Near Threatened C2a(ii)	291
Purple-crowned Fairy-wren *Malurus coronatus* Purple-crowned Fairy-wren (western) *M. c. coronatus* Purple-crowned Fairy-wren (eastern) *M. c. macgillivrayi*	Least Concern Endangered B2ab(ii,iii,iv,v) Near Threatened C2a(ii)	 292 294
White-winged Fairy-wren *Malurus leucopterus* White-winged Fairy-wren (Barrow Island) *M. l. edouardi* White-winged Fairy-wren (Dirk Hartog Island) *M. l. leucopterus*	Least Concern Vulnerable D2 Vulnerable D2	 296 298
Variegated Fairy-wren *Malurus lamberti* Variegated Fairy-wren (Shark Bay) *M. l. bernieri*	Least Concern Vulnerable D2	 299

Taxon name	IUCN Red List Status of Australian population	Page no.
Southern Emu-wren *Stipiturus malachurus*	Least Concern	
Southern Emu-wren (Dirk Hartog Island) *S. m. hartogi*	Vulnerable D2	301
Southern Emu-wren (Fleurieu Peninsula) *S. m. intermedius*	Endangered B1ab(i,ii,iii,iv,v)+2ab(i,ii,iii,iv,v); C2a(i)	302
Southern Emu-wren (Eyre Peninsula) *S. m. parimeda*	Endangered B1ab(ii,iii,iv,v)+2ab(ii,iii,iv,v), C2a(i)	304
Mallee Emu-wren *Stipiturus mallee*	Endangered A2c	306
Grey Grasswren *Amytornis barbatus*	Least Concern	
Grey Grasswren (Bulloo) *A. b. barbatus*	Endangered B1ab(i,ii,iii,v)+2ab(i,ii,iii,v)	308
Striated Grasswren *Amytornis striatus*	Least Concern	
Striated Grasswren (sandplain) *A. s. striatus*	Near Threatened A2bc+3c+4bc	310
Short-tailed Grasswren *Amytornis merrotsyi*	Vulnerable B2ab(ii,iii,v)	311
Short-tailed Grasswren (Flinders Ranges) *A. m. merrotsyi*	Vulnerable B1ab(ii,iii,v)+2ab(ii,iii,v)	313
Short-tailed Grasswren (Gawler Ranges) *A. m. pedleri*	Vulnerable D1	314
White-throated Grasswren *Amytornis woodwardi*	Vulnerable B1ab(ii,iii,iv,v)+2ab(ii,iii,iv,v), C2a(i)	316
Carpentarian Grasswren *Amytornis dorotheae*	Near Threatened A2c, C2a(ii)	318
Thick-billed Grasswren *Amytornis modestus*	Least Concern	320
Thick-billed Grasswren (MacDonnell Ranges) *A. m. modestus*	Extinct	320
Thick-billed Grasswren (NW NSW) *A. m. obscurior*	Critically Endangered D	321
Thick-billed Grasswren (central NSW) *A. m. inexpectatus*	Extinct	322
Thick-billed Grasswren (Flinders Ranges) *A. m. raglessi*	Vulnerable B1ab(i,ii,iii,iv)+2ab(i,ii,iii,iv)	323
Thick-billed Grasswren (Lake Frome Basin) *A. m. curnamona*	Near Threatened D1	325
Western Grasswren *Amytornis textilis*	Least Concern	
Western Grasswren (Gawler Ranges) *A. t. myall*	Vulnerable C2a(ii)	326
Black Grasswren *Amytornis housei*	Near Threatened B1ab(i,ii,iii,iv,v)+2ab(i,ii,iii,iv,v)	328
Eastern Bristlebird *Dasyornis brachypterus*	Endangered B1ab(i,ii,iii,iv)+2ab(i,ii,iii,iv)	330
Eastern Bristlebird (southern) *D. b. brachypterus*	Endangered B1ab(ii)+2ab(ii)	332
Eastern Bristlebird (northern) *D. b. monoides*	Critically Endangered B2ab(ii,iii,iv,v), C2a(i), D	334
Western Bristlebird *Dasyornis longirostris*	Endangered B1ab(iii)	336
Rufous Bristlebird *Dasyornis broadbenti*	Least Concern	
Rufous Bristlebird (western) *D. b. litoralis*	Extinct	338
Scrubtit *Acanthornis magnus*	Least Concern	
Scrubtit (King Island) *A. m. greenianus*	Critically Endangered B1ab(i,ii,iii,iv,v)+2ab(i,ii,iii,iv,v), C2a(i,ii), D	339
Chestnut-rumped Heathwren *Hylacola pyrrhopygia*	Least Concern	340
Chestnut-rumped Heathwren (Mount Lofty Ranges) *H. p. parkeri*	Endangered B1ab(i,ii,iii,iv,v)+2ab(i,ii,iii,iv,v)	342
Chestnut-rumped Heathwren (Flinders Ranges) *H. p. pedleri*	Vulnerable D1	
Rufous Fieldwren *Calamanthus campestris*	Least Concern	
Rufous Fieldwren (Dorre Island) *C. c. dorrie*	Vulnerable D2	344
Rufous Fieldwren (Dirk Hartog Island) *C. c. hartogi*	Vulnerable D2	345
Norfolk Island Gerygone *Gerygone modesta*	Near Threatened D2	347
Lord Howe Gerygone *Gerygone insularis*	Extinct	348
Slender-billed Thornbill *Acanthiza iredalei*	Least Concern	
Slender-billed Thornbill (Gulf St Vincent) *A. i. rosinae*	Vulnerable B1ab(i,ii,iii,iv,v)+2ab(i,ii,iii,iv,v)	349
Brown Thornbill *Acanthiza pusilla*	Least Concern	
Brown Thornbill (King Island) *A. p. archibaldi*	Critically Endangered D	351
Chestnut-breasted Whiteface *Aphelocephala pectoralis*	Near Threatened C2a(i)	352
Forty-spotted Pardalote *Pardalotus quadragintus*	Endangered A2a, B2ab(ii,iii,iv,v)	354
Yellow-tufted Honeyeater *Lichenostomus melanops*	Least Concern	
Yellow-tufted Honeyeater (Helmeted) *L. m. cassidix*	Critically Endangered B1ab(iii,v)+2ab(iii,v), C2a(i), D	356
Black-eared Miner *Manorina melanotis*	Endangered A2ce, B2ab(ii,iii,iv,v); C2a(i)	358
Regent Honeyeater *Anthochaera phrygia*	Critically Endangered A2b	360
Yellow Wattlebird *Anthochaera paradoxa*	Least Concern	
Yellow Wattlebird (King Island) *A. p. kingi*	Near Threatened D2	362
Yellow Chat *Epthianura crocea*	Least Concern	
Yellow Chat (Capricorn) *E. c. macgregori*	Endangered B2ab(i,ii,iv,v), C2a(i), D	363
Yellow Chat (Alligator Rivers) *E. c. tunneyi*	Endangered B2ab(ii,iii,iv,v), C2a(ii)	365

TAXA PROFILES

Subspecies Conservation Summary

Southern Cassowary (Australian)

Casuarius casuarius johnsonii F. Müller, 1866
Casuariidae

Conservation status
Vulnerable A2c

Reasons for listing
Historical declines in habitat over the last 3 generations are thought to have caused a loss >30% of the population in the last 3 generations (44 years), but the is population now almost stable

Status 2000
Endangered A2c
Reason for change in 2010: habitat loss has largely ceased through effective conservation of habitat

Status 1990
Endangered A2c

Taxonomy
Seven extralimital subspecies; the species is Vulnerable
Taxonomic uniqueness: high (2 genera/family, 3 species/genus, 8 subspecies/species)

Range
Confined to eastern north Queensland (Department of Sustainability, Environment, Water, Population and Communities 2011). The largest subpopulation occurs between Cooktown and Townsville, west to the extent of the rainforest, with most records from the coastal lowlands between Ingham and Mossman, and the upland areas incorporating Bluewater, Paluma, Seaview and Kirrama Ranges, the Atherton Tablelands, the Lamb Range and the Carbine, Finnegan and Thornton uplands (Department of Environment and Resource Management 2009). Also occurs on the eastern side of Cape York Peninsula, from the Jardine River south to the Stewart River, with a probable gap between Harmer Creek and the Pascoe River. There are also no recent records from north of Bamaga at the tip of Cape York Peninsula (Marchant and Higgins 1990). About 65% of the wet tropics lowland habitat has been cleared for agriculture (Department of Environment and Resource Management 2009), but less of the uplands. This compares with a probable reduction in the total area of rainforest of 53% at the last glacial maximum (Hilbert *et al.* 2007). The area of occupancy of 2100 km^2 is based on all records held by Birds Australia, including survey records from James Cook University, and is likely to be an underestimate given not all of the rainforests have been sampled; the total area of suitable habitat is 8000 km^2 with a further 4900 km^2 on Cape York Peninsula (Latch 2007).

Abundance
Estimates of population size are based on scant data, much of questionable validity (Westcott 1999). In the Wet Tropics bioregion there are thought to be about 1500 individuals, based on extrapolations from surveys (Moore 1997) and on Cape York Peninsula, 300–500 based primarily on the area of available habitat (S. Goosem in Garnett and Crowley 2000). Upper estimates of the population in the Wet Tropics were said to be 1500 in 2002, 1200 in 2007 (Wet Tropics Management Authority 2002, 2007). One population counted at Mission Beach, using a combination of sightings and foot measurements, contained 49 adults from 102 km^2 with an average home range of each female of 5.36 km^2 (Moore 2007). Because of the uncertainty, the population size was estimated using a structured elicitation process with an expert panel (14 members) to assess all available information using a 2-stage Delphi technique to minimise biases (Burgman 2005). The median view of the panel was that the population was likely to be 2500 mature individuals and that, while there had been a loss of 30–50% of the population over the last 3 generations as a result of land clearing and subsequent extirpation from fragmented remnants, future declines were likely to be small: the median estimate for the population size in 3 generations time (44 years from now) was 2200 mature individuals. More recently females tracked over 3 months using GPS-based telemetry were found to have home ranges of 0.25–0.43 km^2 (H. Campbell *in litt.*), a density 12–21 times that of Moore (2007). Corroboration of this data may result in a reappraisal of population size.

Ecology
Lives in rainforest and associated habitats that can provide a year-round supply of fleshy fruit. All nests have been recorded in rainforest or woodland mosaics with rainforest elements, and cassowaries appear to be the only bird capable of dispersing some rainforest fruits. Other habitats – eucalypt woodlands and savannas, mangroves, exotic fruit plantations and open ground – are used intermittently (Crome 1975; Bentrupperbaumer 1998). Birds will sometimes walk across wide expanses of open land and savanna woodland to feed in rainforest patches, gardens and exotic fruit plantations. Sexes maintain independent, but overlapping, territories. Cassowar-

Current eligibility against IUCN Red List Criteria

IUCN category	Criteria eligibility
A	Vulnerable: population decline of 30–49% over the last 3 generations (44 years) suspected from deterioration in habitat quality
B	Not applicable: >10 locations, not severely fragmented, population fluctuations not extreme
C	Not applicable: population 500–2500 mature individuals, continuing decline not observed or estimated, 1 subpopulation >1000 mature individuals, <95% in largest subpopulation
D	Not applicable: population >1000 mature individuals, >5 locations
E	Not applicable: no population viability analysis undertaken

IUCN Red List assessment data

	Estimate	Reliability
Extent of occurrence trend	23 400 km^2 stable	medium high
Area of occupancy trend	2100 km^2 stable	low medium
No. of mature individuals trend	2500 declining	low low
No. subpopulations	2	low
No. locations	>10	high
Largest subpopulation	1750	low
Generation time	14.5 years	low
Global population share	100%	high

ies lay 3–5 eggs on the ground which the male incubates before raising the young (Bentrupperbaumer 1998). A generation time of 14.5 years is estimated, based on age at first breeding of 4.0 years (value elicited from expert committee) and a maximum longevity in the wild of 21.0 years (Moore 2007).

Threats

Clearance for agriculture is responsible for the loss of most prime habitat of the southern part of the species range, but clearance largely ceased during the 1970s. There has been active rehabilitation of habitat for the last 20 years at multiple locations. Much habitat remains fragmented or connected by only narrow corridors, but other habitat patches have been reconnected. Cassowaries also sometimes travel through woodland and farmland, and can swim across streams. Nevertheless many small remnants no longer contain cassowaries. One locally extirpated subpopulation was on Mt Whitfield within Cairns, but this is now thought to have been derived from captive birds released from the Flecker gardens in the 1950s (B. Venables pers. comm.) as a result of road accidents and attacks by dogs (Moore 1998). Other causes of death include disease (tuberculosis-related disease, aspergillosis) and possible competition with feral pigs *Sus scrofa* for food (Crome and Moore 1990; Bentrupperbaumer 1998). Cyclones can cause temporary local food shortages (Latch 2007), with some populations appearing more vulnerable than others. The younger cohorts are particularly affected and breeding can be delayed for up to 2 years following large cyclones (P. Buosi *in litt.*).

Conservation objectives

1. Maintain existing populations
2. Documented positive or stable population trends at multiple sites

Information required

1. Population densities at multiple locations
2. Demographic trends
3. Impact of dogs, traffic, disease and fragmentation on persistence of small populations
4. Impact of cyclones on survivorship, demography and behaviour and the impact of post-cyclone supplementary feeding
5. The merits of translocation as part of rescue, rehabilitation and release
6. Survivorship and dispersal range of juveniles once they leave the parental home range
7. Home range size and placement of adult males and females
8. Rate of displacement of adult birds from their home areas
9. The social and environmental conditions underpinning demographic processes

Management actions required

1. Monitor key sites
2. Develop and implement population surveys using faecal DNA
3. Improve understanding of density and movements
4. Describe the demography and genetic structure of the Mission Beach cassowary population
5. Identify areas and corridors to protect, restore, manage, develop and implement Cassowary Conservation Local Area Plans as part of local planning
6. Minimise cassowary road deaths and dog attacks, and assess impact of pigs
7. If appropriate implement a translocation plan as part of rescue, rehabilitation and release
8. Develop and implement post-cyclone response plans that maximise persistence of cassowaries and minimise harmful interactions with people
9. Involve community in cassowary conservation and management

Bibliography

Bentrupperbaumer J (1998) Reciprocal ecosystem impact and behavioural interactions between cassowaries, *Casuarius casuarius* and humans, *Homo sapiens* exploring the natural-human environment interface and its implications for endangered species recovery in north Queensland, Australia. PhD thesis, James Cook University of North Queensland, Townsville.

Burgman MA (2005) *Risks and Decisions for Conservation and Environmental Management.* Cambridge University Press, Cambridge.

Crome FHJ (1975) Some observations on the biology of the Cassowary in northern Queensland. *Emu* **76**, 8–14.

Crome FHJ, Moore LA (1990) Cassowaries in north-eastern Queensland: report of a survey and a review and assessment of their status and conservation and management needs. *Australian Wildlife Research* **17**, 369–385.

Department of Environment and Resource Management (2009) *Distribution of Cassowary Habitat in the Wet Tropics Bioregion, Queensland*. 3rd edn. DERM, Queensland.

Department of Sustainability, Environment, Water, Population and Communities (2011) *Casuarius casuarius johnsonii*. In 'Species Profile and Threats Database'. Department of Sustainability, Environment, Water, Population and Communities, Canberra. Retrieved 24 February, 2011 from <http://www.environment.gov.au/sprat>.

Garnett S, Crowley GM (2000) *The Action Plan for Australian Birds 2000*. Environment Australia, Canberra.

Hilbert DW, Graham A, Hopkins MS (2007) Glacial and interglacial refugia within a long-term rainforest refugium: the Wet Tropics Bioregion of NE Queensland, Australia. *Palaeogeography, Palaeoclimatology, Palaeoecology* **251**, 104–118.

Latch P (2007) 'National recovery plan for the southern cassowary *Casuarius casuarius johnsonii*'. Report to Department of the Environment, Water, Heritage and the Arts, Canberra.

Marchant S, Higgins PJ (Eds) (1990) *Handbook of Australian, New Zealand and Antarctic Birds. Volume 1: Ratites to Ducks*. Oxford University Press, Melbourne.

Moore LA (1997) 'Daintree Cassowary Monitoring Stage 2 (November/December 1996)'. Report to the Wet Tropics Management Authority.

Moore LA (1998) 'Cassowary conservation roads: a management strategy and road upgrade assessment for El Arish and Tully-Mission Beach Roads, Mission Beach'. Report to the Queensland Department of Main Roads.

Moore LA (2007) Population ecology of the southern cassowary *Casuarius casuarius johnsonii*, Mission Beach north Queensland. *Journal of Ornithology* **148**, 357–366.

Westcott DA (1999) Counting cassowaries: what does cassowary sign reveal about their abundance? *Wildlife Research* **26**, 61–67.

Wet Tropics Management Authority (2002) 'Appendix 4. State of the Wet Tropics Report 2001–2002'. Wet Tropics Management Authority, Cairns.

Wet Tropics Management Authority (2009) 'State of the Wet Tropics Report 2008–2009'. Wet Tropics Management Authority, Cairns.

Comments received from

Barry Baker, Peter Buosi, Allan Burbidge, Andrew Burbidge, Stuart Butchart, Cassowary Recovery Team, Hamish Campbell, Les Christidis, Hugh Ford, Anneke de Graf, Ian Gynther, Graham Harrington, Tim Holmes, Richard Loyn, Peter Menkhorst, Charles Meredith, Penny Olsen, Scott Sullivan, Andy Symes, David Watson, David Westcott

Subspecies Conservation Summary

Emu (Tasmanian)

Dromaius novaehollandiae diemenensis Le Souef, 1907
Casuariidae

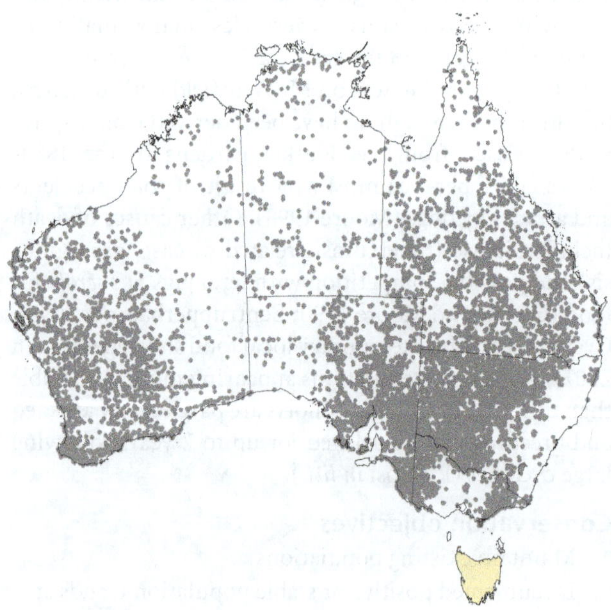

Conservation status
Extinct

Reasons for listing
This subspecies was last seen in the wild in 1865

Status 2000
Extinct

Status 1990
Extinct

Taxonomy
The nominate subspecies, *D. n. novaehollandiae* (mainland Australia) is Least Concern
Taxonomic uniqueness: high (2 genera/family, 3 species/genus, 2 subspecies/species)

Former range
This subspecies of Emu was endemic to Tasmania, most records being from the east and near the north coast (Dove 1926).

Former abundance
No population estimate has ever been made. In articles published in the early 20th century it was described as rare as well as abundant (Department of Sustainability, Environment, Water, Population and Communities 2011). It certainly became rare soon after settlement. Birds were locally extinct by the 1830s (Dove 1924) and the last speci-

mens were collected in 1845 (Dickison 1926). The subspecies was last seen in the wild in about 1865.

Ecology

It appears to have lived in dry forest and grasslands and similar 'park-like' habitats (Dove 1924), and possibly also in upland plains and marshes (Legge 1907).

Reasons for extinction

The subspecies was hunted for its meat and its eggs taken (Legge 1907; Dove 1924), which are assumed to be the cause of its extinction (Dove 1926). They were often kept in captivity, especially birds taken from the wild as young birds, or those raised from chicks that hatched from eggs collected in the wild. One a captive bird is reported to have survived until 1873 (Department of Sustainability, Environment, Water, Population and Communities 2011).

Bibliography

Department of Sustainability, Environment, Water, Population and Communities (2011) *Dromaius novaehollandiae diemenensis*. In 'Species Profile and Threats Database'. Department of Sustainability, Environment, Water, Population and Communities, Canberra. Retrieved 24 February, 2011 from <http://www.environment.gov.au/sprat>.

Dickison D (1926) When did the Tasmanian Emu become extinct? *Emu* **25**, 213.

Dove HS (1924) Notes on the Tasmanian emu. *Emu* **23**, 221–222.

Dove HS (1926) How Tasmania lost the emu. *Emu* **25**, 213.

Legge WV (1907) The Emus of Tasmania and King Island. *Emu* **3**, 113–119.

Comments received from

Eric Woehler

Species Conservation Summary

King Island Emu

Dromaius ater Vieillot, 1817

Casuariidae

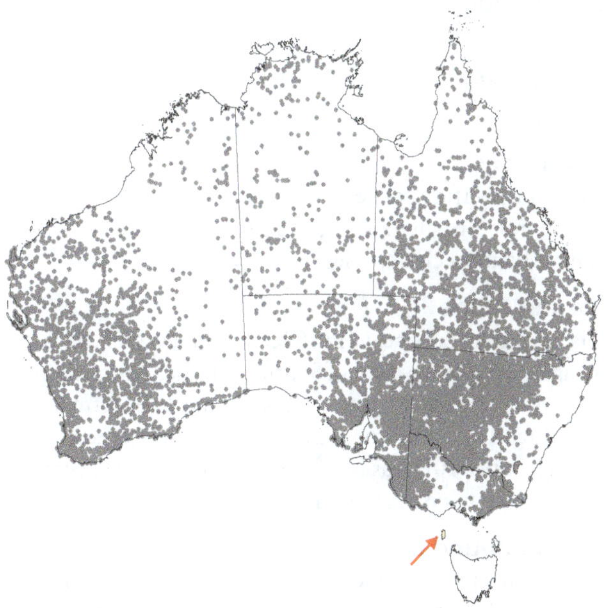

Conservation status

Extinct

Reasons for listing

This species was last seen in the wild in about 1805

Status 2000

Extinct

Status 1990

Extinct

Taxonomy

No infraspecific taxa described

Taxonomic uniqueness: high (2 genera/family, 3 species/genus, 1 subspecies/species)

Former range

Endemic to King I. in Bass Strait (Department of Sustainability, Environment, Water, Population and Communities 2011).

Former abundance

It was considered common at the beginning of the 19th century, but no population estimate was ever made (Department of Sustainability, Environment, Water, Population and Communities 2011).

Ecology

None of the few descriptions of the species' appearance or its behaviour are first hand. Even when alive, there was much confusion about this Emu species (Pfennigwerth 2010). While emus were recorded along the shores and near lagoons on the island, feeding on berries and seaweed (Brasil 1914), it was also described as inhabiting dense forest (Alexander 1921).

Reasons for extinction

Discovered by the Baudin expedition in 1802, the King Island Emu became extinct in the wild in about 1805, probably as a result of hunting with dogs by seal hunters (Brasil 1914). Two or 3 individuals were brought back to France in 1804 and were kept in captivity, the last one dying in 1822 (Pfennigwerth 2010).

Bibliography

Alexander WB (1921) Notes on the fauna of King Island from the logbooks of the 'Lady Nelson'. *Emu* **21**, 318–319.

Brasil L (1914) The Emu of King Island. *Emu* **14**, 88–97.

Department of Sustainability, Environment, Water, Population and Communities (2011) *Dromaius ater.* In 'Species Profile and Threats Database'. Department of Sustainability, Environment, Water, Population and Communities, Canberra. Retrieved 24 February, 2011 from <http://www.environment.gov.au/sprat>.

Pfennigwerth S (2010) 'The mighty cassowary': the discovery and demise of the King Island emu. *Archives of Natural History* **37**, 74–90.

Species Conservation Summary

Kangaroo Island Emu

Dromaius baudinianus S. A. Parker, 1984
Casuariidae

Conservation status
Extinct

Reasons for listing
This species was probably extinct by 1827

Status 2000
Extinct

Status 1990
Extinct

Taxonomy
No infraspecific taxa described
Taxonomic uniqueness: high (2 genera/family, 3 species/genus, 1 subspecies/species)

Former range
Discovered by Europeans in 1802, this emu was endemic to Kangaroo I., off the coast of South Australia. It is presumed to have occurred across the whole island (Department of Sustainability, Environment, Water, Population and Communities 2011), as is reflected in modern place names (Emu Bay, Ravine des Casoars).

Former abundance
When found it was described as numerous (Morgan and Sutton 1928). The emus taken to Paris by Captain Baudin in 1803, of which one surviving until 1822 (Howchin 1926) are now thought to have been King Island Emu *D. ater* (Pfennigwerth 2010).

Ecology
Little is known about the species' habitat preferences. It was mostly recorded on beaches but also occurred in forest (Howchin 1926). Reports of extensive grasslands on Kangaroo I. appear to have been for the purpose of attracting settlers, and were later found to be grossly inaccurate (Hope 1968). Presence of emus in such grasslands (Parker *et al.* 1979) must also be viewed sceptically. There are no records of the species' diet or reproduction.

Reasons for extinction
The last published record of the Kangaroo Island Emu was in 1819, it was probably extinct by 1827 and certainly extinct by 1836 (Department of Sustainability, Environment, Water, Population and Communities 2011). Its demise has been attributed to hunting and excessive burning by settlers in order to create pasture (Ashby 1924; Morgan and Sutton 1928).

Bibliography
Ashby E (1924) Notes on extinct or rare Australian birds, with suggestions as to some of the causes of their disappearance. *Emu* **23**, 294–298.

Department of Sustainability, Environment, Water, Population and Communities (2011) *Dromaius baudinianus.* In 'Species Profile and Threats Database'. Department of Sustainability, Environment, Water, Population and Communities, Canberra. Retrieved 24 February, 2011 from <http://www.environment.gov.au/sprat>.

Hope P (1968) *The Voyage of the Africaine.* Heinemann Educational Australia, South Yarra.

Howchin W (1926) Some references to the literature concerning the extinct emus of Kangaroo Island and elsewhere. *South Australian Ornithologist* **8**, 244–253.

Morgan AM, Sutton J (1928) A critical description of some recently discovered bones of the extinct Kangaroo Island Emu (*Dromaius diemenianus*). *Emu* **28**, 1–19.

Parker SA, Eckert HJ, Ragless GB, Cox JB, Reid NCH (1979) *An Annotated Checklist of the Birds of South Australia. Part One: Emus to Spoonbills.* South Australian Ornithological Association, Adelaide.

Pfennigwerth S (2010) 'The mighty cassowary': the discovery and demise of the King Island emu. *Archives of Natural History* **37**, 74–90.

Species Conservation Summary

Malleefowl

Leipoa ocellata Gould, 1840

Megapodiidae

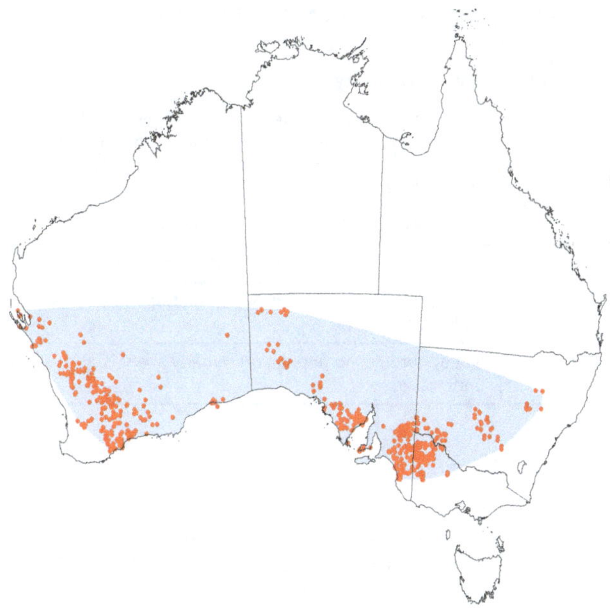

Conservation status

Vulnerable A2bce

Reasons for listing

Estimated population decline of 30–50% over the last 3 generations (48 years) based on counts of active mounds and habitat availability

Status 2000

Vulnerable A2bce+3ce+4bce

Status 1990

Vulnerable A2bce+3ce+4bce

Taxonomy

No infraspecific taxa recognised

Taxonomic uniqueness: high (6 genera/family, 1 species/genus, 1 subspecies/species)

Range

Occurs in scattered locations across much of southern mainland Australia from the west coast to the Great Dividing Range in the east (Department of Sustainability Environment Water Population and Communities 2011). The largest contiguous area of habitat stretches east from the Western Australian wheatbelt but almost all the highest quality habitat in both eastern and western Australia has been cleared for agriculture. Since 1981, the species' range contracted by 22% in eastern Australia (New South Wales and Victoria), 26% in South Australia and around 30% in WA (Benshemesh 2007; Parsons *et al.* 2008). The species may be extinct in the Northern Territory where no confirmed records since the 1950s (Pavey 2006). The species still occurs in the eastern half of the WA Wheatbelt, but there may be a 70-year time lag before the effects of agricultural development result in their complete disappearance from an area.

Abundance

There are no accurate estimates of population size. Garnett and Crowley (2000) estimated the population size at 100 000 based on a density of 1–2 pairs/km^2 over an AOO of 40 000 km^2. Over the past decade there has been a general increase in Vic and south-west NSW and numbers have leveled out in SA, especially since 2007 following the breaking of a long drought. In Vic, numbers in 2010 were higher than they had been for 15 years at 34 sites that have been monitored for that long (J. Benshemesh *in litt.*). However, the long generation time and historical losses means the species still qualifies for Vulnerable.

Ecology

Inhabits semi-arid shrublands and low woodlands dominated by mallee eucalypts and/or acacias (Benshemesh 2007), requiring a sandy substrate and abundant leaf litter for breeding. Malleefowl are highly productive, but rainfall has an important influence on their fecundity and survival (Benshemesh 2007). Clutch size is highly variable, consisting of 3 to 35 eggs, with an average of 15 to 20 (Priddel and Wheeler 2005; Department of Sustainability, Environment, Water, Population and Communities 2011). Malleefowl are generalist and opportunistic, feeding on seeds, flowers and fruits of shrubs (especially legumes), herbs, invertebrates, tubers and fungi (Benshemesh 2007) and their diet is locality specific (Reichelt and Jones 2008). A generation time of 16.0 years (BirdLife International 2011) is derived from an age at first breeding of 3.5 years (Jones *et al.* 1995) and a maximum longevity of 28.5 years (Benshemesh 2007).

Threats

Clearance for agriculture has eliminated and fragmented much of the Malleefowl habitat, resulting in localised extinctions and fragmented populations. The remaining isolated populations are now mostly in suboptimal habitat, since the mallee on the best soil has been cleared, and are vulnerable to catastrophic events, such as bushfires. Habitat degradation continues with overgrazing by both domestic and feral herbivores, and in some areas, kangaroos *Macropus* spp., sometimes affecting food availability, especially for the precocial young. Breeding is usually reduced for about 40 years after a fire event, especially those that leave few patches. While foxes *Vulpes vulpes* kill Malleefowl at all stages of their life cycle (Priddel and Wheeler 2003), fox control generally has little impact on Malleefowl numbers (Benshemesh *et al.* 2007). Malleefowl are also exposed to predation by feral cats *Felis catus* and raptors, especially in open habitats. Cat numbers can increase after fox baiting by meso-predator release (Short 2004). Feeding on spilt grain or road-

Current eligibility against IUCN Red List Criteria

IUCN category	Criteria eligibility
A	Vulnerable: population decline of 30–49% over the last 3 generations (48 years) inferred from monitoring and suspected from deteriorating habitat quality and effects of introduced taxa
B	Not applicable: EOO>20 000 km^2, AOO >2000 km^2
C	Not applicable: population >10 000 mature individuals
D	Not applicable: population >1000 mature individuals, >5 locations
E	Not applicable: no population viability analysis undertaken

IUCN Red List assessment data

	Estimate	Reliability
Extent of occurrence trend	2 250 000 km^2 stable	high high
Area of occupancy trend	40 000 km^2 decreasing	medium high
No. of mature individuals trend	95 000 decreasing	low high
No. subpopulations	100	low
No. locations	>10	low
Largest subpopulation	30 000	low
Generation time	16.0 years	low
Global population share	100%	low

sides make them prone to collision with vehicles (Department of Sustainability, Environment, Water, Population and Communities 2011).

Conservation objectives

1. Major existing populations retained
2. Breeding densities increased at selected sites
3. Area of occupancy expanded

Information required

1. Circumstances under which fox control is necessary for maintenance of Malleefowl populations
2. Population and demographic trends at key sites
3. Longevity, recruitment and turnover of breeding Malleefowl
4. Minimum population size for subpopulation persistence
5. Effectiveness of translocation, captive-rearing and breeding programs

Management actions required

1. Retain areas that support Malleefowl, and those that support Malleefowl habitat, and protect them from clearing
2. Encourage land-holders to undertake conservation covenants and similar agreements
3. Remove goats and sheep from conservation reserves, and keep them at low numbers
4. Close or fence artificial sources of water in conservation reserves
5. Erect stock exclusion fencing around Malleefowl habitat
6. Reduce rabbits where abundant in or near Malleefowl habitat
7. Work with graziers to reduce grazing by domestic stock in Malleefowl habitat
8. Reduce the incidence of large fires, and promote fire patchiness, in reserves and other areas containing Malleefowl
9. Record and centralise details of fox control that is undertaken in areas where Malleefowl abundance being monitored
10. Concentrate fox control on areas where foxes suspected of causing declines
11. Maintain or establish habitat corridors between patches of habitat suitable for Malleefowl
12. Work with farmers near Malleefowl habitat remnants to ensure some grain is grown annually
13. Minimise the amount of grain spilt on roadsides through Malleefowl habitat and erect warning signs where Malleefowl road fatalities are likely to occur
14. Monitor population and demographic trends at key sites
15. Assemble and analyse all existing trend data
16. Coordinate management through a national conservation team

Bibliography

Benshemesh J (2007) 'National recovery plan for Malleefowl'. Department for Environment and Heritage, South Australia.

Benshemesh J, Barker R, Macfarlane R (2007) 'Trend analysis of Malleefowl monitoring data (revised version)'. Report to the Victorian Malleefowl Recovery Group and the Mallee Catchment Management Authority, Mildura.

BirdLife International (2011) 'Species factsheet: *Leipoa ocellata*'. Retrieved 17 January, 2011 from <http://www.birdlife.org/>.

Department of Sustainability, Environment, Water, Population and Communities (2011) *Leipoa ocellata*. In 'Species Profile and Threats Database'. Department of Sustainability, Environment, Water, Population and Communities, Canberra. Retrieved 24 February, 2011 from <http://www.environment.gov.au/sprat>.

Garnett S, Crowley GM (2000) *The Action Plan for Australian Birds 2000*. Environment Australia, Canberra.

Jones DN, Dekker RWRJ, Roselaar CS (1995) *The Megapodes*. Oxford University Press, Oxford.

Parsons BC, Short JC, Roberts JD (2008) Contraction in the range of Malleefowl (*Leipoa ocellata*) in Western Australia: a comparative assessment using presence-only and presence–absence datasets. *Emu* **108**, 221–231.

Pavey CR (2006) 'Malleefowl'. Department of Natural Resources, Environment and the Arts, Darwin. Retrieved 10 April, 2011 from <http://www.nt.gov.au/nreta/wildlife/animals/threatened/pdf/birds/malleefowl_cr.pdf>.

Priddel D, Wheeler R (2003) Nesting activity and demography of an isolated population of malleefowl (*Leipoa ocellata*). *Wildlife Research* **30**, 451–464.

Priddel D, Wheeler R (2005) Fecundity, egg size and the influence of rainfall in an isolated population of malleefowl (*Leipoa ocellata*). *Wildlife Research* **32**, 639–648.

Reichelt RC, Jones DN (2008) Long-term observations of the diet of the Malleefowl *Leipoa ocellata* near the Little Desert, Western Victoria. *Australian Field Ornithology* **25**, 22–30.

Short J (2004) Conservation of the Malleefowl: are there lessons from the successful conservation of native mammals by intensive fox control? *Proceedings of the National Malleefowl Forum 2004.*

Comments received from

Joe Benshemesh, Rod Kavanagh, David Priddel

Subspecies Conservation Summary

Cape Barren Goose (south-western)

Cereopsis novaehollandiae grisea (Storr 1980)

Anatidae

Conservation status

Vulnerable D1

Reasons for listing

Population <1000 mature individuals at more than 5 locations and not declining

Status 2000

Vulnerable D1

Status 1990

Vulnerable D1

Taxonomy

C. n. novaehollandiae (south-eastern Australia, Tasmania, Kangaroo I. and Bass Strait Is) is Least Concern, as is the species

Taxonomic uniqueness: medium (50 genera/family, 1 species/genus, 2 subspecies/species)

Range

Southern Western Australia, breeding on many islands in the Archipelago of the Recherche, but found in small numbers on the mainland from Busselton to the Nullarbor Plain. A survey in 1993 found them on 79 of the 232 islands and rocks surveyed in the Archipelago of the Recherche, 2 small flocks on the mainland opposite and a small number on Red I. 200 km to the west (Halse *et al.* 1995).

Abundance

The single comprehensive survey in 1993 found 631 birds, 612 on the Archipelago of the Recherche, 15 birds on the mainland and 4 on Red I. (Halse *et al.* 1995). There is no new information since 2000 (Garnett and Crowley 2000).

Ecology

Lives primarily in grassland on rocky islands where apparently mostly sedentary. Occasionally visit pastures and beaches on the mainland (Halse *et al.* 1995). They lay 4–5 eggs in nests among grass tussocks (Marchant and Higgins 1990). A generation time of 16.1 years (BirdLife International 2011) is derived from an age at first breeding of 3.0 years, extrapolated from congeners, and a maximum longevity of 29.1 years, based on banding data (Australian Bird and Bat Banding Scheme).

Threats

The small population is vulnerable to extremes of weather, particularly hot summers (Halse *et al.* 1995). Large numbers of birds were hunted for food before 1937, when this was prohibited (Garnett 1992). In 1991, many geese died of starvation or heat stress during a drought and exceptionally high temperatures (Shaughnessy and Haberley 1994; Halse *et al.* 1995). Should the climate in south-western Australia become hotter and drier, the Archipelago may become less suitable for Cape Barren Geese.

Conservation objectives

1. Population retained

Information required

1. Population trends, particularly in counts undertaken following heatwaves

Management actions required

1. Monitor population following heatwaves

Bibliography

BirdLife International (2011) 'Species factsheet: *Cereopsis novaehollandiae*'. Retrieved 17 January, 2011 from <http://www.birdlife.org/>.

Garnett S, Crowley GM (2000) *The Action Plan for Australian Birds 2000.* Environment Australia, Canberra.

Current eligibility against IUCN Red List Criteria

IUCN category	Criteria eligibility
A	Not applicable: past, current or future population declines are thought unlikely to exceed 20% in any 3-generation period
B	Not applicable: >10 locations, not severely fragmented, population fluctuations not extreme
C	Not applicable: no continuing population decline or extreme fluctuations
D	Vulnerable: population 250–1000 mature individuals
E	Not applicable: no population viability analysis undertaken

IUCN Red List assessment data

	Estimate	Reliability
Extent of occurrence trend	9300 km^2 stable	high high
Area of occupancy trend	100 km^2 stable	medium high
No. of mature individuals trend	650 stable	medium medium
No. subpopulations	1	medium
No. locations	1	medium
Generation time	16.1 years	low
Global population share	100%	high

Garnett ST (1992) *The Action Plan for Australian Birds.* Australian National Parks and Wildlife Service, Canberra.

Halse SA, Burbidge AA, Lane JAK, Haberley B, Pearson GB, Clarke A (1995) Size of the Cape Barren goose population in Western Australia. *Emu* **95**, 77–83.

Marchant S, Higgins PJ (Eds) (1990) *Handbook of Australian, New Zealand and Antarctic Birds. Volume 1: Ratites to Ducks.* Oxford University Press, Melbourne.

Shaughnessy PD, Haberley B (1994) Surveys of Cape Barren Geese *Cereopsis novaehollandiae* in Western Australia, 1987–1992. *Corella* **18**, 8–13.

Comments received from

Barry Baker, Andrew Burbidge, Sarah Comer, Stuart Halse

Species Conservation Summary

Blue-billed Duck

Oxyura australis Gould, 1836

Anatidae

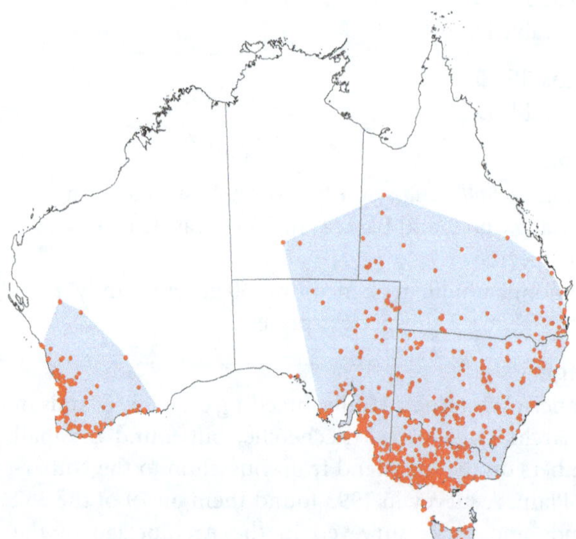

Conservation status

Near Threatened C1

Reasons for listing

Population may be as low as 10 000 and monitoring suggests it may be declining by as much as 10% within 3 generations (18 years)

Status 2000

Least Concern

Reason for change in 2010: declines becoming apparent

Status 1990

Least Concern

Taxonomy

No infraspecific taxa described

Taxonomic uniqueness: medium (50 genera/family, 6 species/genus, 1 subspecies/species)

Range

Across much of south-eastern Australia from southern Queensland to eastern South Australia including entire Murray-Darling basin and through wetter parts of south-western Australia (Barrett *et al.* 2003). The scatter of inland records suggest at least some interchange each generation, if infrequent, so 1 subpopulation.

Abundance

Maximum counts at coastal Victorian drought refuges include 13 000 at Western Treatment Plant, Werribee in 2000–2006 (ARI database on behalf of Melbourne Water Corporation), 3100 at Devilbend Reservoir, Mornington in 2006 (IBA database) with older records of 2000 at Lake Bindegolly in Qld in 1997 (QPWS unpublished data) and 1200 at Peel-Harvey Estuary in WA in 1983 (Jaensch *et al.*

Current eligibility against IUCN Red List Criteria

IUCN category	Criteria eligibility
A	Not applicable: past, current or future population declines are thought unlikely to exceed 20% in any 3-generation period
B	Not applicable: EOO >20 000 km^2, AOO >2000 km^2
C	Near Threatened: population near 10 000 mature individuals, estimated continuing decline near 10% in 3 generations (18 years)
D	Not applicable: population >1000 mature individuals, >5 locations
E	Not applicable: no population viability analysis undertaken

IUCN Red List assessment data

	Estimate	Reliability
Extent of occurrence trend	3 300 000 km^2 stable	high high
Area of occupancy trend	5000 km^2 decreasing	low medium
No. of mature individuals trend	15 000 decreasing	medium low
No. subpopulations	1	high
No. locations	>10	high
Generation time	6.0 years	low
Global population share	100%	high

1988). The National Waterbird Survey in 2008 recorded a national total of only 56 birds and a weakly significant decline 1983–2010 but this species is readily overlooked in aerial surveys (R. Kingsford & J. Porter *in litt.* 2011). Population estimated here follows Wetlands International (Wetlands International 2006): 10 000 in east Australia and 5000 in Western Australia.

Ecology

Blue-billed Ducks are found on temperate, fresh to saline, terrestrial wetlands, including reservoirs, sewerage ponds, rivers, salt lakes and saltpans. They prefer deep, permanent open water, within or near dense vegetation and eat aquatic insect larvae, and seeds and leaves of freshwater plants (Marchant and Higgins 1990). Many birds appear to nest on inland wetlands, such as in the Channel Country of west Queensland and north-east South Australia, when flooded after rain and move to permanent waterbodies, often artificial wetlands such as reservoirs, in dry periods (Costelloe *et al.* 2004; Jaensch *et al.* 2010). A generation time of 6.0 years (BirdLife International 2011) is derived from an age at first breeding of 1.3 years, a maximum longevity of 13.6 years, both extrapolated from congeners, and an adult annual survival of 70%, extrapolated from mean values for other genera in Anatidae.

Threats

Threatened by diversion of water for irrigation, drainage of swamps for agriculture or urban development, habitat degradation by introduced fish, and reduced water inflows and salinisation because of land use and climate change (Marchant and Higgins 1990; Kingsford 2000). A small number are probably shot by accident during the duck hunting season (Loyn 1991), which continues in Victoria.

Conservation objectives

1. Stable numbers of birds recorded at key sites
2. Key breeding sites protected

Information required

1. Population trends, particularly from drought to drought to assess sustainability of post-drought recovery

Management actions required

1. Maintain survey and monitoring activities
2. Identify principal breeding wetlands
3. Protect key wetlands against degradation
4. Advocate for continued natural flows of rivers in Lake Eyre Basin

Bibliography

Barrett G, Silcocks A, Barry S, Cunningham R, Poulter R (2003) *The New Atlas of Australian Birds.* Royal Australian Ornithologists Union, Hawthorn East.

BirdLife International (2011) 'Species factsheet: *Oxyura australis*'. Retrieved 17 January, 2011 from <http://www.birdlife.org/>.

Costelloe JF, Hudson PJ, Pritchard JC, Puckridge JT, Reid JRW (2004) 'ARIDFLO scientific report: environmental flow requirements of arid zone rivers with particular reference to the Lake Eyre Drainage Basin'. School of Earth and Environmental Sciences, University of Adelaide, Adelaide. Final Report to South Australian Department of Water, Land and Biodiversity Conservation and Commonwealth Department of Environment and Heritage.

Jaensch R, Reid J, Kingsford R (2010) Free-flowing rivers of the Channel Country. Conservation implications of the 2009 Georgina-Diamantina flood. *Wetlands Australia* **18**, 18–19.

Jaensch RP, Vervest RM, Hewish MJ (1988) Waterbirds in nature reserves of south-western Australia, 1981–1985: reserve accounts. *Royal Australasian Ornithologists Union Report* **30**, 1–290.

Kingsford RT (2000) Ecological impacts of dams, water diversions and river management on floodplain wetlands in Australia. *Austral Ecology* **25**, 109–127.

Loyn RH (1991) Assessing and managing the impact of duck hunting in Victoria – a new approach. *Wildfowl* **42**, 155–161.

Marchant S, Higgins PJ (Eds) (1990) *Handbook of Australian, New Zealand and Antarctic Birds. Volume 1: Ratites to Ducks.* Oxford University Press, Melbourne.

Wetlands International (2006) *Waterbird Population Estimates.* Wetlands International, Wageningen, The Netherlands.

Comments received from

Ian Gynther, Richard Kingsford, John Porter

Species Conservation Summary

Red-tailed Tropicbird

Phaethon rubricauda Boddaert, 1783
Phaethontidae

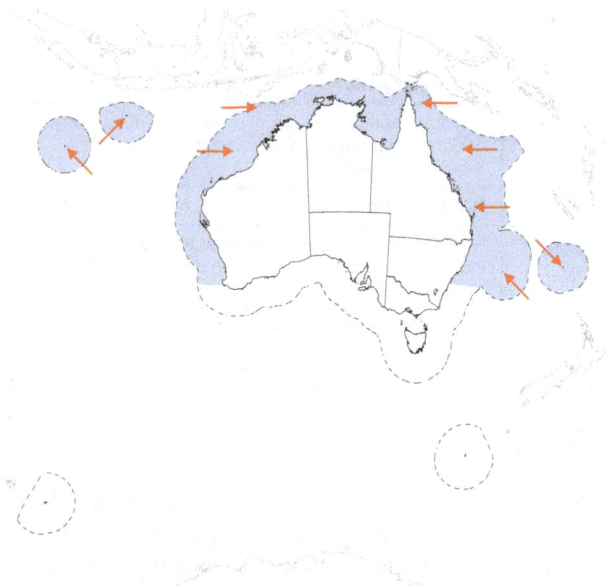

Conservation status

Australian breeding population: Near Threatened A3be+4be
Population visiting Australian territory: Least Concern

Reasons for listing

Australian breeding population: Declines likely in largest population on Christmas I. from cat and Black Rat predation, suspected declines on Raine I. of 20–29% in 3 generations (39 years)

Status 2000

Australian breeding population: Near Threatened A3e
Population visiting Australian territory: Least Concern

Status 1990

Australian breeding population: Least Concern
Reason for change in 2000: evidence of threats on Christmas I.
Population visiting Australian territory: Least Concern

Taxonomy

No infraspecific taxa described
Taxonomic uniqueness: high (1 genus/family, 3 species/genus, 1 subspecies/species)

Range

Breeding on numerous islands in Pacific and Indian Oceans. In Australian territory, breeds on Ashmore Reef, Rowley Shoals, Western Australia on Raine I., Moulter and MacLennan Cays and on the North Reed to Lady Elliot Is. in Queensland, North-east and South-west Herald Cays, Coringa I. and Diamond I. in the Coral Sea, Lord Howe I. and Ball's Pyramid, Cocos-Keeling Is., Christmas I., Norfolk I. and Phillip I. (Marchant and Higgins 1990; Burbidge and Fuller 1996; Driscoll 2010). Formerly bred on Rat I., Rottnest I. off south-west Australia, as well as on the mainland at Busselton (Burbidge and Fuller 1996) and no recent records from Sugarloaf Rock or Houtman Abrolhos, where last seen on Pelsaert I. in 2001 (Surman and Nicholson 2009). Non-breeding birds occur throughout the central Pacific and Indian Oceans, being seen regularly off western and northern coast between Cape Naturaliste, WA and Cape Weymouth, Qld, and east of the Great Barrier Reef, and infrequently elsewhere along the coast (Marchant and Higgins 1990).

Abundance

Largest subpopulations estimated at 1400 pairs at Christmas I., 500–1000 pairs in the Lord Howe I. group, about 200 pairs on Norfolk I. and 500 pairs on the Herald Cays. Small numbers elsewhere (Marchant and Higgins 1990; McAllan *et al.* 2004; Baker and Holdsworth 2009). Numbers on Raine I. have declined by 38% over a 20-year period to 64 pairs (Batianoff and Cornelius 2005) and are assumed to be declining rapidly on Christmas I. based on high rates of nest failure (Beeton *et al.* 2010). The removal of feral and domestic cats *Felis catus* from Lord Howe I., however, appears to have resulted in an increase from about 300 pairs in 1974 (McAllan *et al.* 2004) and the population on North-east Herald Cay has fluctuated in annual monitoring in 1992–2008, but the overall trend is stable (Baker and Holdsworth 2009).

Ecology

Nests alone or in loose colonies on inaccessible islands, stacks, atolls, cays or coastal cliffs. It is a pelagic feeder, preferring waters of intermediate salinity and temperature, mainly taking fish and squid (Marchant and Higgins 1990). A generation time of 13.0 years (BirdLife International 2011) is derived from an age at first breeding of 3.0 years and an annual survival of adults of 90.0% (Schreiber and Burger 2002).

Threats

Predation by cats appears to be the main cause of decline on Christmas I., where nearly all chicks in monitored nests were killed (Ishii 2006; Beeton *et al.* 2010). Introduced Yellow Crazy Ants *Anoplolepis gracilipes* were considered a threat to the Christmas I. population (Garnett and Crowley 2000) but losses from ants have not been demonstrated, and have now been temporarily reduced. However, changes in the ecology of the island, particularly the prevalence of introduced predators, could be detrimental to the tropicbirds should ant control be relaxed. Inflated numbers of gulls, human interference, and mining disturbance have also been implicated in individual nest failures and/or desertion of breeding colonies elsewhere (Marchant and Higgins 1990). The decline in south-western Australia may be related to variations in the Leeuwin Current and those on Raine I. may be related to sea level rise or changes in local food availability (Garnett and Crowley 2000;

Current eligibility against IUCN Red List Criteria

IUCN category	Criteria eligibility
Australian breeding population	
A	Near Threatened: current and anticipated population declines of 20–29% over periods of 3 generations (39 years) inferred from monitoring and effects of introduced taxa
B	Not applicable: >10 locations, not severely fragmented, population fluctuations not extreme
C	Not applicable: population 2500–10 000 mature individuals, continuing decline not observed or estimated, largest subpopulation >1000 mature individuals, >1 subpopulation
D	Not applicable: population >1000 mature individuals, >5 locations
E	Not applicable: no population viability analysis undertaken

IUCN Red List assessment data

	Estimate	Reliability
Australian breeding population		
Extent of occurrence trend	2 500 000 km^2 stable	high medium
Area of occupancy trend	50 km^2 stable	medium medium
No. of mature individuals trend	6000 decreasing	low low
No. subpopulations	>10	medium
No. locations	>10	high
Largest subpopulation	2800	low
Generation time	13.0 years	medium
Global population share	1–10%	low
Level of genetic exchange	low	medium

Batianoff and Cornelius 2005). Reproductive success and nesting habitat are also affected by cyclones (Hennicke and Flachsbarth 2009), though such events are rare at most breeding sites.

Conservation objectives

1. Stable breeding populations on Christmas and Raine Is.
2. Cats controlled

Information required

1. Cat control techniques for Christmas I.
2. Cause of decline on Raine I.

Management actions required

1. Continue monitoring major breeding colonies at Christmas, Lord Howe, Norfolk and Raine Is., and North-east Herald Cay
2. Maintain or establish strict quarantine at all major nesting islands
3. Control or eradicate cats on Christmas I.
4. Create a multi-stakeholder team with responsibility for ensuring threatened species are conserved on Christmas I.

Bibliography

Baker B, Holdsworth M (2009) 'Seabird monitoring study at Coringa Herald National Nature Reserve'. Report to the Department of the Environment, Water, Heritage and the Arts, Canberra.

Batianoff GN, Cornelius NJ (2005) Birds of Raine Island: population trends breeding behaviour and nesting habitats. *Proceedings of the Royal Society of Queensland* **112**, 1–29.

Beeton B, Burbidge AA, Grigg G, Harrison P, How RA, Humphries B, McKenzie N, Woinarski J (2010) 'Final report, Christmas Island Expert Working Group to Minister for the Environment, Heritage and the Arts'. Retrieved 1 April, 2010 from <http://www.environment.gov.au/parks/publications/christmas/pubs/final-report.pdf>.

BirdLife International (2011) 'Species factsheet: *Phaethon rubricauda*'. Retrieved 17 January, 2011 from <http://www.birdlife.org/>.

Burbidge AA, Fuller PJ (1996) The Western Australian Department of Conservation and Land Management seabird breeding islands database. In *Proceedings of the Status of Australia's Seabirds. Proceedings of the National Seabird Workshop*, Canberra. (Eds GJB Ross, K Weaver and JC Greig) pp. 73–137. Biodiversity Group, Environment Australia.

Driscoll PV (2010) 'Phase 1 analysis of coastal bird atlas data'. Great Barrier Reef Marine Park Authority.

Garnett S, Crowley GM (2000) *The Action Plan for Australian Birds 2000*. Environment Australia, Canberra.

Hennicke JC, Flachsbarth K (2009) Effects of Cyclone Rosie on breeding Red-tailed Tropicbirds *Phaethon rubricauda* on Christmas Island, Indian Ocean. *Marine Ornithology* **37**, 175–178.

Ishii N (2006) 'A survey of Red-Tailed Tropicbird *Phaethon rubricauda* at the Sitting Room and Rumah inggi, Christmas Island, April–July, 2006'. Report to Parks Australia North, Christmas Island Biodiversity Monitoring Programme.

Marchant S, Higgins PJ (Eds) (1990) *Handbook of Australian, New Zealand and Antarctic Birds. Volume 1: Ratites to Ducks.* Oxford University Press, Melbourne.

McAllan IAW, Curtis BR, Hutton I, Cooper RM (2004) The birds of the Lord Howe Island Group: a review of records. *Australian Field Ornithology* **21**, 1–82.

Schreiber EA, Burger J (2002) *Biology of Marine Birds*. CRC Press, Boca Raton, FL.

Surman CA, Nicholson LW (2009) A survey of the breeding seabirds and migratory shorebirds of the Houtman Abrolhos, Western Australia. *Corella* **33**, 81–98.

Comments received from

Barry Baker, Andrew Burbidge, Janos Hennicke, David James, Mike Smith, John Woinarski

Subspecies Conservation Summary

White-tailed Tropicbird (Christmas Island)

Phaethon lepturus fulvus Brandt, 1840
Phaethontidae

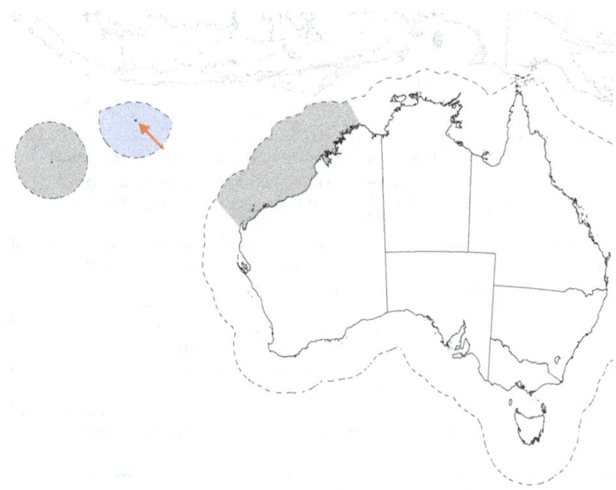

Conservation status

Endangered B2ab(v)

Reasons for listing

Single endemic population breeds in an area of 10–500 km^2; predation by Black Rats and cats are probably reducing the population, which may also be being depleted by hybridization

Status 2000

Endangered B2ab(v)

Status 1990

Vulnerable D2
Reason for change in 2000: declines first observed

Taxonomy

P. l. lepturus (Indian Ocean) is Endangered in Australian territory (Cocos-Keeling and Ashmore Reefs), *P. l. dorotheae* (tropical Pacific Ocean) is a non-breeding visitor to eastern Australian waters. There are 3 extralimital subspecies. The species is Least Concern.
Taxonomic uniqueness: high (1 genus/family, 3 species/genus, 6 subspecies/species)

Range

Endemic to Christmas I. where it breeds throughout the forested area and on the cliffs, presumably foraging over the Indian Ocean when away from the island (Dunlop *et al.* 2001).

Abundance

Population estimates in the last 25 years vary from 600–12 000 pairs but there have been no definitive estimates (Beeton *et al.* 2010) and nor are there estimates of trends. The figure here is the average of the estimate of Dunlop (1988) of 12 000–24 000. However, unpublished data suggest that numbers have declined substantially since 2000 (J.Hennicke pers. comm.).

Ecology

Nests both in tree hollows in rainforest, which are scarce, and rock crevices on the terraces. It forages singly, in warm waters, feeding on fish and cephalopods (Dunlop *et al.* 2001). A generation time of 11.1 years (BirdLife International 2011) is derived from an age at first breeding of 4.0 years and an annual survival of adults of 86.0% (Schreiber and Burger 2002).

Threats

Predation by feral cats *Felis catus* and, to a lesser extent, Black Rats *Rattus rattus*, is likely to be the main cause of any decline on Christmas I. (J. Hennicke pers. comm.; Beeton *et al.* 2010). Introduced Yellow Crazy Ants *Anoplolepis gracilipes* were considered a threat to Christmas I. (Garnett and Crowley 2000) but losses from ants have not been demonstrated, and ants have now been temporarily reduced, though changes in the ecology of the island, particularly the prevalence of introduced predators, could be detrimental to the tropicbirds should ant control be relaxed. The presence of all white birds at Christmas I. may also be increasing and could represent either a shift in diet, if this is the cause of the yellow wash to the plumage, or an incursion of *P. l. lepturus.*

Conservation objectives

1. Stable or increasing population
2. Black Rats and feral cats eradicated or controlled

Information required

1. Cat control techniques for Christmas I.
2. Relationship to *P. l. lepturus* and the cause of the distinctive colouring of *P. l. fulvus*

Management actions required

1. Increase intensity and frequency of monitoring
2. Upgrade quarantine standards for Christmas I.
3. Control or eradicate feral cats
4. Eradicate Black Rats
5. Create a multi-stakeholder team with responsibility for ensuring threatened species are conserved on Christmas I.

Bibliography

Beeton B, Burbidge AA, Grigg G, Harrison P, How RA, Humphries B, McKenzie N, Woinarski J (2010) 'Final report, Christmas Island Expert Working Group to Minister for the Environment, Heritage and the Arts'. Retrieved 1 April, 2010 from <http://www.environment.gov.au/parks/publications/christmas/pubs/final-report.pdf>.

BirdLife International (2011) 'Species factsheet: *Phaethon lepturus*'. Retrieved 17 January, 2011 from <http://www.birdlife.org/>.

Dunlop JN (1988) 'The status and biology of the Golden Bosunbirds *Phaethon lepturus fulvus*.' Unpublished report. Australian National Parks and Wildlife Service, Christmas Island.

Current eligibility against IUCN Red List Criteria

IUCN category	Criteria eligibility
A	Near Threatened: current and anticipated population declines of 20–29% over periods of 3 generations (33 years) suspected from informal observations, effects of introduced taxa and hybridisation
B	Endangered: AOO <500 km^2, at ≤5 locations with plausible future threat, suspected continuing decline in no. mature individuals
C	Vulnerable: population 2500–10 000 mature individuals, suspected continuing decline, 100% in 1 subpopulation
D	Vulnerable: ≤5 locations with plausible future threat
E	Not applicable: no population viability analysis undertaken

Dunlop JN, Surman CA, Wooller RD (2001) The marine distribution of seabirds from Christmas Island, Indian Ocean. *Emu* **101**, 19–24.

Garnett S, Crowley GM (2000) *The Action Plan for Australian Birds 2000.* Environment Australia, Canberra.

Schreiber EA, Burger J (2002) *Biology of Marine Birds.* CRC Press, Boca Raton, FL.

IUCN Red List assessment data

	Estimate	Reliability
Extent of occurrence trend	140 000 km^2 stable	high high
Area of occupancy trend	134 km^2 stable	medium medium
No. of mature individuals trend	18 000 decreasing	low low
No. subpopulations	1	high
No. locations	1	high
Generation time	11.1 years	low
Global population share	100%	high

Comments received from

Barry Baker, Andrew Burbidge, Janos Hennicke, David James, Mike Smith, John Woinarski

Subspecies Conservation Summary

White-tailed Tropicbird (Indian Ocean)

Phaethon lepturus lepturus Daudin, 1802

Phaethontidae

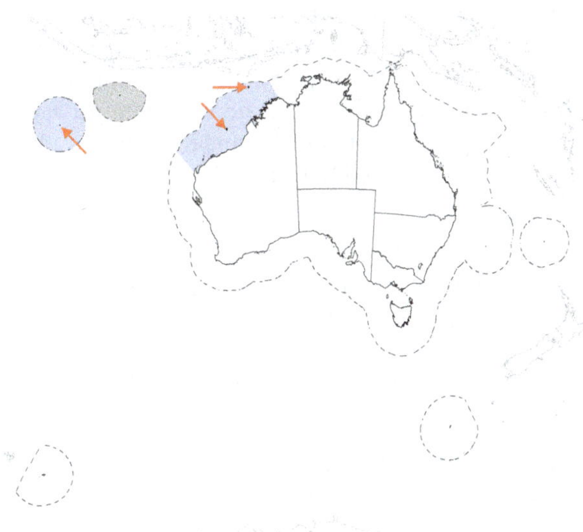

Conservation status

Australian breeding population: Endangered D

Populations visiting Australian territories: Least Concern

Reasons for listing

Australian breeding population: Population 50–250 mature individuals

Status 2000

Australian breeding population: Endangered D

Population visiting Australian territory: Least Concern

Status 1990

Australian breeding population: Endangered D

Population visiting Australian territory: Least Concern

Taxonomy

P. l. fulvus (Christmas I.) is Endangered, *P. l. dorotheae* (tropical Pacific Ocean) is a non-breeding visitor to eastern Australian waters; 3 extralimital subspecies. The species is Least Concern.

Taxonomic uniqueness: high (1 genus/family, 3 species/genus, 6 subspecies/species)

Range

In Australian territory, the principal breeding site is Cocos (Keeling) Is., mostly North Keeling I., but occasionally Home I (Stokes *et al.* 1984; Hadden 2006). Also breeds on 3 islands of Ashmore Reef, on Rowley Shoals, Western Australia and may visit Christmas I. (Marchant and Higgins 1990; Burbidge and Fuller 1996). Also breeds on islands across the northern Indian Ocean; foraging in oceanic water throughout the region.

Current eligibility against IUCN Red List Criteria

IUCN category	Criteria eligibility
Australian breeding population	
A	Not applicable: past, current or future population declines are thought unlikely to exceed 20% in any 3-generation period
B	Not applicable: EOO, AOO, habitat, locations or population not declining, population fluctuations not extreme
C	Not applicable: no continuing population decline or extreme fluctuations
D	Endangered: population 50–250 mature individuals
E	Not applicable: no population viability analysis undertaken

IUCN Red List assessment data

	Estimate	Reliability
Australian breeding population		
Extent of occurrence trend	1 400 000 km^2 stable	medium high
Area of occupancy trend	4 km^2 stable	high high
No. of mature individuals trend	120 stable	low high
No. subpopulations	3	high
No. locations	3	high
Largest subpopulation	120	low
Generation time	11.1 years	medium
Global population share	1–10%	medium
Level of genetic exchange	medium	medium

Abundance

Total population is about 120 breeding individuals including 40–50 pairs North Keeling I., occasionally a pair on Home I., up to 10 pairs on Ashmore Reef and 1 pair on Rowley Shoals (Stokes *et al.* 1984; Marchant and Higgins 1990; Burbidge and Fuller 1996). However, only 25 birds have been seen in the air at any one time on North Keeling I. (J. Reid *in litt.*), so the population there may be smaller.

Ecology

Nests both in tree hollows and in rock crevices. It forages singly, in warm waters, feeding on fish and cephalopods (Dunlop *et al.* 2001). A generation time of 11.1 years (BirdLife International 2011) is derived from an age at first breeding of 4.0 years and an annual survival of adults of 86.0% (Schreiber and Burger 2002).

Threats

The main potential threat to White-tailed Tropicbirds on North Keeling is the accidental introduction of exotic predators. Predation by feral cats *Felis catus* and Black Rats *Rattus rattus* has been implicated in nest failures and/or desertion of breeding colonies (BirdLife International 2011).

Conservation objectives

1. A viable and secure population on North Keeling I

Information required

1. Population census and trends on North Keeling I

Management actions required

1. Continue to protect North Keeling I as outlined in management plan
2. Monitor other populations when possible
3. Monitor by transect every 6 months for 3 years, then review

Bibliography

BirdLife International (2011) 'Species factsheet: *Phaethon lepturus*'. Retrieved 17 January, 2011 from <http://www.birdlife.org/>.

Burbidge AA, Fuller PJ (1996) The Western Australian Department of Conservation and Land Management seabird breeding islands database. In *Proceedings of The Status of Australia's Seabirds. Proceedings of the National Seabird Workshop*, Canberra. (Eds GJB Ross, K Weaver and JC Greig) pp. 73–137. Biodiversity Group, Environment Australia.

Dunlop JN, Surman CA, Wooller RD (2001) The marine distribution of seabirds from Christmas Island, Indian Ocean. *Emu* **101**, 19–24.

Hadden D (2006) Cocos (Keeling) Island birds. *Wingspan* **16**, 34–37.

Marchant S, Higgins PJ (Eds) (1990) *Handbook of Australian, New Zealand and Antarctic Birds. Volume 1: Ratites to Ducks.* Oxford University Press, Melbourne.

Schreiber EA, Burger J (2002) *Biology of Marine Birds.* CRC Press, Boca Raton, FL.

Stokes AS, Shiels W, Dunn K (1984) Birds of the Cocos (Keeling) Islands, Indian Ocean. *Emu* **84**, 23–28.

Comments received from

Andrew Burbidge, Rohan Clarke, Julian Reid

Subspecies Conservation Summary

White-throated Pigeon (Lord Howe Island)

Columba vitiensis godmanae (Mathews, 1915)
Columbidae

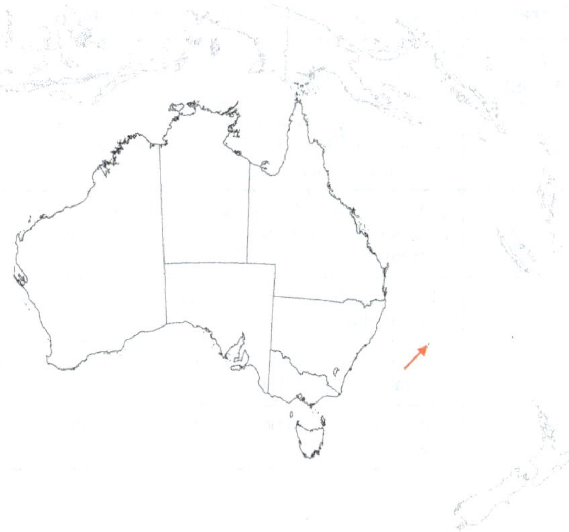

Conservation status
Extinct

Reasons for listing
The last individual of this subspecies was recorded in 1853

Status 2000
Extinct

Status 1990
Extinct

Taxonomy
Seven other subspecies occur in South-East Asia and on other Pacific islands. The species is Least Concern. Taxonomic uniqueness: low (42 genera/family, 35 species/genus, 8 subspecies/species)

Former range
This subspecies of the White-throated Pigeon was endemic to Lord Howe I. (Hindwood 1940).

Former abundance
Some of the first visitors to the island described it as abundant (Hindwood 1940). Others, even as early as 1791, had difficulty obtaining live specimens (McKean and Hindwood 1965). Hindwood (1940) suggests thousands were killed for food, but this would have to have been over many years as it is unlikely the island could have sustained such a large population at any particular time. The subspecies was certainly very rare by 1853 and there are no subsequent records (Hindwood 1940). It was probably extinct by 1869 (McAllan *et al.* 2004). The subspecies is known only from paintings; there are no museum specimens.

Ecology
The White-throated Pigeon probably occurred in all wooded habitats throughout the lowlands of Lord Howe I. It is likely that the pigeons fed on fruits and seeds, as do the other subspecies (Hutton 1991).

Reasons for extinction
Large numbers of White-throated Pigeons were killed for food, which probably caused the taxon's extinction (Hindwood 1940).

Bibliography

Hindwood KA (1940) The birds of Lord Howe Island. *Emu* **40**, 1–86.

Hutton I (1991) *Birds of Lord Howe Island, Past and Present*. The Author, Coffs Harbour.

McAllan IAW, Curtis BR, Hutton I, Cooper RM (2004) The birds of the Lord Howe Island Group: a review of records. *Australian Field Ornithology* **21**, 1–82.

McKean JL, Hindwood KA (1965) Additional notes on the birds of Lord Howe Island. *Emu* **64**, 79–97.

Comments received from
David Priddel

Subspecies Conservation Summary

Emerald Dove (Christmas Island)

Chalcophaps indica natalis Lister, 1889
Columbidae

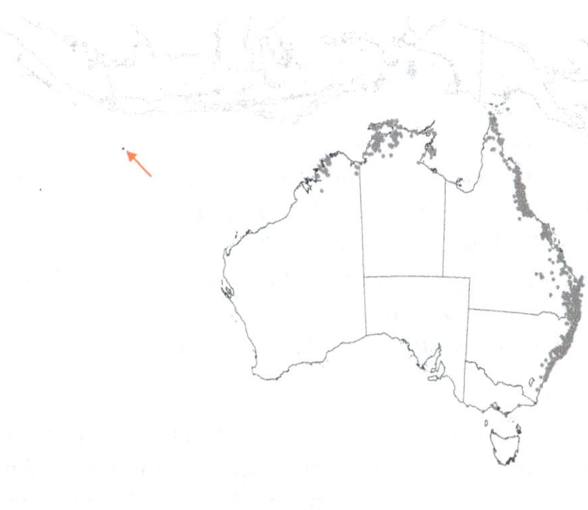

Conservation status
Near Threatened D2

Reasons for listing
Occurs at a single location where susceptible to introduction of predators, competitors or disease, although the risk is now considered low

Current eligibility against IUCN Red List Criteria

IUCN category	Criteria eligibility
A	Not applicable: past, current or future population declines are thought unlikely to exceed 20% in any 3-generation period
B	Not applicable: EOO, AOO, habitat, locations or population not declining, population fluctuations not extreme
C	Not applicable: no continuing population decline or extreme fluctuations
D	Near Threatened: ≤5 locations with plausible but low probability future threat
E	Not applicable: no population viability analysis undertaken

IUCN Red List assessment data

	Estimate	Reliability
Extent of occurrence trend	134 km^2 stable	high high
Area of occupancy trend	134 km^2 stable	high high
No. of mature individuals trend	2000 stable	low low
No. subpopulations	1	high
No. locations	1	high
Generation time	6.5 years	low
Global population share	100%	high

Status 2000

Near Threatened A3ce, D2

Status 1990

Near Threatened D2

Taxonomy

C. i. rogersi (eastern Australia, Lord Howe and Norfolk Is.), *C. i. longirostris* (north-west Australia and Northern Territory) and 7 extralimital subspecies are all Least Concern, as is the species

Taxonomic uniqueness: low (41 genera/family, 2 species/genus, 10 subspecies/species)

Range

Endemic to Christmas I., Indian Ocean where it occurs throughout the island (Department of Sustainability Environment Water Population and Communities 2011). Area of occupancy may have been reduced by clearance for phosphate mining from 1890s to 1980s but it remains relatively abundant in remaining forest and in regrowth sites.

Abundance

There are no good estimates of the population size. The species was recorded in 78% of 123 sites surveyed by James and Retallick (2007). The most recent population estimates, of which the average is used here, are 900–3500 individuals and 1000 pairs (Beeton *et al.* 2010). There is no information on trends, although the species is 9–14 times less numerous in ant-invaded forest than areas where there are no ants (Davis *et al.* 2008).

Ecology

Occupies most forested habitats, including settlements, regrowth forest and terraces, nesting in low, flimsy stick nests. They usually feed on the ground, mostly on seeds (Department of Sustainability, Environment, Water, Population and Communities 2011). A generation time of 6.5 years (BirdLife International 2011) is derived from an average age at first breeding of 1.0 years (extrapolated from other Columbidae) and a maximum longevity in the wild of 12.0 years (Australian Bird and Bat Banding Scheme).

Threats

Like other Christmas I. species, Emerald Doves are potentially threatened by the introduced Yellow Crazy Ant *Anoplolepis gracilipes*, which can alter the whole ecology of sites it occupies by killing the super-numerous Red Crab *Gecaroidea natalis*, as well as by tending scale insects that damage trees (O'Dowd *et al.* 2003; Abbott and Green 2007). The ants have been observed killing Emerald Dove nestlings (Garnett and Crowley 2000) and the species is less numerous in areas with ants than without (Davis *et al.* 2008). Faltering of ant control and infestation of all rainforest could reduce productivity below sustainable levels. Cats *Felis catus* and Black Rats *Rattus rattus* are also likely to kill birds and reduce nesting success, although the species has persisted on other islands where both predators occur. The species may also be affected by the pesticide Fipronil, used for ant control, but this remains unproven (Beeton *et al.* 2010). Quarantine measures on the island are not currently sufficient to prevent invasion by additional unspecified predators. Despite these threats, overall extinction risk is considered low.

Conservation objectives

1. Existing population maintained
2. Yellow Crazy Ant populations constrained to levels that do not threaten subspecies
3. Black Rats and feral cats eradicated or controlled

Information required

1. Alternatives to Fipronil for controlling Yellow Crazy Ants
2. Population trends
3. Levels of ant abundance that will threaten subspecies
4. Effects of rodenticides

Management actions required

1. Continue population monitoring
2. Upgrade quarantine standards for Christmas I.
3. Continue control of Yellow Crazy Ants
4. Eradicate Black Rats
5. Control or eradicate feral cats
6. Create a multi-stakeholder team with responsibility for ensuring threatened species are conserved on Christmas I.

Bibliography

Abbott KL, Green PT (2007) Collapse of an ant-scale mutualism in a rainforest on Christmas Island. *Oikos* **116**, 1238–1246.

Beeton B, Burbidge AA, Grigg G, Harrison P, How RA, Humphries B, McKenzie N, Woinarski J (2010) 'Final report,

Christmas Island Expert Working Group to Minister for the Environment, Heritage and the Arts'. Retrieved 1 April, 2010 from <http://www.environment.gov.au/parks/publications/christmas/pubs/final-report.pdf>.

BirdLife International (2011) 'Species factsheet: *Chalcophaps indica*'. Retrieved 17 January, 2011 from <http://www.birdlife.org/>.

Davis NE, O'Dowd DJ, Green PT, Mac Nally R (2008) Effects of an alien ant invasion on abundance, behavior, and reproductive success of endemic island birds. *Conservation Biology* **22**, 1165–1176.

Department of Sustainability Environment Water Population and Communities (2011) *Chalcophaps indica natalis*. In 'Species Profile and Threats Database'. Department of Sustainability, Environment, Water, Population and Communities, Canberra. Retrieved 24 February, 2011 from <http://www.environment.gov.au/sprat>.

Garnett S, Crowley GM (2000) *The Action Plan for Australian Birds 2000*. Environment Australia, Canberra.

James DJ, Retallick K (2007) 'Forest birds of Christmas Island: a baseline survey of abundance'. Parks Australia North Christmas Island Biodiversity Monitoring Programme: report to the Department of Finance and Administration and the Department of the Environment and Water Resources, Canberra.

O'Dowd DJ, Green PT, Lake PS (2003) Invasional 'meltdown' on an oceanic island. *Ecology Letters* **6**, 812–817.

Comments received from

Andrew Burbidge, David James, Mike Smith, John Woinarski

Species Conservation Summary

Partridge Pigeon

Geophaps smithii (Jardine & Selby, 1830)
Columbidae

Conservation status

Vulnerable A2bc+3c+4c

Reasons for listing

Decline >30% at several monitored sites, probably as a result of change in fire regime, possibly also from predation by feral cats

Status 2000

Near Threatened A2bc+3c+4c
Reason for change in 2010: evidence of severe declines in recent surveys

Status 1990

Near Threatened A2bc+3c+4c

Taxonomy

Both *G. s. smithii* (northern Northern Territory) and *G. s. blaauwi* (Kimberley Region, north-west Western Australia) are Vulnerable
Taxonomic uniqueness: medium (41 genera/family, 3 species/genus, 2 subspecies/species)

Range

G. s. smithii occurs in sub-coastal Northern Territory, from Yinberrie Hills (about 50 km north of Katherine) in the south and Litchfield National Park in the west to western Arnhem Land in the east, with a separate subpopulation on the Tiwi Is. (Melville and Bathurst Is.) with smaller scattered populations elsewhere in the Top End (Woinarski 2004; Woinarski *et al.* 2007). *G. s. blaauwi* occurs in remote areas within 100 km of the coast in the north-west of the Kimberley region, northern Western Australia. Historically the distribution of *G. s. smithii* extended from Kununurra, WA, east to McArthur R., Northern Territory while that of *G. s. blaauwi* extended east to the Durack R. (Higgins and Davies 1996; Woinarski *et al.* 2007). The AOO is estimated here at about 5% of the EOO.

Abundance

Population densities have been estimated at 2 locations where it is comparatively common: 0.33 birds/ha on Melville I. and 0.2 birds/ha in Kakadu National Park (Department of Sustainability, Environment, Water, Population and Communities 2011). The density in the Kimberley is thought to be lower than in the east. Based on the AOO there are over 100 000 birds remaining. However, populations in Kakadu have declined substantially over the last decade (Woinarski 2009), the population on the Tiwi Is. may have declined as a result of clearance for forestry and *G. s. blaauwi* is now extremely scarce in the eastern half of its former range.

Ecology

G. s. smithii lives primarily in open forest and woodland dominated by Darwin Stringybark *Eucalyptus tetrodonta* and Darwin Woollybutt *E. miniata* that has a structurally diverse understorey, usually in areas with a fire regime that promotes a mosaic of fire ages, including wet season burns, which promote grass diversity and year-round seed availability (Fraser *et al.* 2003). *G. s. blaauwi* occurs primarily in open woodland, particularly on the ecotone between the rugged King Leopold Sandstones and alluvial flats. The ground cover is mostly tall grasses, though the pigeons are usually seen feeding in recently burnt areas, by roads and in short grass (Johnstone 1981). Usually lays 2 eggs in nests made on the ground, most often at the base of a clump of grass. Feeds on seeds taken from bare ground, often amidst sparse grasses, but commonly where the ground layer has been removed by fire (Frith 1982). It appears to rely on perennial grass species which set seed relatively early (Fraser 2001). A generation time of 5.6 years (BirdLife International 2011) is derived from an age at first breeding of 1.0 years and a maximum longevity in the wild of 10.1 years, both extrapolated from other Columbidae.

Current eligibility against IUCN Red List Criteria

IUCN category	Criteria eligibility
A	Vulnerable: past, current and anticipated population declines of 30–49% over periods of 3 generations (17 years) estimated from monitoring and inferred from deteriorating habitat quality
B	Not applicable: EOO >20 000 km^2, AOO >2000 km^2
C	Not applicable: population >10 000 mature individuals
D	Not applicable: population >1000 mature individuals, >5 locations
E	Not applicable: no population viability analysis undertaken

IUCN Red List assessment data

	Estimate	Reliability
Extent of occurrence trend	129 000 km^2 stable	high medium
Area of occupancy trend	6600 km^2 decreasing	low medium
No. of mature individuals trend	126 000 decreasing	low medium
No. subpopulations	3	low
No. locations	>10	high
Largest subpopulation	115 000	low
Generation time	5.6 years	low
Global population share	100%	high

Threats

Both subspecies are threatened by a change in fire regime to one where fires extend over large areas and the mosaic of fire ages is erased (Fraser 2001; Fraser *et al.* 2003; Woinarski *et al.* 2007). Two further threats are associated with fire. Predation by feral cats *Felis catus* may increase when extensive fires reduce cover (Woinarski 2004) and, in the Northern Territory, exotic pasture grasses, particularly Gamba Grass *Andropogon gayanus*, are invading habitat and change the vegetation structures because the fires they generate are so intense (Rossiter *et al.* 2003). Also some suitable habitat has been cleared on the Tiwi Is. and in the Darwin-Daly area (Woinarski 2004).

Conservation objectives

1. Stable population at key sites
2. Effective fire management across region

Information required

1. Population trends at key sites
2. Relative significance of grazing and cat predation

Management actions required

1. Monitor at key sites
2. Assess impacts of grazing, introduced grasses and cat predation
3. Continue to undertake annual fire planning and management
4. Control populations of introduced herbivores and cats as appropriate

Bibliography

BirdLife International (2011) 'Species factsheet: *Geophaps smithii*'. Retrieved 17 January, 2011 from <http://www.birdlife.org/>.

Department of Sustainability, Environment, Water, Population and Communities (2011) *Geophaps smithii smithii*. In 'Species Profile and Threats Database'. Department of Sustainability, Environment, Water, Population and Communities, Canberra. Retrieved 24 February, 2011 from <http://www.environment.gov.au/sprat>.

Fraser F (2001) The impacts of fire and grazing on the Partridge Pigeon: the ecological requirements of a declining tropical granivore. PhD thesis, CRES, The Australian National University, Canberra.

Fraser F, Lawson V, Morrison S, Christophersen P, McGreggor S, Rawlinson M (2003) Fire management experiment for the declining partridge pigeon, Kakadu National Park. *Ecological Management and Restoration* **4**, 94–102.

Frith HJ (1982) *Pigeons and Doves of Australia*. Rigby, Adelaide.

Higgins PJ, Davies SJJF (Eds) (1996) *Handbook of Australian, New Zealand and Antarctic Birds. Volume 3: Snipe to Pigeons*. Oxford University Press, Melbourne.

Johnstone RE (1981) Notes on the distribution, ecology and taxonomy of the Partridge Pigeon (*Geophaps smithii*) and Spinifex Pigeon (*Geophaps plumifera*) in Western Australia. *Records of the Western Australian Museum* **9**, 49–63.

Rossiter NA, Setterfield SA, Douglas MM, Hutley LB (2003) Testing the grass-fire cycle: alien grass invasion in the tropical savannas of northern Australia. *Diversity and Distributions* **9**, 169–176.

Woinarski J, Pavey C, Kerrigan R, Cowie I, Ward S (Eds) (2007) *Lost from Our Landscape: Threatened Species of the Northern Territory*. Northern Territory Government, Darwin.

Woinarski JCZ (2004) 'National multi-species recovery plan for the Partridge Pigeon [eastern subspecies] *Geophaps smithii smithii*, Crested Shrike-tit [northern (sub)species] *Falcunculus (frontatus) whitei*, Masked Owl [north Australian mainland subspecies] *Tyto novaehollandiae kimberli*; and Masked Owl [Tiwi Islands subspecies] *Tyto novaehollandiae melvillensis*, 2004–2009'. Northern Territory Department of Infrastructure Planning and Environment, Darwin.

Woinarski JCZ (2009) 'Fauna monitoring at fire plots in Kakadu National Park, 2008–09'. Interim report to Parks Australia, NRETAS, Darwin.

Comments received from

Don Franklin, Simon Ward, John Woinarski

Subspecies Conservation Summary

Partridge Pigeon (western)

Geophaps smithii blaauwi Mathews, 1912
Columbidae

Conservation status

Vulnerable C2a(ii)

Reasons for listing

Population <10 000 mature individuals and suspected continuing decline, probably as a result of a change in fire regime, possibly combined with introduced predators

Status 2000

Vulnerable C2a(ii)

Status 1990

Vulnerable C2a(ii)

Taxonomy

G. s. smithii (northern Northern Territory) is also Vulnerable, as is the species
Taxonomic uniqueness: medium (41 genera/family, 3 species/genus, 2 subspecies/species)

Range

Occurs in remote areas within 100 km of the coast in the north-west of the Kimberley region, northern Western Australia (Storr 1980; Johnstone 1981). Recorded from at least 6 locations since 1998 but probably under-recorded in inaccessible locations (Department of Sustainability, Environment, Water, Population and Communities 2011). Historically occurred east to the Durack River (House 1902). The AOO is nominally estimated at about 5% of the EOO.

Abundance

The population size and trends are poorly known but suggested as c.5000 mature individuals in a single subpopulation, with a gradual ongoing decline (Garnett and Crowley 2000). Historically, it was very numerous west of Durack R. (House 1902) and at Kalumburu (Hill 1911). It has been considered to be uncommon to common (Johnstone and Storr 1998). At Kalumburu only 1 pair was found in searches during 1999 (Garnett and Crowley 2000) and it was reported to have declined dramatically in the Mitchell Plateau region (Department of Sustainability, Environment, Water, Population and Communities 2011). Anecdotal reports suggest lower population densities than the eastern subspecies, which may be up to 0.3 birds/hectare in favoured locations (D. Baker-Gabb *in litt.* 2006). The population size is here estimated as c.6000 birds, or about 0.1 birds/ha in occupied habitat.

Ecology

Occurs primarily in open woodland, particularly on the ecotone between the rugged King Leopold Sandstones and alluvial flats. The ground cover is mostly tall grasses, though the pigeons are usually seen feeding in recently burnt areas, by roads and on short grass (Johnstone 1981; Garnett and Crowley 2000). It is probably similar to the eastern subspecies in being more common in intricately burnt mosaics and dependent on perennial grasses (Fraser 2001; Fraser *et al.* 2003). The species generally lays 2 eggs in nests made on the ground, most often at the base of a clump of grass (Frith 1982). A generation time of 5.6 years (BirdLife International 2011) is derived from an age at first breeding of 1.0 years and a maximum longevity in the wild of 10.1 years (extrapolated from other Columbidae).

Threats

As with the better-known eastern subspecies, it is probably threatened by broad-scale environmental changes, fuelled by altered fire regimes, grazing by livestock and feral animals, and invasion of native woodlands by exotic plants, particularly introduced pasture grasses (Woinarski 2004). Frequent extensive late dry season fires promote a uniform vegetation of tall annual grasses, which restrict food resources and support further extensive hot fires (Fraser 2001; Fraser *et al.* 2003; Woinarski *et al.* 2005). Other possible threats to the subspecies include grazing and trampling by feral animals and predation by feral cats *Felis catus* (Australian Government 2008; Department of Sustainability, Environment, Water, Population and Communities 2011).

Conservation objectives

1. Stable population at key sites
2. Effective fire management across region

Information required

1. Population trends at key sites
2. Relative significance of grazing and cat predation

Management actions required

1. Monitor at key sites
2. Assess impacts of grazing and cat predation
3. Implement annual fire management

Current eligibility against IUCN Red List Criteria

IUCN category	Criteria eligibility
A	Not applicable: past, current or future population declines are thought unlikely to exceed 20% in any 3-generation period
B	Not applicable: >10 locations, not severely fragmented, population fluctuations not extreme
C	Vulnerable: population 2500–10 000 mature individuals, suspected continuing decline, 100% in 1 subpopulation
D	Not applicable: population >1000 mature individuals, >5 locations
E	Not applicable: no population viability analysis undertaken

IUCN Red List assessment data

	Estimate	Reliability
Extent of occurrence trend	12 400 km^2 stable	medium low
Area of occupancy trend	600 km^2 decreasing	low low
No. of mature individuals trend	6000 decreasing	low medium
No. subpopulations	1	high
No. locations	>10	high
Generation time	5.6 years	low
Global population share	100%	high

4. Control populations of introduced herbivores and cats as appropriate

Bibliography

Australian Government (2008) 'Approved conservation advice for *Geophaps smithii blaauwi* (Partridge Pigeon (western))'. Retrieved 18 March, 2011 from <http://www.environment.gov.au/biodiversity/threatened/species/pubs/66501-conservation-advice.pdf>.

BirdLife International (2011) 'Species factsheet: *Geophaps smithii*'. Retrieved 17 January, 2011 from <http://www.birdlife.org/>.

Department of Sustainability, Environment, Water, Population and Communities (2011) *Geophaps smithii blaauwi*. In 'Species Profile and Threats Database'. Department of Sustainability, Environment, Water, Population and Communities, Canberra. Retrieved 24 February, 2011 from <http://www.environment.gov.au/sprat>.

Fraser F (2001) The impacts of fire and grazing on the Partridge Pigeon: the ecological requirements of a declining tropical granivore. PhD thesis, CRES, The Australian National University, Canberra.

Fraser F, Lawson V, Morrison S, Christophersen P, McGreggor S, Rawlinson M (2003) Fire management experiment for the declining partridge pigeon, Kakadu National Park. *Ecological Management and Restoration* **4**, 94–102.

Frith HJ (1982) *Pigeons and Doves of Australia*. Rigby, Adelaide.

Garnett S, Crowley GM (2000) *The Action Plan for Australian Birds 2000*. Environment Australia, Canberra.

Hill GF (1911) Field notes on the birds of Kimberley, North-West Australia. *Emu* **10**, 258–290.

House FM (1902) Kimberley exploring expedition. In 'Report on exploration of north-west Kimberley, 1901'. (Ed. FS Brockman) pp. 18–19. Western Australian Government, Perth.

Johnstone RE (1981) Notes on the distribution, ecology and taxonomy of the Partridge Pigeon (*Geophaps smithii*) and Spinifex Pigeon (*Geophaps plumifera*) in Western Australia. *Records of the Western Australian Museum* **9**, 49–63.

Johnstone RE, Storr GM (1998) *Handbook of Western Australian Birds. Volume 1: Non-passerines (Emu to Dollarbird)*. Western Australian Museum, Perth.

Storr GM (1980) *Birds of the Kimberley Division, Western Australia*. Western Australian Museum, Perth.

Woinarski JCZ (2004) 'National multi-species recovery plan for the Partridge Pigeon [eastern subspecies] *Geophaps smithii smithii*, Crested Shrike-tit [northern (sub)species] *Falcunculus (frontatus) whitei*, Masked Owl [north Australian mainland subspecies] *Tyto novaehollandiae kimberli*; and Masked Owl [Tiwi Islands subspecies] *Tyto novaehollandiae melvillensis*, 2004–2009'. Northern Territory Department of Infrastructure Planning and Environment, Darwin.

Woinarski JCZ, Williams RJ, Price JO, Bankmore B (2005) Landscapes without boundaries: wildlife and their environments in northern Australia. *Wildlife Research* **25**, 377–388.

Comments received from

Don Franklin, Sarah Legge, Simon Ward, John Woinarski

Subspecies Conservation Summary

Partridge Pigeon (eastern)

Geophaps smithii smithii (Jardine & Selby, 1830)
Columbidae

Conservation status
Vulnerable A2bc+3c+4c

Reasons for listing
Decline >30% at several monitored sites, probably as a result of change in fire regime, possibly also from predation by feral cats

Status 2000
Near Threatened A2bc+3c+4c
Reason for change in 2010: evidence of severe declines in recent surveys

Status 1990
Near Threatened A2bc+3c+4c

Taxonomy
G. s. blaauwi (Kimberley Region, north-west Western Australia) is Vulnerable, as is the species
Taxonomic uniqueness: medium (41 genera/family, 3 species/genus, 2 subspecies/species)

Range
Now recorded only in sub-coastal Northern Territory, from Yinberrie Hills (about 50 km north of Katherine) in the south and Litchfield National Park in the west to western Arnhem Land in the east with a separate subpopulation on the Tiwi Is (Melville and Bathurst Is.). Smaller scattered populations elsewhere in the Top End (Woinarski 2004; Woinarski *et al.* 2007) could be isolated subpopulations, particularly that on Coburg Peninsula. Historically the distribution extended from Kununurra, Western Australia, east to McArthur R., NT, with severe range contractions from western, eastern and southern parts of the former distribution, mostly occurring early in the 20th century (Higgins and Davies 1996; Woinarski *et al.* 2007). The AOO is estimated at about 5% of the EOO.

Abundance
The population size and trends are poorly known. Population densities have been estimated at 2 locations where it is comparatively common: 0.33±0.07 birds per hectare on Melville I. in 2001–2002 (based on 156 birds on 192 km of transects and extrapolated to 1464 birds in 4436 ha of woodlands on Melville I. (D. Baker-Gabb *in litt.*) and 0.2 birds per hectare in surveyed areas of Kakadu National Park (Department of Sustainability, Environment, Water, Population and Communities 2011). Assuming an AOO of about 6000 km^2 and density of about 0.2 birds/ha, the total population may be about 120 000 mature individuals. Apart from the historical contraction in range, recent surveys in Kakadu National Park recorded a significant decline in relative abundance from 0.24 in 2001–2004 to 0.05 in 2007–2009. The species was also scarce, being detected at only 15 out of 272 plots (Woinarski 2009). The population on the Tiwi Is. may also have declined as a result of clearance for forestry.

Ecology
Lives primarily in open forest and woodland dominated by Darwin Stringybark *Eucalyptus tetrodonta* and Darwin Woollybutt *E. miniata* that has a structurally diverse understorey, usually in areas with a fire regime that promotes a mosaic of fire ages, including wet season burns, which promote grass diversity and year-round seed availability (Fraser *et al.* 2003). Usually lays 2 eggs in nests made on the ground, most often at the base of a clump of grass. Feeds on seeds taken from bare ground, often amidst sparse grasses, but commonly where the ground layer has been removed by fire (Frith 1982). It appears to rely on perennial grass species which set seed relatively early (Fraser 2001). A generation time of 5.6 years (BirdLife International 2011) is derived from an age at first breeding of 1.0 years and a maximum longevity in the wild of 10.1 years, both extrapolated from other Columbidae.

Threats
The biggest threat is change in fire regime to one where fires extend over large areas and the mosaic of fire ages is erased (Fraser 2001; Fraser *et al.* 2003; Woinarski *et al.* 2007). Two further threats are associated with fire. Predation by feral cats *Felis catus* may increase when extensive fires reduce cover (Woinarski 2004) and exotic pasture grasses, particularly Gamba Grass *Andropogon gayanus*, which are invading habitat in and around Litchfield National Park are likely to disadvantage the pigeons because their fuel loads are so large that, when they burn, they change the vegetation structure (Rossiter *et al.* 2003). Also some suitable habitat has been cleared on the Tiwi Is. and in the Darwin-Daly area (Woinarski 2004).

Conservation objectives
1. Stable population at key sites

Current eligibility against IUCN Red List Criteria

IUCN category	Criteria eligibility
A	Vulnerable: past, current and anticipated population declines of 30–49% over periods of 3 generations (17 years) estimated from monitoring and inferred from deteriorating habitat quality
B	Not applicable: EOO >20 000 km^2, AOO >2000 km^2
C	Not applicable: population >10 000 mature individuals
D	Not applicable: population >1000 mature individuals, >5 locations
E	Not applicable: no population viability analysis undertaken

IUCN Red List assessment data

	Estimate	Reliability
Extent of occurrence trend	117 000 km^2 stable	high medium
Area of occupancy trend	6000 km^2 decreasing	low medium
No. of mature individuals trend	120 000 decreasing	low high
No. subpopulations	2	low
No. locations	>10	high
Largest subpopulation	115 000	low
Generation time	5.6 years	low
Global population share	100%	high

2. Effective fire management across region

Information required

1. Population trends at key sites
2. Relative significance of grazing and cat predation

Management actions required

1. Monitor at key sites
2. Assess impacts of grazing, introduced grasses and cat predation
3. Continue to undertake annual fire planning and management
4. Control populations of introduced herbivores and cats as appropriate

Bibliography

BirdLife International (2011) 'Species factsheet: *Geophaps smithii*'. Retrieved 17 January, 2011 from <http://www.birdlife.org/>.

Department of Sustainability, Environment, Water, Population and Communities (2011) *Geophaps smithii smithii*. In 'Species Profile and Threats Database'. Department of Sustainability, Environment, Water, Population and Communities, Canberra. Retrieved 24 February, 2011 from <http://www.environment.gov.au/sprat>.

Fraser F (2001) The impacts of fire and grazing on the Partridge Pigeon: the ecological requirements of a declining tropical granivore. PhD thesis, CRES, The Australian National University, Canberra.

Fraser F, Lawson V, Morrison S, Christophersen P, McGreggor S, Rawlinson M (2003) Fire management experiment for the declining partridge pigeon, Kakadu National Park. *Ecological Management and Restoration* **4**, 94–102.

Frith HJ (1982) *Pigeons and Doves of Australia*. Rigby, Adelaide.

Higgins PJ, Davies SJJF (Eds) (1996) *Handbook of Australian, New Zealand and Antarctic Birds. Volume 3: Snipe to Pigeons*. Oxford University Press, Melbourne.

Rossiter NA, Setterfield SA, Douglas MM, Hutley LB (2003) Testing the grass-fire cycle: alien grass invasion in the tropical savannas of northern Australia. *Diversity and Distributions* **9**, 169–176.

Woinarski J, Pavey C, Kerrigan R, Cowie I, Ward S (Eds) (2007) *Lost from Our Landscape: Threatened Species of the Northern Territory*. Northern Territory Government, Darwin.

Woinarski JCZ (2004) 'National multi-species recovery plan for the Partridge Pigeon [eastern subspecies] *Geophaps smithii smithii*, Crested Shrike-tit [northern (sub)species] *Falcunculus (frontatus) whitei*, Masked Owl [north Australian mainland subspecies] *Tyto novaehollandiae kimberli*; and Masked Owl [Tiwi Islands subspecies] *Tyto novaehollandiae melvillensis*, 2004–2009'. Northern Territory Department of Infrastructure Planning and Environment, Darwin.

Woinarski JCZ (2009) 'Fauna monitoring at fire plots in Kakadu National Park, 2008–09'. Interim report to Parks Australia, NRETAS, Darwin.

Comments received from

Don Franklin, Simon Ward, John Woinarski

Subspecies Conservation Summary

White-quilled Rock-Pigeon (Stokes Range)

Petrophassa albipennis boothi Goodwin, 1969
Columbidae

Conservation status

Near Threatened B1ab(ii,iii,v)+2ab(ii,iii,v), C2a(ii)

Reasons for listing

Increased fire across its very small range is suspected to be causing a continuing decline in area occupied, habitat quality, number of mature individuals and the number of locations may approach 10

Status 2000

Near Threatened B1ab(ii,iii,v)+2ab(ii,iii,v), C2a(ii)

Status 1990

Near Threatened B1ab(ii,iii,v)+2ab(ii,iii,v), C2a(ii)

Taxonomy

P. a. albipennis (Kimberley region and Northern Territory north of Baines River is Least Concern, as is the species
Taxonomic uniqueness: medium (41 genera/family, 2 species/genus, 2 subspecies/species)

Range

North-west Northern Territory, upper Victoria and Baines River catchments, north to Stokes Range, west to at least Waterloo Station and east to Delamere and south to Jasper Gorge (Higgins and Davies 1996; Schodde and Mason 1997). Recorded in the east of Gregory National Park, which is mostly suitable habitat (A. Fisher pers. comm.) and at the Bradshaw Field Training Area (Fisher and Woinarski 2002). The AOO is assumed to be about 10% of the EOO.

Abundance

Population size and trends very poorly known in this little-surveyed region. In the 8710 km^2 Bradshaw Field Training Area, only 10 records at 7 locations in a 6-day survey in 2002 and known from 15 other records (Fisher and Woinarski 2002). Given the low encounter rate and the small, although under-recorded, area of occupancy, the population is estimated here as near 10 000 mature individuals with about 0.5 birds/ha of AOO. The population structure is even less well known. Although judged here to comprise a single subpopulation across geographically isolated rocky outcrops, there might be multiple subpopulations. A continuing decline is inferred from the scarcity of this subspecies in apparently suitable habitat and what appears to have been an unfavourable fire regime in recent decades.

Ecology

The nominate subspecies inhabits rocky plateaux, rock faces and adjacent alluvium of low ranges and tumbled breakaways with patches of variable vegetation, often open woodland and spinifex *Triodia* (Schodde and Mason 1997). Feeds on the ground on seeds of herbs, shrubs and sedges, in pairs or small groups (Frith *et al.* 1976), and nests on rock ledges and crevices with a clutch size of 2 (Frith 1972). A generation time of 5.6 years (BirdLife International 2011) is derived from an age at first breeding of 1.0 years and a maximum longevity in the wild of 10.2 years, both extrapolated from other Columbidae.

Threats

The primary threat appears to be changed fire regimes from a patchwork of early and late burns to more frequent, large late dry season fires. An average of 40% of Bradshaw Station burnt annually and 89% of the area had a fire return interval of <4 years in 1990–1999 (Yates and Russell-Smith 2003) and an average of 29% of the rocky areas of Gregory National Park burnt annually and 80% had a fire return interval of <3 years in 1998–2008 (Russell-Smith *et al.* 2010). Just how this affects the rock-pigeon is unknown. Feral cats *Felis catus* are common across the species' range and may be a threat.

Conservation objectives

1. Stable numbers

Information required

1. Appropriate methods and locations for long-term monitoring
2. Trends in population
3. Effect of fire

Management actions required

1. Determine the relationship between pigeon demography and fire regime
2. Reduce fire frequency and extent within range of the pigeon

Bibliography

BirdLife International (2011) 'Species factsheet: *Petrophassa albipennis*'. Retrieved 17 January, 2011 from <http://www.birdlife.org/>.

Current eligibility against IUCN Red List Criteria

IUCN category	Criteria eligibility
A	Not applicable: past, current or future population declines are thought unlikely to exceed 20% in any 3-generation period
B	Near Threatened: EOO <20 000 km^2, AOO <2000 km^2, near 10 locations with plausible future threat, suspected continuing decline in AOO habitat quality, no. mature individuals
C	Near Threatened: population near 10 000 mature individuals, suspected continuing decline, 100% in 1 subpopulation
D	Not applicable: population >1000 mature individuals, >5 locations
E	Not applicable: no population viability analysis undertaken

IUCN Red List assessment data

	Estimate	Reliability
Extent of occurrence trend	2300 km^2 stable	low low
Area of occupancy trend	230 km^2 declining	low low
No. of mature individuals trend	12 000 declining	low low
No. subpopulations	1	low
No. locations	>10	medium
Generation time	5.6 years	low
Global population share	100%	high

Fisher A, Woinarski J (2002) 'Assessment of the vertebrate fauna of the Bradshaw (Juliki) Field Training Area, Northern Territory'. Parks and Wildlife Commission of the Northern Territory.

Frith HJ (1972) Nesting of the Black-banded pigeon and the Australian rock pigeons. *Emu* **72**, 13–16.

Frith HJ, Wove TO, Barker RD (1976) Food of eight species of Columbidae, in the genera *Geopelia, Phaps, Geophaps* and *Petrophassa*. *Australian Wildlife Research* **3**, 159–171.

Higgins PJ, Davies SJJF (Eds) (1996) *Handbook of Australian, New Zealand and Antarctic Birds. Volume 3: Snipe to Pigeons.* Oxford University Press, Melbourne.

Russell-Smith J, Yates CP, Brock C, Westcott VC (2010) Fire regimes and interval-sensitive vegetation in semiarid Gregory National Park, northern Australia. *Australian Journal of Botany* **58**, 300–317.

Schodde R, Mason IJ (Eds) (1997) *Zoological Catalogue of Australia 37.2: Aves (Columbidae to Coraciidae)*. CSIRO Publishing, Melbourne.

Yates C, Russell-Smith J (2003) Fire regimes and vegetation sensitivity analysis: an example from Bradshaw Station, monsoonal northern Australia. *International Journal of Wildland Fire* **12**, 349–358.

Comments received from

Alaric Fisher, Don Franklin, Simon Ward, John Woinarski

Species Conservation Summary

Norfolk Island Ground-Dove

Gallicolumba norfolciensis (Latham, 1801)

Columbidae

Conservation status
Extinct

Reasons for listing
The last individual of this species was recorded in 1800

Status 2000
Extinct

Status 1990
Extinct

Taxonomy
No infraspecific taxa described

Taxonomic uniqueness: low (42 genera/family, 19 species/genus, 1 subspecies/species)

Former range
Endemic to Norfolk I. (Hindwood 1965; Schodde *et al.* 1983).

Former abundance
No estimates of its population size are available. The species was last recorded in 1800 (Schodde *et al.* 1983).

Ecology

Presumably the Norfolk Island Ground-Dove was a species of the coastal forest. It may have utilised forest habitat throughout the island, but no bones have been found in the higher parts of the island (M. Christian).

Reasons for extinction

Hunting and predation by introduced cats *Felis catus* and pigs *Sus scrofa* are the likely causes of the Norfolk Island Ground-Dove's extinction (Schodde *et al.* 1983). Numbers may already have been suppressed by the introduction of the Pacific Rat *Rattus exulans* with Polynesians 800 years ago (Holdaway and Anderson 1998).

Bibliography

Hindwood KA (1965) John Hunter: a naturalist and artist of the First Fleet. *Emu* **65**, 83–96.

Holdaway RN, Anderson AJ (1998) 14C AMS dates on *Rattus exulans* bones from natural and archaeological contexts on Norfolk Island, South-west Pacific. *Archaeology in New Zealand* **41**, 195–198.

Schodde R, Fullagar P, Hermes N (1983) *A Review of the Status of Norfolk Island Birds: Past and Present.* Australian National Parks and Wildlife Service, Canberra.

Comments received from

Margaret Christian, Richard Holdaway

Species Conservation Summary

Christmas Island Imperial-Pigeon

Ducula whartoni (Sharpe, 1887)

Columbidae

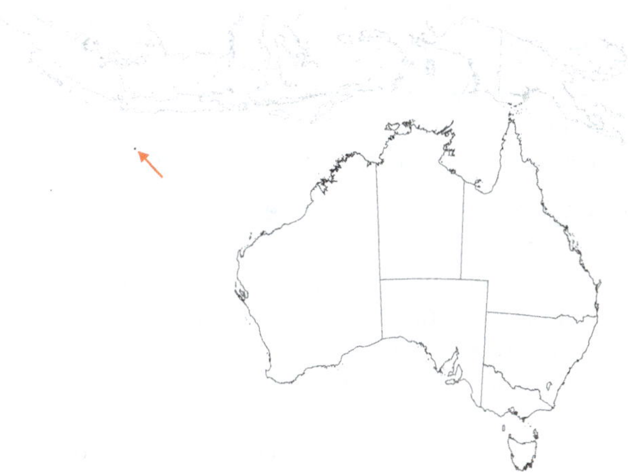

Conservation status

Near Threatened D2

Reasons for listing

Present at a single location with a plausible but low probability threat

Status 2000

Near Threatened D2

Status 1990

Near Threatened D2

Taxonomy

No infraspecific taxa described

Taxonomic uniqueness: low (41 genera/family, 35 species/genus, 1 subspecies/species)

Range

Endemic to Christmas I., Indian Ocean where it occurs throughout the island. The area occupied was reduced by clearance for phosphate mining from 1890s to 1980s but at least some of this area is now rehabilitated and used by the pigeons. Introduction to Cocos-Keeling Is. at end of 19th century failed because of hunting and/or lack of suitable food trees.

Abundance

The most recent population estimate, of which the average is used here, was of 1000–10 000 individuals (Beeton *et al.* 2010). The species was recorded in 98% of 123 sites surveyed by James and Retallick (2007). There is no information on trends.

Ecology

Occurs more commonly in evergreen forest than in either semi-deciduous forests or cleared areas (James and Retallick 2007). It nests in the tops of rainforest trees and other dense vegetation, and feeds on rainforest and other seeds, as well as buds and leaves (BirdLife International 2011). A generation time of 7.8 years (BirdLife International 2011) is derived from an average age at first breeding of 1.5 years and a maximum longevity in the wild of 14.0 years, both extrapolated from other Columbidae.

Threats

Like other Christmas I. species, Christmas Island Imperial-Pigeons are potentially threatened by the introduced Yellow Crazy Ant *Anoplolepis gracilipes*, which can alter the whole ecology of sites it occupies by killing the super-numerous Red Crab *Gecaroidea natalis*, as well as by tending scale insects that damage trees (O'Dowd *et al.* 2003; Abbott and Green 2007). However, there are fewer ants in the canopy where the pigeons feed than on the ground, and threats from the ant are considered very low given they have been temporarily controlled. Black Rats *Rattus rattus* may also reduce nesting success. Quarantine measures on the island are not currently sufficient to prevent invasion by additional unspecified predators.

Conservation objectives

1. Existing population maintained

Current eligibility against IUCN Red List Criteria

IUCN category	Criteria eligibility
A	Not applicable: past, current or future population declines are thought unlikely to exceed 20% in any 3-generation period
B	Not applicable: EOO, AOO, habitat, locations or population not declining, population fluctuations not extreme
C	Not applicable: no continuing population decline or extreme fluctuations
D	Near Threatened: ≤5 locations with plausible but low probability future threat
E	Not applicable: no population viability analysis undertaken

IUCN Red List assessment data

	Estimate	Reliability
Extent of occurrence trend	134 km^2 stable	high high
Area of occupancy trend	134 km^2 stable	high medium
No. of mature individuals trend	5000 stable	low medium
No. subpopulations	1	high
No. locations	1	high
Generation time	7 years	low
Global population share	100%	high

2. Yellow Crazy Ant populations constrained to levels that do not threaten species
3. Black Rats and feral cats eradicated or controlled

Information required

1. Population trends
2. Levels of ant abundance that will threaten species

Management actions required

1. Continue population monitoring
2. Upgrade quarantine standards for Christmas I.
3. Continue control of Yellow Crazy Ants
4. Eradicate Black Rats
5. Control or eradicate feral cats
6. Create a multi-stakeholder team with responsibility for ensuring threatened species are conserved on Christmas I.

Bibliography

Abbott KL, Green PT (2007) Collapse of an ant-scale mutualism in a rainforest on Christmas Island. *Oikos* **116**, 1238–1246.

Beeton B, Burbidge AA, Grigg G, Harrison P, How RA, Humphries B, McKenzie N, Woinarski J (2010) 'Final report, Christmas Island Expert Working Group to Minister for the Environment, Heritage and the Arts'. Retrieved 1 April, 2010 from <http://www.environment.gov.au/parks/publications/christmas/pubs/final-report.pdf>.

BirdLife International (2011) 'Species factsheet: *Ducula whartoni*'. Retrieved 17 January, 2011 from <http://www.birdlife.org/>.

James DJ, Retallick K (2007) 'Forest birds of Christmas Island: a baseline survey of abundance'. Parks Australia North Christmas Island Biodiversity Monitoring Programme: report to the Department of Finance and Administration and the Department of the Environment and Water Resources, Canberra.

O'Dowd DJ, Green PT, Lake PS (2003) Invasional 'meltdown' on an oceanic island. *Ecology Letters* **6**, 812–817.

Comments received from

Andrew Burbidge, David James, Mike Smith, John Woinarski

Subspecies Conservation Summary

New Zealand Pigeon (Norfolk Island)

Hemiphaga novaeseelandiae spadicea (Latham, 1802)
Columbidae

Conservation status

Extinct

Reasons for listing

Last individual recorded in 1900

Status 2000

Extinct

Status 1990
Extinct

Taxonomy
Other subspecies are found on the North, South and off-shore islands of New Zealand and on the Chatham Is. The status of the species is Near Threatened
Taxonomic uniqueness: low (42 genera/family, 1 species/genus, 3 subspecies/species)

Former range
Endemic to Norfolk I. (Schodde *et al.* 1983).

Former abundance
There are no estimates of its abundance; though recorded by early visitors to the island, no more than the subspecies' presence was noted. It has not been recorded since 1900 (Schodde *et al.* 1983).

Ecology
The nominate subspecies of the New Zealand Pigeon in New Zealand relies on the availability of a wide range of fruits in order to breed successfully, as well as the availability of nutritious leaves in the non-fruiting season (Clout 1990). On Norfolk I., the subspecies probably relied on a variety of fruiting trees to survive.

Reasons for extinction
Hunting was almost certainly the main reason for the subspecies' extinction (Schodde *et al.* 1983). Known as a Wood Quest, it was extremely tame and was already scarce by the 1850s (McAllan 1999).

Bibliography
Clout M (1990) The Kereru and its forests. *Birds International* **2**, 10–19.

McAllan IAW (1999) A common name for the Norfolk Island Pigeon. *Australian Bird Watcher* **18**, 179–180.

Schodde R, Fullagar P, Hermes N (1983) *A Review of the Status of Norfolk Island Birds: Past and Present*. Australian National Parks and Wildlife Service, Canberra.

Comments received from
Margaret Christian, Richard Holdaway

Subspecies Conservation Summary

Glossy Swiftlet (Christmas Island)

Collocalia esculenta natalis (Lister, 1889)

Apodidae

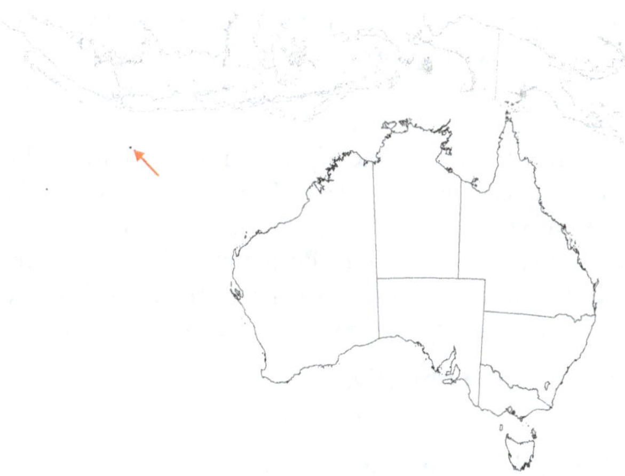

Conservation status
Near Threatened D2

Reasons for listing
Occurs at a single location with a plausible but low probability threat

Status 2000
Near Threatened D2

Status 1990
Near Threatened D2

Taxonomy
There are 31 other subspecies in South-East Asia and the south-west Pacific; the species is Least Concern
Taxonomic uniqueness: low (19 genera/family, 3 species/genus, 32 subspecies/species)

Range
Endemic to Christmas I., Indian Ocean where it occurs throughout the island.

Abundance
No real estimates. Population estimated at 100 000–1 000 000 by van Tets (1975). The figure of 5000 cited by Garnett and Crowley (2000) appears to be an error. Swiftlets were frequently encountered in surveys by James and Retallick (2007), recorded in 302 out of 527 counts. Given this frequency the figure listed here is at the lower end of the earlier estimate, pending accurate counts. Concerns that the swiftlet might decline rapidly following the spread of the introduced Yellow Crazy Ant *Anoplolepis gracilipes* (Garnett and Crowley 2000) have been forestalled by control of the ant; there is no information on trends.

Ecology
Nests in caves and feeds on small flying insects over most habitats, including settlements, regrowth, primary for-

Current eligibility against IUCN Red List Criteria

IUCN category	Criteria eligibility
A	Not applicable: past, current or future population declines are thought unlikely to exceed 20% in any 3-generation period
B	Not applicable: EOO, AOO, habitat, locations or population not declining, population fluctuations not extreme
C	Not applicable: no continuing population decline or extreme fluctuations
D	Near Threatened: ≤5 locations with plausible but low probability future threat
E	Not applicable: no population viability analysis undertaken

IUCN Red List assessment data

	Estimate	Reliability
Extent of occurrence trend	134 km^2 stable	high high
Area of occupancy trend	134 km^2 stable	high high
No. of mature individuals trend	100 000 stable	low low
No. subpopulations	1	high
No. locations	1	high
Generation time	5.0 years	low
Global population share	100%	high

ests and on the terraces (Stokes 1988). A generation time of 5.0 years (BirdLife International 2011) is derived from an average age at first breeding of 1.0 years, an annual survival of adults of 45.0%, and a maximum longevity in the wild of 13.4 years, all extrapolated from other Apodidae.

Threats

Unlike other Christmas I. species, swiftlets are not potentially threatened by the introduced Yellow Crazy Ant *Anoplolepis gracilipes*, unless they so alter the ecology of the island by killing the super-numerous Red Crab *Gecaroidea natalis*, as well as by tending scale insects that damage trees (O'Dowd *et al.* 2003; Abbott and Green 2007) that the abundance of flying insects declines. This is considered highly unlikely. The species may also be affected by the pesticide Fipronil, used for ant control, but this remains unproven (Beeton *et al.* 2010). Quarantine measures on the island are not currently sufficient to prevent invasion by additional unspecified predators.

Conservation objectives

1. Existing population maintained

Information required

1. Population trends
2. Susceptibility of nesting colonies to any threats

Management actions required

1. Initiate population monitoring
2. Upgrade quarantine standards for Christmas I.
3. Create a multi-stakeholder team with responsibility for ensuring threatened species are conserved on Christmas I.

Bibliography

Abbott KL, Green PT (2007) Collapse of an ant-scale mutualism in a rainforest on Christmas Island. *Oikos* **116**, 1238–1246.

Beeton B, Burbidge AA, Grigg G, Harrison P, How RA, Humphries B, McKenzie N, Woinarski J (2010) 'Final report, Christmas Island Expert Working Group to Minister for the Environment, Heritage and the Arts'. Retrieved 1 April, 2010 from <http://www.environment.gov.au/parks/publications/christmas/pubs/final-report.pdf>.

BirdLife International (2011) 'Species factsheet: *Collocalia esculenta*'. Retrieved 17 January, 2011 from <http://www.birdlife.org/>.

Garnett S, Crowley GM (2000) *The Action Plan for Australian Birds 2000.* Environment Australia, Canberra.

James DJ, Retallick K (2007) 'Forest birds of Christmas Island: a baseline survey of abundance'. Parks Australia North Christmas Island Biodiversity Monitoring Programme: report to the Department of Finance and Administration and the Department of the Environment and Water Resources, Canberra.

O'Dowd DJ, Green PT, Lake PS (2003) Invasional 'meltdown' on an oceanic island. *Ecology Letters* **6**, 812–817.

Stokes T (1988) 'A review of the birds of Christmas Island, Indian Ocean'. Australian National Parks and Wildlife Service Occasional Paper.

van Tets GF (1975) A report on the conservation of resident birds on Christmas Island. *Bulletin of the International Council for Bird Preservation* **12**, 238–242.

Comments received from

Andrew Burbidge, David James, Mike Smith, John Woinarski

Species Conservation Summary

Matsudaira's Storm-Petrel

Hydrobates matsudairae (Kuroda, 1922)
Hydrobatidae

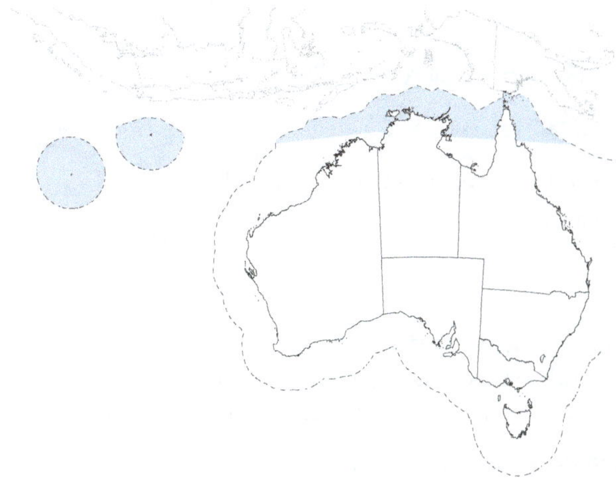

Conservation status

Vulnerable D2

Reasons for listing

Present at 2 sites (outside Australia) where introduction of exotic predators represents a plausible threat

Status 2000

Vulnerable D2

Status 1990

Vulnerable D2

Taxonomy

No infraspecific taxa described
Taxonomic uniqueness: high (1 genus/family, 4 species/genus, 1 subspecies/species)

Range

Breeds only on the Iwo Is. and possibly the Ogasawara Is., southern Japan. After breeding, they appear to move south through the Timor Sea off north-west Australia to the Indian Ocean, as well as to seas off north-east New Guinea (BirdLife International 2011).

Abundance

Population figures are nominal, given the lack of surveys and could be wrong by an order of magnitude (BirdLife International 2011).

Ecology

A colonial species that nests in burrows on high ground on offshore islands. In the non-breeding season it is usually seen well out to sea. A generation time of 14.6 years (BirdLife International 2011) is derived from an age at first breeding of 4.8 years, a maximum age 28.0 years and an annual survival of adults of 87.5%, all extrapolated from congeners.

Threats

It is listed by BirdLife International (2011) as Data Deficient because of lack of knowledge of the condition of the breeding islands. However, the main Iwo Jima islands are long settled so the risk that breeding sites will be colonised by mammalian predators seems feasible. The major threat at sea is pollution. The main migratory pathway appears to pass near the site of the Montara oil spill, though no mortalities were recorded at the time and such events are infrequent.

Conservation objectives

1. At-sea threats maintained at acceptable levels
2. Global agreement on conservation measures

Information required

1. Population trends or changes in adult and juvenile survival

Management actions required

1. None in Australia

Bibliography

BirdLife International (2011) 'Species factsheet: *Oceanodroma matsudairae*'. Retrieved 17 January, 2011 from <http://www.birdlife.org/>.

Comments received from

Barry Baker, Rosemary Gales, Tim Reid, Ben Sullivan

Current eligibility against IUCN Red List Criteria

IUCN category	Criteria eligibility
Population visiting Australian territory	
A	Not applicable: past, current or future population declines are thought unlikely to exceed 20% in any 3-generation period
B	Not applicable: EOO, AOO, habitat, locations or population not declining, population fluctuations not extreme
C	Not applicable: population >10 000 mature individuals
D	Vulnerable: ≤5 locations with plausible future threat
E	Not applicable: no population viability analysis undertaken

IUCN Red List assessment data

	Estimate	Reliability
Population visiting Australian territory		
Extent of occurrence trend	1 800 000 km^2 stable	high high
Area of occupancy trend	1 800 000 km^2 stable	high high
No. of mature individuals trend	10 000 stable	low low
No. subpopulations	1	high
No. locations	2	medium
Generation time	14.6 years	low
Global population share	50–100%	medium
Level of genetic exchange	high	high

Subspecies Conservation Summary

Wilson's Storm-Petrel (subantarctic)

Oceanites oceanicus oceanicus (Kuhl, 1820)
Oceanitidae

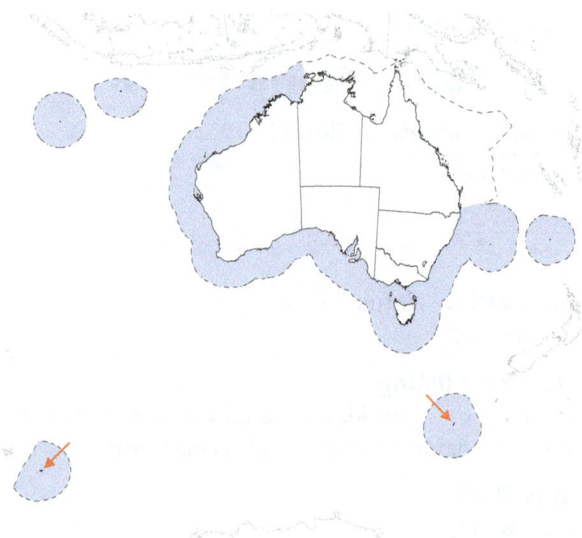

Conservation status

Australian breeding population: Vulnerable D2
Population visiting Australian territory: Least Concern

Reasons for listing

Australian breeding population: Present at 2 locations where there is a plausible threat of introduction of predators

Status 2000

Australian breeding population: Vulnerable D2
Population visiting Australian territory: Least Concern

Status 1990

Australian breeding population: Vulnerable D2
Population visiting Australian territory: Least Concern

Taxonomy

O. o. exasperatus (Antarctic mainland and Scotia Sea islands) is extralimital. The species is Least Concern. Taxonomic uniqueness: high (6 genera/family, 3 species/genus, 2 subspecies/species)

Range

Breeds in Australian territory at Heard I. (Downes *et al.* 1959; Woehler 1991, 2006, 2010) and on Bishop I. off Macquarie (Garnett and Crowley 2000). Extralimital populations breed south of Indian and Atlantic Oceans. Migrates north after breeding, reaching well north of the equator (Marchant and Higgins 1990).

Abundance

There have been no attempts to assess population size on Heard I as the birds are widely dispersed in vegetated and unvegetated areas of the island, including lava fields. Given that breeding data were obtained for the species by Downes *et al.* (1959), numbers are likely to be substantial. A nominal figure of 5000 pairs cited here may be wrong by several orders of magnitude. One pair recorded breeding on Bishop I. (N. Brothers pers. comm.). The global population for species estimated at 12–30 million birds (BirdLife International 2011).

Ecology

Nests in burrows or crevices among rocks, and in vegetated and unvegetated areas, and feed on small crustaceans, particularly euphausiids, fish and cephalopods plucked from the ocean surface (Marchant and Higgins 1990). A generation length of 15.6 years (BirdLife International 2011) is derived from an age at first breeding of 4.5 years and an annual adult survival of 91.0% (Schreiber and Burger 2002).

Threats

There is a plausible risk of predators being introduced to Heard I. The absence of Wilson's Storm-petrels from Macquarie I. may be because introduced cats *Felis catus* and Black Rats *Rattus rattus* eliminated them before they were recorded. There are no known threats at sea, but the potential impact of plastic ingestion is unknown.

Conservation objectives

1. Populations maintained at Heard and Bishop Is.

Information required

1. Population size, trends and demographic parameters

Management actions required

1. Maintain strict quarantine provisions at Heard and Bishop Is.

Bibliography

BirdLife International (2011) 'Species factsheet: *Oceanites oceanicus*'. Retrieved 17 January, 2011 from <http://www.birdlife.org/>.

Downes MC, Ealey EHM, Gwynne AM, Young PS (1959) 'The birds of Heard Island'. Australian National Antarctic Research Reports (Series B).

Garnett S, Crowley GM (2000) *The Action Plan for Australian Birds 2000*. Environment Australia, Canberra.

Marchant S, Higgins PJ (Eds) (1990) *Handbook of Australian, New Zealand and Antarctic Birds. Volume 1: Ratites to Ducks*. Oxford University Press, Melbourne.

Schreiber EA, Burger J (2002) *Biology of Marine Birds*. CRC Press, Boca Raton, FL.

Woehler EJ (1991) The status and conservation of seabirds of Heard Island and the McDonald Islands. In 'Seabird Status and Conservation. ICBP Tech. Publ. No 11.'. (Ed. JP Croxall) pp. 263–277. ICBP, Cambridge, UK.

Woehler EJ (2006) Status and conservation of the seabirds of Heard Island and the McDonald Islands. In *Heard Island: Southern Ocean Sentinel*. (Eds K Green and EJ Woehler) pp. 128–165. Surrey Beatty & Sons, Chipping Norton, NSW.

Current eligibility against IUCN Red List Criteria

IUCN category	Criteria eligibility
Australian breeding population	
A	Not applicable: past, current or future population declines are thought unlikely to exceed 20% in any 3-generation period
B	Not applicable: EOO, AOO, habitat, locations or population not declining, population fluctuations not extreme
C	Not applicable: no continuing population decline or extreme fluctuations
D	Vulnerable: ≤5 locations with plausible future threat
E	Not applicable: no population viability analysis undertaken

IUCN Red List assessment data

	Estimate	Reliability
Australian breeding population		
Extent of occurrence trend	7 200 000 km^2 stable	high high
Area of occupancy trend	40 km^2 stable	low medium
No. of mature individuals trend	10 000 stable	low high
No. subpopulations	2	high
No. locations	2	high
Largest subpopulation	10 000	medium
Generation time	15.6 years	low
Global population share	<1%	medium
Level of genetic exchange	low	medium

Woehler EJ (2010) Status and conservation of the seabirds of Heard Island. *The State of Australia's Birds 2010: Islands and Birds. Wingspan* **Supplement 20**, 8–9.

Comments received from

Barry Baker, Rosemary Gales, Tim Reid, Ben Sullivan, Eric Woehler

Species Conservation Summary

Grey-backed Storm-Petrel

Garrodia nereis (Gould, 1840)

Oceanitidae

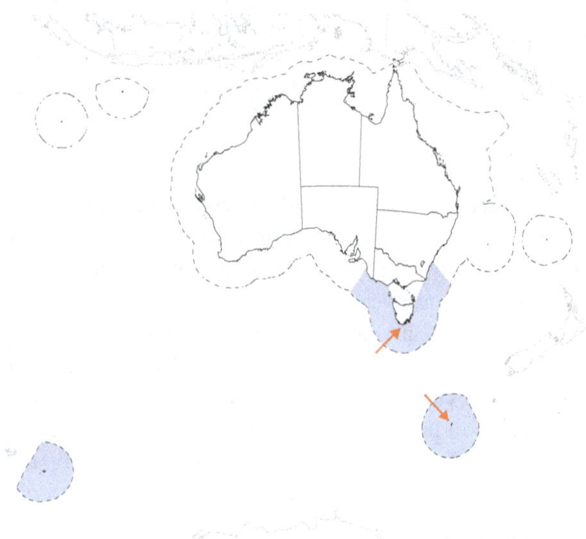

Conservation status

Australian breeding population: Endangered B2ab(iii), C2a(i), D

Population visiting Australian territory: Least Concern

Reasons for listing

Australian breeding population: AOO <500 km^2, at ≤5 locations and habitat declining in quality; Population <2500, largest subpopulation <250 and probably declining; population <250

Status 2000

Australian breeding population: Endangered D

Reason for change in 2010: criteria changed because of new threats on Macquarie I.

Population visiting Australian territory: Least Concern

Status 1990

Australian breeding population: Endangered D

Population visiting Australian territory: Least Concern

Taxonomy

No infraspecific taxa described

Taxonomic uniqueness: high (6 genera/family, 1 species/genus, 1 subspecies/species)

Range

Probably currently (2011) breeds in Australian territory on Macquarie I. (Parks and Wildlife Service 2006) and on islands off Tasmania (Brothers *et al.* 1996). Extralimital populations breed on islands throughout the Southern Ocean. During winter, southern birds migrate northwards after breeding, and subspecies becomes more numerous near the Australian coast (Marchant and Higgins 1990). They generally occur over the edge of the continental shelf and may only be pelagic when dispersing (BirdLife International 2011). Woehler (2006) reports them from Heard I. EEZ, but they are not a breeding species on Heard I.

Abundance

The population on Macquarie I was estimated at 100 pairs in the early 1980s (Rounsevell and Brothers 1984).

Current eligibility against IUCN Red List Criteria

IUCN category	Criteria eligibility
Australian breeding population	
A	Not applicable: past, current or future population declines are thought unlikely to exceed 20% in any 3-generation period
B	Endangered: AOO <500 km^2, at ≤5 locations with plausible future threat, inferred continuing decline in habitat quality
C	Endangered: population 250–2500 mature individuals, suspected continuing decline, largest subpopulation <250 mature individuals
D	Endangered: population 50–250 mature individuals
E	Not applicable: no population viability analysis undertaken

IUCN Red List assessment data

	Estimate	Reliability
Australian breeding population		
Extent of occurrence trend	2 000 000 km^2 stable	high high
Area of occupancy trend	20 km^2 declining	low low
No. of mature individuals trend	250 stable	low low
No. subpopulations	2	medium
No. locations	5	low
Largest subpopulation	200	low
Generation time	15.2 years	medium
Global population share	<1%	medium
Level of genetic exchange	medium	low

Ecology

They breed in loose colonies among rocks or low vegetation. They feed on small crustaceans, including immature barnacles, euphausiids, copepods and amphipods, as well as fish and squid, mostly plucked from the ocean surface but occasionally captured by plunge-diving (Marchant and Higgins 1990). They occasionally follow ships. A generation time of 15.2 years (BirdLife International 2011) is derived from an age at first breeding of 4.7 years, a maximum age 30.8 years and an annual survival of adults of 87.8%, all extrapolated from other Hydrobatidae.

Threats

The increase in rabbits *Oryctolagus cuniculus* numbers on Macquarie I. since feral cats *Felis catus* were eradicated in 1999 has led to an extensive destruction of habitat and soil erosion from overgrazing at nesting sites and exposure of nests to predation by Brown Skua *Stercorarius lonnbergi*. Black Rats *Rattus rattus* are also likely to take eggs and kill chicks. These threats may continue, unless an eradication program planned in 2011–14 is successful. There are no known threats at sea, but potential impacts of plastic ingestion are unknown.

Conservation objectives

1. Stable or increasing population on Macquarie I.

Information required

1. Population size, trends and demographic parameters

Management actions required

1. Eradicate rabbits and Black Rats on Macquarie I.
2. Monitor at least samples of the population

Bibliography

BirdLife International (2011) 'Species factsheet: *Garrodia nereis*'. Retrieved 17 January, 2011 from <http://www.birdlife.org/>.

Brothers N, Pemberton D, Gales R, Skira I (1996) The status of seabirds in Tasmania. In *The Status of Australia's Seabirds: Proceedings of the National Seabird Workshop*, Canberra, 1–2 November 1993'. (Eds GJB Ross, K Weaver and JC Greig) pp. 181–184. Biodiversity Group, Environment Australia, Canberra.

Marchant S, Higgins PJ (Eds) (1990) *Handbook of Australian, New Zealand and Antarctic Birds. Volume 1: Ratites to Ducks.* Oxford University Press, Melbourne.

Parks and Wildlife Service (2006) 'Macquarie Island Nature Reserve and World Heritage Area Management Plan'. Parks and Wildlife Service, Department of Tourism, Arts and the Environment, Hobart.

Rounsevell DE, Brothers NP (1984) The status of seabirds on Macquarie Island. In 'Status and Conservation of the World's Seabirds. ICBP Tech. Publ. 2.' (Eds JP Croxall, PGH Evans and RW Schreiber) pp. 587–592. ICBP, Cambridge, UK.

Woehler EJ (2006) Status and conservation of the seabirds of Heard Island and the McDonald Islands. In *Heard Island: Southern Ocean Sentinel*. (Eds K Green and EJ Woehler) pp. 128–165. Surrey Beatty & Sons, Chipping Norton, NSW.

Comments received from

Barry Baker, Rosemary Gales, Tim Reid, Ben Sullivan, Eric Woehler

Subspecies Conservation Summary

White-bellied Storm-Petrel (Tasman Sea)

Fregetta grallaria grallaria (Vieilott, 1817)
Oceanitidae

Conservation status

Australian breeding population: Vulnerable D2
Population visiting Australian territory: Near Threatened D2

Reasons for listing

Australian breeding population: small population at 5 locations with plausible future threat
Population visiting Australian territory: 7 locations with plausible future threat

Status 2000

Australian breeding population: Vulnerable D2
Population visiting Australian territory: Near Threatened D2

Status 1990

Australian breeding population: Vulnerable D2
Population visiting Australian territory: Near Threatened D2

Taxonomy

Three extralimital subspecies. The species is Least Concern.
Taxonomic uniqueness: medium (6 genera/family, 2 species/genus, 4 subspecies/species).

Range

In Australian territory, breeding recorded on Roach I., Ball's Pyramid, Muttonbird I., Soldiers Cap and South I. in the Lord Howe I. group (Hindwood 1940; Hutton 1991; N. Carlile *in litt.*). Extralimitally, it breeds on Macauley and Curtis Is. on the Kermadec Is. group (Veitch *et al.* 2004). It is highly pelagic and rarely approaches land except when near colonies (Department of Sustainability, Environment, Water, Population and Communities 2011).

Abundance

About 100 pairs are estimated to nest on Soldiers Cap, Roach, Muttonbird and South Is. (N. Carlile *in litt.*). Numbers on Ball's Pyramid are unknown. Locally extinct from Lord Howe I. since 1913 (Hindwood 1940). Fewer than 700 pairs are estimated to breed on the Kermadec Is. (Veitch *et al.* 2004). Number of breeding birds of all subspecies is around 300 000 (BirdLife International 2011).

Ecology

Breeds in loose colonies on offshore islands, in rock crevices or on grassy slopes. Forages pelagically on squid and crustaceans which are caught on the wing by pattering and dipping. The species has been recorded feeding in the company of other seabirds and following ships (Marchant and Higgins 1990). A generation time of 15.2 years (BirdLife International 2011) is derived from an average age at first breeding of 4.7 years, an annual survival of adults of 87.8% and a maximum longevity of 30.4 years, all extrapolated from congeners.

Threats

In Australian territory, breeding colonies are confined to 5 islets, and their small numbers are vulnerable to catastrophe, such as the accidental introduction of the Black Rat *Rattus rattus* such as has occurred on the larger island nearby. Black Rats caused the Storm-petrel's extinction on Lord Howe I. (Hindwood 1940; McAllan *et al.* 2004) and has probably reduced the population in the Kermadec Is. (Veitch *et al.* 2004).

Conservation objectives

1. Maintain current breeding population
2. Expansion of population to main islands

Information required

1. Population size and trends

Management actions required

1. Maintain strict quarantine for islets within the Lord Howe I. group
2. Eradicate rats from Lord Howe I.

Bibliography

BirdLife International (2011) 'Species factsheet: *Fregetta grallaria*'. Retrieved 17 January, 2011 from <http://www.birdlife.org/>.

Department of Sustainability, Environment, Water, Population and Communities (2011) *Fregetta grallaria grallaria*. In 'Species Profile and Threats Database'. Department of Sustainability, Environment, Water, Population and Communities, Canberra. Retrieved 24 February, 2011 from <http://www.environment.gov.au/sprat>.

Hindwood KA (1940) The birds of Lord Howe Island. *Emu* **40**, 1–86.

Hutton I (1991) *Birds of Lord Howe Island, Past and Present*. The Author, Coffs Harbour.

Marchant S, Higgins PJ (Eds) (1990) *Handbook of Australian, New Zealand and Antarctic Birds. Volume 1: Ratites to Ducks*. Oxford University Press, Melbourne.

McAllan IAW, Curtis BR, Hutton I, Cooper RM (2004) The birds of the Lord Howe Island Group: a review of records. *Australian Field Ornithology* **21**, 1–82.

Current eligibility against IUCN Red List Criteria

IUCN category	Criteria eligibility
Australian breeding population	
A	Not applicable: past, current or future population declines are thought unlikely to exceed 20% in any 3-generation period
B	Not applicable: EOO, AOO, habitat, locations or population not declining, population fluctuations not extreme
C	Not applicable: no continuing population decline or extreme fluctuations
D	Vulnerable: population 250–1000 mature individuals; ≤5 locations with plausible future threat
E	Not applicable: no population viability analysis undertaken
Population visiting Australian territory	
A	Not applicable: past, current or future population declines are thought unlikely to exceed 20% in any 3-generation period
B	Not applicable: EOO, AOO, habitat, locations or population not declining, population fluctuations not extreme
C	Not applicable: no continuing population decline or extreme fluctuations
D	Near Threatened: near locations with plausible future threat
E	Not applicable: no population viability analysis undertaken

IUCN Red List assessment data

	Estimate	Reliability
Australian breeding population		
Extent of occurrence trend	1 500 000 km^2 stable	medium high
Area of occupancy trend	1 km^2 stable	high high
No. of mature individuals trend	300 stable	low low
No. subpopulations	1	low
No. locations	5	high
Generation time	15.2 years	low
Global population share	15%	low
Level of genetic exchange	low	low
Population visiting Australian territory		
Extent of occurrence trend	1 500 000 km^2 stable	medium medium
Area of occupancy trend	1 500 000 km^2 stable	high medium
No. of mature individuals trend	2000 stable	low low
No. subpopulations	1	high
No. locations	7	medium
Global population share	<1%	low

Veitch CR, Miskelly CM, Harper GA, Taylor G, Tennyson AJD (2004) Birds of the Kermadec Islands, south-west Pacific. *Notornis* **51**, 61–90.

Comments received from

Barry Baker, Nicholas Carlile, David Priddel

Species Conservation Summary

Wandering Albatross

Diomedea exulans Linnaeus, 1758

Diomedeidae

Conservation status

Australian breeding population: Critically Endangered B2ab(ii,iii), D

Population visiting Australian territory: Vulnerable A4bd

Reasons for listing

Australian breeding population: Population occurs at a single location where habitat quality deteriorating and the area occupied possibly contracting as a result of rabbit grazing; population <50 mature individuals

Population visiting Australian territory: Population decline 30–49% over 3 generations (70 years) currently under way

Status 2000

Australian breeding population: Critically Endangered D

Current eligibility against IUCN Red List Criteria

IUCN category	Criteria eligibility
Australian breeding population	
A	Not applicable: past, current or future population declines are thought unlikely to exceed 20% in any 3-generation period
B	Critically Endangered: AOO<10 km^2, 1 location with plausible future threat, inferred continuing decline in AOO, habitat quality
C	Not applicable: no continuing population decline or extreme fluctuations
D	Critically Endangered: population <50 mature individuals
E	Not applicable: no population viability analysis undertaken
Population visiting Australian territory	
A	Vulnerable: current population decline of 30–49% over 3 generations (70 years) estimated and projected from monitoring and levels of exploitation
B	Not applicable: >10 locations, not severely fragmented, population fluctuations not extreme
C	Not applicable: population >10 000 mature individuals
D	Not applicable: population >1000 mature individuals, >5 locations
E	Not applicable: no population viability analysis undertaken

IUCN Red List assessment data

	Estimate	Reliability
Australian breeding population		
Extent of occurrence trend	4 000 000 km^2 stable	high high
Area of occupancy trend	4 km^2 decreasing	high medium
No. of mature individuals trend	36 stable	medium low
No. subpopulations	1	high
No. locations	1	high
Generation time	23.3 years	medium
Global population share	<1%	high
Level of genetic exchange	low	medium
Population visiting Australian territory		
Extent of occurrence trend	4 000 000 km^2 stable	high high
Area of occupancy trend	4 000 000 km^2 stable	high high
No. of mature individuals trend	26 000 decreasing	medium medium
No. subpopulations	5	high
No. locations	>10	high
Largest subpopulation	6000	medium
Global population share	50–100%	high

Population visiting Australian territory: Vulnerable A4bd

Status 1990

Australian breeding population: Critically Endangered D

Population visiting Australian territory: Vulnerable A4bd

Taxonomy

No infraspecific taxa described; sometimes combined with Amsterdam Albatross *D. amsterdamensis*, Antipodean Albatross *D. antipodensis* and Tristan Albatross *D. dabbenena*

Taxonomic uniqueness: high (4 genera/family, 6 species/genus, 1 subspecies/species)

Range

Breeds on Macquarie I. (Department of Sustainability, Environment, Water, Population and Communities 2011). A pair observed with a large chick on Heard I. (Woehler 1991). Extralimitally breeds South Georgia, Prince Edward Is., Crozet Is. and Kerguelen Is. Non-breeding and juvenile birds remain between subantarctic and subtropical waters north of 50°S with a significant proportion wintering around the southern and eastern coast of Australia (Woehler *et al.* 1990; Reid *et al.* 2002).

Abundance

The status of the extremely small population (5 pairs in 2008) on Macquarie I. is currently uncertain, with trends in numbers and survival in the past similar to those observed in the Indian Ocean populations (Delord *et al.* 2008). Numbers declined by 9.3% 1965–1984 (28 pairs 1968 to 2 pairs 1985), increased by 6.3% 1984–1999, possibly decreased by 4.8% 1999–2008. Adult survival is high at 95.3% but juvenile survival only 46%. The total annual breeding population is estimated at approximately 8050 pairs, representing about 28 000 mature individuals and a total population of 55 000. All populations decreased at some stage over the last 25 years (Agreement on the Conservation of Albatrosses and Petrels 2009).

Ecology

Breeds biennially on the ground on isolated subantarctic islands. Feeds mostly on fish and cephalopods, seizing them from the surface during the day. Well known for following fishing vessels, sometimes in large groups, and competing vigorously for fisheries discards (van den Hoff 2001; Xavier *et al.* 2003). A generation time of 23.3 years (BirdLife International 2011) is derived from an age at first breeding of 9.0 years and an annual survival of adults of 93.0% (Schreiber and Burger 2002).

Threats

On Macquarie I. the increase in rabbit *Oryctolagus cuniculus* numbers since feral cats *Felis catus* were eradicated in 1999 has led to overgrazing of native vegetation and the extensive destruction of habitat and soil erosion around nesting sites. This threat may continue unless an eradication program planned in 2011–14 is successful (Agreement on the Conservation of Albatrosses and Petrels 2009). At sea incidental mortality in longline fishing operations is a major threat. Chicks also ingest marine debris and hooks fed to them by their parents (BirdLife International 2011).

Conservation objectives

1. Stabilise nesting sites on Macquarie I.
2. At-sea threats reduced to acceptable levels
3. Global agreement on conservation measures
4. Increased public awareness of the conservation needs of albatrosses

Information required

1. Population trends or changes in adult and juvenile survival
2. Distribution of immature birds
3. Overlap with fishing operations and documented levels of bycatch so that mitigation measures can be targeted
4. Status of breeding population on Heard I.

Management actions required

1. Secure nest sites on Macquarie I. by eliminating rabbits
2. Continue existing monitoring and research programs
3. Conduct complete censuses at all sites at regular intervals
4. Continue tracking studies to determine spatial and temporal overlap with fisheries for populations and life stages where these data do not exist
5. Improve bycatch assessment and identification of taxa within the 'wandering albatross' group in distant-water and South American fisheries
6. Maintain mitigation strategies for fisheries, including via intergovernmental mechanisms under the auspices of the Agreement on the Conservation of Albatrosses and Petrels, the Food and Agriculture Organisation and Regional Fisheries Management Organisations

Bibliography

Agreement on the Conservation of Albatrosses and Petrels (2009) 'ACAP species assessment: Wandering Albatross *Diomedea exulans*'. Retrieved 10 September, 2009 from <http://www.acap.aq/>.

BirdLife International (2011) 'Species factsheet: *Diomedea exulans*'. Retrieved 17 January, 2011 from <http://www.birdlife.org/>.

Delord K, Besson D, Barbraud C, Weimerskirch H (2008) Population trends in a community of large Procellariiforms of Indian Ocean: potential effects of environment and fisheries interactions. *Biological Conservation* **141**, 1840–1856.

Department of Sustainability, Environment, Water, Population and Communities (2011) *Diomedea exulans exulans*. In 'Species Profile and Threats Database'. Department of Sustainability, Environment, Water, Population and Communities, Canberra. Retrieved 24 February, 2011 from <http://www.environment.gov.au/sprat>.

Reid TA, Hindell MA, Eades DW, Newman M (2002) *Seabird Atlas of South-eastern Australian Waters*. Birds Australia Monograph 4.

Schreiber EA, Burger J (2002) *Biology of Marine Birds*. CRC Press, Boca Raton, FL.

van den Hoff J (2001) Further observations on the cephalopod diet of Wandering Albatrosses (*Diomedea exulans* L.) at Macquarie Island. *Emu* **101**, 169–172.

Woehler EJ (1991) The status and conservation of seabirds of Heard Island and the McDonald Islands. In 'Seabird Status and Conservation. ICBP Tech. Publ. No 11'. (Ed. JP Croxall) pp. 263–277. ICBP, Cambridge, UK.

Woehler EJ, Hodges CL, Watts DJ (1990) 'An atlas of the pelagic distribution and abundance of seabirds in the southern Indian Ocean 1981 to 1990'. ANARE Research Notes 77.

Xavier JC, Croxall JP, Reid K (2003) Interannual variation in the diets of two albatross species breeding at South Georgia: implications for breeding performance. *Ibis* **145**, 593–610.

Comments received from

Barry Baker, Rosemary Gales, Wieslawa Misiak, Tim Reid, Ben Sullivan, Eric Woehler

Species Conservation Summary

Tristan Albatross

Diomedea dabbenena Mathews, 1929
Diomedeidae

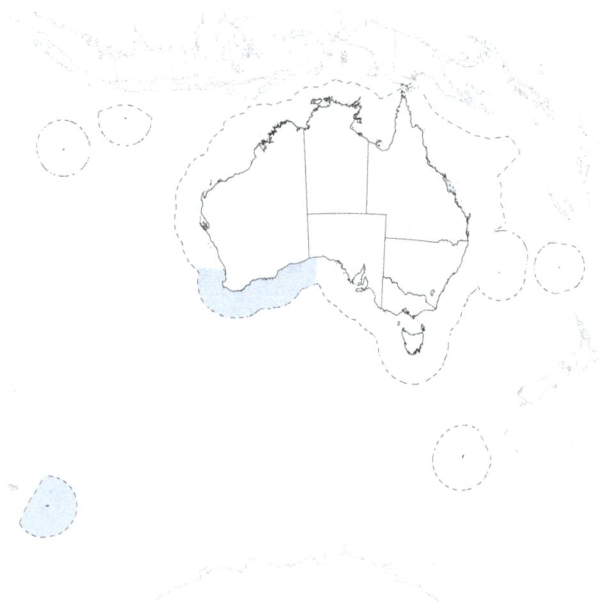

Conservation status

Population visiting Australian territory: Critically Endangered A4ade

Reasons for listing

Population decline >80% in 3 generations (75 years) occurring as a result of at-sea mortality associated with fisheries, particularly longline fishing vessels, as well as predation at its breeding islands

Status 2000

Vulnerable A4ade
Reason for change: initial declines intensifying

Status 1990

Near Threatened A4ade
Reason for change: initial declines intensifying

Taxonomy

This species is sometimes included with Wandering Albatross *D. exulans*, Amsterdam Albatross *D. amsterdamensis* and Antipodean Albatross *D. antipodensis*
Taxonomic uniqueness: high (4 genera/family, 6 species/genus, 1 subspecies/species)

Range

Breeds on Gough and Inaccessible Is. Although only one confirmed Australian record, satellite-tracking studies indicate that non-breeding birds disperse to the east coast of South America and the south west of Western Australia, as well as moving north to waters off Namibia and Angola (Agreement on the Conservation of Albatrosses and Petrels 2009; Department of Sustainability, Environment, Water, Population and Communities 2011).

Abundance

There are about 2700 pairs, all but 1 on Gough I., with a rate of decline of 2.9–5.3% p.a., translating to a 79% decline over a 70-year period (Agreement on the Conservation of Albatrosses and Petrels 2009).

Ecology

Colonial species that breeds biennially. Mostly feeds by surface seizing, possibly largely by night, on squid, fish and crustaceans (Agreement on the Conservation of Albatrosses and Petrels 2009). A generation time of 28.7 years (BirdLife International 2011) is derived from an age at first breeding of 9.7 years and an annual survival of adults of 94.8% (Schreiber and Burger 2002).

Threats

On land introduced House Mice *Mus musculus* cause major mortality of chicks. At sea there is ongoing mortality from longline fishing, particularly off South America (BirdLife International 2011).

Conservation objectives

1. At-sea threats reduced to acceptable levels
2. Global agreement on conservation measures
3. Increased public awareness of the conservation needs of albatrosses

Information required

1. Genetic profiles to determine provenance of birds caught as bycatch
2. Population trends or changes in adult and juvenile survival
3. Distribution of immature and non-breeding birds
4. Overlap with fishing operations and documented levels of bycatch so that mitigation measures can be targeted

Management actions required

1. Improve bycatch assessment and identification of taxa within the 'wandering albatross' group in distant-water and South American fisheries
2. Maintain mitigation strategies for fisheries, including via intergovernmental mechanisms under the auspices of the Agreement on the Conservation of Albatrosses and Petrels, the Food and Agriculture Organisation and Regional Fisheries Management Organisations

Bibliography

Agreement on the Conservation of Albatrosses and Petrels (2009) 'ACAP species assessment: Tristan Albatross *Diomedea dabbenena*'. Retrieved 4 September, 2009 from <http://www.acap.aq/>.

BirdLife International (2011) 'Species factsheet: *Diomedea dabbenena*'. Retrieved 17 January, 2011 from <http://www.birdlife.org/>.

Department of Sustainability, Environment, Water, Population and Communities (2011) *Diomedea exulans exulans*. In 'Species Profile and Threats Database'. Department of Sustainability, Environment, Water, Population and Communities, Canberra. Retrieved 24 February, 2011 from <http://www.environment.gov.au/sprat>.

Current eligibility against IUCN Red List Criteria

IUCN category	Criteria eligibility
Population visiting Australian territory	
A	Critically Endangered: current population decline of >80% over 3 generations (75 years) based on direct observation and inferred from levels of exploitation and effects of introduced taxa
B	Endangered: AOO (breeding) <500 km^2, at ≤5 locations with plausible future threat, observed continuing decline in habitat quality, no. mature individuals
C	Not applicable: population >10 000 mature individuals
D	Vulnerable: ≤5 locations with plausible future threat
E	Not applicable: no population viability analysis undertaken

IUCN Red List assessment data

	Estimate	Reliability
Population visiting Australian territory		
Extent of occurrence trend	1 340 000 km^2 stable	high high
Area of occupancy trend	1 340 000 km^2 stable	high high
No. of mature individuals trend	11 300 decreasing	medium medium
No. subpopulations	1	high
No. locations	2	high
Generation time	28.7 years	medium
Global population share	50–100%	low

Schreiber EA, Burger J (2002) *Biology of Marine Birds*. CRC Press, Boca Raton, FL.

Comments received from

Barry Baker, Rosemary Gales, Wieslawa Misiak, Tim Reid, Ben Sullivan, Eric Woehler

Subspecies Conservation Summary

Antipodean Albatross (Auckland Islands)

Diomedea antipodensis gibsoni Robertson & Warham, 1992

Diomedeidae

Conservation status

Population visiting Australian territory: Endangered A4ad

Reasons for listing

Population decline 50–79% in 3 generations (82 years) occurring as a result of at-sea mortality associated with fisheries, particularly longline fishing vessels

Status 2000

Vulnerable A4ad

Reason for change: initial declines intensifying

Status 1990

Near Threatened A4ad

Reason for change: initial declines intensifying

Taxonomy

D. a. antipodensis (Antipodes and Campbell Is.) is Vulnerable. Species sometimes included with Wandering Albatross *D. exulans*, Amsterdam Albatross *D. amsterdamensis* and Tristan Albatross *D. dabbenena*

Taxonomic uniqueness: high (4 genera/family, 6 species/genus, 2 subspecies/species)

Range

Breeds on Adams, Disappointment and Auckland Is. south of New Zealand. Circumpolar distribution when not breeding. Auckland Is. birds mostly foraged in the Tasman Sea. Satellite-tracking studies indicate that non-breeding birds spend over 50% of their time in the Exclusive Economic Zones of Australia, New Zealand and Chile (Agreement on the Conservation of Albatrosses and Petrels 2009; Department of Sustainability, Environment, Water, Population and Communities 2011).

Abundance

The total breeding population on Adams I. was estimated at approximately 9682 pairs in 2006–09, with an average of 3277 pairs breeding annually over the four seasons.

Current eligibility against IUCN Red List Criteria

IUCN category	Criteria eligibility
Population visiting Australian territory	
A	Endangered: current population decline of 50–79% over 3 generations (82 years) observed directly and estimated and projected from levels of exploitation
B	Not applicable: at ≤5 locations but no plausible future threat to locations, population fluctuations not extreme
C	Not applicable: population >10 000 mature individuals
D	Not applicable: population >1000 mature individuals, >5 locations
E	Not applicable: no population viability analysis undertaken

IUCN Red List assessment data

	Estimate	Reliability
Population visiting Australian territory		
Extent of occurrence trend	4 700 000 km^2 stable	high high
Area of occupancy trend	4 700 000 km^2 stable	high high
No. of mature individuals trend	20 000 decreasing	medium medium
No. subpopulations	1	high
No. locations	4	high
Generation time	27.4 years	medium
Global population share	50–100%	high

Earlier surveys recorded 352 pairs on Disappointment I. and 72 pairs on Auckland I. Breeding success has declined since 1991, and in 2006 was the lowest ever recorded (25%). Adult survival in 2005 and 2006 was also the lowest ever recorded. The total Adams I. breeding population has decreased by roughly 40% between counts in 1997 and 2009 (Agreement on the Conservation of Albatrosses and Petrels 2009).

Ecology

Colonial species that breeds biennially. Mostly feeds by surface seizing. Diet is poorly known but cephalopods are followed by fish during chick rearing. In winter feeds on moribund Giant Cuttlefish *Sepia apama* off the New South Wales coast of eastern Australia (Agreement on the Conservation of Albatrosses and Petrels 2009; Department of Sustainability, Environment, Water, Population and Communities 2011). A generation time of 27.4 years (BirdLife International 2011) is derived from an age at first breeding of 9.2 years and an annual survival of adults of 94.5% (Schreiber and Burger 2002).

Threats

On land introduced feral cats *Felis catus* and pigs *Sus scrofa* cause some mortality on Auckland I. Principal reason for decline is mortality from longline fishing, particularly of females. It is also possible that changes in water temperature have reduced local productivity and underpin some of the declines on Adams I. (BirdLife International 2011).

Conservation objectives

1. At-sea threats reduced to acceptable levels
2. Global agreement on conservation measures
3. Increased public awareness of the conservation needs of albatrosses

Information required

1. Determine impacts of increased ocean temperatures and changes in productivity in the Tasman Sea
2. Population trends or changes in adult and juvenile survival
3. Distribution of immature and non-breeding birds
4. Overlap with fishing operations and documented levels of bycatch so that mitigation measures can be targeted
5. Extent of incidental mortality, especially in unregulated longline fisheries operating in the central and south east Pacific and the central Tasman Sea

Management actions required

1. Improve bycatch assessment and identification of taxa within the 'wandering albatross' group in distant-water and South American fisheries
2. Maintain mitigation strategies for fisheries, including via intergovernmental mechanisms under the auspices of the Agreement on the Conservation of Albatrosses and Petrels, the Food and Agriculture Organisation and Regional Fisheries Management Organisations

Bibliography

Agreement on the Conservation of Albatrosses and Petrels (2009) 'ACAP species assessment: Antipodean Albatross *Diomedea antipodensis*'. Retrieved 18 September, 2009 from <http://www.acap.aq/>.

BirdLife International (2011) 'Species factsheet: *Diomedea antipodensis*'. Retrieved 17 January, 2011 from <http://www.birdlife.org/>.

Department of Sustainability, Environment, Water, Population and Communities (2011) *Diomedea exulans antipodensis*. In 'Species Profile and Threats Database'. Department of Sustainability, Environment, Water, Population and Communities, Canberra. Retrieved 24 February, 2011 from <http://www.environment.gov.au/sprat>.

Schreiber EA, Burger J (2002) *Biology of Marine Birds*. CRC Press, Boca Raton, FL.

Comments received from

Barry Baker, Rosemary Gales, Wieslawa Misiak, Tim Reid, Ben Sullivan

Species Conservation Summary

Northern Royal Albatross

Diomedea sanfordi Murphy, 1917

Diomedeidae

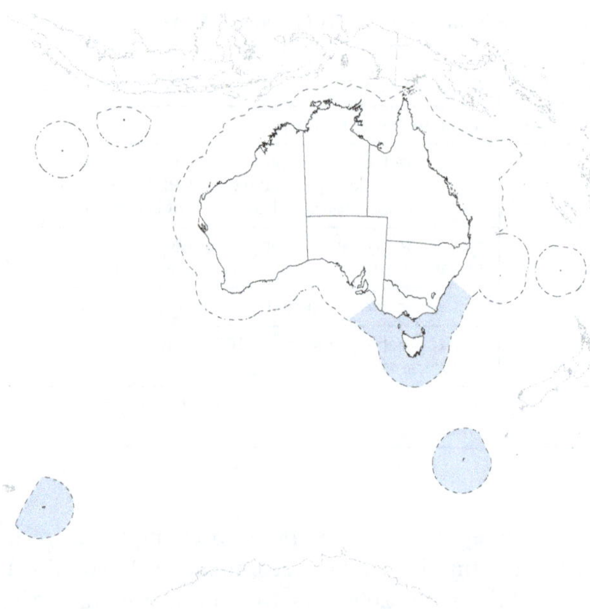

Conservation status

Population visiting Australian territory: Endangered A4bc; B2a+b(iii,v)

Reasons for listing

Population decline 50–79% in 3 generations (81 years) thought to be occurring as a result of reduced breeding success following storm damage to principal breeding site

Status 2000

Endangered A4bc; B2a+b(iii,v)

Status 1990

Endangered A4bc; B2a+b(iii,v)

Taxonomy

Species sometimes included with Southern Royal Albatross *D. epomophora*

Taxonomic uniqueness: high (4 genera/family, 6 species/genus, 1 subspecies/species)

Range

Breeds on Forty-Fours, Big and Little Sister Is. (Chatham Is.) and Taiaroa Head (Otago Peninsula, South I., New Zealand). Non-breeding birds have a circumpolar migration, flying downwind generally between 30° and 45°S (Agreement on the Conservation of Albatrosses and Petrels 2009; Department of Sustainability, Environment, Water, Population and Communities 2011).

Abundance

There are 5800 pairs on the Chatham Is. with a further 32 pairs on Taiaroa Head. There are 8 hybrid individuals of *D. epomophora* x *D. sanfordi* at Taiaroa Head and another 2 hybrid pairs on Enderby I. (Auckland Is.). A major storm damaged nesting habitat on Chatham Is. in 1985 that greatly reduced subsequent productivity to the extent that a major decline in the population is expected, though evidence of the decline has yet to become apparent (Agreement on the Conservation of Albatrosses and Petrels 2009).

Ecology

Colonial species that breeds biennially. Mostly feeds by surface seizing, largely taking moribund cephalopods, fish, tunicates and crustaceans (Agreement on the Conservation of Albatrosses and Petrels 2009; Department of Sustainability Environment Water Population and Communities 2011). A generation time of 27.0 years (BirdLife International 2011) is derived from an age at first breeding of 8.0 years and an annual survival of adults of 94.8% (Schreiber and Burger 2002).

Threats

On land a cyclonic storm washed away much of the breeding habitat on the Chatham Is., reducing subsequent juvenile recruitment. Longline fishing is not thought to be having a major effect on this albatross species (BirdLife International 2011).

Conservation objectives

1. At-sea threats maintained at acceptable levels
2. Global agreement on conservation measures
3. Increased public awareness of the conservation needs of albatrosses

Information required

1. Population trends or changes in adult and juvenile survival
2. Distribution of immature and non-breeding birds

Management actions required

1. Maintain mitigation strategies for fisheries, including via intergovernmental mechanisms under the auspices of the Agreement on the Conservation of Albatrosses and Petrels, the Food and Agriculture Organisation and Regional Fisheries Management Organisations

Bibliography

Agreement on the Conservation of Albatrosses and Petrels (2009) 'ACAP species assessment: Northern Royal Albatross *Diomedea sanfordi*'. Retrieved 31 August, 2009 from <http://www.acap.aq/>.

BirdLife International (2011) 'Species factsheet: *Diomedea sanfordi*'. Retrieved 17 January, 2011 from <http://www.birdlife.org/>.

Department of Sustainability, Environment, Water, Population and Communities (2011) *Diomedea epomophora sanfordi*. In 'Species Profile and Threats Database'. Department of Sustainability, Environment, Water, Population and Communities, Canberra. Retrieved 24 February, 2011 from <http://www.environment.gov.au/sprat>.

Schreiber EA, Burger J (2002) *Biology of Marine Birds*. CRC Press, Boca Raton, FL.

Current eligibility against IUCN Red List Criteria

IUCN category	Criteria eligibility
Population visiting Australian territory	
A	Endangered: current population decline of 50–79% over 3 generations (81 years) estimated and projected from monitoring and inferred from deteriorating habitat quality
B	Endangered: AOO (breeding) <500 km^2, at ≤5 locations with plausible future threat, observed continuing decline in habitat quality, no. mature individuals
C	Not applicable: population >10 000 mature individuals
D	Vulnerable: ≤5 locations with plausible future threat
E	Not applicable: no population viability analysis undertaken

IUCN Red List assessment data

	Estimate	Reliability
Population visiting Australian territory		
Extent of occurrence trend	2 300 000 km^2 stable	high high
Area of occupancy trend	2 300 000 km^2 stable	high high
No. of mature individuals trend	17 000 decreasing	medium medium
No. subpopulations	2	high
No. locations	2	high
Largest subpopulation	11 300	medium
Generation time	27.0 years	medium
Global population share	50–100%	low

Comments received from

Barry Baker, Rosemary Gales, Wieslawa Misiak, Tim Reid, Ben Sullivan, Eric Woehler

Species Conservation Summary

Southern Royal Albatross

Diomedea epomophora Lesson, 1825

Diomedeidae

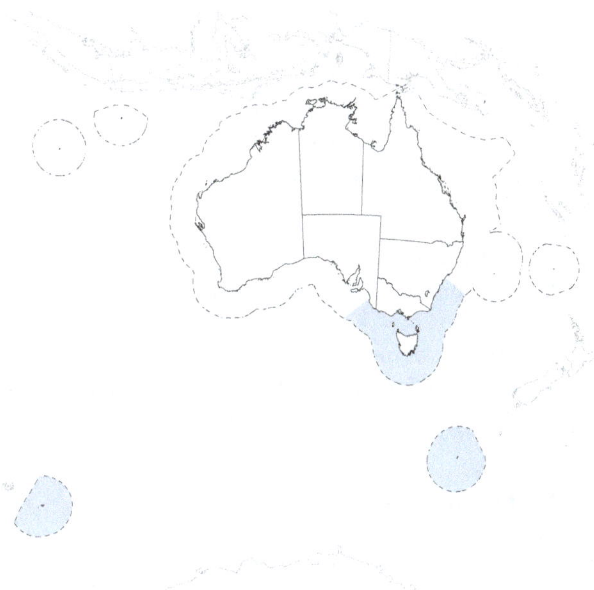

Conservation status

Population visiting Australian territory: Vulnerable D2

Reasons for listing

Occurs at ≤5 locations with plausible future threat

Status 2000

Vulnerable D2

Status 1990

Vulnerable D2

Taxonomy

Species sometimes included with Northern Royal Albatross *D. sanfordi*

Taxonomic uniqueness: high (4 genera/family, 6 species/genus, 1 subspecies/species)

Range

New Zealand endemic, 99% breeding on Campbell I. with the remainder on the Auckland Is. Circumpolar distribution when not breeding, with juveniles mostly found off western South America, immatures off eastern South America, and adults near New Zealand (Department of Sustainability, Environment, Water, Population and Communities 2011).

Abundance

There were 7800 annual breeding pairs on Campbell I. in 2008, 15 on Adams I. in 1991, 2 on Auckland I. in 1989 and 69 on Enderby I. in 2001. Numbers on Campbell and Enderby Is. have increased over recent decades (Agreement on the Conservation of Albatrosses and Petrels 2009).

Ecology

Breeds biennially in colonies among grass tussocks. Feeds pelagically, largely surface seizing moribund squid, fish and tunicates. It also follows fishing boats to retrieve offal (Agreement on the Conservation of Albatrosses and Petrels 2009; Department of Sustainability, Environment, Water, Population and Communities 2011). A generation time of 42.3 years (BirdLife International 2011) is derived from an age at first breeding of 9.0 years and an annual survival of adults of 97.0% (Schreiber and Burger 2002).

Current eligibility against IUCN Red List Criteria

IUCN category	Criteria eligibility
Population visiting Australian territory	
A	Not applicable: past, current or future population declines are thought unlikely to exceed 20% in any 3-generation period
B	Not applicable: EOO, AOO, habitat, locations or population not declining, population fluctuations not extreme
C	Not applicable: population >10 000 mature individuals
D	Vulnerable: ≤5 locations with possible future threat
E	Not applicable: no population viability analysis undertaken

IUCN Red List assessment data

	Estimate	Reliability
Population visiting Australian territory		
Extent of occurrence trend	2 300 000 km^2 stable	high high
Area of occupancy trend	2 300 000 km^2 stable	high high
No. of mature individuals trend	29 000 stable	high high
No. subpopulations	1	high
No. locations	2	high
Generation time	42.3 years	medium
Global population share	50–100%	medium

Threats

There are currently no known substantial land-based threats to Southern Royal Albatross and the population appears to be recovering from hunting and predation by feral animals during the 20th century. At sea the primary threat is drowning in longline fishing gear, but these losses appear to have peaked in the 1970s and 1980s (BirdLife International 2011) and have not reduced the population as much as was feared in 2000 (Garnett and Crowley 2000).

Conservation objectives

1. At-sea threats maintained at acceptable levels
2. Global agreement on conservation measures
3. Increased public awareness of the conservation needs of albatrosses

Information required

1. Population trends or changes in adult and juvenile survival
2. Distribution of immature and non-breeding birds
3. Overlap with fishing operations and documented levels of bycatch so that mitigation measures can be targeted

Management actions required

1. Assess the impact of trawl fisheries bycatch
2. Maintain mitigation strategies for fisheries, including via intergovernmental mechanisms under the auspices of the Agreement on the Conservation of Albatrosses and Petrels, the Food and Agriculture Organisation and Regional Fisheries Management Organisations

Bibliography

Agreement on the Conservation of Albatrosses and Petrels (2009) 'ACAP species assessment: Southern Royal Albatross *Diomedea epomophora*'. Retrieved 2 September, 2009 from <http://www.acap.aq/>.

BirdLife International (2011) 'Species factsheet: *Diomedea epomophora*'. Retrieved 17 January, 2011 from <http://www.birdlife.org/>.

Department of Sustainability, Environment, Water, Population and Communities (2011) *Diomedea epomophora epomophora*. In 'Species Profile and Threats Database'. Department of Sustainability, Environment, Water, Population and Communities, Canberra. Retrieved 24 February, 2011 from <http://www.environment.gov.au/sprat>.

Garnett S, Crowley GM (2000) *The Action Plan for Australian Birds 2000*. Environment Australia, Canberra.

Schreiber EA, Burger J (2002) *Biology of Marine Birds*. CRC Press, Boca Raton, FL.

Comments received from

Barry Baker, Rosemary Gales, Wieslawa Misiak, Tim Reid, Ben Sullivan

Species Conservation Summary

Black-browed Albatross

Thalassarche melanophrys (Temminck, 1828)
Diomedeidae

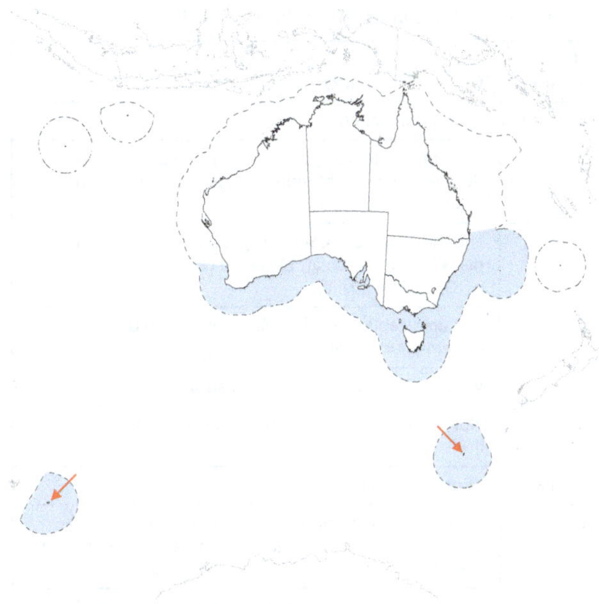

Conservation status

Australian breeding population: Endangered B2ab(ii,iii)
Population visiting Australian territory: Endangered A4bd

Reasons for listing

Australian breeding population: Population occurs at only 4 locations where the habitat quality is declining as a result of rabbit grazing and the area occupied is probably contracting
Population visiting Australian territory: Population decline 50–79% over three generations (65 years) currently under way

Status 2000

Australian breeding population: Vulnerable D2
Reason for change in 2010: active threats to Macquarie I. breeding sites
Population visiting Australian territory: Near Threatened A4bd
Reason for change in 2010: intensification of declines

Status 1990

Australian breeding population: Vulnerable D2
Population visiting Australian territory: Least Concern
Reason for change in 2000: declines detected

Taxonomy

No infraspecific taxa described; sometimes included with Campbell Albatross *T. impavida*
Taxonomic uniqueness: high (4 genera/family, 11 species/genus, 1 subspecies/species)

Range

Breeding occurs on Heard, McDonald, Bishop and Clerk and Macquarie Is. (Department of Sustainability Environment Water Population and Communities 2011). Extralimitally breeds South Georgia, Crozet, Kerguelen, Antipodes, and Campbell Is., as well as on Falkland Is., and 4 island groups off southern Chile (Agreement on the Conservation of Albatrosses and Petrels 2009). At sea it has a circumpolar distribution, ranging from subtropical to polar waters, with the majority of breeding birds found in the subantarctic zone, and the majority of non-breeding birds occurring in subtropical as well as subantarctic waters. Prefers shallow (<1000 m) waters but they also frequently forage in deeper waters in association with the polar front or other oceanic frontal systems (Terauds *et al.* 2006; Agreement on the Conservation of Albatrosses and Petrels 2009). Satellite tracking suggests that many individuals remain close to breeding islands all year but some or all birds from islands near South America may visit Australian waters in addition to those that breed there.

Abundance

Australian breeding population is 600 pairs on Heard I. (last census 2001), 90 pairs McDonald I. (1980), 141 pairs Bishop and Clerk Is. (1993) and 46 pairs Macquarie I. (2007; Woehler 2006; Department of Sustainability, Environment, Water, Population and Communities 2011). Globally about 1.2 million mature individuals (Agreement on the Conservation of Albatrosses and Petrels 2009). Increasing trends reported for Heard I. (Woehler *et al.* 2002; Woehler 2006). Population on Macquarie I. was stable 1995–2007 (Terauds *et al.* 2005). Population trend on Bishop and Clerk Is. is unknown. Global population is declining rapidly (Agreement on the Conservation of Albatrosses and Petrels 2009).

Ecology

Nests colonially on subantarctic islands, usually annually although only 75% of successful breeders and 67% of failed breeders breed the following year. Feeds primarily on fish, cephalopods, and in some areas crustaceans, mostly by surface seizing but sometimes by diving up to 4.5 m deep (BirdLife International 2011). Diet varies between years depending on food availability (Xavier *et al.* 2003). A generation time of 21.5 years (BirdLife International 2011) is derived from an age at first breeding of 9.0 years and an annual survival of adults of 92.0% (Schreiber and Burger 2002).

Threats

At breeding sites recent volcanic eruptions at McDonald Is. (1993–2005) appear not to have affected Maxwell Hill and Meyer Rock, the sites where the albatrosses have been recorded breeding (Stephenson *et al.* 2005). Eruptions on Heard I. (1985–2008) are unlikely to have affected breeding sites (contra. Agreement on the Conservation of Albatrosses and Petrels 2009; BirdLife International 2011). On Macquarie I. the increase in rabbit *Oryctolagus cuniculus* numbers since feral cats *Felis catus* were eradicated in 1999 has led to overgrazing of native vegetation and extensive destruction of habitat and soil erosion at nesting

Current eligibility against IUCN Red List Criteria

IUCN category	Criteria eligibility
Australian breeding population	
A	Not applicable: past, current or future population declines are thought unlikely to exceed 20% in any 3-generation period
B	Endangered: AOO <500 km^2, at ≤5 locations with plausible future threat, observed continuing decline in AOO, habitat quality
C	Not applicable: no continuing population decline or extreme fluctuations
D	Vulnerable: ≤5 locations with plausible future threat
E	Not applicable: no population viability analysis undertaken
Population visiting Australian territory	
A	Endangered: current population decline of 50–79% over 3 generations (65 years) estimated and projected from monitoring and levels of exploitation
B	Vulnerable: AOO <2000 km^2, at ≤10 locations with plausible future threat, estimated continuing decline in AOO, habitat quality, no. mature individuals
C	Not applicable: population >10 000 mature individuals
D	Not applicable: population >1000 mature individuals, >5 locations
E	Not applicable: no population viability analysis undertaken

IUCN Red List assessment data

	Estimate	Reliability
Australian breeding population		
Extent of occurrence trend	4 400 000 km^2 stable	high high
Area of occupancy trend	6 km^2 decreasing	high medium
No. of mature individuals trend	2000 stable	medium high
No. subpopulations	2	high
No. locations	4	high
Largest subpopulation	1300	high
Generation time	21.5 years	medium
Global population share	<1%	high
Level of genetic exchange	low	medium
Population visiting Australian territory		
Extent of occurrence trend	4 400 000 km^2 stable	high high
Area of occupancy trend	4 400 000 km^2 stable	high high
No. of mature individuals trend	1 200 000 decreasing	medium high
No. subpopulations	10	high
No. locations	>10	high
Largest subpopulation	400 000	high
Global population share	26–50%	low

sites. This threat may continue unless an eradication program planned in 2011–14 is successful (Agreement on the Conservation of Albatrosses and Petrels 2009).

At sea this species frequently associates in large numbers with commercial fishing operations and is one of the most common bycatch species in many fishery operations throughout its range (BirdLife International 2011). Most birds breeding on Macquarie and Heard Is. feed locally where risk of being caught is low, though some feed in unregulated waters in the southern Indian Ocean (Terauds *et al.* 2006; Lawton *et al.* 2008). The overlap between the distribution of Black-browed Albatrosses and trawl fishing may be significant (Phillips *et al.* 2010) and although, in Commonwealth waters, a Threat Abatement Plan under the EPBC Act has effectively reduced bycatch from longline fishing, trawl fisheries require greater attention (Baker and Finley 2010).

Conservation objectives

1. Nesting sites on Macquarie I. stabilised
2. At-sea threats reduced to acceptable levels
3. Global agreement on conservation measures
4. Increased public awareness of the conservation needs of albatrosses

Information required

1. Information on population trends and demographic parameters for most sites, particularly Heard I.
2. Population trends or changes in adult and juvenile survival
3. Distribution of immature and non-breeding birds
4. Overlap with fishing operations and documented levels of bycatch so that mitigation measures can be targeted

Management actions required

1. Secure nest sites on Macquarie I. by eliminating rabbits
2. Continue existing monitoring and research programs
3. Conduct complete censuses at all sites at regular intervals
4. Assess the impact of trawl fisheries bycatch
5. Maintain mitigation strategies for fisheries, including via intergovernmental mechanisms under the auspices of the Agreement on the Conservation of Albatrosses and Petrels, the Food and Agriculture Organisation and Regional Fisheries Management Organisations

Bibliography

Agreement on the Conservation of Albatrosses and Petrels (2009) 'ACAP species assessment: Black-browed Albatross *Thalassarche melanophrys*'. Retrieved 1 September, 2009 from <http://www.acap.aq/>.

Baker GB, Finley LA (2010) '2008 National Assessment Report for reducing the incidental catch of seabirds in longline fisheries'. Bureau of Rural Sciences, Canberra.

BirdLife International (2011) 'Species factsheet: *Thalassarche melanophrys*'. Retrieved 17 January, 2011 from <http://www.birdlife.org/>.

Department of Sustainability, Environment, Water, Population and Communities (2011) *Thalassarche melanophris.* In 'Species Profile and Threats Database'. Department of Sustainability, Environment, Water, Population and Communities, Canberra. Retrieved 24 February, 2011 from <http://www.environment.gov.au/sprat>.

Lawton K, Kirkwood R, Robertson G, Raymond B (2008) Preferred foraging areas of Heard Island albatrosses during chick raising and implications for the management of incidental mortality in fisheries. *Aquatic Conservation: Marine and Freshwater Ecosystems* **18**, 309–320.

Phillips K, Giannini F, Lawrence E, Bensley N (2010) 'Cumulative assessment of the catch of non-target species in Commonwealth fisheries: a scoping study'. Bureau of Rural Sciences, Canberra.

Schreiber EA, Burger J (2002) *Biology of Marine Birds*. CRC Press, Boca Raton, FL.

Stephenson J, Budd GM, Manning J, Hansbro P (2005) Major eruption-induced changes to the McDonald Islands, southern Indian Ocean. *Antarctic Science* **17**, 259–266.

Terauds A, Gales R, Alderman R (2005) Trends in numbers and survival of Black-browed (*Thalassarche melanophrys*) and Grey-headed (*T. chrysostoma*) Albatrosses breeding on Macquarie Island. *Emu* **105**, 159–167.

Terauds A, Gales R, Baker GB, Alderman R (2006) Foraging areas of black-browed and grey-headed albatrosses breeding on Macquarie Island in relation to marine protected areas. *Aquatic Conservation: Marine and Freshwater Ecosystems* **16**, 133–146.

Woehler EJ (2006) Status and conservation of the seabirds of Heard Island and the McDonald Islands. In *Heard Island: Southern Ocean Sentinel*. (Eds K Green and EJ Woehler) pp. 128–165. Surrey Beatty & Sons, Chipping Norton, NSW.

Woehler EJ, Auman HJ, Riddle MJ (2002) Long-term population increase of black-browed albatrosses *Thalassarche melanophrys* at Heard Island, 1947/1948–2000/2001. *Polar Biology* **25**, 921–927.

Xavier JC, Croxall JP, Reid K (2003) Interannual variation in the diets of two albatross species breeding at South Georgia: implications for breeding performance. *Ibis* **145**, 593–610.

Comments received from

Barry Baker, Rosemary Gales, Wieslawa Misiak, Tim Reid, Ben Sullivan, Eric Woehler

Species Conservation Summary

Campbell Albatross

Thalassarche impavida (Mathews, 1912)
Diomedeidae

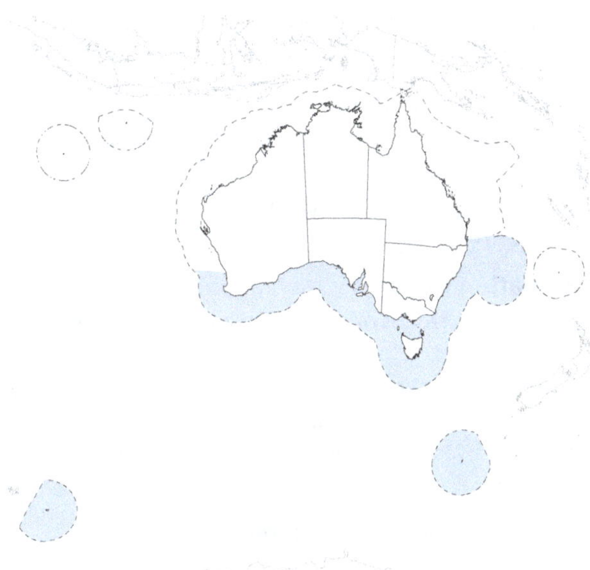

Conservation status

Population visiting Australian territory: Vulnerable D2

Reasons for listing

Occurs at ≤5 locations with plausible future threat

Status 2000

Vulnerable D2

Status 1990

Vulnerable D2

Taxonomy

No infraspecific taxa described. Species sometimes included with Black-browed Albatross *T. melanophrys*
Taxonomic uniqueness: high (4 genera/family, 11 species/genus, 1 subspecies/species)

Range

Breeds only on Campbell I. and the adjacent Isle de Jeanette Marie south of New Zealand. Juvenile and non-breeding individuals forage over temperate shelf waters of New Zealand, Australia and the central and western Pacific (Woehler *et al.* 1990; Reid *et al.* 2002; Agreement on the Conservation of Albatrosses and Petrels 2009).

Abundance

The last population estimate for Campbell I. was of an average of 24 600 pairs annually from 1995–1997, down from 31 300 pairs in the 1940s. Numbers decreased steeply in the 1970s and 1980s, attributed to bycatch from longline fisheries, but were either stable or increasing from 1984 to 1997. There is no more recent census (Agreement on the Conservation of Albatrosses and Petrels 2009).

Ecology

Breeds annually in colonies among grass tussocks, mostly in mixed colonies with the Grey-headed Albatross *T. chrysostoma*. It feeds pelagically over near-shore waters, surface seizing or undertaking shallow dives to catch squid, fish and crustaceans. It also follows fishing boats to retrieve offal (Agreement on the Conservation of Albatrosses and Petrels 2009). A generation time of 28.2 years

Current eligibility against IUCN Red List Criteria

IUCN category	Criteria eligibility
Population visiting Australian territory	
A	Not applicable: past, current or future population declines are thought unlikely to exceed 20% in any 3-generation period
B	Not applicable: EOO, AOO, habitat, locations or population not declining, population fluctuations not extreme
C	Not applicable: population >10 000 mature individuals
D	Vulnerable: ≤5 locations with plausible future threat
E	Not applicable: no population viability analysis undertaken

IUCN Red List assessment data

	Estimate	Reliability
Population visiting Australian territory		
Extent of occurrence trend	4 400 000 km^2 stable	high high
Area of occupancy trend	4 400 000 km^2 stable	high high
No. of mature individuals trend	50 000 stable	high high
No. subpopulations	1	high
No. locations	2	high
Generation time	28.2 years	medium
Global population share	50–100%	medium

(BirdLife International 2011) is derived from an age at first breeding of 10.0 years and an annual survival of adults of 94.5% (Schreiber and Burger 2002).

Threats

There are currently no known substantial land-based threats to Campbell Albatross. At sea the primary threat is drowning in longline fishing gear, particularly of sub-adults. It is not known whether deaths from this cause are compromising population viability (BirdLife International 2011) but the severe declines suspected in 2000 (Garnett and Crowley 2000) have not been realised.

Conservation objectives

1. At-sea threats maintained at acceptable levels
2. Global agreement on conservation measures
3. Increased public awareness of the conservation needs of albatrosses

Information required

1. Population trends or changes in adult and juvenile survival
2. Distribution of immature and non-breeding birds
3. Overlap with fishing operations and documented levels of bycatch so that mitigation measures can be targeted

Management actions required

1. Assess the impact of trawl fisheries bycatch
2. Maintain mitigation strategies for fisheries, including via intergovernmental mechanisms under the auspices of the Agreement on the Conservation of Albatrosses and Petrels, the Food and Agriculture Organisation and Regional Fisheries Management Organisations

Bibliography

Agreement on the Conservation of Albatrosses and Petrels (2009) 'ACAP species assessment: Campbell Albatross *Thalassarche impavida*'. Retrieved 26 August, 2009 from <http://www.acap.aq/>.

BirdLife International (2011) 'Species factsheet: *Thalassarche impavida*'. Retrieved 17 January, 2011 from <http://www.birdlife.org/>.

Garnett S, Crowley GM (2000) *The Action Plan for Australian Birds 2000*. Environment Australia, Canberra.

Reid TA, Hindell MA, Eades DW, Newman M (2002) *Seabird Atlas of South-eastern Australian Waters*. Birds Australia Monograph 4.

Schreiber EA, Burger J (2002) *Biology of Marine Birds*. CRC Press, Boca Raton, FL.

Woehler EJ, Hodges CL, Watts DJ (1990) 'An atlas of the pelagic distribution and abundance of seabirds in the southern Indian Ocean 1981 to 1990'. ANARE Research Notes 77.

Comments received from

Barry Baker, Rosemary Gales, Wieslawa Misiak, Tim Reid, Ben Sullivan, Eric Woehler

Species Conservation Summary

Shy Albatross

Thalassarche cauta (Gould, 1841)
Diomedeidae

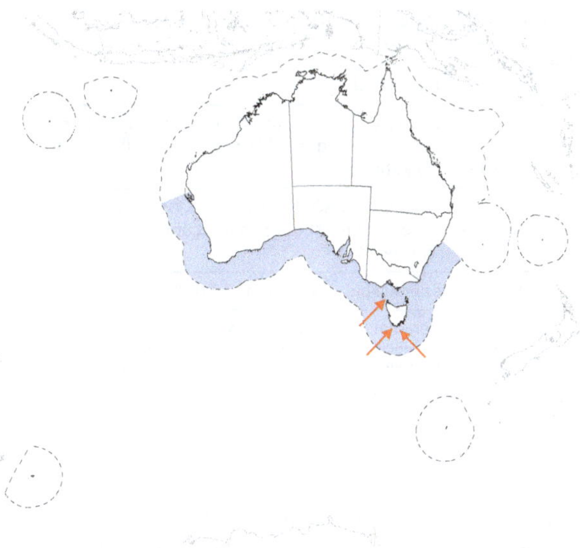

Conservation status

Vulnerable D2

Reasons for listing

Occurs at only 3 locations the smallest of which faces a plausible future threat from competition with Australasian Gannets *Morus serrator* and rough weather. The loss of this breeding location could render the status as Endangered B2ab(iv) but, when no longer occupied, this species would again become Vulnerable D2, or even Least Concern if no plausible future threat to the remaining two islands can be established; therefore it is retained as Vulnerable. Losses from fishing bycatch are not known to be reducing the population

Status 2000

Vulnerable D2

Status 1990

Vulnerable D2

Taxonomy

No infraspecific taxa described. Species sometimes included with White-capped Albatross *T. steadi*, Chatham Island Albatross *T. eremita* and Salvin's Albatross *T. salvini*
Taxonomic uniqueness: high (4 genera/family, 11 species/genus, 1 subspecies/species)

Range

Breeding occurs on Albatross I. in Bass Strait and on the Mewstone and Pedra Branca south-west of Tasmania. At sea adults largely remain in Australian waters, mostly remaining near nesting islands, as do juveniles from Albatross I. Juveniles from Mewstone and Pedra Branca travel further west, sometimes to South Africa where they forage over shelf waters (Brothers *et al.* 1997; Brothers *et al.* 1998; Hedd and Gales 2005; Agreement on the Conservation of Albatrosses and Petrels 2009; Alderman *et al.* 2010; Department of Sustainability, Environment, Water, Population and Communities 2011). There is genetic differentiation among the three subpopulations (Abbott and Double 2003).

Abundance

Population on Mewstone estimated at 7300 pairs in 1996, 5017 pairs at Albatross I. in 2007 and 268 pairs at Pedra Branca in 1996. Numbers on Albatross I. were about 20 000 pairs when first discovered in the late 18th century but this was reduced to as low as 250 pairs by the early 20th century as a result of the feather trade (Johnstone *et al.* 1975). Numbers have been recovering since, increasing at 3% per year from 1999 to 2007 and is now at 25% of estimated original size. On Pedra Branca, chick production between 1993 and 2007 decreased by about 9% per year from over 100 to 31 (Agreement on the Conservation of Albatrosses and Petrels 2009). Trends on Mewstone are unknown. However, concerns held in 2000 that the population as a whole would fall rapidly (Garnett and Crowley 2000) have not been realised.

Ecology

Breeds annually on the ground on small isolated islands. Feeds mostly on fish with some cephalopods and <1% tunicates and crustaceans. Most food is thought to be obtained by seizing from the surface during the day but may dive down to 7 m and feed by moonlight. Commonly follows fishing boats (Hedd *et al.* 1997; Hedd and Gales 2001; Hedd *et al.* 2001; Agreement on the Conservation of Albatrosses and Petrels 2009). A generation time of 23.1 years (BirdLife International 2011) is derived from an age at first breeding of 8.4 years, extrapolated from congeners, and an annual survival of adults of 95.8% (B. Baker *in litt.*).

Threats

The small population breeding on Pedra Branca is being competitively excluded from the limited space by Australasian Gannets *Morus serrator*, the regional population of which was increasing at about 4% per year from 1985 to 2000 (Bunce *et al.* 2002). Pedra Branca is particularly vulnerable to wave action in rough weather. Avian pox virus, probably transmitted by parasitic fleas and ticks, also kills an unknown number of birds on Albatross I. with infection rates varying among years (Johnstone *et al.* 1975; Woods 2004). At sea, entanglement in fishing warps when feeding on fishery waste is the major threat (Baker *et al.* 2007). However, nearly all the estimated 7000 (3000–12 500) individuals killed this way off South Africa each year (Ryan *et al.* 2002; Watkins *et al.* 2008) are likely to be White-capped Albatross *T. steadi* (Abbott *et al.* 2006). The spatial and temporal overlap between the distribution of adult Shy Albatrosses and Australian longline fishing effort is small (BirdLife International

Current eligibility against IUCN Red List Criteria

IUCN category	Criteria eligibility
A	Not applicable: past, current or future population declines are thought unlikely to exceed 20% in any 3-generation period
B	Not applicable: AOO <250 km^2, 3 locations one of which could be lost but is still occupied so this criterion not triggered
C	Not applicable: population >10 000 mature individuals
D	Vulnerable: ≤5 locations with plausible future threat
E	Not applicable: no population viability analysis undertaken

IUCN Red List assessment data

	Estimate	Reliability
Extent of occurrence trend	2 780 000 km^2 stable	high high
Area of occupancy trend	4 km^2 stable	high high
No. of mature individuals trend	26 000 stable	medium low
No. subpopulations	3	high
No. locations	3	high
Largest subpopulation	15 000	medium
Generation time	23.1 years	poor
Global population share	100%	high

2011) but may be significant (Phillips *et al.* 2010). In Commonwealth waters a Threat Abatement Plan under the EPBC Act has effectively reduced bycatch from longline fishing but trawl fisheries require greater attention (Baker and Finley 2010).

Conservation objectives

1. Secure status of Albatross I. and the Mewstone maintained
2. Decision on conservation of Pedra Branca population implemented
3. At-sea threats reduced to acceptable levels
4. Global agreement on conservation measures
5. Increased public awareness of the conservation needs of albatrosses

Information required

1. Extent of seabird bycatch in State and Territory trawl fisheries
2. Improved seabird bycatch mitigation measures for pelagic longline gear

Management actions required

1. Active decision on whether to attempt conservation on Pedra Branca
2. Implement decision on Pedra Branca if active management considered necessary
3. Conduct complete censuses at all sites at regular intervals
4. Continue monitoring of survival and productivity on Albatross I.
5. Increase levels of observer coverage required by the Threat Abatement Plan for all Commonwealth fisheries
6. Maintain mitigation strategies for fisheries, including via the Threat Abatement Plan and intergovernmental mechanisms under the auspices of the Agreement on the Conservation of Albatrosses and Petrels, the Food and Agriculture Organisation and Regional Fisheries Management Organisations

Bibliography

Abbott CL, Double MC, Gales R, Baker GB, Lashko A, Robertson CJR, Ryan PG (2006) Molecular provenance analysis for shy and white-capped albatrosses killed by fisheries interactions in Australia, New Zealand, and South Africa. *Conservation Genetics* 7, 531–542.

Abbott CL, Double ML (2003) Genetic structure, conservation genetics and evidence of speciation by range expansion in shy and white-capped albatrosses. *Molecular Ecology* **12**, 2953–2962.

Agreement on the Conservation of Albatrosses and Petrels (2009) 'ACAP species assessment: Shy Albatross *Thalassarche cauta*'. Retrieved 31 August, 2009 from <http://www.acap.aq/>.

Alderman R, Gales R, Hobday AJ, Candy SG (2010) Post-fledging survival and dispersal of shy albatross from three breeding colonies in Tasmania. *Marine Ecology Progress Series* **405**, 271–285.

Baker GB, Double MC, Gales R, Tuck GN, Abbott CL, Ryan PG, Petersen SL, Robertson CJR, Alderman R (2007) A global assessment of the impact of fisheries-related mortality on Shy and White-capped Albatrosses: conservation implications. *Biological Conservation* **137**, 319–333.

Baker GB, Finley LA (2010) '2008 National Assessment Report for reducing the incidental catch of seabirds in longline fisheries'. Bureau of Rural Sciences, Canberra.

BirdLife International (2011) 'Species factsheet: *Thalassarche cauta*'. Retrieved 17 January, 2011 from <http://www.birdlife.org/>.

Brothers N, Gales R, Hedd A, Robertson G (1998) Foraging movements of the Shy Albatross *Diomedea cauta* breeding in Australia; implications for interactions with longline fisheries. *Ibis* **140**, 446–457.

Brothers NP, Reid T, Gales R (1997) At sea distribution of Shy Albatrosses *Diomedea cauta cauta* derived from records of band recoveries and colour-marked birds. *Emu* **97**, 231–239.

Bunce A, Norman FI, Brothers N, Gales R (2002) Long-term trends in the Australasian gannet (*Morus serrator*) population in Australia: the effect of climate change and commercial fisheries. *Marine Biology* **141**, 263–269.

Department of Sustainability, Environment, Water, Population and Communities (2011) *Thalassarche cauta cauta*. In 'Species Profile and Threats Database'. Department of Sustainability, Environment, Water, Population and Communities, Canberra. Retrieved 24 February, 2011 from <http://www.environment.gov.au/sprat>.

Garnett S, Crowley GM (2000) *The Action Plan for Australian Birds 2000*. Environment Australia, Canberra.

Hedd A, Gales R (2001) The diet of shy albatrosses (*Thalassarche cauta*) at Albatross Island, Tasmania. *Journal of The Zoological Society of London* **253**, 69–90.

Hedd A, Gales R (2005) Breeding and overwintering ecology of Shy Albatrosses in southern Australia: year-round patterns of colony attendance and foraging-trip durations. *Condor* **107**, 375–387.

Hedd A, Gales R, Brothers N (2001) Foraging strategies of Shy Albatross *Thalassarche cauta* breeding at Albatross Island, Tasmania, Australia. *Marine Ecology Progress Series* **224**, 267–282.

Hedd A, Gales R, Brothers N, Robertson G (1997) Diving behaviour of the shy albatross *Diomedea cauta* in Tasmania – initial findings and dive recorder assessment. *Ibis* **139**, 452–460.

Johnstone GW, Milledge D, Dorwood DF (1975) The White-capped Albatross of Albatross Island: numbers and breeding behaviour. *Emu* **75**, 1–11.

Phillips K, Giannini F, Lawrence E, Bensley N (2010) 'Cumulative assessment of the catch of non-target species in Commonwealth fisheries: a scoping study'. Bureau of Rural Sciences, Canberra.

Ryan PG, Keith DG, Kroese M (2002) Seabird bycatch by longline fisheries off southern Africa, 1998–2000. *South African Journal of Marine Science* **24**, 103–110.

Watkins BP, Petersen SL, Ryan PG (2008) Interactions between seabirds and deep-water hake trawl gear: an assessment of impacts in South African waters. *Animal Conservation* **11**, 247–254.

Woods R (2004) Result of a preliminary disease survey in Shy Albatross (*Thalassarche cauta* Gould 1841) chicks at Albatross Island, Bass Strait Tasmania. In *Proceedings of the Annual Conference of the Australian Association of Veterinary Conservation Biologists*. Canberra.

Comments received from

Barry Baker, Rosemary Gales, Wieslawa Misiak, Tim Reid, Ben Sullivan, Eric Woehler

Species Conservation Summary

White-capped Albatross

Thalassarche steadi (Falla, 1933)
Diomedeidae

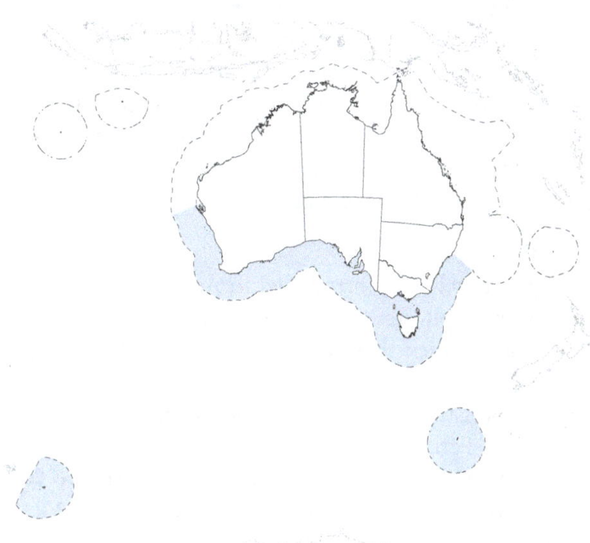

Conservation status

Population visiting Australian territory: Vulnerable D2

Reasons for listing

Occurs at ≤5 locations with pig predation a plausible threat to one location

Status 2000

Vulnerable D2

Status 1990

Vulnerable D2

Taxonomy

No infraspecific taxa described. Sometimes included with Shy Albatross *T. cauta*, Chatham Island Albatross *T. eremita* and Salvin's Albatross *T. salvini*
Taxonomic uniqueness: high (4 genera/family, 11 species/genus, 1 subspecies/species)

Range

Virtually the entire population breeds in the Auckland Is., mostly on Disappointment I. with smaller numbers on nearby Auckland and Adams Is., very low numbers on Bollons I. near Antipodes I. and the Forty-fours in the Chatham Is., all near New Zealand. Juveniles and non-breeding adults are thought to range throughout the waters off southern Australia and South Africa, breeding adults forage mainly within New Zealand's EEZ, including around the Chatham Is. and south of Auckland Is. Chick-rearing birds utilise areas off the south-east coast of Australia and around Tasmania (Reid *et al.* 2002; Agreement on the Conservation of Albatrosses and Petrels 2009).

Abundance

The population of about 200 000 is a minimum as it is now thought that this species breeds biennially not annually as first thought. The population on Disappointment I. was 91 694 pairs in 2008 with 5264 on Auckland I. and 131 on Adams I. Elsewhere 20 pairs bred on Bollons I. and 3 on the Forty-fours. A male hybrid has also paired with a Black-browed Albatross *T. melanophrys* on South Georgia in the south Atlantic Ocean but has not reared young. At least 8000 are known to be caught on longlines each year and counts over the last 4 years suggest declines but these are not confirmed (Baker

Current eligibility against IUCN Red List Criteria

IUCN category	Criteria eligibility
Population visiting Australian territory	
A	Near Threatened: current and anticipated population declines of 20–29% over periods of 3 generations (69 years) estimated and projected from rates of exploitation
B	Not applicable: at ≤5 locations but no plausible future threat to locations, population fluctuations not extreme
C	Not applicable: population >10 000 mature individuals
D	Vulnerable: ≤5 locations with plausible future threat
E	Not applicable: no population viability analysis undertaken

IUCN Red List assessment data

	Estimate	Reliability
Population visiting Australian territory		
Extent of occurrence trend	3 900 000 km^2 stable	high high
Area of occupancy trend	3 900 000 km^2 stable	high high
No. of mature individuals trend	200 000 stable	low high
No. subpopulations	3	high
No. locations	5	high
Largest subpopulation	200 000	high
Generation time	23.1 years	medium
Global population share	50–100%	medium

et al. 2009, 2010; Agreement on the Conservation of Albatrosses and Petrels 2009).

Ecology

Had been thought to breed annually but recent research suggests biennial breeding. Nests in colonies among grass tussocks. It feeds largely by surface seizing or undertaking shallow dives to catch squid or fish but there are no detailed studies. It often follows fishing boats to retrieve offal (Agreement on the Conservation of Albatrosses and Petrels 2009). A generation time of 23.1 years (BirdLife International 2011) is derived from an age at first breeding of 8.4 years and an annual survival of adults of 93.2%, both extrapolated from congeners.

Threats

Feral pigs *Sus scrofa*, introduced to Auckland I. in 1807, have been implicated in reduction of nesting colonies. Predation of chicks has been recorded to such an extent that pigs were considered to have a 'serious impact', at least periodically, except in inaccessible nesting areas on steep cliffs. Feral cats *Felis catus* are also thought to kill chicks on Auckland I., but all remaining areas are free from introduced mammals (BirdLife International 2011). At sea an estimated 8000 White-capped Albatrosses are killed annually as a result of trawl and longline fisheries throughout the Southern Ocean. Both adult and juvenile birds are killed by drowning in longline fishing gear and collision with cables and warps used on fishing trawlers. The scale of the loss is such that trends from census data at breeding islands are considered unreliable and a rapid decline is suspected (Baker *et al.* 2007; BirdLife International 2011). However, the severe declines suspected likely in 2000 (Garnett and Crowley 2000) have not been proven.

Conservation objectives

1. At-sea threats reduced to acceptable levels
2. Global agreement on conservation measures
3. Increased public awareness of the conservation needs of albatrosses

Information required

1. Population trends or changes in adult and juvenile survival
2. Distribution of immature and non-breeding birds
3. Overlap with fishing operations and documented levels of bycatch so that mitigation measures can be targeted

Management actions required

1. Assess the impact of trawl fisheries bycatch
2. Maintain and strengthen mitigation strategies for fisheries, including via intergovernmental mechanisms under the auspices of the Agreement on the Conservation of Albatrosses and Petrels, the Food and Agriculture Organisation and Regional Fisheries Management Organisations

Bibliography

Agreement on the Conservation of Albatrosses and Petrels (2009) 'ACAP species assessment: White-capped Albatross *Thalassarche steadi*'. Retrieved 17 September, 2009 from <http://www.acap.aq/>.

Baker GB, Double MC, Gales R, Tuck GN, Abbott CL, Ryan PG, Petersen SL, Robertson CJR, Alderman R (2007) A global assessment of the impact of fisheries-related mortality on Shy and White-capped Albatrosses: conservation implications. *Biological Conservation* **137**, 319–333.

Baker GB, Jensz K, Cunningham R (2009) 'Data collection of demographic, distributional and trophic information on the white-capped albatross to allow estimation of effects of fishing on population viability – 2008 field season'. Report to the New Zealand Ministry of Fisheries.

Baker GB, Jensz K, Cunningham R (2010) 'Data collection of demographic, distributional and trophic information on the white-capped albatross to allow estimation of effects of fishing on population viability – 2009 field season'. Report to the New Zealand Ministry of Fisheries.

BirdLife International (2011) 'Species factsheet: *Thalassarche steadi*'. Retrieved 17 January, 2011 from <http://www.birdlife.org/>.

Garnett S, Crowley GM (2000) *The Action Plan for Australian Birds 2000*. Environment Australia, Canberra.

Reid TA, Hindell MA, Eades DW, Newman M (2002) *Seabird Atlas of South-eastern Australian Waters*. Birds Australia Monograph 4.

Comments received from

Barry Baker, Rosemary Gales, Wieslawa Misiak, Tim Reid, Ben Sullivan, Eric Woehler

Species Conservation Summary

Salvin's Albatross

Thalassarche salvini (Rothschild, 1893)
Diomedeidae

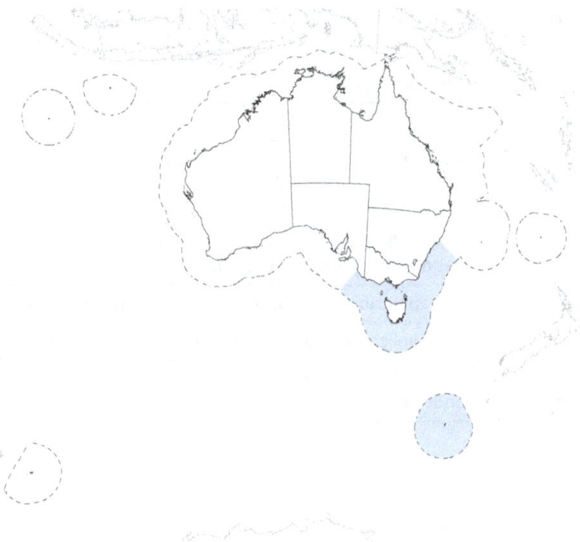

Conservation status
Population visiting Australian territory: Vulnerable A4bd

Reasons for listing
Population decline of 30–49% possibly occurring as a result of fishing-related mortality

Status 2000
Vulnerable A3bd+4bd

Status 1990
Vulnerable A3bd

Taxonomy
No infraspecific taxa described. Sometimes included with Shy Albatross *T. cauta*, White-capped Albatross *T. steadi* and Chatham Island Albatross *T. eremita*
Taxonomic uniqueness: high (4 genera/family, 11 species/genus, 1 subspecies/species)

Range
Almost the entire population breeds on the Bounty Is. (9 islands and islets) and the Western Chain islets, Snares Is. (2 islets). Three nests have been recorded from the Chatham Is. (The Pyramid 1995, 2006, the Forty-Fours 2007) and 4 from Ile des Pingouins, Crozet Is. 1986. It ranges widely through the south Pacific with large numbers of birds occurring off South America, including the South Atlantic (Agreement on the Conservation of Albatrosses and Petrels 2009) and, infrequently, the waters off eastern Australia and Macquarie I. (Baker *et al.* 2007). Aberrant records of breeding away from the Bounty and Snares Is. suggests that there is one subpopulation.

Abundance
In 2010, the population on the Bounty Is. was estimated at 41 000 pairs. A further 1200 pairs breed on the Snares Is. (Baker *et al.* 2011; P. Sagar *in litt.*). At sea the maximum population off Chile has been estimated at 133 000 (confidence interval 83 000–184 000) of which 114 000 were estimated to be adults (Spear *et al.* 2003). The number of nests on the Snares Is. is thought to have been stable between 1984 and 2010 (P. Sagar *in litt.*). In 1978, 76 000 breeding pairs were estimated to breed on the Bounty Is. and 30 750 pairs were counted in 1998 but methods have varied so changes may not be significant (Agreement on the Conservation of Albatrosses and Petrels 2009).

Ecology
Nests annually in colonies and feeds largely by surface seizing or undertaking shallow dives to catch squid or fish but there are no detailed studies. It often follows fishing boats to retrieve offal (Agreement on the Conservation of Albatrosses and Petrels 2009). A generation time of 23.1 years (BirdLife International 2011) is derived from an age at first breeding of 8.4 years and an annual survival of adults of 93.2%, both extrapolated from congeners.

Threats
Breeding sites are currently not threatened (Agreement on the Conservation of Albatrosses and Petrels 2009) and while some islands could be overwhelmed by introduced predators or particularly inclement weather (BirdLife International 2011), the species breeds on at least 11 islands. Many were being killed in demersal longlines and trawling operations off New Zealand 1996–2002, but numbers killed were declining as mitigation techniques were applied (Robertson *et al.* 2004). The situation is less clear in the fisheries off Chile (Spear *et al.* 2003).

Conservation objectives
1. At-sea threats reduced to acceptable levels
2. Global agreement on conservation measures
3. Increased public awareness of the conservation needs of albatrosses

Information required
1. Population trends or changes in adult and juvenile survival
2. Distribution of immature and non-breeding birds
3. Overlap with fishing operations and documented levels of bycatch so that mitigation measures can be targeted

Management actions required
1. Assess the impact of trawl fisheries bycatch
2. Maintain and strengthen mitigation strategies for fisheries, including via intergovernmental mechanisms under the auspices of the Agreement on the Conservation of Albatrosses and Petrels, the Food and Agriculture Organisation and Regional Fisheries Management Organisations

Bibliography
Agreement on the Conservation of Albatrosses and Petrels (2009) 'ACAP Species assessment: Salvin's Albatross *Thalas-*

Current eligibility against IUCN Red List Criteria

IUCN category	Criteria eligibility
Population visiting Australian territory	
A	Vulnerable: current and anticipated population declines of 30–49% over periods of 3 generations (69 years) estimated and projected from levels of exploitation
B	Not applicable: at >5 locations, not severely fragmented, population fluctuations not extreme
C	Not applicable: population >10 000 mature individuals
D	Not applicable: >1000 mature individuals, >5 locations
E	Not applicable: no population viability analysis undertaken

IUCN Red List assessment data

Population visiting Australian territory	Estimate	Reliability
Extent of occurrence trend	3 900 000 km^2 stable	high high
Area of occupancy trend	3 900 000 km^2 stable	high high
No. of mature individuals trend	133 000 decreasing	low low
No. subpopulations	1	low
No. locations	>10	high
Generation time	23.1 years	medium
Global population share	11-25%	Low

sarche salvini'. Retrieved 17 September, 2009 from <http://www.acap.aq/>.

Baker GB, Double MC, Gales R, Tuck GN, Abbott CL, Ryan PG, Petersen SL, Robertson CJR, Alderman R (2007) A global assessment of the impact of fisheries-related mortality on Shy and White-capped Albatrosses: conservation implications. *Biological Conservation* 137, 319–333.

Baker GB, Jensz K, Sagar P (2011) 'Data collection of demographic, distributional and trophic information on Salvin's Albatross to allow estimation of effects of fishing on population viability'. Report to the New Zealand Ministry of Fisheries.

BirdLife International (2011) 'Species factsheet: *Thalassarche salvini*'. Retrieved 17 January, 2011 from <http://www.birdlife.org/>.

Robertson CJR, Bell E, Scofield P (2004) 'Autopsy report for seabirds killed and returned from New Zealand fisheries, 1 October 2001 to 30 September 2002'. DOC Science Internal Series 155. Department of Conservation, Wellington.

Spear LB, Ainley DG, Webb SW (2003) Distribution, abundance and behaviour of Buller's, Chatham Island and Salvin's Albatrosses off Chile and Peru. *Ibis* **145**, 253–269.

Comments received from

Barry Baker, Rosemary Gales, Wieslawa Misiak, Tim Reid, Ben Sullivan, Eric Woehler

Species Conservation Summary

Grey-headed Albatross

Thalassarche chrysostoma (J. R. Forster, 1785)

Diomedeidae

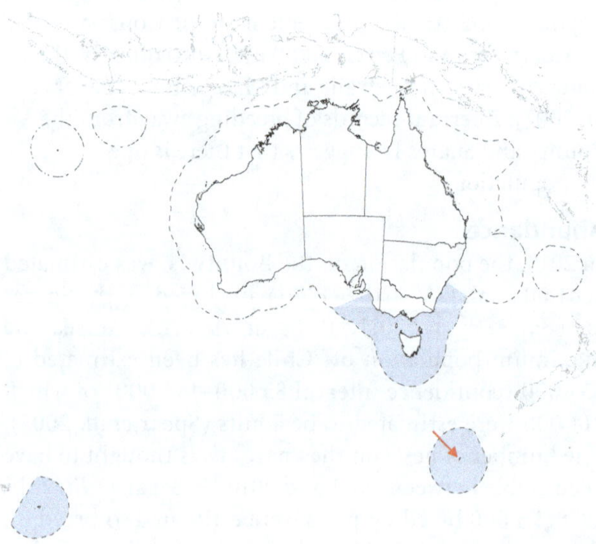

Conservation status

Australian breeding population: Critically Endangered B2ab(ii,iii)

Population visiting Australian territory: Vulnerable A4bd

Reasons for listing

Australian breeding population: population occurs at a single location where the habitat quality is deteriorating as a result of rabbit grazing and the minute area occupied is probably contracting

Population visiting Australian territory: Population decline 20–29% over 3 generations (90 years) currently under way

Status 2000

Australian breeding population: Endangered D

Reason for change in 2010: active threats to Macquarie I. breeding sites

Current eligibility against IUCN Red List Criteria

IUCN category	Criteria eligibility
Australian breeding population	
A	Not applicable: past, current or future population declines are thought unlikely to exceed 20% in any 3-generation period
B	Critically Endangered: AOO <10 km², 1 location, observed continuing decline in AOO, habitat quality
C	Not applicable: no continuing population decline or extreme fluctuations
D	Vulnerable: ≤5 locations with plausible future threat
E	Not applicable: no population viability analysis undertaken
Population visiting Australian territory	
A	Vulnerable: current population decline of 30–49% in 3 generations (90 years) estimated and projected from monitoring and levels of exploitation
B	Not applicable: >10 locations, not severely fragmented, population fluctuations not extreme
C	Not applicable: population >10 000 mature individuals
D	Not applicable: population >1000 mature individuals, >5 locations
E	Not applicable: no population viability analysis undertaken

IUCN Red List assessment data

	Estimate	Reliability
Australian breeding population		
Extent of occurrence trend	2 400 000 km² stable	high high
Area of occupancy trend	0.12 km² decreasing	high medium
No. of mature individuals trend	250 stable	high high
No. subpopulations	1	high
No. locations	1	high
Generation time	30.0 years	medium
Global population share	<1%	high
Level of genetic exchange	low	medium
Population visiting Australian territory		
Extent of occurrence trend	2 400 000 km² stable	high high
Area of occupancy trend	2 400 000 km² stable	high high
No. of mature individuals trend	250 000 decreasing	medium high
No. subpopulations	7	high
No. locations	>10	high
Largest subpopulation	100 000	medium
Global population share	50–100%	high

Population visiting Australian territory: Vulnerable A4bd

Status 1990

Australian breeding population: Endangered D

Population visiting Australian territory: Near Threatened A4bd

Reason for change in 2000: initial declines intensifying

Taxonomy

No infraspecific taxa described

Taxonomic uniqueness: high (4 genera/family, 11 species/genus, 1 subspecies/species)

Range

Has bred on the southern and western flanks of Petrel Peak, Macquarie I. for at least the past 30 years (Terauds *et al.* 2005). Extralimitally breeds on subantarctic islands in the Pacific, Indian and Atlantic Oceans, and on Cape Horn (Agreement on the Conservation of Albatrosses and Petrels 2009). At sea occurs throughout subantarctic and Antarctic waters (Woehler *et al.* 1990), satellite tracking from South Georgia shows that males are the most likely to complete circumpolar migration (Terauds *et al.* 2006; Agreement on the Conservation of Albatrosses and Petrels 2009).

Abundance

The small population on Macquarie I. (about 94 pairs breeding annually) was stable between 1995 and 2007, and is likely to have been since the mid-1970s (Terauds *et al.* 2005; Agreement on the Conservation of Albatrosses and Petrels 2009). It had a higher mean breeding success and juvenile and adult survival rates than larger populations, although the study periods were not always directly comparable between islands. Of the global population of about 250 000 individuals, about 96 000 pairs breed annually (Agreement on the Conservation of Albatrosses and Petrels 2009). This species may be declining at >30% in 3 generations based on direct estimates of trends (Agreement on the Conservation of Albatrosses and Petrels 2009).

Ecology

Nests biennially in dispersed colonies. Its diet includes squid, fish and crustaceans, the composition varying between years and sites. Most food is taken by surface-seizing, but can dive down to 6 m, mostly by day, but sometimes by night. Although they infrequently follow ships, presence of large skeletal remains and tails of Patagonian Toothfish *Dissostichus eleginoides* in the diet at some, but not all sites, suggests some degree of interaction with longline fisheries operating in the vicinity (BirdLife International 2011; Department of Sustainability, Environment, Water, Population and Communities 2011). A generation time of 30.0 years (BirdLife International 2011) is derived from an age at first breeding of 10.0 years and an annual survival of adults of 95.0% (Schreiber and Burger 2002).

Threats

On Macquarie I. the increase in rabbit *Oryctolagus cuniculus* numbers since feral cats *Felis catus* were eradicated in 1999 has led to overgrazing of native vegetation and the extensive destruction of habitat and soil erosion at nesting sites. This threat may continue unless an eradication program planned for 2011–14 is successful (Agreement on the Conservation of Albatrosses and Petrels 2009). At sea large numbers were killed by tuna fisheries south of Africa in the early 1990s and around Australia 1989–1995 (up to 409 per year), with immatures and adults also being killed in New Zealand waters. Much of this loss has been controlled within territorial waters but the species occurs in all southern fishery zones and declines in non-Australian colonies are continuing (Agreement on the Conservation of Albatrosses and Petrels 2009).

Conservation objectives

1. Nesting sites on Macquarie I. stabilised
2. At-sea threats reduced to acceptable levels
3. Global agreement on conservation measures
4. Increased public awareness of the conservation needs of albatrosses

Information required

1. Information on population trends and demographic parameters for most sites
2. Trends in adult and juvenile survival
3. Distribution of immature and non-breeding birds
4. Overlap with fishing operations and documented levels of bycatch so that mitigation measures can be targeted

Management actions required

1. Secure nest sites on Macquarie I. by eliminating rabbits
2. Continue existing monitoring and research programs
3. Assess the impact of trawl fisheries bycatch
4. Maintain mitigation strategies for fisheries, including via intergovernmental mechanisms under the auspices of the Agreement on the Conservation of Albatrosses and Petrels, the Food and Agriculture Organisation and Regional Fisheries Management Organisations

Bibliography

Agreement on the Conservation of Albatrosses and Petrels (2009) 'ACAP species assessment: Grey-headed Albatross *Thalassarche chrysostoma*'. Retrieved 24 September, 2009 from <http://www.acap.aq/>.

BirdLife International (2011) 'Species factsheet: *Thalassarche chrysostoma*'. Retrieved 17 January, 2011 from <http://www.birdlife.org/>.

Department of Sustainability, Environment, Water, Population and Communities (2011) *Thalassarche chrysostoma*. In 'Species Profile and Threats Database'. Department of Sustainability, Environment, Water, Population and Communities, Canberra. Retrieved 24 February, 2011 from <http://www.environment.gov.au/sprat>.

Schreiber EA, Burger J (2002) *Biology of Marine Birds*. CRC Press, Boca Raton, FL.

Terauds A, Gales R, Alderman R (2005) Trends in numbers and survival of Black-browed (*Thalassarche melanophrys*) and Grey-headed (*T. chrysostoma*) Albatrosses breeding on Macquarie Island. *Emu* **105**, 159–167.

Terauds A, Gales R, Baker GB, Alderman R (2006) Foraging areas of black-browed and grey-headed albatrosses breeding on Macquarie Island in relation to marine protected areas. *Aquatic Conservation: Marine and Freshwater Ecosystems* **16**, 133–146.

Woehler EJ, Hodges CL, Watts DJ (1990) 'An atlas of the pelagic distribution and abundance of seabirds in the southern Indian Ocean 1981 to 1990'. ANARE Research Notes 77.

Comments received from

Barry Baker, Rosemary Gales, Wieslawa Misiak, Tim Reid, Ben Sullivan, Eric Woehler

Species Conservation Summary

Indian Yellow-nosed Albatross

Thalassarche carteri (Mathews, 1912)
Diomedeidae

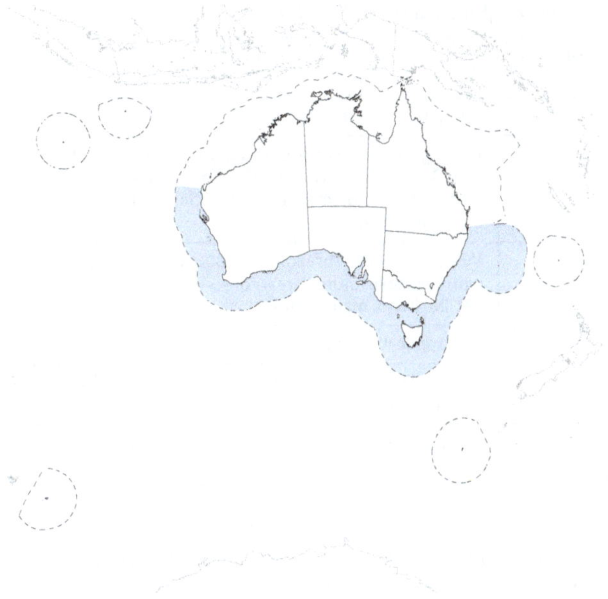

Conservation status
Population visiting Australian territory: Endangered A4bde

Reasons for listing
Population declines 50–79% in three generations (72 years) occurring as a result of at-sea mortality associated with fisheries, particularly longline fishing vessels

Status 2000
Vulnerable A4bde, B2ab(v)
Reason for change in 2010: initial declines intensifying

Status 1990
Near Threatened A4bde, B2ab(v)

Taxonomy
No infraspecific taxa described. Sometimes included with Atlantic Yellow-nosed Albatross *D. chlororhynchos*
Taxonomic uniqueness: high (4 genera/family, 11 species/genus, 1 subspecies/species)

Range
Breeds on Prince Edward, Kerguelen, Crozet, Amsterdam and Saint Paul Is. During winter birds disperse from their breeding islands and commonly occur off southern Africa and Australia (Woehler *et al.* 1990; Reid *et al.* 2002). Also known to occur through the Tasman Sea and as far east as the Chatham Is. (Agreement on the Conservation of Albatrosses and Petrels 2009).

Abundance
The population is estimated as 36 500 breeding pairs, corresponding to a total population of between 160 000 and 180 000 individuals of all ages. Between 1982 and 2006, the total population on Amsterdam I. is estimated to have decreased from 37 000 to 27 000 pairs (Agreement on the Conservation of Albatrosses and Petrels 2009; BirdLife International 2011).

Ecology
Nests annually, solitarily or in loose colonies on slopes or cliffs, typically in bare, rocky areas but sometimes in tussock-grass and ferns. Feeds on squid and fish by surface seizing and shallow dives, with crustaceans being taken less frequently. It attends fishing boats (Agreement on the Conservation of Albatrosses and Petrels 2009; BirdLife International 2011; Department of Sustainability, Environment, Water, Population and Communities 2011). A generation time of 23.7 years (BirdLife International 2011) is derived from an age at first breeding of 9.0 years and an annual survival of adults of 93.2% (Schreiber and Burger 2002).

Threats
On land introduced disease, Black Rats *Rattus rattus* and feral cats *Felis catus* cause mortality on Amsterdam I. in particular. Principal reason for declines is thought to be mortality as bycatch from longline fishing (BirdLife International 2011).

Conservation objectives
1. At-sea threats reduced to acceptable levels
2. Global agreement on conservation measures
3. Increased public awareness of the conservation needs of albatrosses

Information required
1. Population trends or changes in adult and juvenile survival
2. Distribution of immature and non-breeding birds
3. Overlap with fishing operations and documented levels of bycatch so that mitigation measures can be targeted

Management actions required
1. Maintain mitigation strategies for fisheries, including via intergovernmental mechanisms under the auspices of the Agreement on the Conservation of Albatrosses and Petrels, the Food and Agriculture Organisation and Regional Fisheries Management Organisations

Bibliography

Agreement on the Conservation of Albatrosses and Petrels (2009) 'ACAP species assessment: Indian Yellow-nosed Albatross *Thalassarche carteri*'. Retrieved 27 August, 2009 from <http://www.acap.aq/>.

BirdLife International (2011) 'Species factsheet: *Thalassarche carteri*'. Retrieved 17 January, 2011 from <http://www.birdlife.org/>.

Department of Sustainability, Environment, Water, Population and Communities (2011) *Thalassarche chlororhynchos*. In 'Species Profile and Threats Database'. Department of Sustainability, Environment, Water, Population and Commu-

Current eligibility against IUCN Red List Criteria

IUCN category	Criteria eligibility
Population visiting Australian territory	
A	Endangered: current population decline of 50–79% over 3 generations (72 years) estimated and projected from monitoring, levels of exploitation and effects of introduced taxa
B	Vulnerable: AOO (breeding) <2000 km^2, ≤10 locations with plausible future threat, observed continuing decline in no. mature individuals
C	Not applicable: population >10 000 mature individuals
D	Not applicable: population >1000 mature individuals, >5 locations
E	Not applicable: no population viability analysis undertaken

IUCN Red List assessment data

	Estimate	Reliability
Population visiting Australian territory		
Extent of occurrence trend	3 700 000 km^2 stable	high high
Area of occupancy trend	3 700 000 km^2 stable	high high
No. of mature individuals trend	85 000 decreasing	medium high
No. subpopulations	4	high
No. locations	10	high
Largest subpopulation	36 000	medium
Generation time	23.7 years	medium
Global population share	50–100%	high

nities, Canberra. Retrieved 24 February, 2011 from <http://www.environment.gov.au/sprat>.

Reid TA, Hindell MA, Eades DW, Newman M (2002) *Seabird Atlas of South-eastern Australian Waters*. Birds Australia Monograph 4.

Schreiber EA, Burger J (2002) *Biology of Marine Birds*. CRC Press, Boca Raton, FL.

Woehler EJ, Hodges CL, Watts DJ (1990) 'An atlas of the pelagic distribution and abundance of seabirds in the southern Indian Ocean 1981 to 1990'. ANARE Research Notes 77.

Comments received from

Barry Baker, Rosemary Gales, Wieslawa Misiak, Tim Reid, Ben Sullivan, Eric Woehler

---

Species Conservation Summary

Buller's Albatross

Thalassarche bulleri (Rothschild, 1893)
Diomedeidae

Conservation status

Population visiting Australian territory: Near Threatened D2

Reasons for listing

Breeds at 4 locations (outside Australian territory) which have been damaged by storms historically but there are no current threats, and future human or stochastic events are unlikely to make the species Extinct or Critically Endangered in a short time

Status 2000

Near Threatened D2

Status 1990

Near Threatened D2

Taxonomy

Two subspecies: *T. b. bulleri* and *T. b. platei*, both Near Threatened

Taxonomic uniqueness: high (4 genera/family, 11 species/genus, 2 subspecies/species)

Range

New Zealand breeding endemic with colonies on only 4 island groups: the Snares Is., Solander Is., the Sisters and Forty-Fours in the Chatham Is., and Rosemary Rock in the Three Kings Is. During incubation birds range along the shelf slope off the east and west coasts of the South I., New Zealand and into the Tasman Sea; during and after chick-rearing around the southern island of New Zealand. After breeding, birds of all ages (including fledglings) migrate to slopes off Chile and Peru (Agreement on the Conservation of Albatrosses and Petrels 2009).

Abundance

About 30 500 pairs breed annually at all sites, about half that number on the Forty-Fours alone. Numbers on Snares Is. and on Solander Is. have been increasing in last few decades; numbers on Chatham and Three Kings Is. are thought to be stable (Agreement on the Conservation of Albatrosses and Petrels 2009).

Ecology

Breeds annually in colonies in a variety of habitats including grassy meadows, tussock-covered slopes and cliffs, scrub and under forest canopy. Although usually forages singly, large numbers may gather to feed at concentrated food sources. Feeds using surface plunges or shallow dives for fish (mostly fishery discards), squid, tunicates, octopuses and crustacea (Agreement on the Conservation of Albatrosses and Petrels 2009; Department of Sustainability, Environment, Water, Population and Communities 2011). A generation time of 16.5 years (BirdLife International 2011) is derived from an age at first breeding of 5.0 years and an annual survival of adults of 91.3% (Schreiber and Burger 2002).

Current eligibility against IUCN Red List Criteria

IUCN category	Criteria eligibility
Population visiting Australian territory	
A	Not applicable: past, current or future population declines are thought unlikely to exceed 20% in any 3-generation period
B	Not applicable: EOO, AOO, habitat, locations or population not declining, population fluctuations not extreme
C	Not applicable: population >10 000 mature individuals
D	Near Threatened: ≤5 locations but plausible threats not severe
E	Not applicable: no population viability analysis undertaken

IUCN Red List assessment data

	Estimate	Reliability
Population visiting Australian territory		
Extent of occurrence trend	1 700 000 km^2 stable	high high
Area of occupancy trend	1 700 000 km^2 stable	high high
No. of mature individuals trend	64 000 stable	high high
No. subpopulations	3	high
No. locations	4	high
Largest subpopulation	32 000	medium
Generation time	16.5 years	medium
Global population share	50–100%	high

Threats

At sea it is often caught as bycatch in fisheries off Chile, though this appears not to have affected trends of any population. Large storms have degraded nesting habitat in the past and could do so further in the future (BirdLife International 2011).

Conservation objectives

1. At-sea threats maintained at acceptable levels
2. Global agreement on conservation measures
3. Increased public awareness of the conservation needs of albatrosses

Information required

1. Population trends or changes in adult and juvenile survival
2. Distribution of immature and non-breeding birds
3. Overlap with fishing operations and documented levels of bycatch so that mitigation measures can be targeted

Management actions required

1. Maintain mitigation strategies for fisheries, including via intergovernmental mechanisms under the auspices of the Agreement on the Conservation of Albatrosses and Petrels, the Food and Agriculture Organisation and Regional Fisheries Management Organisations

Bibliography

Agreement on the Conservation of Albatrosses and Petrels (2009) 'ACAP species assessment: Buller's Albatross *Thalassarche bulleri*'. Retrieved 21 September, 2009 from <http://www.acap.aq/>.

BirdLife International (2011) 'Species factsheet: *Thalassarche bulleri*'. Retrieved 17 January, 2011 from <http://www.birdlife.org/>.

Department of Sustainability, Environment, Water, Population and Communities (2011) *Thalassarche bulleri*. In 'Species Profile and Threats Database'. Department of Sustainability, Environment, Water, Population and Communities, Canberra. Retrieved 24 February, 2011 from <http://www.environment.gov.au/sprat>.

Schreiber EA, Burger J (2002) *Biology of Marine Birds*. CRC Press, Boca Raton, FL.

Comments received from

Barry Baker, Rosemary Gales, Wieslawa Misiak, Tim Reid, Ben Sullivan

Subspecies Conservation Summary

Buller's Albatross (southern)

Thalassarche bulleri bulleri (Rothschild, 1893)
Diomedeidae

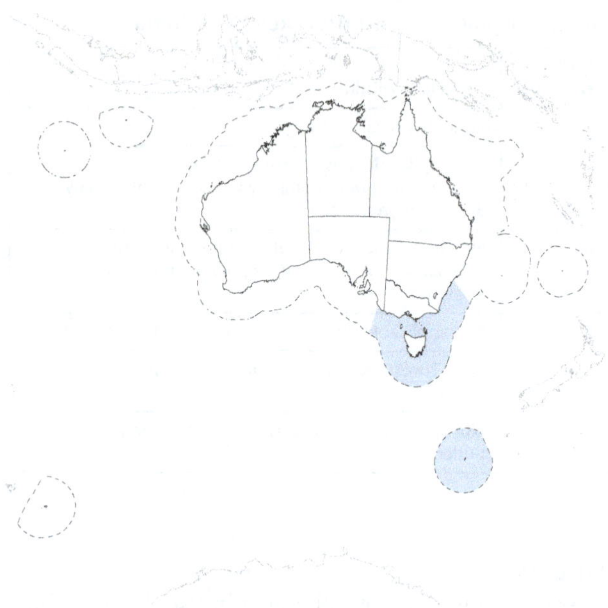

Conservation status
Population visiting Australian territory: Near Threatened D2

Reasons for listing
Breeds at 2 locations (outside Australian territory), which have been damaged by storms historically, but there are no current threats, and future human or stochastic events are unlikely to make the species Extinct or Critically Endangered in a short time

Status 2000
Near Threatened D2

Status 1990
Near Threatened D2

Taxonomy
Buller's Albatross (northern) *T. b. platei* is also Near Threatened
Taxonomic uniqueness: high (4 genera/family, 11 species/genus, 2 subspecies/species)

Range
New Zealand breeding endemic with colonies on only 2 island groups: the Snares Is. and the Solander Is. During incubation birds range along the shelf slope off the east and west coasts of the South I., New Zealand and into the Tasman Sea; during and after chick-rearing around the southern island of New Zealand. After breeding, birds of all ages (including fledglings) migrate to slopes off Chile and Peru (Agreement on the Conservation of Albatrosses and Petrels 2009).

Abundance
About 13 700 pairs breed annually at the 2 sites, 8900 on Snares Is. and 4900 on Solander Is. Populations have been increasing in the last few decades on both island groups (Agreement on the Conservation of Albatrosses and Petrels 2009).

Ecology
Breeds annually in colonies in a variety of habitats including grassy meadows, tussock-covered slopes and cliffs, scrub and under forest canopy. Although usually forages singly, large numbers may gather to feed at concentrated food sources. Feeds using surface plunges or shallow dives for fish (mostly fishery discards), squid, tunicates, octopuses and crustacea (Agreement on the Conservation of Albatrosses and Petrels 2009; Department of Sustainability, Environment, Water, Population and Communities 2011). A generation time of 16.5 years (BirdLife International 2011) is derived from an age at first breeding of 5.0 years and an annual survival of adults of 91.3% (Schreiber and Burger 2002).

Threats
At sea it is often caught as bycatch in fisheries off Chile, though this appears not to have affected populations trends. A storm at the breeding site at Solander I. stripped most vegetation from the nesting area and recovery is slow (BirdLife International 2011).

Conservation objectives
1. At-sea threats maintained at acceptable levels
2. Global agreement on conservation measures
3. Increased public awareness of the conservation needs of albatrosses

Information required
1. Changes in adult and juvenile survival
2. Distribution of immature and non-breeding birds
3. Overlap with fishing operations and documented levels of bycatch so that mitigation measures can be targeted

Management actions required
1. Maintain mitigation strategies for fisheries, including via intergovernmental mechanisms under the auspices of the Agreement on the Conservation of Albatrosses and Petrels, the Food and Agriculture Organisation and Regional Fisheries Management Organisations

Bibliography

Agreement on the Conservation of Albatrosses and Petrels (2009) 'ACAP species assessment: Buller's Albatross *Thalassarche bulleri*'. Retrieved 21 September, 2009 from <http://www.acap.aq/>.

BirdLife International (2011) 'Species factsheet: *Thalassarche bulleri*'. Retrieved 17 January, 2011 from <http://www.birdlife.org/>.

Department of Sustainability, Environment, Water, Population and Communities (2011) *Thalassarche bulleri*. In 'Species Profile and Threats Database'. Department of Sustainabil-

Current eligibility against IUCN Red List Criteria

IUCN category	Criteria eligibility
Population visiting Australian territory	
A	Not applicable: past, current or future population declines are thought unlikely to exceed 20% in any 3-generation period
B	Not applicable: EOO, AOO, habitat, locations or population not declining, population fluctuations not extreme
C	Not applicable: population >10 000 mature individuals
D	Near Threatened: ≤5 locations but plausible threats not severe
E	Not applicable: no population viability analysis undertaken

ity, Environment, Water, Population and Communities, Canberra. Retrieved 24 February, 2011 from <http://www.environment.gov.au/sprat>.

Schreiber EA, Burger J (2002) *Biology of Marine Birds.* CRC Press, Boca Raton, FL.

IUCN Red List assessment data

	Estimate	Reliability
Population visiting Australian territory		
Extent of occurrence trend	1 700 000 km^2 stable	high high
Area of occupancy trend	1 700 000 km^2 stable	high high
No. of mature individuals trend	28 000 stable	high high
No. subpopulations	1	high
No. locations	2	high
Generation time	16.5 years	medium
Global population share	50–100%	high

Comments received from

Barry Baker, Rosemary Gales, Wieslawa Misiak, Tim Reid, Ben Sullivan, Eric Woehler

Subspecies Conservation Summary

Buller's Albatross (northern)

Thalassarche bulleri platei (Reichenow, 1898)
Diomedeidae

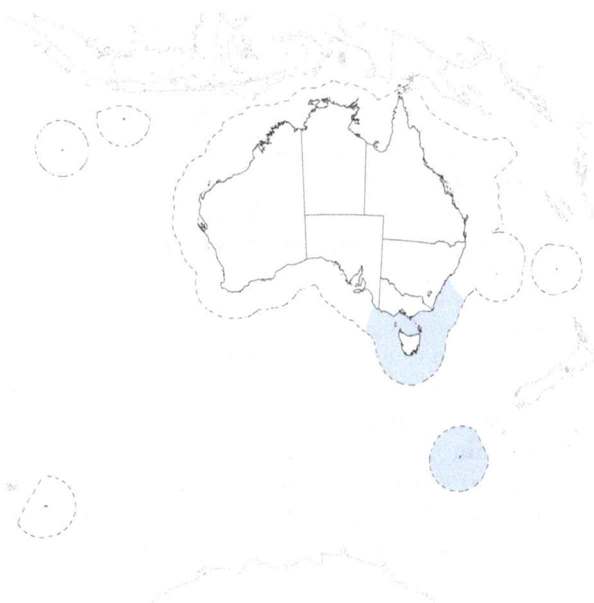

Conservation status

Population visiting Australian territory: Near Threatened D2

Reasons for listing

Breeds at 3 locations (outside Australian territory), 1 very small. There are no current threats, and future human or stochastic events are unlikely to make the species Extinct or Critically Endangered in a short time

Status 2000

Near Threatened D2

Status 1990

Near Threatened D2

Taxonomy

Buller's Albatross (southern) *T. b. bulleri* and the species are also Near Threatened
Taxonomic uniqueness: high (4 genera/family, 11 species/genus, 2 subspecies/species)

Range

New Zealand breeding endemic with colonies on only 3 island groups: the Sisters and Forty-Fours in the Chatham Is., and Rosemary Rock in the Three Kings Is. During incubation it is assumed that individuals of this subspecies, like those of the southern subspecies, range along the shelf slope off the east and west coasts of the South I., New Zealand and into the Tasman Sea; during and after chick-rearing around the southern island of New Zealand. After breeding, birds of all ages (including fledglings) migrate to slopes off Chile and Peru (Agreement on the Conservation of Albatrosses and Petrels 2009). However, there has been no satellite tracking of this subspecies.

Abundance

About 18 000 pairs breed annually at all sites, 16 000 on the Forty-Fours, 2130 pairs on Big and Little Sister Is. in the Chatham Is. group, and 20 pairs on Rosemary Rock,

Current eligibility against IUCN Red List Criteria

IUCN category	Criteria eligibility
Population visiting Australian territory	
A	Not applicable: past, current or future population declines are thought unlikely to exceed 20% in any 3-generation period
B	Not applicable: EOO, AOO, habitat, locations or population not declining, population fluctuations not extreme
C	Not applicable: population >10 000 mature individuals
D	Near Threatened: ≤5 locations but plausible threats not severe
E	Not applicable: no population viability analysis undertaken

IUCN Red List assessment data

	Estimate	Reliability
Population visiting Australian territory		
Extent of occurrence trend	1 700 000 km^2 stable	high high
Area of occupancy trend	1 700 000 km^2 stable	high high
No. of mature individuals trend	36 000 stable	high high
No. subpopulations	2	high
No. locations	3	high
Largest subpopulation	32 000	medium
Generation time	16.5 years	medium
Global population share	50–100%	low

Three Kings Is. off North I. Numbers on Chatham Is. are thought to be stable and the tiny colony on Three Kings Is. had 5 nests in 1983 and 13 in 1985 (Agreement on the Conservation of Albatrosses and Petrels 2009).

Ecology

Breeds annually in colonies in a variety of habitats including grassy meadows, tussock-covered slopes and cliffs, scrub and under forest canopy. Although usually forages singly, large numbers may gather to feed at concentrated food sources. Feeds using surface plunges or shallow dives for fish (mostly fishery discards), squid, tunicates, octopuses and crustacea (Agreement on the Conservation of Albatrosses and Petrels 2009; Department of Sustainability, Environment, Water, Population and Communities 2011). A generation time of 16.5 years (BirdLife International 2011) is derived from an age at first breeding of 5.0 years and an annual survival of adults of 91.3% (Schreiber and Burger 2002).

Threats

At sea it is often caught as bycatch in fisheries off Chile, though this appears not to have affected population trends (BirdLife International 2011). Concerns that longline fishing has been causing a rapid decline (Garnett and Crowley (2000) listed taxon at Vulnerable A2d) have not eventuated. The tiny breeding site on Rosemary Rock is probably susceptible to large storms or other natural catastrophes.

Conservation objectives

1. At-sea threats maintained at acceptable levels
2. Global agreement on conservation measures
3. Increased public awareness of the conservation needs of albatrosses

Information required

1. Population trends or changes in adult and juvenile survival
2. Distribution of immature and non-breeding birds
3. Overlap with fishing operations and documented levels of bycatch so that mitigation measures can be targeted

Management actions required

1. Maintain mitigation strategies for fisheries, including via intergovernmental mechanisms under the auspices of the Agreement on the Conservation of Albatrosses and Petrels, the Food and Agriculture Organisation and Regional Fisheries Management Organisations

Bibliography

Agreement on the Conservation of Albatrosses and Petrels (2009) 'ACAP species assessment: Buller's Albatross *Thalassarche bulleri*'. Retrieved 21 September, 2009 from <http://www.acap.aq/>.

BirdLife International (2011) 'Species factsheet: *Thalassarche bulleri*'. Retrieved 17 January, 2011 from <http://www.birdlife.org/>.

Department of Sustainability, Environment, Water, Population and Communities (2011) *Thalassarche bulleri*. In 'Species Profile and Threats Database'. Department of Sustainability, Environment, Water, Population and Communities, Canberra. Retrieved 24 February, 2011 from <http://www.environment.gov.au/sprat>.

Garnett S, Crowley GM (2000) *The Action Plan for Australian Birds 2000*. Environment Australia, Canberra.

Schreiber EA, Burger J (2002) *Biology of Marine Birds*. CRC Press, Boca Raton, FL.

Comments received from

Barry Baker, Rosemary Gales, Wieslawa Misiak, Tim Reid, Ben Sullivan, Eric Woehler

Species Conservation Summary

Sooty Albatross

Phoebetria fusca (Hilsenberg, 1822)
Diomedeidae

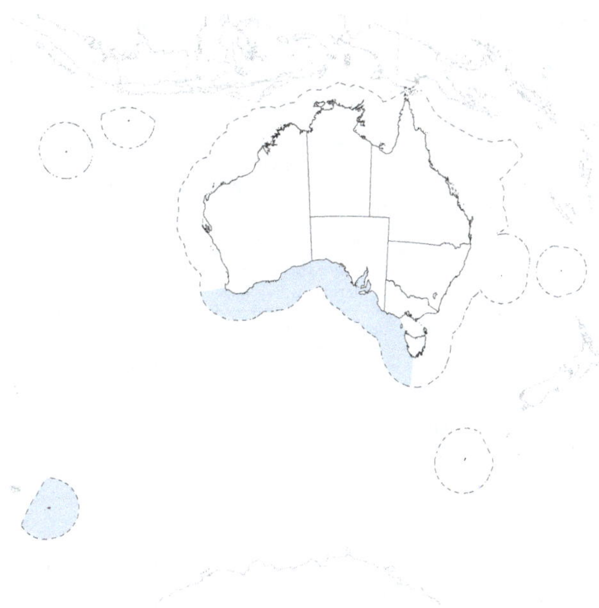

Conservation status

Population visiting Australian territory: Endangered A4bd

Reasons for listing

Population declines 50–79% in three generations (90 years) occurring as a result of at-sea mortality associated with fisheries, particularly longline fishing vessels

Status 2000

Vulnerable A4bd
Reason for change in 2010: initial declines intensifying

Status 1990

Near Threatened A4bd
Reason for change in 2000: initial declines intensifying

Taxonomy

No infraspecific taxa described
Taxonomic uniqueness: high (4 genera/family, 2 species/genus, 1 subspecies/species)

Range

Breeds on Prince Edward, Marion, Kerguelen, Crozet, Amsterdam, Saint Paul, Gough and Tristan da Cunha Is. At sea occurs mainly between 30°S and 60°S in the southern Indian and Atlantic Oceans (Woehler *et al.* 1990; Reid *et al.* 2002), with limits between c. 65°S and c. 20°S. Adults move north in winter from sub-Antarctic to subtropical seas, whereas immature birds tend to remain in subtropical seas year round. Infrequently disperses eastward to the Tasman Sea (Agreement on the Conservation of Albatrosses and Petrels 2009).

Abundance

About 15 700 pairs breed on 15 islands with declines reported from all islands where there have been repeated surveys (Agreement on the Conservation of Albatrosses and Petrels 2009).

Ecology

Breeds biennially, laying a single egg. Nests are solitary or in colonies of up to 60 nests. Feeds mainly by surface-seizing on squid, fish, crustaceans and carrion, proportions varying between years and locations. This species takes seabird carrion more than other albatrosses, including dead penguins and small petrels, and follows fishing boats to scavenge offal at least occasionally (Agreement on the Conservation of Albatrosses and Petrels 2009; Department of Sustainability, Environment, Water, Population and Communities 2011). A generation time of 30.0 years (BirdLife International 2011) is derived from an age at first breeding of 10.0 years and an annual survival of adults of 95.0% (Schreiber and Burger 2002).

Threats

On land introduced Black Rats *Rattus rattus* and feral cats *Felis catus* cause mortality on some islands. Principal reason for declines is thought to be mortality as bycatch from longline fishing (BirdLife International 2011).

Conservation objectives

1. At-sea threats reduced to acceptable levels
2. Global agreement on conservation measures
3. Increased public awareness of the conservation needs of albatrosses

Information required

1. Population trends or changes in adult and juvenile survival
2. Distribution of immature and non-breeding birds
3. Overlap with fishing operations and documented levels of bycatch so that mitigation measures can be targeted

Management actions required

1. Maintain mitigation strategies for fisheries, including via intergovernmental mechanisms under the auspices of the Agreement on the Conservation of Albatrosses and Petrels, the Food and Agriculture Organisation and Regional Fisheries Management Organisations

Bibliography

Agreement on the Conservation of Albatrosses and Petrels (2009) 'ACAP species assessment: Sooty Albatross *Phoebetria fusca*'. Retrieved 1 September, 2009 from <http://www.acap.aq/>.

BirdLife International (2011) 'Species factsheet: *Phoebetria fusca*'. Retrieved 17 January, 2011 from <http://www.birdlife.org/>.

Department of Sustainability, Environment, Water, Population and Communities (2011) *Phoebetria fusca*. In 'Species Profile and Threats Database'. Department of Sustainability, Environment, Water, Population and Communities, Can-

Current eligibility against IUCN Red List Criteria

IUCN category	Criteria eligibility
Population visiting Australian territory	
A	Endangered: current population decline of 50–79% over 3 generations (90 years) estimated and projected from monitoring and levels of exploitation
B	Vulnerable: AOO (breeding) <2000 km^2, ≤10 locations with plausible future threat, observed continuing decline in no. mature individuals
C	Not applicable: population >10 000 mature individuals
D	Not applicable: population >1000 mature individuals, >5 locations
E	Not applicable: no population viability analysis undertaken

IUCN Red List assessment data

	Estimate	Reliability
Population visiting Australian territory		
Extent of occurrence trend	2 140 000 km^2 stable	high high
Area of occupancy trend	2 140 000 km^2 stable	high high
No. of mature individuals trend	42 000 decreasing	medium high
No. subpopulations	5	high
No. locations	10	high
Largest subpopulation	10 000	medium
Generation time	30.0 years	medium
Global population share	50–100%	high

berra. Retrieved 24 February, 2011 from <http://www.environment.gov.au/sprat>.

Reid TA, Hindell MA, Eades DW, Newman M (2002) *Seabird Atlas of South-eastern Australian Waters*. Birds Australia Monograph 4.

Schreiber EA, Burger J (2002) *Biology of Marine Birds*. CRC Press, Boca Raton, FL.

Woehler EJ, Hodges CL, Watts DJ (1990) 'An atlas of the pelagic distribution and abundance of seabirds in the southern Indian Ocean 1981 to 1990'. ANARE Research Notes 77.

Comments received from

Barry Baker, Rosemary Gales, Wieslawa Misiak, Tim Reid, Ben Sullivan, Eric Woehler

Species Conservation Summary

Light-mantled Albatross

Phoebetria palpebrata (J. R. Forster, 1785)

Diomedeidae

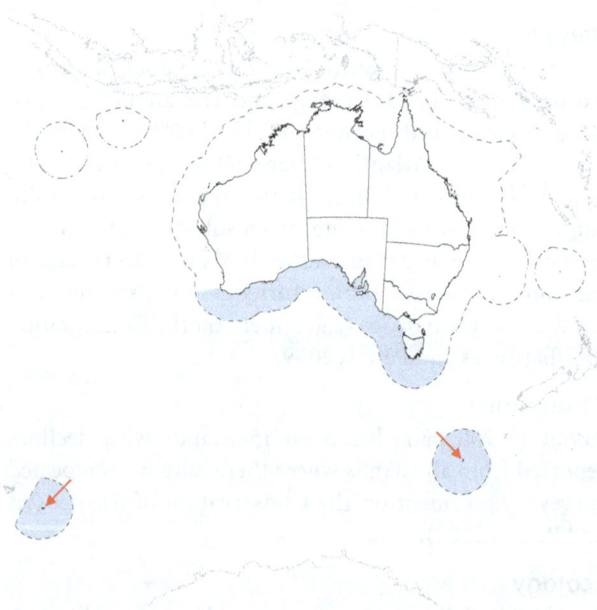

Conservation status

Australian breeding population: Endangered B2ab(ii,iii)

Population visiting Australian territory: Near Threatened A4bd

Reasons for listing

Australian breeding population: Population occurs at just 2 locations; habitat quality on Macquarie I. is deteriorating and the area occupied is probably contracting

Population visiting Australian territory: Decline 20–29% over the next 100 years possible based on counts and known levels of fishing mortality (3 generations >100 years)

Status 2000

Australian breeding population: Vulnerable D2

Reason for change in 2010: active threats to Macquarie I. breeding sites

Population visiting Australian territory: Near Threatened A4bd

Reason for change in 2000: intensification of fishing in feeding zone

Status 1990

Australian breeding population: Vulnerable D2

Population visiting Australian territory: Least Concern

Taxonomy

No infraspecific taxa described

Taxonomic uniqueness: high (4 genera/family, 2 species/genus, 1 subspecies/species)

Range

Breeds on Heard and Macquarie Is. Extralimitally breeds South Georgia, Prince Edward, Marion, Crozet, Kerguelen, Auckland, Antipodes and Campbell Is. At sea, has a circumpolar distribution over the Southern Ocean and

Current eligibility against IUCN Red List Criteria

IUCN category	Criteria eligibility
Australian breeding population	
A	Not applicable: past, current or future population declines are thought unlikely to exceed 20% in any 3-generation period
B	Endangered: AOO <500 km^2, at ≤5 locations with plausible future threat, inferred continuing decline in AOO, habitat quality
C	Not applicable: no continuing population decline or extreme fluctuations
D	Vulnerable: ≤5 locations with plausible future threat
E	Not applicable: no population viability analysis undertaken
Population visiting Australian territory	
A	Near Threatened: current population decline of 20–29% in 100 years estimated and projected from monitoring and levels exploitation
B	Not applicable: >10 locations, not severely fragmented, population fluctuations not extreme
C	Not applicable: population >10 000 mature individuals
D	Not applicable: population >1000 mature individuals, >5 locations
E	Not applicable: no population viability analysis undertaken

IUCN Red List assessment data

	Estimate	Reliability
Australian breeding population		
Extent of occurrence trend	2 900 000 km^2 stable	high high
Area of occupancy trend	40 km^2 decreasing	high medium
No. of mature individuals trend	4600 stable	low medium
No. subpopulations	2	high
No. locations	2	high
Largest subpopulation	3600	medium
Generation time	44.0 years	medium
Global population share	8%	high
Level of genetic exchange	low	medium
Population visiting Australian territory		
Extent of occurrence trend	2 900 000 km^2 stable	high high
Area of occupancy trend	2 900 000 km^2 stable	high high
No. of mature individuals trend	58 000 decreasing	low medium
No. subpopulations	6	high
No. locations	>10	high
Largest subpopulation	12 500	medium
Global population share	11–25%	low

exhibits the most southerly distribution of all the albatross species (Woehler *et al.* 1990; Agreement on the Conservation of Albatrosses and Petrels 2009).

Abundance

Estimates of the Australian population are, for Heard I., 200–500 (last census 1954; Agreement on the Conservation of Albatrosses and Petrels 2009) or at least 500 (Woehler 2006) pairs and 1110 (Terauds and Gales 2006) or 1281 pairs (2007; Agreement on the Conservation of Albatrosses and Petrels 2009) at Macquarie I. Globally about 140 000 mature individuals (Agreement on the Conservation of Albatrosses and Petrels 2009). Little change in population on Macquarie I. from 1994 and 2005 and population Heard I. may be increasing (Woehler 2006). Global population is probably declining, though negative trends in the 1990s have been reversed at most sites in the 2000s. Little monitoring overall (Agreement on the Conservation of Albatrosses and Petrels 2009).

Ecology

Nests solitarily or in small colonies, building pedestal nests along sheltered, vegetated cliff edges. Although considered a biennial breeder when successful, unsuccessful breeders may breed the following year with an average success rate of 59%. Major food items are squid, fish, crustaceans and carrion which are taken using a variety of feeding strategies including surface seizing, surface filtering and plunging to up to 12.4 m underwater in pursuit of prey. This species does not follow ships as often as other albatross species (Green *et al.* 1998; Terauds and Gales 2006; BirdLife International 2011; Department of Sustainability, Environment, Water, Population and Communities 2011). A generation time of 44.0 years (BirdLife International 2011) is derived from an age at first breeding of 7.0 years and an annual survival of adults of 97.3% (Schreiber and Burger 2002).

Threats

On Macquarie I. the increase in rabbit *Oryctolagus cuniculus* numbers since feral cats *Felis catus* were eradicated in 1999 has led to overgrazing of native vegetation and the extensive destruction of habitat and soil erosion at nesting sites. This threat may continue unless an eradication program planned in 2011–14 is successful (Agreement on the Conservation of Albatrosses and Petrels 2009). Recent eruptions on Heard I. (1985–2008; Seach 2011) and those on McDonald I. (1993–2005; Stephenson *et al.* 2005; Woehler 2006) have not affected breeding sites (contra Agreement on the Conservation of Albatrosses and Petrels 2009; BirdLife International 2011). At sea mortality is associated with fisheries, specifically longlining. While follows fishing vessels less often than most other albatrosses, their more agile diving behavior allows them greater access to baited hooks. They overlap with fishing activity most often during the non breeding period when they are in the northern part of their range (BirdLife International 2011).

Conservation objectives

1. Stabilise nesting sites on Macquarie I.
2. At-sea threats reduced to acceptable levels
3. Global agreement on conservation measures

4. Increased public awareness of the conservation needs of albatrosses

Information required

1. Information on population trends and demographic parameters for most sites, particularly Heard I.
2. Population trends or changes in adult and juvenile survival
3. Distribution of immature and non-breeding birds
4. Overlap with fishing operations and documented levels of bycatch so that mitigation measures can be targeted

Management actions required

1. Secure nest sites on Macquarie I. by eliminating rabbits
2. Continue existing monitoring and research programs
3. Conduct complete censuses at all sites at regular intervals
4. Assess the impact of trawl fisheries bycatch
5. Maintain mitigation strategies for fisheries, including via intergovernmental mechanisms under the auspices of the Agreement on the Conservation of Albatrosses and Petrels, the Food and Agriculture Organisation and Regional Fisheries Management Organisations

Bibliography

Agreement on the Conservation of Albatrosses and Petrels (2009) 'ACAP species assessment: Light-mantled Albatross *Phoebetria palpebrata*'. Retrieved 31 August, 2009 from <http://www.acap.aq/>.

BirdLife International (2011) 'Species factsheet: *Phoebetria palpebrata*'. Retrieved 17 January, 2011 from <http://www.birdlife.org/>.

Department of Sustainability, Environment, Water, Population and Communities (2011) *Phoebetria palpebrata*. In 'Species Profile and Threats Database'. Department of Sustainability, Environment, Water, Population and Communities, Canberra. Retrieved 24 February, 2011 from <http://www.environment.gov.au/sprat>.

Green K, Kerry K, Disney T, Clarke M (1998) Dietary studies of Light-mantled Sooty Albatrosses *Phoebetria palpebrata* from Macquarie and Heard Islands. *Marine Ornithology* **26**, 19–26.

Schreiber EA, Burger J (2002) *Biology of Marine Birds*. CRC Press, Boca Raton, FL.

Seach J (2011) 'Heard Island Volcano'. Retrieved 26 January, 2011 from <http://www.volcanolive.com/heardisland.html>.

Stephenson J, Budd GM, Manning J, Hansbro P (2005) Major eruption-induced changes to the McDonald Islands, southern Indian Ocean. *Antarctic Science* **17**, 259–266.

Terauds A, Gales R (2006) Provisioning strategies and growth patterns of Light-mantled Sooty Albatrosses *Phoebetria palpebrata* on Macquarie Island. *Polar Biology* **29**, 917–926.

Woehler EJ (2006) Status and conservation of the seabirds of Heard Island and the McDonald Islands. In *Heard Island: Southern Ocean Sentinel*. (Eds K Green and EJ Woehler) pp. 128–165. Surrey Beatty & Sons, Chipping Norton, NSW.

Woehler EJ, Hodges CL, Watts DJ (1990) 'An atlas of the pelagic distribution and abundance of seabirds in the southern Indian Ocean 1981 to 1990'. ANARE Research Notes 77.

Comments received from

Barry Baker, Rosemary Gales, Wieslawa Misiak, Tim Reid, Ben Sullivan, Eric Woehler

---

Species Conservation Summary

Blue Petrel

Halobaena caerulea (Gmelin, 1789)

Procellariidae

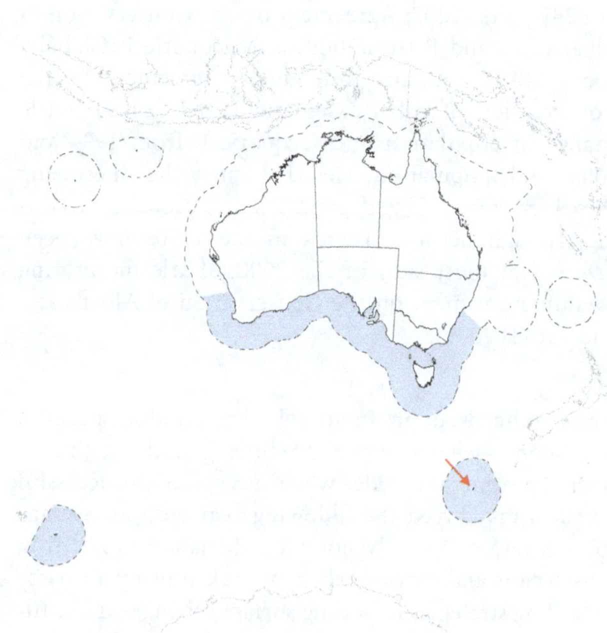

Conservation status

Australian breeding population: Critically Endangered B2ab(iii,v)

Population visiting Australian territory: Least Concern

Reasons for listing

Australian breeding population: Australian population breeds in area of <10 km^2 at 1 location and continuing decline likely until rabbits and predators removed

Status 2000

Australian breeding population: Critically Endangered B2ab(iii,v)

Current eligibility against IUCN Red List Criteria

IUCN category	Criteria eligibility
Australian breeding population	
A	Not applicable: past, current or future population declines are thought unlikely to exceed 20% in any 3-generation period
B	Critically Endangered: AOO <10 km^2, 1 location with plausible future threat, suspected continuing decline in habitat quality, no. mature individuals
C	Endangered: population 250–2500 mature individuals, suspected continuing decline, >95% mature individuals in 1 subpopulation
D	Vulnerable: population 250–1000 mature individuals, ≤5 locations with plausible future threat
E	Not applicable: no population viability analysis undertaken

IUCN Red List assessment data

	Estimate	Reliability
Australian breeding population		
Extent of occurrence trend	3 600 000 km^2 stable	high high
Area of occupancy trend	2 km^2 stable	medium medium
No. of mature individuals trend	1000 decreasing	low low
No. subpopulations	1	high
No. locations	1	high
Generation time	11.7 years	low
Global population share	<1%	medium
Level of genetic exchange	medium	low

Population visiting Australian territory: Least Concern

Status 1990

Australian breeding population: Critically Endangered B2ab(iii,v)

Population visiting Australian territory: Least Concern

Taxonomy

No infraspecific taxa described

Taxonomic uniqueness: high (14 genera/family, 1 species/genus, 1 subspecies/species)

Range

Breeding known only from a few rock stacks off Macquarie I., although may once have been more widespread on the main island. Extralimitally breeds on numerous other subantarctic islands in Indian and Atlantic Oceans including Crozet, Kerguelen, Marion, Prince Edward Is. and South Georgia. Probably foraging around breeding colonies throughout year, but also found throughout Southern Ocean throughout the year (Marchant and Higgins 1990).

Abundance

Population at Macquarie I. was estimated at 500–600 pairs in 1979 (Brothers 1984). Persistence of the species over several decades is the only evidence that some successful breeding must occur (Garnett and Crowley 2000). The global population is estimated at 80 000 mature individuals (BirdLife International 2011).

Ecology

Blue Petrels nest in colonies, laying a single egg in rock crevices or burrows dug among rocks or tussock grasses. They forage in Antarctic and subantarctic waters for pelagic crustaceans, fish, cephalopods and insects (Marchant and Higgins 1990). A generation time of 11.7 years (BirdLife International 2011) is derived from an average age at first breeding of 5.4 years and an annual survival of adults of 84.0%, both extrapolated from closely-related *Pachyptila* spp.

Threats

All nests found in the vicinity of Macquarie I. have failed because of nest destruction or predation. Nests are frequently deserted because of trampling of burrows by Imperial Shags *Leucocarbo atriceps* (Garnett and Crowley 2000), flooding of burrows or exceptionally high waves that washed part of 1 colony away. Many adults are taken by Brown Skua *Stercorarius lonnbergi* or Black Rats *Rattus rattus*. Their presence on the island during winter, when many other species of seabirds are absent, makes Blue Petrels particularly vulnerable to predators (Brothers 1984). Extinction from Macquarie I. itself may have been hastened by introduced Wekas *Gallirallus australis*, cats *Felis catus* and rats. Rats are either resident on, or occasionally visit, all but 1 of the breeding stacks (Brothers 1984) and, on the main island, are most common in the tussock grassland where the petrels were reported to breed (Garnett and Crowley 2000). These threats may continue, unless a rat eradication program planned for 2011–14 is successful.

Conservation objectives

1. Ongoing presence at existing sites
2. Nesting on Macquarie I.

Information required

1. Numbers and trends

Management actions required

1. Eradicate Black Rats and rabbits on Macquarie I.
2. Monitor samples of the population

Bibliography

BirdLife International (2011) 'Species factsheet: *Halobaena caerulea*'. Retrieved 17 January, 2011 from <http://www.birdlife.org/>.

Brothers NP (1984) Breeding, distribution and status of burrow-nesting petrels at Maqaurie Island. *Australian Wildlife Research* **11**, 113–131.

Garnett S, Crowley GM (2000) *The Action Plan for Australian Birds 2000*. Environment Australia, Canberra.

Marchant S, Higgins PJ (Eds) (1990) *Handbook of Australian, New Zealand and Antarctic Birds. Volume 1: Ratites to Ducks*. Oxford University Press, Melbourne.

Comments received from

Barry Baker, Rosemary Gales, Tim Reid, Ben Sullivan, Eric Woehler

Species Conservation Summary

Antarctic Prion

Pachyptila desolata (Gmelin, 1789)
Procellariidae

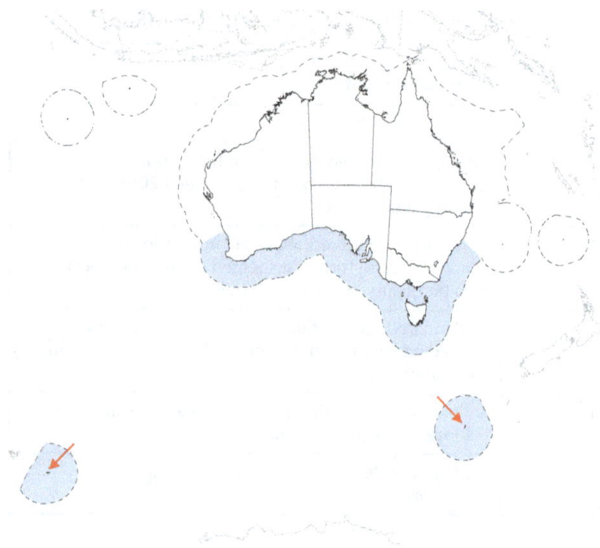

Conservation status

Australian breeding population: Endangered B2ab(iii,v)
Population visiting Australian territory: Least Concern

Reasons for listing

Australian breeding population: Present at only 2 locations at 1 of which rapid decline likely because of predators and rabbits

Status 2000

Australian breeding population: Vulnerable D2
Reason for change in 2010: declines on Macquarie I. after cats removed and rabbit populations increased
Population visiting Australian territory: Least Concern

Status 1990

Australian breeding population: Vulnerable D2
Population visiting Australian territory: Least Concern

Taxonomy

No infraspecific taxa described
Taxonomic uniqueness: medium (14 genera/family, 6 species/genus, 1 subspecies/species)

Range

Breeding occurs on Heard and Macquarie Is. (Marchant and Higgins 1990; Woehler 2006, 2010). Extralimitally breeds South Georgia, Scotia Archipelago and Crozet, Kerguelen, Auckland, South Sandwich and Scott Is. After breeding disperses from pack ice in Antarctica to as far north as Australia, South Africa and off South America as far north as Peru.

Abundance

On Macquarie I. was estimated at 49 000 pairs in 1984 (Brothers 1984), now 40 000 pairs (Parks and Wildlife Service 2006). On Heard I. the population is conservatively estimated at 100 000 pairs (Woehler 2006, 2010). The global population estimated at 50 million individuals, but this is a very coarse estimate (BirdLife International 2011).

Ecology

Digs burrows among rocks or under low vegetation in which to nest. Feeds on small crustaceans, particularly euphausiids, copepods and amphipods, as well as fish and squid plucked from the ocean surface (Marchant and Higgins 1990). A generation time of 11.3 years (BirdLife International 2011) is derived from an average age at first breeding is 5.0 years (Schreiber and Burger 2002) and an annual survival of adults of 84.0%, extrapolated from congeners.

Threats

On Macquarie I. the increase in rabbit *Oryctolagus cuniculus* numbers since feral cats *Felis catus* were eradicated in 1999 has led to an extensive destruction of habitat and soil erosion at nesting sites from overgrazing, and exposure of nests to predation by Brown Skua *Stercorarius lonnbergi*. Black Rats *Rattus rattus* are also likely to take eggs and chicks. These threats may continue unless an eradication program planned for 2011–14 is successful. The accidental introduction of mammalian predators to Heard I. is a plausible threat. There are no known threats at sea, although the impact of plastic ingestion is unknown.

Conservation objectives

1. Nesting sites on Macquarie I. stabilised

Information required

1. Information on population size, trends and demographic parameters

Management actions required

1. Eradicate Black Rats and rabbits on Macquarie I.
2. Monitor at least samples of the population
3. Maintain strict quarantine for Heard I.

Bibliography

BirdLife International (2011) 'Species factsheet: *Pachyptila desolata*'. Retrieved 17 January, 2011 from <http://www.birdlife.org/>.

Brothers NP (1984) Breeding, distribution and status of burrow-nesting petrels at Maqaurie Island. *Australian Wildlife Research* **11**, 113–131.

Marchant S, Higgins PJ (Eds) (1990) *Handbook of Australian, New Zealand and Antarctic Birds. Volume 1: Ratites to Ducks.* Oxford University Press, Melbourne.

Parks and Wildlife Service (2006) 'Macquarie Island Nature Reserve and World Heritage Area Management Plan'. Parks and Wildlife Service, Department of Tourism, Arts and the Environment, Hobart.

Schreiber EA, Burger J (2002) *Biology of Marine Birds.* CRC Press, Boca Raton, FL.

Woehler EJ (2006) Status and conservation of the seabirds of Heard Island and the McDonald Islands. In *Heard Island: Southern Ocean Sentinel.* (Eds K Green and EJ Woehler) pp. 128–165. Surrey Beatty & Sons, Chipping Norton, NSW.

Current eligibility against IUCN Red List Criteria

IUCN category	Criteria eligibility
Australian breeding population	
A	Not applicable: past, current or future population declines are thought unlikely to exceed 20% in any 3-generation period
B	Endangered: AOO <500 km², at ≤5 locations with plausible future threat, inferred continuing decline in habitat quality, no. mature individuals
C	Not applicable: population >10 000 mature individuals
D	Vulnerable: ≤5 locations with plausible future threat
E	Not applicable: no population viability analysis undertaken

Woehler EJ (2010) Status and conservation of the seabirds of Heard Island. *The State of Australia's Birds 2010: Islands and Birds. Wingspan* **Supplement 20**, 8–9.

IUCN Red List assessment data

	Estimate	Reliability
Australian breeding population		
Extent of occurrence trend	3 600 000 km² stable	high high
Area of occupancy trend	40 km² decreasing	low medium
No. of mature individuals trend	240 000 decreasing	low high
No. subpopulations	2	high
No. locations	2	high
Largest subpopulation	200 000	medium
Generation time	11.3 years	medium
Global population share	<1%	medium
Level of genetic exchange	low	medium

Comments received from

Barry Baker, Rosemary Gales, Tim Reid, Ben Sullivan, Eric Woehler

Subspecies Conservation Summary

Fairy Prion (southern)

Pachyptila turtur subantarctica Oliver, 1955
Procellariidae

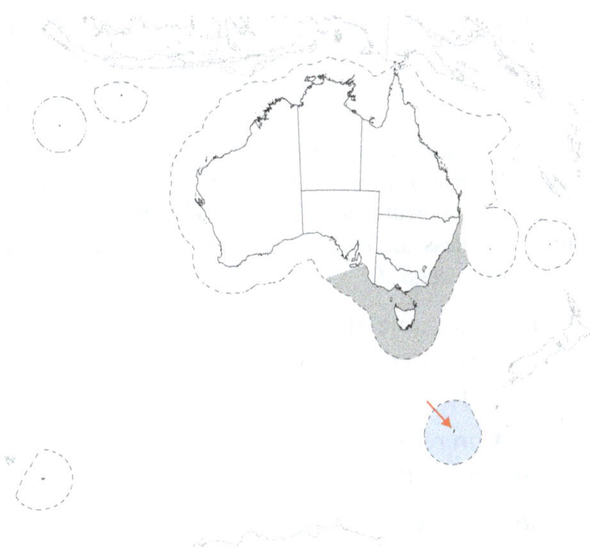

Conservation status

Australian breeding population: Endangered D
Population visiting Australian territory: Least Concern

Reasons for listing

Australian breeding population: Population <250 mature individuals

Status 2000

Australian breeding population: Endangered D
Population visiting Australian territory: Least Concern

Status 1990

Australian breeding population: Endangered D
Population visiting Australian territory: Least Concern

Taxonomy

P. t. turtur (Bass Strait islands, New Zealand offshore islands, Chatham Is, Falkland Is, Indian Ocean subantarctic islands) is Least Concern, as is the species
Taxonomic uniqueness: medium (14 genera/family, 6 species/genus, 2 subspecies/species)

Range

Breeding known only from 2 rock stacks off Macquarie I., 1 near Langdon Point, the other near Davis Point (Brothers 1984) with a second location on Bishop and Clerk Is. nearby (Garnett and Crowley 2000). Outside Australian territory, breeding on Antipodes I., Big South Cape and Snares I. with some individuals migrating north towards New Zealand and southern Australia in winter (Marchant and Higgins 1990).

Abundance

Two very small populations of little over 50 pairs (Brothers 1984), though may have been larger before Black Rats *Rattus rattus* arrived on Macquarie I. The global population is estimated at 80 000 birds (Blaber *et al.* 1996).

Ecology

Digs burrows among rocks or low vegetation in which to nest. Feeds on small crustaceans, particularly euphausiids, copepods and amphipods, as well as fish and squid plucked from the ocean surface (Marchant and Higgins 1990). A generation time of 10.8 years (BirdLife International 2011) is derived from an average age at first breed-

Current eligibility against IUCN Red List Criteria

IUCN category	Criteria eligibility
Australian breeding population	
A	Not applicable: past, current or future population declines are thought unlikely to exceed 20% in any 3-generation period
B	Not applicable: EOO, AOO, habitat, locations or population not declining, population fluctuations not extreme
C	Not applicable: no continuing population decline or extreme fluctuations
D	Endangered: population 50–250 mature individuals
E	Not applicable: no population viability analysis undertaken

IUCN Red List assessment data

	Estimate	Reliability
Australian breeding population		
Extent of occurrence trend	610 000 km^2 stable	high high
Area of occupancy trend	3 km^2 stable	medium medium
No. of mature individuals trend	110 stable	low high
No. subpopulations	1	high
No. locations	3	high
Generation time	10.8 years	medium
Global population share	<1%	medium
Level of genetic exchange	low	medium

ing of 4.5 years (Schreiber and Burger 2002) and an annual survival of adults of 84.0%, extrapolated from congeners.

Threats

Only threat to existing range is accidental introduction of rats or other vertebrate predators, which is unlikely. The current distribution, however, may reflect historic loss from Macquarie I. the colonisation of which is probably constrained by the presence of introduced Black Rats *Rattus rattus*. These constraints on nesting away from rock stacks are even stronger following the increase in rabbit *Oryctolagus cuniculus* numbers that has occurred since feral cats *Felis catus* were eradicated in 1999, resulting in increased erosion from over grazing, exposing potential nest sites to predation by Brown Skua *Stercorarius lonnbergi*. These threats may continue unless an eradication program planned for 2011–14 is successful.

Conservation objectives

1. Ongoing presence at existing sites
2. Nesting on Macquarie I.

Information required

1. Numbers and trends

Management actions required

1. Eradicate Black Rats and rabbits on Macquarie I.
2. Monitor population to test success of management

Bibliography

BirdLife International (2011) 'Species factsheet: *Pachyptila turtur*'. Retrieved 17 January, 2011 from <http://www.birdlife.org/>.

Blaber S, Battam H, Brothers N, Garnett ST (1996) Threatened and migratory seabird species in Australia: an overview of status, conservation and management. In *The Status of Australia's Seabirds: Proceedings of the National Seabird Workshop*, Canberra, 1–2 November 1993. (Eds GJB Ross, K Weaver and JC Greig) pp. 13–27. Biodiversity Group, Environment Australia, Canberra.

Brothers NP (1984) Breeding, distribution and status of Burrow-nesting Petrels at Maqaurie Island. *Australian Wildlife Research* **11**, 113–131.

Garnett S, Crowley GM (2000) *The Action Plan for Australian Birds 2000*. Environment Australia, Canberra.

Marchant S, Higgins PJ (Eds) (1990) *Handbook of Australian, New Zealand and Antarctic Birds. Volume 1: Ratites to Ducks*. Oxford University Press, Melbourne.

Schreiber EA, Burger J (2002) *Biology of Marine Birds*. CRC Press, Boca Raton, FL.

Comments received from

Barry Baker, Rosemary Gales, Tim Reid, Ben Sullivan, Eric Woehler

Subspecies Conservation Summary

Fulmar Prion (southern)

Pachyptila crassirostris eatoni (Mathews, 1912)
Procellariidae

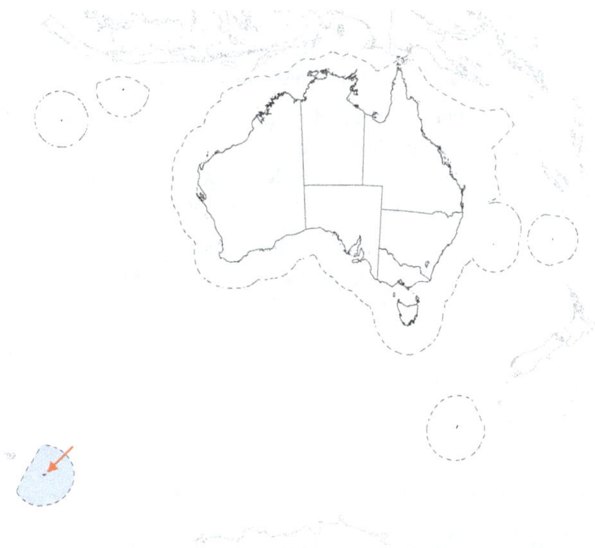

Conservation status

Australian breeding population: Vulnerable D2
Population visiting Australian territory: Least Concern

Reasons for listing

Australian breeding population: Present at 1 location where the introduction of predators is a plausible future threat

Status 2000

Australian breeding population: Vulnerable D2
Population visiting Australian territory: Least Concern

Status 1990

Australian breeding population: Vulnerable D2
Population visiting Australian territory: Least Concern

Taxonomy

P. c. crassirostris (Chatham, Bounty and Snares Is. off New Zealand) is Least Concern
Taxonomic uniqueness: medium (14 genera/family, 6 species/genus, 2 subspecies/species)

Range

Breeding occurs on Heard I. in Australian Territory. Extralimitally breeds Rose, Ocean and Ewing Is. in the Auckland I. group (New Zealand). Non-breeding birds disperse well out to sea (Woehler *et al.* 1990).

Abundance

The population on Heard I. is estimated to be at least 10 000 breeding pairs (Woehler 2006, 2010). The global population of the species is estimated at 150 000–300 000 individuals (BirdLife International 2011).

Ecology

Individuals dig burrows among rocks, on unstable coastal lava cliffs, on ledges of coastal headlands or low vegetation in which to nest (Woehler 2006). They feed on small crustaceans, particularly euphausiids, copepods and amphipods, as well as fish and squid plucked from the ocean surface (Marchant and Higgins 1990). A generation time of 11.7 years (BirdLife International 2011) is derived from an average age at first breeding of 5.4 years and an annual survival of adults of 84.0%, both extrapolated from congeners.

Threats

The accidental introduction of mammalian predators to Heard I. is a plausible threat. The only possible threat at sea is ingestion of plastic particles.

Conservation objectives

1. Population on Heard I. remains secure

Information required

1. Precise population size estimate

Management actions required

1. Maintain strict quarantine for Heard I.

Bibliography

BirdLife International (2011) 'Species factsheet: *Pachyptila crassirostris*'. Retrieved 17 January, 2011 from <http://www.birdlife.org/>

Current eligibility against IUCN Red List Criteria

IUCN category	Criteria eligibility
Australian breeding population	
A	Not applicable: past, current or future population declines are thought unlikely to exceed 20% in any 3-generation period
B	Not applicable: EOO, AOO, habitat, locations or population not declining, population fluctuations not extreme
C	Not applicable: no continuing population decline or extreme fluctuations
D	Vulnerable: ≤5 locations with plausible future threat
E	Not applicable: no population viability analysis undertaken

IUCN Red List assessment data

	Estimate	Reliability
Australian breeding population		
Extent of occurrence trend	520 000 km^2 stable	high high
Area of occupancy trend	20 km^2 stable	low medium
No. of mature individuals trend	20 000 stable	low high
No. subpopulations	1	high
No. locations	1	high
Generation time	11.7 years	medium
Global population share	<1%	medium
Level of genetic exchange	low	medium

Marchant S, Higgins PJ (Eds) (1990) Handbook of Australian, New Zealand and Antarctic Birds. Volume 1: Ratites to Ducks. Oxford University Press, Melbourne.

Woehler EJ (2006) Status and conservation of the seabirds of Heard Island and the McDonald Islands. In *Heard Island: Southern Ocean Sentinel*. (Eds K Green and EJ Woehler) pp. 128–165. Surrey Beatty & Sons, Chipping Norton, NSW.

Woehler EJ (2010) Status and conservation of the seabirds of Heard Island. *The State of Australia's Birds 2010: Islands and Birds. Wingspan* **Supplement 20**, 8–9.

Woehler EJ, Hodges CL, Watts DJ (1990) 'An atlas of the pelagic distribution and abundance of seabirds in the southern Indian Ocean 1981 to 1990'. ANARE Research Notes 77.

Comments received from

Barry Baker, Rosemary Gales, Tim Reid, Ben Sullivan, Eric Woehler

Species Conservation Summary

White-chinned Petrel

Procellaria aequinoctialis Linnaeus, 1758
Procellariidae

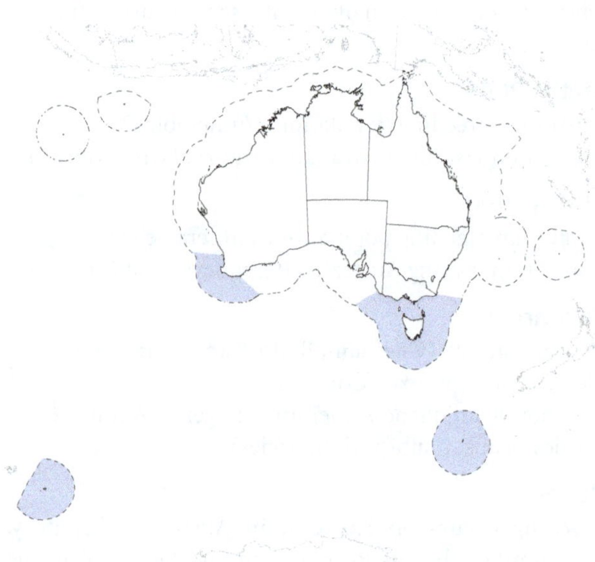

Conservation status

Population visiting Australian territory: Vulnerable A4bcde

Reasons for listing

Ongoing population decline of 30–49% in 3 generations (74 years)

Status 2000

Vulnerable A4bcde
Reason for change in 2010: initial declines intensifying

Status 1990

Near Threatened A4bcde
Reason for change in 2000: initial declines detected

Taxonomy

No infraspecific taxa described
Taxonomic uniqueness: medium (14 genera/family, 4 species/genus, 1 subspecies/species)

Range

Breeds on South Georgia, Prince Edward, Crozet, Kerguelen, Auckland, Campbell, Antipodes and Falkland Is. At sea birds occur widely across southern oceans from the subtropics to the edge of Antarctica (Agreement on the Conservation of Albatrosses and Petrels 2010) including Australian waters.

Abundance

The population is estimated to be 1 196 000 to 1 334 000 pairs, corresponding to a total population of about 3 500 000 individuals. Declines of 28% over 20 years recorded on Bird I., South Georgia, 86% from 1981–1993 Prydz Bay, Antarctica, 37% decline 1996–2000 on Marion I., and a 37% decline 1983–2004 on Crozet archipelago (Agreement on the Conservation of Albatrosses and Petrels 2010; BirdLife International 2011).

Ecology

A colonial species that breeds annually and it is the largest petrel to nest in burrows. Birds feed by surface seizing, but can dive to 15 m. They feed mainly on krill and fish with some squid and amphipods. They follow ships and feed on fisheries discards or dives on baited hooks, making them highly vulnerable to accidental death in longline fisheries (Agreement on the Conservation of Albatrosses and Petrels 2010). A generation time of 24.7 years (BirdLife International 2011) is derived from an average age at first breeding of 6.5 years and an annual survival of adults of 94.5% (Schreiber and Burger 2002).

Threats

On land Black Rats *Rattus rattus*, Brown Rats *Rattus norvegicus*, feral cats *Felis catus*, reindeer *Rangifer tarandus* and Antarctic Fur Seal *Arctocephalus gazella* cause mortality or damage breeding habitat. Principal reason for declines is thought to be mortality as bycatch from longline fishing (BirdLife International 2011) although some effective mitigation techniques are now available (Robertson *et al.* 2006).

Conservation objectives

1. At-sea threats reduced to acceptable levels
2. Global agreement on conservation measures

Current eligibility against IUCN Red List Criteria

IUCN category	Criteria eligibility
Population visiting Australian territory	
A	Vulnerable: current population decline of 30–49% over 3 generations (74 years) estimated and projected from monitoring and inferred from deteriorating habitat quality, levels of exploitation and effects of introduced taxa
B	Not applicable: >10 locations, not severely fragmented, population fluctuations not extreme
C	Not applicable: population >10 000 mature individuals
D	Not applicable: population >1000 mature individuals, >5 locations
E	Not applicable: no population viability analysis undertaken

IUCN Red List assessment data

	Estimate	Reliability
Population visiting Australian territory		
Extent of occurrence trend	2 500 000 km^2 stable	high high
Area of occupancy trend	900 000 km^2 stable	high high
No. of mature individuals trend	3 500 000 decreasing	medium high
No. subpopulations	2	high
No. locations	>10	high
Largest subpopulation	1 800 000	medium
Generation time	24.7 years	medium
Global population share	1–10%	high

3. Increased public awareness of the conservation needs of albatrosses

Information required

1. Population trends or changes in adult and juvenile survival
2. Distribution of immature and non-breeding birds
3. Overlap with fishing operations and documented levels of bycatch so that mitigation measures can be targeted

Management actions required

1. Maintain mitigation strategies for fisheries, including via intergovernmental mechanisms under the auspices of the Agreement on the Conservation of Albatrosses and Petrels, the Food and Agriculture Organisation and Regional Fisheries Management Organisations

Bibliography

Agreement on the Conservation of Albatrosses and Petrels (2010) 'ACAP species assessment: White-chinned Petrel *Procellaria aequinoctialis*'. Retrieved 14 September, 2009 from <http://www.acap.aq/>.

BirdLife International (2011) 'Species factsheet: *Procellaria aequinoctialis*'. Retrieved 17 January, 2011 from <http://www.birdlife.org/>.

Robertson G, McNeill M, Smith N, Wienecke B, Candy S, Olivier F (2006) Fast sinking (integrated weight) longlines reduce mortality of white-chinned petrels (*Procellaria aequinoctialis*) and sooty shearwaters (*Puffinus griseus*) in demersal longline fisheries. *Biological Conservation* **132**, 458–471.

Schreiber EA, Burger J (2002) *Biology of Marine Birds*. CRC Press, Boca Raton, FL.

Comments received from

Barry Baker, Rosemary Gales, Wieslawa Misiak, Tim Reid, Ben Sullivan, Eric Woehler

Species Conservation Summary

Westland Petrel

Procellaria westlandica Falla, 1946

Procellariidae

Conservation status

Population visiting Australian territory: Vulnerable D2

Reasons for listing

Present at 1 site (outside Australia) where development nearby represents a plausible future threat

Status 2000

Vulnerable D2

Current eligibility against IUCN Red List Criteria

IUCN category	Criteria eligibility
Population visiting Australian territory	
A	Not applicable: past, current or future population declines are thought unlikely to exceed 20% in any 3-generation period
B	Not applicable: EOO, AOO, habitat, locations or population not declining, population fluctuations not extreme
C	Not applicable: population >10 000 mature individuals
D	Vulnerable: ≤5 locations with plausible future threat
E	Not applicable: no population viability analysis undertaken

IUCN Red List assessment data

	Estimate	Reliability
Population visiting Australian territory		
Extent of occurrence trend	570 000 km^2 stable	high high
Area of occupancy trend	570 000 km^2 stable	high high
No. of mature individuals trend	20 000 stable	medium medium
No. subpopulations	1	high
No. locations	1	high
Generation time	40.6 years	high
Global population share	1–10%	high

Status 1990

Vulnerable D2

Taxonomy

No infraspecific taxa described

Taxonomic uniqueness: medium (14 genera/family, 4 species/genus, 1 subspecies/species)

Range

Breeds only at Punakaiki, South Island, New Zealand, migrating in summer to the central Pacific and eastern New Zealand waters as well as to waters off South America (BirdLife International 2011). Recorded infrequently but year round in Australian waters (Agreement on the Conservation of Albatrosses and Petrels 2009).

Abundance

The total population is about 4000 pairs which translates to a total population of 20 000 individuals. Although modelling foretold a decline, this appears not to have eventuated (BirdLife International 2011).

Ecology

Colonial burrow-nesting, annually-breeding species. Feeds from the surface and by shallow diving in groups of up to 300, probably at night, as well as scavenging around fishing vessels and cetaceans. Its diet is dominated by squid and supplemented by fish, tunicates, crustaceans and cyclostomes (BirdLife International 2011). A generation time of 40.6 years (BirdLife International 2011) is derived from an average age at first breeding of 12.0 years (Schreiber and Burger 2002) and an annual survival of adults of 96.5% (Agreement on the Conservation of Albatrosses and Petrels 2009).

Threats

At sea the principal threat is as bycatch from pelagic longlines and trawl vessels. It is threatened at breeding sites by inappropriate development for tourism or mining (Agreement on the Conservation of Albatrosses and Petrels 2009).

Conservation objectives

1. At-sea threats maintained at acceptable levels
2. Global agreement on conservation measures
3. Increased public awareness of the conservation needs of large petrels

Information required

1. Population trends or changes in adult and juvenile survival
2. Distribution of immature and non-breeding birds
3. Overlap with fishing operations and documented levels of bycatch so that mitigation measures can be targeted

Management actions required

1. Assess the impact of trawl fisheries bycatch
2. Maintain mitigation strategies for fisheries, including via intergovernmental mechanisms under the auspices of the Agreement on the Conservation of Albatrosses and Petrels, the Food and Agriculture Organisation and Regional Fisheries Management Organisations

Bibliography

Agreement on the Conservation of Albatrosses and Petrels (2009) 'ACAP species assessment: Westland Petrel *Procellaria westlandica*'. Retrieved 11 September, 2009 from <http://www.acap.aq/>.

BirdLife International (2011) 'Species factsheet: *Procellaria westlandica*'. Retrieved 17 January, 2011 from <http://www.birdlife.org/>.

Schreiber EA, Burger J (2002) *Biology of Marine Birds*. CRC Press, Boca Raton, FL.

Comments received from

Barry Baker, Rosemary Gales, Wieslawa Misiak, Tim Reid, Ben Sullivan

Species Conservation Summary

Black Petrel

Procellaria parkinsoni Gray, 1862
Procellariidae

Conservation status
Population visiting Australian territory: Vulnerable D2

Reasons for listing
Present at 2 sites (outside Australia) where introduction of exotic predators represents a credible threat

Status 2000
Vulnerable D2

Status 1990
Vulnerable D2

Taxonomy
No infraspecific taxa described
Taxonomic uniqueness: medium (14 genera/family, 4 species/genus, 1 subspecies/species)

Range
Formerly widespread across the North Island and northern South Island of New Zealand, now restricted to Little and Great Barrier Is. in the Hauraki Gulf, in northern New Zealand. Non-breeding birds feed in the Pacific Ocean, and regularly visit the east coast of Australia. During the breeding season found mainly east of breeding islands in subtropical waters.

Abundance
The total population is c.1300 pairs on Great Barrier I. and c.100 breeding pairs on Little Barrier Island (Bell *et al.* 2009), with an estimated total of c.5000 individuals including non-breeding birds (BirdLife International 2011).

Ecology
Colonial burrow-nesting, annually-breeding species. They feed on the surface and by shallow diving in groups of up to 300, probably at night. They also frequently scavenge around fishing vessels and cetaceans. Their diet is dominated by squid and supplemented by fish, tunicates, crustaceans and cyclostomes (BirdLife International 2011). A generation time of 23.7 years (BirdLife International 2011) is derived from an average age at first breeding of 7.0 years and an annual survival of adults of 94.0% (Schreiber and Burger 2002).

Threats
There are few records of the marine threats; however, as scavengers and known boat followers, this species is vulnerable to interactions with fishing vessels. Information from New Zealand fisheries indicates that this species interacts with, and is killed on, pelagic longlines in New Zealand waters (Agreement on the Conservation of Albatrosses and Petrels 2009). Both breeding islands are vulnerable to the accidental introduction of invasive predators and to fire.

Conservation objectives
1. At-sea threats maintained at acceptable levels
2. Global agreement on conservation measures
3. Increased public awareness of the conservation needs of large petrels

Current eligibility against IUCN Red List Criteria

IUCN category	Criteria eligibility
Population visiting Australian territory	
A	Not applicable: past, current or future population declines are thought unlikely to exceed 20% in any 3-generation period
B	Not applicable: EOO, AOO, habitat, locations or population not declining, population fluctuations not extreme
C	Not applicable: population >10 000 mature individuals
D	Vulnerable: ≤5 locations with plausible future threat
E	Not applicable: no population viability analysis undertaken

IUCN Red List assessment data

	Estimate	Reliability
Population visiting Australian territory		
Extent of occurrence trend	850 000 km^2 stable	high high
Area of occupancy trend	850 000 km^2 stable	high high
No. of mature individuals trend	5000 stable	medium medium
No. subpopulations	1	high
No. locations	2	high
Generation time	23.4 years	medium
Global population share	<1%	high

Information required
1. Population trends or changes in adult and juvenile survival
2. Distribution of immature and non-breeding birds
3. Overlap with fishing operations and documented levels of bycatch so that mitigation measures can be targeted

Management actions required
1. Assess the impact of trawl fisheries bycatch
2. Maintain mitigation strategies for fisheries, including via intergovernmental mechanisms under the auspices of the Agreement on the Conservation of Albatrosses and Petrels, the Food and Agriculture Organisation and Regional Fisheries Management Organisations

Bibliography
Agreement on the Conservation of Albatrosses and Petrels (2009) 'ACAP species assessment: Black Petrel *Procellaria parkinsoni*'. Retrieved 20 August, 2009 from <http://www.acap.aq/>.

Bell EA, Sim JL, Scofield P (2009) 'Population parameters and distribution of the black petrel (*Procellaria parkinsoni*), 2005/06'. DOC Research and Development Series 307. Department of Conservation, Wellington.

BirdLife International (2011) 'Species factsheet: *Procellaria parkinsoni*'. Retrieved 17 January, 2011 from <http://www.birdlife.org/>.

Schreiber EA, Burger J (2002) *Biology of Marine Birds.* CRC Press, Boca Raton, FL.

Comments received from
Barry Baker, Rosemary Gales, Wieslawa Misiak, Tim Reid, Ben Sullivan

Species Conservation Summary

Grey Petrel

Procellaria cinerea (Gmelin, 1789)
Procellariidae

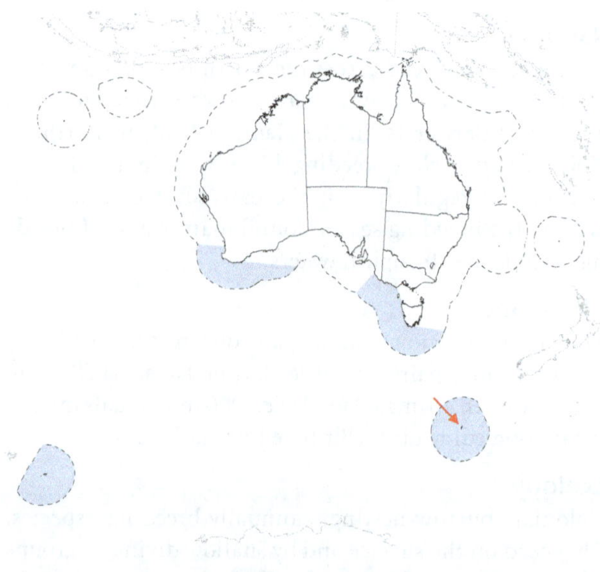

Taxonomic uniqueness: medium (14 genera/family, 4 species/genus, 1 subspecies/species)

Conservation status
Australian breeding population: Endangered B3ab(iii)
Population visiting Australian territory: Near Threatened A2cde+3cde+4cde

Reasons for listing
Australian breeding population: population occurs at a single location where the habitat quality is declining and the area occupancy is probably contracting because of grazing by rabbits and subsequent predation; could be Critically Endangered but some evidence of genetic exchange with the extralimital Near Threatened population so downlisted 1 level
Population visiting Australian territory: recent and ongoing declines approaching 30% in 3 generations (75 years)

Status 2000
Australian breeding population: Endangered D
Reason for change in 2010: active threats to Macquarie I. breeding sites
Population visiting Australian territory: Near Threatened A2cde+3cde+4cde

Status 1990
Australian breeding population: Endangered D
Population visiting Australian territory: Near Threatened A2cde+3cde+4cde

Taxonomy
No infraspecific taxa described

Range
Breeds on Macquarie I. as well as Gough, St Helena, Prince Edward, Marion, Crozet, Kerguelen, Amsterdam, Campbell and the Antipodes Is. Its total population size is poorly known. At sea has circumpolar distribution over subantarctic and subtropical waters, mostly between 32–58°S, but farther north off the west coast of South America. Females forage farther north than males, in waters north of the Subtropical Convergence (BirdLife International 2011).

Abundance
No breeding records on Macquarie I. for 80 years to 1993 (Brothers 1984; Rounsevell and Brothers 1984), though possibly for lack of researchers searching steep slopes in winter. From 1993 to 2003, population rose to as high as 80 pairs in 26 colonies (Schulz *et al.* 2005) before decreasing to 74 pairs in 2007 (Agreement on the Conservation

Current eligibility against IUCN Red List Criteria

IUCN category	Criteria eligibility
Australian breeding population	
A	Not applicable: population up and down within 1 generation, so no trend
B	Critically Endangered: AOO <10 km^2, 1 location with plausible future threat, inferred continuing decline in AOO, habitat quality (but downgraded to Endangered as global status Near Threatened)
C	Not applicable: although population <250; 100% in 1 subpopulation, population up and down within 1 generation
D	Vulnerable: population 250–1000 mature individuals, ≤5 locations with plausible future threat
E	Not applicable: no population viability analysis undertaken
Population visiting Australian territory	
A	Near Threatened: past, current and anticipated population declines of 20–29% over periods of 3 generations (75 years) inferred from deteriorating habitat quality, levels of exploitation and effects of introduced taxa
B	Not applicable: >10 locations, not severely fragmented, population fluctuations not extreme
C	Not applicable: population >10 000 mature individuals
D	Not applicable: population >1000 mature individuals, >5 locations
E	Not applicable: no population viability analysis undertaken

IUCN Red List assessment data

	Estimate	Reliability
Australian breeding population		
Extent of occurrence trend	2 200 000 km^2 stable	high high
Area of occupancy trend	2 km^2 contracting	high medium
No. of mature individuals trend	140 declining	high high
No. subpopulations	1	high
No. locations	1	high
Generation time	25.2 years	medium
Global population share	<1%	high
Level of genetic exchange	high	medium
Population visiting Australian territory		
Extent of occurrence trend	2 200 000 km^2 stable	high high
Area of occupancy trend	2 200 000 km^2 stable	high high
No. of mature individuals trend	400 000 decreasing	medium high
No. subpopulations	3	high
No. locations	10	high
Largest subpopulation	106 000	medium
Global population share	26–50%	high

of Albatrosses and Petrels 2009) and 32 pairs in 2010 as nest sites were exposed by rabbit *Oryctolagus cuniculus* grazing (R. Gales pers. comm.). Global population is estimated tentatively at 400 000 individuals, a figure that could be incorrect by a factor of 2–3, with the largest population probably on the Antipodes Is., with 53 000 pairs estimated in 2001 (BirdLife International 2011). There is no population trend data for most sites, but based on a population model and data from fisheries, global population is probably declining due to oceanic fisheries bycatch.

Ecology

Grey Petrels nest annually in colonies in burrows during winter on well-drained areas. On Macquarie I. most nests are in tussock grassland on steep, seaward-facing or near-coastal slopes in deep soil with little surface rock, particularly in deeper soils or on westerly spur crests (Schulz *et al.* 2005). They usually feed alone or in small groups, plunging from up to 10 m above the surface then swimming underwater to catch cephalopods and some fish (Marchant and Higgins 1990; BirdLife International 2011). A generation time of 25.2 years (BirdLife International 2011) is derived from an average age at first breeding of 7.0 years (Schreiber and Burger 2002) and an annual survival of adults of 94.5%, extrapolated from congeners.

Threats

On Macquarie I. the increase in rabbit numbers, since feral cats *Felis catus* were eradicated in 1999, has led to overgrazing of native vegetation and the extensive destruction of habitat and soil erosion of nesting sites. Burrow entrances are usually concealed by foliage; when this is removed exercising chicks are more vulnerable to predation by Brown Skua *Catharacta lonnbergi*. This threat may continue unless a rabbit eradication program planned for 2011–14 is successful (Agreement on the Conservation of Albatrosses and Petrels 2009). Predation by cats and introduced Wekas *Gallirallus australis* probably caused the temporary extinction of the small breeding colony of Grey Petrels on Macquarie I. (Rounsevell and Brothers 1984); predator eradication then allowed recolonisation by petrels. However, Black Rats *Rattus rattus* remain on the island and continue preying on eggs and chicks (Schulz *et al.* 2005). At sea, the key threat is from commercial longline fisheries, this species was the fourth most common observed killed in New Zealand fisheries between 1998–2004, particularly by bottom long-lining vessels. Most of the killed birds were adult females. Substantial incidental mortality has also been recorded in fisheries off Australia in the 1990s and continues in other legal and illegal fisheries in the south Indian Ocean and off South America (Agreement on the Conservation of Albatrosses and Petrels 2009).

Conservation objectives

1. Nesting sites on Macquarie I. stabilised
2. At-sea threats reduced to acceptable levels
3. Global agreement on conservation measures
4. Increased public awareness of the conservation needs of large petrels

Information required

1. Information on population trends and demographic parameters

2. Trends or changes in adult and juvenile survival
3. Distribution of immature and non-breeding birds
4. Overlap with fishing operations and documented levels of bycatch so that mitigation measures can be targeted

Management actions required

1. Eradicate Black Rats and rabbits from Macquarie I.
2. Continue existing monitoring
3. Assess the impact of fisheries bycatch
4. Maintain mitigation strategies for fisheries, including via intergovernmental mechanisms under the auspices of the Agreement on the Conservation of Albatrosses and Petrels, the Food and Agriculture Organisation and Regional Fisheries Management Organisations

Bibliography

Agreement on the Conservation of Albatrosses and Petrels (2009) 'ACAP species assessment: Grey Petrel *Procellaria cinerea*'. Retrieved 27 August, 2009 from <http://www.acap.aq/>.

BirdLife International (2011) 'Species factsheet: *Procellaria cinerea*'. Retrieved 17 January, 2011 from <http://www.birdlife.org/>.

Brothers NP (1984) Breeding, distribution and status of Burrow-nesting Petrels at Maqaurie Island. *Australian Wildlife Research* **11**, 113–131.

Marchant S, Higgins PJ (Eds) (1990) *Handbook of Australian, New Zealand and Antarctic Birds. Volume 1: Ratites to Ducks.* Oxford University Press, Melbourne.

Rounsevell DE, Brothers NP (1984) The status of seabirds on Macquarie Island. In 'Status and Conservation of the World's Seabirds. ICBP Tech. Publ. 2.' (Eds JP Croxall, PGH Evans and RW Schreiber) pp. 587–592. ICBP, Cambridge, UK.

Schreiber EA, Burger J (2002) *Biology of Marine Birds*. CRC Press, Boca Raton, FL.

Schulz M, Robinson S, Gales R (2005) Breeding of the Grey Petrel (*Procellaria cinerea*) on Macquarie Island: population size and nesting habitat. *Emu* **105**, 323–329.

Comments received from

Barry Baker, Rosemary Gales, Wieslawa Misiak, Tim Reid, Ben Sullivan, Eric Woehler

---

Species Conservation Summary

Buller's Shearwater

Ardenna bulleri (Salvin, 1888)

Procellariidae

Conservation status

Population visiting Australian territory: Vulnerable D2

Reasons for listing

Present at 2 locations (outside Australia) where introduction of exotic predators represents a plausible future threat

Status 2000

Vulnerable D2

Status 1990

Vulnerable D2

Taxonomy

No infraspecific taxa described

Taxonomic uniqueness: medium (14 genera/family, 7 species/genus, 1 subspecies/species)

Range

Breeds only on the 2 main islands, Aorangi and Tawhiti Rahi, and 5 other islets and stacks in the Poor Knights group as well as 1 pair on Simmonds Is. off New Zealand. When not breeding migrates to the north Pacific Ocean. Regularly seen off the east coast of Australia during the breeding season (Marchant and Higgins 1990).

Abundance

The total population is estimated at 2 500 000 individuals, which is still increasing following removal of feral pigs *Sus scrofa* from Aorangi I. (BirdLife International 2011).

Ecology

It nests in burrows or on rock ledges, often under dense vegetation, and feeds on krill, small fish, salps and jellyfish (Marchant and Higgins 1990). A generation time of 18.3 years (BirdLife International 2011) is derived from an average age at first breeding of 5.8 years and an annual survival of adults of 92.0%, both extrapolated from congeners.

Current eligibility against IUCN Red List Criteria

IUCN category	Criteria eligibility
Population visiting Australian territory	
A	Not applicable: past, current or future population declines are thought unlikely to exceed 20% in any 3-generation period
B	Not applicable: EOO, AOO, habitat, locations or population not declining, population fluctuations not extreme
C	Not applicable: population >10 000 mature individuals
D	Vulnerable: ≤5 locations with plausible future threat
E	Not applicable: no population viability analysis undertaken

IUCN Red List assessment data

	Estimate	Reliability
Population visiting Australian territory		
Extent of occurrence trend	1 500 000 km^2 stable	high high
Area of occupancy trend	1 500 000 km^2 stable	high high
No. of mature individuals trend	2 500 000 increasing	medium medium
No. subpopulations	1	high
No. locations	3	high
Generation time	18.3 years	medium
Global population share	<1%	high

Threats

Population on Aorangi increased after removal of feral pigs *Sus scrofa* 1936. Buller's Shearwater were previously caught in drift-nets in the north Pacific, and are still potentially at risk from fishing (BirdLife International 2011).

Conservation objectives

1. At-sea threats maintained at acceptable levels
2. Global agreement on conservation measures

Information required

1. Population trends or changes in adult and juvenile survival
2. Overlap with fishing operations and documented levels of bycatch so that mitigation measures can be targeted

Management actions required

1. Assess the impact of trawl fisheries bycatch
2. Continue to maintain mitigation strategies for trawl fisheries, including via intergovernmental mechanisms under the auspices of the Agreement on the Conservation of Albatrosses and Petrels the Food and Agriculture Organisation and Regional Fisheries Management Organisations

Bibliography

BirdLife International (2011) 'Species factsheet: *Puffinus bulleri*'. Retrieved 17 January, 2011 from <http://www.birdlife.org/>.

Marchant S, Higgins PJ (Eds) (1990) *Handbook of Australian, New Zealand and Antarctic Birds. Volume 1: Ratites to Ducks.* Oxford University Press, Melbourne.

Species Conservation Summary

Flesh-footed Shearwater

Ardenna carneipes (Gould, 1844)

Procellariidae

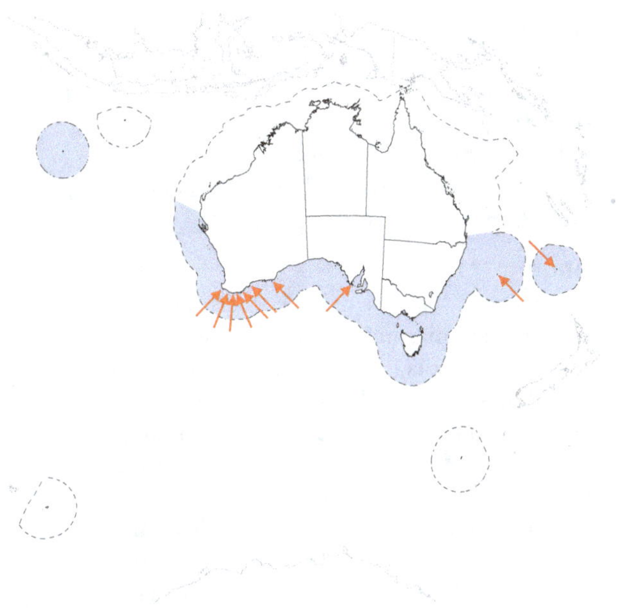

Conservation status

Australian breeding population: Near Threatened A2bc

Population visiting Australian territory: Near Threatened A2bc

Reasons for listing

Australian breeding population: Population decline may be approaching 30% in 3 generations (55 years) from loss of breeding habitat, because often caught as bycatch by longline fishing and may be affected by pollution

Population visiting Australian territory: Population decline may be approaching 30% in 3 generations (55 years) because often caught as bycatch by longline fishing

Current eligibility against IUCN Red List Criteria

IUCN category	Criteria eligibility
Australian breeding population	
A	Near Threatened: population decline of 20–29% over the last 3 generations (55 years) inferred from monitoring and deteriorating habitat quality
B	Not applicable: >10 locations, not severely fragmented, population fluctuations not extreme
C	Not applicable: population >10 000 mature individuals
D	Not applicable: population >1000 mature individuals, >5 locations
E	Not applicable: no population viability analysis undertaken
Population visiting Australian territory	
A	Near Threatened: population decline of 20–29% over the last 3 generations (55 years) inferred from monitoring and deteriorating habitat quality
B	Not applicable: >10 locations, not severely fragmented, population fluctuations not extreme
C	Not applicable: population >10 000 mature individuals
D	Not applicable: population >1000 mature individuals, >5 locations
E	Not applicable: no population viability analysis undertaken

IUCN Red List assessment data

	Estimate	Reliability
Australian breeding population		
Extent of occurrence trend	4 600 000 km² stable	low high
Area of occupancy trend	50 km² decreasing	high high
No. of mature individuals trend	620 000 decreasing	low low
No. subpopulations	>10	low
No. locations	>10	high
Largest subpopulation	400 000	low
Generation time	18.3 years	low
Global population share	90%	low
Level of genetic exchange	medium	low
Population visiting Australian territory		
Extent of occurrence trend	4 600 000 km² stable	medium medium
Area of occupancy trend	4 600 000 km² stable	high medium
No. of mature individuals trend	80 000 decreasing	low low
No. subpopulations	>10	low
No. locations	>10	high
Largest subpopulation	20 000	low
Global population share	10%	low

Status 2000

Australian breeding population: Least Concern
Reason for change in 2010: evidence for decline in population
Population visiting Australian territory: Least Concern
Reason for change in 2010: evidence for decline in population

Status 1990

Australian breeding population: Least Concern
Population visiting Australian territory: Least Concern

Taxonomy

No infraspecific taxa described
Taxonomic uniqueness: medium (14 genera/family, 7 species/genus, 1 subspecies/species)

Range

Breeds on 41 islands off the coast of south-western Western Australia, on Smith I. off the south-eastern coast of Eyre Peninsula, South Australia, on Phillip I. off Norfolk I. and on Lord Howe I. Extralimitally breeds St Paul I. in the Indian Ocean and on islands off New Zealand of which 10 are important (Marchant and Higgins 1990; Priddel *et al.* 2010). On Lord Howe I. at least 5 ha of breeding habitat has been lost through clearing (Fullagar and Disney 1981) and the remaining breeding grounds have contracted from 37.8 ha in 1978 to 24.3 ha in 2002 (Priddel *et al.* 2006). In WA the species no longer breeds in the Torbay region on the mainland WA (last recorded in 1937), Mistake I. (1906) and Green I. (Department of Sustainability, Environment, Water, Population and Communities 2011). At sea western birds are thought to migrate through the Indian Ocean and Indonesia to join birds from the east in the north-west Pacific Ocean (BirdLife International 2011).

Abundance

Population estimated at about 350 000 pairs, mostly in WA where population was estimated at 100 000–310 000 pairs in 1996 (Burbidge and Fuller 1996). Population on Lord Howe I. 17 500 pairs (Priddel *et al.* 2006), in SA about 150 pairs (Copley 1996), on St Paul I. and around New Zealand 10 000 pairs (Baker *et al.* 2011). The population on Lord Howe I. declined by around 19% from 1978 to 2002–2003 (Priddel *et al.* 2006) with further declines subsequently (Reid 2010). The fishing effort has been sufficient to cause a decline of 50% within 55 years (Baker and Wise 2005).

Ecology

Nests in colonies in burrows under trees or shrubs. On Lord Howe I. it favours the flatter areas in the central lowlands (Priddel *et al.* 2006). Most feeding is undertaken offshore over continental shelves where it feeds on fish and squid, mostly caught by pursuit-plunging (Marchant and Higgins 1990). Readily takes baits from longlines (Baker and Wise 2005). A generation time of 18.3 years (BirdLife International 2011) is derived from an average age at first breeding of 5.8 years and an annual survival of adults of 92.0%, both extrapolated from congeners.

Threats

Nesting habitat has been lost on Lord Howe Island partly from clearing of forest 50 years ago and currently as a result of Kikuyu Grass *Pennisetum clandestinum* smothering nests (Priddel *et al.* 2006) and many birds are killed by cars. At sea driftnets and on longline fishing operations are the major threat (Thalman *et al.* 2009) with 5000–6000 individuals being killed annually off eastern Australia (Baker and Wise 2005) and an unknown number off WA where fishing effort has increased in recent years, with unknown effects on this species (Priddel *et al.* 2006). It may be substantial given that the shearwaters were among the 2 most commonly caught species in the Indian Ocean and the Coral Sea, and may also be under-estimated by at least half using current monitoring methods (Brothers *et al.* 2010). Mortality is also likely in the northern Pacific Ocean (McAllan *et al.* 2004). They are also known to ingest considerable quantities of plastics (Hutton *et al.* 2008) which may lead to exceptional levels of heavy metal contamination of feathers (Bond and Lavers 2010).

Conservation objectives

1. Population on Lord Howe I. stabilised
2. Population in WA maintained
3. At-sea threats reduced to acceptable levels

Information required

1. Trends in numbers or changes in adult and juvenile survival on Lord Howe I. and in selected WA colonies
2. Effect of plastic ingestion on survivorship and population stability
3. Distribution of immature and non-breeding birds
4. Overlap with fishing operations, particularly for WA populations, and documented levels of bycatch so that mitigation measures can be targeted
5. Extent of incidental mortality, especially in unregulated longline fisheries operating in the central and south east Pacific and the central Tasman Sea

Management actions required

1. Control Kikuyu Grass in Lord Howe I. colonies
2. Reduce roadkill on Lord Howe I.
3. Undertake a baseline census at selected WA colonies
4. Investigate bycatch in WA fisheries and apply appropriate mitigation techniques
5. Maintain mitigation strategies for fisheries, including via intergovernmental mechanisms under the auspices of the Agreement on the Conservation of Albatrosses and Petrels, the Food and Agriculture Organisation and Regional Fisheries Management Organisations

Bibliography

Baker GB, Hedley G, Cunningham RB (2011) 'Data collection of demographic, distributional, and trophic information on the Flesh-footed Shearwater to allow estimation of effects of fishing on population viability. 2010–2011 field season'. Latitude 42 Environmental Consulting Pty. Ltd., Hobart.

Baker GB, Wise BS (2005) The impact of pelagic longline fishing on the flesh-footed shearwater *Puffinus carneipes* in Eastern Australia. *Biological Conservation* **126**, 306–316.

BirdLife International (2011) 'Species factsheet: *Puffinus carneipes*'. Retrieved 17 January, 2011 from <http://www.birdlife.org/>.

Bond AL, Lavers JL (2010) Trace element concentrations in feathers of Flesh-footed Shearwaters (*Puffinus carneipes*) from across their breeding range. *Archives of Environmental Contamination and Toxicology* 09/2010.

Brothers N, Duckworth AR, Safina C, Gilman EL (2010) Seabird bycatch in pelagic longline fisheries is grossly underestimated when using only haul data. *PLoS ONE* **5**, e12491.

Burbidge AA, Fuller PJ (1996) The Western Australian Department of Conservation and Land Management seabird breeding islands database. In *Proceedings of the Status of Australia's Seabirds. Proceedings of the National Seabird Workshop*, Canberra. (Eds GJB Ross, K Weaver and JC Greig) pp. 73–137. Biodiversity Group, Environment Australia.

Copley PB (1996) The status of seabirds in South Australia. In *Proceedings of the Status of Australia's Seabirds. Proceedings of the National Seabird Workshop*, Canberra. (Eds GJB Ross, K Weaver and JC Grieg) pp. 139–180. Biodiversity Group, Environment Australia.

Department of Sustainability, Environment, Water, Population and Communities (2011) *Ardenna carneipes*. In 'Species Profile and Threats Database'. Department of Sustainability, Environment, Water, Population and Communities, Canberra. Retrieved 24 February, 2011 from <http://www.environment.gov.au/sprat>.

Fullagar PJ, Disney HJ (1981) Studies on the Flesh-footed Shearwaters, *Puffinus carneipes*. *Occasional Reports of the Australian Museum* **1**, 31–32.

Hutton I, Carlile N, Priddel D (2008) Plastic ingestion by Flesh-footed (*Puffinus carneipes*) and Wedge-tailed (*P. pacificus*) shearwaters. *Papers and Proceedings of the Royal Society of Tasmania* **142**, 1–6.

Marchant S, Higgins PJ (Eds) (1990) *Handbook of Australian, New Zealand and Antarctic Birds. Volume 1: Ratites to Ducks.* Oxford University Press, Melbourne.

McAllan IAW, Curtis BR, Hutton I, Cooper RM (2004) The birds of the Lord Howe Island Group: a review of records. *Australian Field Ornithology* **21**, 1–82.

Priddel D, Carlile N, Evans O, Evans B, McCoy H (2010) A review of the seabirds of Phillip Island in the Norfolk Island Group. *Notornis* **57**, 113–127.

Priddel D, Carlile N, Fullagar P, Hutton I, O'Neill L (2006) Decline in the distribution and abundance of flesh-footed shearwaters (*Puffinus carneipes*) on Lord Howe Island, Australia. *Biological Conservation* **128**, 412–424.

Reid T (2010) Modelling the foraging ecology of the Flesh-footed Shearwater *Puffinus carneipes* in relation to fisheries and oceanography. PhD thesis, University of Tasmania, Hobart.

Thalman SJ, Baker GB, Hindell M, Tuck GN (2009) Longline fisheries and foraging distribution of Flesh-Footed Shearwaters in eastern Australia. *Journal of Wildlife Management* **73**, 399–406.

Comments received from

Barry Baker, Nicholas Carlile, Jennifer Lavers, David Priddel

Species Conservation Summary

Sooty Shearwater

Ardenna grisea (Gmelin, 1789)
Procellariidae

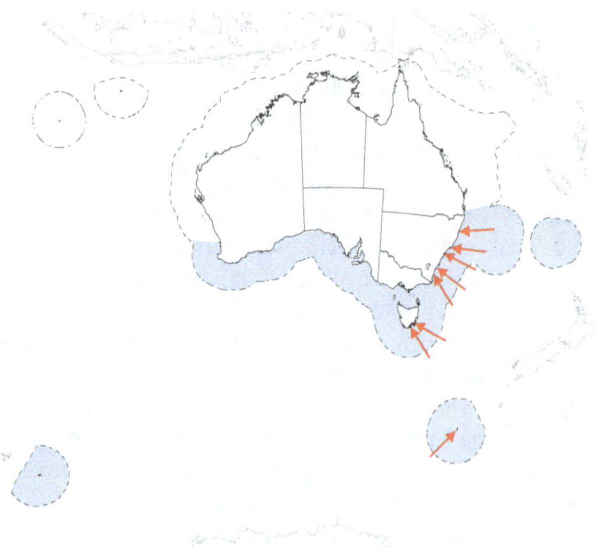

Conservation status
Australian breeding population: Near Threatened A2de+3de+4de
Population visiting Australian territory: Near Threatened A2de+3de+4de

Reasons for listing
Australian breeding population: Population decline may be following global trend of close to 30% in 3 generations (61 years)
Population visiting Australian territory: Population decline has been approaching 30% in 3 generations (61 years) and is thought likely to continue

Status 2000
Australian breeding population: Least Concern
Reason for change in 2010: Evidence for decline in population
Population visiting Australian territory: Least Concern
Reason for change in 2010: Evidence for decline in population

Status 1990
Australian breeding population: Least Concern
Population visiting Australian territory: Least Concern

Taxonomy
No infraspecific taxa described
Taxonomic uniqueness: medium (14 genera/family, 7 species/genus, 1 subspecies/species)

Range
Breeds on islands off eastern Australia (17 islands), and on Macquarie I., as well as off New Zealand, Chile, and the Falkland Is. While breeding commonly forages along the Antarctic Polar Front far to the south of breeding grounds (Woehler, 2006, 2010; Raymond *et al.*, 2011). Migrates to the northern Pacific Ocean in the non-breeding season (BirdLife International, 2011).

Abundance
In Australia total population is estimated <1000 pairs but over 2 million off Chile, up to 20 000 in the Falklands and about 0.5 million pairs off New Zealand (BirdLife International, 2011). Trends in Australia are unknown and unmonitored but largest New Zealand colony declined by 37% between 1969 and 2000 and numbers seen in the California Current fell by 90% in the last 20 years, though this could be a distributional shift (Veit *et al.*, 1997; BirdLife International, 2011).

Ecology
Nests on islands and headlands in burrows dug beneath tussock grass and low scrub. Feeds on fish, crustacea and cephalopods, caught while diving (Marchant and Higgins, 1990). A generation time of 21.2 years (BirdLife International, 2011) is derived from an average age at first breeding of 6.0 years and an annual survival of adults of 93.4% (Schreiber and Burger, 2002).

Threats
There are some threats on breeding sites including an annual harvest in New Zealand of a quarter of a million birds. Black Rats *Rattus rattus* and Brown Rats *R. norvegicus* kill some eggs and chicks. For both sources of mortality the decline seems larger and more rapid than the estimated rates of mortality while breeding. Pelagic drift-nets formerly drowned up to 350 000 birds annually and longline fisheries are responsible for a high number of deaths. Climate change may affect food availability but corroborative data are lacking (BirdLife International, 2011).

Conservation objectives
1. Australian populations remain secure
2. At-sea threats eliminated

Information required
1. Trends in numbers or changes in adult and juvenile survival
2. Distribution of immature and non-breeding birds
3. Overlap with fishing operations and documented levels of bycatch so that mitigation measures can be targeted
4. Extent of incidental mortality, especially in unregulated longline fisheries

Management actions required
1. Undertake a baseline census at selected colonies
2. Investigate bycatch in fisheries and apply appropriate mitigation techniques
3. Continue to maintain mitigation strategies for trawl fisheries, including via intergovernmental mechanisms under the auspices of the Agreement on the Conservation of Albatrosses and Petrels the Food and Agriculture Organisation and Regional Fisheries Management Organisations

Current eligibility against IUCN Red List Criteria

IUCN category	Criteria eligibility
Australian breeding population	
A	Near Threatened: past, current and anticipated population declines of 20–29% over periods of 3 generations (61 years) inferred from levels of exploitation and effects of introduced taxa
B	Not applicable: >10 locations, not severely fragmented, population fluctuations not extreme
C	Not applicable: population 500–2500 mature individuals, continuing decline not observed or estimated, largest subpopulation >1000 mature individuals but <95% of total population
D	Not applicable: population >1000 mature individuals, >5 locations
E	Not applicable: no population viability analysis undertaken
Population visiting Australian territory	
A	Near Threatened: past, current and anticipated population declines of 20–29% over periods of 3 generations (61 years) inferred from levels of exploitation and effects of introduced taxa
B	Not applicable: >10 locations, not severely fragmented, population fluctuations not extreme
C	Not applicable: population 500–2500 mature individuals, continuing decline not observed or estimated, largest subpopulation >1000 mature individuals but <95% of total population
D	Not applicable: population >1000 mature individuals, >5 locations
E	Not applicable: no population viability analysis undertaken

IUCN Red List assessment data

	Estimate	Reliability
Australian breeding population		
Extent of occurrence trend	4 800 000 km^2 stable	low high
Area of occupancy trend	50 km^2 stable	high medium
No. of mature individuals trend	2000 decreasing	low low
No. subpopulations	>10	low
No. locations	>10	high
Largest subpopulation	1000	low
Generation time	21.2 years	low
Global population share	<1%	low
Level of genetic exchange	medium	low
Population visiting Australian territory		
Extent of occurrence trend	4 800 000 km^2 stable	medium medium
Area of occupancy trend	4 800 000 km^2 stable	high medium
No. of mature individuals trend	20 000 000 decreasing	low high
No. subpopulations	>10	low
No. locations	>10	high
Global population share	1–10%	low

Bibliography

BirdLife International (2011) 'Species factsheet: *Puffinus griseus*'.

Marchant S, Higgins PJ (Eds) (1990) *Handbook of Australian, New Zealand and Antarctic Birds*. Oxford University Press, Melbourne.

Raymond B, McInnes J, Dambacher JM, Way S, Bergstrom DM (2011) Qualitative modelling of invasive species eradication on subantarctic Macquarie Island. *Journal of Applied Ecology* **48**, 181–191.

Schreiber EA, Burger J (2002) *Biology of Marine Birds*. CRC Press, Boca Raton, FL.

Veit RR, McGowan JA, Ainley DG, Wahl TR, Pyle P (1997) Apex marine predator declines 90% in association with changing oceanic climate. *Global Change Biology* **3**, 23–28.

Woehler EJ (2006) Status and conservation of the seabirds of Heard Island and the McDonald Islands. In *Heard Island: Southern Ocean Sentinel*. (Eds K Green, EJ Woehler) pp. 128–165. Surrey Beatty & Sons, Chipping Norton, NSW.

Woehler EJ (2010) Status and conservation of the seabirds of Heard Island. *The State of Australia's Birds 2010: Islands and Birds*. *Wingspan* **Supplement 20**, 8–9.

Comments received from

Eric Woehler

Species Conservation Summary

Hutton's Shearwater

Puffinus huttoni Mathews, 1912
Procellariidae

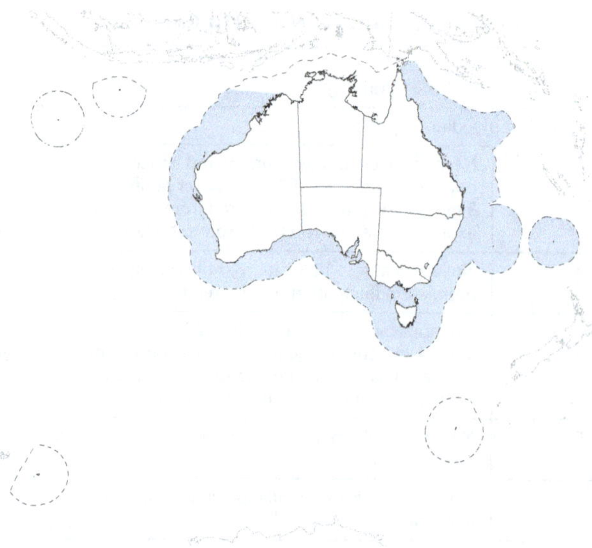

Conservation status

Population visiting Australian territory: Endangered B2ab(ii,iii)

Reasons for listing

Present at 2 locations (outside Australia) where the breeding sites are threatened by trampling by introduced herbivores

Status 2000

Endangered B2ab(ii,iii)

Status 1990

Endangered B2ab(ii,iii)

Taxonomy

No infraspecific taxa described
Taxonomic uniqueness: medium (14 genera/family, 12 species/genus, 1 subspecies/species)

Range

Breeds only at 2 locations on the Seaward Kaikoura Range, north-east South Island, New Zealand, with a third subpopulation on the Kaikoura peninsula. Numbers and distribution within the Kaikoura Ranges have decreased with 8 of 10 known colonies having been extirpated this century. When not breeding, migrates to waters off the south, east and north-western coasts of Australia (Marchant and Higgins 1990).

Abundance

The total population is estimated at 300 000–350 000 individuals, which, though stable from 1990–2000, declined substantially over the previous century, with 8 of 10 known colonies having been extirpated (BirdLife International 2011).

Ecology

Nesting is in burrows on mountain slopes from 1200–1800 m above sea level, usually under grass tussocks or low alpine shrubs. They feed mostly on small fish and krill (Marchant and Higgins 1990). A generation time of 19.5 years (BirdLife International 2011) is derived from an average age at first breeding of 5.0 years and an annual survival of adults of 93.1% (Schreiber and Burger 2002).

Threats

Introduced herbivores have caused heavy erosion at some former colonies and local fishing could compete for food (BirdLife International 2011), but few of the threats seem valid for the population in Australian waters.

Conservation objectives

1. Population maintained

Information required

1. Population trends

Management actions required

1. No actions currently required in Australian territory

Bibliography

BirdLife International (2011) 'Species factsheet: *Puffinus huttoni*'. Retrieved 17 January, 2011 from <http://www.birdlife.org/>.

Marchant S, Higgins PJ (Eds) (1990) *Handbook of Australian, New Zealand and Antarctic Birds. Volume 1: Ratites to Ducks.* Oxford University Press, Melbourne.

Schreiber EA, Burger J (2002) *Biology of Marine Birds.* CRC Press, Boca Raton, FL.

Current eligibility against IUCN Red List Criteria

IUCN category	Criteria eligibility
Population visiting Australian territory	
A	Not applicable: past, current or future population declines are thought unlikely to exceed 20% in any 3-generation period
B	Endangered: AOO (when breeding) <500 km^2, at <5 locations, with plausible future threat, inferred continuing decline in AOO, habitat quality
C	Not applicable: population >10 000 mature individuals
D	Vulnerable: ≤5 locations with plausible future threat
E	Not applicable: no population viability analysis undertaken

IUCN Red List assessment data

	Estimate	Reliability
Population visiting Australian territory		
Extent of occurrence trend	6 500 000 km^2 stable	high high
Area of occupancy trend	6 500 000 km^2 stable	high high
No. of mature individuals trend	300 000 stable	low medium
No. subpopulations	3	high
No. locations	3	high
Largest subpopulation	300 000	low
Generation time	19.5 years	medium
Global population share	>90%	high

Subspecies Conservation Summary

Little Shearwater (Tasman Sea)

Puffinis assimilis assimilis Gould, 1838
Procellariidae

Conservation status

Vulnerable D2

Reasons for listing

Present at 4 locations, introduction of exotic predators represents a plausible future threat

Status 2000

Vulnerable D2

Status 1990

Vulnerable D2

Taxonomy

P. a. tunneyi (south-west Western Australia) is Least Concern. There are 6 extralimital subspecies
Taxonomic uniqueness: medium (14 genera/family, 12 species/genus, 8 subspecies/species)

Range

Endemic to Lord Howe I. and Norfolk I. groups (contra Garnett and Crowley 2000; Department of Sustainability, Environment, Water, Population and Communities 2011). After breeding, apparently disperses over the Tasman Sea and possibly the Coral Sea. Disappeared from Lord Howe I. after 1913 (Hindwood 1940) and was confined to Roach I. until the 1980s. Recolonised the main island following removal of feral and domestic cats *Felis catus* (Priddel *et al.* 2003). Was present at Anson Point on Norfolk I. in the 1970s (Director of National Parks 2010) but now breeding confined to Phillip and Nepean Is.

Abundance

Population on Roach I. estimated at 4000 pairs with a further 3–50 pairs on the main island (McAllan *et al.* 2004). Population in the Norfolk I. group estimated at 100–1000 pairs (Schodde *et al.* 1983; Priddel *et al.* 2010). Total here is double the Roach I. population and the average of the estimate for the Norfolk I. group.

Ecology

Breeds in burrows dug in soft soil among mats of succulents or among loose rocks. For feeding frequents warm tropical and subtropical waters. Feeds on squid, fish and crustaceans relatively close to breeding islands (BirdLife International 2011). Most populations are considered fairly sedentary with some dispersal when immature (BirdLife International 2011). A generation time of 18.0 years (BirdLife International 2011) is derived from an average age at first breeding of 5.5 years, an annual adult survival of 92.0% and a maximum longevity of 30.2 years, all extrapolated from congeners.

Threats

Little Shearwaters appear to have been eliminated from Lord Howe I. by Black Rats *Rattus rattus* (Hindwood 1940), which is also the likely cause of their apparent disappearance from Norfolk I. However, the elimination of feral cats *Felis catus* from Lord Howe I. was followed by recolonisation, and cats may have contributed to loss of the breeding birds from Norfolk I. The accidental introduction of either predator to their refuges could result in their local extinction.

Conservation objectives

1. Maintain current breeding subpopulations
2. Expansion of population to main islands

Current eligibility against IUCN Red List Criteria

IUCN category	Criteria eligibility
A	Not applicable: past, current or future population declines are thought unlikely to exceed 20% in any 3-generation period
B	Not applicable: EOO, AOO, habitat, locations or population not declining, population fluctuations not extreme
C	Not applicable: no continuing population decline or extreme fluctuations
D	Vulnerable: ≤5 locations with plausible future threat
E	Not applicable: no population viability analysis undertaken

IUCN Red List assessment data

	Estimate	Reliability
Extent of occurrence trend	1 800 000 km^2 stable	high high
Area of occupancy trend	8 km^2 stable	medium high
No. of mature individuals trend	9000 stable	low medium
No. subpopulations	2	low
No. locations	4	high
Largest subpopulation	8000	medium
Generation time	18.0 years	low
Global population share	100%	low

Information required

1. None

Management actions required

1. Maintain strict quarantine for Roach, Nepean and Phillip Is.
2. Establish cooperative rodent control programs throughout Norfolk I, with a view to rat eradication
3. Eradicate rats from Lord Howe I.
4. Enhance rat baiting and cat trapping on Norfolk I. and monitor their efficacy

Bibliography

BirdLife International (2011) 'Species factsheet: *Puffinus assimilis*'. Retrieved 17 January, 2011 from <http://www.birdlife.org/>.

Department of Sustainability Environment Water Population and Communities (2011) *Puffinus assimilis assimilis*. In 'Species Profile and Threats Database'. Department of Sustainability, Environment, Water, Population and Communities, Canberra. Retrieved 24 February, 2011 from <http://www.environment.gov.au/sprat>.

Director of National Parks (2010) 'Draft Norfolk Island Region Threatened Species Recovery Plan'. Department of the Environment, Water, Heritage and the Arts, Canberra.

Garnett S, Crowley GM (2000) *The Action Plan for Australian Birds 2000*. Environment Australia, Canberra.

Hindwood KA (1940) The birds of Lord Howe Island. *Emu* **40**, 1–86.

McAllan IAW, Curtis BR, Hutton I, Cooper RM (2004) The birds of the Lord Howe Island Group: a review of records. *Australian Field Ornithology* **21**, 1–82.

Priddel D, Hutton I, Carlile N, Bester A (2003) Little Shearwaters *Puffinus assimilis assimilis* breeding on Lord Howe Island. *Emu* **103**, 67–70.

Priddel D, Carlile N, Evans O, Evans B, McCoy H (2010) A review of the seabirds of Phillip Island in the Norfolk Island Group. *Notornis* **57**, 113–127.

Schodde R, Fullagar P, Hermes N (1983) *A Review of the Status of Norfolk Island Birds: Past and Present*. Australian National Parks and Wildlife Service, Canberra.

Comments received from

Nicholas Carlile, Margaret Christian, David Priddel

Species Conservation Summary

Tahiti Petrel

Pseudobulweria rostrata (Peale, 1848)
Procellariidae

Conservation status

Population visiting Australian territory: Near Threatened A2bde+3bde+4bde; B2ab(i,ii,iii,iv,v)

Reasons for listing

Present at multiple locations in south Pacific Ocean (outside Australia), but totalling a small area, where exotic predators and development probably causing declines approaching 30% in 3 generations (53 years)

Status 2000

Near Threatened A2bde+3bde+4bde; B2ab(i,ii,iii,iv,v)

Status 1990

Least Concern

Reason for change in 2000: declines apparent at sea and on land

Taxonomy

Pseudobulweria rostrata trouessarti (New Caledonia) and *P. r. rostrata* (Pacific) are both Near Threatened

Taxonomic uniqueness: medium (14 genera/family, 4 species/genus, 2 subspecies/species)

Range

Breeds on Marquesas, Society and Gambier Is. (*rostrata*) and New Caledonia (*trouessarti*). Populations on Fiji and American Samoa may be of either subspecies (Bretagnolle *et al.* 1998). When not breeding, it disperses widely (BirdLife International 2011). Regularly seen off the east coast of Australia during the breeding season (Marchant and Higgins 1990).

Abundance

The total population of the species is estimated to be 20 000 mature individuals (BirdLife International 2011), but there were estimated to be on average 437 000 in the eastern Pacific during surveys 1988–1990 and 282 000 in 1998–2000, a significant decline (Balance *et al.* 2002). These figures also suggest that many sites remain undiscovered. There are indications of decline in known colonies on Tahiti, Moorea and New Caledonia (BirdLife International 2011).

Ecology

Breeds on steep hills or almost inaccessible cliffs (Villard *et al.* 2006). Feeds mostly alone and is tuna-dependent, in that it feeds exclusively by scavenging on squid left floating on the surface by feeding tuna (Balance *et al.* 2002). A generation time of 17.7 years (BirdLife International 2011) is derived from an average age at first breeding is 6.4 years and an annual survival of adults of 91.4%, both extrapolated from congeners.

Threats

Although they have coexisted with introduced Black Rats *Rattus rattus* on nesting islands for decades, the extent to which rats are a threat is unknown. Cats *Felis catus*, dogs *Canis familiaris* and pigs *Sus scrofa* are also probable predators (Villard *et al.* 2006). Young birds are attracted by lights at night at mines and in the cities. Some nest

Current eligibility against IUCN Red List Criteria

IUCN category	Criteria eligibility
Population visiting Australian territory	
A	Near Threatened: past, current and anticipated population declines of 20–29% over periods of 3 generations (53 years) suspected from monitoring, deteriorating habitat quality, levels exploitation and effects of introduced taxa
B	Near Threatened: AOO near 2000 km^2: near 10 locations with plausible future threat, suspected continuing decline in EOO, AOO, habitat quality, no. locations, no. mature individuals
C	Not applicable: population >10 000 mature individuals
D	Not applicable: population >1000 mature individuals, >5 locations
E	Not applicable: no population viability analysis undertaken

IUCN Red List assessment data

	Estimate	Reliability
Population visiting Australian territory		
Extent of occurrence trend	2 400 000 km^2 stable	high high
Area of occupancy trend	2 400 000 km^2 stable	high high
No. of mature individuals trend	20 000 decreasing	low medium
No. subpopulations	>10	high
No. locations	>10	high
Largest subpopulation	10 000	low
Generation time	17.7 years	medium
Global population share	<1%	high

sites on New Caledonia are also threatened by nickel mining (BirdLife International 2011). Given the declines recorded in the marine surveys and their dependence on dead squid left over by tuna (Balance *et al.* 2002), there must also be concern for threats at sea from over-fishing of tuna.

Conservation objectives

1. At-sea threats maintained at acceptable levels
2. Global agreement on conservation measures

Information required

1. Population trends
2. Effects of fishing on food availability

Management actions required

1. Work in international fora to ensure tuna fishing is sustainable

Bibliography

Balance LT, Pitman RL, Spear LB, Fiedler PC (2002) 'Investigations into temporal patterns in distribution, abundance and habitat relationships within seabird communities of the eastern tropical Pacific'. Southwest Fisheries Science Center, National Marine Fisheries Service Administrative Report LJ-02-17.

BirdLife International (2011) 'Species factsheet: *Pseudobulweria rostrata*'. Retrieved 17 January, 2011 from <http://www.birdlife.org/>.

Bretagnolle V, Attie C, Pasquet E (1998) Cytochrome-B evidence for validity and phylogenetic relationships of *Pseudobulweria* and *Bulweria* (Procellariidae). *Auk* **115**, 188–195.

Marchant S, Higgins PJ (Eds) (1990) *Handbook of Australian, New Zealand and Antarctic Birds. Volume 1: Ratites to Ducks.* Oxford University Press, Melbourne.

Villard P, Dano S, Bretagnolle V (2006) Morphometrics and the breeding biology of the Tahiti Petrel *Pseudobulweria rostrata*. *Ibis* **148**, 285–291.

Comments received from

David Priddel

Subspecies Conservation Summary

Tahiti Petrel (Pacific)

Pseudobulweria rostrata rostrata (Peale, 1848)

Procellariidae

Conservation status

Population visiting Australian territory: Near Threatened A2bde+3bde+4bde; B2ab(i,ii,iii,iv,v)

Reasons for listing

Present at multiple locations near New Caledonia (outside Australia), but totalling a small area, where exotic predators and development probably causing declines approaching 30% in 3 generations (53 years)

Current eligibility against IUCN Red List Criteria

IUCN category	Criteria eligibility
Population visiting Australian territory	
A	Near Threatened: past, current and anticipated population declines of 20–29% over periods of 3 generations (53 years) suspected from monitoring, deteriorating habitat quality, levels of exploitation and effects of introduced taxa
B	Near Threatened: AOO near 2000 km^2: near 10 locations with plausible future threat, suspected continuing decline in EOO, AOO, habitat quality, no. locations, no. mature individuals
C	Not applicable: population >10 000 mature individuals
D	Not applicable: population >1000 mature individuals, >5 locations
E	Not applicable: no population viability analysis undertaken

IUCN Red List assessment data

	Estimate	Reliability
Population visiting Australian territory		
Extent of occurrence trend	2 400 000 km^2 stable	high high
Area of occupancy trend	2 400 000 km^2 stable	high high
No. of mature individuals trend	10 000 decreasing	low medium
No. subpopulations	>10	medium
No. locations	>10	high
Generation time	17.7 years	medium
Global population share	<1%	high

Status 2000
Near Threatened A2bde+3bde+4bde; B2ab(i,ii,iii,iv,v)

Status 1990
Least Concern

Reasons for change in 2000: declines and threats becoming apparent

Taxonomy
Pseudobulweria rostrata trovessanti (New Caledonia) is Near Threatened, as is the species

Taxonomic uniqueness: medium (14 genera/family, 4 species/genus, 2 subspecies/species)

Range
Breeds on Marquesas, Society and Gambier Is. Populations on Fiji and American Samoa may be of other subspecies (Bretagnolle *et al.* 1998). When not breeding disperses widely (BirdLife International 2011) and is regularly seen off the east coast of Australia during the breeding season (Marchant and Higgins 1990).

Abundance
Population data is sparse but 1000–5000 pairs are estimated but there are indicators of decline in known colonies on Tahiti and Moorea (BirdLife International 2011).

Ecology
Breeds on steep hills or almost inaccessible cliffs (Villard *et al.* 2006). Feeds mostly alone and feeds almost exclusively by scavenging on squid killed by tuna and left floating on the surface (Balance *et al.* 2002). A generation time of 17.7 years (BirdLife International 2011) is derived from an average age at first breeding is 6.4 years and an annual survival of adults of 91.4%, both extrapolated from congeners.

Threats
Although they have coexisted with introduced Black Rats *Rattus rattus* on nesting islands for decades, the extent to which they are a threat is unknown. Young birds are attracted by lights at night in the cities, particularly Papeete (BirdLife International 2011). Given the declines recorded in the marine surveys and their dependence on dead squid left over by tuna (Balance *et al.* 2002), there must also be concern for threats at sea from over-fishing of tuna.

Conservation objectives
1. At-sea threats maintained at acceptable levels
2. Global agreement on conservation measures

Information required
1. Population trends

Management actions required
1. Work in international fora to ensure tuna fishing is sustainable

Bibliography

Balance LT, Pitman RL, Spear LB, Fiedler PC (2002) 'Investigations into temporal patterns in distribution, abundance and habitat relationships within seabird communities of the eastern tropical Pacific'. Southwest Fisheries Science Center, National Marine Fisheries Service Administrative Report LJ-02-17.

BirdLife International (2011) 'Species factsheet: *Pseudobulweria rostrata*'. Retrieved 17 January, 2011 from <http://www.birdlife.org/>.

Bretagnolle V, Attie C, Pasquet E (1998) Cytochrome-B evidence for validity and phylogenetic relationships of *Pseudobulweria* and *Bulweria* (Procellariidae). *Auk* **115**, 188–195.

Marchant S, Higgins PJ (Eds) (1990) *Handbook of Australian, New Zealand and Antarctic Birds. Volume 1: Ratites to Ducks.* Oxford University Press, Melbourne.

Villard P, Dano S, Bretagnolle V (2006) Morphometrics and the breeding biology of the Tahiti Petrel *Pseudobulweria rostrata*. *Ibis* **148**, 285–291.

Comments received from
David Priddel

Subspecies Conservation Summary

Tahiti Petrel (New Caledonia)

Pseudobulweria rostrata trouessarti (Brazil, 1917)
Procellariidae

Conservation status

Population visiting Australian territory: Near Threatened A2bde+3bde+4bde; B2ab(i,ii,iii,iv,v)

Reasons for listing

Present at multiple locations near New Caledonia (outside Australia), but totalling a small area, where exotic predators and development probably causing declines approaching 30% in 3 generations (53 years)

Status 2000

Near Threatened A2bde+3bde+4bde; B2ab(i,ii,iii,iv,v)

Status 1990

Least Concern

Reasons for change in 2000: declines and threats becoming apparent

Taxonomy

Pseudobulweria rostrata rostrata (Pacific) is Near Threatened, as is the species

Taxonomic uniqueness: medium (14 genera/family, 4 species/genus, 2 subspecies/species)

Range

Breeds on New Caledonia on small islets in the southern lagoon and in the Grant Terre mountains such as Massif du Koniambo (Villard *et al.* 2006). When not breeding disperses widely (BirdLife International 2011) and regularly seen off the east coast of Australia during the breeding season (Marchant and Higgins 1990).

Abundance

Population data are sparse but 1000–5000 pairs are estimated to breed on Grand Terre (Villard *et al.* 2006) while there are probably fewer than 100 pairs on the islets in the southern lagoon, with a severe reduction known to have occurred on 1 of them (BirdLife International 2011).

Ecology

Nests in scattered colonies in burrows on rocky slopes, in open upland forest or sandy islands. Feeds mostly alone, almost exclusively on squid left floating on the surface by feeding tuna (Balance *et al.* 2002). A generation time of 17.7 years (BirdLife International 2011) is derived from an average age at first breeding is 6.4 years and an annual survival of adults of 91.4%, both extrapolated from congeners.

Threats

Although they have coexisted with introduced Black Rats *Rattus rattus* on nesting islands for decades, the extent to which rats are a threat is unknown. Cats *Felis catus*, dogs *Canis familiaris* and pigs *Sus scrofa* are also probable predators (Villard *et al.* 2006). Young birds are attracted by lights at night at mines and in the cities. Some nest sites are also threatened by nickel mining (BirdLife International 2011). Given the declines recorded in the marine surveys and their dependence on dead squid left over by tuna (Balance *et al.* 2002), there must also be concern for threats at sea from over-fishing of tuna.

Current eligibility against IUCN Red List Criteria

IUCN category	Criteria eligibility
Population visiting Australian territory	
A	Near Threatened: past, current and anticipated population declines of 20–29% over periods of 3 generations (53 years) suspected from monitoring, deteriorating habitat quality, levels of exploitation and effects of introduced taxa
B	Near Threatened: AOO near 2000 km^2: near 10 locations with plausible future threat, suspected continuing decline in EOO, AOO, habitat quality, no. locations, no. mature individuals
C	Not applicable: population >10 000 mature individuals
D	Not applicable: population >1000 mature individuals, >5 locations
E	Not applicable: no population viability analysis undertaken

IUCN Red List assessment data

	Estimate	Reliability
Population visiting Australian territory		
Extent of occurrence trend	2 400 000 km^2 stable	high high
Area of occupancy trend	2 400 000 km^2 stable	high high
No. of mature individuals trend	10 000 decreasing	low medium
No. subpopulations	1	medium
No. locations	>10	high
Generation time	17.7 years	medium
Global population share	<1%	high

Conservation objectives

1. At-sea threats maintained at acceptable levels
2. Global agreement on conservation measures

Information required

1. Population trends

Management actions required

1. Work in international fora to ensure tuna fishing is sustainable

Bibliography

Balance LT, Pitman RL, Spear LB, Fiedler PC (2002) 'Investigations into temporal patterns in distribution, abundance and habitat relationships within seabird communities of the eastern tropical Pacific'. Southwest Fisheries Science Center, National Marine Fisheries Service Administrative Report LJ-02-17.

BirdLife International (2011) 'Species factsheet: *Pseudobulweria rostrata*'. Retrieved 17 January, 2011 from <http://www.birdlife.org/>.

Marchant S, Higgins PJ (Eds) (1990) *Handbook of Australian, New Zealand and Antarctic Birds. Volume 1: Ratites to Ducks.* Oxford University Press, Melbourne.

Villard P, Dano S, Bretagnolle V (2006) Morphometrics and the breeding biology of the Tahiti Petrel *Pseudobulweria rostrata. Ibis* **148**, 285–291.

Comments received from

David Priddel

Subspecies Conservation Summary

Kermadec Petrel (western)

Pterodroma neglecta neglecta (Schlegel, 1863)
Procellariidae

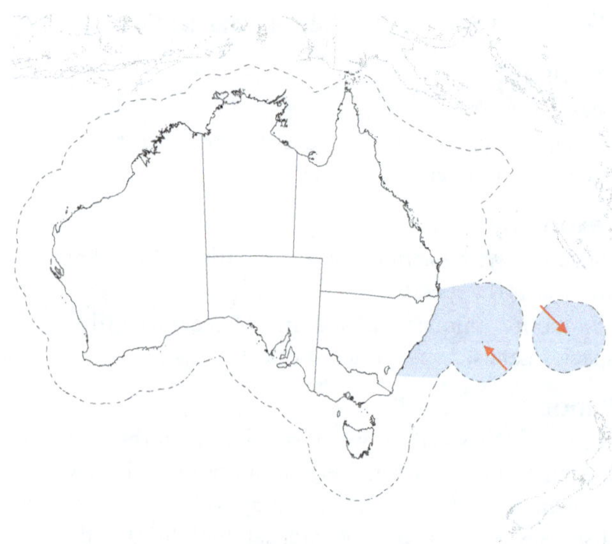

Conservation status

Australian breeding population: Endangered D
Population visiting Australian territory: Least Concern

Reasons for listing

Australian breeding population: Population 50–250 mature individuals

Status 2000

Australian breeding population: Endangered D
Population visiting Australian territory: Least Concern

Status 1990

Australian breeding population: Endangered D
Population visiting Australian territory: Least Concern

Taxonomy

P. n. juana (eastern Pacific) is extralimital. The species is Least Concern

Taxonomic uniqueness: medium (14 genera/family, 30 species/genus, 2 subspecies/species)

Range

Breeds on islands across southern Pacific Ocean from Easter Island west to the Norfolk and Lord Howe I. groups.

Abundance

Small numbers (10–100 pairs) nest on Ball's Pyramid, south of Lord Howe I. and on Phillip I. near Norfolk I. (10–100 pairs; Priddel *et al.* 2010). Last recorded nesting on Lord Howe I. in 1913 (McAllan *et al.* 2004). Foraging distribution covers the tropical and subtropical Pacific Ocean, both north and south of equator (BirdLife International 2011).

Ecology

Typically nests on the surface in loose colonies among rocks and vegetation. On Ball's Pyramid it nests only on steep cliffs above 400 m (Brown 1979). On Phillip I. it nests under stands of African Olive *Olea europea cuspidata* (Priddel *et al.* 2010). This species is marine and highly pelagic, rarely approaching land except at colonies. Little is known about its diet, though squid and crustaceans have been recorded as prey (Marchant and Higgins 1990). A generation time of 15.9 years (BirdLife International 2011) is derived from an average age at first breeding of 6.1 years and an annual survival of adults of 89.5%, both extrapolated from congeners.

Threats

In Australian territory, breeding colonies are confined to 2 islets, and their small numbers are vulnerable to catastrophes. Accidental introduction of the Black Rat *Rattus*

Current eligibility against IUCN Red List Criteria

IUCN category	Criteria eligibility
Australian breeding population	
A	Not applicable: past, current or future population declines are thought unlikely to exceed 20% in any 3-generation period
B	Not applicable: EOO, AOO, habitat, locations or population not declining, population fluctuations not extreme
C	Not applicable: no continuing population decline or extreme fluctuations
D	Endangered: population 50–250 mature individuals
E	Not applicable: no population viability analysis undertaken

IUCN Red List assessment data

	Estimate	Reliability
Australian breeding population		
Extent of occurrence trend	1 300 000 km^2 stable	low high
Area of occupancy trend	2 km^2 stable	high high
No. of mature individuals trend	200 stable	low low
No. subpopulations	2	low
No. locations	2	high
Largest subpopulation	100	medium
Generation time	15.9 years	low
Global population share	<1%	low
Level of genetic exchange	low	low

rattus caused the petrel's extinction on Lord Howe I. (McAllan *et al.* 2004) and probably prevents colonisation of Norfolk I. Brown Rats *Rattus norvegicus* and feral cats *Felis catus* reduced the population on Raoul I., in the Kermadec Is., from half a million to 2 pairs between 1908 and 1966 (Veitch *et al.* 2004). On Phillip I., Nankeen Kestrels *Falco cenchroides* have been reported preying on adult petrels and Purple Swamphens *Porphyio porphyrio* may prey on eggs and chicks (Director of National Parks 2010; Priddel *et al.* 2010), but the extent to which this threatens the population is unknown given the prevalence of other potential prey. Further information is required to determine whether the species is affected by longline fishing.

Conservation objectives

1. Maintain current breeding population
2. Enable expansion of population to main islands

Information required

1. Factors affecting breeding success
2. Impact on breeding of Purple Swamphens

Management actions required

1. Maintain strict quarantine for Ball's Pyramid and Phillip I.
2. Determine threats on Phillip I. and manage as required
3. Eradicate rats from Lord Howe I.

Bibliography

BirdLife International (2011) 'Species factsheet: *Pterodroma neglecta*'. Retrieved 17 January, 2011 from <http://www.birdlife.org/>.

Brown I (1979) The birds of Ball's Pyramid, Lord Howe Island. *Australian Birds* **13**, 41–42.

Director of National Parks (2010) 'Draft Norfolk Island Region Threatened Species Recovery Plan'. Department of the Environment, Water, Heritage and the Arts, Canberra.

Marchant S, Higgins PJ (Eds) (1990) *Handbook of Australian, New Zealand and Antarctic Birds. Volume 1: Ratites to Ducks.* Oxford University Press, Melbourne.

McAllan IAW, Curtis BR, Hutton I, Cooper RM (2004) The birds of the Lord Howe Island Group: a review of records. *Australian Field Ornithology* **21**, 1–82.

Priddel D, Carlile N, Evans O, Evans B, McCoy H (2010) A review of the seabirds of Phillip Island in the Norfolk Island Group. *Notornis* **57**, 113–127.

Veitch CR, Miskelly CM, Harper GA, Taylor G, Tennyson AJD (2004) Birds of the Kermadec Islands, south-west Pacific. *Notornis* **51**, 61–90.

Comments received from

Margaret Christian, Coral Rowston, Ron Ward

Species Conservation Summary

Herald Petrel

Pterodroma heraldica (Salvin, 1888)
Procellariidae

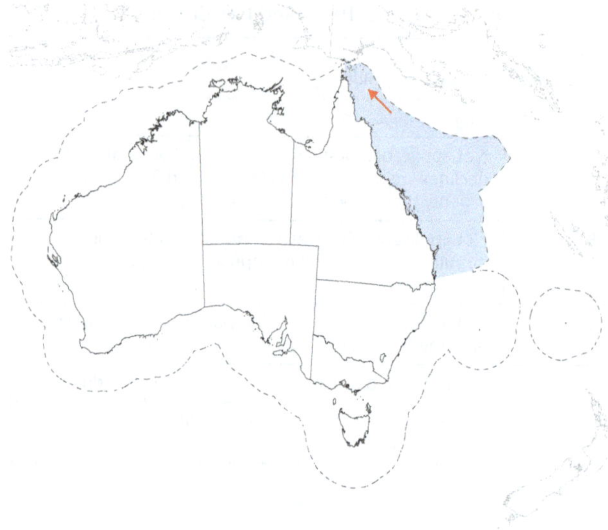

Conservation status

Australian breeding population: Vulnerable D1+2
Population visiting Australian territory: Least Concern

Reasons for listing

Australian breeding population: Almost certainly contains fewer than 50 mature individuals but, because not genetically isolated from a much larger Least Concern global population, so downgraded 2 categories to Vulnerable (following IUCN 2003)

Status 2000

Australian breeding population: Vulnerable D1+2
Population visiting Australian territory: Least Concern

Status 1990

Australian breeding population: Vulnerable D1+2
Population visiting Australian territory: Least Concern

Taxonomy

No infraspecific taxa described. A bird found on Cocos (Keeling) I., dubbed the 'Round Island' Petrel and identified to *P. arminjoniana* by Garnett and Crowley (2000), is likely to have been *P. heraldica*
Taxonomic uniqueness: low (14 genera/family, 30 species/genus, 1 subspecies/species)

Range

In Australian territory breeds only on Raine I., where it was last recorded in 2002 (Batianoff and Cornelius 2005), and possibly on other small cays in the Coral Sea. Globally it occurs primarily in the central and western Pacific, with colonies in the Indian (Round I.) and the Atlantic (Trindade and Martim Vaz Is., Brazil) Oceans (BirdLife International 2011). Despite their separation, the Round and Raine Is. birds are in the same subpopulation: a Herald Petrel banded on Raine I. as a breeding female in 1984 was recaptured nesting successfully on Round I., Mauritius in 2006 (M.Brooke *in litt.*). Breeding of petrels on Cocos (Keeling) has never been confirmed. Feeds pelagically, mostly south of the equator (BirdLife International 2011) across the Pacific and, though seen rarely, the Indian Oceans.

Abundance

Very small numbers breed on Raine I. Since its discovery in 1959, was recorded breeding irregularly in small numbers until 1987, once more in June 2002 when only 1 breeding pair was observed, and not again, though winter visits to the island have been few (Batianoff and Cornelius 2005). A total of 22 birds have been banded, but rarely are more than a few pairs found on any 1 visit. The global populations is estimated at 150 000 mature individuals (BirdLife International 2011).

Ecology

In Australia, they breed in winter in burrows on a vegetated coral cay. Elsewhere they mostly breed on high islands in loose colonies among rocks and on the ground

Current eligibility against IUCN Red List Criteria

IUCN category	Criteria eligibility
Australian breeding population	
A	Not applicable: past, current or future population declines are thought unlikely to exceed 20% in any 3-generation period
B	Not applicable: EOO, AOO, habitat, locations or population not declining, population fluctuations not extreme
C	Not applicable: no continuing population decline or extreme fluctuations
D	Critically Endangered: population <50 mature individuals (but downgraded to Vulnerable as global status Least Concern)
E	Not applicable: no population viability analysis undertaken

IUCN Red List assessment data

	Estimate	Reliability
Australian breeding population		
Extent of occurrence trend	1 500 000 km^2 stable	medium low
Area of occupancy trend	2 km^2 stable	low low
No. of mature individuals trend	10 decreasing	low low
No. subpopulations	1	high
No. locations	1	medium
Generation time	15.6 years	low
Global population share	<1%	High
Level of genetic exchange	high	medium

up to 1000 m above sea level. This species is marine and highly pelagic, rarely approaching land except at colonies. Little is known about its diet, but is known to feed on squid (Marchant and Higgins 1990). A generation time of 15.6 years (BirdLife International 2011) is derived from an age at first breeding of 6.1 years and an annual survival of adults of 89.5% (Schreiber and Burger 2002).

Threats

The tiny population is vulnerable to catastrophes, such as the accidental introduction of predators.

Conservation objectives

1. Current breeding population maintained

Information required

1. Regularity of breeding on Raine I.

Management actions required

1. Maintain strict quarantine for Raine I.
2. Check for presence of petrels, and introduced predators, during visits to the island

Bibliography

Batianoff GN, Cornelius NJ (2005) Birds of Raine Island: population trends breeding behaviour and nesting habitats. *Proceedings of the Royal Society of Queensland* **112**, 1–29.

BirdLife International (2011) 'Species factsheet: *Pterodroma heraldica*'. Retrieved 17 January, 2011 from <http://www.birdlife.org/>.

Garnett S, Crowley GM (2000) *The Action Plan for Australian Birds 2000*. Environment Australia, Canberra.

Marchant S, Higgins PJ (Eds) (1990) *Handbook of Australian, New Zealand and Antarctic Birds. Volume 1: Ratites to Ducks*. Oxford University Press, Melbourne.

Schreiber EA, Burger J (2002) *Biology of Marine Birds*. CRC Press, Boca Raton, FL.

Comments received from

Barry Baker, Mike Brooke, John Cornelius

Species Conservation Summary

Soft-plumaged Petrel

Pterodroma mollis (Gould, 1844)

Procellariidae

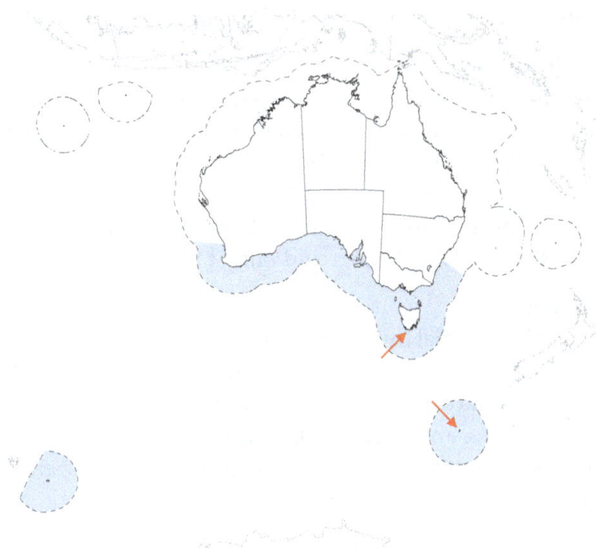

Conservation status

Australian breeding population: Critically Endangered D

Population visiting Australian territory: Least Concern

Reasons for listing

Australian breeding population: population <50 mature individuals

Status 2000

Australian breeding population: Critically Endangered D

Population visiting Australian territory: Least Concern

Status 1990

Australian breeding population: Critically Endangered D

Population visiting Australian territory: Least Concern

Taxonomy

Two subspecies, *P. m. mollis* (Gough and Antipodes Is.) and *P. m. dubia* (Amsterdam, Crozet, Kerguelen, Prince Edward and Marion Is.) are widely accepted. A third form *P. m. deceptornis* (Amsterdam I.), which has a similar call to the Maatsuyker I. birds (Bretagnolle 1995), is rarely recognised. Although Garnett & Crowley (2000) tentatively ascribed records of the species breeding in Australia to *P. m. deceptornis*, here we eschew any such attribution following the inability of Wiltshire et al. (2004) to ascribe birds from Maatsuyker I. held in the hand to any subspecies. Thus this account considers the whole species as 1 entity.

Taxonomic uniqueness: medium (14 genera/family, 30 species/genus, 2 subspecies/species)

Range

In Australian territory, breeds at 2 sites: on Maatsuyker I. off Tasmania (Wiltshire *et al.* 2004) and on Macquarie I. (Way *et al.* 2009). Elsewhere, breeding on Gough, Antipodes, Amsterdam, Crozet, Kerguelen, Prince Edward and Marion Is. It disperses outside the breeding season to seas throughout the Southern Ocean (Woehler *et al.* 1990; Reid *et al.* 2002).

Abundance

The population on Maatsuyker I. is very small: 6 nests had eggs in 2001–2002 of which 4 had fledglings at the

Current eligibility against IUCN Red List Criteria

IUCN category	Criteria eligibility
Australian breeding population	
A	Not applicable: past, current or future population declines are thought unlikely to exceed 20% in any 3-generation period
B	Not applicable: EOO, AOO, habitat, locations or population not declining, population fluctuations not extreme
C	Not applicable: no continuing population decline or extreme fluctuations
D	Critically Endangered: population <50 mature individuals
E	Not applicable: no population viability analysis undertaken

IUCN Red List assessment data

	Estimate	Reliability
Australian breeding population		
Extent of occurrence trend	3 500 000 km^2 stable	low low
Area of occupancy trend	2 km^2 stable	low low
No. of mature individuals trend	16 stable	medium low
No. subpopulations	2	low
No. locations	2	low
Generation time	15.6 years	low
Global population share	<1%	low
Level of genetic exchange	medium	low

end of the season, and 5 nests had eggs in 2002–2003 (1 in a different burrow from the 2001–2002 season) of which 3 had fledglings. There were thus at least 7 breeding burrows, so 14 individuals (Wiltshire *et al.* 2004). On Macquarie I. a single chick was found in 2006–2007 (Way *et al.* 2009). The global population is estimated at 5 million mature individuals (BirdLife International 2011).

Ecology

Breeds in burrows among rocks and tussocks, and forages for squid, crustaceans and fish, mainly by surface seizing (BirdLife International 2011). A generation time of 15.6 years (BirdLife International 2011) is derived from an age at first breeding of 6.1 years and an annual survival of adults of 89.5% (Schreiber and Burger 2002).

Threats

The only potential threat on Maatsuyker I. is the accidental introduction of predators (Wiltshire *et al.* 2004). Introduced predators, including cats *Felis catus*, Black Rats *Rattus rattus* and Wekas *Gallirallus australis*, may have been been responsible for the modern absence of the petrel from Macquarie I. (Brothers 1984).

Conservation objectives

1. Population stable or increasing

Information required

1. Population trends

Management actions required in Australia

1. Monitor numbers on Maatsuyker I.
2. Monitor for species on Macquarie I. once rats and rabbits have been eradicated
3. Maintain strict quarantine on Maatsuyker I.

Bibliography

BirdLife International (2011) 'Species factsheet: *Pterodroma mollis*'. Retrieved 17 January, 2011 from <http://www.birdlife.org/>.

Bretagnolle V (1995) Systematics of the Soft-plumaged Petrel *Pterodroma mollis* (Procellariidae): new insight from the study of vocalizations. *Ibis* **137**, 207–218.

Brothers NP (1984) Breeding, distribution and status of Burrow-nesting Petrels at Maqaurie Island. *Australian Wildlife Research* **11**, 113–131.

Garnett S, Crowley GM (2000) *The Action Plan for Australian Birds 2000.* Environment Australia, Canberra.

Reid TA, Hindell MA, Eades DW, Newman M (2002) *Seabird Atlas of South-eastern Australian Waters.* Birds Australia Monograph 4.

Schreiber EA, Burger J (2002) *Biology of Marine Birds.* CRC Press, Boca Raton, FL.

Way S, McInnes J, Derry A (2009) 'Baseline population monitoring of key seabird species, Macquarie Island Marine Park: Report on the 2008/09 season'. Unpublished report produced for the Tasmanian Parks and Wildlife Service, Hobart.

Wiltshire A, Hamilton S, Brothers N (2004) First record of Soft-plumaged Petrels, *Pterodroma mollis*, breeding in Australia. *Emu* **104**, 363–368.

Woehler EJ, Hodges CL, Watts DJ (1990) 'An atlas of the pelagic distribution and abundance of seabirds in the southern Indian Ocean 1981 to 1990'. ANARE Research Notes 77.

Comments received from

Barry Baker, Sheryl Hamilton, Eric Woehler

Species Conservation Summary

White-headed Petrel

Pterodroma lessonii (Garnot, 1826)
Procellariidae

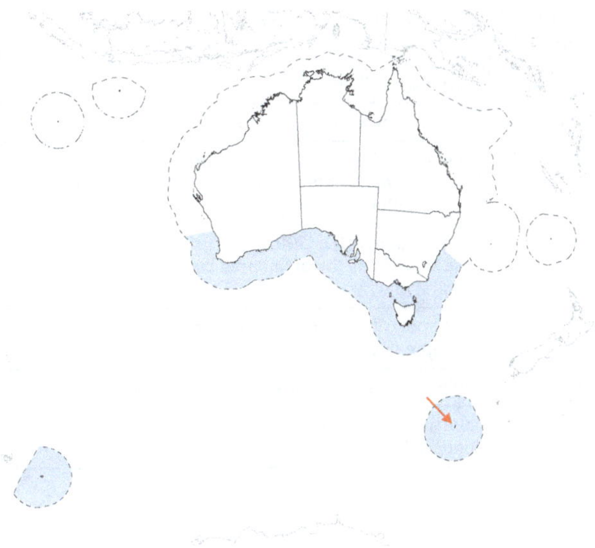

Conservation status

Australian breeding population: Endangered B2ab(iii,v)
Population visiting Australian territory: Least Concern

Reasons for listing

Australian breeding population: Present only at 1 Australian site that is affected by rabbit grazing

Status 2000

Australian breeding population: Vulnerable D2
Reason for change in 2010: increase in rabbit grazing degrading breeding habitat
Population visiting Australian territory: Least Concern

Status 1990

Australian breeding population: Vulnerable D2
Population visiting Australian territory: Least Concern

Taxonomy

No infraspecific taxa described
Taxonomic uniqueness: medium (14 genera/family, 30 species/genus, 1 subspecies/species)

Range

In Australian territory, breeds on Macquarie I. Extralimitally breeds on other subantarctic islands, including Auckland, Antipodes, Crozet, Kerguelen and possibly Prince Edward Is. (BirdLife International 2011). At sea has a circumpolar range and oceanic distribution throughout the Southern Oceans.

Abundance

This was the most abundant burrowing petrel breeding on Macquarie I. in 1960 (Warham 1967) but was thought to have declined greatly by the time a survey had estimated the population at 8000 pairs in the late 1970s. It was little different in 1993 (Garnett and Crowley 2000), but was estimated at 9000 pairs in 2006 (Parks and Wildlife Service 2006), the figure used here without allowing for non-breeding mature birds. However, this total may have declined substantially since, because heavy grazing by rabbits *Oryctolagus cuniculus* in recent years has exposed nests to predation from Brown Skua *Catharacta stercorarius*. The global population is estimated at 600 000 individuals (BirdLife International 2011).

Ecology

Breeds singly or in loose colonies in burrows dug among tussocks and herbfields on subantarctic islands. They feed on oceanic squid and crustaceans, hunting mostly at night (Marchant and Higgins 1990). A generation time of 12.6 years (BirdLife International 2011) is derived from an average age at first breeding of 5.5 years and an annual survival of adults of 86.0% (Schreiber and Burger 2002).

Threats

On Macquarie I. the rapid increase in rabbit numbers since feral cats *Felis catus* were eradicated in 1999 has led to overgrazing of native vegetation, extensive destruction of habitat and soil erosion at nesting sites. Burrow entrances are usually concealed by foliage; when this is removed exercising chicks are more vulnerable to predation by Brown Skua. This threat may continue unless a rabbit eradication program planned in 2011–2014 is successful. Historically the population at Macquarie I. probably declined because of predation by introduced Wekas *Gallirallus australis*, cats, and Black Rats *Rattus rattus*. Appears not to be threatened at sea, even though it often follows fishing boats (Brothers *et al.* 2010).

Conservation objectives

1. Nesting sites on Macquarie I. stabilised

Information required

1. Information on population trends and demographic parameters

Management actions required

1. Eradicate Black Rats and rabbits from Macquarie I.
2. Monitor a sample of nest sites

Bibliography

BirdLife International (2011) 'Species factsheet: *Pterodroma lessonii*'. Retrieved 17 January, 2011 from <http://www.birdlife.org/>.

Brothers N, Duckworth AR, Safina C, Gilman EL (2010) Seabird bycatch in pelagic longline fisheries is grossly underestimated when using only haul data. *PLoS ONE* **5**, e12491.

Garnett S, Crowley GM (2000) *The Action Plan for Australian Birds 2000.* Environment Australia, Canberra.

Marchant S, Higgins PJ (Eds) (1990) *Handbook of Australian, New Zealand and Antarctic Birds. Volume 1: Ratites to Ducks.* Oxford University Press, Melbourne.

Parks and Wildlife Service (2006) 'Macquarie Island Nature Reserve and World Heritage Area Management Plan'. Parks and Wildlife Service, Department of Tourism, Arts and the Environment, Hobart.

Current eligibility against IUCN Red List Criteria

IUCN category	Criteria eligibility
Australian breeding population	
A	Not applicable: past, current or future population declines are thought unlikely to exceed 20% in any 3-generation period
B	Endangered: AOO <500 km2, at ≤5 locations with plausible future threat, inferred continuing decline in habitat quality
C	Not applicable: no continuing population decline or extreme fluctuations
D	Vulnerable: ≤5 locations with plausible future threat
E	Not applicable: no population viability analysis undertaken

Schreiber EA, Burger J (2002) *Biology of Marine Birds*. CRC Press, Boca Raton, FL.

Warham J (1967) The White-headed Petrel *Pterodroma lessonii* at Macquarie Island. *Emu* **67**, 1–22.

IUCN Red List assessment data

	Estimate	Reliability
Australian breeding population		
Extent of occurrence trend	3 600 000 km^2 stable	medium high
Area of occupancy trend	40 km^2 stable	medium high
No. of mature individuals trend	18 000 decreasing	medium medium
No. subpopulations	1	high
No. locations	1	high
Generation time	12.6 years	medium
Global population share	1–10%	high
Level of genetic exchange	Low	low

Comments received from

Barry Baker, Rosemary Gales, Tim Reid, Ben Sullivan, Eric Woehler

Species Conservation Summary

Providence Petrel

Pterodroma solandri (Gould, 1844)

Procellariidae

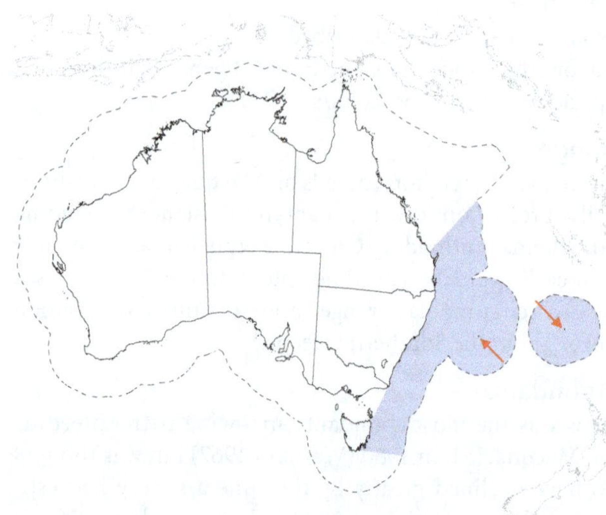

Conservation status

Vulnerable D2

Reasons for listing

Present only at only two Australian sites with a plausible threat of the loss of 1 location

Status 2000

Vulnerable D2

Status 1990

Vulnerable D2

Taxonomy

No infraspecific taxa described

Taxonomic uniqueness: medium (14 genera/family, 30 species/genus, 1 subspecies/species)

Range

Breeds only on islands in the Tasman Sea on Lord Howe I., and on Phillip I. near Norfolk I. Norfolk I. itself was also known briefly as a third breeding location before the bird's extirpation in the early 1800s (Medway 2002). The birds on Phillip I., while only discovered in 1985 (Hermes *et al.* 1986), may have been there much longer (Coyne 2010; Priddel *et al.* 2010). Birds are present on Lord Howe I. throughout the year, breeding occurs in winter and birds are often seen foraging off eastern Australia (BirdLife International 2011) as well as other sites distant from the island (Bester *et al.* 2002). They also migrate to the north Pacific Ocean east of Japan (BirdLife International 2011).

Abundance

Approximately 32 000 pairs nest on Mt Gower and Mt Lidgbird (Bester 2003) with some nesting on lower slopes and now also on Northern Hills (D. Priddel *in litt.*). Only 10–100 pairs breed on Phillip I. (Priddel *et al.* 2010). The population on Norfolk I. was once very large: from 1790 to 1793 about 1 million adults and young were harvested (Medway 2002).

Current eligibility against IUCN Red List Criteria

IUCN category	Criteria eligibility
A	Not applicable: past, current or future population declines are thought unlikely to exceed 20% in any 3-generation period
B	Not applicable: EOO, AOO, habitat, locations or population not declining, population fluctuations not extreme
C	Not applicable: no continuing population decline or extreme fluctuations
D	Vulnerable: ≤5 locations with a plausible future threat
E	Not applicable: no population viability analysis undertaken

IUCN Red List assessment data

	Estimate	Reliability
Extent of occurrence trend	1 300 000 km^2 stable	low high
Area of occupancy trend	3 km^2 stable	high high
No. of mature individuals trend	100 000 stable	medium low
No. subpopulations	2	low
No. locations	2	high
Largest subpopulation	100 000	medium
Generation time	15.9 years	low
Global population share	100%	high

Ecology

They breed in burrows in the ground, often within rainforest. At sea, they generally prefer warm waters for foraging, taking fish, cephalopods, crustaceans and offal, and will forage near fishing boats (Marchant and Higgins 1990). A generation time of 15.9 years (BirdLife International 2011) is derived from an average age at first breeding of 6.1 years and an annual survival of adults of 89.5%, both extrapolated from congeners.

Threats

The petrels were hunted to extinction, or killed by pigs *Sus scrofa*, on Norfolk I. in the early 1800s (Medway 2002), although subject to Australia's first species protection law (Bonyhardy 2000). On Phillip I., Nankeen Kestrels *Falco cenchroides* and possibly Purple Swamphens *Porphyio porphyrio* take some seabirds (Director of National Parks 2010), but the Providence Petrels are well protected in their burrows. However, Wedge-tailed Shearwater *Puffinus pacificus* sometimes evict young Providence Petrels from burrows (Priddel *et al.* 2010). On Lord Howe I. they have thrived despite numerous threats, even though some nests on the lower mountain slopes may have been lost to feral pigs or cats *Felis catus* before their eradication in the 1980s (McAllan *et al.* 2004). Currently the main causes of nest failure are inundation of burrows and predation of eggs and chicks by the endemic Endangered Lord Howe Woodhen *Gallirallus sylvestris*, but neither of these threats is significant (Bester *et al.* 2002). The only potential threats are the introduction of a new disease or predator to either location, and further information is required to determine whether the species is affected by longline fishing. While all threats are plausible, they have a low probability given the history of the species. It is listed as Near Threatened both for these reasons and because the very small Phillip I. population may be lost simply through stochastic variation.

Conservation objectives

1. Maintain current breeding population

Information required

1. Trends on Phillip I.

Management actions required

1. Maintain quarantine for Lord Howe and Phillip I.
2. Determine threats on Phillip I. and manage as required

Bibliography

Bester A (2003) The breeding, foraging ecology and conservation of the Providence Petrel *Pterodroma solandri* on Lord Howe Island, Australia. PhD thesis, Charles Sturt University, Albury.

Bester A, Klomp NI, Priddel D, Carlile N (2002) Chick-provisioning behaviour of the Providence Petrel, *Pterodroma solandri*. *Emu* **102**, 297–304.

BirdLife International (2011) 'Species factsheet: *Pterodroma solandri*'. Retrieved 17 January, 2011 from <http://www.birdlife.org/>.

Bonyhardy T (2000) *The Colonial Earth*. Melbourne University Press, Melbourne.

Coyne P (2010) Ecological rebound on Phillip Island, South Pacific. *Ecological Management and Restoration* **11**, 4–15.

Director of National Parks (2010) 'Draft Norfolk Island Region Threatened Species Recovery Plan'. Department of the Environment, Water, Heritage and the Arts, Canberra.

Hermes N, Evans O, Evans B (1986) Norfolk Island birds: a review. *Notornis* **33**, 141–149.

Marchant S, Higgins PJ (Eds) (1990) *Handbook of Australian, New Zealand and Antarctic Birds. Volume 1: Ratites to Ducks.* Oxford University Press, Melbourne.

McAllan IAW, Curtis BR, Hutton I, Cooper RM (2004) The birds of the Lord Howe Island Group: a review of records. *Australian Field Ornithology* **21**, 1–82.

Medway DG (2002) History and causes of extinction of the Providence Petrel (*Pterodroma solandri*) on Nordfolk Island. *Notornis* **49**, 246–258.

Priddel D, Carlile N, Evans O, Evans B, McCoy H (2010) A review of the seabirds of Phillip Island in the Norfolk Island Group. *Notornis* **57**, 113–127.

Comments received from

Nicholas Carlile, Margaret Christian, David Priddel, Coral Rowston, Ron Ward

Species Conservation Summary

Mottled Petrel

Pterodroma inexpectata (J. R. Forster, 1844)
Procellariidae

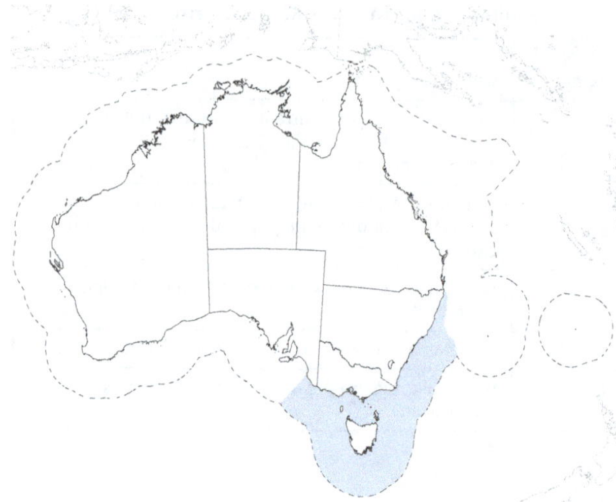

Conservation status

Population visiting Australian territory: Near Threatened A2e+3e+4e; B2ab(iv)

Reasons for listing

Population declining at 20–30% in 3 generations (47 years) because of predation by exotic predators at some of the small number of breeding sites near New Zealand

Status 2000

Near Threatened A2e+3e+4e; B2ab(iv)

Status 1990

Near Threatened A2e+3e+4e; B2ab(iv)

Taxonomy

No infraspecific taxa described
Taxonomic uniqueness: medium (14 genera/family, 30 species/genus, 1 subspecies/species)

Range

Formerly widespread across the North and South I. of New Zealand, now restricted to islands off the South and Stewart Is. as well as the Solander and Snares Is. Mostly frequents Antarctic and subantarctic waters, migrating rapidly through the tropics to far northern waters to spend the non-breeding season in the northern hemisphere, mostly in the Gulf of Alaska and the Aleutian Islands. Regularly seen off the east coast of Australia (BirdLife International 2011).

Abundance

The total population is estimated at 1 500 000 individuals of which up to 400 000 pairs estimated to be on Codfish I. (BirdLife International 2011). Only a small proportion are likely to visit Australian waters in any particular year.

Ecology

Nests in burrows or on rocky surfaces and feeds on cephalopods, crustaceans and fish (BirdLife International 2011). A generation time of 15.6 years (BirdLife International 2011) is derived from an average age at first breeding of 6.1 years and an annual survival of adults of 89.5%, both extrapolated from congeners.

Threats

There are no records of marine threats. On land suffers from predation by Weka *Gallirallus australis*, which have been introduced to several colonies, and possibly Black Rats *Rattus rattus*, which are present on Big South Cape I. Some birds may be taken incidentally, and burrows trampled, on islands where Sooty Shearwater *Puffinus griseus* chicks are harvested.

Conservation objectives

1. At-sea threats maintained at acceptable levels

Information required

1. None for Australia

Management actions required

1. Support efforts by New Zealand to eradicate introduced predators

Bibliography

BirdLife International (2011) 'Species factsheet: *Pterodroma inexpectata*'. Retrieved 17 January, 2011 from <http://www.birdlife.org/>.

Comments received from

Barry Baker, David Priddel, Eric Woehler

Current eligibility against IUCN Red List Criteria

IUCN category	Criteria eligibility
Population visiting Australian territory	
A	Near Threatened: past, current and anticipated population declines of 20–29% over periods of 3 generations (47 years) inferred from effects of introduced taxa
B	Near Threatened: AOO (breeding) <500 km², near 10 locations, continuing decline in no. locations
C	Not applicable: population >10 000 mature individuals
D	Not applicable: population >1000 mature individuals, >5 locations
E	Not applicable: no population viability analysis undertaken

IUCN Red List assessment data

	Estimate	Reliability
Population visiting Australian territory		
Extent of occurrence trend	1 300 000 km² stable	high high
Area of occupancy trend	1 300 000 km² stable	high high
No. of mature individuals trend	1 500 000 stable	medium medium
No. subpopulations	1	high
No. locations	12	high
Generation time	5.6 years	medium
Global population share	<1%	high
Level of genetic exchange	high	high

Species Conservation Summary

Pycroft's Petrel

Pterodroma pycrofti Falla, 1933
Procellariidae

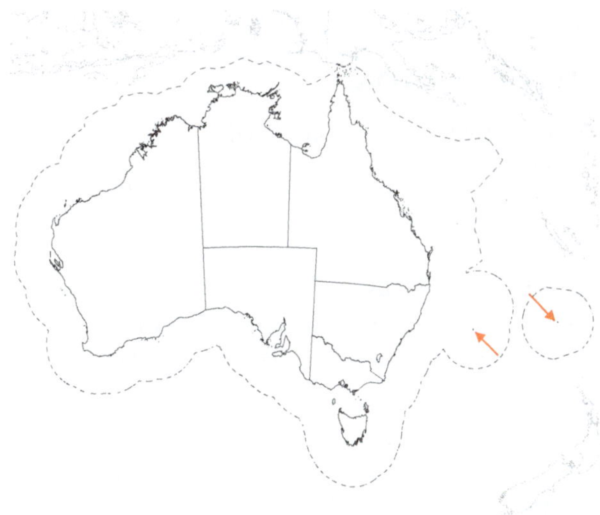

Conservation status
Australian breeding population: Extinct

Reasons for listing
Birds thought to be this species were last recorded on Norfolk I. in about 1800. On Lord Howe I., the species is known only from subfossils

Status 2000
Extinct

Status 1990
Extinct

Taxonomy
No infraspecific taxa described. Globally the status is Vulnerable.
Taxonomic uniqueness: medium (14 genera/family, 30 species/genus, 1 subspecies/species)

Former range
Subfossils have been found on Lord Howe and Norfolk Is., and a painting of a blue-footed petrel from Norfolk I. has been attributed to this species (Holdaway and Anderson 2001). Still breeds in the Poor Knights, Hen and Chicken, Mercury, and Ririwha Is. off the coast of New Zealand, migrating to the North Pacific Ocean (Heather and Robertson 1997) and is a vagrant to Australia (Marchant and Higgins 1990).

Former abundance
No information.

Ecology
Nests in burrows, often within colonies of other petrels. Their diet is little known, but contains cephalopods (BirdLife International 2011).

Reasons for extinction
Predation by introduced cats *Felis catus* and pigs *Sus scrofa* probably caused the extinction of Pycroft's Petrel from Norfolk and Lord Howe Is., although numbers on Norfolk I. may already have been suppressed by Pacific Rats *Rattus exulans* introduced by Polynesians 800 years ago (Holdaway and Anderson 1998). On New Zealand islands, Pacific Rats have a significant impact on breeding success, taking eggs and chicks (Heather and Robertson 1997).

Bibliography
BirdLife International (2011) 'Species factsheet: *Pterodroma pycrofti*'. Retrieved 17 January, 2011 from <http://www.birdlife.org/>.

Heather BD, Robertson HA (1997) *The Field Guide to the Birds of New Zealand*. Oxford University Press, Oxford, UK.

Holdaway RN, Anderson AJ (1998) 14C AMS dates on *Rattus exulans* bones from natural and archaeological contexts on Norfolk Island, South-west Pacific. *Archaeology in New Zealand* **41**, 195–198.

Holdaway RN, Anderson AJ (2001) Avifauna from the Emily Bay Settlement Site, Norfolk Island: a preliminary account. *Records of the Australian Museum, Supplement* **27**, 85–100.

Marchant S, Higgins PJ (Eds) (1990) *Handbook of Australian, New Zealand and Antarctic Birds. Volume 1: Ratites to Ducks.* Oxford University Press, Melbourne.

Comments received from
Richard Holdaway, David Priddel

Species Conservation Summary

Gould's Petrel

Pterodroma leucoptera (Gould, 1844)
Procellariidae

Conservation status

Australian breeding population: Vulnerable D2
Population visiting Australian territory and the species: Vulnerable B2ab(v), D2

Reasons for listing

Australian breeding population: Occurs at ≤5 locations where there are plausible future threats
Population visiting Australian territory and the species: Breeding sites (outside Australia) are small and few in number and there is ongoing loss of birds to introduced predators

Status 2000

Australian breeding population: Vulnerable D2
Population visiting Australian territory and the species: Vulnerable B2ab(v), D2

Status 1990

Australian breeding population: Endangered B1ab(i,ii,iii,iv,v)+B2ab(i,ii,iii,iv,v), C2a(i)
Reason for change in 2000: effective conservation intervention
Population visiting Australian territory and the species: Vulnerable B2ab(v), D2

Current eligibility against IUCN Red List Criteria

IUCN category	Criteria eligibility
Australian breeding population	
A	Not applicable: past, current or future population declines are thought unlikely to exceed 20% in any 3-generation period
B	Not applicable: EOO, AOO, habitat, locations or population not declining, population fluctuations not extreme
C	Not applicable: no continuing population decline or extreme fluctuations
D	Vulnerable: ≤5 locations with plausible future threat
E	Not applicable: no population viability analysis undertaken
Population visiting Australian territory	
A	Not applicable: past, current or future population declines are thought unlikely to exceed 20% in any 3-generation period
B	Vulnerable: AOO (breeding) <2000 km², ≤10 locations with plausible future threat, suspected continuing decline in no. mature individuals
C	Near Threatened: population near 10 000 mature individuals, suspected continuing decline, near 100% in 1 subpopulation
D	Vulnerable: ≤5 locations with plausible future threat
E	Not applicable: no population viability analysis undertaken

Taxonomy

P. l. leucoptera (New South Wales) and *P. l. caledonica* (New Caledonia) are both Vulnerable
Taxonomic uniqueness: medium (14 genera/family, 10 species/genus, 2 subspecies/species)

Range

P. l. leucoptera breeds at 4 locations off the coast of New South Wales, *P. l. caledonica* breeds between Mts Dzumac and Poya on New Caledonia and at Ravavae in French Polynesia. When not breeding disperses throughout the Tasman Sea and across to the eastern Pacific Ocean (BirdLife International 2011).

Abundance

P. l. leucoptera has >1000 breeding pairs after successful conservation management recovered population from 250 pairs in 1989–1991 (Priddel and Carlile 2009). The 3 main breeding sites of *P. l. caledonica* support 1000–2000 pairs and the total population may be as much as 10 000 pairs (BirdLife International 2011), but is declining overall.

Ecology

In Australia breeds colonially in natural cavities in rock scree. In New Caledonia breeds in small, loose colonies of 10–50 pairs in burrows on steep, vegetated slopes. Feeds at sea, probably largely on cephalopods (Marchant and Higgins 1990). A generation time of 15.6 years (BirdLife International 2011) is derived from an average age at first breeding is 6.1 years and an annual survival of adults of 89.5%, both extrapolated from congeners.

Threats

The main threats to *P. l. leucoptera* are introduction of rats *Rattus* spp. or cats *Felis catus* from the mainland. On

IUCN Red List assessment data

	Estimate	Reliability
Australian breeding population		
Extent of occurrence trend	2 400 000 km² stable	high high
Area of occupancy trend	4 km² stable	high high
No. of mature individuals trend	2500 increasing	low medium
No. subpopulations	1	high
No. locations	4	high
Generation time	15.6 years	medium
Global population share	100%	high
Population visiting Australian territory		
Extent of occurrence trend	2 400 000 km² stable	high high
Area of occupancy trend	2 400 000 km² stable	high low
No. of mature individuals trend	10 000 decreasing	low medium
No. subpopulations	2	low
No. locations	5	low
Largest subpopulation	10 000	medium

New Caledonia, they breed in the presence of both the introduced pig *Sus scrofa* and Black Rat *Rattus rattus* which kill adults and young and damage nesting habitat (BirdLife International 2011).

Conservation objectives

1. Stable or increasing population
2. At-sea threats maintained at acceptable levels

Information required

1. Demographic and life history data
2. Foraging locations and diet

Management actions required

1. Continue monitoring
2. Initiate dietary and foraging studies
3. Determine size and security of Broughton Is. populations
4. Continue control of Pied Currawongs and Birdlime Trees as required on Cabbage Tree Island

Bibliography

BirdLife International (2011) 'Species factsheet: *Pterodroma leucoptera*'. Retrieved 17 January, 2011 from <http://www.birdlife.org/>.

Marchant S, Higgins PJ (Eds) (1990) *Handbook of Australian, New Zealand and Antarctic Birds. Volume 1: Ratites to Ducks.* Oxford University Press, Melbourne.

Priddel D, Carlile N (2009) Key elements in achieving a successful recovery programme: a discussion illustrated by the Gould's Petrel case study. *Ecological Management and Restoration* **10**, S97–S102.

Comments received from

Nicholas Carlile, David Priddel

Subspecies Conservation Summary

Gould's Petrel (New Caledonian)

Pterodroma leucoptera caledonica (Gould, 1844)
Procellariidae

Conservation status

Population visiting Australian territory: Vulnerable B2ab(v), D2

Reasons for listing

Breeding sites (outside Australia) are small and few in number and there is ongoing loss of birds to introduced predators

Status 2000

Vulnerable B2ab(v), D2

Status 1990

Vulnerable B2ab(v), D2

Taxonomy

P. l. leucoptera (New South Wales) is Vulnerable, as is the species
Taxonomic uniqueness: medium (14 genera/family, 10 species/genus, 2 subspecies/species)

Range

Breeds in New Caledonia occur between Mts Dzumac and Poya but may breed on other mountain massifs. A small colony breeds on Ravavae, in French Polynesia. When not breeding disperses throughout the Tasman Sea and across to the eastern Pacific Ocean (BirdLife International 2011).

Abundance

In New Caledonia, the 3 main breeding sites support 1000–2000 pairs but total population may approach 10 000 pairs (BirdLife International 2011). The population size used here is the average of 1000–10 000 pairs.

Ecology

Breeds in small, loose colonies of 10–50 pairs in burrows on steep, vegetated slopes and feeds well out to sea, coming ashore at night. A generation time of 15.6 years (BirdLife International 2011) is derived from an average age at first breeding is 6.1 years and an annual survival of adults of 89.5%, both extrapolated from congeners.

Threats

Major threat is introduced predators at nests: introduced pigs *Sus scrofa* excavate burrows, Black Rats *Rattus rattus* take eggs and adults. Birds also die when they fly into lights in Noumea. Threats from fishing appear minimal (BirdLife International 2011).

Conservation objectives

1. At-sea threats maintained at acceptable levels
2. Global agreement on conservation measures

Information required

1. Population trends or changes in adult and juvenile survival

Current eligibility against IUCN Red List Criteria

IUCN category	Criteria eligibility
Population visiting Australian territory	
A	Not applicable: past, current or future population declines are thought unlikely to exceed 20% in any 3-generation period
B	Vulnerable: AOO (breeding) <2000 km^2, ≤10 locations with plausible future threat, suspected continuing decline in no. mature individuals
C	Near Threatened: population near 10 000 mature individuals, suspected continuing decline, near 100% in 1 subpopulation
D	Vulnerable: ≤5 locations with plausible future threat
E	Not applicable: no population viability analysis undertaken

IUCN Red List assessment data

	Estimate	Reliability
Population visiting Australian territory		
Extent of occurrence trend	2 400 000 km^2 stable	high high
Area of occupancy trend	2 400 000 km^2 stable	high low
No. of mature individuals trend	10 000 decreasing	low medium
No. subpopulations	2	low
No. locations	5	low
Largest subpopulation	10 000	medium
Generation time	15.6 years	medium
Global population share	50–100%	high

Management actions required

1. Assist New Caledonia with conservation management as requested

Comments received from

Nicholas Carlile, David Priddel

Bibliography

BirdLife International (2011) 'Species factsheet: *Pterodroma leucoptera*'. Retrieved 17 January, 2011 from <http://www.birdlife.org/>.

Subspecies Conservation Summary

Gould's Petrel (Australian)

Pterodroma leucoptera leucoptera (Gould, 1844)
Procellariidae

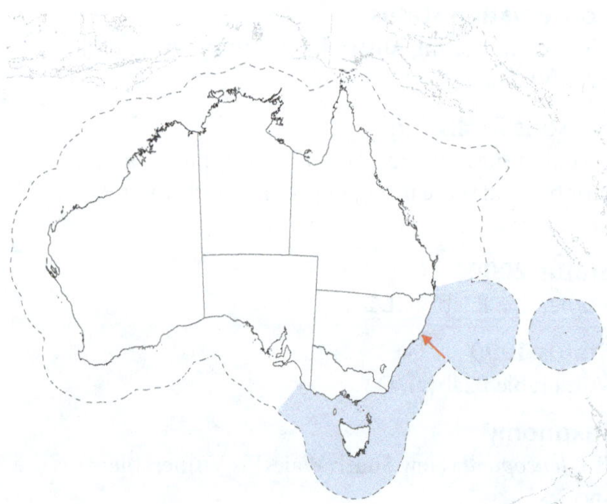

Conservation status

Vulnerable D2

Reasons for listing

Occurs at ≤5 locations where there are plausible future threats

Status 2000

Vulnerable D2

Status 1990

Endangered B1ab(i,ii,iii,iv,v)+B2ab(i,ii,iii,iv,v), C2a(i)
Reason for change in 2000: effective conservation intervention

Taxonomy

P. l. caledonica (New Caledonia) is Vulnerable, as is the species
Taxonomic uniqueness: medium (14 genera/family, 10 species/genus, 2 subspecies/species)

Range

Breeds at 4 locations off the coast of New South Wales: on Cabbage Tree I.; following transfer of 200 nestlings in 1999–2000, on nearby Boondelbah I. (Priddel *et al.* 2006); and on Broughton and Little Broughton Is. off Port Stephens where small populations were found in 2009 and 2010. When not breeding, disperses throughout the Tasman Sea and across to the eastern Pacific Ocean (BirdLife International 2011).

Abundance

Following recovery actions, breeding success increased from less than 20% to about 50%, and the population from fewer than 250 breeding pairs during 1989–1991 to more than 1000 breeding pairs. The number of fledglings

Current eligibility against IUCN Red List Criteria

IUCN category	Criteria eligibility
A	Not applicable: past, current or future population declines are thought unlikely to exceed 20% in any 3-generation period
B	Not applicable: EOO, AOO, habitat, locations or population not declining, population fluctuations not extreme
C	Not applicable: no continuing population decline or extreme fluctuations
D	Vulnerable: ≤5 locations with plausible future threat
E	Not applicable: no population viability analysis undertaken

IUCN Red List assessment data

	Estimate	Reliability
Extent of occurrence trend	2 400 000 km^2 stable	high high
Area of occupancy trend	4 km^2 stable	high high
No. of mature individuals trend	2500 increasing	low medium
No. subpopulations	1	high
No. locations	4	high
Generation time	15.6 years	medium
Global population share	100%	high

produced per annum increased from fewer than 50 to more than 450 (Priddel and Carlile 2009). There are 70+ breeding individuals on Boondelbah I. as a result of the introduction (Priddel *et al.* 2006). Currently only small numbers are known from the Broughton Is.

Ecology

Breeds colonially in natural cavities in rock scree. Feeds on cephalopods and fish. Though diet not well known (Marchant and Higgins 1990), variation in foraging success is substantial (O'Dwyer *et al.* 2007). A generation time of 15.6 years (BirdLife International 2011) is derived from an average age at first breeding is 6.1 years and an annual survival of adults of 89.5%, both extrapolated from congeners.

Threats

The main threats are introduction of Black Rats *Rattus rattus* or cats *Felis catus* from the mainland. Since 1992 threats to the population on Cabbage Tree I. that developed as a result of grazing by rabbits *Oryctolagus cuniculus*, including proliferation of the Birdlime Tree *Pisonia umbellifera* and predation by Pied Currawongs *Strepera graculina*, have been largely remedied, though some control continues (Priddel *et al.* 2000). As the understorey returns, the need to manage currawongs and the Birdlime Tree should diminish, hopefully to the extent where Gould's Petrel is no longer dependent on management intervention for its survival.

Conservation objectives

1. Population increase maintained

Information required

1. Demographic and life history data
2. Foraging behaviour and marine food resources

Management actions required

1. Continue annual monitoring
2. Initiate dietary and foraging studies
3. Determine size and security of Broughton Is. populations
4. Continue control of Pied Currawongs and Birdlime Trees as required

Bibliography

BirdLife International (2011) 'Species factsheet: *Pterodroma leucoptera*'. Retrieved 17 January, 2011 from <http://www.birdlife.org/>.

Marchant S, Higgins PJ (Eds) (1990) *Handbook of Australian, New Zealand and Antarctic Birds. Volume 1: Ratites to Ducks.* Oxford University Press, Melbourne.

O'Dwyer TW, Buttemer W, Priddel DM (2007) Differential rates of offspring provisioning in Gould's Petrels: are better feeders better breeders? *Australian Journal of Zoology* **55**, 155–160.

Priddel D, Carlile N (2009) Key elements in achieving a successful recovery programme: a discussion illustrated by the Gould's Petrel case study. *Ecological Management and Restoration* **10**, S97–S102.

Priddel D, Carlile N, Wheeler R (2000) Eradication of European rabbits (*Oryctolagus cuniculus*) from Cabbage Tree Island, NSW, Australia, to protect the breeding habitat of Gould's petrel (*Pterodroma leucoptera leucoptera*). *Biological Conservation* **94**, 115–125.

Priddel D, Carlile N, Wheeler R (2006) Establishment of a new breeding colony of Gould's petrel (*Pterodroma leucoptera leucoptera*) through the creation of artificial nesting habitat and the translocation of nestlings. *Biological Conservation* **128**, 553–563.

Comments received from

Nicholas Carlile, David Priddel

Subspecies Conservation Summary

White-necked Petrel (southern)

Pterodroma cervicalis cervicalis (Salvin, 1891)
Procellariidae

Conservation status

Australian breeding population: Endangered D
Population visiting Australian territory: Vulnerable D2

Reasons for listing

Australian breeding population: Population 50–250 mature individuals; some genetic mixing so could be downgraded but global population also threatened
Population visiting Australian territory: Breeds on a few small islands where vulnerability to predators represents a plausible future threat

Status 2000

Australian breeding population: Endangered D
Population visiting Australian territory: Vulnerable D2

Status 1990

Australian breeding population: Endangered D
Population visiting Australian territory: Vulnerable D2

Taxonomy

P. c. occulta (Vanuatu) is sometimes considered a separate species: Vanuatu Petrel. The species is Vulnerable.
Taxonomic uniqueness: medium (14 genera/family, 30 species/genus, 2 subspecies/species)

Range

A small colony present on Phillip I., off Norfolk I. (Priddel *et al.* 2010). Extralimitally breeds on Macauley I. in the Kermadec Is., New Zealand, with about 50 000 pairs in 1988 and possibly increasing (Veitch *et al.* 2004). It migrates to the north Pacific Ocean when not breeding (BirdLife International 2011).

Abundance

The population on Phillip I. is estimated at 10–100 pairs (Priddel *et al.* 2010) with a recent incomplete survey finding 18 nests (D. Priddel *in litt.*).

Ecology

Nests in burrows excavated in gently sloping grassy areas, and feeds on squid and crustaceans taken from the open

Current eligibility against IUCN Red List Criteria

IUCN category	Criteria eligibility
Australian breeding population	
A	Not applicable: past, current or future population declines are thought unlikely to exceed 20% in any 3-generation period
B	Not applicable: EOO, AOO, habitat, locations or population not declining, population fluctuations not extreme
C	Not applicable: no continuing population decline or extreme fluctuations
D	Endangered: population 50–250 mature individuals
E	Not applicable: no population viability analysis undertaken
Population visiting Australian territory	
A	Not applicable: past, current or future population declines are thought unlikely to exceed 20% in any 3-generation period
B	Not applicable: EOO, AOO, habitat, locations or population not declining, population fluctuations not extreme
C	Not applicable: no continuing population decline or extreme fluctuations
D	Vulnerable: ≤5 locations with plausible future threat
E	Not applicable: no population viability analysis undertaken

IUCN Red List assessment data

	Estimate	Reliability
Australian breeding population		
Extent of occurrence trend	1 340 000 km^2 stable	high high
Area of occupancy trend	1 km^2 stable	medium low
No. of mature individuals trend	60 stable	low low
No. subpopulations	1	high
No. locations	1	high
Generation time	15.6 years	low
Global population share	<1%	high
Level of genetic exchange	low	low
Population visiting Australian territory		
Extent of occurrence trend	1 340 000 km^2 stable	medium medium
Area of occupancy trend	1 340 000 km^2 stable	high medium
No. of mature individuals trend	100 000 increasing	medium low
No. subpopulations	1	medium
No. locations	1	high
Global population share	1–10%	low

ocean (BirdLife International 2011). On Phillip I., most nests are in rock cavities under dense low trees. A generation time of 15.6 years (BirdLife International 2011) is derived from an average age at first breeding of 6.1 years and an annual survival of adults of 89.5%, both extrapolated from congeners.

Threats

As breeding is confined to 1 islet, the species is vulnerable to catastrophe, such as the accidental introduction of the Black Rat *Rattus rattus*, which occurs on the adjacent Norfolk I. Brown Rats *Rattus norvegicus* are thought to have caused their extinction from Raoul I. (Veitch *et al.* 2004). Purple Swamphens *Porphyio porphyrio* may prey on eggs and chicks (Director of National Parks 2010), but the extent to which this threatens the population is unknown given the prevalence of other potential prey.

Conservation objectives

1. Maintain current breeding population
2. Enable expansion of population to main islands

Information required

1. Factors affecting breeding success

Management actions required

1. Maintain strict quarantine for Phillip I.
2. Determine threats on Phillip I. and manage as required

Bibliography

BirdLife International (2011) 'Species factsheet: *Pterodroma cervicalis*'. Retrieved 17 January, 2011 from <http://www.birdlife.org/>.

Director of National Parks (2010) 'Draft Norfolk Island Region Threatened Species Recovery Plan'. Department of the Environment, Water, Heritage and the Arts, Canberra.

Priddel D, Carlile N, Evans O, Evans B, McCoy H (2010) A review of the seabirds of Phillip Island in the Norfolk Island Group. *Notornis* **57**, 113–127.

Veitch CR, Miskelly CM, Harper GA, Taylor G, Tennyson AJD (2004) Birds of the Kermadec Islands, south-west Pacific. *Notornis* **51**, 61–90.

Comments received from

Nicholas Carlile, Margaret Christian, David Priddel, Coral Rowston, Ron Ward

---

Subspecies Conservation Summary

Common Diving-Petrel (southern)

Pelecanoides urinatrix exsul (Salvin, 1896)
Procellariidae

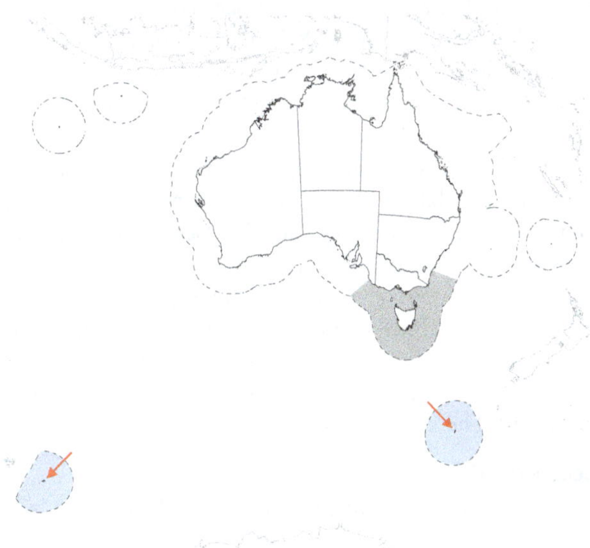

Conservation status

Australian breeding population: Vulnerable D2
Population visiting Australian territory: Least Concern

Reasons for listing

Australian breeding population: Present at 2 sites; introduction of exotic predators represents a plausible future threat

Status 2000

Australian breeding population: Vulnerable D2
Population visiting Australian territory: Least Concern

Status 1990

Australian breeding population: Vulnerable D2
Population visiting Australian territory: Least Concern

Taxonomy

P. u. urinatrix (islands of south-east Australia, Tasmania, New Zealand and Cook Strait) is Least Concern, as is the species. There are 4 extralimital subspecies.
Taxonomic uniqueness: medium (14 genera/family, 4 species/genus, 6 subspecies/species)

Range

Circumpolar distribution, breeding on Heard and Macquarie Is., as well as on South Georgia, Marion, Crozet, Kerguelen, Auckland and Antipodes Is. (BirdLife International 2011). When not breeding, they are widespread across the Southern Ocean (Woehler *et al.* 1990).

Abundance

Population on Heard I. is conservatively estimated at 1000–10 000 individuals (Woehler 2010), with about 20 pairs at Macquarie I. (Brothers 1984). The global population is estimated at 16 million of which a substantial proportion is likely to be this subspecies (BirdLife International 2011).

Current eligibility against IUCN Red List Criteria

IUCN category	Criteria eligibility
Australian breeding population	
A	Not applicable: past, current or future population declines are thought unlikely to exceed 20% in any 3-generation period
B	Not applicable: EOO, AOO, habitat, locations or population not declining, population fluctuations not extreme
C	Not applicable: no continuing population decline or extreme fluctuations
D	Vulnerable: ≤5 locations with plausible future threat
E	Not applicable: no population viability analysis undertaken

IUCN Red List assessment data

	Estimate	Reliability
Australian breeding population		
Extent of occurrence trend	1 100 000 km^2 stable	low high
Area of occupancy trend	20 km^2 stable	medium high
No. of mature individuals trend	5000 stable	low medium
No. subpopulations	2	high
No. locations	2	high
Largest subpopulation	5000	low
Generation time	7.8 years	low
Global population share	1–10%	high
Level of genetic exchange	low	low

Ecology

Nests in colonies among boulders in soil or peat, or under vegetation. Feeds on planktonic crustaceans and squid (Marchant and Higgins 1990). A generation time of 7.8 years (BirdLife International 2011) is derived from an average age at first breeding of 4.5 years and an annual survival of adults of 91.0% (Schreiber and Burger 2002).

Threats

There is a plausible risk of predators being introduced to Heard I. Their scarcity on Macquarie I. may be because introduced cats *Felis catus*, Black Rats *Rattus rattus* and Weka *Gallirallus australis* reduced their numbers before surveys occurred. All wekas and cats have been eradicated on Macquarie I. but, following the elimination of cats, an increase in rabbit *Oryctolagus cuniculus* numbers resulted in over-grazing of native vegetation and leading to extensive destruction of habitat and soil erosion at nesting sites and exposure of nests to predation by Brown Skua *Stercorarius lonnbergi*. Black Rats are also likely to take eggs and kills chicks and possibly adults. These threats may continue unless an eradication program planned for 2011–2014 is successful. There are no known threats at sea.

Conservation objectives

1. Populations maintained at Heard and Macquarie Is.

Information required

1. Population size, trends and demographic parameters

Management actions required

1. Maintain strict quarantine provisions at Heard I.
2. Eradicate Black Rats and rabbits from Macquarie I.

Bibliography

BirdLife International (2011) 'Species factsheet: *Pelecanoides urinatrix*'. Retrieved 17 January, 2011 from <http://www.birdlife.org/>.

Brothers NP (1984) Breeding, distribution and status of Burrow-nesting Petrels at Maqaurie Island. *Australian Wildlife Research* **11**, 113–131.

Schreiber EA, Burger J (2002) *Biology of Marine Birds*. CRC Press, Boca Raton, FL.

Woehler EJ (2010) Status and conservation of the seabirds of Heard Island. *The State of Australia's Birds 2010: Islands and Birds. Wingspan* **Supplement 20**, 8–9.

Woehler EJ, Hodges CL, Watts DJ (1990) 'An atlas of the pelagic distribution and abundance of seabirds in the southern Indian Ocean 1981 to 1990'. ANARE Research Notes 77.

Comments received from

Barry Baker, Rosemary Gales, Tim Reid, Ben Sullivan, Eric Woehler

Species Conservation Summary

South Georgian Diving-Petrel

Pelecanoides georgicus Murphy and Harper, 1916
Procellariidae

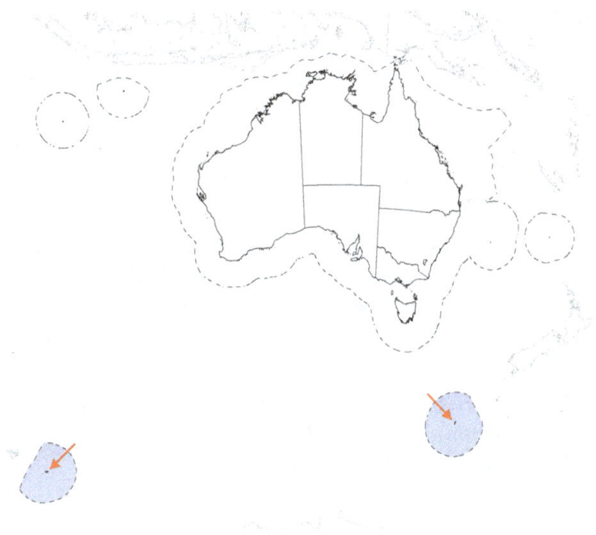

Conservation status

Australian breeding population: Vulnerable D2
Population visiting Australian territory: Least Concern

Reasons for listing

Australian breeding population: Present at 2 sites; introduction of exotic predators represents a plausible future threat

Status 2000

Australian breeding population: Vulnerable D2
Population visiting Australian territory: Least Concern

Status 1990

Australian breeding population: Vulnerable D2
Population visiting Australian territory: Least Concern

Taxonomy

No infraspecific taxa described
Taxonomic uniqueness: medium (14 genera/family, 4 species/genus, 6 subspecies/species)

Range

Circumpolar distribution, breeding on Heard and Macquarie Is. in Australian waters and extralimitally on South Georgia, Prince Edward, Crozet and Kerguelen Is. and Codfish I. off New Zealand (BirdLife International 2011). When not breeding they are widespread across the Southern Ocean (Woehler *et al.* 1990).

Abundance

Population on Heard I. is estimated at 10 000–100 000 individuals (Woehler 2010), with 3–4 pairs found breeding on Bishop and Clerk Is. near Macquarie I. (Brothers 1984). The global population is estimated at 15 million individuals (BirdLife International 2011).

Ecology

Nests in colonies in burrows dug in sandy areas, plains and hillsides. Feeds on crustaceans, particularly krill, but also on small fish and cephalopods (Marchant and Higgins 1990). A generation time of 7.8 years (BirdLife International 2011) is derived from an average age at first breeding of 4.5 years and an annual survival of adults of 91.0% (Schreiber and Burger 2002).

Threats

There is a plausible risk of predators being introduced to Heard I., which would be devastating (Holdaway *et al.* 2003). Their absence from Macquarie I. may be because introduced cats *Felis catus* and Black Rats *Rattus rattus* eliminated them before they were recorded. There are no major threats to the species at sea.

Conservation objectives

1. Populations maintained at Heard and Macquarie Is.

Information required

1. Population size, trends and demographic parameters

Management actions required

1. Maintain strict quarantine provisions at Heard I. and Bishop and Clerk Is.

Current eligibility against IUCN Red List Criteria

IUCN category	Criteria eligibility
Australian breeding population	
A	Not applicable: past, current or future population declines are thought unlikely to exceed 20% in any 3-generation period
B	Not applicable: EOO, AOO, habitat, locations or population not declining, population fluctuations not extreme
C	Not applicable: no continuing population decline or extreme fluctuations
D	Vulnerable: ≤5 locations with plausible future threat
E	Not applicable: no population viability analysis undertaken

IUCN Red List assessment data

	Estimate	Reliability
Australian breeding population		
Extent of occurrence trend	1 100 000 km^2 stable	low high
Area of occupancy trend	20 km^2 stable	medium high
No. of mature individuals trend	50 000 stable	low medium
No. subpopulations	2	high
No. locations	2	high
Largest subpopulation	50 000	low
Generation time	7.8 years	low
Global population share	1–10%	high
Level of genetic exchange	low	low

Bibliography

BirdLife International (2011) 'Species factsheet: *Pelecanoides georgicus*'. Retrieved 17 January, 2011 from <http://www.birdlife.org/>.

Brothers NP (1984) Breeding, distribution and status of Burrow-nesting Petrels at Maqaurie Island. *Australian Wildlife Research* **11**, 113–131.

Holdaway RN, Jones MD, Beavan Athfield NR (2003) Establishment and extinction of a population of South Georgian diving petrel (*Pelecanoides georgicus*) at Mason Bay, Stewart Island, New Zealand, during the late Holocene. *Journal of the Royal Society of New Zealand* **33**, 601–622.

Marchant S, Higgins PJ (Eds) (1990) *Handbook of Australian, New Zealand and Antarctic Birds. Volume 1: Ratites to Ducks.* Oxford University Press, Melbourne.

Schreiber EA, Burger J (2002) *Biology of Marine Birds.* CRC Press, Boca Raton, FL.

Woehler EJ (2010) Status and conservation of the seabirds of Heard Island. *The State of Australia's Birds 2010: Islands and Birds. Wingspan* **Supplement 20**, 8–9.

Woehler EJ, Hodges CL, Watts DJ (1990) 'An atlas of the pelagic distribution and abundance of seabirds in the southern Indian Ocean 1981 to 1990'. ANARE Research Notes 77.

Comments received from

Barry Baker, Rosemary Gales, Tim Reid, Ben Sullivan, Eric Woehler

---

Subspecies Conservation Summary

Gentoo Penguin (subantarctic)

Pygoscelis papua papua (J. R. Forster, 1781)
Spheniscidae

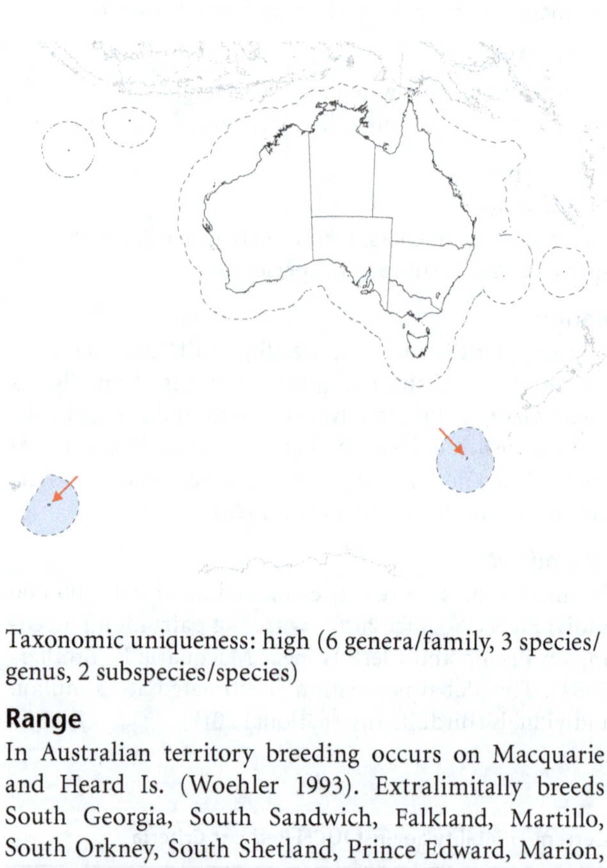

Conservation status

Australian breeding population: Near Threatened A2bc+3bc+4bc
Population visiting Australian territory: Near Threatened A2bc+3bc+4bc

Reasons for listing

Australian breeding population: Australian population not monitored but assumed to be following trend of global population: occurs at only 2 locations but no known current site-based threats
Population visiting Australian territory: Rapid declines globally approaching 30% in 3 generations (21 years) in the past and anticipated

Status 2000

Australian breeding population: Near Threatened A2bc+3bc+4bc
Population visiting Australian territory: Near Threatened A2bc+3bc+4bc

Status 1990

Australian breeding population: Least Concern
Reason for change in 2000: declines detected in global population
Population visiting Australian territory: Least Concern
Reason for change in 2000: declines detected in global population

Taxonomy

P. p. ellsworthii (South Sandwich, South Orkney and South Shetland Is., Antarctic Peninsula) is also Near Threatened, as is the global status of the species

Taxonomic uniqueness: high (6 genera/family, 3 species/genus, 2 subspecies/species)

Range

In Australian territory breeding occurs on Macquarie and Heard Is. (Woehler 1993). Extralimitally breeds South Georgia, South Sandwich, Falkland, Martillo, South Orkney, South Shetland, Prince Edward, Marion, Crozet, Kerguelen Is. and mainland Argentina (BirdLife International 2011).

Abundance

About 16 000 pairs were breeding on Heard I. in 1987–88 (Woehler 1991, 2006). This is possibly higher than the 10 000–12 000 pairs occurring in 16 colonies in 1955 (Downes *et al.* 1959), although recent surveys included areas not visited during the 1950s due to glacier retreats. More of Heard I. is ice free than in 1955, so a greater area is suitable for breeding by Gentoo Penguins (and other species) but inter-annual variation can be high. However, real declines in the area of occupancy may have occurred in the Atlas Cove region between the 1980s and 2000–01, and at the western end of Heard I. more generally

Current eligibility against IUCN Red List Criteria

IUCN category	Criteria eligibility
Australian breeding population	
A	Near Threatened: past, current and anticipated population declines of 20–29% over periods 3 generations (21 years) suspected from monitoring the global population and deteriorating habitat quality
B	Not applicable: at ≤5 locations but no plausible future threat to locations, population fluctuations not extreme
C	Not applicable: population >10 000 mature individuals
D	Not applicable: ≤5 locations but no plausible future threat
E	Not applicable: no population viability analysis undertaken
Population visiting Australian territory	
A	Near Threatened: past, current and anticipated population declines of 20–29% over periods 3 generations (21 years) inferred from monitoring, deteriorating habitat quality and levels exploitation
B	Not applicable: >10 locations, not severely fragmented, population fluctuations not extreme
C	Not applicable: population >10 000 mature individuals
D	Not applicable: population >1000 mature individuals, >5 locations
E	Not applicable: no population viability analysis undertaken

IUCN Red List assessment data

	Estimate	Reliability
Australian breeding population		
Extent of occurrence trend	1 130 000 km^2 stable	high high
Area of occupancy trend	40 km^2 stable	medium medium
No. of mature individuals trend	40 000 decreasing	low low
No. subpopulations	2	high
No. locations	2	high
Largest subpopulation	30 000	low
Generation time	7.0 years	medium
Global population share	<1%	high
Level of genetic exchange	low	medium
Population visiting Australian territory		
Extent of occurrence trend	1 130 000 km^2 stable	high high
Area of occupancy trend	1 130 000 km^2 stable	high high
No. of mature individuals trend	386 000 decreasing	medium medium
No. subpopulations	>10	high
No. locations	>10	high
Largest subpopulation	170 000	medium
Global population share	8%	low

between the 1950s and 1980s, and between the 1980s and 2003–2004 (Woehler 2006). On Macquarie I. population estimates vary from 4700 pairs in 1984 (Robertson 1986), 6820 pairs in 1992–2003 (Ellis *et al.* 1998) and 3800 pairs in 2002 (Parks and Wildlife Service 2006). Elsewhere there have been substantial declines of this subspecies at Falkland, Bird I., South Georgia, Marion and Kerguelen Is. and, while *P. p. ellsworthii* has increased, there have been no confirmed increases of *P. p. papua* (Lescroël and Bost 2006).

Ecology

Nest are built on a wide range of vegetated and non-vegetated habitats in ice-free areas, breeding site possibly being determined by access to shallow near-shore feeding grounds. They feed principally on fish and some euphausiids (BirdLife International 2011). Although they appear to be more flexible in both choice of food and feeding behaviours than congeners (Miller *et al.* 2009), subantarctic populations may be at the limit of their physical capabilities for gathering food and raising young as they lay in winter, have a lower provisioning rate, lower growth rate, a longer rearing period and lower breeding success than southern populations (Lescroël *et al.* 2009). A generation time of 7.0 years (BirdLife International 2011) is derived from an average age at first breeding of 2.0 years and an annual survival of adults of 80.0% (Schreiber and Burger 2002).

Threats

There are no major land-based threats to Gentoo Penguins on either Heard or Macquarie Is. While cats *Felis catus* may have eaten some chicks in the past and Black Rats *Rattus rattus* may occasionally take eggs, these are unlikely to affect the population as a whole. Tourists may have adversely affected some breeding success in the past but are now controlled (Parks and Wildlife Service 2006). No reasons for the declines occurring elsewhere have been proven. The effect of climate change on food supply is plausible (Ellis *et al.* 1998), however, at other sites competition with commercial fishing may also have had a direct effect before it was controlled (Ainley and Blight 2009).

Conservation objectives

1. A stable or increasing population over a period of decades

Information required

1. Rates and direction of change in sea temperature and other oceanographic variables, and their effect on prey distributions, in the southern Indian Ocean
2. Population trends or changes in adult and juvenile survival
3. Ecology and movements when not breeding

Management actions required

1. Maintain quarantine status of Heard I.
2. Update the population estimates for Heard and Macquarie Is.
3. Monitor rates and effects of marine debris ingestion
4. Establish demographic parameters, particularly survival of different age classes
5. Study the potential impacts of climate change
6. Eradicate rats from Macquarie I.

Bibliography

Ainley DG, Blight LK (2009) Ecological repercussions of historical fish extraction from the Southern Ocean. *Fish and Fisheries* **10**, 13–38.

BirdLife International (2011) 'Species factsheet: *Pygoscelis papua*'. Retrieved 17 January, 2011 from <http://www.birdlife.org/>.

Downes MC, Ealey EHM, Gwynne AM, Young PS (1959) 'The birds of Heard Island'. Australian National Antarctic Research Reports (Series B).

Ellis S, Croxall JP, Cooper J (Eds) (1998) 'Penguin conservation assessment and management plan'. IUCN/SSC Conservation Breeding Specialist Group, Apple Valley.

Lescroël A, Bajzak C, Bost C-A (2009) Recent decrease in gentoo penguin populations at Iles Kerguelen. *Polar Biology* **32**, 1495–1505.

Lescroël A, Bost C-A (2006) Recent decrease in gentoo penguin populations at Iles Kerguelen. *Antarctic Science* **18**, 171–174.

Miller AK, Karnovsky NJ, Trivelpiece WZ (2009) Flexible foraging strategies of gentoo penguins *Pygoscelis papua* over 5 years in the South Shetland Islands, Antarctica. *Marine Biology* **156**, 2527–2537.

Parks and Wildlife Service (2006) 'Macquarie Island Nature Reserve and World Heritage Area Management Plan'. Parks and Wildlife Service, Department of Tourism, Arts and the Environment, Hobart.

Robertson G (1986) Population size and breeding success of the Gentoo Penguin, *Pygoscelis papua*, at Macquarie Island. *Australian Wildlife Research* **13**, 583–587.

Schreiber EA, Burger J (2002) *Biology of Marine Birds*. CRC Press, Boca Raton, FL.

Woehler EJ (1991) The status and conservation of seabirds of Heard Island and the McDonald Islands. In 'Seabird Status and Conservation. ICBP Tech. Publ. No 11.' (Ed. JP Croxall) pp. 263–277. ICBP, Cambridge, UK.

Woehler EJ (1993) *The Distribution and Abundance of Antarctic and Subantarctic Penguins*. Scientific Committee on Antarctic Research, Cambridge, UK.

Woehler EJ (2006) Status and conservation of the seabirds of Heard Island and the McDonald Islands. In *Heard Island: Southern Ocean Sentinel*. (Eds K Green and EJ Woehler) pp. 128–165. Surrey Beatty & Sons, Chipping Norton, NSW.

Comments received from

Rosemary Gales, Eric Woehler

Subspecies Conservation Summary

Southern Rockhopper Penguin (eastern)

Eudyptes chrysocome filholi Hutton, 1879

Spheniscidae

Conservation status

Australian breeding population: Vulnerable A2bc+3bc+4bc

Population visiting Australian territory: Vulnerable A2bc+3bc+4bc

Reasons for listing

Australian breeding population: Australian population likely to be declining along with the global population: global declines of >30% within 3 generations (34 years) have occurred and are thought likely to continue, it is assumed because of changes in habitat

Population visiting Australian territory: Most monitored populations have declined by 30–49% over the last 3 generations (34 years) and the declines appear to be continuing

Status 2000

Australian breeding population: Vulnerable A2bc+3bc+4bc

Population visiting Australian territory: Vulnerable A2bc+3bc+4bc

Status 1990

Australian breeding population: Vulnerable A2bc+3bc+4bc

Population visiting Australian territory: Vulnerable A2bc+3bc+4bc

Taxonomy

E. c. chrysocome (Falkland Is and off Cape Horn) is extralimital. Species status is Vulnerable. Occasionally interbreeds with Macaroni Penguins *E. chrysolophus* (Woehler and Gilbert 1990).

Taxonomic uniqueness: high (6 genera/family, 7 species/genus, 2 subspecies/species

Current eligibility against IUCN Red List Criteria

IUCN category	Criteria eligibility
Australian breeding population	
A	Vulnerable: past, current and anticipated population declines of 30–49% over periods of 3 generations (34 years) suspected from monitoring the global population and deteriorating habitat quality
B	Not applicable: at ≤5 locations but no plausible future threat to locations, population fluctuations not extreme
C	Not applicable: population >10 000 mature individuals
D	Not applicable: ≤5 locations but no plausible future threat
E	Not applicable: no population viability analysis undertaken
Population visiting Australian territory	
A	Vulnerable: past, current and anticipated population declines of 30–49% over periods of 3 generations (34 years) inferred from monitoring and deteriorating habitat quality
B	Not applicable: >10 locations, not severely fragmented, population fluctuations not extreme
C	Not applicable: population >10 000 mature individuals
D	Not applicable: population >1000 mature individuals, >5 locations
E	Not applicable: no population viability analysis undertaken

IUCN Red List assessment data

	Estimate	Reliability
Australian breeding population		
Extent of occurrence trend	1 130 000 km^2 stable	high high
Area of occupancy trend	20 km^2 stable	medium medium
No. of mature individuals trend	170 000 decreasing	low low
No. subpopulations	2	high
No. locations	2	high
Largest subpopulation	150 000	low
Generation time	11.4 years	medium
Global population share	11–25 %	high
Level of genetic exchange	low	medium
Population visiting Australian territory		
Extent of occurrence trend	1 130 000 km^2 stable	high high
Area of occupancy trend	1 130 000 km^2 stable	high high
No. of mature individuals trend	1 200 000 decreasing	medium medium
No. subpopulations	9	high
No. locations	>10	high
Largest subpopulation	530 000	medium
Global population share	<1%	low

Range

In Australian territory breeding occurs on Macquarie, Bishop and Clerk, Heard and McDonald Is. (Woehler 1993). Extralimitally breeds on Prince Edward, Marion, Crozet, Kerguelen, Campbell, Auckland and Antipodes Is. (Woehler 1993; Jouventin *et al.* 2006; U.S. Fish and Wildlife Service 2008; BirdLife International 2011). In winter and spring, when not breeding, presumed to move widely across the Southern Ocean, being regular visitors to Tasmania (Woehler 1993).

Abundance

In 1990 population was estimated at 23 colonies and 100 000–300 000 pairs on Macquarie I., 2 colonies and 20 pairs on Bishop and Clerk Is., 12 colonies and 10 000 pairs on Heard I., and 4 colonies and at least 10 pairs on McDonald I. (Woehler 1991, 1993; Ellis *et al.* 1998; Woehler 2006). Numbers on Macquarie I. were estimated as 32 000–43 000 pairs in 2006–2007 (BirdLife International 2011). There are no recent data from Bishop and Clerk, Heard or McDonald Is. (Woehler 2006). It is not known whether the big difference between estimates on Macquarie I. in the 1990s and in 2006–2007 are from a change in survey methods or real declines but, given the difference between the minimum estimate from Woehler (1993) and the maximum quoted by BirdLife International (2011), and the declines on Campbell I., a decline on Macquarie I. is assumed. Extralimitally there have been decreases recorded on Marion I. (Crawford *et al.* 2003; Crawford *et al.* 2008), Campbell I., Antipodes I. and Auckland Is. (Ellis *et al.* 1998) and of *E. c. chrysocome* on the Falkland Is. (Ellis *et al.* 1998). Differences in counts for Kerguelen and Crozet Is. between Jouventin *et al.* (Jouventin *et al.* 2006) and Woehler (1993), which were used to dismiss claims of a decline in the Indian Ocean (U.S. Fish and Wildlife Service 2008), remain unexplained.

Ecology

Breeds on subantarctic islands in colonies from sea-level to cliff-tops, and sometimes inland, with nests being built among rocks and tussock grass. Major food items are krill and other crustaceans, squid, octopus and fish which, when breeding, penguins obtain near the nesting island (Hull 1999b, 1999a; BirdLife International 2011). A generation time of 11.5 years (BirdLife International 2011) is derived from an age at first breeding of 5.4 years and an annual survival of adults of 83.7%, both extrapolated from congeners.

Threats

Although Black Rats *Rattus rattus* take eggs and young on Macquarie I., it is not known if this is at significant levels. Some were also taken by feral cats *Felis catus* before these were eradicated. Diseases brought by human visitors is another potential terrestrial threat. The largest population decline on Campbell I. occurred from 1946–1956 and was associated with a rise in sea surface temperature causing movement of food away from shore (Cunningham and Moors 1994). Other declines are not understood but also appear to be associated with food availability since the weights of penguins returning to breed on Marion I. was associated with breeding success, with both weights and success have been declining (Crawford *et al.* 2008). Fishing is currently unlikely to affect the birds on Macquarie or Heard Is. given that both are protected with controlled fisheries within the Exclusive Economic Zone, but they could have been affected by fishing historically, from which fish stocks may not have recovered (Ainley and Blight 2009). The extent of the threat from ingested plastic debris (Ellis *et al.* 1998) remains unknown.

Conservation objectives

1. A stable or increasing population over a period of decades

Information required

1. Rates and direction of change in sea temperature and other oceanographic variables, and their effect on prey distributions, in the southern Indian Ocean
2. Population trends or changes in adult and juvenile survival
3. Ecology and movements when not breeding

Management actions required

1. Maintain quarantine status of Heard I.
2. Update the population estimates for Heard and Macquarie Is.
3. Monitor rates and effects of marine debris ingestion
4. Establish demographic parameters, particularly survival of different age classes
5. Study the potential impacts of climate change
6. Eradicate rats from Macquarie I.

Bibliography

Ainley DG, Blight LK (2009) Ecological repercussions of historical fish extraction from the Southern Ocean. *Fish and Fisheries* **10**, 13–38.

BirdLife International (2011) 'Species factsheet: *Eudyptes chrysocome*'. Retrieved 17 January, 2011 from <http://www.birdlife.org/>.

Crawford RJM, Cooper J, Dyer BM, Greyling MD, Klages NTW, Nel DC, Peterson SL, Wolfaardt AC, (2003) Decrease in numbers of the eastern Rockhopper penguin *Eudyptes chrysocome filholi* at Marion Island, 1994/95–2002/03. *African Journal of Marine Science* **25**, 487–498.

Crawford RJM, Makhado AB, Upfold L, Dyer BM (2008) Mass on arrival of rockhopper penguins at Marion Island correlated with breeding success. *African Journal of Marine Science* **30**, 185–188.

Cunningham DM, Moors PJ (1994) The decline of Rockhopper penguins *Eudyptes chrysocome* at Campbell Island, Southern Ocean and the influence of rising sea temperatures. *Emu* **94**, 27–36.

Ellis S, Croxall JP, Cooper J (Eds) (1998) 'Penguin Conservation Assessment and Management Plan'. IUCN/SSC Conservation Breeding Specialist Group, Apple Valley.

Hull CL (1999a) Comparison of the diets of breeding Royal (*Eudyptes schlegeli*) and Rockhopper (*Eudyptes chrysocome*) Penguins on Macquarie Island over three years. *Journal of Zoology* **247**, 507–530.

Hull CL (1999b) The foraging zones of breeding royal (*Eudyptes schlegeli*) and rockhopper (*E. chrysocome*) penguins: an assessment of techniques and species comparison. *Wildlife Research* **26**, 789–803.

Jouventin P, Cuthbert RJ, Ottvall R (2006) Genetic isolation and divergence in sexual traits: evidence for the Northern Rockhopper Penguin *Eudyptes moseleyi* being a sibling species. *Molecular Ecology* **15**, 3413–3423.

U.S. Fish and Wildlife Service (2008) 'Endangered and threatened wildlife and plants; 12-month finding on a petition to list four penguin species as Threatened or Endangered under the Endangered Species Act and proposed rule to list the Southern Rockhopper Penguin in the Campbell Plateau Portion of its range. Federal Register'. Retrieved 17 February, 2011 from <http://www.federalregister.gov/articles/2008/12/18/E8-29673/endangered-and-threatened-wildlife-and-plants-12-month-finding-on-a-petition-to-list-four-penguin#p-39>.

Woehler EJ (1991) The status and conservation of seabirds of Heard Island and the McDonald Islands. In 'Seabird Status and Conservation. ICBP Tech. Publ. No 11.' (Ed. JP Croxall) pp. 263–277. ICBP, Cambridge, UK.

Woehler EJ (1993) *The Distribution and Abundance of Antarctic and Subantarctic Penguins.* Scientific Committee on Antarctic Research, Cambridge, UK.

Woehler EJ (2006) Status and conservation of the seabirds of Heard Island and the McDonald Islands. In *Heard Island: Southern Ocean Sentinel.* (Eds K Green and EJ Woehler) pp. 128–165. Surrey Beatty & Sons, Chipping Norton, NSW.

Woehler EJ, Gilbert CA (1990) Hybrid Rockhopper-Macaroni Penguins, interbreeding and mixed species pairs at Heard and Marion Islands. *Emu* **90**, 198–210.

Comments received from

Rosemary Gales, Eric Woehler

Species Conservation Summary

Macaroni Penguin

Eudyptes chrysolophus (Brandt, 1837)
Spheniscidae

Conservation status

Australian breeding population: Near Threatened A2bc+3bc+4bc
Population visiting Australian territory: Near Threatened A2bc+3bc+4bc

Reasons for listing

Australian breeding population: Australian population assumed to be following global trends; occurs at only 2 Australian locations but no plausible future threat to that location
Population visiting Australian territory: Declines approaching 30% within 3 generations (34 years) thought to have occurred and likely to be continuing

Status 2000

Australian breeding population: Near Threatened A2bc+3bc+4bc
Population visiting Australian territory: Near Threatened A2bc+3bc+4bc

Status 1990

Australian breeding population: Least Concern
Reason for change in 2000: Australian population not monitored but assumed to follow global trend
Population visiting Australian territory: Least Concern
Reasons for change in 2000: declines recorded at major breeding locations in South Atlantic Ocean

Taxonomy

No infraspecific taxa described; sometimes thought to include Royal Penguin *E. schlegeli* of Macquarie I. Occasionally interbreeds with Rockhopper Penguins *E. chrysocome* (Woehler and Gilbert 1990).
Taxonomic uniqueness: high (6 genera/family, 7 species/genus, 1 subspecies/species)

Range

Australian populations breed on Heard and McDonald Is.; extralimitially they breed on the Antarctic Peninsula, southern Chile, South Georgia, and the Falkland Is., South Sandwich, South Orkney, South Shetland, Bouvetøya, Prince Edward, Marion, the Crozet Islands and the Kerguelen Is. When breeding they feed usually feed within 40 km of breeding sites (Trathan *et al.* 1998;, but see Deagle *et al.* (2008) for Heard I. study). Non-breeding birds from Kerguelen I. foraged during the non-breeding season in a narrow latitudinal band (47–49°S) within the central Indian Ocean (70–110°E), corresponding oceanographically to the Polar Frontal Zone (Bost *et al.* 2009). Heard I. and McDonald I. birds may do the same.

Abundance

About 1 million pairs were estimated to breed on each of Heard and McDonald Is. (Woehler 1991, 1993, 2006). Satellite imagery suggested that, in 2005, there were no birds breeding on McDonald I. following extensive volcanic activity (Woehler 2006; Department of Sustainability, Environment, Water, Population and Communities 2011), but there have been no surveys, the island area is now larger and beaches appear to be more extensive. In 2002, after which the island has changed little in shape, what had been Flat I. was deserted but high numbers of birds were observed ashore on the north-eastern slopes and summit of Maxwell Hill and on terraces on the debris cliffs of the east coast (Stephenson *et al.* 2005); whether these are breeding colonies not known. The area occupied on Heard I. appears to have decreased in the Atlas Cove region of Heard I. between the 1980s and 2000–2001 but it is unclear whether these changes are real or simply gradual shifts in the areas occupied (Woehler 2006). There have been substantial decreases in the numbers of birds breeding on islands in the South Atlantic Ocean and in Chile including at Bird I., South Georgia, Marion I. and Bouvet I., but the populations at Kerguelen are stable or increasing (BirdLife International 2011) and overall decreases have been questioned (U.S. Fish and Wildlife Service 2008).

Ecology

Nests in dense colonies on level to steep ground, often walking hundreds of metres across steep landscapes to nest-sites. Breeding areas usually have little or no vegetation due to erosion by birds. Individuals feed primarily on crustaceans but will also take small fish and squid, diving to over 100 m deep (Green *et al.* 1998). For Heard I. fish are, at least sometimes, particularly important, contributing 46% by weight to the diet (Deagle *et al.* 2008). Unlike in the South Atlantic Ocean, non-breeding birds from Kerguelen Is. fed on crustaceans other than krill during the non-breeding season (Bost *et al.* 2009). A generation time of 11.4 years (BirdLife International 2011) is derived from an age at first breeding of 6.0 years and an annual survival of adults of 81.5% (Schreiber and Burger 2002).

Threats

Recent volcanic eruptions at McDonald I. (1993–2005) have destroyed some breeding habitat, but may not have

Current eligibility against IUCN Red List Criteria

IUCN category	Criteria eligibility
Australian breeding population	
A	Near Threatened: past, current and anticipated population declines of 20–29% over periods of 3 generations (34 years) suspected from monitoring the global population and deteriorating habitat quality
B	Not applicable: at ≤5 locations but no plausible future threat to locations, population fluctuations not extreme
C	Not applicable: population >10 000 mature individuals
D	Not applicable: ≤5 locations but no plausible future threat
E	Not applicable: no population viability analysis undertaken
Population visiting Australian territory	
A	Near Threatened: past, current and anticipated population declines of 20–29% over periods of 3 generations (34 years) inferred from monitoring and deteriorating habitat quality
B	Not applicable: >10 locations, not severely fragmented, population fluctuations not extreme
C	Not applicable: population >10 000 mature individuals
D	Not applicable: population >1000 mature individuals, >5 locations
E	Not applicable: no population viability analysis undertaken

IUCN Red List assessment data

	Estimate	Reliability
Australian breeding population		
Extent of occurrence trend	520 000 km^2 stable	high high
Area of occupancy trend	20 km^2 stable	medium medium
No. of mature individuals trend	2 000 000 decreasing	low low
No. subpopulations	1	high
No. locations	2	high
Generation time	11.4 years	medium
Global population share	11–25%	high
Level of genetic exchange	low	medium
Population visiting Australian territory		
Extent of occurrence trend	520 000 km^2 stable	high high
Area of occupancy trend	520 000 km^2 stable	high high
No. of mature individuals trend	18 000 000 decreasing	low medium
No. subpopulations	>10	high
No. locations	>10	high
Largest subpopulation	5 000 000	low
Global population share	<1%	low

resulted in loss of birds. While some areas were deserted in 2002, others may have been re-colonised. Eruptions on Heard I. (1985–2008) are unlikely to have affected breeding sites. Otherwise there are no plausible threats to the breeding populations. At sea there is concern that climatic shifts in prey availability are causing the declines seen around South America and the Antarctic Peninsula (Cresswell *et al.* 2008). Marine pollution and fishing may pose threats but remain unconfirmed (Ellis *et al.* 1998), but overfishing in particular is an alternative explanation for declines in the species, particularly in areas where the species relies on fish (Ainley and Blight 2009).

Conservation objectives

1. Stable or increasing population

Information required

1. Population trends or changes in adult and juvenile survival
2. Ecology and movements when not breeding
3. Rates and direction of change in sea temperature and other oceanographic variables, and their effect on prey distributions, in the southern Indian Ocean

Management actions required

1. Update and increase precision of the population estimate.
2. Maintain quarantine status of Heard and McDonald Is.

Bibliography

Ainley DG, Blight LK (2009) Ecological repercussions of historical fish extraction from the Southern Ocean. *Fish and Fisheries* **10**, 13–38.

BirdLife International (2011) 'Species factsheet: *Eudyptes chrysolophus*'. Retrieved 17 January, 2011 from <http://www.birdlife.org/>.

Bost CQ, Thiebot JB, Pinaud D, Cherel Y, Trathan PN (2009) Where do penguins go during the inter-breeding period? Using geolocation to track the winter dispersion of the macaroni penguin. *Biology Letters* **5**, 473–476.

Cresswell KA, Wiedenmann J, Mangel M (2008) Can macaroni penguins keep up with climate- and fishing-induced changes in krill? *Polar Biology* **31**, 641–649.

Deagle BE, Gales NJ, Hindell MA (2008) Variability in foraging behaviour of chick-rearing macaroni penguins *Eudyptes chrysolophus* and its relation to diet. *Marine Ecology Progress Series* **359**, 295–309.

Department of Sustainability, Environment, Water, Population and Communities (2011) *Eudyptes chrysolophus sensu lato.* In 'Species Profile and Threats Database'. Department of Sustainability, Environment, Water, Population and Communities, Canberra. Retrieved 24 February, 2011 from <http://www.environment.gov.au/sprat>.

Ellis S, Croxall JP, Cooper J (Eds) (1998) 'Penguin Conservation Assessment and Management Plan'. IUCN/SSC Conservation Breeding Specialist Group, Apple Valley.

Green K, Williams R, Green MG (1998) Foraging ecology and diving behaviour of Macaroni Penguins *Eudyptes chrysolophus* at Heard Island. *Marine Ornithology* **26**, 27–34.

Schreiber EA, Burger J (2002) *Biology of Marine Birds.* CRC Press, Boca Raton, FL.

Stephenson J, Budd GM, Manning J, Hansbro P (2005) Major eruption-induced changes to the McDonald Islands, southern Indian Ocean. *Antarctic Science* **17**, 259–266.

Trathan PN, Croxall JP, Murphy EJ, Everson I (1998) Use of at-sea distribution data to derive potential foraging ranges of macaroni penguins during the breeding season. *Marine Ecology Progress Series* **169**, 263–275.

U.S. Fish and Wildlife Service (2008) 'Endangered and threatened wildlife and plants; 12-month finding on a petition to list four penguin species as Threatened or Endangered under the Endangered Species Act and proposed rule to list the Southern Rockhopper Penguin in the Campbell Plateau Portion of its range. Federal Register'. Retrieved 17 February, 2011 from <http://www.federalregister.gov/articles/2008/12/18/E8-29673/endangered-and-threatened-wildlife-and-plants-12-month-finding-on-a-petition-to-list-four-penguin#p-39>.

Woehler EJ (1991) The status and conservation of seabirds of Heard Island and the McDonald Islands. In 'Seabird Status and Conservation. ICBP Tech. Publ. No 11.' (Ed. JP Croxall) pp. 263–277. ICBP, Cambridge, UK.

Woehler EJ (1993) *The Distribution and Abundance of Antarctic and Subantarctic Penguins*. Scientific Committee on Antarctic Research, Cambridge, UK.

Woehler EJ (2006) Status and conservation of the seabirds of Heard Island and the McDonald Islands. In *Heard Island: Southern Ocean Sentinel*. (Eds K Green and EJ Woehler) pp. 128–165. Surrey Beatty & Sons, Chipping Norton, NSW.

Woehler EJ, Gilbert CA (1990) Hybrid Rockhopper-Macaroni Penguins, interbreeding and mixed species pairs at Heard and Marion Islands. *Emu* **90**, 198–210.

Comments received from

Eric Woehler

Species Conservation Summary

Royal Penguin

Eudyptes schlegeli (Finsch, 1876)
Spheniscidae

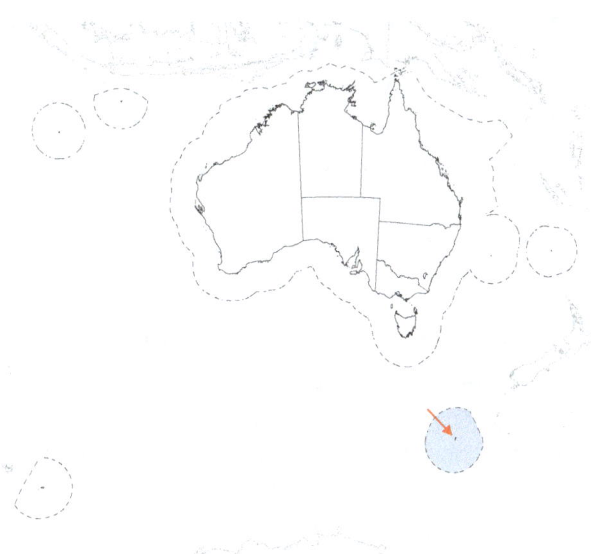

Conservation status

Near Threatened D2

Reasons for listing

Present only at only 1 location with a plausible but low probability threat from marine pollution

Status 2000

Near Threatened D2

Status 1990

Near Threatened D2

Taxonomy

No infraspcific taxa described; sometimes combined with Macaroni Penguin *E. chrysolophus*
Taxonomic uniqueness: high (6 genera/family, 7 species/genus, 1 subspecies/species)

Range

Confined to Macquarie and nearby Bishop and Clerk Is., travelling to the polar frontal zone to feed (Hull 1999a). Non-breeding and juvenile birds are assumed to stay near their breeding islands but small numbers of similar looking birds are sometimes seen at other subantarctic islands (Marchant and Higgins 1990) and in Tasmania (Woehler 1992).

Abundance

Populations are not monitored but about 850 000 pairs were estimated to nest on Macquarie I. in 1984–1985 and over 1000 pairs on Bishop and Clerk Is. (Marchant and Higgins 1990) but these estimates are now dated and considered unreliable. Populations are assumed to be stable but with no data to support this assertion. Assumed to have fully recovered after heavy harvesting in the 19th century (Rounsevell and Brothers 1984).

Ecology

Nests in large colonies on bare, level, pebbly, rocky or sandy ground (Hull and Wilson 1996). Breeding birds eat euphausiids, fish and squid (Hindell 1988; Hull 1999b) but their ecology during the winter, when birds are away from the islands, is unknown (Marchant and Higgins 1990). A generation time of 14.1 years (BirdLife International 2011) is derived from an age at first breeding of 7.0 years and an annual survival of adults of 86.0% (Schreiber and Burger 2002).

Threats

On Macquarie I. no significant threats though introduced Black Rats *Rattus rattus* eat some eggs and young, and tourists and researchers sometimes cause disturbance at colonies (Garnett and Crowley 2000). The species was heavily hunted for oil in the 19th and early 20th century (Rounsevell and Brothers 1984). At sea marine pollution,

Current eligibility against IUCN Red List Criteria

IUCN category	Criteria eligibility
A	Not applicable: past, current or future population declines are thought unlikely to exceed 20% in any 3-generation period
B	Not applicable: EOO, AOO, habitat, locations or population not declining, population fluctuations not extreme
C	Not applicable: no continuing population decline or extreme fluctuations
D	Near Threatened: ≤5 locations with a plausible but low probability threat
E	Not applicable: no population viability analysis undertaken

IUCN Red List assessment data

	Estimate	Reliability
Extent of occurrence trend	600 000 km^2 stable	high high
Area of occupancy trend	130 km^2 stable	high low
No. of mature individuals trend	1 700 000 stable	low low
No. subpopulations	1	high
No. locations	2	medium
Generation time	14.1 years	medium
Global population share	100%	high

particularly ingested plastics, kills some birds and fishing may adversely affect the species but the significance of neither threat is known. Marine pollution and fishing may pose threats but remain unconfirmed (Ellis *et al.* 1998), with historical overfishing sometimes having long-lasting effects (Ainley and Blight 2009). Although occurring at only 1 location, the probability of a plausible threat is low so listed as Near Threatened.

Conservation objectives

1. Stable or increasing population

Information required

1. Population trends or changes in adult and juvenile survival
2. Ecology and movements when not breeding

Management actions required

1. Update the population estimate
2. Monitor rates and effects of marine debris ingestion
3. Monitor the effects of fishing
4. Establish demographic parameters, particularly survival of different age classes
5. Study the potential impacts of climate change
6. Eradicate rats from Macquarie I.

Bibliography

Ainley DG, Blight LK (2009) Ecological repercussions of historical fish extraction from the Southern Ocean. *Fish and Fisheries* **10**, 13–38.

BirdLife International (2011) 'Species factsheet: *Eudyptes schlegeli*'. Retrieved 26 January, 2011 from <http://www.birdlife.org>.

Ellis S, Croxall JP and Cooper J (Eds) (1998) 'Penguin Conservation Assessment and Management Plan'. IUCN/SSC Conservation Breeding Specialist Group, Apple Valley.

Garnett ST and Crowley GM (2000) *Action Plan for Australian Birds 2000*. Environment Australia, Canberra.

Hindell MA (1988) The diet of the Royal Penguin *Eudyptes schlegeli* at Macquarie Island. *Emu* **88**, 219–226.

Hull CL (1999a) The foraging zones of breeding Royal (*Eudyptes schlegeli*) and Rockhopper (*Eudyptes chrysocome*) Penguins: an assessment of techniques and species comparison. *Wildlife Research* **26**, 789–803.

Hull CL (1999b) Comparison of the diets of breeding Royal (*Eudyptes schlegeli*) and Rockhopper (*Eudyptes chrysocome*) Penguins on Macquarie Island over three years. *Journal of Zoology* **247**, 507–530.

Hull CL and Wilson J (1996) Location of colonies of Royal Penguins *Eudyptes schlegeli*: potential costs and consequences for breeding success. *Emu* **96**, 135–143.

Marchant S and Higgins PJ (Eds) (1990) *The Handbook of Australian, New Zealand and Antarctic Birds*. Oxford University Press, Melbourne.

Rounsevell DE and Brothers NP (1984) The status of seabirds on Macquarie Island. In 'Status and Conservation of the World's Seabirds. ICBP Tech. Publ. 2'. (Eds JP Croxall, PGH Evans, RW Schreiber) pp. 587–592. ICBP, Cambridge, UK.

Schreiber EA and Burger J (2002) *Biology of Marine Birds*. CRC Press, Boca Raton, FL.

Woehler EJ (1992) Records of vagrant penguins from Tasmania. *Marine Ornithology* **20**, 61–73.

Comments received from

Barry Baker, Rosemary Gales

Species Conservation Summary

Christmas Island Frigatebird

Fregata andrewsi Mathews, 1914
Fregatidae

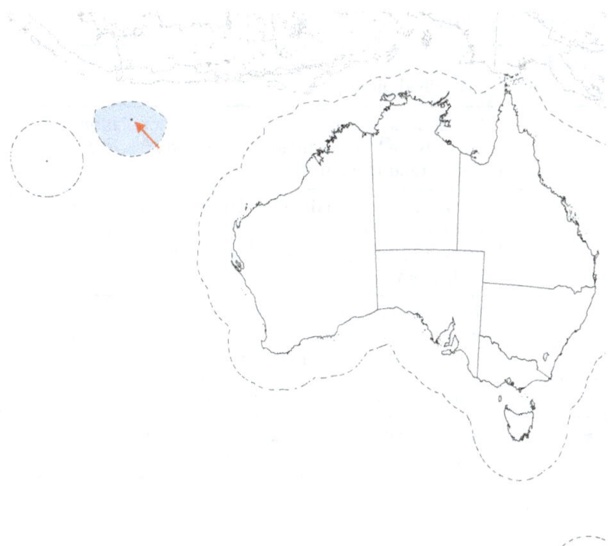

Conservation status
Critically Endangered B2ab(ii,iii,v)

Reasons for listing
Breeding colonies occupy 1.7 km^2 at 1 location, and there is a continuing decline likely in area of occupancy, area and quality of habitat, and number of mature individuals as a result of storms and/or weed invasion

Status 2000
Critically Endangered B2ab(ii,iii,v)

Status 1990
Critically Endangered B2ab(ii,iii,v)

Taxonomy
No infraspecific taxa described
Taxonomic uniqueness: high (1 genus/family, 5 species/genus, 1 subspecies/species)

Range
Endemic to Christmas I. where it breeds in 3 small colonies with a total area of 1.7 km^2. Breeding birds frequently forage hundreds or even thousands of kilometres from the colony over the Indian Ocean and throughout the Indo-Malay Archipelago. When not breeding, the species ranges widely across the seas of South-East Asia to Indochina and south to northern Australia, but its status in the Indian Ocean to the west is less well known (Hill and Dunn 2004).

Abundance
The pre-settlement population was thought to have been about 6300 breeding pairs each year with the population declining by around 66% over the last 3 generations (James 2003), apparently owing to habitat clearance and dust fallout from phosphate mining, marine pollution, over-fishing and bycatch in fishing gear. Total population was estimated in 1985 as 1600 pairs, with 850 nest at Golf Course, 370 nests at Cemetery, and 100 nest at Dryers (Stokes 1988). In 2003 there were 4 subcolonies (since reduced to 3) covering an area of c.49 ha: Flying Fish Cove (c.10 ha; 2 nests), Dryers (c.62 ha; 20 nests); Golf Course (c.25 ha; 828±42 nests) and Cemetery (46 ha; 321±15 nests; James 2003). This is a reduction of 3–16% from 1985. Currently it is estimated that about 1200 pairs breed each year (Beeton *et al.* 2010), making the population 4800 mature individuals allowing for biennial breeding.

Ecology
Nests in tall forest trees, particularly Indian Almond *Terminalia catappa* and Stinkwood *Celtis timorensis*, that are sheltered from the prevailing south-eastern trade winds. Raises 1 fledgling every 2 years. The species forages over relatively warm, low salinity waters, taking flying fish, squid and other marine creatures, and stealing food from other birds on the wing (BirdLife International 2011). A generation time of 15.5 years (BirdLife International 2011) is derived from an age at first breeding of 5.3 years and a maximum longevity in the wild of 27.6 years (Schreiber and Burger 2002).

Threats
Most nests are now located in a single colony and thus the species is more vulnerable to cyclones than when there were several colonies (Hill and Dunn 2004) and weed invasion also a potential problem (Beeton *et al.* 2010). The number and size of colonies on Christmas I. have decreased because of human disturbance. About 90 ha of breeding colony were cleared before 1946 for phosphate mining and the major colony at Flying Fish Cove was virtually abandoned because of continuing dust fall-out from phosphate dryers. Less specific threats include over-fishing and marine pollution, plus clearance of vegetation and hunting on non-breeding roost islands. Approximately 10% of the population nests outside the national park and does not have any formal protection (BirdLife International 2011).

Conservation objectives
1. Stable or increasing population
2. Some former breeding habitat rehabilitated and reoccupied
3. Bird-safe fishing practiced in the Christmas I. Exclusive Economic Zone

Information required
1. Appropriate techniques to monitor the total breeding population size
2. Breeding biology and success
3. Impact of longline fishing

Management actions required
1. Upgrade quarantine standards for Christmas I.
2. Increase intensity and frequency of monitoring
3. Control weeds in and around colonies
4. Identify feeding habitat of breeding adults and juveniles

Current eligibility against IUCN Red List Criteria

IUCN category	Criteria eligibility
A	Not applicable: past, current or future population declines are thought unlikely to exceed 20% in any 3-generation period
B	Critically Endangered: AOO <10 km^2, 1 location with plausible future threat, inferred continuing decline in AOO, area and quality of habitat, no. mature individuals
C	Vulnerable: population 2500–10 000 mature individuals, inferred continuing decline, >95% mature individuals in 1 subpopulation
D	Vulnerable: ≤5 locations with plausible future threat
E	Not applicable: no population viability analysis undertaken

IUCN Red List assessment data

	Estimate	Reliability
Extent of occurrence trend	140 000 km^2 stable	high high
Area of occupancy trend	0.5 km^2 decreasing	high medium
No. of mature individuals trend	4800 decreasing	medium low
No. subpopulations	1	high
No. locations	1	high
Generation time	15.5 years	low
Global population share	100%	high

5. Monitor likely impact of developing fisheries around Christmas I.
6. Create a multi-stakeholder team with responsibility for ensuring threatened species are conserved on Christmas I.

Bibliography

Beeton B, Burbidge AA, Grigg G, Harrison P, How RA, Humphries B, McKenzie N, Woinarski J (2010) 'Final Report, Christmas Island Expert Working Group to Minister for the Environment, Heritage and the Arts'. Retrieved 1 April, 2010 from <http://www.environment.gov.au/parks/publications/christmas/pubs/final-report.pdf>.

BirdLife International (2011) 'Species factsheet: *Fregata andrewsi*'. Retrieved 17 January, 2011 from <http://www.birdlife.org/>.

Hill R, Dunn A (2004) 'National Recovery Plan for the Christmas Island Frigatebird *Fregata andrewsi*'. Commonwealth of Australia, Canberra.

James DJ (2003) 'A survey of Christmas Island frigatebird nests in 2003'. Unpublished report to Parks Australia. Tropical Ecology Consulting Services.

Schreiber EA, Burger J (2002) *Biology of Marine Birds*. CRC Press, Boca Raton, FL.

Stokes T (1988) 'A review of the birds of Christmas Island, Indian Ocean'. Australian National Parks and Wildlife Service Occasional Paper.

Comments received from

Andrew Burbidge, Janos Hennicke, Mike Smith, John Woinarski

Species Conservation Summary

Abbott's Booby

Papasula abbotti (Ridgway, 1893)
Sulidae

Conservation status

Endangered B2ab(ii,iii,v)

Reasons for listing

Occupies only 25 km^2 at a single location with a continuing decline likely in area of occupancy, area and quality of habitat, and number of mature individuals, largely from the ongoing effects of past clearing for mining

Status 2000

Critically Endangered A3ce
Reason for change in 2010: effective ant control applied

Status 1990

Endangered B2ab(ii,iii,v)
Reason for change in 2000: proliferation of invasive ants

Taxonomy

No infraspecific taxa described
Taxonomic uniqueness: high (3 genera/family, 1 species/genus, 1 subspecies/species)

Current eligibility against IUCN Red List Criteria

IUCN category	Criteria eligibility
A	Vulnerable: population decline of 30–49% over the last 3 generations (49 years) inferred from monitoring and deteriorating habitat quality
B	Endangered: AOO <500 km^2, at ≤5 locations with plausible future threat, inferred continuing decline in AOO, area and quality of habitat, no. mature individuals
C	Vulnerable: population 2500–10 000 mature individuals, suspected continuing decline, 100% in 1 subpopulation
D	Vulnerable: ≤5 locations with plausible future threat
E	Not applicable: no population viability analysis undertaken

IUCN Red List assessment data

	Estimate	Reliability
Extent of occurrence trend	140 000 km^2 stable	high high
Area of occupancy trend	25 km^2 decreasing	high medium
No. of mature individuals trend	7000 decreasing	low low
No. subpopulations	1	high
No. locations	1	high
Generation time	16.3 years	medium
Global population share	100%	medium

Range

Now breeds only on Christmas I. where it breeds on the central and western areas of plateau and in upper terrace forest of north coast (Yorkston and Green 1997). The distribution appears to have shifted since clearing of the plateau and most nests are between the cleared land, even if rehabilitated, and the coastal terraces (M. Smith, C. Boland, B. Tiernan, D. Maple and F. Napier *in litt.*). Until the early 19th century, the species was found on several islands in the Indian and Pacific Oceans, persisting on Chagas I. (BirdLife International 2011). Tracking studies suggest that they forage within 40–100 km of Christmas I. and show no clear association with any oceanographic features (BirdLife International 2011). The presence of an adult female in the Mariana Is. indicates either an extensive foraging range or another breeding colony nearby (Pratt *et al.* 2009).

Abundance

A partial survey found 2300 active breeding pairs in 1967 (Nelson 1971), with 2500 active pairs recorded in 1992 (Yorkston and Green 1997), of which 1900 pairs were within the area surveyed by Nelson in 1967. Population was estimated at 1500–2500 breeding pairs in 2002 (Beeton *et al.* 2010) and 3000–4000 breeding birds in 2005 (Olsen 2005). The population is estimated here as 7000 mature individuals, the midpoint of the most recent estimate, doubled because birds breed biennially.

Ecology

Abbott's Booby nests in tall, emergent rainforest trees, mostly *Planchonella nitida* and *Syzygium nervosum* (Yorkston and Green 1997). Most nest trees are associated with uneven terrain created by gullies, hill-sides or cliffs (Department of the Environment and Heritage 2004). Successful pairs nest once every 2 years whilst unsuccessful pairs may breed in successive years (Nelson and Powell 1986). The species feeds in warm, low salinity waters, taking squid and fish (Department of Sustainability, Environment, Water, Population and Communities 2011). A generation time of 16.3 years (BirdLife International 2011) is derived from an age at first breeding of 3.0 years and an annual survival of adults of 92.5%, both extrapolated from its largest relative, *Sula dactylatra*.

Threats

The principal reason for the decline is clearance of about a third of the former nesting habitat for phosphate mining from 1965 to 1987 with deleterious effects on the forest nearby through wind shear, and on remaining breeding boobies through disruptive competition for nest sites (Hicks and Powell 1983), leading to a shift in major nesting areas to some which may be of lower quality (M. Smith *et al. in litt.*). This may also increase sensitivity to extreme weather events: in 1988 winds associated with the tail of a cyclone destroyed a third of monitored fledglings and nest-sites (Reville *et al.* 1990). Like other Christmas I. species, the boobies are potentially threatened by the introduced Yellow Crazy Ant *Anoplolepis gracilipes*, which can alter the whole ecology of sites it occupies by killing the super-numerous Red Crab *Gecaroidea natalis*, as well as by tending scale insects that damage trees (O'Dowd *et al.* 2003; Abbott and Green 2007). Such changes may affect nesting habitat over the next 3 generations (49 years). An Immigration Reception and Processing Centre built adjacent to prime Abbott's Booby habitat, and in an area with a high priority for rehabilitation (Department of the Environment and Heritage 2004), could have a localised impact. The species could also be threatened in its marine habitat by overfishing and entanglement in longline fishing gear, as well as by marine pollution. Quarantine is insufficient to prevent invasion by other unspecified exotic pests.

Conservation objectives

1. A stable population
2. All breeding habitat protected
3. Yellow Crazy Ant populations constrained to levels that do not threaten subspecies
4. Bird-safe fishing practiced in the Christmas I. Exclusive Economic Zone

Information required

1. At-sea distribution
2. Refined techniques for controlling Yellow Crazy Ants

3. Improved techniques for monitoring population trends and breeding success
4. Feeding habitat of breeding adults
5. Impact of longline fishing

Management actions required

1. Upgrade quarantine standards for Christmas I.
2. Continue control of Yellow Crazy Ants
3. Continue population monitoring
4. Continue rehabilitation of cleared areas, and monitor the impact of rehabilitation efforts on areas affected by wind turbulence
5. Identify feeding habitat of breeding adults and juveniles
6. Monitor likely impact of developing fisheries around Christmas I.
7. Ensure activities associated with operation of Immigration Reception and Processing Centre do not impede recovery
8. Create a multi-stakeholder team with responsibility for ensuring threatened species are conserved on Christmas I.

Bibliography

Abbott KL, Green PT (2007) Collapse of an ant-scale mutualism in a rainforest on Christmas Island. *Oikos* **116**, 1238–1246.

Beeton B, Burbidge AA, Grigg G, Harrison P, How RA, Humphries B, McKenzie N, Woinarski J (2010) 'Final Report, Christmas Island Expert Working Group to Minister for the Environment, Heritage and the Arts'. Retrieved 1 April, 2010 from <http://www.environment.gov.au/parks/publications/christmas/pubs/final-report.pdf>.

BirdLife International (2011) 'Species factsheet: *Papasula abbotti*'. Retrieved 17 January, 2011 from <http://www.birdlife.org/>.

Department of Sustainability, Environment, Water, Population and Communities (2011) *Papasula abbotti*. In 'Species Profile and Threats Database'. Department of Sustainability, Environment, Water, Population and Communities, Canberra. Retrieved 24 February, 2011 from <http://www.environment.gov.au/sprat>.

Department of the Environment and Heritage (2004) 'National Recovery Plan for the Abbott's Booby *Papasula abbotti*'. Department of the Environment and Heritage, Canberra.

Hicks J, Powell D (1983) 'Forest margin survey on Abbott's Booby site protection'. Unpublished report, Australian National Parks and Wildlife Service, Canberra, and Phosphate Mining Company, Christmas Island.

Nelson JB (1971) The biology of Abbott's Booby. *Ibis* **113**, 429–467.

Nelson JB, Powell DA (1986) The breeding ecology of Abbott's Booby. *Emu* **86**, 33–46.

O'Dowd DJ, Green PT, Lake PS (2003) Invasional 'meltdown' on an oceanic island. *Ecology Letters* **6**, 812–817.

Olsen P (2005) 'National Recovery Plan for the Abbott's Booby *Papasula abbotti*'. Department of the Environment and Heritage, Canberra.

Pratt HD, Retter MLP, Chapman D, Ord WM, Pisano P (2009) An Abbott's Booby *Papasula abbotti* on Rota, Mariana Islands: first historical record for the Pacific Ocean. *Bulletin of the British Ornithological Club* **129**, 87–91.

Reville BJ, Tranter JD, Yorkston HD (1990) Impact of forest clearing on the endangered seabird *Sula abbotti*. *Biological Conservation* **51**, 23–38.

Yorkston HD, Green PT (1997) The breeding distribution and status of Abbott's Booby (Sulidae: *Papasula abbotti*) on Christmas Island, Indian Ocean. *Biological Conservation* **79**, 293–301.

Comments received from

Andrew Burbidge, Janos Hennicke, Mike Smith, John Woinarski

Subspecies Conservation Summary

Masked Booby (Tasman Sea)

Sula dactylatra tasmani O'Brien and Davies, 1990

Sulidae

Conservation status

Near Threatened D2

Reasons for listing

Present at only 2 locations with plausible but low probability future threats

Status 2000

Near Threatened D2

Current eligibility against IUCN Red List Criteria

IUCN category	Criteria eligibility
Australian breeding population	
A	Not applicable: past, current or future population declines are thought unlikely to exceed 20% in any 3-generation period
B	Not applicable: EOO, AOO, habitat, locations or population not declining, population fluctuations not extreme
C	Not applicable: no continuing population decline or extreme fluctuations
D	Near Threatened: ≤5 locations with a plausible but low probability threat
E	Not applicable: no population viability analysis undertaken

IUCN Red List assessment data

	Estimate	Reliability
Australian breeding population		
Extent of occurrence trend	1 400 000 km^2 stable	high high
Area of occupancy trend	10 km^2 stable	high high
No. of mature individuals trend	6200 stable	medium high
No. subpopulations	2	high
No. locations	2	high
Largest subpopulation	5000	medium
Generation time	16.3 years	medium
Global population share	90%	high
Level of genetic exchange	medium	high

Status 1990

Near Threatened D2

Infraspecific taxa

Previously, this taxon was considered to be *S. dactylatra fullagari* (Steeves *et al.* 2010). *S. d. personata* (now assumed to include *S. d. bedouti*) breeds on islands of the central and western Pacific and eastern Indian Ocean and is Least Concern. Two extralimital subspecies breed in the Caribbean, some Atlantic islands and in the western Indian Ocean. The species is Least Concern. Taxonomic uniqueness: medium (3 genera/family, 8 species/genus, 4 subspecies/species)

Range

This subspecies breeds on Lord Howe I. at Mutton Bird and King Points as well as Roach, South, Sugarloaf, Mutton Bird and Gower Is., Sail Rock and Ball's Pyramid (Priddel *et al.* 2005). It also breeds on Phillip and Nepean Is. off Norfolk I. as well as offshore stacks. Extralimitally occurs on Herald, Macauley, Hazard, Curtis and Cheeseman Is. in the Kermadec Is. (Veitch *et al.* 2004) but these birds are not assumed to come to Australian waters in significant numbers. Banded birds have been recovered primarily to the north including New Caledonia, Vanuatu and the Coral Sea as well as a few from Tonga and the mainland Australian coast (McAllan *et al.* 2004).

Abundance

Up to 2500 pairs breed in the Lord Howe I. group (D. Priddel *in litt.*) and another 600 pairs off Norfolk I. (Garnett and Crowley 2000; Priddel *et al.* 2010). Extralimitally up to 250 pairs breed in the Kermadec I. (Veitch *et al.* 2004). There is no data on trends but no suggestion of declines.

Ecology

Masked Boobies nest on the ground, on rocks, bare ground or a flattened area of grass. They feed on fish, principally flying fish and squid (Marchant and Higgins 1990). A generation time of 16.3 years (BirdLife International 2011) is derived from an age of first breeding of 3.0 years (Schreiber and Burger 2002) and an annual survival of adults of 92.5% extrapolated from congeners (BirdLife International 2011).

Threats

It is known that this species was eaten by Polynesians (Holdaway and Anderson 2001) and later by Europeans on Norfolk Island and was almost driven to extinction on Lord Howe I. by hunting for food (Marchant and Higgins 1990). Birds that attempted to breed on Norfolk I. in the 1980s failed after a few years when some birds were shot and the remainder disturbed by dogs and people (Director of National Parks 2010). Black Rats *Rattus rattus* are present on Lord Howe I., but do not seem to be a significant predator (Priddel *et al.* 2005). Bycatch in line fisheries is a plausible future threat (Priddel *et al.* 2005).

Conservation objectives

1. Ensure persistence of population

Information required

1. Extent of interaction with fisheries and mortality as bycatch

Management actions required

1. Assess extent of interaction with fisheries
2. Census subpopulation on Lord Howe I. group at least once every 5 years
3. Census subpopulation on Phillip, Nepean and Norfolk Is, including offshore stacks, every 5 years
4. Maintain quarantine on Phillip and Nepean Is.

Bibliography

BirdLife International (2011) 'Species factsheet: *Sula dactylatra*'. Retrieved 17 January, 2011 from <http://www.birdlife.org/>.

Director of National Parks (2010) 'Draft Norfolk Island Region Threatened Species Recovery Plan'. Department of the Environment, Water, Heritage and the Arts, Canberra.

Garnett S, Crowley GM (2000) *The Action Plan for Australian Birds 2000.* Environment Australia, Canberra.

Holdaway RN, Anderson A (2001) Avifauna from the Emily Bay settlement site, Norfolk Island: a preliminary account. In 'The Prehistoric Archaeology of Norfolk Island, Southwest Pacific'. (Eds A Anderson and P White) *Records of the Austral-*

ian Museum, Supplement **27**, 85–100. Australian Museum, Sydney.

Marchant S, Higgins PJ (Eds) (1990) *Handbook of Australian, New Zealand and Antarctic Birds. Volume 1: Ratites to Ducks.* Oxford University Press, Melbourne.

McAllan IAW, Curtis BR, Hutton I, Cooper RM (2004) The birds of the Lord Howe Island Group: a review of records. *Australian Field Ornithology* **21**, 1–82.

Priddel D, Carlile N, Evans O, Evans B, McCoy H (2010) A review of the seabirds of Phillip Island in the Norfolk Island Group. *Notornis* **57**, 113–127.

Priddel D, Hutton I, Olson S, Wheeler R (2005) Breeding biology of Masked Boobies (*Sula dactylatra tasmani*) on Lord Howe Island, Australia. *Emu* **105**, 105–113.

Schreiber EA, Burger J (2002) *Biology of Marine Birds.* CRC Press, Boca Raton, FL.

Steeves TE, Holdaway RN, Hale ML, McLay E, McAllan IAW, Christian M, Hauber ME, Bunce M (2010) Merging ancient and modern DNA: extinct seabird taxon rediscovered in the North Tasman Sea. *Biology Letters* **6**, 94–97.

Veitch CR, Miskelly CM, Harper GA, Taylor G, Tennyson AJD (2004) Birds of the Kermadec Islands, south-west Pacific. *Notornis* **51**, 61–90.

Comments received from

Barry Baker, Margaret Christian, David Priddel

Subspecies Conservation Summary

Imperial Shag (Heard Island)

Leucocarbo atriceps nivalis Falla, 1937
Phalacrocoracidae

Conservation status

Near Threatened D2

Reasons for listing

Present only at only 1 site with a plausible but low probability threat from marine pollution

Status 2000

Near Threatened D2

Status 1990

Near Threatened D2

Taxonomy

L. a. purpurascens (Macquarie I.) is Vulnerable; 6 extralimital subspecies on remote subantarctic islands including *L. a. verrucosus*, a vagrant to Australian waters. The species is Least Concern.
Taxonomic uniqueness: high (2 genera/family, 2 species/genus, 8 subspecies/species)

Range

Endemic to Heard I., breeding at 3 sites on the north-western coast: Stephenson Lagoon, Saddle Point and Sydney Cove (Marchant and Higgins 1990), and at Cape Pillar on the south-west coast on the western periphery of a Macaroni Penguin *Eudyptes chrysolophus* colony (Woehler *et al.* 2001).

Abundance

The population is currently estimated at 1100 pairs (Woehler 2010).

Ecology

Breeding colonies are on cliff tops or boulder beaches with birds roosting on offshore stacks, boulder and sand beaches, areas of volcanic sand and cliff tops. They feed on polychaete worms and benthic fish caught off the island's coast (Green *et al.* 1990; Green and Williams 1997). A generation length of 11.3 years (BirdLife International 2011) is derived from an age of first breeding of 3.0 years (Schreiber and Burger 2002) and an annual survival of adults of 88.0% (extrapolated from congeners).

Threats

Fluctuations in population size and breeding success can be attributed to frequent inclement weather (Pemberton and Gales 1987) or predator-prey cycles. While plausible marine pollution is unlikely to pose a threat in the next 3 generations so the subspecies listed as Near Threatened not Vulnerable.

Conservation objectives

1. Persistence of population

Information required

1. Population trends

Management actions required

1. Maintain quarantine provisions

Current eligibility against IUCN Red List Criteria

IUCN category	Criteria eligibility
A	Not applicable: past, current or future population declines are thought unlikely to exceed 20% in any 3-generation period
B	Not applicable: EOO, AOO, habitat, locations or population not declining, population fluctuations not extreme
C	Not applicable: no continuing population decline or extreme fluctuations
D	Near Threatened: ≤5 locations with a plausible but low probability threat
E	Not applicable: no population viability analysis undertaken

IUCN Red List assessment data

	Estimate	Reliability
Extent of occurrence trend	320 km^2 stable	high high
Area of occupancy trend	3 km^2 stable	high high
No. of mature individuals trend	2200 stable	medium high
No. subpopulations	1	high
No. locations	1	high
Generation time	11.3 years	medium
Global population share	100%	high
Level of genetic exchange	low	high

Bibliography

BirdLife International (2011) 'Species factsheet: *Phalacrocorax atriceps*'. Retrieved 17 January, 2011 from <http://www.birdlife.org/>.

Green K, Williams R (1997) Biology of the Heard Island Shag *Phalacrocorax nivalis*. 3. Foraging, diet and diving behaviour. *Emu* **97**, 76–83.

Green K, Williams R, Woehler EJ, Burton HR, Gales NJ, Jones RT (1990) Diet of the Heard Island Cormorant *Phalacrocorax atriceps nivalis*. *Antarctic Science* **2**, 139–141.

Marchant S, Higgins PJ (Eds) (1990) *Handbook of Australian, New Zealand and Antarctic Birds. Volume 1: Ratites to Ducks.* Oxford University Press, Melbourne.

Pemberton D, Gales RP (1987) Notes on the status and breeding of the Imperial Cormorant *P. atriceps* at Heard Island. *Cormorant* **15**, 33–40.

Schreiber EA, Burger J (2002) *Biology of Marine Birds.* CRC Press, Boca Raton, FL.

Woehler EJ (2010) Status and conservation of the seabirds of Heard Island. *The State of Australia's Birds 2010: Islands and Birds. Wingspan* **Supplement 20**, 8–9.

Woehler EJ, Auman HJ, Riddle MJ (2001) 'Heard Island's seabirds under scrutiny. Australian Antarctic Magazine 1'. Retrieved 31 March, 2011 from <http://www.antarctica.gov.au/about-antarctica/australian-antarctic-magazine/issue-1-autumn-2001/science/heard-islands-seabirds-under-scrutiny>.

Comments received from

Eric Woehler

Subspecies Conservation Summary

Imperial Shag (Macquarie Island)

Leucocarbo atriceps purpurascens (Brandt, 1837)

Phalacrocoracidae

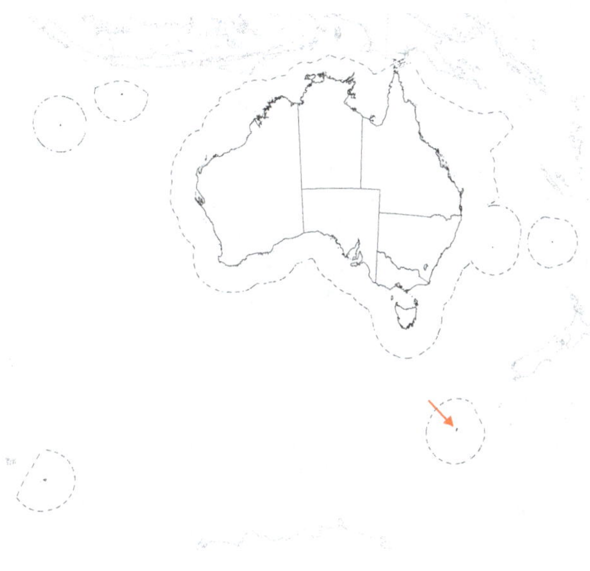

Conservation status

Vulnerable D1

Reasons for listing

Population 250–1000 mature individuals

Status 2000

Vulnerable D1

Status 1990

Vulnerable D1

Taxonomy

L. a. nivalis (Heard I.) is Near Threatened; 6 extralimital subspecies on remote subantarctic islands including *L. a. verrucosus*, a vagrant to Australian waters. The species is Least Concern.

Taxonomic uniqueness: high (2 genera/family, 2 species/genus, 8 subspecies/species)

Range

Restricted to Macquarie Island and the adjacent Bishop Island and Clerk Island, 33 km to the south (Marchant

Current eligibility against IUCN Red List Criteria

IUCN category	Criteria eligibility
A	Not applicable: past, current or future population declines are thought unlikely to exceed 20% in any 3-generation period
B	Not applicable: EOO, AOO, habitat, locations or population not declining, population fluctuations not extreme
C	Not applicable: no continuing population decline or extreme fluctuations
D	Vulnerable: 250–1000 mature individuals
E	Not applicable: no population viability analysis undertaken

IUCN Red List assessment data

	Estimate	Reliability
Extent of occurrence trend	75 km^2 stable	high high
Area of occupancy trend	50 km^2 stable	high high
No. of mature individuals trend	940 stable	medium medium
No. subpopulations	1	high
No. locations	2	high
Generation time	11.3 years	medium
Global population share	100%	high
Level of genetic exchange	low	high

and Higgins 1990). There are 23 recorded breeding sites for the shags, with all but 2 sites restricted to offshore stacks or islets, or stacks attached to the shore (Brothers 1985). Not all sites are used each year, and birds may interchange between colonies (Department of Sustainability, Environment, Water, Population and Communities 2011).

Abundance

Eleven colonies comprising 472 nesting pairs were located in October 2003 (Schulz and Lynn 2003). By comparison the population in 1975–1979 was estimated at 760 pairs, including 100 pairs on the Bishop and Clerk Islets. Of 23 breeding colonies recorded, were 19 active with between 3 and 320 pairs in each colony. Numbers of breeding birds fluctuate between years and pairs probably move between breeding sites (Brothers 1985). Although it is difficult to detect trends from these data, this figure represents 30% fewer than the 660 pairs observed in 19 sites in the 1970s (Schulz and Lynn 2003).

Ecology

Breeding colonies are on cliff tops or boulder beaches with birds roosting on offshore stacks, boulder and sand beaches, areas of volcanic sand and cliff tops. They forage in the subantarctic inshore waters of Macquarie I. and its outlying islands, largely on fish, some crustaceans and other benthic invertebrates (Brothers 1985; Marchant and Higgins 1990). A generation length of 11.3 years (BirdLife International 2011) is derived from an age of first breeding of 3.0 years (Schreiber and Burger 2002) and an annual survival of adults of 88.0% (extrapolated from congeners).

Threats

In the past, cats *Felis catus* were identified as a possible predator of chicks; since their eradication the only other introduced predators are Black Rats *Rattus rattus* (Schulz and Lynn 2003). Native predators, such as Brown Skuas *Catharacta lonnbergi* also kill chicks (Brothers 1985). The effects of adverse weather on breeding success may be manifest not only in the physical destruction of nests, but also by restricting access to available prey. On Macquarie I. up to 90% mortality of Shag nestlings has been observed after a single storm (Pemberton and Gales 1987). Storms on Macquarie Island have also been directly implicated in the death of adults when colonies are wave-washed and in severe winter storm events (Schulz and Lynn 2003). Human mediated deaths have also been recorded with fatal strikes with radio masts quite common (Brothers 1985). The shallow and inshore foraging behaviour of the Shags makes them unlikely candidates for interactions with commercial fishing operations around Macquarie I. although starvation kills many chicks (Brothers 1985).

Conservation objectives

1. Persistence of population

Information required

1. Population trends

Management actions required

1. Repeat surveys every 5 years
2. Maintain quarantine provisions

Bibliography

BirdLife International (2011) 'Species factsheet: *Phalacrocorax atriceps*'. Retrieved 17 January, 2011 from <http://www.birdlife.org/>.

Brothers NP (1985) Breeding biology, diet and morphometrics of the King Shag, *Phalacrocorax albiventer purpurascens* at Macquarie Island. *Australian Wildlife Research* **12**, 81–94.

Department of Sustainability, Environment, Water, Population and Communities (2011) *Leucocarbo atriceps purpurascens*. In 'Species Profile and Threats Database'. Department of Sustainability, Environment, Water, Population and Communities, Canberra. Retrieved 24 February, 2011 from <http://www.environment.gov.au/sprat>.

Garnett S, Crowley GM (2000) *The Action Plan for Australian Birds 2000*. Environment Australia, Canberra.

Marchant S, Higgins PJ (Eds) (1990) *Handbook of Australian, New Zealand and Antarctic Birds. Volume 1: Ratites to Ducks*. Oxford University Press, Melbourne.

Pemberton D, Gales RP (1987) Notes on the status and breeding of the Imperial Cormorant *P. atriceps* at Heard Island. *Cormorant* **15**, 33–40.

Schreiber EA, Burger J (2002) *Biology of Marine Birds*. CRC Press, Boca Raton, FL.

Schulz M, Lynn J (2003) 'Burrowing petrels on Macquarie Island, late March to early December 2003'. Report to

Nature Conservation Branch, Department of Primary Industries, Water and Environment, Hobart, Tasmania. Department of Primary Industries, Water and Environment, Hobart, Tasmania.

Comments received from

Eric Woehler

Species Conservation Summary

Australasian Bittern

Botaurus poiciloptilus (Wagler, 1827)
Ardeidae

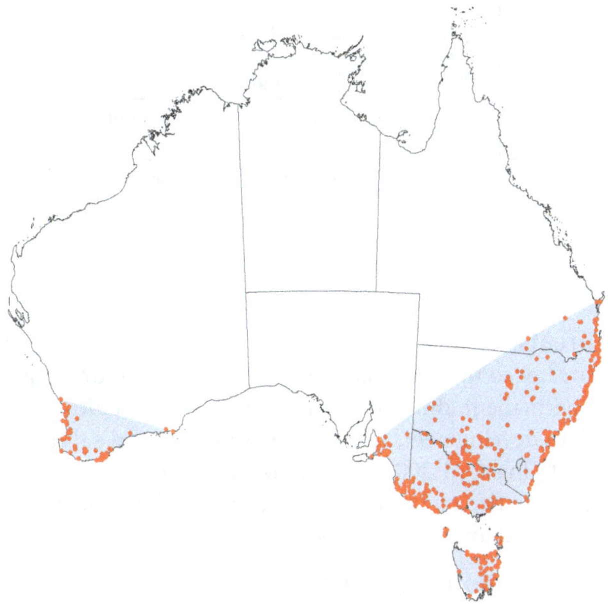

Conservation status

Australian breeding population: Endangered C1

Reasons for listing

Population <2500 and surveys indicate a decline of >20% within 2 generations (11 years)

Status 2000

Vulnerable C2a(ii)
Reason for change in 2010: declines during since 2000 more pronounced

Status 1990

Vulnerable C2a(ii)

Taxonomy

No infraspecific taxa described. The species is Endangered C1.
Taxonomic uniqueness: medium (20 genera/family, 4 species/genus, 1 subspecies/species)

Range

Two subpopulations; in south-eastern Australia from south-east Queensland to south-eastern South Australia including Tasmania, and, in south-western Australia, between Moora and Cape Arid (Marchant and Higgins 1990). In recent decades, numbers in the previous strongholds in the Murray-Darling Basin have declined, the Western Australian subpopulation has become more restricted with few records away from the south coast and Lake Muir wetlands, breeding in South Australia confined to coast from Adelaide to Victorian border, and few recent Queensland records (Barrett *et al.* 2003). There have been few confirmed records from the Swan Coastal Plain since 1992 and surveys in 2007 and 2008 found that half the wetlands that supported the species in 1980 now retained no suitable habitat (Pickering and Gole 2008).

Abundance

Based on 2 years of surveys at all wetlands where Australasian Bitterns had been reported between 1998 and 2008, Birds Australia estimated the number of adult birds in 2009–2010 to be 3–16 in Qld, 82–162 in NSW, 86–248 in Vic, 12–100 in Tas, 26–116 in SA and 38–154 in WA; a national total of 247–796 (Birds Australia *in litt.*). The reporting rate in Atlas surveys has decreased from being recorded in 260 10-minute grid squares in 1977–1981, to 142 squares in 1998–2003, and 61 in 2003–2008 (Birds Australia *in litt.* 2008). The reporting rate declined by >90% in Tas and WA, and by 63% in the Riverina. The long-term rate of decline is estimated to be 20–30% over 2 generations (11 years). Extralimitally there were c.580–725 in New Zealand in 1980 (Heather and Robertson 1997) but numbers may be greater given the lack of targeted surveys and the size of suitable swamps, and probably none in New Caledonia (Wetlands International 2006). The distribution in New Zealand declined between the 2 Atlases in 1969–1979 and 1999–2004 (Wilson 2008).

Ecology

Inhabits shallow, vegetated freshwater or brackish swamps, favouring those dominated by sedges, rushes and/or reeds (Marchant and Higgins 1990). Recorded in rice paddies in the Murray Darling basin, but it is not thought to use such habitats for breeding. It feeds on fish, eels, frogs, freshwater crayfish and aquatic insects (Heather and Robertson 1997). The species appears to disperse widely, including to coastal wetlands during periods of drought and to ephemeral wetlands when wet. During years of extensive flooding, there may be a short to medium term increase in population (Department of Sustainability, Environment, Water, Population and Communities 2011). A generation time of 5.5 years (BirdLife International 2011) is derived from a mean age at first breeding of 2.0 years, an annual survival of adults of 68.0%, and a maximum longevity in the wild of 9.8 years, all values extrapolated from mean values for other genera in Ardeidae.

Current eligibility against IUCN Red List Criteria

IUCN category	Criteria eligibility
Australian breeding population	
A	Not applicable: past, current or future population declines are thought unlikely to exceed 20% in any 3-generation period
B	Not applicable: >10 locations, not severely fragmented, population fluctuations not extreme
C	Endangered: population 250–2500 individuals, estimated continuing decline between 20% in 2 generations and 25% in 1 generation
D	Not applicable: population >1000 mature individuals, >5 locations
E	Not applicable: no population viability analysis undertaken

IUCN Red List assessment data

	Estimate	Reliability
Australian breeding population		
Extent of occurrence trend	1 270 000 stable	high high
Area of occupancy trend	2580 decreasing	high medium
No. of mature individuals trend	1000 decreasing	medium medium
No. subpopulations	2	high
No. locations	>10	high
Largest subpopulation	900	medium
Generation time	5.5 years	low
Global population share	26–50%	high
Level of genetic exchange	low	medium

Threats

Threatened by diversion of water for irrigation, drainage of permanent and ephemeral swamps for agriculture or urban development, reduced water inflows and salinisation (Kingsford 2000). Overgrazing by stock and inappropriate fire regimes can also reduce habitat suitability (Marchant and Higgins 1990).

Conservation objectives

1. Stable or increasing numbers of birds recorded in standardised surveys at key sites
2. Former breeding sites rehabilitated and reoccupied

Information required

1. Greater understanding of movements
2. Principal breeding and non-breeding wetlands
3. Exact habitat needs by collating detailed information on habitat before, during and after use

Management actions required

1. Continue survey and monitoring activities
2. Protect and manage principal breeding wetlands
3. Rehabilitate selected former breeding wetlands as part of Landcare
4. Protect and manage drought refuges

Bibliography

Barrett G, Silcocks A, Barry S, Cunningham R, Poulter R (2003) *The New Atlas of Australian Birds.* Royal Australian Ornithologists Union, Hawthorn East.

BirdLife International (2011) 'Species factsheet: *Botaurus poiciloptilus*'. Retrieved 17 January, 2011 from <http://www.birdlife.org/>.

Department of Sustainability, Environment, Water, Population and Communities (2011) *Botaurus poiciloptilus.* In 'Species Profile and Threats Database'. Department of Sustainability, Environment, Water, Population and Communities, Canberra. Retrieved 24 February, 2011 from <http://www.environment.gov.au/sprat>.

Heather BD, Robertson HA (1997) *The Field Guide to the Birds of New Zealand.* Oxford University Press, Oxford, UK.

Kingsford RT (2000) Ecological impacts of dams, water diversions and river management on floodplain wetlands in Australia. *Austral Ecology* **25**, 109–127.

Marchant S, Higgins PJ (Eds) (1990) *Handbook of Australian, New Zealand and Antarctic Birds. Volume 1: Ratites to Ducks.* Oxford University Press, Melbourne.

Pickering R, Gole C (2008) 'Swan coastal plain Australasian Bittern surveys 2007–2008'. Unpublished report.

Wetlands International (2006) 'Waterbird population estimates'. Wetlands International, Wageningen, The Netherlands.

Wilson KJ (2008) *The State of New Zealand's Birds 2008.* Ornithological Society of New Zealand.

Comments received from

Cheryl Gole, Roger Jaensch, Max Maddock, Colin O'Donnell, Robyn Pickering, Andrew Silcocks

Species Conservation Summary

Letter-winged Kite

Elanus scriptus Gould, 1842
Accipitridae

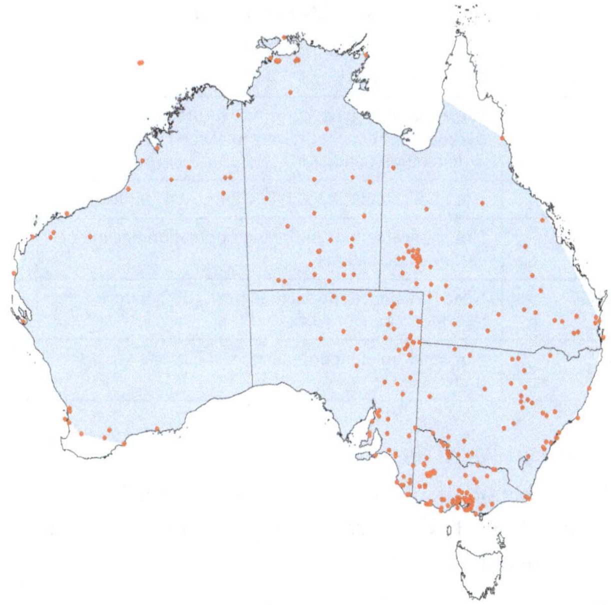

Conservation status
Near Threatened D1

Reasons for listing
Fluctuating population that may approach 1000 individuals in the poorest years

Status 2000
Near Threatened D1

Status 1990
Near Threatened D1

Taxonomy
No infraspecific taxa described
Taxonomic uniqueness: medium (64 genera/family, 4 species/genus, 1 subspecies/species)

Range
Fluctuating range in eastern Australian arid zone, but occasionally irrupts across the continent. In lean years, when prey populations are low, persists in small, poorly-known core areas, notably the Cooper Creek drainage system. In the years following rain, if prey numbers increase, the kite breeds rapidly. When dry times return and prey populations crash, kites disperse, usually across the Barkly Tablelands in the eastern Northern Territory and river systems in south-western Queensland, north-eastern South Australia and north-western New South Wales, but can reach all parts of the continent. The area occupied is nominally put at 0.1% of the EOO but is essentially unknown. There is no evidence of any decline in range (Marchant and Higgins 1993; Olsen 1998).

Abundance
The population size fluctuates. When prey is numerous, the kite can be locally common but little is known about the intervening times when the species is rarely seen and populations fall (Marchant and Higgins 1993; Olsen 1998; Aumann and Bellchambers 1999). The reporting rate in the second Birds Australia Atlas (Barrett *et al.* 2003) was the lowest of all the raptors: just 3% of 365 one-degree grid-blocks. Minimum population size during these lean times is estimated here, on a precautionary basis, as near to 1000 mature individuals.

Ecology
Nocturnal inhabitant of open grassy or sparsely wooded country, often along dry timbered watercourses. They roost, nest and sometimes hunt in groups, with some breeding colonies containing up to a hundred individuals. Irruptive: population cycles are linked to those of the principal prey, the Long-haired or Plague Rat *Rattus villosissimus*, which has population explosions following high rainfall. During a rat plague, kites can produce several clutches in succession and the population can rapidly increase 10-fold. After rat populations crash, birds disperse, often far from their core range although no subpopulation based on irruptive birds has persisted for long (Marchant and Higgins 1993; Olsen 1998). Irruptive House Mice *Mus domesticus*, native mice, reptiles and invertebrates are also important foods (Baker-Gabb and Pettigrew 1982; Pavey *et al.* 2008b). A generation time of 6.6 years (BirdLife International 2011) is derived from an age at first breeding of 2.0 years and an annual survival of 78.0%, both values extrapolated from other Accipitridae genera.

Threats
There are no known major threats, although competition for prey with introduced predators like fox *Vulpes vulpes* and feral cat *Felis catus* may reduce breeding productivity (Pavey *et al.* 2008a; Pavey *et al.* 2008b). Rat populations are otherwise thought to be fairly secure, even in extremely dry years, although may be affected by intensification of grazing in the arid zone (MacMillen and MacMillen 2009) and the frequency of irruptive years seems to have declined (Woinarski and Aplin 2008). Cats may take significant number of nestlings, but the demographic impact of this predation remains unquantified (Olsen 1995).

Conservation objectives
1. Stable populations in core refugia

Information required
1. Size of by core population and area occupied
2. Population fluctuations better understood from regular surveys and analysis of ad-hoc sightings, particularly during years when prey are scarce
3. Impact of cattle grazing on numbers of rats and other native prey mammals
4. Impact of nest predation by feral cats

Current eligibility against IUCN Red List Criteria

IUCN category	Criteria eligibility
A	Not applicable: past, current or future population declines are thought unlikely to exceed 20% in any 3-generation period
B	Not applicable: EOO >20 000 km², AOO >2000 km²
C	Not applicable: no continuing population decline or extreme fluctuations
D	Near Threatened: population near 1000 mature individuals in poor years
E	Not applicable: no population viability analysis undertaken

IUCN Red List assessment data

	Estimate	Reliability
Extent of occurrence trend	7 300 000 km² stable	high medium
Area of occupancy trend	7000 km² fluctuating	low medium
No. of mature individuals trend	1200 fluctuating	low medium
No. subpopulations	1	high
No. locations	>10	high
Generation time	6.6 years	low
Global population share	100%	high

5. Impact of cat and fox predation and grazing on amplitude and occurrence of population irruptions of main prey species

Management actions required

1. Identify and protect core refuge sites
2. Control cats at breeding sites if necessary

Bibliography

Aumann T, Bellchambers K (1999) Letter-winged Kites *Elanus scriptus* in the south-west of the Northern Territory, 1994–95. *Australian Bird Watcher* **18**, 68–74.

Baker-Gabb DJ, Pettigrew JD (1982) Non-breeding diet of the Letter-winged Kite in north-eastern South Australia. *Corella* **6**, 19–20.

Barrett G, Silcocks A, Barry S, Cunningham R, Poulter R (2003) *The New Atlas of Australian Birds.* Royal Australian Ornithologists Union, Hawthorn East.

BirdLife International (2011) 'Species factsheet: *Elanus scriptus*'. Retrieved 17 January, 2011 from <http://www.birdlife.org/>.

MacMillen RE, MacMillen BJ (2009) *Meanderings in the Bush.* 2nd edn. CSIRO Publishing, Melbourne.

Marchant S, Higgins PJ (Eds) (1993) *Handbook of Australian, New Zealand and Antarctic Birds. Volume 2: Raptors to Lapwings.* Oxford University Press, Melbourne.

Olsen P (1995) *Australian Birds of Prey.* Johns Hopkins University Press, Baltimore, Maryland.

Olsen P (1998) 'Australia's raptors: diurnal birds of prey and owls'. Birds Australia Conservation Statement no. 3.

Pavey CR, Eldridge SR, Heywood M (2008a) Population dynamics and prey selection of native and introduced predators during a rodent outbreak in arid Australia. *Journal of Mammalogy* **89**, 674–683.

Pavey CR, Gorman J, Heywood M (2008b) Dietary overlap between the nocturnal letter-winged kite *Elanus scriptus* and barn owl *Tyto alba* during a rodent outbreak in arid Australia. *Journal of Arid Environments* **72**, 2282–2286.

Woinarski J, Aplin K (2008) *Rattus villosissimus.* In 'IUCN 2010. IUCN Red List of Threatened Species. Version 2010.4'. Retrieved 17 May, 2011 from <www.iucnredlist.org/>.

Comments received from

Greg Czechura, Stephen Debus, Rod Hobson, Penny Olsen, Chris Pavey, Julian Reid

Subspecies Conservation Summary

Brown Goshawk (Papuan)

Accipiter fasciatus dogwa Rand, 1941
Accipitridae

Conservation status

Near Threatened D1

Reasons for listing

Australian population probably small enough for Endangered D but not isolated from Least Concern main population in New Guinea, so downgraded 2 categories to Near Threatened

Status 2000

Near Threatened D1

Current eligibility against IUCN Red List Criteria

IUCN category	Criteria eligibility
A	Not applicable: past, current or future population declines are thought unlikely to exceed 20% in any 3-generation period
B	Not applicable: EOO, AOO, habitat, locations or population not declining, population fluctuations not extreme
C	Not applicable: no continuing population decline or extreme fluctuations
D	Endangered: population 50–250 mature individuals (but downgraded to Near Threatened as global status Least Concern)
E	Not applicable: no population viability analysis undertaken

IUCN Red List assessment data

	Estimate	Reliability
Extent of occurrence trend	400 km^2 stable	high high
Area of occupancy trend	168 km^2 stable	high high
No. of mature individuals trend	70 stable	low low
No. subpopulations	2	high
No. locations	2	high
Largest subpopulation	40	low
Generation time	7.2 years	poor
Global population share	<1%	high
Level of genetic exchange	high	medium

Status 1990

Near Threatened D1

Taxonomy

A. f. didimus (northern Australia) and *A. f. fasciatus* (southern Australia) are Least Concern, as is the species. There are 10 extralimital taxa. The species is Least Concern.

Taxonomic uniqueness: low (64 genera/family, 51 species/genus, 12 subspecies/species)

Range

There have been a recent series of records of this subspecies from the north Torres Strait islands of Boigu (72 km^2) and Saibai (96 km^2). Records from the southern Torres Strait islands (e.g. Draffan *et al.* 1983) are probably the subspecies *A. f. didimus.*

Abundance

At least 10 sightings of 1 or 2 individuals have been reported on birdwatching visits to the islands (Clarke 2004). Population densities of *A. f. fasciatus* recorded in southern Australia were 0.02–0.30 pairs/km^2 (Baker-Gabb 1984; Aumann 1989). Much of Boigu and Saibai is probably unsuitable habitat, being mangroves and freshwater swamps, but there are few other predators so all habitats may be used. Thus the population is estimated here as probably about 70 birds (0.20 pairs/km^2). Outside Australia, this subspecies is relatively common and widespread in south New Guinea (BirdLife International 2011).

Ecology

In Papua New Guinea, inhabits savanna, forest edge, open forest and suburban gardens from sea-level to mid-mountains, where it feeds mostly on birds but also reptiles and large invertebrates (Coates 1985). Builds a large stick nest in a tall tree in forest edge (Marchant and Higgins 1993). Given the lack of subspecific variation between many relatively distant islands in the species' range (e.g. *A. f. fasciatus* on Australia, Timor and Rennell), it is considered to be a dispersive species with regular exchange between the Torres Strait islands and New Guinea. A generation time of 7.2 years (BirdLife International 2011) derived from an average age at first breeding of 2.0 years and maximum longevity of 12.4 years, both values extrapolated from congeners.

Threats

The subspecies is listed as threatened because it has a very small population. There are no plausible threats; though Saibai and Boigu are low-lying they are too far north for catastrophic cyclones. The species is sometimes killed in New Guinea because it may kill domestic chickens, but few if any are likely to be killed on the Torres Strait islands.

Conservation objectives

1. Population size maintained

Information required

1. Population density and size estimated

Management actions required

1. Initiate community-based ecotourism business promoting the islands' unique avifauna. Monitoring of subpopulations and habitat could then be undertaken by visiting birdwatchers and local rangers

Bibliography

Aumann T (1989) Breeding parameters of the Brown Goshawk *Accipiter fasciatus* in south-eastern Australia. *Emu* **89**, 112–118.

Baker-Gabb DJ (1984) The breeding ecology of twelve species of diurnal raptor in north-western Victoria. *Australian Wildlife Research* **11**, 145–160.

BirdLife International (2011) 'Species factsheet: *Accipiter fasciatus*'. Retrieved 17 January, 2011 from <http://www.birdlife.org/>.

Clarke RH (2004) The avifauna of Northern Torres Strait: notes on a wet-season visit. *Australian Field Ornithology* **21**, 49–66.

Draffan RDW, Garnett ST, Malone GJ (1983) Birds of the Torres Strait: an annotated list and biogeographical analysis. *Emu* **83**, 207–234.

Marchant S, Higgins PJ (Eds) (1993) *Handbook of Australian, New Zealand and Antarctic Birds. Volume 2: Raptors to Lapwings.* Oxford University Press, Melbourne.

Comments received from

Rohan Clarke

Subspecies Conservation Summary

Variable Goshawk (Christmas Island)

Accipiter hiogaster natalis (Lister, 1889)
Accipitridae

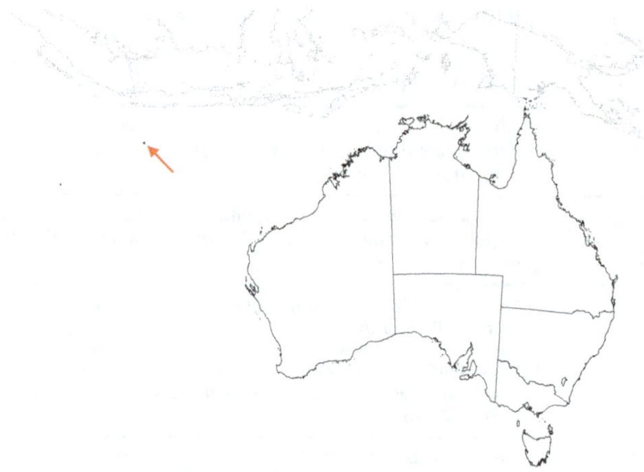

Conservation status
Endangered D

Reasons for listing
Population 50–250 mature individuals

Status 2000
Endangered D

Status 1990
Endangered D

Taxonomy
16 extralimital subspecies. The species is Least Concern. Sometimes included with the Brown Goshawk *Accipiter fasciatus*.

Taxonomic uniqueness: low (64 genera/family, 51 species/genus, 17 subspecies/species)

Range
Endemic to Christmas I., Indian Ocean. Forest clearance for phosphate mining has reduced the area of habitat suitable for nesting by about one-third (Department of Sustainability Environment Water Population and Communities 2011), but it still occurs over the entire island (M. Smith pers. comm.).

Abundance
The population size was estimated at 100 breeding adults (Hill 2004) but colour-banding studies suggest there may be as many as 250 mature individuals (Beeton *et al.* 2010). It was the least frequently encountered of diurnal land birds surveyed by James and Retallick (2007), occurring in only 6 out of 527 counts. The breeding population has probably declined in proportion to the area cleared for phosphate mining, although it does use rehabilitated areas (Hill 2004; Beeton *et al.* 2010). Reports of historical declines and recovery (Stokes 1988) are backed by little evidence and cannot be confirmed. Concerns that the Goshawk might decline rapidly following the spread of the introduced Yellow Crazy Ant *Anoplolepis gracilipes* (Garnett and Crowley 2000) were not realised, and the ants are currently controlled. Thus the population is likely to be stable.

Ecology
Occurs in all forest types on the island, but is uncommonly seen in secondary forest, settlements and rehabilitated habitat. Adults tend to occupy the primary forest with subadults more common in disturbed areas (Hurley 2005). Eats a wide range of vertebrates including birds, small mammals and invertebrates such as centipedes and large insects (Hill 2004). Builds a stick nest, breeding in both wet and dry seasons, but usually fledge only 1 young (Hill 2004; Hurley 2005). A generation time of 8.4 years (BirdLife International 2011) is derived from an age at first breeding of 2.0 years (extrapolated from congeners) and a maximum longevity in the wild of 14.9 years (extrapolated from *A. f. fasciatus*; Australian Bird and Bat Banding Scheme).

Threats
Like other Christmas I. species, a proliferation of scale insects and a number of introduced ants which tend them, particularly the Yellow Crazy Ant, have the potential to change vegetation structure profoundly (O'Dowd *et al.* 2003; Abbott and Green 2007). Thus any faltering of ant control and infestation of all rainforest could eventually affect the goshawk and its prey. The species may also be affected by the pesticide Fipronil used for ant control (V. Hurley pers. comm.), but this remains unproven (Beeton *et al.* 2010). However, the ants have been controlled with application of pesticide by helicopter and there is no evidence of decline in the goshawk, even where ants are present. Quarantine measures on the island are not currently sufficient to prevent invasion by additional unspecified predators. In the past forest clearance has destroyed about 25% of the habitat, but these areas are recovering and are used by at least immature goshawks (Hurley 2005). Hunting around settlements (Department of Sustainability, Environment, Water, Population and Communities 2011) may have killed birds in the past but is no longer a threat.

Conservation objectives
1. Existing population maintained

Information required
1. Population trends
2. Alternatives to Fipronil for controlling Yellow Crazy Ants

Management actions required
1. Standardise long-term population monitoring
2. Upgrade quarantine standards for Christmas I.
3. Continue control of Yellow Crazy Ants

Current eligibility against IUCN Red List Criteria

IUCN category	Criteria eligibility
A	Not applicable: past, current or future population declines are thought unlikely to exceed 20% in any 3-generation period
B	Not applicable: EOO, AOO, habitat, locations or population not declining, population fluctuations not extreme
C	Not applicable: no continuing population decline or extreme fluctuations
D	Endangered: population 50–250 mature individuals
E	Not applicable: no population viability analysis undertaken

IUCN Red List assessment data

	Estimate	Reliability
Extent of occurrence trend	134 km^2 stable	high high
Area of occupancy trend	134 km^2 stable	medium low
No. of mature individuals trend	250 stable	low low
No. subpopulations	1	high
No. locations	1	high
Generation time	8.4 years	medium
Global population share	100%	high
Level of genetic exchange	low	high

4. Create a multi-stakeholder team with responsibility for ensuring threatened species are conserved on Christmas I.

Bibliography

Abbott KL, Green PT (2007) Collapse of an ant-scale mutualism in a rainforest on Christmas Island. *Oikos* **116**, 1238–1246.

Beeton B, Burbidge AA, Grigg G, Harrison P, How RA, Humphries B, McKenzie N, Woinarski J (2010) 'Final Report, Christmas Island Expert Working Group to Minister for the Environment, Heritage and the Arts'. Retrieved 1 April, 2010 from <http://www.environment.gov.au/parks/publications/christmas/pubs/final-report.pdf>.

BirdLife International (2011) 'Species factsheet: *Accipiter fasciatus*'. Retrieved 17 January, 2011 from <http://www.birdlife.org/>.

Department of Sustainability, Environment, Water, Population and Communities (2011) *Accipiter hiogaster natalis*. In 'Species Profile and Threats Database'. Department of Sustainability, Environment, Water, Population and Communities, Canberra. Retrieved 24 February, 2011 from <http://www.environment.gov.au/sprat>.

Garnett S, Crowley GM (2000) *The Action Plan for Australian Birds 2000*. Environment Australia, Canberra.

Hill R (2004) 'National Recovery Plan for the Christmas Island Goshawk *Accipiter fasciatus natalis*'. Commonwealth of Australia, Canberra.

Hurley VG (2005) 'Colour banding Christmas Island Goshawks, a first stage report'. A report by Birds Australia to Parks Australia North (Christmas Island), Christmas Island, Indian Ocean, and Department of Environment and Heritage, Canberra.

James DJ, Retallick K (2007) 'Forest birds of Christmas Island: a baseline survey of abundance'. Parks Australia North Christmas Island Biodiversity Monitoring Programme: report to the Department of Finance & Administration and the Department of the Environment & Water Resources, Canberra.

O'Dowd DJ, Green PT, Lake PS (2003) Invasional 'meltdown' on an oceanic island. *Ecology Letters* **6**, 812–817.

Stokes T (1988) 'A review of the birds of Christmas Island, Indian Ocean'. Australian National Parks and Wildlife Service Occasional Paper.

Comments received from

Stephen Debus, Mark Holdsworth, Victor Hurley, Penny Olsen, Mike Smith, John Woinarski

Species Conservation Summary

Red Goshawk

Erythrotriorchis radiatus (Latham, 1802)
Accipitridae

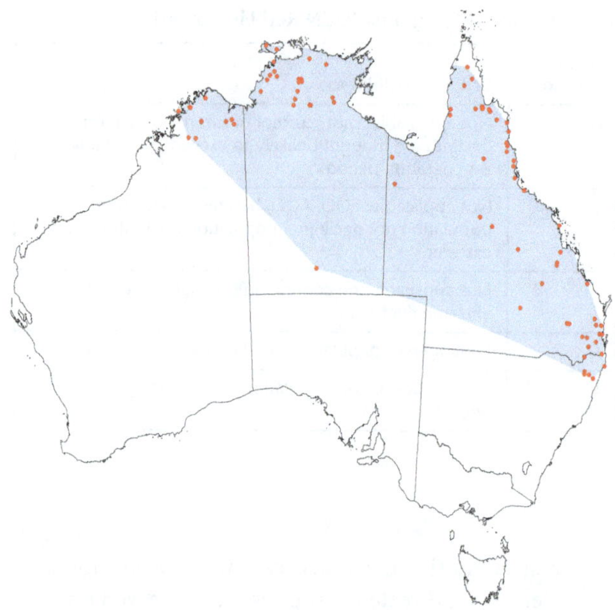

Conservation status
Near Threatened C2a(i), D1

Reasons for listing
Population estimated at about 1400 mature individuals in at least 2 subpopulations and possibly declining

Status 2000
Near Threatened C2a(i), D1

Status 1990
Near Threatened C2a(i), D1

Taxonomy
No infraspecific taxa described

Taxonomic uniqueness: medium (64 genera/family, 2 species/genus, 1 subspecies/species)

Range
Northern and eastern Australia, north of 28°S in the east, a contraction of about 500 km from the historical range south to 33°S, and north of about 19°S in the west (Marchant and Higgins 1993). Birds seen in central Australia may be resident, but are possibly dispersive individuals (Czechura and Hobson 2000; Aumann 2001; Department of Sustainability, Environment, Water, Population and Communities 2011). Virtually extinct in New South Wales (NSW National Parks and Wildlife Service 2002; Barrett *et al.* 2003), but found along the length of eastern Queensland (Czechura and Hobson 2000). Patchy distribution in subcoastal areas of the Gulf of Carpentaria and western Cape York Peninsula, but historical records and recent surveys suggest that the breeding distribution is continuous (Blakers *et al.* 1984; Czechura and Hobson 2000; Barrett *et al.* 2003).

Abundance
Because of its inconspicuousness, the population size is probably historically under-estimated. The total population was previously estimated at about 350 pairs (Aumann and Baker-Gabb 1991), but is now estimated at about 700 pairs. This includes 100 pairs on Tiwi Islands, probably a separate subpopulation (Department of Environment and Resource Management 2009), where 5 nest sites are rendered unsuitable (D. Baker-Gabb *in litt.*). In Qld, the population includes 10–30 pairs in the South Eastern Queensland bioregion, 25–30 pairs in the Wet Tropics bioregion, 65–70 pairs in the Cape York Peninsula bioregion, about 5 pairs in the Mt Isa Inlier bioregion with 3 bioregions (Central Mackay Coast, Brigalow Belt and Gulf Plains) currently unsurveyed (Czechura *et al.* 2011). Given the habitat loss, a gradual decline in eastern Qld may still be occurring.

Ecology
Inhabits coastal and subcoastal tall open forests and woodlands, tropical savannas traversed by wooded or forested rivers, and along the edges of rainforest, usually on fertile soils (Marchant and Higgins 1993). In partly cleared parts of eastern Qld, it is associated with gorge and escarpment country (Czechura and Hobson 2000; Czechura *et al.* 2009). It rarely breeds in areas with fragmented native vegetation (Aumann and Baker-Gabb 1991; Czechura 2001). The stick nests, in which 1–2 eggs are laid, are restricted to trees that are taller than 20 m within 1 km of a watercourse or wetland (Aumann and Baker-Gabb 1991). They hunt within a home range of up to 200 km^2 in open forests and gallery forests, taking mostly medium to large birds (Czechura and Hobson 2000). In winter in eastern Australia, they appear to move from their nest sites in the ranges to coastal plains, where they are associated with permanent wetlands and often feed on waterbirds (Czechura and Hobson 2000). A generation time of 8.3 years (BirdLife International 2011) is derived from an age at first breeding of 3.0 years and an annual survival of adults of 78.0%, both extrapolated from mean values for Accipitridae, and longevity of 14.9 years (extrapolated from mean values for Australian *Accipiter* ssp.; Australian Bird and Bat Banding scheme).

Threats
Clearance for agriculture is thought to have caused the historical decline in NSW and southern Qld (Czechura and Hobson 2000; Czechura *et al.* 2011). Small numbers have been lost from forestry operations on Melville I., NT (Woinarski *et al.* 2003; Woinarski *et al.* 2007). Clearance for sub-coastal urban development in Qld may cause ongoing declines (Czechura *et al.* 2011). Fragmentation of natural habitat may be a threat as Red Goshawks in Qld are most scarce where lowland forests have been cleared for agriculture. Nests are particularly vulnerable as they are usually in the tallest trees that are valuable for timber, or are exposed to storm damage and other disturbance,

Current eligibility against IUCN Red List Criteria

IUCN category	Criteria eligibility
A	Not applicable: past, current or future population declines are thought unlikely to exceed 20% in any 3-generation period
B	Not applicable: EOO >20 000 km^2, AOO >2000 km^2
C	Near Threatened: population 500–2500 mature individuals, suspected continuing decline, largest subpopulation near 1000 mature individuals
D	Near Threatened: population near 1000 mature individuals
E	Not applicable: no population viability analysis undertaken

IUCN Red List assessment data

	Estimate	Reliability
Extent of occurrence trend	2 900 000 km^2 stable	high high
Area of occupancy trend	290 000 km^2 stable	low medium
No. of mature individuals trend	1400 declining	medium low
No. subpopulations	2	high
No. locations	>10	high
Largest subpopulation	1200	medium
Generation time	8.3 years	low
Global population share	100%	high

even if left intact after logging and clearance. Prey abundance may be reduced by loss or degradation of freshwater wetlands, loss of hollow-bearing trees in which prey breed, over-grazing by livestock and feral herbivores, and changed fire regimes (Czechura and Hobson 2000; Franklin *et al.* 2005; Czechura *et al.* 2011).

Conservation objectives

1. Populations maintained or increased across their range
2. Threatening processes reduced around known nest sites

Information required

1. Demography, especially adult survivorship
2. Nest localities for monitoring
3. Impact of habitat fragmentation on nesting pairs

Management actions required

1. Protect habitat through purchase or voluntary conservation agreements
2. Encourage landholders to protect and manage Red Goshawk territories
3. Promote information used to identify and protect nesting habitat
4. Produce habitat descriptions and maps for management purposes
5. Produce educational materials that promote the recovery process
6. Limit access to known nest sites

Bibliography

Aumann T (2001) Habitat use, temporal activity patterns and foraging behaviour of raptors in the south-west of the Northern Territory, Australia. *Wildlife Research* **28**, 365–378.

Aumann T, Baker-Gabb DJ (1991) 'A management plan for the Red Goshawk'. RAOU Report 75.

Barrett G, Silcocks A, Barry S, Cunningham R, Poulter R (2003) *The New Atlas of Australian Birds.* Royal Australian Ornithologists Union, Hawthorn East.

BirdLife International (2011) 'Species factsheet: *Erythrotriorchis radiatus*'. Retrieved 17 January, 2011 from <http://www.birdlife.org/>.

Blakers MS, Davies JJF, Reilly PN (1984) *The Atlas of Australian Birds.* Melbourne University Press, Melbourne.

Czechura GV (2001) 'The status and distribution of the Red Goshawk *Erythrotriorchis radiatus* on Cape York Peninsula, Queensland'. Unpublished report to Birds Australia.

Czechura GV, Hobson RG (2000) 'The Red Goshawk *Erythrotriorchis radiatus* in northern Queensland: status and distribution'. Report to Queensland Parks and Wildlife Service.

Czechura GV, Hobson RG, Stewart DA (2009) Observations on the biology of the Red Goshawk *Erythrotriorchis radiatus* in Queensland. *Australian Field Ornithology* **26**, 148–156.

Czechura GV, Hobson RG, Stewart DA (2011) Distribution, status and habitat of the Red Goshawk *Erythrotriorchis radiatus* in Queensland. *Corella* **35**, 1–10.

Department of Environment and Resource Management (2009) 'National recovery plan for the red goshawk *Erythrotriorchis radiatus*'. Report to the Department of the Environment, Water, Heritage and the Arts, Canberra. Queensland Department of Environment and Resource Management, Brisbane.

Department of Sustainability Environment Water Population and Communities (2011) *Erythrotriorchis radiatus.* In 'Species Profile and Threats Database'. Department of Sustainability, Environment, Water, Population and Communities, Canberra. Retrieved 24 February, 2011 from <http://www.environment.gov.au/sprat>.

Franklin DC, Whitehead PJ, Pardon G, Matthews J, McMahon P, McIntyre D (2005) Geographic patterns and correlates of decline of granivorous birds in northern Australia. *Wildlife Research* **32**, 399–408.

Marchant S, Higgins PJ (Eds) (1993) *Handbook of Australian, New Zealand and Antarctic Birds. Volume 2: Raptors to Lapwings.* Oxford University Press, Melbourne.

NSW National Parks and Wildlife Service (2002) 'Approved recovery plan for the Red Goshawk (*Erythrotriorchis radiatus*)'. NSW National Parks and Wildlife Service, Hurstville.

Woinarski J, Brennan K, Hempel C, Armstrong M, Milne D, Chatto R (2003) 'Biodiversity conservation on the Tiwi islands, Northern Territory. Part 2. Fauna'. Department of Infrastructure Planning and Environment, Darwin.

Woinarski J, Pavey C, Kerrigan R, Cowie I, Ward S (Eds) (2007) *Lost from Our Landscape: Threatened Species of the Northern Territory.* Northern Territory Government, Darwin.

Comments received from

David Baker-Gabb, Greg Czechura, Stephen Debus, Rod Hobson, David Stewart

Subspecies Conservation Summary

Wedge-tailed Eagle (Tasmanian)

Aquila audax fleayi Condon & Amadon, 1954
Accipitridae

Conservation status

Vulnerable D1

Reasons for listing

Population 250–1000 mature individuals

Status 2000

Endangered C2a(ii)
Reason for change in 2010: declines arrested

Status 1990

Endangered C2a(ii)

Taxonomy

The nominate subspecies, *A. a. audax* of mainland Australia and southern New Guinea is Least Concern, as is the species
Taxonomic uniqueness: low (64 genera/family, 12 species/genus, 2 subspecies/species)

Range

Breeds throughout Tasmania but not King I. or the smaller islands of Bass Strait, although they visit occasionally. Territories are widely dispersed in very large home ranges (Department of Sustainability Environment Water Population and Communities 2011). Area of Occupancy derived from no. territories × 4 km^2 to allow for alternative nest sites. There is no known movement between Tasmania and the mainland (Threatened Species Section 2006).

Abundance

The number of breeding adults is estimated at between 466 and 766 birds (average 600), derived from approximately 426 territories and a 54–90% territory occupancy (Threatened Species Section 2006). A projected 65% decline over the next 160 years in north-east Tasmania (Bekessy *et al.* 2009) is equivalent to a projected 38% in 3 generations, but mitigation measures are in place to reduce the major sources of mortality that were driving that decline (B. Brown pers. comm.).

Ecology

This subspecies forages in all habitats from coastal dunes to mountain peaks. Nesting habitat is generally restricted to predominantly mature forest with nests almost always being in large eucalypts on sheltered aspects. They breed solitarily in monogamous pairs and pair bonds last a lifetime. They are carnivorous, feeding on both live prey (mainly on rabbits *Oryctolagus cuniculus* and Hares *Lepus europaeus*, but a wide variety of other vertebrates as well) and carrion (Department of Sustainability, Environment, Water, Population and Communities 2011). A generation time of 25.0 years (BirdLife International 2011) is derived from an age at first breeding of 5.0 years (Marchant and Higgins 1993) and an annual survival of adults of 95% (Bell and Mooney 1998).

Threats

Major threats to the subspecies are loss of habitat, specifically nesting habitat, nest disturbance and collision with wind turbines and powerlines, causing injury or electrocution. Shooting, poisoning, vehicle strikes and trapping still kill some birds. These sources of mortality were causing, or were predicted to cause, a decline in the mean age of the population and decline in recruitment (Threatened Species Section 2006) but mitigation measures are now in place to protect nesting habitat, reduce nest disturbance and reduce losses from wind turbines, or compensate for losses through protection of nest sites (B. Brown pers. comm.).

Conservation objectives

1. Maintain a stable or increasing population

Information required

1. Effectiveness of current management prescriptions
2. Trends and spatial patterns of breeding success

Management actions required

1. Locate and protect nests on private land and in State Forests
2. Demonstrably reduce eagle mortality on electrical infrastructure and wind turbines
3. Monitor breeding success and abundance
4. Promote the conservation of the subspecies to land owners and managers

Bibliography

Bekessy SA, Wintle BA, Gordon A, Fox JC, Chisholm R, Brown B, Regan T, Mooney N, Read SM, Burgman MA (2009) Modelling human impacts on the Tasmanian wedge-tailed eagle (*Aquila audax fleayi*). *Biological Conservation* **142**, 2438–2448.

Bell P, Mooney N (1998) 'Recovery plan for the Wedge-tailed Eagle: 1998–2003'. TAS DPIWE. Retrieved 28 April, 2011

Current eligibility against IUCN Red List Criteria

IUCN category	Criteria eligibility
A	Not applicable: past, current or future population declines are thought unlikely to exceed 20% in any 3-generation period
B	Not applicable: EOO, AOO, habitat, locations or population not declining, population fluctuations not extreme
C	Not applicable: no continuing population decline or extreme fluctuations
D	Vulnerable: population 250–1000 mature individuals
E	Not applicable: population viability analysis does not indicate >10% probability of extinction in 100 years

IUCN Red List assessment data

	Estimate	Reliability
Extent of occurrence trend	70 000 km^2 stable	high high
Area of occupancy trend	700 km^2 stable	high high
No. of mature individuals trend	600 stable	medium high
No. subpopulations	1	high
No. locations	>10	high
Generation time	25 years	medium
Global population share	100%	high

from <http://www.environment.gov.au/biodiversity/threatened/publications/recovery/wedge-tail/index.html>.

BirdLife International (2011) 'Species factsheet: *Aquila audax*'. Retrieved 17 January, 2011 from <http://www.birdlife.org/>.

Department of Sustainability, Environment, Water, Population and Communities (2011) *Aquila audax fleayi*. In 'Species Profile and Threats Database'. Department of Sustainability, Environment, Water, Population and Communities, Canberra. Retrieved 24 February, 2011 from <http://www.environment.gov.au/sprat>.

Marchant S, Higgins PJ (Eds) (1993) *Handbook of Australian, New Zealand and Antarctic Birds. Volume 2: Raptors to Lapwings*. Oxford University Press, Melbourne.

Threatened Species Section (2006) 'Threatened Tasmanian Eagles Recovery Plan 2006–2010'. Department of Primary Industries and Water, Hobart.

Comments received from

Phil Bell, Bill Brown, Stephen Debus, Nick Mooney, Eric Woehler

Species Conservation Summary

Grey Falcon

Falco hypoleucos Gould, 1841

Falconidae

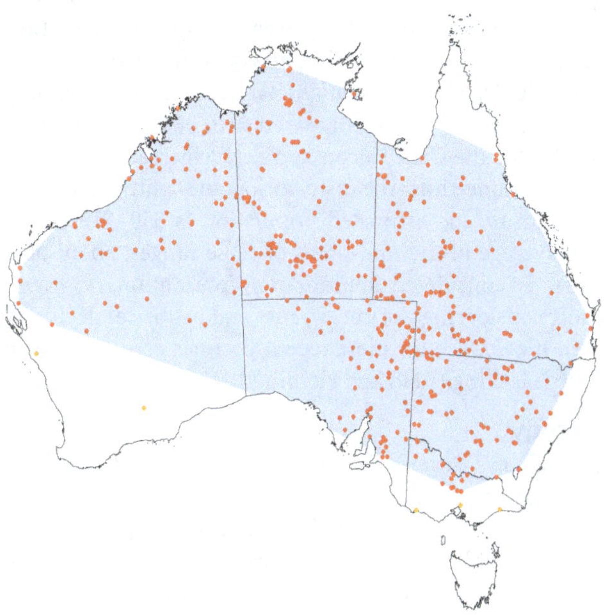

Conservation status

Vulnerable D1

Reasons for listing

Population may be fewer than 1000 individuals

Status 2000

Vulnerable D1

Status 1990

Vulnerable D1

Taxonomy

No infraspecific taxa described

Taxonomic uniqueness: medium (11 genera/family, 38 species/genus, 1 subspecies/species)

Range

Arid and semi-arid Australia, including the Murray-Darling basin, Eyre basin, central Australia and Western Australia (Marchant and Higgins 1993). Mainly found where annual rainfall <500 mm, except when wet years are followed by drought, when more widespread. Some evidence of regular movements, probably mostly of immatures, during the non-breeding season toward northern and coastal areas (Olsen and Olsen 1986; Martin and Royal 2000). May have been eliminated early in 20th century from some former breeding areas, particularly those with >500 mm annual rainfall in New South Wales (Olsen and Olsen 1986; Olsen 1998), now largely absent from WA south of 26°S (Johnstone and

Current eligibility against IUCN Red List Criteria

IUCN category	Criteria eligibility
A	Not applicable: past, current or future population declines are thought unlikely to exceed 20% in any 3-generation period
B	Not applicable: EOO >20 000 km^2, AOO >2000 km^2
C	Not applicable: no continuing population decline or extreme fluctuations
D	Vulnerable: population 250–1000 mature individuals
E	Not applicable: no population viability analysis undertaken

IUCN Red List assessment data

	Estimate	Reliability
Extent of occurrence trend	6 100 000 km^2 stable	high high
Area of occupancy trend	6000 km^2 stable	low high
No. of mature individuals trend	1000 stable	low medium
No. subpopulations	1	high
No. locations	>10	medium
Generation time	6.2 years	medium
Global population share	100%	high

Storr 1998) it is now an extremely rare vagrant to Victoria (Hurley *in litt.*).

Abundance

Always occurs at low densities and other grey raptors are often misreported as Grey Falcons (Schoenjahn 2010); the AOO is nominally estimated at 0.1% of the EOO. Two nests were found 23 km apart (Sutton 2011) but this is exceptional: the species has been encountered very infrequently during extensive dedicated searching in many parts of arid Australia over the last decade (Schoenjahn 2011). By comparing the range and number of sightings per 1 degree block in the first Atlas (Blakers *et al.* 1984), it is estimated that the Grey Falcon occupies about 0.27× the area occupied by the Peregrine Falcon *F. peregrinus* (99 compared to 365 grid blocks) at an average of one-quarter its density. Given an estimated 3000–5000 pairs of Peregrines in Australia (Olsen and Olsen 1988), this suggests a total of 200 to 350 pairs of Grey Falcon (Schoenjahn 2011). The second atlas (Barrett *et al.* 2003) reports sightings in 118 (14%) compared with 384 (47%) of grid blocks, for the Grey Falcon and Peregrine Falcon respectively. At one-third the distribution and a little over half the density, the estimated population is 550–915 pairs. The average of the mid-point of these ranges, about 500 pairs, is considered appropriately precautionary, especially considering the uncertainty and historical declines, even if evidence for more recent declines across the arid zone is lacking (Reid and Fleming 1992).

Ecology

The falcon's distribution is centred on inland drainage systems where the average annual rainfall is <500 mm. There it frequents timbered lowland plains, particularly acacia shrublands that are crossed by tree-lined watercourses. It also hunts far out into treeless areas and frequents tussock grassland and open woodland, especially in winter (Olsen and Olsen 1986). It preys on birds, mostly granivorous parrots and pigeons, and sometimes mammals, reptiles and insects (Olsen and Olsen 1986; Marchant and Higgins 1993; Falkenberg 2011; Sutton 2011). One to 4 eggs are laid in the old nests of other birds, particularly those of other raptors or corvids. The nests chosen are usually in the tallest trees along watercourses, particularly River Red Gum *Eucalyptus camaldulensis* and Coolibah *E. coolabah*, but also on telecommunication towers (Marchant and Higgins 1993; Schoenjahn 2009; Falkenburg 2010). In any 1 area, nesting may be confined to years of above average rainfall, with the falcons leaving during drier times (Aumann 2001). A generation time of 6.2 years (BirdLife International 2011) is derived from an age at first breeding of 1.6 years, a maximum longevity of 13.6 years and mean adult annual survival of 67%, all values extrapolated from congeners.

Threats

All threats to the Grey Falcon are speculative and are concentrated at the wetter margins of the species' range (Olsen and Olsen 1986). In particular, the contraction in its breeding distribution in the 20th century is attributed to habitat degradation due to over grazing and ongoing clearing for agriculture, which has reduced the suitability of some semi-arid habitat and appears to have restricted the species to the arid zone (Olsen 1998, Falkenburg 2011). The provision of water for stock may have favoured the more mesic-adapted Peregrine Falcon over this arid-adapted species (Hollands 1984). In the very long term, nest site availability, particularly in sparsely-treed parts of the inland, could become limiting, especially where grazing by exotic herbivores is preventing regeneration of trees (Marchant and Higgins 1993; Duncan *et al.* 2007; Cheal 2009). Even in the remotest parts of the arid zone there is grazing by camels *Camelus dromerdarius.*

Conservation objectives

1. Stable populations in monitored regions
2. Successful breeding over the long term at monitored nests

Information required

1. Methods for assessing population trends
2. Base-line ecological and demographic information
3. Replacement rate of trees in nesting areas

Management actions required

1. Institute standardised long-term monitoring in selected parts of range, including both arid and semi-arid zones

Bibliography

Aumann T (2001) Breeding biology of raptors in the south-west of the Northern Territory, Australia. *Emu* **101**, 305–315.

Barrett G, Silcocks A, Barry S, Cunningham R, Poulter R (2003) *The New Atlas of Australian Birds*. Royal Australian Ornithologists Union, Hawthorn East.

BirdLife International (2011) 'Species factsheet: *Falco hypoleucos*'. Retrieved 17 January, 2011 from <http://www.birdlife.org/>.

Blakers MS, Davies JJF, Reilly PN (1984) *The Atlas of Australian Birds*. Melbourne University Press, Melbourne.

Cheal D (2009) Twenty years of grazing reduction in semi-arid woodlands. *Pacific Conservation Biology* **15**, 268–277.

Duncan D, Moxham C, Read C (2007) 'Effect of stock removal on woodlands in the Murray Mallee and Wimmera Bioregions of Victoria'. A report by the Arthur Rylah Institute for Environmental Research, Heidelberg for the Wimmera and Mallee Catchment Management Authorities.

Falkenburg ID (2010) Aspects of the ecology of the Grey Falcon *Falco hypoleucos* in the South Australian arid zone. *Corella* **35**, 23–28.

Hollands D (1984) *Eagles, Hawks and Falcons of Australia*. Thomas Nelson, Melbourne.

Johnstone RE, Storr GM (1998) *Handbook of Western Australian Birds. Vol. 1. Non-passerines (Emu to Dollarbird)*. Western Australian Museum, Perth.

Marchant S, Higgins PJ (Eds) (1993) *Handbook of Australian, New Zealand and Antarctic Birds. Volume 2: Raptors to Lapwings*. Oxford University Press, Melbourne.

Martin WK, Royal MJ (2000) Easterly records of the Grey Falcon in New South Wales. *Australian Bird Watcher* **18**, 132–134.

Olsen P (1998) 'Australia's raptors: diurnal birds of prey and owls'. Birds Australia Conservation Statement no. 3.

Olsen PD, Olsen J (1986) Distribution, status, movements and breeding of the Grey Falcon *Falco hypoleucos*. *Emu* **86**, 47–51.

Olsen PD, Olsen J (1988) Population trends, distribution and status of the Peregrine Falcon in Australia. In *Peregrine Falcon Populations: Their Management and Recovery*. (Eds TJ Cade, JH Enderson, CG Thelander, CM White) pp. 258–274. The Peregrine Fund, New York.

Reid J, Fleming M (1992) The conservation status of birds in arid Australia. *Rangeland Journal* **14**, 65–91.

Schoenjahn J (2009) Movements and genetics of Grey Falcons. *Boobook* **26**, 53–54.

Schoenjahn J (2010) Field Identification of the Grey Falcon *Falco hypoleucos*. *Australian Field Ornithology* **27**, 49–58.

Schoenjahn J (2011) How scarce is the Grey Falcon? *Boobook* **29**, 24–25.

Sutton AJG (2011) Aspects of the biology of the Grey Falcon *Falco hypoleucos* in the Pilbara region of Western Australia. *Corella* **35**, 11–15.

Comments received from

Stephen Debus, Victor Hurley, Penny Olsen, Chris Pavey, Julian Reid, Jonny Schoenjahn

Species Conservation Summary

White Gallinule

Porphyrio albus (Shaw, 1790)
Rallidae

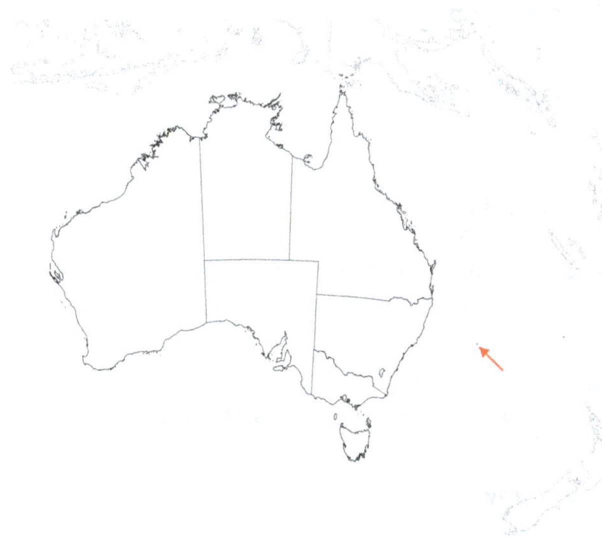

Conservation status
Extinct

Reasons for listing
Last recorded in 1788

Status 2000
Extinct

Status 1990
Extinct

Taxonomy
No infraspecific taxa described. Large and flightless with massive skull, long legs and short toes, the White Gallinule is most closely related to the Takahe *Notornis mantelli* (Hutton 1991).
Taxonomic uniqueness: medium (33 genera/family, 10 species/genus, 1 subspecies/species)

Former range
The White Gallinule was endemic to Lord Howe Island. The species was noted when the island was discovered in 1788 (Hutton 1991). As the first settlers on the island in the 1830s did not comment on the species, it had presumably already become Extinct (Hindwood 1940). Records from Ball's Pyramid and Norfolk Island are considered spurious (Hindwood 1940). The species is known from 2 skins, several paintings, and some subfossil bones (BirdLife International 2011).

Former abundance
It is thought to have been numerous (Hutton 1991), though, if the carrying capacity of the Lord Howe

Woodhen *Gallirallus sylvestris* is only 200 (Brook *et al.* 1997), the carrying capacity of this larger bird may not have been all that great.

Ecology

Nothing is known of the species, habits or breeding (McAllan *et al.* 2004). It is thought to have been a ground-feeding omnivore, feeding on plants and invertebrates (Hutton 1991).

Reasons for extinction

The White Gallinule was hunted with enthusiasm by members of the First Fleet. Almost certainly, hunting by the sailors and subsequent itinerant visitors caused the species' extinction (Hindwood 1940).

Bibliography

BirdLife International (2011) 'Species factsheet: *Porphyrio albus*'. Retrieved 17 January, 2011 from <http://www.birdlife.org/>.

Brook BW, Lim L, Harden R, Frankham R (1997) How secure is the Lord Howe Island Woodhen? A population viability analysis using VORTEX. *Pacific Conservation Biology* **3**, 125–133.

Hindwood KA (1940) The birds of Lord Howe Island. *Emu* **40**, 1–86.

Hutton I (1991) *Birds of Lord Howe Island, Past and Present*. The Author, Coffs Harbour.

McAllan IAW, Curtis BR, Hutton I, Cooper RM (2004) The birds of the Lord Howe Island Group: a review of records. *Australian Field Ornithology* **21**, 1–82.

Comments received from

David Priddel

Subspecies Conservation Summary

Lewin's Rail (Tasmanian)

Lewinia pectoralis brachipus (Swainson, 1838)
Rallidae

Conservation status

Vulnerable C2a(ii)

Reasons for listing

Population probably <10 000 mature individuals, in a single population and declining

Status 2000

Vulnerable A2b
Reason for change in 2010: rapid declines thought to have occurred during previous 2 decades

Status 1990

Near Threatened A4b
Reason for change in 2010: rapid declines were probably occurring

Taxonomy

R. p. clelandi (Western Australia) is Extinct, *R. p. pectoralis* (mainland eastern Australia) is Least Concern. There are 7 other subspecies. The species is Least Concern.
Taxonomic uniqueness: medium (33 genera/family, 3 species/genus, 9 subspecies/species)

Range

Scattered distribution through near-coastal environments around Tasmania and some offshore islands including Furneaux Is, King I. and many small islands (Marchant and Higgins 1993; Barrett *et al.* 2003). Although recorded on King I. in the first Birds Australia Atlas in 1977–1981, there have been no Atlas records there since 1998 (Barrett *et al.* 2003; Birds Australia *in litt.*).

Abundance

Assessing population size and trends is difficult because the rail is secretive and calls infrequently. However, whereas the reporting rate in the first Birds Australia Atlas in 1977–1981 was 5 times higher in the 4 Tasmanian bioregions where it was recorded than in the 15 mainland bioregions, the number of records from Tasmania were too few to analyse for the period 1998–2002 (S. Garnett pers. obs.). Subsequent reports suggest no sign of recovery. An estimate of 5000 birds that appears still to be declining is offered here pending more accurate estimates.

Ecology

Inhabits a wide range of habitats but are most common in coastal or near-coastal permanent to ephemeral, fresh to

Current eligibility against IUCN Red List Criteria

IUCN category	Criteria eligibility
A	Not applicable: past, current or future population declines are thought unlikely to exceed 20% in any 3-generation period
B	Not applicable: >10 locations, not severely fragmented, population fluctuations not extreme
C	Vulnerable: population 2500–10 000 mature individuals, suspected continuing decline, 100% in 1 subpopulation
D	Not applicable: population >1000 mature individuals, >5 locations
E	Not applicable: no population viability analysis undertaken

IUCN Red List assessment data

	Estimate	Reliability
Extent of occurrence trend	70 000 km^2 decreasing	high low
Area of occupancy trend	240 km^2 decreasing	high low
No. of mature individuals trend	5000 decreasing	low low
No. subpopulations	1	medium
No. locations	>10	high
Generation time	5.5 years	low
Global population share	100%	high

saline wetlands that have dense emergent or fringing vegetation. They sometimes also use artificial habitats with similar structural features, rainforest wetlands, riverine forest, thickets of wetland shrubs and wet heathlands, or even dry habitats such as parks, agricultural land, grass or ferns under scrub or thickets of brambles or *Lantana*. They mostly eat invertebrates, rarely birds eggs and frogs (Marchant and Higgins 1993). A generation time of 3.7 years (BirdLife International 2011) is derived from an average age at first breeding of 1.0 years and maximum longevity of 6.3 years, both extrapolated from mean values for other genera in the Rallidae.

Threats

Although extended drought has meant that many wetlands have been unsuitable for extended periods, the rapid decline in Tas is still puzzling as recording frequency has not increased following rainfall, suggesting that drought may have had a more permanent impact. Lewin's Rails are likely to abandon wetlands where the fringing vegetation has been degraded by grazing, inappropriate burning, or trampling by stock (Marchant and Higgins 1993). Extinction of the south-western subspecies suggests that the species is more vulnerable to habitat change than many other wetland birds. They may also be vulnerable to predation by fox *Vulpes vulpes*, which has recently become established in Tas, and feral cats *Felis catus*.

Conservation objectives

1. Stable numbers of birds recorded in standardised surveys at key sites

Information required

1. Location and trends in populations
2. Improved understanding of movements
3. Principal breeding and non-breeding wetlands

Management actions required

1. Undertake a dedicated survey to determine distribution and relative abundance
2. Collaborate with land owners and managers to conserve swamps and other wetlands where subspecies occurs

Bibliography

Barrett G, Silcocks A, Barry S, Cunningham R, Poulter R (2003) *The New Atlas of Australian Birds*. Royal Australian Ornithologists Union, Hawthorn East.

BirdLife International (2011) 'Species factsheet: *Lewinia pectoralis*'. Retrieved 17 January, 2011 from <http://www.birdlife.org/>.

Marchant S, Higgins PJ (Eds) (1993) *Handbook of Australian, New Zealand and Antarctic Birds. Volume 2: Raptors to Lapwings*. Oxford University Press, Melbourne.

Comments received from

Richard Donaghey, Mark Holdsworth, Eric Woehler

Subspecies Conservation Summary

Lewin's Rail (Western Australian)

Rallus pectoralis clelandi (Mathews, 1911)
Rallidae

Conservation status
Extinct

Reasons for listing
There have been no records of this subspecies since 1932, despite intensive surveys

Status 2000
Extinct

Status 1990
Extinct

Taxonomy
R. p. brachipus (Tasmanian) is Vulnerable, *R. p. pectoralis* (mainland eastern Australia) is Least Concern. There are 7 other subspecies. The species is Least Concern.
Taxonomic uniqueness: medium (33 genera/family, 3 species/genus, 10 subspecies/species)

Former range
Occurred only in the south-western corner of Western Australia (BirdLife International 2011; Department of Sustainability, Environment, Water, Population and Communities 2011).

Former abundance
Although Whittell (1933) implied that it was reasonably common in the early 1930s, he procured only 1 specimen, near Bridgetown in December 1931, and saw another bird 10 months later. His records are the last to appear in the literature.

Ecology
Occupied swamps (Whittell 1933), but no more detailed description of its habitat or ecology exists.

Reasons for Extinction
The subspecies would have been particularly vulnerable at times of drought, when it is likely to have been confined to a few refugia. Many swamps in south-western Australia were drained for agriculture early this century (Sedgwick 1940; Serventy 1947) and other swamps in the south-west were burnt regularly by graziers (A. H. Burbidge pers. comm., R. E. Johnstone pers. comm.).

Bibliography

BirdLife International (2011) 'Species factsheet: *Lewinia pectoralis*'. Retrieved 17 January, 2011 from <http://www.birdlife.org/>.

Department of Sustainability, Environment, Water, Population and Communities (2011) *Lewinia pectoralis clelandi*. In 'Species Profile and Threats Database'. Department of Sustainability, Environment, Water, Population and Communities, Canberra. Retrieved 24 February, 2011 from <http://www.environment.gov.au/sprat>.

Sedgwick EH (1940) The birds of the Rockingham district. *Emu* **40**, 129–152.

Serventy DL (1947) Birds of the Swan River district, Western Australia. *Emu* **47**, 241–286.

Whittell HM (1933) The birds of Bridgetown district, south-west Australia. *Emu* **32**, 182–189.

Subspecies Conservation Summary

Buff-banded Rail (Cocos (Keeling) Islands)

Gallirallus philippensis andrewsi (Mathews, 1911)
Rallidae

Conservation status
Vulnerable D1+2

Reasons for listing
Population 250–1000 mature individuals in 1 location; exotic predator introduction is a plausible future threat

Status 2000
Vulnerable D1+2

Status 1990
Vulnerable D1+2

Taxonomy
G. p. mellori (Tasmania and mainland Australia) is Least Concern and *G. p. macquariensis* (Macquarie I.) is Extinct. There are 6 extant extralimital subspecies. The species global status is Least Concern.
Taxonomic uniqueness: low (33 genera/family, 16 species/genus, 20 subspecies/species)

Range
Widespread and common on all islands of Cocos (Keeling) Is. until early 1900s (Stokes *et al.* 1984). Now confined to North Keeling I., but persisted on South I. (Pulu Atas) and West I. (Pulu Panjang) of the main atoll until the early 1980s (Stokes *et al.* 1984; Marchant and Higgins 1993) but could not be found in 2005 despite earlier anecdotal reports (Reid and Hill 2005; Department of Sustainability, Environment, Water, Population and Communities 2011).

Abundance
Population in 1999 estimated at 750–800 individuals (Reid 2000). More accurate transect counts in 2005 estimated a population of c.850–1000, suggesting the population is stable (Reid and Hill 2005). The population nominated here is the average for the most recent estimate.

Ecology
Uses all habitats on North Keeling, foraging along the lagoon shore and nesting in Birdlime Tree *Pisonia grandis* debris, grass tussocks or similar ground layer vegetation, the forks of Birdlime Trees, and the bases of coconut palms. Up to 40 birds at a time have been seen foraging on the lagoon shore, and this seems to be a preferred feeding habitat (Reid and Hill 2005). On West I., they were seen feeding on the airstrip and sheltering in *Scaevola taccada* thickets. The rails are omnivorous, taking mostly crustaceans, molluscs, worms and insects, but also plants, seeds, fruits, frogs, eggs, carrion and refuse (Stokes *et al.* 1984; Marchant and Higgins 1993; Reid 2000). A generation time of 3.4 years (BirdLife International 2011) is derived from an age at first breeding of 1.0 years and an annual survival of adults of 58.0%, both extrapolated from mean values for *Rallus* spp.

Threats
Conversion of diverse *Pisonia* forests to coconut plantations, predation by feral cats *Felis catus*, Black Rats *Rattus rattus* and perhaps exploitation by people, have probably caused its extinction on the main atoll. The greatest threat now is the accidental introduction of additional predators (Reid 2000; Commonwealth of Australia 2005).

Conservation objectives
1. A viable and secure population on North Keeling I.
2. All major threats on the island targeted for reintroduction in the southern atoll securely abated

Current eligibility against IUCN Red List Criteria

IUCN category	Criteria eligibility
A	Not applicable: past, current or future population declines are thought unlikely to exceed 20% in any 3-generation period
B	Not applicable: EOO, AOO, habitat, locations or population not declining, population fluctuations not extreme
C	Not applicable: no continuing population decline or extreme fluctuations
D	Vulnerable: population 250–1000 mature individuals, ≤5 locations with plausible future threat
E	Not applicable: no population viability analysis undertaken

IUCN Red List assessment data

	Estimate	Reliability
Extent of occurrence trend	4 km^2 stable	high high
Area of occupancy trend	2 km^2 stable	high high
No. of mature individuals trend	930 stable	medium low
No. subpopulations	1	high
No. locations	1	medium
Generation time	3.4 years	low
Global population share	100%	high

3. A second viable population
4. Community support for and involvement in the rail's recovery developed
5. Significant threat reduction over other islands in the southern atoll for dispersing rails

Information required

1. Population trends North Keeling I.
2. Predator levels on southern atoll

Management actions required

1. Continue to protect North Keeling I as outlined in management plan
2. Monitor by transect every 6 months for 3 years, then review
3. Translocate rails to the southern atoll
4. Ensure engagement and support from local community

Bibliography

BirdLife International (2011) 'Species factsheet: *Gallirallus philippensis*'. Retrieved 17 January, 2011 from <http://www.birdlife.org/>.

Commonwealth of Australia (2005) 'National recovery plan for the Buff-banded Rail (Cocos (Keeling) Islands) *Gallirallus philippensis andrewsi*'. Department of the Environment and Heritage, Canberra.

Department of Sustainability, Environment, Water, Population and Communities (2011) *Gallirallus philippensis andrewsi.* In 'Species Profile and Threats Database'. Department of Sustainability, Environment, Water, Population and Communities, Canberra. Retrieved 24 February, 2011 from <http://www.environment.gov.au/sprat>.

Marchant S, Higgins PJ (Eds) (1993) *Handbook of Australian, New Zealand and Antarctic Birds. Volume 2: Raptors to Lapwings.* Oxford University Press, Melbourne.

Reid JRW (2000) 'Survey of the Buff-banded Rail (*Rallus philippensis andrewsi*) in Pulu Keeling National Park, Cocos Islands, Indian Ocean'. Report to Parks Australia North.

Reid JRW, Hill BM (2005) 'Recent surveys of the Cocos Buff-banded Rail'. CRES, The Australian National University, Canberra.

Stokes AS, Shiels W, Dunn K (1984) Birds of the Cocos (Keeling) Islands, Indian Ocean. *Emu* **84**, 23–28.

Comments received from

Andrew Burbidge, Julian Reid

Subspecies Conservation Summary

Buff-banded Rail (Macquarie Island)

Gallirallus philippensis macquariensis (Hutton, 1879)
Rallidae

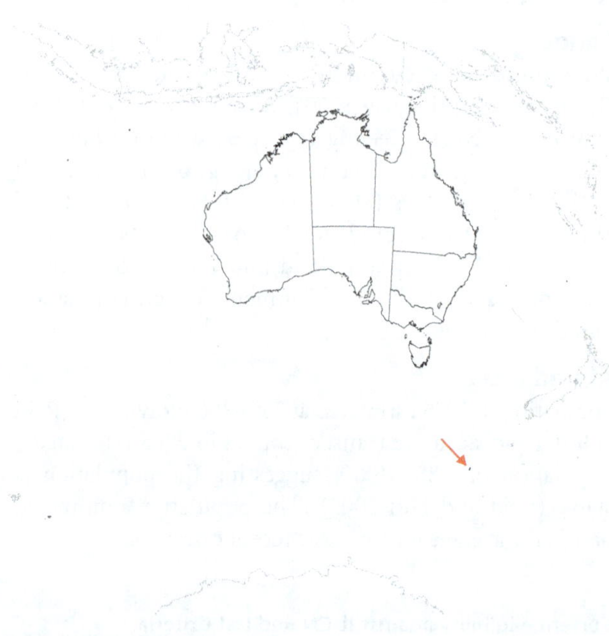

Conservation status

Extinct

Reasons for listing

This subspecies was last recorded in 1879

Status 2000

Extinct

Status 1990

Extinct

Taxonomy

G. p. andrewsi (Cocos-Keeling Is) is Vulnerable and *G. p. mellori* (Tasmania and mainland Australia) is Least Concern. There are 16 other extant subspecies through South-East Asia and the Pacific. The species is Least Concern.

Taxonomic uniqueness: low (33 genera/family, 16 species/genus, 20 subspecies/species)

Former range

This subspecies was endemic to Macquarie I.

Former abundance

It was formerly considered common (Scott 1882), although no population estimate was ever made (Vestjens 1963). There were so few records that it is unknown whether the subspecies occurred over the whole island (Department of Sustainability, Environment, Water, Population and Communities 2011).

Ecology

The subspecies occupied tussock grassland (Hamilton 1894), but nothing else is known of its ecology.

Reasons for extinction

This rail co-existed with feral cats *Felis catus* on Macquarie I. for at least 70 years, but Wekas *Gallirallus australis scotti* were introduced in 1872, and may have also preyed on eggs and young (Brothers and Skira 1983). The decline of the rails was particularly rapid after 1878, when rabbits *Oryctolagus cuniculus* became established. It is thought that the rabbits reduced the area of suitable habitat by grazing tussocks and enabled the population of cats to increase. Rabbits are scarce in winter (Jones 1977), and the cats may then have hunted rails (Hamilton 1894).

Bibliography

Brothers NP, Skira IJ (1983) The Weka on Macquarie Island. *Notornis* **31**, 145–154.

Department of Sustainability, Environment, Water, Population and Communities (2011) *Gallirallus philippensis macquariensis*. In 'Species Profile and Threats Database'. Department of Sustainability, Environment, Water, Population and Communities, Canberra. Retrieved 24 February, 2011 from <http://www.environment.gov.au/sprat>.

Hamilton A (1894) Notes on a visit to Macquarie Island. *Transactions and Proceedings of the New Zealand Institute* **27**, 559–579.

Jones E (1977) Ecology of the Feral Cat *Felis catus* (L.), (Carnivora: Felidae) on Macquarie Island. *Australian Wildlife Research* **4**, 249–262.

Scott JH (1882) Macquarie Island. *Transactions and Proceedings of the New Zealand Institute* **15**, 484–493.

Vestjens WJM (1963) Remains of the extinct Banded Rail at Macquarie Island. *Emu* **62**, 249–250.

Species Conservation Summary

Lord Howe Woodhen

Gallirallus sylvestris (Sclater, 1869)

Rallidae

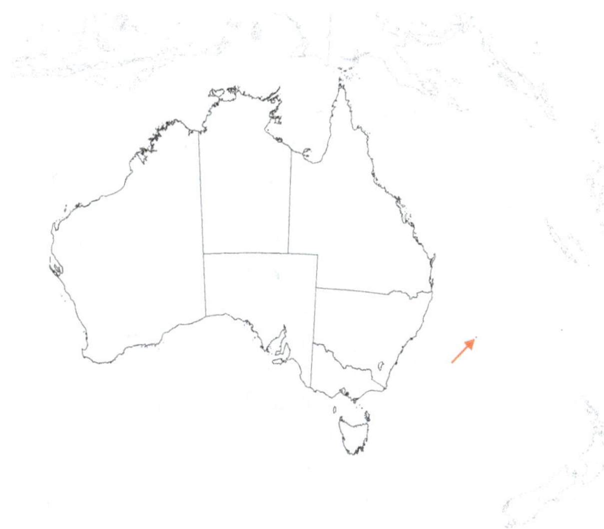

Conservation status

Endangered D

Reasons for listing

Population <250 mature individuals

Status 2000

Endangered D

Status 1990

Endangered D

Taxonomy

No infraspecific taxa described

Taxonomic uniqueness: medium (33 genera/family, 16 species/genus, 1 subspecies/species)

Range

The Lord Howe Woodhen is endemic to Lord Howe Island. When discovered in 1788 it was described as common and distributed from sea-level to the tops of the 2 mountains. From the mid-19th century Woodhens were confined to summit regions of Mt Gower and Mt Lidgbird (Hutton 1991). At its nadir in the 1970s the population comprised fewer than 10 breeding pairs (Brook *et al.* 1997). Following control of threats, captive-bred birds were reintroduced to lowland sites in the 1980s (Miller and Mullette 1985). At present, there are birds on the summit of Mt Gower and at Big and Little Slopes, Grey face and Far Flats with about half the population in the settlement area (Department of Environment and Climate Change (NSW) 2007).

Abundance

The total population is estimated to be 220–230 birds and 71–74 breeding pairs and stable, having reached carrying capacity within 5 years of threats having been removed (NSW National Parks and Wildlife Service 2002). The island's total carrying capacity has been estimated at 220 individuals (Brook *et al.* 1997).

Ecology

The Lord Howe Woodhen is sedentary and flightless. On Mt Gower they occur in gnarled mossy forest that is characteristic of the mountain summits. At lower altitudes, most territories are in megaphyllous broad sclerophyll forest, particularly Kentia Palm *Howea forsterana* forests on, or immediately downslope of, areas of igneous geology. Territories in other settings have been associated with residences where supplementary food has been

Current eligibility against IUCN Red List Criteria

IUCN category	Criteria eligibility
A	Not applicable: past, current or future population declines are thought unlikely to exceed 20% in any 3-generation period
B	Not applicable: EOO, AOO, habitat, locations or population not declining, population fluctuations not extreme
C	Not applicable: no continuing population decline or extreme fluctuations
D	Endangered: population 50–250 mature individuals
E	Vulnerable: population viability analysis suggests a 10% probability of extinction in 100 years

IUCN Red List assessment data

	Estimate	Reliability
Extent of occurrence trend	15 km^2 stable	high high
Area of occupancy trend	15 km^2 stable	high high
No. of mature individuals trend	150 stable	high high
No. subpopulations	1	high
No. locations	1	high
Generation time	8 years	medium
Global population share	100%	high

available (Harden and Robertshaw 1988, 1989). They are rarely found in the rainforest that covers most of the island. They forage amongst litter on the forest floor, feeding on worms, molluscs and invertebrates. They nest on the ground under thick vegetation or in petrel burrows (Department of Sustainability, Environment, Water, Population and Communities 2011).

Threats

Reported to be the principal food of the early settlers (Hutton 1991), Woodhens were eliminated from the lowlands in the 19th century as a result of hunting by people, and predation by feral cats *Felis catus* and feral pigs *Sus scrofa*, as well as habitat degradation caused by pigs and goats *Capra hircus* (Miller and Mullette 1985). These threats were eliminated in the 1980s. Masked Owls *Tyto novaehollandiae castanops*, introduced to control introduced Black Rats *Rattus rattus* in the 1920s, are thought to be responsible for a major decline in the number of birds using Little Slope in 1989, and remain a threat. Buff-banded Rails *Gallirallus philippensis* apparently co-existed with Woodhens before rats were introduced, after which the Buff-banded Rails disappeared. However, since Buff-banded Rails re-colonised the island since the 1970s (McAllan *et al.* 2004) they may outcompete the Lord Howe Woodhen: they have temporarily expelled Woodhens from settlement territories and are often in conflict with Woodhens, especially during the breeding season (Department of Environment and Climate Change (NSW) 2007). Woodhens have successfully persisted alongside rats since 1918 but are likely to be in danger of poisoning from brodifacoum, the bait favoured for planned eradication of Black Rats from the Lord Howe I. group, which is why it is planned to take as many Woodhens as possible into captivity while the baits are toxic (Lord Howe Island Board 2009). Population Viability Analysis using VORTEX suggests that the species remains highly susceptible to changes in survival or fecundity (Brook *et al.* 1997).

Conservation objectives

1. Woodhen population on Lord Howe I. maintained
2. Lord Howe I. community involved in monitoring, management, habitat rehabilitation and threat abatement
3. Viable captive populations established at sites other than Lord Howe I.

Information required

1. Long-term impact of Buff-banded Rail on Lord Howe Woodhen

Management actions required

1. Capture and retain in captivity on the island as many individuals as possible during the rat eradication period, releasing back to the wild as soon as possible after poisoned rodents and baits have decayed
2. Establish a captive population off the island
3. Eradicate Masked Owls
4. Assess the potential impact of Buff-banded Rail and, if necessary, develop and implement a strategy for control
5. Continue annual population monitoring
6. Maintain high quarantine standards

Bibliography

Brook BW, Lim L, Harden R, Frankham R (1997) How secure is the Lord Howe Island Woodhen? A population viability analysis using VORTEX. *Pacific Conservation Biology* **3**, 125–133.

Department of Environment and Climate Change (NSW) (2007) 'Lord Howe Island Biodiversity Management Plan'. Department of Environment and Climate Change (NSW), Sydney.

Department of Sustainability, Environment, Water, Population and Communities (2011) *Gallirallus sylvestris.* In 'Species Profile and Threats Database'. Department of Sustainability, Environment, Water, Population and Communities, Canberra. Retrieved 24 February, 2011 from <http://www.environment.gov.au/sprat>.

Harden R, Robertshaw J (1988) 'Lord Howe Island Woodhen Census 1987'. Unpublished report. Environmental Survey and Research Branch, National Parks and Wildlife Service.

Harden R, Robertshaw J (1989) 'Lord Howe Island Woodhen Census 1988'. Unpublished report. Environmental Survey and Research Branch, National Parks and Wildlife Service.

Hutton I (1991) *Birds of Lord Howe Island, Past and Present.* The Author, Coffs Harbour.

Lord Howe Island Board (2009) 'Draft Lord Howe Island Rodent Eradication Plan'. Lord Howe Island Board, Lord Howe Island.

McAllan IAW, Curtis BR, Hutton I, Cooper RM (2004) The birds of the Lord Howe Island Group: a review of records. *Australian Field Ornithology* **21**, 1–82.

Miller BJ, Mullette KJ (1985) Rehabilitation of an endangered Australian bird: the Lord Howe Island Woodhen *Tricholimnas sylvestris* (Selater). *Biological Conservation* **34**, 55–95.

NSW National Parks and Wildlife Service (2002) 'Approved recovery plan for the Lord Howe Woodhen'. NSW National Parks and Wildlife Service, Hurstville, NSW.

Comments received from

Robert Harden, David Priddel

Subspecies Conservation Summary

Black-faced Sheathbill (Heard Island)

Chionis minor nasicornis Reichenow, 1904

Chionididae

Conservation status

Vulnerable D1

Reasons for listing

Population probably <1000 mature individuals

Status 2000

Vulnerable D1

Status 1990

Vulnerable D1

Taxonomy

C. m. minor (Kerguelen I.), *C. m. crozettensis* (Crozet I.) and *C m. marionensis* (Prince Edward I.) are extralimital. Species status is Least Concern.

Taxonomic uniqueness: high (1 genus/family, 2 species/genus, 4 subspecies/species)

Range

Endemic to Heard I., foraging along the shoreline, believed to remain on the island year-round (BirdLife International 2011).

Abundance

The population is estimated at 100–1000 (Woehler 1991) and <1000 pairs (Woehler 2010).

Ecology

Omnivorous scavengers found on subantarctic islands (Marchant and Higgins 1993; Jouventin *et al.* 1996). In the warmer months, they mainly scavenge from seals and birds, taking meat, blood, eggs, excreta and/or milk. During winter, when few other vertebrates remain on the islands, algae are taken from rocks. They also feed on a range of invertebrates and, on other islands, scavenge among human refuse. Breeding pairs are highly territorial, and make their nests under rocks on scree slopes (Marchant and Higgins 1993). Breeding coincides with that of other seabirds, especially penguins, on whose eggs and chicks they prey to provide food for nestlings (Marchant and Higgins 1993; Jouventin *et al.* 1996). A generation time of 9.0 years (BirdLife International 2011) is derived from an age of first breeding of 4.0 years and a maximum longevity in the wild of 14.0 years (Marchant and Higgins 1993).

Threats

Currently no plausible threats but any small, isolated population is vulnerable to catastrophes. Decreases in the number of either seabirds or seals could have an adverse impact on them. Their naivety to predators may mean they could be affected should any predators reach the breeding grounds.

Conservation objectives

1. Persistence of population

Information required

1. Population trends

Management actions required

1. Maintain quarantine provisions

Bibliography

BirdLife International (2011) 'Species factsheet: *Chionis minor*'. Retrieved 17 January, 2011 from <http://www.birdlife.org/>.

Jouventin P, Bried J, Ausilio E (1996) Life-history variations of the Lesser Sheathbill *Chionis minor* in contrasting situations. *Ibis* **138**, 732–741.

Marchant S, Higgins PJ (Eds) (1993) *Handbook of Australian, New Zealand and Antarctic Birds. Volume 2: Raptors to Lapwings.* Oxford University Press, Melbourne.

Woehler EJ (1991) The status and conservation of seabirds of Heard Island and the McDonald Islands. In 'Seabird Status

Current eligibility against IUCN Red List Criteria

IUCN category	Criteria eligibility
A	Not applicable: past, current or future population declines are thought unlikely to exceed 20% in any 3-generation period
B	Not applicable: EOO, AOO, habitat, locations or population not declining, population fluctuations not extreme
C	Not applicable: no continuing population decline or extreme fluctuations
D	Vulnerable: population 250–1000 mature individuals, ≤5 locations but no plausible future threat
E	Not applicable: no population viability analysis undertaken

IUCN Red List assessment data

	Estimate	Reliability
Extent of occurrence trend	150 km^2 stable	high high
Area of occupancy trend	30 km^2 stable	high high
No. of mature individuals trend	1000 stable	medium high
No. subpopulations	1	high
No. locations	1	high
Generation time	9.0 years	medium
Global population share	100%	high

and Conservation. ICBP Tech. Publ. No 11.' (Ed. JP Croxall) pp. 263–277. ICBP, Cambridge, UK.

Woehler EJ (2010) Status and conservation of the seabirds of Heard Island. *The State of Australia's Birds 2010: Islands and Birds. Wingspan* **Supplement 20**, 8–9.

Comments received from

Eric Woehler

Species Conservation Summary

Grey Plover

Pluvialis squatarola (Linnaeus, 1758)

Charadriidae

Conservation status

Population visiting Australia: Near Threatened A2bc+3c+4bc

Reasons for listing

Past, recent and ongoing decline of 30–49% in 3 generations (22 years) based on survey data and habitat loss. However, non-declining populations in other flyways are not phenotypically differentiated, suggesting genetic interchange, so downgraded to Near Threatened

Status 2000

Least Concern

Reason for change in 2010: decline in population and loss of habitat at staging posts

Status 1990

Least Concern

Taxonomy

No infraspecific taxa described

Taxonomic uniqueness: high (11 genera/family, 4 species/genus, 1 subspecies/species)

Range

Breeds around the Arctic, wintering worldwide along temperate and tropical coasts (Bamford *et al.* 2008). Non-breeding birds occur around coastal Australia, with large numbers at sites in both the north and south (Department of Sustainability, Environment, Water, Population and Communities 2011).

Abundance

The global population has been estimated at 692 000 individuals and the flyway population at 125 000 (Wetlands International 2006; Bamford *et al.* 2008) including 12 000 in Australia (Stewart *et al.* 2007). Numbers declined on Eighty-mile Beach and Bush Point, Western Australia, by c.28% between 2000 and 2008 (Rogers *et al.* 2009), at Gulf St Vincent, South Australia, by c.52% between c.1981 and c.2004 (Close 2008), by 3% per year at sites across Queensland up to 2008 (Fuller *et al.* 2009) and by c.40% across 49 Australian sites between c.1983 and c.2007 (AWSG database: Birds Australia *in litt.*). Populations in Japan have declined in general, and by about 38% in spring between

Current eligibility against IUCN Red List Criteria

IUCN category	Criteria eligibility
Population visiting Australia	
A	Vulnerable: past, current and anticipated population declines of 30–49% over periods of 3 generations (22 years) estimated and projected from monitoring and from deteriorating habitat quality (downgraded to Near Threatened)
B	Not applicable: EOO >20 000 km^2; AOO >2000 km^2
C	Not applicable: population >10 000 mature individuals
D	Not applicable: population >1000 mature individuals, >5 locations
E	Not applicable: no population viability analysis undertaken

IUCN Red List assessment data

	Estimate	Reliability
Population visiting Australia		
Extent of occurrence trend	36 200 km^2 stable	high high
Area of occupancy trend	3000 km^2 stable	low medium
No. of mature individuals trend	12 000 decreasing	medium high
No. subpopulations	1	high
No. locations	>10	high
Generation time	7.2 years	low
Global population share	1–10%	high

1998 and 2008 (Amano *et al.* 2010). These suggest an overall 30–49% decline over 22 years across the flyway. Declines are likely to continue given ongoing threats to its migratory staging sites in East Asia (Barter 2002; Kelin and Qiang 2006; Moores 2006). Populations in other flyways are stable or declining slowly and the species is not considered threatened (Wetlands International 2006).

Ecology

Breeds in tundra, and non-breeding birds migrate to the temperate and tropical coasts worldwide, where they inhabit intertidal mudflats, saltmarshes, sandflats and beaches and feed on polychaete worms, molluscs and crustaceans (Marchant and Higgins 1993). A generation time of 7.2 years (BirdLife International 2011) is derived from an average age at first breeding of 2.5 years (Cramp *et al.* 1983), an annual adult survival of 64% (extrapolated from congeners) and a maximum longevity of 15.6 years (Australian Bird and Bat Banding Scheme).

Threats

Threatened by wetland degradation in East Asia, especially of intertidal flats in the Yellow Sea where c.80% of the flyway population stages on northward migration (Barter 2002; Bamford *et al.* 2008). General threats at migratory staging sites include environmental pollution, reduced river flows, human disturbance, hunting and reclamation for tidal power plants and barrages, industrial use, urban expansion (Barter *et al.* 1997; Barter 2002; Kelin and Qiang 2006; Moores 2006). Threats in Australia, especially eastern and southern Australia, include ongoing human disturbance, habitat loss and degradation from pollution, changes to the water regime and invasive plants (Rogers *et al.* 2006; Australian Government 2009).

Conservation objectives

1. Stable Australian non-breeding populations
2. Disturbance at key roosting and feeding sites reduced

Information required

1. Trends at more key sites for the Australian non-breeding population
2. Improved knowledge of dependence on key migratory staging sites
3. Improved knowledge of impacts of disturbance

Management actions required

1. Work with governments along the flyway at all levels, and local communities, to prevent destruction of key migratory staging sites
2. Negotiate with the Chinese government in particular to protect remaining tidal flats in Bohai Bay
3. Advocate for restoration of alienated wetland sites
4. Maintain and improve protection of roosting and feeding sites in Australia

Bibliography

Amano T, Székely T, Koyama K, Amano H, Sutherland WJ (2010) A framework for monitoring the status of populations: an example from wader populations in the East Asian–Australasian flyway. *Biological Conservation* **143**, 2238–2247.

Australian Government (2009) 'Draft significant impact guidelines for 36 migratory shorebirds.' Draft EPBC Act Policy Statement 3.21, Canberra. Retrieved 25 February, 2011 from <http://www.environment.gov.au/epbc/publications/migratory-shorebirds.html>.

Bamford M, Watkins D, Bancroft W, Tischler G, Wahl J (2008) 'Migratory shorebirds of the East Asian – Australasian flyway: population estimates and internationally important sites'. Wetlands International – Oceania, Canberra.

Barter M, Fawen Q, Sixian T, Xiao Y, Tonkinson D (1997) Hunting of migratory waders on Chongming Dao: a declining occupation? *Stilt* **31**, 19–22.

Barter MA (2002) 'Shorebirds of the Yellow Sea: importance, threats and conservation status'. Canberra, Australia.

BirdLife International (2011) 'Species factsheet: *Pluvialis squatarola*'. Retrieved 17 January, 2011 from <http://www.birdlife.org/>.

Close DH (2008) Changes in wader numbers in the Gulf St Vincent, South Australia, 1979–2008. *Stilt* **54**, 24–27.

Cramp S, Simmons KEL, Brooks DC, Collar NJ, Dunn E, Gillmor R, Hollom PAD, Hudson R, Nicholson EM, Ogilvie MA, Olney PJS, Roselaar CS, Voous KH, Wallace DIM, Wattel J, Wilson MG (Eds) (1983) *Handbook of the Birds of Europe, the Middle East and North Africa. The Birds of the Western Palearctic: 3. Waders to Gulls.* Oxford University Press, Oxford, UK.

Department of Sustainability, Environment, Water, Population and Communities (2011) *Pluvialis squatarola.* In 'Species Profile and Threats Database'. Department of Sustainability, Environment, Water, Population and Communities, Canberra. Retrieved 24 February, 2011 from <http://www.environment.gov.au/sprat>.

Fuller RA, Wilson HB, Kendall BE, Possingham HP (2009) 'Monitoring shorebirds using counts by the Queensland Wader Study Group'. Report to the Queensland Wader Study Group and the Department of Environment and Resource Management, Brisbane.

Kelin C, Qiang X (2006) Conserving migratory shorebirds in the Yellow Sea region. In *Waterbirds Around the World.* (Eds G Boere, C Galbraith and D Stroud) p. 319. The Stationery Office, Edinburgh, UK.

Marchant S, Higgins PJ (Eds) (1993) *Handbook of Australian, New Zealand and Antarctic Birds. Volume 2: Raptors to Lapwings.* Oxford University Press, Melbourne.

Moores N (2006) South Korea's shorebirds: a review of abundance, distribution, threats and conservation status. *Stilt* **50**, 62–72.

Rogers D, Hassell C, Oldland J, Clemens R, Boyle A, Rogers K (2009) 'Monitoring Yellow Sea migrants in Australia (MYSMA): north-western Australian shorebird surveys and workshops, December 2008.'

Rogers DI, Piersma T, Hassell CJ (2006) Roost availability may constrain shorebird distribution: exploring the energetic costs of roosting and disturbance around a tropical bay. *Biological Conservation* **133**, 225–235.

Stewart DA, Rogers A, Rogers DI (2007) Species description. In *Shorebirds of Australia.* (Eds A Geering, L Agnew and S Harding) pp. 75–196. CSIRO Publishing, Melbourne.

Wetlands International (2006) 'Waterbird population estimates'. Wetlands International, Wageningen, The Netherlands.

Comments received from

Rob Clemens, Ken Gosbell, Roger Jaensch, David Milton, Clive Minton, Danny Rogers

Subspecies Conservation Summary

Lesser Sand Plover (Mongolian)

Charadrius mongolus mongolus Pallas, 1776
Charadriidae

Conservation status

Population visiting Australia: Endangered A2bc+3c+4bc

Reasons for listing

Past, current and anticipated declines of 50–79% over periods of 3 generations (24 years) based on survey data and habitat loss

Status 2000

Near Threatened A3c

Reason for change in 2010: decline in population and ongoing loss of habitat at staging posts

Status 1990

Least Concern

Reason for change in 2000: decline in population occurring

Taxonomy

C. m. stegmanni (Kamchatkan) is Endangered in Australia, 3 other subspecies are extralimital. The species is Least Concern.

Taxonomic uniqueness: medium (11 genera/family, 35 species/genus, 5 subspecies/species)

Range

Breeds in Mongolia and Russia, wintering along coasts from Taiwan to Australasia (Bamford *et al.* 2008). Non-breeding birds occur along Australian coasts, especially in east Australia (Stewart *et al.* 2007; Department of Sustainability, Environment, Water, Population and Communities 2011).

Abundance

The global population was estimated at 40 000 individuals (Wetlands International 2006). There are no estimates of the Australian population, which is here assumed to be 50% of the Australian species population of 20 000–25 000, which also includes *C. m. stegmanni* (Stewart *et al.* 2007; Bamford *et al.* 2008). Numbers of the 2 subspecies combined declined by c.84% across 49 Australian sites between c.1983 and c.2007 (AWSG database: Birds Australia *in litt.*) but there are few data from northern Australia (e.g. Rogers *et al.* 2009). Trends in the sub-

Current eligibility against IUCN Red List Criteria

IUCN category	Criteria eligibility
Population visiting Australia	
A	Endangered: past, current and anticipated population declines of 50–79% over periods of 3 generations (24 years) estimated and projected from monitoring and from deteriorating habitat quality
B	Not applicable: EOO >20 000 km^2; AOO >2000 km^2
C	Not applicable: population >10 000 mature individuals
D	Not applicable: population >1000 mature individuals, >5 locations
E	Not applicable: no population viability analysis undertaken

IUCN Red List assessment data

	Estimate	Reliability
Population visiting Australia		
Extent of occurrence trend	35 300 km^2 stable	high high
Area of occupancy trend	2600 km^2 stable	low medium
No. of mature individuals trend	12 000 decreasing	medium high
No. subpopulations	1	high
No. locations	>10	high
Generation time	8.0 years	low
Global population share	26–50%	high

population outside Australia are poorly known but, in Japan, the 2 subspecies combined have declined in general, and by about 61% in autumn between 1978 and 2008 (Amano *et al.* 2010). These suggest an overall 50–79% decline over 16 years across the flyway, which is inferred to continue given likely ongoing threats to its migratory staging sites in East Asia (Barter 2002; Kelin and Qiang 2006; Moores 2006). Population trends in other flyways are poorly known and the species as a whole is not considered threatened (Wetlands International 2006; BirdLife International 2011).

Ecology

Breeds in barren valleys and basins in tundra and mountain steppe. Non-breeding birds migrate to the coast, where they inhabit mudflats of coastal bays and estuaries, sand-flats and dunes, and feed on insects, crustaceans, molluscs and polycheate worms (Marchant and Higgins 1993). A generation time of 8.0 years (BirdLife International 2011) is derived from an average age at first breeding of 2.0 years, an annual adult survival of 56% and a maximum longevity of 12.6 years, all extrapolated from *C. leschenaultii.*

Threats

Threatened by wetland degradation in East Asia, especially intertidal flats in the Yellow Sea where 50% of the estimated combined populations of *C. m. mongolus* and *C. m. stegmanni* stage during their northward migration (Barter 2002). General threats at migratory staging sites include environmental pollution, reduced river flows, human disturbance and reclamation for tidal power plants and barrages, industrial use, urban expansion and hunting (Barter *et al.* 1997; Barter 2002; Kelin and Qiang 2006; Moores 2006). Threats in Australia, especially eastern and southern Australia, include ongoing human disturbance, habitat loss and degradation from pollution, changes to the water regime and invasive plants (Rogers *et al.* 2006; Australian Government 2009).

Conservation objectives

1. Stable Australian non-breeding populations
2. Disturbance at key roosting and feeding sites reduced

Information required

1. Numbers and trends of Australian non-breeding population
2. Improved knowledge of dependence on key migratory staging sites
3. Improved knowledge of impacts of disturbance

Management actions required

1. Work with governments along the flyway at all levels, and local communities, to prevent destruction of key migratory staging sites
2. Negotiate with the Chinese and South Korean governments in particular to protect remaining tidal flats beside the Yellow Sea
3. Advocate for restoration of alienated wetland sites
4. Maintain and improve protection of roosting and feeding sites in Australia

Bibliography

Amano T, Székely T, Koyama K, Amano H, Sutherland WJ (2010) A framework for monitoring the status of populations: an example from wader populations in the East Asian–Australasian flyway. *Biological Conservation* **143**, 2238–2247.

Australian Government (2009) 'Draft significant impact guidelines for 36 migratory shorebirds'. Draft EPBC Act Policy Statement 3.21, Canberra. Retrieved 25 February, 2011 from <http://www.environment.gov.au/epbc/publications/migratory-shorebirds.html>.

Bamford M, Watkins D, Bancroft W, Tischler G, Wahl J (2008) 'Migratory shorebirds of the East Asian – Australasian flyway: Population estimates and internationally important sites'. Wetlands International – Oceania, Canberra.

Barter M, Fawen Q, Sixian T, Xiao Y, Tonkinson D (1997) Hunting of migratory waders on Chongming Dao: a declining occupation? *Stilt* **31**, 19–22.

Barter MA (2002) 'Shorebirds of the Yellow Sea: importance, threats and conservation status'. Canberra, Australia.

BirdLife International (2011) 'Species factsheet: *Charadrius mongolus*'. Retrieved 17 January, 2011 from <http://www.birdlife.org/>.

Department of Sustainability, Environment, Water, Population and Communities (2011) *Charadrius mongolus*. In 'Species Profile and Threats Database'. Department of Sustainability, Environment, Water, Population and Communities, Canberra. Retrieved 24 February, 2011 from <http://www.environment.gov.au/sprat>.

Kelin C, Qiang X (2006) Conserving migratory shorebirds in the Yellow Sea region. In *Waterbirds Around the World*. (Eds G Boere, C Galbraith and D Stroud) p. 319. The Stationery Office, Edinburgh, UK.

Marchant S, Higgins PJ (Eds) (1993) *Handbook of Australian, New Zealand and Antarctic Birds. Volume 2: Raptors to Lapwings*. Oxford University Press, Melbourne.

Moores N (2006) South Korea's shorebirds: a review of abundance, distribution, threats and conservation status. *Stilt* **50**, 62–72.

Rogers D, Hassell C, Oldland J, Clemens R, Boyle A, Rogers K (2009) 'Monitoring Yellow Sea migrants in Australia (MYSMA): north-western Australian shorebird surveys and workshops, December 2008'.

Rogers DI, Piersma T, Hassell CJ (2006) Roost availability may constrain shorebird distribution: exploring the energetic costs of roosting and disturbance around a tropical bay. *Biological Conservation* **133**, 225–235.

Stewart DA, Rogers A, Rogers DI (2007) Species description. In *Shorebirds of Australia*. (Eds A Geering, L Agnew and S Harding) pp. 75–196. CSIRO Publishing, Melbourne.

Wetlands International (2006) 'Waterbird population estimates'. Wetlands International, Wageningen, The Netherlands.

Comments received from

Rob Clemens, Ken Gosbell, Roger Jaensch, David Milton, Clive Minton, Danny Rogers

Subspecies Conservation Summary

Lesser Sand Plover (Kamchatkan)

Charadrius mongolus stegmanni Portenko, 1939

Charadriidae

Conservation status

Population visiting Australia: Endangered A2bc+3c+4bc

Reasons for listing

Past, current and anticipated declines of 50–79% over periods of 3 generations (24 years) based on survey data and habitat loss

Status 2000

Near Threatened A3c

Reason for change in 2010: decline in population and ongoing loss of habitat at staging posts

Status 1990

Least Concern

Reason for change in 2000: decline in population occurring

Taxonomy

C. m. mongolus (Mongolian) is Endangered in Australia, 3 other subspecies are extralimital. The species is Least Concern.

Taxonomic uniqueness: medium (11 genera/family, 35 species/genus, 5 subspecies/species)

Range

Breeds around Kamchatka, Kuril and Commander Is. and Chukotski, Russia, wintering along coasts from Japan and Taiwan to Australasia (Bamford *et al.* 2008). Non-breeding birds occur along north Australian coasts (Stewart *et al.* 2007; Department of Sustainability, Environment, Water, Population and Communities 2011).

Abundance

The global population was estimated at 20 000 individuals (Wetlands International 2006). There are no estimates of the Australian population, which is here assumed to be 50% of the Australian species population of 20 000–25 000, which also includes *C. m. mongolus* (Stewart *et al.* 2007; Bamford *et al.* 2008). Numbers of the 2 subspecies combined declined by c.84% across 49 Australian sites between c.1983 and c.2007 (AWSG database: Birds Australia *in litt.*) but there are few data from northern Australia where species more numerous (e.g. Rogers *et al.* 2009). Trends in the subpopulation outside Australia are poorly known but, in Japan, the 2 subspecies combined have declined in general, and by about 61% in autumn between 1978 and 2008 (Amano *et al.* 2010). These suggest an overall decline of 50–79% over 16 years across the flyway, which is inferred to continue

Current eligibility against IUCN Red List Criteria

IUCN category	Criteria eligibility
Population visiting Australia	
A	Endangered: past, current and anticipated population declines of 50–79% over periods of 3 generations (24 years) estimated and projected from monitoring and from deteriorating habitat quality
B	Not applicable: EOO >20 000 km^2; AOO >2000 km^2
C	Not applicable: population >10 000 mature individuals
D	Not applicable: population >1000 mature individuals, >5 locations
E	Not applicable: no population viability analysis undertaken

IUCN Red List assessment data

	Estimate	Reliability
Population visiting Australia		
Extent of occurrence trend	35 300 km^2 stable	high high
Area of occupancy trend	2200 km^2 stable	low medium
No. of mature individuals trend	12 000 decreasing	medium high
No. subpopulations	1	high
No. locations	>10	high
Generation time	8.0 years	low
Global population share	51–100%	high

given likely ongoing threats to its migratory staging sites in East Asia (Barter 2002; Kelin and Qiang 2006; Moores 2006). Population trends in other flyways are poorly known and the species as a whole is not considered threatened (Wetlands International 2006; BirdLife International 2011).

Ecology

Breeds in sand dunes, shingle and open habitats. Non-breeding birds migrate to coasts, where they inhabit mudflats of coastal bays and estuaries, sand-flats and dunes, and feed on insects, crustaceans, molluscs and polycheate worms (Marchant and Higgins 1993). A generation time of 8.0 years (BirdLife International 2011) is derived from an average age at first breeding of 2.0 years, an annual adult survival of 56% and a maximum longevity of 12.6 years, all extrapolated from *C. leschenaultii*.

Threats

Threatened by wetland degradation in East Asia, especially intertidal flats in the Yellow Sea where 50% of the estimated combined populations of *C. m. mongolus* and *C. m. stegmanni* stage during their northern migration (Barter 2002). General threats at migratory staging sites include environmental pollution, reduced river flows, human disturbance and reclamation for tidal power plants and barrages, industrial use, urban expansion and hunting (Barter *et al.* 1997; Barter 2002; Kelin and Qiang 2006; Moores 2006). Generalised threats in Australia include ongoing human disturbance, habitat loss and degradation from pollution, changes to the water regime and invasive plants (Rogers *et al.* 2006; Australian Government 2009) but most birds overwinter along coasts that have little disturbance.

Conservation objectives

1. Stable Australian non-breeding populations

Information required

1. Numbers and trends of Australian non-breeding population
2. Improved knowledge of dependence on key migratory staging sites

Management actions required

1. Work with governments along the flyway at all levels, and local communities, to prevent destruction of key migratory staging sites
2. Negotiate with the Chinese and South Korean governments in particular to protect remaining tidal flats beside the Yellow Sea
3. Advocate for restoration of alienated wetland sites

Bibliography

Amano T, Székely T, Koyama K, Amano H, Sutherland WJ (2010) A framework for monitoring the status of populations: an example from wader populations in the East Asian–Australasian flyway. *Biological Conservation* **143**, 2238–2247.

Australian Government (2009) 'Draft significant impact guidelines for 36 migratory shorebirds'. Draft EPBC Act Policy Statement 3.21, Canberra. Retrieved 25 February, 2011 from <http://www.environment.gov.au/epbc/publications/migratory-shorebirds.html>.

Bamford M, Watkins D, Bancroft W, Tischler G, Wahl J (2008) 'Migratory shorebirds of the East Asian – Australasian flyway: population estimates and internationally important sites'. Wetlands International – Oceania, Canberra.

Barter M, Fawen Q, Sixian T, Xiao Y, Tonkinson D (1997) Hunting of migratory waders on Chongming Dao: a declining occupation? *Stilt* **31**, 19–22.

Barter MA (2002) 'Shorebirds of the Yellow Sea: importance, threats and conservation status'. Canberra, Australia.

BirdLife International (2011) 'Species factsheet: *Charadrius mongolus*'. Retrieved 17 January, 2011 from <http://www.birdlife.org/>.

Department of Sustainability, Environment, Water, Population and Communities (2011) *Charadrius mongolus*. In 'Species Profile and Threats Database'. Department of Sustainability, Environment, Water, Population and Communities, Canberra. Retrieved 24 February, 2011 from <http://www.environment.gov.au/sprat>.

Kelin C, Qiang X (2006) Conserving migratory shorebirds in the Yellow Sea region. In *Waterbirds Around the World*. (Eds G Boere, C Galbraith and D Stroud) p. 319. The Stationery Office, Edinburgh, UK.

Marchant S, Higgins PJ (Eds) (1993) *Handbook of Australian, New Zealand and Antarctic Birds. Volume 2: Raptors to Lapwings.* Oxford University Press, Melbourne.

Moores N (2006) South Korea's shorebirds: a review of abundance, distribution, threats and conservation status. *Stilt* **50**, 62–72.

Rogers D, Hassell C, Oldland J, Clemens R, Boyle A, Rogers K (2009) 'Monitoring Yellow Sea migrants in Australia (MYSMA): north-western Australian shorebird surveys and workshops, December 2008'.

Rogers DI, Piersma T, Hassell CJ (2006) Roost availability may constrain shorebird distribution: exploring the energetic costs of roosting and disturbance around a tropical bay. *Biological Conservation* **133**, 225–235.

Stewart DA, Rogers A, Rogers DI (2007) Species description. In *Shorebirds of Australia.* (Eds A Geering, L Agnew and S Harding) pp. 75–196. CSIRO Publishing, Melbourne.

Wetlands International (2006) 'Waterbird population estimates'. Wetlands International, Wageningen, The Netherlands.

Comments received from

Rob Clemens, Ken Gosbell, Roger Jaensch, David Milton, Clive Minton, Danny Rogers

Subspecies Conservation Summary

Greater Sand Plover (Mongolian)

Charadrius leschenaultii leschenaultii Lesson, 1826

Charadriidae

Conservation status

Population visiting Australia: Vulnerable A2bc+3c+4bc

Reasons for listing

Past, recent and ongoing declines of 30–49% in 3 generations (24 years) based on survey data and habitat loss

Status 2000

Near Threatened A3c

Reason for change in 2010: decline in population and ongoing loss of habitat at staging posts

Status 1990

Least Concern

Reason for change in 2000: habitat loss in staging posts accelerating

Taxonomy

C. m. columbinus and *C. m. crassirostris* are extralimital. The species is Least Concern.

Taxonomic uniqueness: low (11 genera/family, 35 species/genus, 3 subspecies/species)

Range

Breeds in China, Mongolia and nearby parts of Russia, wintering along coasts from Japan through South-East Asia to Australasia (Bamford *et al.* 2008). Non-breeding birds occur along all Australian coasts, especially in the north (Department of Sustainability, Environment, Water, Population and Communities 2011).

Abundance

The global population has been estimated at 270 000 individuals (Wetlands International 2006; BirdLife International 2011), including 100 000 in the flyway, and 75 000 in Australia (Stewart *et al.* 2007; Bamford *et al.* 2008). Numbers declined at Moreton Bay by 60% between 1998 and 2008 (Fuller *et al.* 2009), on Eighty-mile Beach by c.65% between 2000 and 2008, whereas similar numbers at Bush Point have fluctuated between 2004 and 2008 (Rogers *et al.* 2009) and numbers elsewhere were too small to measure trends (AWSG database: Birds Australia *in litt.*). Trends outside Australia are poorly known but numbers in Japan have generally slightly increased between 1978 and 2008 (Amano *et al.* 2010). These suggest an overall 30–49% decline over 17 years across the flyway, averaging some contradictory trends, which is inferred as likely to continue given ongoing threats to its migratory staging sites in East Asia (Barter 2002; Kelin and Qiang 2006; Moores 2006). Population trends in other flyways are poorly known and the species is not considered threatened (Wetlands International 2006; BirdLife International 2011).

Ecology

Breeds in open, dry, treeless, uncultivated areas up to 3000 m. Non-breeding birds migrate to sheltered sandy,

Current eligibility against IUCN Red List Criteria

IUCN category	Criteria eligibility
Population visiting Australia	
A	Vulnerable: past, current and anticipated population declines of 30–49% over periods of 3 generations (24 years) estimated and projected from monitoring and from deteriorating habitat quality
B	Not applicable: EOO >20 000 km^2; AOO >2000 km^2
C	Not applicable: population >10 000 mature individuals
D	Not applicable: population >1000 mature individuals, >5 locations
E	Not applicable: no population viability analysis undertaken

IUCN Red List assessment data

	Estimate	Reliability
Population visiting Australia		
Extent of occurrence trend	35 700 km^2 stable	high high
Area of occupancy trend	2600 km^2 stable	low medium
No. of mature individuals trend	75 000 decreasing	medium high
No. subpopulations	1	high
No. locations	>10	high
Generation time	8.0 years	low
Global population share	26–50%	high

shelly or muddy, rocky beaches or salt-marshes, and feed on marine invertebrates such as molluscs, worms and crustaceans (Marchant and Higgins 1993). A generation time of 8.0 years (BirdLife International 2011) is derived from an age at first breeding of 2.0 years (Cramp *et al.* 1983), an annual adult survival of 56% (extrapolated from congeners) and a maximum longevity of 12.6 years (Australian Bird and Bat Banding Scheme).

Threats

Threatened by wetland degradation in East Asia, where it uses migration staging sites south of the Yellow Sea (Barter 2002). General threats at migratory staging sites include environmental pollution, reduced river flows, human disturbance, reclamation for tidal power plants, barrages, aquaculture ponds and agriculture, industrial use, urban expansion and hunting (Barter *et al.* 1997; Barter 2002; Ma and Cai 2002; Kelin and Qiang 2006; Moores 2006). Threats in Australia, especially eastern and southern Australia, include ongoing human disturbance, habitat loss and degradation from pollution, changes to the water regime and invasive plants (Rogers *et al.* 2006; Australian Government 2009).

Conservation objectives

1. Stable Australian non-breeding populations
2. Disturbance at key roosting and feeding sites reduced

Information required

1. Numbers and trends of Australian non-breeding population
2. Improved knowledge of dependence on key migratory staging sites
3. Improved knowledge of impacts of disturbance

Management actions required

1. Work with governments along flyway, especially Vietnam and China, to prevent destruction of key migratory staging sites
2. Maintain and improve protection of roosting and feeding sites in Australia
3. Support surveys of breeding grounds for potential threats

Bibliography

Amano T, Székely T, Koyama K, Amano H, Sutherland WJ (2010) A framework for monitoring the status of populations: an example from wader populations in the East Asian–Australasian flyway. *Biological Conservation* **143**, 2238–2247.

Australian Government (2009) 'Draft significant impact guidelines for 36 migratory shorebirds'. Draft EPBC Act Policy Statement 3.21, Canberra. Retrieved 25 February, 2011 from <http://www.environment.gov.au/epbc/publications/migratory-shorebirds.html>.

Bamford M, Watkins D, Bancroft W, Tischler G, Wahl J (2008) 'Migratory shorebirds of the East Asian – Australasian flyway: population estimates and internationally important sites'. Wetlands International – Oceania, Canberra.

Barter M, Fawen Q, Sixian T, Xiao Y, Tonkinson D (1997) Hunting of migratory waders on Chongming Dao: a declining occupation? *Stilt* **31**, 19–22.

Barter MA (2002) 'Shorebirds of the Yellow Sea: importance, threats and conservation status'. Canberra, Australia.

BirdLife International (2011) 'Species factsheet: *Charadrius leschenaultii*'. Retrieved 17 January, 2011 from <http://www.birdlife.org/>.

Cramp S, Simmons KEL, Brooks DC, Collar NJ, Dunn E, Gillmor R, Hollom PAD, Hudson R, Nicholson EM, Ogilvie MA, Olney PJS, Roselaar CS, Voous KH, Wallace DIM, Wattel J, Wilson MG (Eds) (1983) *Handbook of the Birds of Europe, the Middle East and North Africa. The Birds of the Western Palearctic: 3. Waders to Gulls.* Oxford University Press, Oxford, UK.

Department of Sustainability, Environment, Water, Population and Communities (2011) *Charadrius leschenaultii.* In 'Species Profile and Threats Database'. Department of Sustainability, Environment, Water, Population and Communities, Canberra. Retrieved 24 February, 2011 from <http://www.environment.gov.au/sprat>.

Fuller RA, Wilson HB, Kendall BE, Possingham HP (2009) 'Monitoring shorebirds using counts by the Queensland Wader Study Group'. Report to the Queensland Wader Study Group and the Department of Environment and Resource Management, Brisbane.

Kelin C, Qiang X (2006) Conserving migratory shorebirds in the Yellow Sea region. In *Waterbirds Around the World*. (Eds G Boere, C Galbraith and D Stroud) p. 319. The Stationery Office, Edinburgh, UK.

Ma M, Cai D (2002) Threats to Whooper Swans in Xinjiang, China. *Waterbirds* **25**, 331–333.

Marchant S, Higgins PJ (Eds) (1993) *Handbook of Australian, New Zealand and Antarctic Birds. Volume 2: Raptors to Lapwings*. Oxford University Press, Melbourne.

Moores N (2006) South Korea's shorebirds: a review of abundance, distribution, threats and conservation status. *Stilt* **50**, 62–72.

Rogers D, Hassell C, Oldland J, Clemens R, Boyle A, Rogers K (2009) 'Monitoring Yellow Sea migrants in Australia (MYSMA): north-western Australian shorebird surveys and workshops, December 2008'.

Rogers DI, Piersma T, Hassell CJ (2006) Roost availability may constrain shorebird distribution: exploring the energetic costs of roosting and disturbance around a tropical bay. *Biological Conservation* **133**, 225–235.

Stewart DA, Rogers A, Rogers DI (2007) Species description. In *Shorebirds of Australia*. (Eds A Geering, L Agnew and S Harding) pp. 75–196. CSIRO Publishing, Melbourne.

Wetlands International (2006) 'Waterbird population estimates'. Wetlands International, Wageningen, The Netherlands.

Comments received from

Rob Clemens, Ken Gosbell, Roger Jaensch, David Milton, Clive Minton, Danny Rogers

---

Species Conservation Summary

Hooded Plover

Thinornis rubricollis (Gmelin, 1789)
Charadriidae

Conservation status

Vulnerable C1

Reasons for listing

Population <10 000 individuals with an estimated ongoing decline of 10–20% in 3 generations (39 years)

Status 2000

Near Threatened C1
Reason for change in 2010: declines detected in western subspecies

Status 1990

Least Concern
Reason for change in 2000: substantial declines detected in eastern subspecies

Taxonomy

T. r. tregellasi (Western Australia) and *T. r. rubricollis* (eastern Australia) are both Vulnerable
Taxonomic uniqueness: high (11 genera/family, 2 species/genus, 2 subspecies/species)

Range

In the east breeds on the coast from Jervis Bay in New South Wales to Fowlers Bay, South Australia and including Tasmania, Kangaroo I., King I., Flinders I. and other nearby islands (Cameron and Weston 1999). Breeds on the Western Australian coast from Cape Naturaliste to Eyre, and on inland lakes as far north as L. Cowan and L. Moore and Yalgorup Lakes, south of Perth. After breeding, inland birds appear to move to lakes near the west coast or shores of southern lakes (Blakers *et al.* 1984; Johnstone and Storr 1998).

Abundance

The 2 subpopulations are estimated to contain about 5500 mature individuals including about 2500 in the west and 3000 in the east. Ongoing population declines of 10–25% in 3 generations have been recorded in all eastern states based on monitoring (see subspecies text) and, while monitoring in the west has been less rigorous, it suggests ongoing declines which contribute to an overall decline of >10% for the whole species.

Ecology

Inhabits ocean beaches, particularly wide beaches with large amounts of beach-washed seaweed, and creek mouths or inlet entrances, and feeds on polychaetes, molluscs, turions, crustaceans and seeds. In WA, and some parts of SA and Vic also inhabits inland lakes, feeding mostly on *Coxiella* spp. Nests on the upper levels of the beach, adjacent sand dunes and, in Western Australia, on lake shores (Marchant and Higgins 1993). A generation time of 12.9 years is derived from an age at first breeding of 1.7 years (Baird and Dann 2003) and an annual survival of adults of 90.7% (Weston 2003).

Threats

Hooded Plovers on beaches are threatened by a wide array of human-associated disturbances, particularly pedestrian beach traffic, dogs, horses and vehicles. Near settlements, high numbers of Silver Gulls *Larus novaehollandiae* and ravens *Corvus* spp. reduce breeding success, and on the mainland, foxes *Vulpes vulpes* take eggs and chicks. Though largely secure on islands, and where active nest protection operates, a continued decline is likely in the absence of appropriate beach management, which includes active nest protection.

Current eligibility against IUCN Red List Criteria

IUCN category	Criteria eligibility
A	Near Threatened A2: decline has approached 30% in 3 generations (39 years) based on direct observation, decline in AOO and habitat quality and effects of introduced taxa
B	Not applicable: EOO >20 000 km^2; AOO >2000 km^2; at >10 locations; fluctuations not extreme
C	Vulnerable: population <10 000 mature individuals with a continuing decline >10% in 3 generations
D	Not applicable: population >1000 mature individuals, >5 locations
E	Not applicable: no population viability analysis undertaken

IUCN Red List assessment data

	Estimate	Reliability
Extent of occurrence trend	590 000 decreasing	high high
Area of occupancy trend	5600 decreasing	low high
No. of mature individuals trend	5500 decreasing	medium high
No. subpopulations	2	high
No. locations	>10	high
Largest subpopulation	3000	medium
Generation time	12.9 years	medium
Global population share	100%	high

Conservation objectives

1. Stable numbers of birds recorded in standardised surveys
2. Improved breeding success at monitored nests

Information required

1. Demographic trends including population size, breeding success and growth rate
2. Levels of nest predation and breeding success, in areas with and without predator and stock control programs
3. Habitat availability and risk of habitat loss due to weed invasion, rising sea levels and dune morphology changes
4. Relative impacts of different threats so that a threat index can be formulated and beaches ranked and prioritised for management
5. Location of key breeding lakes and winter flocking sites

Management actions required

1. Manage, and in some cases discourage and actively police, the use of key beaches for recreation, and consider bans on vehicle access, horse riding and dog walking in the breeding season
2. Where necessary reduce predation by cats and foxes and manage native predators
3. Advocate for suitable coastal planning and erosion control activities
4. Adequately fence livestock from key breeding sites
5. Identify and manage key breeding lakes and winter flocking sites
6. Engage the public in research, monitoring, management and advocacy

Bibliography

Baird B, Dann P (2003) The breeding biology of Hooded Plovers, *Thinornis rubricollis*, on Phillip Island, Victoria. *Emu* **103**, 323–328.

Blakers MS, Davies JJF, Reilly PN (1984) *The Atlas of Australian Birds*. Melbourne University Press, Melbourne.

Cameron D, Weston MA (1999) The Hooded Plover: first confirmed record in Queensland, the longest movement yet recorded, and a discussion of the range contraction in eastern Australia. *Australian Bird Watcher* **18**, 8–18.

Johnstone RE, Storr GM (1998) *Handbook of Western Australian Birds. Volume 1: Non-passerines (Emu to Dollarbird)*. Western Australian Museum, Perth.

Marchant S, Higgins PJ (Eds) (1993) *Handbook of Australian, New Zealand and Antarctic Birds. Volume 2: Raptors to Lapwings*. Oxford University Press, Melbourne.

Weston MA (2003) 'Managing the Hooded Plover in Victoria: a review of existing information'. Parks Victoria Technical Series 4. Parks Victoria, Melbourne.

Comments received from

Allan Burbidge, Glenn Ehmke, Cheryl Gole, Grainne Maguire, Mike Weston, Eric Woehler

Subspecies Conservation Summary

Hooded Plover (eastern)

Thinornis rubricollis rubricollis (Gmelin, 1789)
Charadriidae

Conservation status

Vulnerable C1+2a(ii)

Reasons for listing

Population <10 000 individuals in a single subpopulation with an estimated ongoing decline of 10–20% in 3 generations (39 years)

Status 2000

Vulnerable C1+2a(ii)

Status 1990

Vulnerable C1+2a(ii)

Taxonomy

T. r. tregellasi (Western Australia) is also Vulnerable, as is the species

Taxonomic uniqueness: high (11 genera/family, 2 species/genus, 2 subspecies/species)

Range

Occurs along coasts of south-east Australia from Jervis Bay in New South Wales, perhaps historically north to Queensland, to Fowlers Bay in South Australia and including Tasmania, Kangaroo I., King I., Flinders I. and other nearby islands (Cameron and Weston 1999; Barrett *et al.* 2003). Some individuals move to traditional aggregation sites during non-breeding season (Weston *et al.* 2009).

Abundance

The population is estimated as 3000 mature individuals, based on counts of 40 in NSW, c.405 in Victoria and 636–800 in SA in 2008 (Birds Australia 2008; Mooney *et al.* 2009), and 600 pairs in Tas, Flinders and King Is. in 2010–2011 (E. J. Woehler *in litt.*). Calculated population declines include 12% across Vic from 2000–2008, and 33% from 600 to 400 from 1980–2008 (Birds Australia 2008); 13% between comprehensive surveys across Vic in 1988 and 1992, but statistically non-significant overall decline 1980–1992 (Weston 1993); 58% on Phillip I. from 1981–1997 (Baird and Dann 2003); c.20% across Tas. from 1982–2006 (Woehler and Park 1997; Bryant 2002) E. J. Woehler *in litt.*); 25% on Kangaroo I. from 1985–2004 (Dennis and Masters 2006); and 55% across NSW from 1988–2008 (Birds Australia 2008; NSW Scientific Committee 2008). The declines in NSW and Kangaroo I. may be biased towards heavily-disturbed beaches. Though largely secure on small islands, it is likely to continue declining wherever appropriate beach management is lacking (Weston 1993). The overall rate of decline is estimated here to be 10–20% in 3 generations (39 years).

Ecology

Hooded Plovers live on ocean beaches, particularly wide beaches backed by dunes with large amounts of beach-washed seaweed, and creek mouths or inlet entrances (Weston 2003), and feed on polychaetes, molluscs, turions, crustaceans and seeds (Marchant and Higgins 1993). Hooded Plovers are highly dispersed nesters with relatively large territory sizes (37 ha) to which they are highly site faithful (Weston *et al.* 2009). A generation time of 12.9 years is derived from an age at first breeding of 1.7 years (Baird and Dann 2003) and an annual survival of adults of 90.7% (Weston 2003).

Threats

The principal threats are disturbance to incubating birds, crushing or disturbance of eggs and chicks by people and vehicles, or predation by dogs, cats *Felis catus*, rats *Rattus* spp. and, on the Australian mainland, foxes *Vulpes vulpes* (Hanisch 1998; Baird and Dann 2003; Weston 2003). Increased human activity also increases numbers of native scavengers and predators such as Silver Gulls *Larus novaehollandiae*, ravens *Corvus* spp. and currawongs *Strepera* spp. (Weston 2003; Burke *et al.* 2004). Up to 81% of nests and 30% of chicks in the Coorong would have been crushed by off-road vehicles (Buick and Paton 1989), and over 30% of nests were trampled by dogs or people at Mornington before controls were introduced (Dowling and Weston 1999). Of 1510 disturbances which could reduce chick foraging and brooding, 81% were from people, 21% dogs and 16% birds, mostly raptors and gulls (Weston and Elgar 2005). Breeding success is also reduced by wandering livestock and beach erosion management which reduces the availability of bare nest sites (Weston 2003). Breeding success is improved by minimising disturbance by people and dogs such as temporary beach closures (Dowling and Weston 1999; Weston *et al.* 2011), and erecting chick shelters as refugia (Baird and Dann 1999; Maguire 2008; Maguire et al. 2010). In Vic, protected nests (signs and fencing) have a significantly greater chance of hatching than unprotected nests (51% vs. 42%; G. Maguire *in litt.*). Furthermore, protected nests on heavily used beaches are significantly more likely to fledge young

Current eligibility against IUCN Red List Criteria

IUCN category	Criteria eligibility
A	Near Threatened: past, current and anticipated population declines of 20–29% in 3 generations (39 years) observed directly and estimated and projected from decline in AOO, habitat quality and effects of introduced taxa
B	Not applicable: >10 locations, not severely fragmented, population fluctuations not extreme
C	Vulnerable: population 2500–10 000 mature individuals, estimated continuing decline between 10% in 3 generations and 20% in 2 generations, 100% in 1 subpopulation
D	Not applicable: population >1000 mature individuals, >5 locations
E	Not applicable: no population viability analysis undertaken

IUCN Red List assessment data

	Estimate	Reliability
Extent of occurrence trend	7900 km^2 stable	high high
Area of occupancy trend	3600 km^2 decreasing	high high
No. of mature individuals trend	3000 decreasing	medium high
No. subpopulations	1	high
No. locations	>10	high
Generation time	12.9 years	medium
Global population share	100%	high

than unprotected nests (38% versus 5% probability of fledging young, n=342 nests), and this is an equivalent fledging rate to remote and isolated beaches (G. Maguire *in litt.*).

Conservation objectives

1. Stable numbers of birds recorded in standardised surveys, especially a stable number of occupied breeding territories
2. Improved breeding success at monitored nests

Information required

1. Demographic trends including population size, breeding success and growth rate
2. Levels of nest and chick predation and overall breeding success, in areas with and without predator control programs
3. Habitat availability and risk of habitat loss due to weed invasion, rising sea levels and dune morphology changes
4. Relative impacts of different threats so that a threat index can be formulated and beaches ranked and prioritised for management

Management actions required

1. Manage, and in some cases discourage and actively police, the use of key beaches for recreation, and consider bans on vehicle access, horse riding and dog walking in the breeding season
2. Where necessary reduce predation by cats and foxes
3. Investigate predator control options for native predators such as ravens, Silver Gulls and kestrels
4. Advocate for suitable coastal planning and erosion control activities
5. Adequately fence livestock from key breeding beaches
6. Seek consistency in coastal policy and legislation for species protection
7. Trial recovery of habitat degraded by Marram Grass
8. Carry out a Population Viability Analysis to set breeding success targets for recovery programs
9. Engage the public in research, monitoring, management and advocacy

Bibliography

Baird B, Dann P (2003) The breeding biology of Hooded Plovers, *Thinornis rubricollis*, on Phillip Island, Victoria. *Emu* **103**, 323–328.

Barrett G, Silcocks A, Barry S, Cunningham R, Poulter R (2003) *The New Atlas of Australian Birds.* Royal Australian Ornithologists Union, Hawthorn East.

Birds Australia (2008) 'November 2008 biennial Hooded Plover count'. Unpublished report. Birds Australia, Melbourne. Retrieved 6 March, 2011 from <http://www.birdsaustralia.com.au/images/stories/current-projects/bnb/Biennial_Report2008.pdf>.

Bryant S (2002) 'Conservation assessment of beach nesting and migratory shorebirds in Tasmania. Report to Birds Australia and Natural Heritage Trust'. Department of Primary Industries, Water and Environment, Hobart.

Buick AM, Paton DC (1989) Impact of off-road vehicles on the nesting success of Hooded Plovers *Charadrius rubricollis* in the Coorong region of South Australia. *Emu* **89**, 159–172.

Burke P, Burke AJ, Weston MA (2004) Silver Gull preys upon Hooded Plover chick. *Wader Study Group Bulletin* **103**, 76–77.

Cameron D, Weston MA (1999) The Hooded Plover: first confirmed record in Queensland, the longest movement yet recorded, and a discussion of the range contraction in eastern Australia. *Australian Bird Watcher* **18**, 8–18.

Dennis TE, Masters P (2006) Long-term trends in the Hooded Plover *Thinornis rubricollis* population on Kangaroo Island, South Australia. *South Australian Ornithologist* **34**, 258–266.

Dowling B, Weston MA (1999) Managing a breeding population of the Hooded Plover *Thinornis rubricollis* in a high-use recreational environment. *Bird Conservation International* **9**, 255–270.

Hanisch DR (1998) The effect of human disturbance on the reproductive performance of the Hooded Plover, *Thinornis rubricollis*, in the south-east of Tasmania. Honours thesis, University of Tasmania.

Maguire GS, Duivenvoorden AK, Weston MA, Adams R (2010) 'Provision of artificial shelter on beaches is associated with improved shorebird fledging success'. Bird Conservation International online early publication.

Maguire GS (2008) 'A practical guide to managing beach-nesting birds in Australia'. Birds Australia report. Retrieved 16 March, 2011 from <http://www.birdsaustralia.com.au/our-projects/management-resources.html>.

Maguire GS, Cullen M, Ehmke GC and Purnell C (2010) 'Supplement to a practical guide to managing beach-nesting birds in Australia'. Birds Australia report. Retrieved 16 March, 2011 from <http://www.birdsaustralia.com.au/our-projects/management-resources.html>.

Marchant S, Higgins PJ (Eds) (1993) *Handbook of Australian, New Zealand and Antarctic Birds. Volume 2: Raptors to Lapwings*. Oxford University Press, Melbourne.

Mooney T, Baxter C, Achurch H, Berris M, Male D (2009) 'Hooded Plover breeding study: Kangaroo Island 2007–2009'. Unpublished report. Department for Environment and Heritage, South Australia.

NSW Scientific Committee (2008) 'Hooded Plover *Thinornis rubricollis*. Review of current information in NSW'. Unpublished report arising from the Review of the Schedules of the Threatened Species Conservation Act 1995. NSW Scientific Committee, Hurstville. Retrieved 16 March, 2011 from <http://www.environment.nsw.gov.au/resources/nature/schedules/HoodedPlover.pdf>.

Weston MA (1993) Twelve years of counting the Hooded Plover in Victoria, Australia. *Stilt* **23**, 15–19.

Weston MA (2003) 'Managing the Hooded Plover in Victoria: a review of existing information'. Parks Victoria Technical Series 4. Parks Victoria, Melbourne.

Weston MA, Ehmke GC, Maguire GS (2009) Manage one beach or two? Movements and space-use of the threatened hooded plover (*Thinornis rubricollis*) in south-eastern Australia. *Wildlife Research* **36**, 289–298.

Weston MA, Ehmke GC, Maguire GS (2011) Nest return times in response to static versus mobile human disturbance. *Journal of Wildlife Management* **75**, 252–255.

Weston MA, Elgar MA (2005) Disturbance to brood-rearing Hooded Plover *Thinornis rubricollis*: responses and consequences. *Bird Conservation International* **15**, 193–209.

Woehler EJ, Park P (1997) 'Interim report on the status of Hooded Plovers *Thinornis rubricollis* in Tasmania'. Report to Tasmanian Parks and Wildlife Service, Hobart.

Comments received from

Terry Dennis, Glenn Ehmke, Grainne Maguire, Mike Weston, Eric Woehler

Subspecies Conservation Summary

Hooded Plover (western)

Thinornis rubricollis tregellasi (Mathews, 1912)
Charadriidae

Conservation status

Vulnerable C2a(ii)

Reasons for listing

Population <10 000 mature individuals a single subpopulation with a suspected ongoing decline

Status 2000

Least Concern
Reason for change in 2010: evidence of decline

Status 1990

Least Concern

Taxonomy

T. r. rubricollis (south-east Australia) is also Vulnerable, as is the species
Taxonomic uniqueness: high (11 genera/family, 2 species/genus, 2 subspecies/species)

Range

Breeds on south-west Western Australian coast, from Cape Naturaliste to Eyre, and on inland lakes as far north-east as L. Cowan and L. Moore and north-west to Yalgorup Lakes, south of Perth. Single birds or non-breeding pairs recorded on L. Arrow, L. Barlee and L. Ballard (Chapman and Lane 1997). After breeding, inland birds appear to move to lakes near the west coast or shores of southern lakes. Flocks and smaller numbers of breeding birds may appear at many remote and under-surveyed lakes when conditions are suitable, e.g. 330 at the Cranbrook salt lake system in 2007 (Elson and Singor 2008).

Abundance

A census in 1995 found 2093 birds, including 1500 on the Esperance Lakes (Newbey 1996). There has been no subsequent comprehensive census. Numbers counted in relatively standardised monitoring of some beaches and lakes between 2002 and 2010 fluctuated around an average of 490 adults. While overall trends were negative (–16%), this trend is influenced by the size of the first count (Singor 2010). However, numbers near Esperance have

Current eligibility against IUCN Red List Criteria

IUCN category	Criteria eligibility
A	Not applicable: past, current or future population declines are thought unlikely to exceed 20% in any 3-generation period
B	Not applicable: EOO >20 000 km^2, AOO >2000 km^2
C	Vulnerable: population 2500–10 000 mature individuals, inferred continuing decline, 100% in 1 subpopulation
D	Not applicable: population >1000 mature individuals, >5 locations
E	Not applicable: no population viability analysis undertaken

IUCN Red List assessment data

	Estimate	Reliability
Extent of occurrence trend	580 000 km^2 stable	high high
Area of occupancy trend	2000 km^2 stable	low medium
No. of mature individuals trend	2500 decreasing	medium low
No. subpopulations	1	high
No. locations	>10	high
Generation time	12.9 years	medium
Global population share	100%	high

never again approached numbers seen in the 1990s and the numbers seen recently are small enough to warrant a precautionary approach. Given that surveys omit some beaches and lakes, the population is assumed here to be 2500 individuals, but declining. Given the recorded movements of banded eastern birds (Weston *et al.* 2009; Department of Sustainability, Environment, Water, Population and Communities 2011), it is assumed here to have a single subpopulation.

Ecology

In WA, Hooded Plovers live on ocean beaches and beside inland lakes. They nest on the upper levels of the beach, in adjacent sand dunes, or on lake shores. They forage at the water's edge mostly for polychaetes, molluscs and crustaceans on the coast, but *Coxiella* spp. are probably the main food around inland lakes (Marchant and Higgins 1993; Singor 1999; Weston and Elgar 2000). A generation time of 12.9 years is calculated based on mean age at first breeding of 1.7 years (Baird and Dann 2003) and mean adult survival of 90.7% (Weston 2003).

Threats

The principal threats are disturbance or crushing by people, vehicles and stock as well as disturbance or predation by dogs *Canis familiaris*, cats *Felis catus*, rats *Rattus* spp. and foxes *Vulpes vulpes* (Buick and Paton 1989; Baird and Dann 2003; Weston 2003), though all these threats except wandering stock are likely to be less important at the more remote beaches and inland lakes, where most Hooded Plover in WA breed, than in eastern Australia. Increased human activity also increases numbers of native scavengers and predators such as Silver Gulls *Larus novaehollandiae* and ravens *Corvus* spp. (Weston 2003; Burke *et al.* 2004). Inland lakes are either dry following extended drought or, as around Esperance, the water level has risen following clearing of nearby vegetation, flooding shoreline habitat.

Conservation objectives

1. Stable numbers of birds recorded in standardised surveys

Information required

1. Demographic trends including population size and growth rate
2. Breeding success, in areas with and without predator and stock control programs

Management actions required

1. Identify and manage key breeding lakes
2. Undertake standardised surveys
3. Prohibit or manage dogs at key breeding sites
4. Where necessary reduce predation by cats and foxes
5. Manage, and in some cases discourage recreation at breeding sites
6. Restrict vehicular access to important breeding sites, and consider bans in the breeding season
7. Actively police breeding sites that have high densities of birds and frequent illegal use of off-road vehicles
8. Adequately fence livestock from key breeding sites
9. Engage the public in research, monitoring, management and advocacy

Bibliography

Baird B, Dann P (2003) The breeding biology of Hooded Plovers, *Thinornis rubricollis*, on Phillip Island, Victoria. *Emu* **103**, 323–328.

Buick AM, Paton DC (1989) Impact of off-road vehicles on the nesting success of Hooded Plovers *Charadrius rubricollis* in the Coorong region of South Australia. *Emu* **89**, 159–172.

Burke P, Burke AJ, Weston MA (2004) Silver Gull preys upon Hooded Plover chick. *Wader Study Group Bulletin* **103**, 76–77.

Chapman A, Lane JAK (1997) Waterfowl usage of wetlands in the south-east arid interior of Western Australia 1992–93. *Emu* **97**, 51–59.

Department of Sustainability, Environment, Water, Population and Communities (2011) *Thinornis rubricollis rubricollis.* In 'Species Profile and Threats Database'. Department of Sustainability, Environment, Water, Population and Communities, Canberra. Retrieved 24 February, 2011 from <http://www.environment.gov.au/sprat>.

Elson SJ, Singor MJC (2008) Hooded Plover *Thinornis rubricollis tregellasi* breeding data from Western Australia. *Stilt* **53**, 6–12.

Marchant S, Higgins PJ (Eds) (1993) *Handbook of Australian, New Zealand and Antarctic Birds. Volume 2: Raptors to Lapwings.* Oxford University Press, Melbourne.

Newbey BJ (1996) Report on the Hooded Plover project. *Supplement to W. A. Bird Notes* **79**.

Singor MJC (1999) Hooded Plover report No. 2, 1996–1999. *Supplement to Western Australian Bird Notes* **90**.

Singor MJC (2010) Hooded Plover report. *Western Australian Bird Notes*.

Weston MA (2003) 'Managing the Hooded Plover in Victoria: a review of existing information'. Parks Victoria Technical Series 4. Parks Victoria, Melbourne.

Weston MA, Ehmke GC, Maguire GS (2009) Manage one beach or two? Movements and space-use of the threatened hooded plover (*Thinornis rubricollis*) in south-eastern Australia. *Wildlife Research* **36**, 289–298.

Weston MA, Elgar MA (2000) The effect of a major rainfall event on Hooded Plovers in a salt-lake in Western Australia. *Emu* **100**, 64–69.

Comments received from

Allan Burbidge, Cheryl Gole, Mike Weston

Species Conservation Summary

Plains-wanderer

Pedionomus torquatus Gould, 1840

Pedionomidae

Conservation status

Endangered C2a(ii)

Reasons for listing

Single declining subpopulation <2500 mature individuals when at its lowest during droughts

Status 2000

Endangered C2a(ii)

Status 1990

Endangered C2a(ii)

Taxonomy

No infraspecific taxa

Taxonomic uniqueness: high (1 genus/family, 1 species/genus, 1 subspecies/species)

Range

Recently recorded from north-central Victoria, eastern South Australia, southern New South Wales around the Riverina and west-central Queensland (Baker-Gabb 2002a; Barrett *et al.* 2003). Recorded historically and possibly no longer extant in south-east SA, eastern NSW, and south-east Qld (Baker-Gabb 2002a). Seven core areas identified in the Riverina, supporting c.72% of the Riverina population in 3400 km^2 (NSW National Parks and Wildlife Service 2002).

Abundance

Difficult to survey and numbers poorly known, but the population is estimated to vary from c.2000 birds during widespread drought to c.5500–7000 birds after several successive seasons with favourable conditions (Baker-Gabb 2002a; Department of Sustainability, Environment, Water, Population and Communities 2011). These numbers have been revised down from c.2500 to c.2000 (Baker-Gabb 1998) because habitat loss has continued (NSW National Parks and Wildlife Service 2001) and birds have been scarcer on recent surveys: 0.13 birds/km in 4286 km of monitoring in the NSW Riverina during the dry years of 2001–2007, compared to 0.3 birds/km in 2121 km of surveys in the same region in the wetter years of 1984–1986 (Baker-Gabb 2002a; Birds Australia 2008). Highest densities are in Riverina region, where the population may be from c.1000 in poor years to c.5500 after good years (Baker-Gabb *et al.* 1990; Baker-Gabb 2002a). The population in west-central Qld is estimated at 1000 birds (Baker-Gabb 2002b), and the north-central Victorian and SA populations are estimated at about 500 birds in each during good years (Maher and Baker-Gabb 1993; Baker-Gabb 2002a). Other important sites may remain undiscovered in south-western and south-central Qld and north-eastern SA (Baker-Gabb 2002a); scattered records outside these areas are probably dispersing individuals and suggest a single subpopulation (contra in Department of Sustainability, Environment, Water, Population and Communities 2011). Declines from localised habitat loss and degradation are likely to be ongoing, in part balanced by improved management in other areas. On balance, it is estimated that there is still a net decline disregarding climate-induced fluctuations. Extreme population fluctuations of an order of magnitude may occur at a local level but are unlikely for the whole population.

Ecology

Inhabits sparse grasslands with c.50% bare ground, most vegetation less than 5 cm in height and some widely spaced plants up to 30 cm high. May occasionally use lower-quality habitat including cereal stubble, but

Current eligibility against IUCN Red List Criteria

IUCN category	Criteria eligibility
A	Not applicable: past, current or future population declines are thought unlikely to exceed 20% in any 3-generation period
B	Not applicable: EOO >20 000 km^2, AOO >2000 km^2
C	Endangered: population 250–2500 mature individuals, inferred continuing decline, >95% mature individuals in 1 subpopulation
D	Not applicable: population >1000 mature individuals, >5 locations
E	Not applicable: no population viability analysis undertaken

IUCN Red List assessment data

	Estimate	Reliability
Extent of occurrence trend	930 000 km^2 stable	high high
Area of occupancy trend	330 km^2 decreasing	low medium
No. of mature individuals trend	2000 decreasing	low low
No. subpopulations	1	medium
No. locations	>10	high
Generation time	7.3 years	low
Global population share	100%	high

cannot persist in an agricultural landscape. Suitable habitat tends to be restricted to small (50–300 ha) patches that do not support dense pasture growth under any seasonal conditions (Baker-Gabb 2002a). Habitat mapping of 2.3 million ha in the Riverina showed that 2.3% was primary habitat suitable all year round and 4.3% was secondary habitat that may be periodically occupied (Roberts and Roberts 2001; NSW National Parks and Wildlife Service 2002). Sedentary for as long as the habitat remains suitable. Feeds on seeds and arthropods throughout the year (Baker-Gabb 1988). The female lays 3–5 eggs in a shallow, grass-lined scrape (Marchant and Higgins 1993; Baker-Gabb 1998). A generation time of 7.3 years (BirdLife International 2011) is derived from an age at first breeding of 1.0 years and a maximum longevity in the wild of 13.5 years, both extrapolated from values for related Charadriiformes of similar size and Ecology.

Threats

The main threat is cultivation of native grassland which, even if left to recover, remains unsuitable for decades. Cultivation has all but eliminated the species from southern SA and Vic and is increasing across the Riverina. Even where patches of habitat remain, they may be too few and dispersed to be effective refuges during drought (Baker-Gabb 1998). Overgrazing causes the species to leave an area when grassland is reduced to a remnant less than 2–3 cm high with 60% or more bare ground (Baker-Gabb *et al.* 1990), probably because the species becomes too vulnerable to predators (Baker-Gabb 1987). Such overgrazing is particularly likely to take place during droughts. Fenitrothion is occasionally used to spray locust plagues, particularly in the Riverina, and may kill Plains-wanderers directly or indirectly through the food chain but a biological control agent, *Metarhizium anisopliae* var. *acridum* (Green Guard®) can be used instead (Story *et al.* 2007). Foxes *Vulpes vulpes* may be significant predators when crops, which support high densities of House Mouse *Mus musculus*, occur within 2 km (Baker-Gabb 1998).

Conservation objectives

1. Extent of Plains-wanderer habitat maintained and quality enhanced
2. Population size increased by improving management of 100 km^2 of suitable habitat in new reserves and refuge areas
3. Recognition as a 'flagship' species for native grasslands

Information required

1. Survey more properties for suitable habitat and identify areas of high conservation value
2. Refine knowledge of numbers and distribution in Queensland and northern South Australia

Management actions required

1. Paddocks with Plains-wanderers should be managed by light to moderate grazing and not overgrazed or cultivated
2. Land-holders interested in managing areas for Plains-wanderers should seek incentive-based land management opportunities from their local catchment management authority
3. Negotiate refuge areas, ensuring Plains-wanderer habitat is not cultivated, has a 2 km buffer from cultivated land wherever possible, is not overgrazed during drought and is integrated into a regional conservation plan
4. Monitor population and habitat condition and provide feedback to land-holders
5. Continue to advocate use of Green Guard® biological control agent for locust control in Plains-wanderer habitat

Bibliography

Baker-Gabb DJ (1987) 'The conservation and management of the Plains-wanderer *Pedionomus torquatus*'. World Wildlife Fund Report 49.

Baker-Gabb DJ (1988) The diet and foraging behaviour of the Plains-wanderer *Pedionomus torquatus*. *Emu* **88**, 115–118.

Baker-Gabb DJ (1998) Native grasslands and the Plains-wanderer. Birds Australia Conservation Statement No. 1. *Wingspan* **8**, 8.

Baker-Gabb DJ (2002a) 'Recovery plan for the Plains-wanderer *Pedionomus torquatus* 2002–2006: conservation of lowland native grassland dependant fauna'. Environment Australia, Canberra.

Baker-Gabb DJ (2002b) 'Surveys for Plains-wanderers *Pedionomus torquatus* on Astrebla Downs National Park, western Queensland'. Unpublished report to Environment Australia.

Baker-Gabb DJ, Benshemes JS, Maher PN (1990) A revision of the distribution, status and management of the Plains-wanderer *Pedionomus torquatus*. *Emu* **90**, 161–168.

Barrett G, Silcocks A, Barry S, Cunningham R, Poulter R (2003) *The New Atlas of Australian Birds*. Royal Australian Ornithologists Union, Hawthorn East.

BirdLife International (2011) 'Species factsheet: *Pedionomus torquatus*'. Retrieved 17 January, 2011 from <http://www.birdlife.org/>.

Birds Australia (2008) 'Review of DECC Plains-wanderer *Pedionomus torquatus* monitoring data 2001–2007'. Birds Australia, Melbourne.

Department of Sustainability, Environment, Water, Population and Communities (2011) *Pedionomus torquatus*. In 'Species Profile and Threats Database'. Department of Sustainability, Environment, Water, Population and Communities, Canberra. Retrieved 24 February, 2011 from <http://www.environment.gov.au/sprat>.

Maher PN, Baker-Gabb D (1993) 'Surveys and conservation of the Plains-wanderer in northern Victoria'. Department of Conservation and Natural Resources, No. 131, Melbourne.

Marchant S, Higgins PJ (Eds) (1993) *Handbook of Australian, New Zealand and Antarctic Birds. Volume 2: Raptors to Lapwings*. Oxford University Press, Melbourne.

NSW National Parks and Wildlife Service (2001) 'Summary of results of Plains-Wanderer API Mapping Project 1999–2001'. Unpublished report to New South Wales Plains-wanderer Recovery Team.

NSW National Parks and Wildlife Service (2002) 'Plains-wanderer (*Pedionomus torquatus*) draft recovery plan'.

Roberts I, Roberts J (2001) 'Plains-wanderer (*Pedionomus torquatus*) habitat mapping including woody vegetation and other landscape features, Riverina Plains, New South Wales'. Unpublished report to New South Wales National Parks and Wildlife Service.

Story P, Oliver D, Deveson T, McCulloch L, Hamilton G, Baker-Gabb D (2007) Estimating and reducing the amount of Plains-wanderer (*Pedionomus torquatus* Gould) habitat sprayed with pesticides for locust control in the New South Wales Riverina. *Emu* **107**, 308–314.

Comments received from

David Baker-Gabb, Damon Oliver, David Parker, Rick Webster

Species Conservation Summary

Australian Painted Snipe

Rostratula australis (Gould, 1838)

Rostratulidae

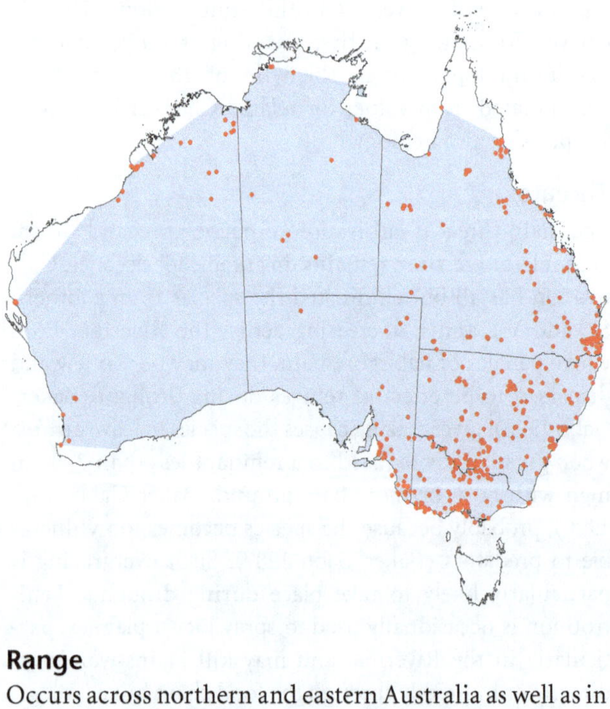

Conservation status

Endangered C2a(ii)

Reasons for listing

Population 250–2500 and occurs as a single subpopulation that is declining owing to ongoing loss and degradation of wetlands

Status 2000

Vulnerable A2b

Reason for change in 2010: evidence that declines intensifying

Status 1990

Near Threatened A2b

Reason for change in 2000: evidence of declines

Taxonomy

No infraspecific taxa

Taxonomic uniqueness: high (2 genera/family, 2 species/genus, 1 subspecies/species)

Range

Occurs across northern and eastern Australia as well as in south-western Australia (Barrett *et al.* 2003), probably as 1 subpopulation. The AOO calculated from the Painted Snipe Database between 1999 and 2008, and the Birds Australia Atlases between 1977 and 2007, decreased by >50% in 3 generations, but comparisons are difficult to quantify because of different survey methods, the snipe's cryptic and dispersive habits, and the lack of extensive surveys in northern Australia (Rogers *et al.* 2005). There

Current eligibility against IUCN Red List Criteria

IUCN category	Criteria eligibility
A	Vulnerable: population decline of 30–49% over the last 3 generations (26 years) estimated from monitoring
B	Not applicable: EOO >20 000 km^2, AOO >2000 km^2
C	Endangered: population 250–2500 mature individuals, inferred continuing decline, >95% mature individuals in 1 subpopulation
D	Not applicable: population >1000 mature individuals, >5 locations
E	Not applicable: no population viability analysis undertaken

IUCN Red List assessment data

	Estimate	Reliability
Extent of occurrence trend	7 100 000 km^2 stable	high high
Area of occupancy trend	2000 km^2 decreasing	low medium
No. of mature individuals trend	1250 decreasing	medium medium
No. subpopulations	1	high
No. locations	>10	high
Generation time	8.6 years	low
Global population share	100%	high

is evidence of partial migration from south-eastern wetlands to coastal central and northern Queensland in autumn and winter (Marchant and Higgins 1993; Black *et al.* 2010).

Abundance

The population is estimated to be between 1000 and 1500, and highly unlikely to exceed 2500 mature individuals, based on counts of 805 birds, including duplicate sightings and juveniles, recorded over *c.*160 sites between 2001 and 2009 (Department of Sustainability, Environment, Water, Population and Communities 2011). A population decline of >30% is estimated for the past 26 years (3 generations), a trend which is likely to continue into the future based on ongoing habitat loss (Department of Sustainability Environment Water Population and Communities 2011).

Ecology

Inhabits shallow, vegetated, temporary or infrequently filled wetlands, sometimes where there are trees or shrubs or samphire, and occasionally brackish wetlands, saltmarsh, claypans, sewage farms, dams, bores and irrigation schemes (Marchant and Higgins 1993). Feed on seeds and invertebrates, including insects, worms, molluscs and crustaceans from the water's edge. A generation time of 8.6 years (BirdLife International 2011) is derived from an average age at first breeding of 1.0 years and a maximum longevity in the wild of 16.2 years, both extrapolated from the mean of other shorebirds.

Threats

Declines have been caused by loss and degradation of wetlands, especially shallow ephemeral wetlands, through drainage and the diversion of water for agriculture and reservoirs. Substantial declines in the reporting rate of the species in the Murray-Darling Basin coincided with major changes in water management, including diversion to irrigated agriculture (Kingsford 2000; Lane and Rogers 2000). The replacement of native wetland vegetation by invasive weeds may also render habitats less suitable or even totally unsuitable for the species (Rogers *et al.* 2005). Grazing and associated trampling of wetland vegetation by cattle may also threaten the snipe in certain regions, particularly where grazing is concentrated around wetlands in the dry season (Johnstone and Storr 1998; Rogers *et al.* 2005). Predation by feral mammals has been suggested as a threat, though there is no evidence that it is a cause of recent declines (Department of Sustainability Environment Water Population and Communities 2011).

Conservation objectives

1. Stable or increasing numbers recorded in standardised surveys

Information required

1. Population trends
2. Improved understanding of movements
3. Principal breeding and non-breeding wetlands, particularly drought refuges
4. Detailed information on habitat before, during and after use

Management actions required

1. Maintain regular counts
2. Protect and manage principal breeding wetlands and wintering grounds. As a precaution, identify and protect habitat used in the last 10 years
3. Rehabilitate selected former breeding wetlands
4. Develop guidelines, in consultation with landholders, for the management of suitable wetlands, including prevention of grazing and trampling by cattle

Bibliography

Barrett G, Silcocks A, Barry S, Cunningham R, Poulter R (2003) *The New Atlas of Australian Birds.* Royal Australian Ornithologists Union, Hawthorn East.

BirdLife International (2011) 'Species factsheet: *Rostratula australis*'. Retrieved 17 January, 2011 from <http://www.birdlife.org/>.

Black R, Houston W, Jaensch R (2010) Evidence of regular seasonal migration by Australian Painted Snipe *Rostratula australis* to the Queensland tropics in autumn and winter. *Stilt* **58**, 1–9.

Department of Sustainability, Environment, Water, Population and Communities (2011) *Rostratula australis.* In 'Species Profile and Threats Database'. Department of Sustainability, Environment, Water, Population and Communities, Canberra. Retrieved 24 February, 2011 from <http://www.environment.gov.au/sprat>.

Johnstone RE, Storr GM (1998) *Handbook of Western Australian Birds. Volume 1: Non-passerines (Emu to Dollarbird).* Western Australian Museum, Perth.

Kingsford RT (2000) Ecological impacts of dams, water diversions and river management on floodplain wetlands in Australia. *Austral Ecology* **25**, 109–127.

Lane BA, Rogers DI (2000) The Australian Painted Snipe, *Rostratula (benghalensis) australis*: an Endangered species? *Stilt* **36**, 26–34.

Marchant S, Higgins PJ (Eds) (1993) *Handbook of Australian, New Zealand and Antarctic Birds. Volume 2: Raptors to Lapwings*. Oxford University Press, Melbourne.

Rogers D, Hance I, Paton S, Tzaros C, Griffioen P, Herring M, Jaensch R, Oring L, Silcocks A, Weston M (2005) The breeding bottleneck: breeding habitat and population decline in the Australian Painted Snipe. In *Status and Conservation of Seabirds in the East Asian-Australasian Flyway: Proceedings of the Australasian Shorebirds Conference*, 13–15 Dec 2003, Canberra, Australia. (Ed. P Straw) pp. 15–23. Wetlands International Global Series 18, International Wader Studies 17, Sydney, Australia.

Comments received from

Dean Ingwersen, Roger Jaensch, Danny Rogers

Subspecies Conservation Summary

Black-tailed Godwit (eastern Siberian)

Limosa limosa melanuroides Gould, 1846

Scolopacidae

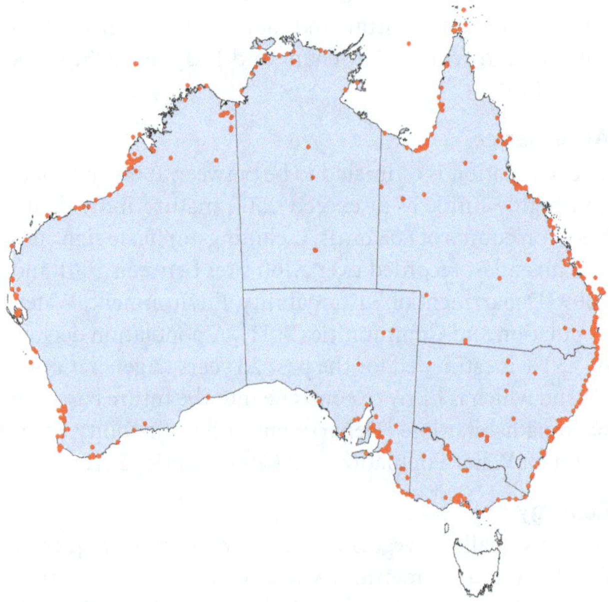

Conservation status

Population visiting Australia: Near Threatened A2b

Reasons for listing

Declines of 20–29% over last 3 generations (26 years) based on survey data

Status 2000

Near Threatened A4b

Reason for change in 2010: Same status but declines assumed to have stopped until better evidence of ongoing declines available

Status 1990

Least Concern

Reason for change in 2000: Population declines must have been under way

Taxonomy

L. l. limosa and *L. l. islandica* are both extralimital. The species status is Near Threatened.

Taxonomic uniqueness: medium (22 genera/family, 4 species/genus, 3 subspecies/species)

Range

Breeds in China, Mongolia and Russia, and winters in south Asia, South-East Asia and Australasia (Bamford *et al.* 2008). Non-breeding birds occur throughout Australia but especially along coasts in the north from Cape York Peninsula west to Arnhem Land (Department of Sustainability, Environment, Water, Population and Communities 2011).

Abundance

The global population has been estimated at 160 000 individuals (Wetlands International 2006; BirdLife International 2011), including 70 000 in Australia (Bamford *et al.* 2008). Numbers declined at northern Roebuck Bay from 6800 in 2001 to 470 in 2008 (Rogers *et al.* 2009), across 49 Australian sites from 830 to 520 between c.1983 and c.2007 (AWSG database: Birds Australia *in litt.*) and at Moreton Bay by 79% between 1993 and 2008 (Fuller *et al.* 2009), but detailed trends are unknown at all other important sites (in the Northern Territory and the Gulf of Carpentaria; Bamford *et al.* 2008). Trends outside Australia are poorly known but numbers in Japan show no clear trends between 1978 and 2008 (Amano *et al.* 2010). Given the variability in trends and inadequate data for most of the population, it is inferred that there may have been an overall decline of 20–29% over the last 26 years. This decline may not continue as its causes are not well known. Globally the species has probably declined by between 14% and 33% over the past 15 years (BirdLife International 2011).

Ecology

Breeds in grassy pastures, hayfields, marshland, bogs, moorland, lake margins and damp grassy depressions in steppe. Non-breeding birds migrate to sheltered bays,

Current eligibility against IUCN Red List Criteria

IUCN category	Criteria eligibility
Population visiting Australia	
A	Near Threatened: population decline of 20–29% over the last 3 generations (26 years) estimated from monitoring
B	Not applicable: EOO >20 000 km^2; AOO >2000 km^2
C	Not applicable: population >10 000 mature individuals
D	Not applicable: population >1000 mature individuals, >5 locations
E	Not applicable: no population viability analysis undertaken

IUCN Red List assessment data

	Estimate	Reliability
Population visiting Australia		
Extent of occurrence trend	7 500 000 km^2 stable	high high
Area of occupancy trend	2500 km^2 stable	low medium
No. of mature individuals trend	70 000 decreasing	medium high
No. subpopulations	1	high
No. locations	>10	high
Generation time	8.6 years	high
Global population share	26–50%	high

estuaries and lagoons, near-coastal wetlands including floodplains and, on migration, rice paddies, and feed on invertebrates such as insects, worms and crustaceans. In Australia the largest numbers are found on prograding coasts backed by mangroves (Higgins and Davies 1996). A generation time of 8.6 years (BirdLife International 2011) is derived from an age at first breeding of 2.0 years (Cramp *et al.* 1983), an adult survival of 63% (Cramp *et al.* 1983) and a maximum longevity of 22.8 years (extrapolated from for congeners).

Threats

Probably threatened by wetland degradation along its migration route in East Asia, where the Yellow Sea supports c.30% of the flyway population and remainder probably migrates through inland China (Barter 2002; Department of Sustainability, Environment, Water, Population and Communities 2011). General threats at migratory staging sites include environmental pollution, reduced river flows, human disturbance, reclamation for tidal power plants, barrages, aquaculture ponds and agriculture, industrial use, urban expansion and agricultural intensification at rice paddies (Barter 2002; Kelin and Qiang 2006; Moores 2006). Possibly threatened on its breeding grounds by drainage and agricultural intensification, as are the other subspecies (BirdLife International 2011). Generalised threats in Australia, especially eastern and southern Australia, include ongoing human disturbance, habitat loss and degradation from pollution, changes to the water regime and invasive plants (Rogers *et al.* 2006; Australian Government 2009) but most of the population overwinters on some of the least disturbed coastline in the world.

Conservation objectives

1. Stable Australian non-breeding populations
2. Disturbance at key roosting and feeding sites reduced

Information required

1. Numbers and trends of Australian non-breeding population
2. Improved knowledge of dependence on key migratory staging sites
3. Improved knowledge of impacts of disturbance

Management actions required

1. Work with governments along the flyway at all levels, and local communities, to prevent destruction of key migratory staging sites
2. Negotiate with the Chinese and South Korean governments in particular to protect remaining tidal flats in the Yellow Sea
3. Advocate for restoration of alienated wetland sites

Bibliography

Amano T, Székely T, Koyama K, Amano H, Sutherland WJ (2010) A framework for monitoring the status of populations: an example from wader populations in the East Asian–Australasian flyway. *Biological Conservation* **143**, 2238–2247.

Australian Government (2009) 'Draft significant impact guidelines for 36 migratory shorebirds'. Draft EPBC Act Policy Statement 3.21, Canberra. Retrieved 25 February, 2011 from <http://www.environment.gov.au/epbc/publications/migratory-shorebirds.html>.

Bamford M, Watkins D, Bancroft W, Tischler G, Wahl J (2008) 'Migratory shorebirds of the East Asian – Australasian flyway: population estimates and internationally important sites'. Wetlands International – Oceania, Canberra.

Barter MA (2002) 'Shorebirds of the Yellow Sea: importance, threats and conservation status'. Canberra, Australia.

BirdLife International (2011) 'Species factsheet: *Limosa limosa*'. Retrieved 17 January, 2011 from <http://www.birdlife.org/>.

Cramp S, Simmons KEL, Brooks DC, Collar NJ, Dunn E, Gillmor R, Hollom PAD, Hudson R, Nicholson EM, Ogilvie MA, Olney PJS, Roselaar CS, Voous KH, Wallace DIM, Wattel J, Wilson MG (Eds) (1983) *Handbook of the Birds of Europe, the Middle East and North Africa. The Birds of the Western Palearctic: 3. Waders to Gulls*. Oxford University Press, Oxford, UK.

Department of Sustainability, Environment, Water, Population and Communities (2011) *Limosa limosa*. In 'Species Profile and Threats Database'. Department of Sustainability, Environment, Water, Population and Communities, Canberra. Retrieved 24 February, 2011 from <http://www.environment.gov.au/sprat>.

Fuller RA, Wilson HB, Kendall BE, Possingham HP (2009) 'Monitoring shorebirds using counts by the Queensland Wader Study Group'. Report to the Queensland Wader Study Group and the Department of Environment and Resource Management, Brisbane.

Higgins PJ, Davies SJJF (Eds) (1996) *Handbook of Australian, New Zealand and Antarctic Birds. Volume 3: Snipe to Pigeons.* Oxford University Press, Melbourne.

Kelin C, Qiang X (2006) Conserving migratory shorebirds in the Yellow Sea region. In *Waterbirds Around the World.* (Eds G Boere, C Galbraith and D Stroud) p. 319. The Stationery Office, Edinburgh, UK.

Moores N (2006) South Korea's shorebirds: a review of abundance, distribution, threats and conservation status. *Stilt* **50**, 62–72.

Rogers D, Hassell C, Oldland J, Clemens R, Boyle A, Rogers K (2009) 'Monitoring Yellow Sea migrants in Australia (MYSMA): north-western Australian shorebird surveys and workshops, December 2008'.

Rogers DI, Piersma T, Hassell CJ (2006) Roost availability may constrain shorebird distribution: exploring the energetic costs of roosting and disturbance around a tropical bay. *Biological Conservation* **133**, 225–235.

Wetlands International (2006) 'Waterbird population estimates'. Wetlands International, Wageningen, The Netherlands.

Comments received from

Rob Clemens, Ken Gosbell, Roger Jaensch, David Milton, Clive Minton, Danny Rogers

Subspecies Conservation Summary

Bar-tailed Godwit (western Alaskan)

Limosa lapponica baueri Naumann, 1836

Scolopacidae

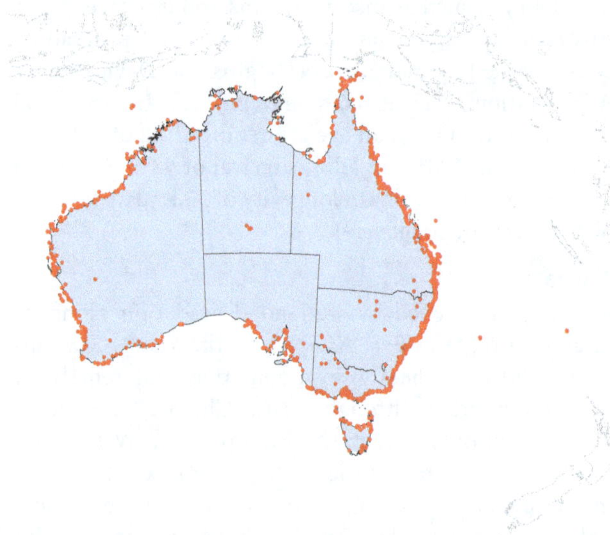

Conservation status

Population visiting Australia: Vulnerable A3c+4bc

Reasons for listing

Suspected recent and ongoing declines of 30–49% in 3 generations (29 years) based on survey data and habitat loss

Status 2000

Near Threatened A3c

Reason for change in 2010: habitat loss intensifying

Status 1990

Least Concern

Reason for change in 2000: population declines first recorded

Taxonomy

L. l. menzbieri is Vulnerable in Australia, *L. l. lapponica* is extralimital. The species is Least Concern.

Taxonomic uniqueness: medium (22 genera/family, 4 species/genus, 3 subspecies/species)

Range

Breeds in north-east Siberia and north-west Alaska, and winters in east Australia and New Zealand (Bamford *et al.* 2008). Non-breeding birds occur along the coasts of Australia especially the north and east (Department of Sustainability, Environment, Water, Population and Communities 2011).

Abundance

The global population has been estimated at 155 000 individuals (Wetlands International 2006; Bamford *et al.* 2008). The population spending its non-breeding period primarily in Australia is assumed to be 61 000 as most of the c.94 000 Bar-tailed Godwits in New Zealand are *L. l. baueri* (Southey 2009). Numbers declined at Moreton Bay by 68% between 1993 and 2008 (Fuller *et al.* 2009), at Corner Inlet by 12% between 1982 and 2009 (C. D. T. Minton *in litt.*), at Gulf St Vincent by 78% between c.1981 and c.2004 (Close 2008), but increased by c.8% across 49 Australian sites between c.1983 and c.2007 (AWSG database: Birds Australia *in litt.*). Populations in Japan have declined in general, and by about 53% in spring between 1998 and 2008 (Amano *et al.* 2010) and in New Zealand declined by 18% (103 000 to 85 000) between 1993 and 2003 (Southey 2009). Given the variability in trends, it is estimated that there may have been an overall decline of 30–49% in Australia over the last 27 years. Given likely ongoing threats to its migratory staging sites in East Asia (Barter 2002; Moores 2006), it is assumed that declines of 30–49% over 27 years will continue. Globally the species is declining gradually and is not considered threatened (BirdLife International 2011).

Current eligibility against IUCN Red List Criteria

IUCN category	Criteria eligibility
Population visiting Australia	
A	Vulnerable: current and anticipated population declines of 30–49% over periods of 3 generations (29 years) estimated and projected from monitoring and from deteriorating habitat quality
B	Not applicable: EOO >20 000 km^2; AOO >2000 km^2
C	Not applicable: population >10 000 mature individuals
D	Not applicable: population >1000 mature individuals, >5 locations
E	Not applicable: no population viability analysis undertaken

IUCN Red List assessment data

	Estimate	Reliability
Population visiting Australia		
Extent of occurrence trend	7 500 000 km^2 stable	high high
Area of occupancy trend	8100 km^2 stable	low medium
No. of mature individuals trend	61 000 decreasing	medium high
No. subpopulations	1	high
No. locations	>10	high
Generation time	9.7 years	high
Global population share	26–50%	high

Ecology

Breeds in marshy areas in lowland tundra. Non-breeding birds migrate to muddy coastlines, estuaries, inlets, mangrove-fringed lagoons and sheltered bays, and feed on annelids, bivalves and crustaceans (Higgins and Davies 1996). A generation time of 9.7 years (BirdLife International 2011) is derived from an average age at first breeding of 2.0 years (Cramp *et al.* 1983), an adult survival of 70.0% (Cramp *et al.* 1983) and a maximum longevity in the wild of 22.8 years (Australian Bird and Bat Banding Scheme).

Threats

Major threat is wetland loss and degradation along its migration route in East Asia, where the Yellow Sea supports c.80% of the flyway population on northward migration (Barter 2002; Department of Sustainability, Environment, Water, Population and Communities 2011). General threats at migratory staging sites include environmental pollution, reduced river flows, human disturbance, reclamation for tidal power plants, barrages, aquaculture ponds and agriculture, industrial use and urban expansion (Barter 2002; Kelin and Qiang 2006; Moores 2006) and hunting (Barter *et al.* 1997; Ming *et al.* 1998). Threats in Australia, especially eastern and southern Australia, include ongoing human disturbance, habitat loss and degradation from pollution, changes to the water regime and invasive plants (Rogers *et al.* 2006; Australian Government 2009).

Conservation objectives

1. Stable Australian non-breeding populations
2. Disturbance at key roosting and feeding sites reduced

Information required

1. Numbers and trends of Australian non-breeding population
2. Improved knowledge of dependence on key migratory staging sites
3. Improved knowledge of impacts of disturbance

Management actions required

1. Work with governments along the flyway at all levels, and local communities, to prevent destruction of key migratory staging sites
2. Negotiate with the Chinese and South Korean governments in particular to protect remaining tidal flats in the Yellow Sea
3. Advocate for restoration of alienated wetland sites
4. Maintain and improve protection of roosting and feeding sites in Australia

Bibliography

Amano T, Székely T, Koyama K, Amano H, Sutherland WJ (2010) A framework for monitoring the status of populations: an example from wader populations in the East Asian–Australasian flyway. *Biological Conservation* **143**, 2238–2247.

Australian Government (2009) 'Draft significant impact guidelines for 36 migratory shorebirds'. Draft EPBC Act Policy Statement 3.21, Canberra. Retrieved 25 February, 2011 from <http://www.environment.gov.au/epbc/publications/migratory-shorebirds.html>.

Bamford M, Watkins D, Bancroft W, Tischler G, Wahl J (2008) 'Migratory shorebirds of the East Asian – Australasian flyway: population estimates and internationally important sites'. Wetlands International – Oceania, Canberra.

Barter M, Fawen Q, Sixian T, Xiao Y, Tonkinson D (1997) Hunting of migratory waders on Chongming Dao: a declining occupation? *Stilt* **31**, 19–22.

Barter MA (2002) 'Shorebirds of the Yellow Sea: importance, threats and conservation status'. Canberra, Australia.

BirdLife International (2011) 'Species factsheet: *Limosa lapponica*'. Retrieved 17 January, 2011 from <http://www.birdlife.org/>.

Close DH (2008) Changes in wader numbers in the Gulf St Vincent, South Australia, 1979–2008. *Stilt* **54**, 24–27.

Cramp S, Simmons KEL, Brooks DC, Collar NJ, Dunn E, Gillmor R, Hollom PAD, Hudson R, Nicholson EM, Ogilvie MA, Olney PJS, Roselaar CS, Voous KH, Wallace DIM, Wattel J, Wilson MG (Eds) (1983) *Handbook of the Birds of Europe, the Middle East and North Africa. The Birds of the Western Palearctic: 3. Waders to Gulls.* Oxford University Press, Oxford, UK.

Department of Sustainability Environment Water Population and Communities (2011) *Limosa lapponica*. In 'Species Profile and Threats Database'. Department of Sustainability, Environment, Water, Population and Communities,

Canberra. Retrieved 24 February, 2011 from <http://www.environment.gov.au/sprat>.

Fuller RA, Wilson HB, Kendall BE, Possingham HP (2009) 'Monitoring shorebirds using counts by the Queensland Wader Study Group'. Report to the Queensland Wader Study Group and the Department of Environment and Resource Management, Brisbane.

Higgins PJ, Davies SJJF (Eds) (1996) *Handbook of Australian, New Zealand and Antarctic Birds. Volume 3: Snipe to Pigeons.* Oxford University Press, Melbourne.

Kelin C, Qiang X (2006) Conserving migratory shorebirds in the Yellow Sea region. In *Waterbirds Around the World.* (Eds G Boere, C Galbraith and D Stroud) p. 319. The Stationery Office, Edinburgh, UK.

Ming M, Jianjian L, Chengjia T, Pingyue S, Wei H (1998) The contribution of shorebirds to the catches of hunters in the Shanghai area, China, during 1997–1998. *Stilt* **33**, 32–36.

Moores N (2006) South Korea's shorebirds: a review of abundance, distribution, threats and conservation status. *Stilt* **50**, 62–72.

Rogers DI, Piersma T, Hassell CJ (2006) Roost availability may constrain shorebird distribution: exploring the energetic costs of roosting and disturbance around a tropical bay. *Biological Conservation* **133**, 225–235.

Southey I (2009) 'Numbers of waders in New Zealand 1994–2003'. DOC Research & Development Series 308. Retrieved 4 May, 2011 from <http://www.doc.govt.nz/upload/documents/science-and-technical/drds308.pdf>.

Wetlands International (2006) 'Waterbird population estimates'. Wetlands International, Wageningen, The Netherlands.

Comments received from

Rob Clemens, Ken Gosbell, Roger Jaensch, David Milton, Clive Minton, Danny Rogers

Subspecies Conservation Summary

Bar-tailed Godwit (northern Siberian)

Limosa lapponica menzbieri Portenko, 1936

Scolopacidae

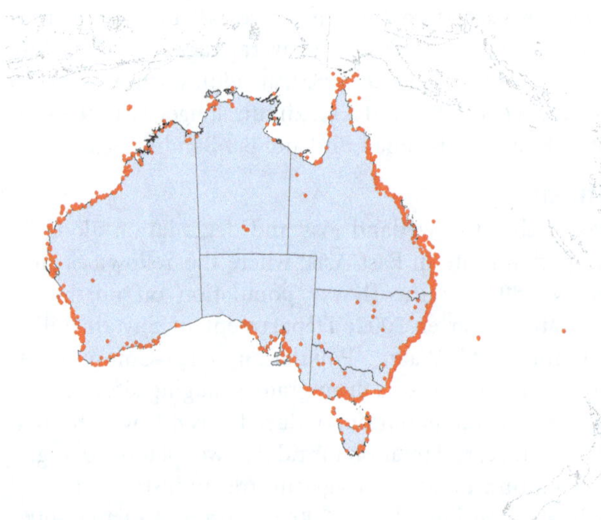

Conservation status

Population visiting Australia: Vulnerable A3c+4bc

Reasons for listing

Suspected recent and ongoing declines of 30–49% in 3 generations (29 years) based on survey data and habitat loss

Status 2000

Near Threatened A3c

Reason for change in 2010: habitat loss intensifying

Status 1990

Least Concern

Reason for change in 2000: population declines first recorded

Taxonomy

L. l. baueri is Vulnerable in Australia, *L. l. lapponica* is extralimital. The species is Least Concern.

Taxonomic uniqueness: medium (22 genera/family, 4 species/genus, 3 subspecies/species)

Range

Breeds in north Siberia east to the Kolyma Delta, and winters in South-East Asia and northern Australia (Bamford *et al.* 2008). Non-breeding birds occur along the coasts of Australia especially the north of Western Australia (Department of Sustainability, Environment, Water, Population and Communities 2011).

Abundance

The global population has been estimated at 170 000 individuals (Wetlands International 2006; Bamford *et al.* 2008). The Australian population estimated here is assumed to be 124 000, based on 185 000 for the species (Bamford *et al.* 2008) minus 61 000 *L. l. baueri*. Numbers of mostly this subspecies declined 2000–2008 on Eighty-mile Beach (110 000–52 000) and at northern Roebuck Bay (12 000–9000 from 2001–2004 to 2005–2008), but with no clear trends at Bush Point 2004–2008 (21 000–34 000; Rogers *et al.* 2009), and numbers of both subspecies increased by c.8% across 49 Australian sites between c.1983 and c.2007 (AWSG database: Birds Australia *in litt.*). Populations of both subspecies in Japan have declined in general, and by about 53% in spring between 1998 and 2008 (Amano *et al.* 2010). Assuming the declines in north-western Australia are typical of the Australian population, it is estimated that there may have been an overall decline of 30–49% over the last 27 years.

Current eligibility against IUCN Red List Criteria

IUCN category	Criteria eligibility
Population visiting Australia	
A	Vulnerable: current and anticipated population declines of 30–49% over periods of 3 generations (29 years) estimated and projected from monitoring and from deteriorating habitat quality
B	Not applicable: EOO >20 000 km^2; AOO >2000 km^2
C	Not applicable: population >10 000 mature individuals
D	Not applicable: population >1000 mature individuals, >5 locations
E	Not applicable: no population viability analysis undertaken

IUCN Red List assessment data

	Estimate	Reliability
Population visiting Australia		
Extent of occurrence trend	7 500 000 km^2 stable	high high
Area of occupancy trend	8100 km^2 stable	low medium
No. of mature individuals trend	124 000 decreasing	medium high
No. subpopulations	1	high
No. locations	>10	high
Generation time	9.7 years	high
Global population share	51–100%	high

Given likely ongoing threats to its migratory staging sites in East Asia (Barter 2002; Moores 2006), it is assumed that these declines will continue. Global the species is declining gradually and is not threatened (BirdLife International 2011).

Ecology

Breeds in marshy areas in lowland tundra. Non-breeding birds migrate to muddy coastlines, estuaries, inlets, mangrove-fringed lagoons and sheltered bays, and feed on annelids, bivalves and crustaceans (Higgins and Davies 1996). A generation time of 8.9 years (BirdLife International 2011) is derived an age at first breeding of 2.0 years (Cramp *et al.* 1983), an adult survival of 70% (Cramp *et al.* 1983) and a maximum longevity of 22.8 years (Australian Bird and Bat Banding Scheme).

Threats

Major threat is wetland degradation in the Yellow Sea where c.80% of the flyway population stages on northward migration (Barter 2002; Bamford *et al.* 2008; Department of Sustainability, Environment, Water, Population and Communities 2011). General threats at migratory staging sites include environmental pollution, reduced river flows, human disturbance and reclamation for tidal power plants and barrages, industrial use and urban expansion (Barter 2002; Kelin and Qiang 2006; Moores 2006). It is also hunted on migration (Barter *et al.* 1997; Ming *et al.* 1998). Threats in Australia, especially eastern and southern Australia, include ongoing human disturbance, habitat loss and degradation from pollution, changes to the water regime and invasive plants (Rogers *et al.* 2006; Australian Government 2009).

Conservation objectives

1. Stable Australian non-breeding populations
2. Disturbance at key roosting and feeding sites reduced

Information required

1. Numbers and trends of Australian non-breeding population
2. Improved knowledge of dependence on key migratory staging sites
3. Improved knowledge of impacts of disturbance

Management actions required

1. Work with governments along the flyway at all levels, and local communities, to prevent destruction of key migratory staging sites
2. Negotiate with the Chinese and South Korean governments in particular to protect remaining tidal flats in the Yellow Sea
3. Advocate for restoration of alienated wetland sites
4. Maintain and improve protection of roosting and feeding sites in Australia

Bibliography

Amano T, Székely T, Koyama K, Amano H, Sutherland WJ (2010) A framework for monitoring the status of populations: an example from wader populations in the East Asian–Australasian flyway. *Biological Conservation* **143**, 2238–2247.

Australian Government (2009) 'Draft significant impact guidelines for 36 migratory shorebirds'. Draft EPBC Act Policy Statement 3.21, Canberra. Retrieved 25 February, 2011 from <http://www.environment.gov.au/epbc/publications/migratory-shorebirds.html>.

Bamford M, Watkins D, Bancroft W, Tischler G, Wahl J (2008) 'Migratory shorebirds of the East Asian – Australasian flyway: population estimates and internationally important sites'. Wetlands International – Oceania, Canberra.

Barter M, Fawen Q, Sixian T, Xiao Y, Tonkinson D (1997) Hunting of migratory waders on Chongming Dao: a declining occupation? *Stilt* **31**, 19–22.

Barter MA (2002) 'Shorebirds of the Yellow Sea: importance, threats and conservation status'. Canberra, Australia.

BirdLife International (2011) 'Species factsheet: *Limosa lapponica*'. Retrieved 17 January, 2011 from <http://www.birdlife.org/>.

Cramp S, Simmons KEL, Brooks DC, Collar NJ, Dunn E, Gillmor R, Hollom PAD, Hudson R, Nicholson EM, Ogilvie MA, Olney PJS, Roselaar CS, Voous KH, Wallace DIM, Wattel J, Wilson MG (Eds) (1983) *Handbook of the Birds of Europe, the Middle East and North Africa. The Birds of the Western Palearctic: 3. Waders to Gulls*. Oxford University Press, Oxford, UK.

Department of Sustainability, Environment, Water, Population and Communities (2011) *Limosa lapponica*. In 'Species Profile and Threats Database'. Department of Sustainability, Environment, Water, Population and Communities, Canberra. Retrieved 24 February, 2011 from <http://www.environment.gov.au/sprat>.

Higgins PJ, Davies SJJF (Eds) (1996) *Handbook of Australian, New Zealand and Antarctic Birds. Volume 3: Snipe to Pigeons.* Oxford University Press, Melbourne.

Kelin C, Qiang X (2006) Conserving migratory shorebirds in the Yellow Sea region. In *Waterbirds Around the World.* (Eds G Boere, C Galbraith and D Stroud) p. 319. The Stationery Office, Edinburgh, UK.

Ming M, Jianjian L, Chengjia T, Pingyue S, Wei H (1998) The contribution of shorebirds to the catches of hunters in the Shanghai area, China, during 1997–1998. *Stilt* **33**, 32–36.

Moores N (2006) South Korea's shorebirds: a review of abundance, distribution, threats and conservation status. *Stilt* **50**, 62–72.

Rogers D, Hassell C, Oldland J, Clemens R, Boyle A, Rogers K (2009) 'Monitoring Yellow Sea migrants in Australia (MYSMA): north-western Australian shorebird surveys and workshops, December 2008'.

Rogers DI, Piersma T, Hassell CJ (2006) Roost availability may constrain shorebird distribution: exploring the energetic costs of roosting and disturbance around a tropical bay. *Biological Conservation* **133**, 225–235.

Wetlands International (2006) 'Waterbird population estimates'. Wetlands International, Wageningen, The Netherlands.

Comments received from

Rob Clemens, Ken Gosbell, Roger Jaensch, David Milton, Clive Minton, Danny Rogers

Subspecies Conservation Summary

Whimbrel (eastern Siberian)

Numenius phaeopus variegatus (Scopoli, 1786)
Scolopacidae

Conservation status

Population visiting Australia: Near Threatened A3c+4bc

Reasons for listing

Suspected recent and ongoing declines of 20–29% over 3 generations (27 years) based on survey data and habitat loss

Status 2000

Least Concern

Reason for change in 2010: decline in population and habitat loss at staging posts

Status 1990

Least Concern

Taxonomy

Five other infraspecific taxa; all vagrant to or absent from Australia. The species is Least Concern.

Taxonomic uniqueness: medium (22 genera/family, 8 species/genus, 6 subspecies/species)

Range

Breeds in east Siberia, and winters in South and South-East Asia and Australasia (Bamford *et al.* 2008). Non-breeding birds occur along the coasts of Australia, especially in Queensland (AWSG database: Birds Australia *in litt.*).

Abundance

The flyway population has been estimated at 100 000 individuals (Wetlands International 2006; Bamford *et al.* 2008) and the Australian population at 10 000 (Stewart *et al.* 2007). Birds are spread out through mangroves so difficult to count but 2 counts at Shoalwater Bay, the most significant Australian site, were 7089 in 1995 (Driscoll 1995) and 2891 in 2007 (Jaensch 2008) and at Moreton Bay numbers declined by 43% between 1993 and 2008 with maximum counts >2000 individuals (Fuller *et al.* 2009). Counts at the 2 next most important sites, Gulf of Carpentaria and Great Sandy Strait, are not available. However, numbers increased from <500 to nearly 2000 on Eighty-mile Beach, Bush Point and northern Roebuck Bay beaches between 2000 and 2008 (Rogers *et al.* 2009), and increased by c.96% across 49 Australian sites between c.1983 and c.2007 (AWSG database: Birds Australia *in litt.*). Populations in Japan have declined consistently in autumn, and by about 80% between 1978 and 2008 (Amano *et al.* 2010). Despite the variability in trends, it is estimated that there may have been an overall decline of 20–29% in Australia over the last 27 years. Given likely ongoing threats to its migratory staging sites in East Asia (Barter 2002; Moores

Current eligibility against IUCN Red List Criteria

IUCN category	Criteria eligibility
Population visiting Australia	
A	Near Threatened: current and anticipated population declines of 20–29% over periods 3 generations (27 years) estimated and projected from monitoring and from deteriorating habitat quality
B	Not applicable: EOO >20 000 km^2; AOO >2000 km^2
C	Not applicable: population >10 000 mature individuals
D	Not applicable: population >1000 mature individuals, >5 locations
E	Not applicable: no population viability analysis undertaken

IUCN Red List assessment data

	Estimate	Reliability
Population visiting Australia		
Extent of occurrence trend	36 000 km^2 Stable	high high
Area of occupancy trend	1400 km^2 stable	low medium
No. of mature individuals trend	10 000 decreasing	medium low
No. subpopulations	1	high
No. locations	>10	high
Generation time	9.1 years	high
Global population share	1–10%	high

2006), it is assumed that declines of 20–30% over 27 years will continue. The species is gradually declining globally but not at a rate to be considered threatened (BirdLife International 2011).

Ecology

Breeds in heathland, tundra, sedge meadows, moorland, bogs and open forest. Non-breeding birds migrate to coastal wetlands including muddy, rocky or sandy beaches, mangroves and short grasslands, and feed on crustaceans, molluscs and polychaete worms (Higgins and Davies 1996). A generation time of 9.1 years (BirdLife International 2011) is derived from an age at first breeding of 2.0 years, an adult survival of 69% and a maximum longevity in the wild of 24.0 years, all extrapolated from congeners.

Threats

Threatened by wetland loss and degradation along its migration route in East Asia, where the Yellow Sea supports c.50% of the flyway population on northern migration (Barter 2002; Department of Sustainability, Environment, Water, Population and Communities 2011). General threats at migratory staging sites include environmental pollution, reduced river flows, human disturbance, reclamation for tidal power plants, barrages, aquaculture ponds and agriculture, industrial use, urban expansion and hunting (Barter *et al.* 1997; Barter 2002; Kelin and Qiang 2006; Moores 2006). Threats in Australia, especially eastern Australia, include ongoing human disturbance, habitat loss and degradation from pollution, changes to the water regime and invasive plants (Rogers *et al.* 2006; Australian Government 2009).

Conservation objectives

1. Stable Australian non-breeding populations
2. Disturbance at key roosting and feeding sites reduced

Information required

1. Numbers and trends of Australian non-breeding population
2. Improved knowledge of dependence on key migratory staging sites
3. Improved knowledge of impacts of disturbance

Management actions required

1. Work with governments along the flyway at all levels, and local communities, to prevent destruction of key migratory staging sites
2. Negotiate with the Chinese and South Korean governments in particular to protect remaining tidal flats in the Yellow Sea
3. Advocate for restoration of alienated wetland sites
4. Maintain and improve protection of roosting and feeding sites in Australia

Bibliography

Amano T, Székely T, Koyama K, Amano H, Sutherland WJ (2010) A framework for monitoring the status of populations: an example from wader populations in the East Asian–Australasian flyway. *Biological Conservation* **143**, 2238–2247.

Australian Government (2009) 'Draft significant impact guidelines for 36 migratory shorebirds'. Draft EPBC Act Policy Statement 3.21, Canberra. Retrieved 25 February, 2011 from <http://www.environment.gov.au/epbc/publications/migratory-shorebirds.html>.

Bamford M, Watkins D, Bancroft W, Tischler G, Wahl J (2008) 'Migratory shorebirds of the East Asian – Australasian flyway: population estimates and internationally important sites'. Wetlands International – Oceania, Canberra.

Barter M, Fawen Q, Sixian T, Xiao Y, Tonkinson D (1997) Hunting of migratory waders on Chongming Dao: a declining occupation? *Stilt* **31**, 19–22.

Barter MA (2002) 'Shorebirds of the Yellow Sea: importance, threats and conservation status'. Canberra, Australia.

BirdLife International (2011) 'Species factsheet: *Numenius phaeopus*'. Retrieved 17 January, 2011 from <http://www.birdlife.org/>.

Department of Sustainability, Environment, Water, Population and Communities (2011) *Numenius phaeopus*. In 'Species Profile and Threats Database'. Department of Sustainability, Environment, Water, Population and Communities, Canberra. Retrieved 24 February, 2011 from <http://www.environment.gov.au/sprat>.

Driscoll PV (1995) 'Survey of shorebird feeding areas and roosts in the Shoalwater Bay area'. Unpublished report.

Fuller RA, Wilson HB, Kendall BE, Possingham HP (2009) 'Monitoring shorebirds using counts by the Queensland Wader Study Group'. Report to the Queensland Wader

Study Group and the Department of Environment and Resource Management, Brisbane.

Higgins PJ, Davies SJJF (Eds) (1996) *Handbook of Australian, New Zealand and Antarctic Birds. Volume 3: Snipe to Pigeons.* Oxford University Press, Melbourne.

Jaensch R (2008) 'A condition assessment of wetlands and waterbirds in the Shoalwater Bay Training Area. Part 4: numbers of migratory shorebirds'. Unpublished report to Australian Government Department of Defence, by Wetlands International, Brisbane.

Kelin C, Qiang X (2006) Conserving migratory shorebirds in the Yellow Sea region. In *Waterbirds Around the World.* (Eds G Boere, C Galbraith and D Stroud) p. 319. The Stationery Office, Edinburgh, UK.

Moores N (2006) South Korea's shorebirds: a review of abundance, distribution, threats and conservation status. *Stilt* **50**, 62–72.

Rogers D, Hassell C, Oldland J, Clemens R, Boyle A, Rogers K (2009) 'Monitoring Yellow Sea migrants in Australia (MYSMA): north-western Australian shorebird surveys and workshops, December 2008'.

Rogers DI, Piersma T, Hassell CJ (2006) Roost availability may constrain shorebird distribution: exploring the energetic costs of roosting and disturbance around a tropical bay. *Biological Conservation* **133**, 225–235.

Stewart DA, Rogers A, Rogers DI (2007) Species description. In *Shorebirds of Australia.* (Eds A Geering, L Agnew and S Harding) pp. 75–196. CSIRO Publishing, Melbourne.

Wetlands International (2006) 'Waterbird population estimates'. Wetlands International, Wageningen, The Netherlands.

Comments received from

Rob Clemens, Ken Gosbell, Roger Jaensch, David Milton, Clive Minton, Danny Rogers

---

Species Conservation Summary

Eastern Curlew

Numenius madagascariensis (Linnaeus, 1766)
Scolopacidae

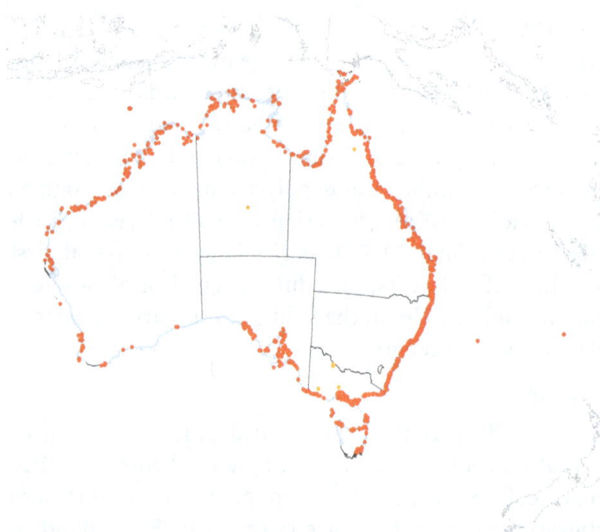

Conservation status

Population visiting Australia: Vulnerable A2bc+3c+4bc

Reasons for listing

Past, recent and ongoing declines of 30–49% in 3 generations (30 years) based on survey data and habitat loss

Status 2000

Near Threatened A3cd

Reason for change in 2010: decline in population and habitat loss at staging posts

Status 1990

Least Concern

Reason for change in 2000: declines in population anticipated

Taxonomy

No infraspecific taxa described

Taxonomic uniqueness: medium (22 genera/family, 8 species/genus, 1 subspecies/species)

Range

Breeds in Siberia, Kamchatka and Mongolia, wintering in coastal East Asia, mostly in Australia, and stopping on migration beside the Yellow Sea. Non-breeding birds occur around coastal Australia, are more common in the north and have disappeared or become much rarer at many sites along the south coast (AWSG database: Birds Australia *in litt.*).

Abundance

The global population has been estimated at *c.*38 000 individuals (Wetlands International 2006) including 28 000 in Australia (Bamford *et al.* 2008), but numbers have subsequently declined including c.1800 which disappeared after reclamation of tidal flats at Saemangeum in the South Korean Yellow Sea in 2006 (Moores 2006). Numbers have declined on Eighty-mile Beach by c.40% between 2000 and 2008, whereas numbers at Roebuck Bay have remained stable (Rogers *et al.* 2009), at Moreton Bay declined by c.5.5% per year between 1998 and 2008 (Fuller *et al.* 2009), in Tasmania by 80% since the 1950s and 2000 (Reid and Park 2003) and by 40% across 49 Australian sites between c.1983 and c.2007 (AWSG database: Birds Australia *in litt.* 2011). Numbers declined less severely elsewhere in the flyway, e.g. no clear trends in Japan between 1978 and 2008 (Amano *et al.* 2010), suggesting a 30–49% decline over 30 years (3 generations) across the flyway. Given that more

Current eligibility against IUCN Red List Criteria

IUCN category	Criteria eligibility
Population visiting Australia	
A	Vulnerable: past, current and anticipated population declines of 30–49% over periods of 3 generations (30 years) estimated and projected from monitoring and from deteriorating habitat quality
B	Not applicable: EOO >20 000 km^2; AOO >2000 km^2
C	Not applicable: population >10 000 mature individuals
D	Not applicable: population >1000 mature individuals, >5 locations
E	Not applicable: no population viability analysis undertaken

IUCN Red List assessment data

	Estimate	Reliability
Population visiting Australia		
Extent of occurrence trend	30 000 km^2 stable	high high
Area of occupancy trend	8500 km^2 decreasing	low medium
No. of mature individuals trend	28 000 decreasing	medium high
No. subpopulations	1	high
No. locations	>10	high
Generation time	10.1 years	medium
Global population share	51–100%	high

reclamation is proposed within the Yellow Sea, it is assumed that declines of 30–49% over 30 years will continue. It may qualify as Endangered if additional data suggests that the whole population has or will decline by >50% in 30 years as suggested by some of the above data.

Ecology

Breeds in bogs and wetlands. Non-breeding birds migrate to estuaries, mangroves, saltmarshes and intertidal flats, particularly those with extensive seagrass (Zosteraceae), where they feed on marine invertebrates, especially crabs and small molluscs (Higgins and Davies 1996). A generation time of 6.8 years (BirdLife International 2011) is derived from an age at first breeding of 2.3 years, an annual adult survival of 79% and a maximum longevity of 24.0 years, all values extrapolated from congeners.

Threats

Threatened by wetland degradation in the Yellow Sea where it stages on migration (Bamford *et al.* 2008; van de Kam *et al.* 2010). Threats include environmental pollution, reduced river flows, human disturbance and reclamation for tidal power plants and barrages, industrial use and urban expansion (Barter 2002; Kelin and Qiang 2006; Moores 2006). Additional threats may include disturbance at nesting and feeding sites and hunting throughout its range (Barter *et al.* 1997). Threats in Australia, especially eastern and southern Australia, include ongoing human disturbance, habitat loss and degradation from pollution, changes to the water regime and invasive plants (Rogers *et al.* 2006; Australian Government 2009).

Conservation objectives

1. Stable Australian non-breeding populations
2. Disturbance at key roosting and feeding sites reduced

Information required

1. Numbers and trends of Australian non-breeding population
2. Improved knowledge of dependence on key migratory staging sites
3. Improved knowledge of impacts of disturbance

Management actions required

1. Work with governments along the flyway at all levels, and local communities, to prevent destruction of key migratory staging sites
2. Negotiate with the Chinese and South Korean governments in particular to protect remaining tidal flats in the Yellow Sea
3. Advocate for restoration of alienated wetland sites
4. Maintain and improve protection of roosting and feeding sites in Australia

Bibliography

Amano T, Székely T, Koyama K, Amano H, Sutherland WJ (2010) A framework for monitoring the status of populations: an example from wader populations in the East Asian–Australasian flyway. *Biological Conservation* **143**, 2238–2247.

Australian Government (2009) 'Draft significant impact guidelines for 36 migratory shorebirds'. Draft EPBC Act Policy Statement 3.21, Canberra. Retrieved 25 February, 2011 from <http://www.environment.gov.au/epbc/publications/migratory-shorebirds.html>.

Bamford M, Watkins D, Bancroft W, Tischler G, Wahl J (2008) 'Migratory shorebirds of the East Asian – Australasian flyway: population estimates and internationally important sites'. Wetlands International – Oceania, Canberra.

Barter M, Fawen Q, Sixian T, Xiao Y, Tonkinson D (1997) Hunting of migratory waders on Chongming Dao: a declining occupation? *Stilt* **31**, 19–22.

Barter MA (2002) 'Shorebirds of the Yellow Sea: importance, threats and conservation status'. Canberra, Australia.

BirdLife International (2011) 'Species factsheet: *Numenius madagascariensis*'. Retrieved 17 January, 2011 from <http://www.birdlife.org/>.

Fuller RA, Wilson HB, Kendall BE, Possingham HP (2009) 'Monitoring shorebirds using counts by the Queensland Wader Study Group'. Report to the Queensland Wader Study Group and the Department of Environment and Resource Management, Brisbane.

Higgins PJ, Davies SJJF (Eds) (1996) *Handbook of Australian, New Zealand and Antarctic Birds. Volume 3: Snipe to Pigeons.* Oxford University Press, Melbourne.

Kelin C, Qiang X (2006) Conserving migratory shorebirds in the Yellow Sea region. In *Waterbirds Around the World*. (Eds G Boere, C Galbraith and D Stroud) p. 319. The Stationery Office, Edinburgh, UK.

Moores N (2006) South Korea's shorebirds: a review of abundance, distribution, threats and conservation status. *Stilt* **50**, 62–72.

Reid T, Park P (2003) Continuing decline of Eastern Curlew, *Numenius madagascariensis*, in Tasmania. *Emu* **103**, 279–283.

Rogers D, Hassell C, Oldland J, Clemens R, Boyle A, Rogers K (2009) 'Monitoring Yellow Sea migrants in Australia (MYSMA): north-western Australian shorebird surveys and workshops, December 2008'.

Rogers DI, Piersma T, Hassell CJ (2006) Roost availability may constrain shorebird distribution: exploring the energetic costs of roosting and disturbance around a tropical bay. *Biological Conservation* **133**, 225–235.

van de Kam J, Battley PF, McCaffery BJ, Rogers DI, Hong J-S, Moores N, Ki J-Y, Lewis J, Piersma T (2010) *Invisible Connections: Why Migrating Shorebirds Need the Yellow Sea.* CSIRO Publishing, Melbourne.

Wetlands International (2006) 'Waterbird population estimates'. Wetlands International, Wageningen, The Netherlands.

Comments received from

Rob Clemens, Ken Gosbell, Roger Jaensch, David Milton, Clive Minton, Danny Rogers

Species Conservation Summary

Grey-tailed Tattler

Tringa brevipes (Vieillot, 1816)

Scolopacidae

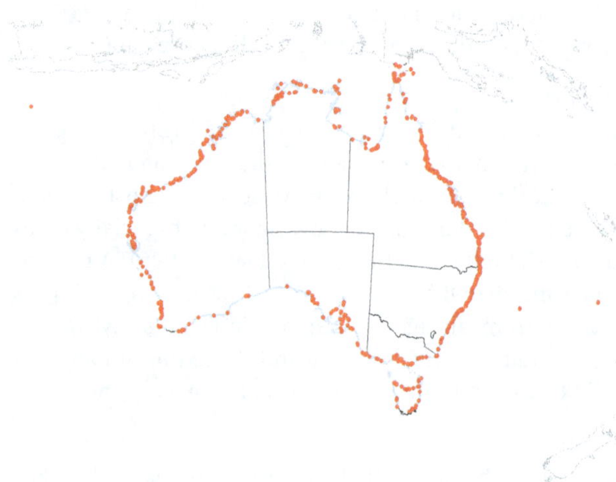

Conservation status

Population visiting Australia: Near Threatened A2b

Reasons for listing

Declines of 20–29% over 3 generations (25 years) based on survey data

Status 2000

Least Concern

Reason for change in 2010: decline in population in the last decade

Status 1990

Least Concern

Taxonomy

No infraspecific taxa described

Taxonomic uniqueness: medium (22 genera/family, 16 species/genus, 1 subspecies/species)

Range

Breeds in north Siberia and Kamchatka, and winters in Indonesia, Philippines and Australasia, with most birds migrating directly between Japan and Australia (Bamford *et al.* 2008; Branson *et al.* 2010). Non-breeding birds occur along the coasts of Australia especially the north (AWSG database: Birds Australia *in litt.*).

Abundance

The global and flyway population has been estimated at 40 000 to 50 000 individuals and the Australian population at 36 000 to 45 000 (Wetlands International 2006; Stewart *et al.* 2007; Bamford *et al.* 2008). Numbers declined at Eighty-mile Beach, Western Australia, by c.46% between 2000 and 2008, increased slightly at northern Roebuck Bay between 2001 and 2008, smaller numbers at Bush Point but declined by c.81% between 2004 and 2008 (Rogers *et al.* 2009), were stable at Moreton Bay, Queensland, between 1998 and 2008 (Fuller *et al.* 2009) and at Shoalwater Bay from 1995 (3014 birds; Driscoll 1995) and 2007 (3831 birds; Jaensch 2008). Much smaller numbers in Gulf St Vincent, South Australia, and across Victoria have declined steeply but stable across 49 Australian sites between c.1983 and c.2007 (Close 2008; R. Clemens *in litt.*; D. Rogers *in litt.*). Populations in Japan in autumn have declined consistently since 1978, and by 57% between 1998 and 2008 (Amano *et al.* 2010). Despite some variability in rates and trends, it is estimated that there may have been an overall decline of 20–29% in Australia over the last 3 generations (25 years).

Ecology

Breeds beside stony riverbeds. Non-breeding birds migrate to sheltered coasts with reefs or intertidal mud-

Current eligibility against IUCN Red List Criteria

IUCN category	Criteria eligibility
Population visiting Australia	
A	Near Threatened: population decline of 20–29% over the last 3 generations (25 years) estimated and projected from monitoring
B	Not applicable: EOO >20 000 km^2; AOO >2000 km^2
C	Not applicable: population >10 000 mature individuals
D	Not applicable: population >1000 mature individuals, >5 locations
E	Not applicable: no population viability analysis undertaken

IUCN Red List assessment data

	Estimate	Reliability
Population visiting Australia		
Extent of occurrence trend	30 200 km^2 stable	high high
Area of occupancy trend	4800 km^2 decreasing	low medium
No. of mature individuals trend	45 000 decreasing	medium high
No. subpopulations	1	high
No. locations	>10	high
Generation time	8.3 years	high
Global population share	50–100%	high

flats, especially those with seagrass or by mangroves, and feed on arthropods, molluscs, crustaceans and marine worms (Higgins and Davies 1996). A generation time of 8.3 years (BirdLife International 2011) is derived from an age at first breeding of 2.5 years (extrapolated from Wandering Tattler *T. incana*), an annual survival of adults of 68.8% (extrapolated from other members of Scolopacidae) and the oldest known banding return of 19.1 years (Australian Bird and Bat Banding Scheme).

Threats

Potentially threatened by wetland loss and degradation along its migration route. Few stop on the Yellow Sea during migration (Barter 2002) with staging in Japan but also southern China, Taiwan and Korea (Branson *et al.* 2010). General threats at migratory staging sites include environmental pollution, reduced river flows, human disturbance, reclamation for tidal power plants, barrages, aquaculture ponds and agriculture, industrial use, urban expansion and hunting (Barter *et al.* 1997; Barter 2002; Kelin and Qiang 2006; Moores 2006). Threats in Australia, especially eastern Australia, include ongoing human disturbance, habitat loss and degradation from pollution, changes to the water regime and invasive plants (Rogers *et al.* 2006; Australian Government 2009).

Conservation objectives

1. Stable Australian non-breeding populations
2. Disturbance at key roosting and feeding sites reduced

Information required

1. Numbers and trends of Australian non-breeding population
2. Improved knowledge of dependence on key migratory staging sites
3. Improved knowledge of impacts of disturbance

Management actions required

1. Work with governments along the flyway at all levels, and local communities, to prevent destruction of key migratory staging sites
2. Advocate for restoration of alienated wetland sites
3. Maintain and improve protection of roosting and feeding sites in Australia

Bibliography

Amano T, Székely T, Koyama K, Amano H, Sutherland WJ (2010) A framework for monitoring the status of populations: an example from wader populations in the East Asian–Australasian flyway. *Biological Conservation* **143**, 2238–2247.

Australian Government (2009) 'Draft significant impact guidelines for 36 migratory shorebirds'. Draft EPBC Act Policy Statement 3.21, Canberra. Retrieved 25 February, 2011 from <http://www.environment.gov.au/epbc/publications/migratory-shorebirds.html>.

Bamford M, Watkins D, Bancroft W, Tischler G, Wahl J (2008) 'Migratory shorebirds of the East Asian – Australasian flyway: population estimates and internationally important sites'. Wetlands International – Oceania, Canberra.

Barter M, Fawen Q, Sixian T, Xiao Y, Tonkinson D (1997) Hunting of migratory waders on Chongming Dao: a declining occupation? *Stilt* **31**, 19–22.

Barter MA (2002) 'Shorebirds of the Yellow Sea: importance, threats and conservation status'. Canberra, Australia.

BirdLife International (2011) 'Species factsheet: *Heteroscelus brevipes*'. Retrieved 17 January, 2011 from <http://www.birdlife.org/>.

Branson NJBA, Shigeta Y, Chiang CY, Minton CDT (2010) Movements of Grey-tailed Tattlers and Terek Sandpipers in the East Asian–Australasian Flyway. *Wader Study Group Bulletin* **117**, 27–34.

Close DH (2008) Changes in Wader Numbers in the Gulf St Vincent, South Australia, 1979–2008. *Stilt* **54**, 24–27.

Driscoll PV (1995) 'Survey of shorebird feeding areas and roosts in the Shoalwater Bay area'. Unpublished report.

Higgins PJ, Davies SJJF (Eds) (1996) *Handbook of Australian, New Zealand and Antarctic Birds. Volume 3: Snipe to Pigeons*. Oxford University Press, Melbourne.

Jaensch R (2008) 'A condition assessment of wetlands and waterbirds in the Shoalwater Bay Training Area. Part 4: numbers of migratory shorebirds'. Unpublished report to Australian Government Department of Defence, by Wetlands International, Brisbane.

Kelin C, Qiang X (2006) Conserving migratory shorebirds in the Yellow Sea region. In *Waterbirds Around the World*. (Eds

G Boere, C Galbraith and D Stroud) p. 319. The Stationery Office, Edinburgh, UK.

Moores N (2006) South Korea's shorebirds: a review of abundance, distribution, threats and conservation status. *Stilt* **50**, 62–72.

Rogers D, Hassell C, Oldland J, Clemens R, Boyle A, Rogers K (2009) 'Monitoring Yellow Sea migrants in Australia (MYSMA): north-western Australian shorebird surveys and workshops, December 2008'.

Rogers DI, Piersma T, Hassell CJ (2006) Roost availability may constrain shorebird distribution: exploring the energetic costs of roosting and disturbance around a tropical bay. *Biological Conservation* **133**, 225–235.

Stewart DA, Rogers A, Rogers DI (2007) Species description. In *Shorebirds of Australia*. (Eds A Geering, L Agnew and S Harding) pp. 75–196. CSIRO Publishing, Melbourne.

Wetlands International (2006) 'Waterbird population estimates'. Wetlands International, Wageningen, The Netherlands.

Comments received from

Rob Clemens, Heather Gibbs, Ken Gosbell, Roger Jaensch, David Milton, Clive Minton, Danny Rogers

Subspecies Conservation Summary

Ruddy Turnstone (Palaearctic)

Arenaria interpres interpres (Linnaeus, 1758)
Scolopacidae

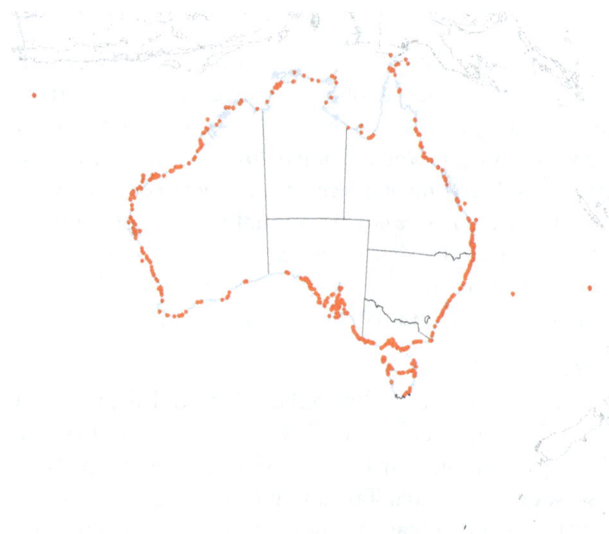

Conservation status

Population visiting Australian territory: Near Threatened A2b

Reasons for listing

Past decline in flyway estimated at 30–49% in 3 generations (25 years) based on survey data. However, non-declining populations in other flyways are not phenotypically differentiated, suggesting genetic interchange, so downgraded to Near Threatened

Status 2000

Least Concern

Reason for change in 2010: decline in population and habitat loss at staging posts

Status 1990

Least Concern

Taxonomy

A. i. morinella is extralimital; the species is Least Concern

Taxonomic uniqueness: medium (22 genera/family, 2 species/genus, 2 subspecies/species)

Range

The flyway population breeds in the Siberian and Alaskan Arctic, and winters on coasts from East Asia through Australasia to Mexico (Bamford *et al.* 2008). Non-breeding birds occur around all Australian coasts (Department of Sustainability, Environment, Water, Population and Communities 2011).

Abundance

The global population has been estimated at c.630 000, the flyway population at 35 000 individuals, and the Australian population at 20 000 (Wetlands International 2006; Bamford *et al.* 2008). Numbers declined at Eighty-mile Beach by 30% between 2000 and 2008 but fluctuated at Bush Point between 2004 and 2008 (Rogers *et al.* 2009), declined at Moreton Bay by 68% between 1993 and 2008 (Fuller *et al.* 2009), across 49 Australian sites by 42% between c.1983 and c.2007 (R. Clemens *in litt.*), and at King I. by a large percentage over the last 25 years (C. D. T. Minton *in litt.*) but fluctuated at Boullanger Bay/Robbins Passage, the second most significant site in Australia, between 1997 and 2005 (E. Woehler *in litt.*), and increased at Gulf St Vincent by 77% between c.1981 and c.2004 (Close 2008). Populations in Japan have declined consistently from 1978, and by about 65% between 1998 and 2008 (Amano *et al.* 2010). Despite the variability in trends, it is estimated that there may have been an overall decline of 30–49% in Australia over the last 22 years. Given likely ongoing threats to its migratory staging sites in East Asia (Barter 2002; Moores 2006), it is assumed that declines will continue. The subspecies is not known to be declining in other flyways and is not considered threatened (BirdLife International 2011).

Ecology

Breeds on coastal plains, marshes and tundra in the high Arctic. Non-breeding birds migrate to rocky and shingle coasts and also sandy beaches with seaweed, saltmarshes, mangroves, reefs and mudflats, and feed on insects, crus-

Current eligibility against IUCN Red List Criteria

IUCN category	Criteria eligibility
Population visiting Australia	
A	Vulnerable: population decline of 30–49% over the last 3 generations (25 years) estimated and projected from deterioration in habitat quality (but downgraded to Near Threatened)
B	Not applicable: EOO >20 000 km^2; AOO >2000 km^2
C	Not applicable: population >10 000 mature individuals
D	Not applicable: population >1000 mature individuals, >5 locations
E	Not applicable: no population viability analysis undertaken

IUCN Red List assessment data

	Estimate	Reliability
Population visiting Australia		
Extent of occurrence trend	36 600 km^2 stable	high high
Area of occupancy trend	4200 km^2 stable	low medium
No. of mature individuals trend	20 000 decreasing	medium high
No. subpopulations	1	high
No. locations	>10	high
Generation time	8.2 years	high
Global population share	1–10%	high

taceans, molluscs and other marine life (Higgins and Davies 1996). A generation time of 8.2 years (BirdLife International 2011) is derived from an age at first breeding of 2.0 years (Evans 1991; Poole 2008), an annual survival of 78.3% (Evans 1991; Poole 2008), and mean longevity of the genus of 17.7 years (Australian Bird and Bat Banding Scheme).

Threats

Potentially threatened by wetland loss and degradation along its migration route especially in the Yellow Sea where c.15% stage during northward migration (Barter 2002). General threats at migratory staging sites include environmental pollution, reduced river flows, human disturbance, reclamation for tidal power plants, barrages, aquaculture ponds and agriculture, industrial use and urban expansion (Barter 2002; Kelin and Qiang 2006; Moores 2006) but may be less affected by coastal reclamation works as less dependent on intertidal habitats. Threats in Australia, especially eastern and southern Australia, include ongoing human disturbance, habitat loss and degradation from pollution, changes to the water regime and invasive plants (Rogers *et al.* 2006; Australian Government 2009).

Conservation objectives

1. Stable Australian non-breeding populations
2. Disturbance at key roosting and feeding sites reduced

Information required

1. Numbers and trends of Australian non-breeding population
2. Improved knowledge of dependence on key migratory staging sites
3. Improved knowledge of impacts of disturbance

Management actions required

1. Work with governments along the flyway at all levels, and local communities, to prevent destruction of key migratory staging sites
2. Negotiate with the Chinese and South Korean governments in particular to protect remaining tidal flats in the Yellow Sea
3. Advocate for restoration of alienated wetland sites
4. Maintain and improve protection of roosting and feeding sites in Australia

Bibliography

Amano T, Székely T, Koyama K, Amano H, Sutherland WJ (2010) A framework for monitoring the status of populations: an example from wader populations in the East Asian–Australasian flyway. *Biological Conservation* **143**, 2238–2247.

Australian Government (2009) 'Draft significant impact guidelines for 36 migratory shorebirds'. Draft EPBC Act Policy Statement 3.21, Canberra. Retrieved 25 February, 2011 from <http://www.environment.gov.au/epbc/publications/migratory-shorebirds.html>.

Bamford M, Watkins D, Bancroft W, Tischler G, Wahl J (2008) 'Migratory shorebirds of the East Asian – Australasian flyway: Population estimates and internationally important sites'. Wetlands International – Oceania, Canberra.

Barter MA (2002) 'Shorebirds of the Yellow Sea: importance, threats and conservation status'. Canberra, Australia.

BirdLife International (2011) 'Species factsheet: *Arenaria interpres*'. Retrieved 17 January, 2011 from <http://www.birdlife.org/>.

Close DH (2008) Changes in wader numbers in the Gulf St Vincent, South Australia, 1979–2008. *Stilt* **54**, 24–27.

Department of Sustainability, Environment, Water, Population and Communities (2011) *Arenaria interpres*. In 'Species Profile and Threats Database'. Department of Sustainability, Environment, Water, Population and Communities, Canberra. Retrieved 24 February, 2011 from <http://www.environment.gov.au/sprat>.

Evans PR (1991) Seasonal and annual patterns of mortality in migratory shorebirds: some conservation implications. In *Bird Populations Studies: Relevance to Conservation and Management*. (Eds CM Perrins, J-D Lebreton and GJM Hirons) pp. 346–359. Oxford University Press, Oxford, UK.

Fuller RA, Wilson HB, Kendall BE, Possingham HP (2009) 'Monitoring shorebirds using counts by the Queensland Wader Study Group'. Report to the Queensland Wader Study Group and the Department of Environment and Resource Management, Brisbane.

Higgins PJ, Davies SJJF (Eds) (1996) *Handbook of Australian, New Zealand and Antarctic Birds. Volume 3: Snipe to Pigeons*. Oxford University Press, Melbourne.

Kelin C, Qiang X (2006) Conserving migratory shorebirds in the Yellow Sea region. In *Waterbirds Around the World*. (Eds

G Boere, C Galbraith and D Stroud) p. 319. The Stationery Office, Edinburgh, UK.

Moores N (2006) South Korea's shorebirds: a review of abundance, distribution, threats and conservation status. *Stilt* **50**, 62–72.

Poole A (Ed.) (2008) *The Birds of North America Online*. Cornell University Press, Ithaca.

Rogers D, Hassell C, Oldland J, Clemens R, Boyle A, Rogers K (2009) 'Monitoring Yellow Sea migrants in Australia (MYSMA): north-western Australian shorebird surveys and workshops, December 2008'.

Rogers DI, Piersma T, Hassell CJ (2006) Roost availability may constrain shorebird distribution: exploring the energetic costs of roosting and disturbance around a tropical bay. *Biological Conservation* **133**, 225–235.

Wetlands International (2006) 'Waterbird population estimates'. Wetlands International, Wageningen, The Netherlands.

Comments received from

Maureen Christie, Rob Clemens, Ken Gosbell, Roger Jaensch, David Milton, Clive Minton, Danny Rogers

Species Conservation Summary

Asian Dowitcher

Limnodromus semipalmatus (Blyth, 1848)
Scolopacidae

Conservation status

Population visiting Australian territory: Near Threatened A3c+4c, C2a(ii), D1

Reasons for listing

Small Australian population unlikely to exceed 1000 nature individuals, possibly declining along with global population which is Near Threatened based on habitat loss. As species is not threatened in Australia, Australian status made consistent with global status

Status 2000

Near Threatened D1
Reason for change in 2010: accelerated habitat loss at staging posts

Status 1990

Near Threatened D1

Taxonomy

No infraspecific taxa described
Taxonomic uniqueness: medium (22 genera/family, 3 species/genus, 1 subspecies/species)

Range

The species breeds in Siberia and adjacent Mongolia and China, and winters on coasts in Asia, especially Indonesia (Bamford *et al.* 2008). Non-breeding birds occur on northern Australian coasts (Department of Sustainability, Environment, Water, Population and Communities 2011). The number of locations in Australia is unknown but probably no more than 10 based on a scatter of records around the generally poorly-surveyed regions of northern Australia (Higgins and Davies 1996). The EOO consists of a coastal strip across coastal north Australia with about 10% assumed to be occupied.

Abundance

The global population has been estimated at 23 000 and the Australian population at 500 based on a maximum count of 414 at Roebuck Bay (Wetlands International 2006; Bamford *et al.* 2008), up to 70 in the Northern Territory (Chatto 2003) and a handful in Queensland. There are no monitoring data from Australia except that small numbers fluctuated at Bush Point between 2004 and 2008 (Rogers *et al.* 2009). There is too little data to determine trends in Australia but the species is believed to be declining globally on the basis of habitat degradation on the flyway (Barter 2002; Moores 2006; BirdLife International 2011).

Ecology

Breeds in extensive freshwater wetlands in the steppe and forest steppe. Non-breeding birds migrate to sheltered coasts, primarily estuarine and intertidal mudflats, and feed on polychaetes, insect larvae and molluscs (Higgins and Davies 1996). A generation time of 5.8 years (BirdLife International 2011) is derived from an age at first breeding of 1.5 years (extrapolated from congeners), an annual survival of 69% (extrapolated from mean values for other genera in Scolopacidae) and a maximum longevity of 12.3 years (extrapolated from congeners).

Threats

Potentially threatened by wetland loss and degradation along its migration route, especially in the Yellow Sea

Current eligibility against IUCN Red List Criteria

IUCN category	Criteria eligibility
Population visiting Australia	
A	Near Threatened: current and anticipated population declines of 20–29% over the last 3 generations (17 years) inferred from deteriorating habitat quality
B	Not applicable: >10 locations, not severely fragmented, population fluctuations not extreme
C	Endangered: population 250–2500 mature individuals, suspected continuing decline, >95% mature individuals in 1 subpopulation (downgraded to Near Threatened)
D	Vulnerable: population 250–1000 mature individuals (downgraded to Near Threatened)
E	Not applicable: no population viability analysis undertaken

IUCN Red List assessment data

	Estimate	Reliability
Population visiting Australia		
Extent of occurrence trend	14 000 km^2 stable	medium medium
Area of occupancy trend	1400 km^2 stable	low medium
No. of mature individuals trend	500 decreasing	low high
No. subpopulations	1	high
No. locations	10	high
Generation time	5.8 years	high
Global population share	1–10%	high

where c.40% stage during northward migration (Barter 2002). General threats at migratory staging sites include environmental pollution, reduced river flows, human disturbance, reclamation for tidal power plants, barrages, aquaculture ponds and agriculture, industrial use and urban expansion (Barter 2002; Kelin and Qiang 2006; Moores 2006). Although general threats to shorebirds in Australia include ongoing human disturbance, habitat loss and degradation from pollution, changes to the water regime and invasive plants (Rogers *et al.* 2006; Australian Government 2009), most Asian Dowitchers occur in areas with almost none of these.

Conservation objectives

1. Stable Australian non-breeding populations

Information required

1. Distribution, numbers and trends of Australian non-breeding population
2. Population trends across flyway
3. Improved knowledge of dependence on key migratory staging sites
4. Improved knowledge of potential threats at key sites

Management actions required

1. Work with governments along the flyway at all levels, and local communities, to prevent destruction of key migratory staging sites
2. Advocate for restoration of alienated wetland sites
3. Support surveys of breeding grounds for potential threats

Bibliography

Australian Government (2009) 'Draft significant impact guidelines for 36 migratory shorebirds'. Draft EPBC Act Policy Statement 3.21, Canberra. Retrieved 25 February, 2011 from <http://www.environment.gov.au/epbc/publications/migratory-shorebirds.html>.

Bamford M, Watkins D, Bancroft W, Tischler G, Wahl J (2008) 'Migratory shorebirds of the East Asian – Australasian flyway: population estimates and internationally important sites'. Wetlands International – Oceania, Canberra.

Barter MA (2002) 'Shorebirds of the Yellow Sea: importance, threats and conservation status'. Canberra, Australia.

BirdLife International (2011) 'Species factsheet: *Limnodromus semipalmatus*'. Retrieved 17 January, 2011 from <http://www.birdlife.org/>.

Chatto R (2003) 'The distribution and status of shorebirds around the coast and coastal wetlands of the Northern Territory'. Technical Report 73. Parks and Wildlife Commission of the Northern Territory, Palmerston.

Department of Sustainability, Environment, Water, Population and Communities (2011) *Limnodromus semipalmatus.* In 'Species Profile and Threats Database'. Department of Sustainability, Environment, Water, Population and Communities, Canberra. Retrieved 24 February, 2011 from <http://www.environment.gov.au/sprat>.

Higgins PJ, Davies SJJF (Eds) (1996) *Handbook of Australian, New Zealand and Antarctic Birds. Volume 3: Snipe to Pigeons.* Oxford University Press, Melbourne.

Kelin C, Qiang X (2006) Conserving migratory shorebirds in the Yellow Sea region. In *Waterbirds Around the World.* (Eds G Boere, C Galbraith and D Stroud) p. 319. The Stationery Office, Edinburgh, UK.

Moores N (2006) South Korea's shorebirds: a review of abundance, distribution, threats and conservation status. *Stilt* **50**, 62–72.

Rogers D, Hassell C, Oldland J, Clemens R, Boyle A, Rogers K (2009) 'Monitoring Yellow Sea migrants in Australia (MYSMA): north-western Australian shorebird surveys and workshops, December 2008'.

Rogers DI, Piersma T, Hassell CJ (2006) Roost availability may constrain shorebird distribution: exploring the energetic costs of roosting and disturbance around a tropical bay. *Biological Conservation* **133**, 225–235.

Wetlands International (2006) 'Waterbird population estimates'. Wetlands International, Wageningen, The Netherlands.

Comments received from

Rob Clemens, Ken Gosbell, Roger Jaensch, David Milton, Clive Minton, Danny Rogers

Species Conservation Summary

Great Knot

Calidris tenuirostris (Horsfield, 1821)
Scolopacidae

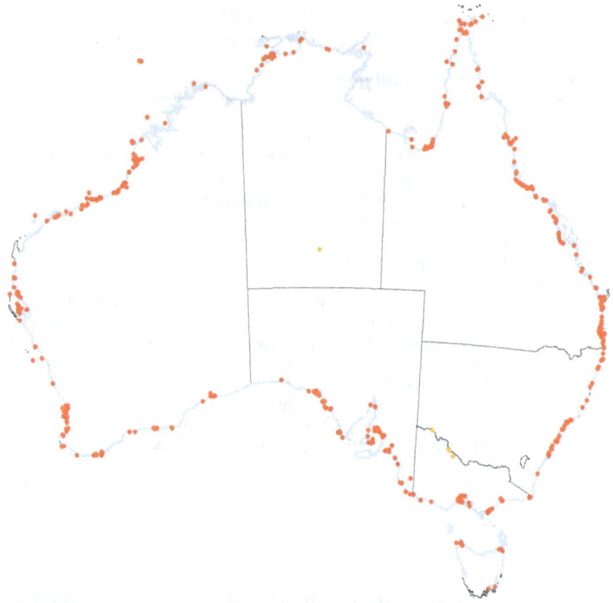

Conservation status

Population visiting Australian territory: Vulnerable A3c+4bc

Reasons for listing

Recent and ongoing declines of 30–49% in 3 generations (26 years) based on survey data and habitat loss

Status 2000

Near Threatened A3c
Reason for change in 2010: decline in population and habitat loss at staging posts

Status 1990

Least Concern
Reason for change in 2000: habitat destruction planned for major staging post

Taxonomy

No infraspecific taxa described
Taxonomic uniqueness: medium (22 genera/family, 20 species/genus, 1 subspecies/species)

Range

Breeds in north-east Siberia, wintering along coasts from Arabia to Australia. Large numbers of non-breeding visitors occur from northern Western Australia and the Northern Territory to the Gulf of Carpentaria, and significant numbers along east Queensland (Department of Sustainability, Environment, Water, Population and Communities 2011). It is no longer regular at some sites along the south coast of Australia which used to support small numbers (AWSG database: Birds Australia *in litt.*).

Abundance

The global population was estimated at *c.*380 000 individuals (Wetlands International 2006) including 360 000 estimated in Australia (Bamford *et al.* 2008), but *c.*90 000 individuals disappeared following the reclamation of tidal flats at Saemangeum in the South Korean Yellow Sea in 2006 (Moores *et al.* 2008). Numbers have declined on Eighty-mile Beach by c.24% between 2000 and 2008 (Rogers *et al.* 2009), at Moreton Bay by 6.0% per year between 1998 and 2008 (Fuller *et al.* 2009) and by 34% across 49 Australian sites between c.1983 and c.2007 (AWSG database: Birds Australia *in litt.*). Given that more reclamation is proposed within the Yellow Sea, it is assumed that declines of 30–49% over 22 years (3 generations) will continue.

Ecology

Breeds on montane tundra in the subarctic. Non-breeding birds migrate to inlets, bays, harbours, estuaries and lagoons with large intertidal mud and sand flats, where they feed on bivalves, gastropods, crustaceans and other invertebrates (Higgins and Davies 1996; Moores 2006). A generation time of 8.6 years (BirdLife International 2011) is derived from an age at first breeding of 1.7 years, an annual adult survival of 79% (both extrapolated from congeners) and a maximum longevity of 19.7 years (Australian Bird and Bat Banding Scheme).

Threats

Threatened by wetland degradation in the Yellow Sea where c.80% of the population stages on northward migration (Barter 2002; Bamford *et al.* 2008; van de Kam *et al.* 2010). Threats include environmental pollution, reduced river flows, human disturbance and reclamation for tidal power plants and barrages, industrial use and urban expansion (Tomkovich 1997; Barter 2002; Kelin and Qiang 2006; Moores 2006). It is also hunted on migration (Barter *et al.* 1997; Ming *et al.* 1998). Threats in Australia, especially eastern and southern Australia, include ongoing human disturbance, habitat loss and degradation from pollution, changes to the water regime and invasive plants (Rogers *et al.* 2006; Australian Government 2009).

Conservation objectives

1. Stable Australian non-breeding populations
2. Disturbance at key roosting and feeding sites reduced

Information required

1. Numbers and trends of Australian non-breeding population
2. Improved knowledge of dependence on key migratory staging sites

Management actions required

1. Work with governments along the flyway at all levels, and local communities, to prevent destruction of key migratory staging sites
2. Negotiate with the Chinese government in particular to protect remaining tidal flats in Bohai Bay
3. Advocate for restoration of alienated wetland sites

Current eligibility against IUCN Red List Criteria

IUCN category	Criteria eligibility
Population visiting Australia	
A	Vulnerable: current and anticipated population declines of 30–49% over periods of 3 generations (26 years) estimated and projected from monitoring and from deteriorating habitat quality
B	Not applicable: EOO >20 000 km^2; AOO >2000 km^2
C	Not applicable: population >10 000 mature individuals
D	Not applicable: population >1000 mature individuals, >5 locations
E	Not applicable: no population viability analysis undertaken

IUCN Red List assessment data

	Estimate	Reliability
Population visiting Australia		
Extent of occurrence trend	35 000 km^2 stable	high high
Area of occupancy trend	2800 km^2 decreasing	low medium
No. of mature individuals trend	290 000 decreasing	medium high
No. subpopulations	1	high
No. locations	>10	high
Generation time	8.6 years	high
Global population share	50–100%	high

4. Maintain and improve protection of roosting and feeding sites in Australia

Bibliography

Australian Government (2009) 'Draft significant impact guidelines for 36 migratory shorebirds'. Draft EPBC Act Policy Statement 3.21. Canberra, Australia. Retrieved 25 February, 2011 from <http://www.environment.gov.au/epbc/publications/migratory-shorebirds.html>.

Bamford M, Watkins D, Bancroft W, Tischler G, Wahl J (2008) 'Migratory shorebirds of the East Asian – Australasian flyway: population estimates and internationally important sites'. Wetlands International – Oceania, Canberra.

Barter M, Fawen Q, Sixian T, Xiao Y, Tonkinson D (1997) Hunting of migratory waders on Chongming Dao: a declining occupation? *Stilt* **31**, 19–22.

Barter MA (2002) 'Shorebirds of the Yellow Sea: importance, threats and conservation status'. Canberra, Australia.

BirdLife International (2011) 'Species factsheet: *Calidris tenuirostris*'. Retrieved 17 January, 2011 from <http://www.birdlife.org/>.

Department of Sustainability, Environment, Water, Population and Communities (2011) *Calidris tenuirostris*. In 'Species Profile and Threats Database'. Department of Sustainability, Environment, Water, Population and Communities, Canberra. Retrieved 24 February, 2011 from <http://www.environment.gov.au/sprat>.

Fuller RA, Wilson HB, Kendall BE, Possingham HP (2009) 'Monitoring shorebirds using counts by the Queensland Wader Study Group'. Report to the Queensland Wader Study Group and the Department of Environment and Resource Management, Brisbane.

Higgins PJ, Davies SJJF (Eds) (1996) *Handbook of Australian, New Zealand and Antarctic Birds. Volume 3: Snipe to Pigeons*. Oxford University Press, Melbourne.

Kelin C, Qiang X (2006) Conserving migratory shorebirds in the Yellow Sea region. In *Waterbirds Around the World*. (Eds G Boere, C Galbraith and D Stroud) p. 319. The Stationery Office, Edinburgh, UK.

Ming M, Jianjian L, Chengjia T, Pingyue S, Wei H (1998) The contribution of shorebirds to the catches of hunters in the Shanghai area, China, during 1997–1998. *Stilt* **33**, 32–36.

Moores N (2006) South Korea's shorebirds: a review of abundance, distribution, threats and conservation status. *Stilt* **50**, 62–72.

Moores N, Rogers DI, Kim R-H, Hassell C, Gosbell K, Kim S-A, Park M-N (2008) 'The 2006–2008 Saemangeum Shorebird Monitoring Program Report'. Birds Korea, Busan.

Rogers D, Hassell C, Oldland J, Clemens R, Boyle A, Rogers K (2009) 'Monitoring Yellow Sea migrants in Australia (MYSMA): north-western Australian shorebird surveys and workshops, December 2008'.

Rogers DI, Piersma T, Hassell CJ (2006) Roost availability may constrain shorebird distribution: exploring the energetic costs of roosting and disturbance around a tropical bay. *Biological Conservation* **133**, 225–235.

Tomkovich PS (1997) Breeding distribution, migrations and conservation status of the Great Knot *Calidris tenuirostris* in Russia. *Emu* **97**, 265–282.

van de Kam J, Battley PF, McCaffery BJ, Rogers DI, Hong J-S, Moores N, Ki J-Y, Lewis J, Piersma T (2010) *Invisible Connections: Why Migrating Shorebirds Need the Yellow Sea*. CSIRO Publishing, Melbourne.

Wetlands International (2006) 'Waterbird population estimates'. Wetlands International, Wageningen, The Netherlands.

Comments received from

Rob Clemens, Ken Gosbell, Roger Jaensch, David Milton, Clive Minton, Danny Rogers

Subspecies Conservation Summary

Red Knot (New Siberian Islands)

Calidris canutus piersmai Tomkovich, 2001
Scolopacidae

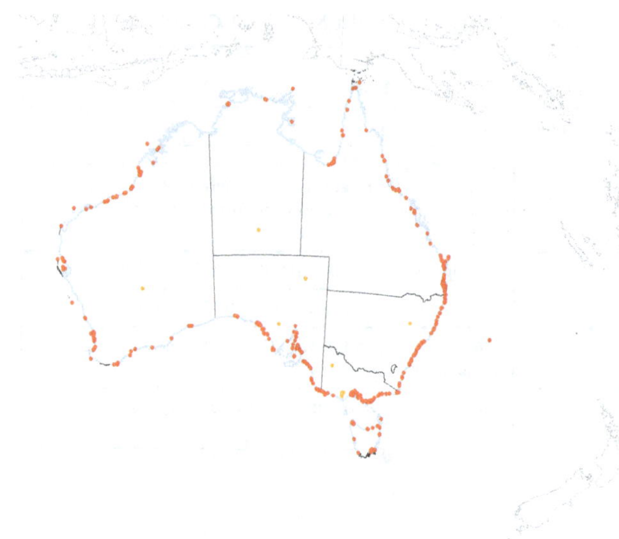

Conservation status

Population visiting Australian territory: Vulnerable A3c+4bc

Reasons for listing

Recent and ongoing declines of 30–49% in 3 generations (23 years) based on survey data and habitat loss

Status 2000

Near Threatened A3c
Reason for change in 2010: decline in population and ongoing loss of habitat at staging posts

Status 1990

Least Concern
Reason for change in 2000: habitat loss in staging posts accelerating

Taxonomy

C. c. rogersi (breeding in north-eastern Siberia) is Vulnerable in Australia, *C. c. canutus* (breeding in central Siberia and vagrant in Australia) is Least Concern. There are 2 extralimital subspecies. The species is Least Concern.
Taxonomic uniqueness: low (22 genera/family, 20 species/genus, 5 subspecies/species)

Range

Breeds in the New Siberian Is., wintering largely in Australia and New Zealand (Rogers *et al.* 2010). Non-breeding birds occur around coastal Australia, mostly in north-west Australia (Rogers *et al.* 2010; Department of Sustainability, Environment, Water, Population and Communities 2011). Birds migrating to New Zealand may stage in Australia, particularly in the Gulf of Carpentaria (Bamford *et al.* 2008; Rogers *et al.* 2010).

Abundance

The global population of *C. c. piersmai* is likely to be between 48 000 and 60 000 birds, mostly in Australia (Rogers *et al.* 2010). As this subspecies is not identifiable in the field in winter, the proportion spending its non-breeding months in Australia is unknown. However, it is estimated at 44 000 as c.25% of Red Knots banded in New Zealand and sighted at Bohai Bay were *C. c. piersmai* (subspecies = 48 000 to 60 000, of which 24.6% of 42 000 are in New Zealand (Rogers *et al.* 2010)). Numbers of all subspecies declined on Eighty-mile Beach by c.22% between 2000 and 2008 (Rogers *et al.* 2009), at Moreton Bay by 75% between 1993 and 2008 (Fuller *et al.* 2009) and by c.27% across 49 Australian sites between c.1983 and c.2007 (AWSG database: Birds Australia *in litt.*). Numbers declined less severely elsewhere in the flyway, e.g. no clear trends in Japan between 1978 and 2008 (Amano *et al.* 2010), suggesting a 30–49% decline over 20 years (3 generations) across the flyway, although this could be higher given the high proportion at Bohai Bay. Given likely ongoing threats to its migratory staging sites in East Asia (Rogers *et al.* 2010), it is assumed that declines of 30–49% over 20 years will continue.

Ecology

Breeds on dry upland tundra in high Arctic. Non-breeding birds migrate to tidal mudflats or sandflats, where they feed on intertidal invertebrates, especially shellfish (Piersma *et al.* 1995; Higgins and Davies 1996; Piersma 2007). A generation time of 6.8 years (BirdLife International 2011) is derived from an average age at first breeding of 2.0 years (Cramp *et al.* 1983), an annual adult survival of 68% (Boyd 1962) and a maximum longevity in the wild of 18.8 years (Australian Bird and Bat Banding Scheme).

Threats

Threatened by wetland degradation in East Asia, where it stages on migration (Bamford *et al.* 2008). Specifically threatened at Bohai Bay where up to 80% of the north-west Australian population stages during the spring migration along 20 km of mudflats which are being reclaimed and developed (Rogers *et al.* 2010). General threats at migratory staging sites include environmental pollution, reduced river flows, human disturbance and reclamation for tidal power plants and barrages, industrial use and urban expansion (Barter 2002; Kelin and Qiang 2006; Moores 2006) and hunting (Ming *et al.* 1998). Threats in Australia, especially in the east and south, include ongoing human disturbance and wetland degradation including habitat loss, pollution, changes to the water regime and invasive plants (Rogers *et al.* 2006; Australian Government 2009).

Conservation objectives

1. Stable Australian non-breeding populations
2. Disturbance at key roosting and feeding sites reduced

Current eligibility against IUCN Red List Criteria

IUCN category	Criteria eligibility
Population visiting Australia	
A	Vulnerable: current and anticipated population declines of 30–49% over periods of 3 generations (23 years) estimated and projected from monitoring and from deteriorating habitat quality
B	Not applicable: EOO >20 000 km^2; AOO >2000 km^2
C	Not applicable: population >10 000 mature individuals
D	Not applicable: population >1000 mature individuals, >5 locations
E	Not applicable: no population viability analysis undertaken

IUCN Red List assessment data

	Estimate	Reliability
Population visiting Australia		
Extent of occurrence trend	36 000 km^2 stable	high high
Area of occupancy trend	2400 km^2 stable	low medium
No. of mature individuals trend	44 000 decreasing	medium high
No. subpopulations	1	high
No. locations	>10	high
Generation time	7.8 years	high
Global population share	50–100%	medium

Information required

1. Numbers and trends of Australian non-breeding population
2. Improved knowledge of dependence on key migratory staging sites
3. Improved knowledge of impacts of disturbance

Management actions required

1. Work with governments along the flyway at all levels, and local communities, to prevent destruction of key migratory staging sites
2. Negotiate with the Chinese government in particular to protect remaining tidal flats in Bohai Bay
3. Advocate for restoration of alienated wetland sites
4. Maintain and improve protection of roosting and feeding sites in Australia

Bibliography

Amano T, Székely T, Koyama K, Amano H, Sutherland WJ (2010) A framework for monitoring the status of populations: an example from wader populations in the East Asian–Australasian flyway. *Biological Conservation* **143**, 2238–2247.

Australian Government (2009) 'Draft Significant impact guidelines for 36 migratory shorebirds'. Draft EPBC Act Policy Statement 3.21, Canberra. Retrieved 25 February, 2011 from <http://www.environment.gov.au/epbc/publications/migratory-shorebirds.html>.

Bamford M, Watkins D, Bancroft W, Tischler G, Wahl J (2008) 'Migratory shorebirds of the East Asian – Australasian flyway: population estimates and internationally important sites'. Wetlands International – Oceania, Canberra.

Barter MA (2002) 'Shorebirds of the Yellow Sea: importance, threats and conservation status'. Canberra, Australia.

BirdLife International (2011) 'Species factsheet: *Calidris canutus*'. Retrieved 17 January, 2011 from <http://www.birdlife.org/>.

Boyd H (1962) Mortality and fertility of European Charadrii. *Ibis* **104**, 368–387.

Cramp S, Simmons KEL, Brooks DC, Collar NJ, Dunn E, Gillmor R, Hollom PAD, Hudson R, Nicholson EM, Ogilvie MA, Olney PJS, Roselaar CS, Voous KH, Wallace DIM, Wattel J, Wilson MG (Eds) (1983) *Handbook of the Birds of Europe, the Middle East and North Africa. The Birds of the Western Palearctic: 3. Waders to Gulls.* Oxford University Press, Oxford, UK.

Department of Sustainability, Environment, Water, Population and Communities (2011) *Calidris canutus.* In 'Species Profile and Threats Database'. Department of Sustainability, Environment, Water, Population and Communities, Canberra. Retrieved 24 February, 2011 from <http://www.environment.gov.au/sprat>.

Fuller RA, Wilson HB, Kendall BE, Possingham HP (2009) 'Monitoring shorebirds using counts by the Queensland Wader Study Group'. Report to the Queensland Wader Study Group and the Department of Environment and Resource Management, Brisbane.

Higgins PJ, Davies SJJF (Eds) (1996) *Handbook of Australian, New Zealand and Antarctic Birds. Volume 3: Snipe to Pigeons.* Oxford University Press, Melbourne.

Kelin C, Qiang X (2006) Conserving migratory shorebirds in the Yellow Sea region. In *Waterbirds Around the World.* (Eds G Boere, C Galbraith and D Stroud) p. 319. The Stationery Office, Edinburgh, UK.

Ming M, Jianjian L, Chengjia T, Pingyue S, Wei H (1998) The contribution of shorebirds to the catches of hunters in the Shanghai area, China, during 1997–1998. *Stilt* **33**, 32–36.

Moores N (2006) South Korea's shorebirds: a review of abundance, distribution, threats and conservation status. *Stilt* **50**, 62–72.

Piersma T (2007) Using the power of comparison to explain habitat use and migration strategies of shorebirds worldwide. *Journal of Ornithology* **148**, 45–59.

Piersma T, Van Gils J, De Goeij P, Van Der Meer J (1995) Holling's Functional Response Model as a tool to link the food-finding mechanism of a probing shorebird with its spatial distribution. *Journal of Animal Ecology* **64**, 493–504.

Rogers D, Hassell C, Oldland J, Clemens R, Boyle A, Rogers K (2009) 'Monitoring Yellow Sea migrants in Australia (MYSMA): north-western Australian shorebird surveys and workshops, December 2008'.

Rogers DI, Piersma T, Hassell CJ (2006) Roost availability may constrain shorebird distribution: exploring the energetic costs of roosting and disturbance around a tropical bay. *Biological Conservation* **133**, 225–235.

Rogers DI, Yang H-Y, Hassell CJ, Boyle AN, Rogers KG, Chen B, Zhang Z-W, Piersma T (2010) Red Knots (*Calidris canutus piersmai* and *C. c. rogersi*) depend on a small threatened staging area in Bohai Bay, China. *Emu* **110**, 307–315.

Comments received from

Rob Clemens, Nicola Crockford, Ken Gosbell, Chris Hassell, Roger Jaensch, David Milton, Clive Minton, Danny Rogers

Subspecies Conservation Summary

Red Knot (north-eastern Siberian)

Calidris canutus rogersi (Mathews, 1913)
Scolopacidae

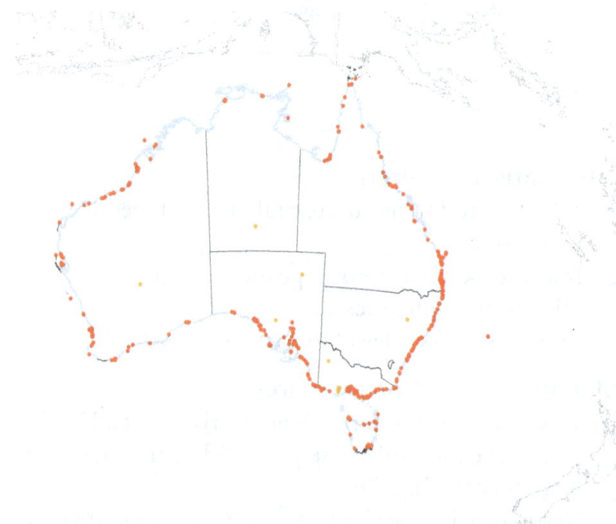

Conservation status

Population visiting Australian territory: Vulnerable A3c+4bc

Reasons for listing

Recent and ongoing declines of 30–49% in 3 generations (23 years) based on survey data and habitat loss

Status 2000

Near Threatened A3c
Reason for change in 2010: decline in population and ongoing loss of habitat at staging posts

Status 1990

Least Concern
Reason for change in 2000: habitat loss in staging posts accelerating

Taxonomy

C. c. piersmai (breeding in the New Siberian Islands) is Vulnerable in Australia, *C. c. canutus* (breeding in central Siberia and vagrant in Australia) is Least Concern. There are 2 extralimital subspecies. The species is Least Concern.
Taxonomic uniqueness: low (22 genera/family, 20 species/genus, 5 subspecies/species)

Range

Breeds in eastern Siberia including Chukotka, wintering largely in Australia and New Zealand (Zöckler and O'Sullivan 2005; Rogers *et al.* 2010). Non-breeding birds occur around coastal Australia, mostly in south-east Australia (Piersma 2007; Rogers *et al.* 2010; Department of Sustainability Environment Water Population and Communities 2011). Birds migrating to New Zealand may stage in Australia, particularly in the Gulf of Carpentaria (Bamford *et al.* 2008; Rogers *et al.* 2010).

Abundance

The global population of *C. c. rogersi* is likely to be between 50 000 and 62 000 birds, mostly in Australia (Rogers *et al.* 2010). As this subspecies is not identifiable in the field in winter, the proportion spending its non-breeding months in Australia is unknown, but estimated at 24 000 as 75% of Red Knots banded in New Zealand and sighted at Bohai Bay were *C. c. rogersi* (subspecies = 50 000 to 62 000, of which 75.4% of 42 000 in New Zealand (Rogers *et al.* 2010)). Numbers of all subspecies declined on Eighty-mile Beach by c.22% between 2000 and 2008 (Rogers *et al.* 2009), at Moreton Bay by 75% 1993–2008 (Fuller *et al.* 2009) and by c.27% across 49 Australian sites between c.1983 and c.2007 (AWSG database: Birds Australia *in litt.*). Numbers declined less severely elsewhere in the flyway, e.g. no clear trends in Japan between 1978 and 2008 (Amano *et al.* 2010), suggesting a 30–49% decline over 20 years (3 generations) across the flyway although this could be higher given the high proportion at Bohai Bay. Given likely ongoing threats to its migratory staging sites in East Asia (Rogers *et al.* 2010), it is assumed that declines of 30–49% over 20 years will continue.

Ecology

Breeds on dry upland tundra in high Arctic. Non-breeding birds migrate to tidal mudflats or sandflats, where they feed on intertidal invertebrates, especially shellfish (Piersma *et al.* 1995; Higgins and Davies 1996; Piersma 2007). A generation time of 7.8 years (BirdLife International 2011) is derived from an average age at first breeding of 2.0 years (Cramp *et al.* 1983), an annual adult

Current eligibility against IUCN Red List Criteria

IUCN category	Criteria eligibility
Population visiting Australia	
A	Vulnerable: current and anticipated population declines of 30–49% over periods of 3 generations (23 years) estimated and projected from monitoring and from deteriorating habitat quality
B	Not applicable: EOO >20 000 km^2; AOO >2000 km^2
C	Not applicable: population >10 000 mature individuals
D	Not applicable: population >1000 mature individuals, >5 locations
E	Not applicable: no population viability analysis undertaken

IUCN Red List assessment data

	Estimate	Reliability
Population visiting Australia		
Extent of occurrence trend	36 000 km^2 stable	high high
Area of occupancy trend	2400 km^2 stable	low medium
No. of mature individuals trend	24 000 decreasing	medium high
No. subpopulations	1	high
No. locations	>10	high
Generation time	7.8 years	high
Global population share	26–50%	medium

survival of 68% (Boyd 1962) and a maximum longevity of 14.8 years (Australian Bird and Bat Banding Scheme).

Threats

Threatened by wetland degradation in East Asia where it stages on migration (Bamford *et al.* 2008). Specifically threatened at Bohai Bay where *c.*54% of the north-western Australian population stages in spring migration along 20 km of mudflats which is being reclaimed and developed (Rogers *et al.* 2010). General threats at migratory staging sites include environmental pollution, reduced river flows, human disturbance and reclamation for tidal power plants and barrages, industrial use and urban expansion (Barter 2002; Kelin and Qiang 2006; Moores 2006) and hunting (Ming *et al.* 1998). Threats in Australia, especially in the east and south, include ongoing human disturbance and wetland degradation including habitat loss, pollution, changes to the water regime and invasive plants (Rogers *et al.* 2006; Australian Government 2009).

Conservation objectives

1. Stable Australian non-breeding populations
2. Disturbance at key roosting and feeding sites reduced

Information required

1. Numbers and trends of Australian non-breeding population
2. Improved knowledge of dependence on key migratory staging sites
3. Improved knowledge of impacts of disturbance

Management actions required

1. Work with governments along the flyway at all levels, and local communities, to prevent destruction of key migratory staging sites
2. Negotiate with the Chinese government in particular to protect remaining tidal flats in Bohai Bay
3. Advocate for restoration of alienated wetland sites
4. Maintain and improve protection of roosting and feeding sites in Australia

Bibliography

Amano T, Székely T, Koyama K, Amano H, Sutherland WJ (2010) A framework for monitoring the status of populations: an example from wader populations in the East Asian–Australasian flyway. *Biological Conservation* **143**, 2238–2247.

Australian Government (2009) 'Draft significant impact guidelines for 36 migratory shorebirds'. Draft EPBC Act Policy Statement 3.21, Canberra. Retrieved 25 February, 2011 from <http://www.environment.gov.au/epbc/publications/migratory-shorebirds.html>.

Bamford M, Watkins D, Bancroft W, Tischler G, Wahl J (2008) 'Migratory shorebirds of the East Asian – Australasian flyway: population estimates and internationally important sites'. Wetlands International – Oceania, Canberra.

Barter MA (2002) 'Shorebirds of the Yellow Sea: importance, threats and conservation status'. Canberra, Australia.

BirdLife International (2011) 'Species factsheet: *Calidris canutus*'. Retrieved 17 January, 2011 from <http://www.birdlife.org/>.

Boyd H (1962) Mortality and fertility of European Charadrii. *Ibis* **104**, 368–387.

Cramp S, Simmons KEL, Brooks DC, Collar NJ, Dunn E, Gillmor R, Hollom PAD, Hudson R, Nicholson EM, Ogilvie MA, Olney PJS, Roselaar CS, Voous KH, Wallace DIM, Wattel J, Wilson MG (Eds) (1983) *Handbook of the Birds of Europe, the Middle East and North Africa. The Birds of the Western Palearctic: 3. Waders to Gulls.* Oxford University Press, Oxford.

Department of Sustainability, Environment, Water, Population and Communities (2011) *Calidris canutus.* In 'Species Profile and Threats Database'. Department of Sustainability, Environment, Water, Population and Communities, Canberra. Retrieved 24 February, 2011 from <http://www.environment.gov.au/sprat>.

Higgins PJ, Davies SJJF (Eds) (1996) *Handbook of Australian, New Zealand and Antarctic Birds. Volume 3: Snipe to Pigeons.* Oxford University Press, Melbourne.

Kelin C, Qiang X (2006) Conserving migratory shorebirds in the Yellow Sea region. In *Waterbirds Around the World.* (Eds G Boere, C Galbraith and D Stroud) p. 319. The Stationery Office, Edinburgh, UK.

Ming M, Jianjian L, Chengjia T, Pingyue S, Wei H (1998) The contribution of shorebirds to the catches of hunters in the Shanghai area, China, during 1997–1998. *Stilt* **33**, 32–36.

Moores N (2006) South Korea's shorebirds: a review of abundance, distribution, threats and conservation status. *Stilt* **50**, 62–72.

Piersma T (2007) Using the power of comparison to explain habitat use and migration strategies of shorebirds worldwide. *Journal of Ornithology* **148**, 45–59.

Piersma T, Van Gils J, De Goeij P, Van Der Meer J (1995) Holling's Functional Response Model as a tool to link the food-finding mechanism of a probing shorebird with its spatial distribution. *Journal of Animal Ecology* **64**, 493–504.

Rogers D, Hassell C, Oldland J, Clemens R, Boyle A, Rogers K (2009) 'Monitoring Yellow Sea migrants in Australia (MYSMA): north-western Australian shorebird surveys and workshops, December 2008'.

Rogers DI, Piersma T, Hassell CJ (2006) Roost availability may constrain shorebird distribution: exploring the energetic costs of roosting and disturbance around a tropical bay. *Biological Conservation* **133**, 225–235.

Rogers DI, Yang H-Y, Hassell CJ, Boyle AN, Rogers KG, Chen B, Zhang Z-W, Piersma T (2010) Red Knots (*Calidris canutus piersmai* and *C. c. rogersi*) depend on a small threatened staging area in Bohai Bay, China. *Emu* **110**, 307–315.

Zöckler C, O'Sullivan J (2005) New Zealand Red Knot breeding in Meinopylgino, Chukotka, NE Russia. *Wader Study Group Bulletin* **108**, 76.

Comments received from

Rob Clemens, Nicola Crockford, Ken Gosbell, Chris Hassell, Roger Jaensch, David Milton, Clive Minton, Danny Rogers

Species Conservation Summary

Curlew Sandpiper

Calidris ferruginea (Pontoppidan, 1763)
Scolopacidae

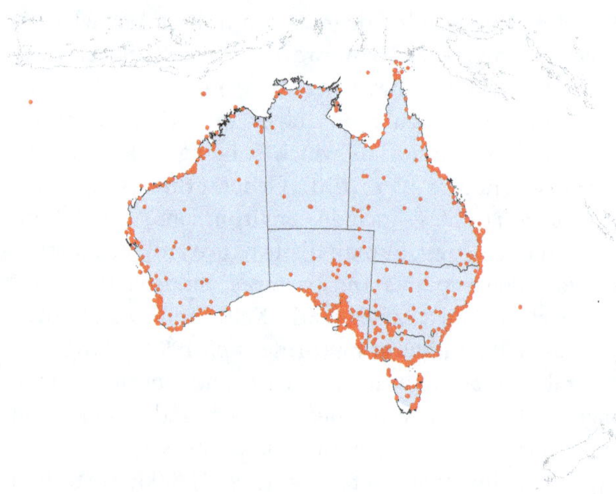

Conservation status

Population visiting Australia: Vulnerable A2bc+3c+4bc

Reasons for listing

Past, recent and ongoing declines of 50–79% in 3 generations (23 years) based on survey data and habitat loss. However, non-declining populations in other flyways are not phenotypically differentiated, suggesting genetic interchange, so downgraded to Vulnerable

Status 2000

Near Threatened A3c+4bc
Reason for change in 2010: decline in population and habitat loss at staging posts

Status 1990

Least Concern
Reason for change in 2000: first declines detected

Taxonomy

No infraspecific taxa described
Taxonomic uniqueness: medium (22 genera/family, 20 species/genus, 1 subspecies/species)

Range

Breeds in north Siberia, wintering from western Africa to Australia (Bamford *et al.* 2008). Non-breeding birds occur around coastal Australia, with large numbers at sites in both the north and south (Department of Sustainability, Environment, Water, Population and Communities 2011).

Abundance

The global population has been estimated at 1 850 000 individuals and the flyway population at 180 000 (Wetlands International 2006) including 115 000 in Australia (Bamford *et al.* 2008). Numbers declined on Eighty-mile Beach by c.59% between 2000 and 2008 (Rogers *et al.* 2009), at the Coorong by 79% between the 1980s and c.2004 (Wainwright and Christie 2008), at sites across Queensland by 6.3% per year between 1998 and 2008 (Fuller *et al.* 2009), at Gulf St Vincent by c.71% between c.1981 and c.2004 (Close 2008), and by c.82% across 49 Australian sites between c.1983 and c.2007 (AWSG database: Birds Australia *in litt.*). Numbers declined less severely elsewhere in the flyway, e.g. no clear trends in Japan between 1978 and 2008 (Amano *et al.* 2010), suggesting >50% decline over 23 years (3 generations) in Australia may not apply across the flyway. Models suggest that this decline is due to reduced adult survival rates (Rogers and Gosbell 2006). Given likely ongoing threats to its migratory staging sites in East Asia (Rogers *et al.* 2010), it is assumed that declines of 30–49% over 23 years will continue. Populations in other flyways are stable and the species is not threatened (Wetlands International 2006; BirdLife International 2011).

Current eligibility against IUCN Red List Criteria

IUCN category	Criteria eligibility
Population visiting Australia	
A	Endangered: past, current and anticipated population declines of 50–79% over periods of 3 generations (23 years) estimated and projected from monitoring and from deteriorating habitat quality (downgraded to Vulnerable)
B	Not applicable: EOO >20 000 km^2; AOO >2000 km^2
C	Not applicable: population >10 000 mature individuals
D	Not applicable: population >1000 mature individuals, >5 locations
E	Not applicable: no population viability analysis undertaken

IUCN Red List assessment data

	Estimate	Reliability
Population visiting Australia		
Extent of occurrence trend	7 600 000 km^2 stable	high high
Area of occupancy trend	6800 km^2 stable	low medium
No. of mature individuals trend	115 000 decreasing	medium high
No. subpopulations	1	high
No. locations	>10	high
Generation time	7.6 years	medium
Global population share	11–25%	high

Ecology

Breeds in open tundra in the lowlands of the high Arctic. Non-breeding birds migrate to coastal brackish lagoons, tidal mud and sand flats, estuaries, saltmarshes and less often inland, where they feed on polycheate worms, molluscs and crustaceans (Higgins and Davies 1996). A generation time of 7.6 years (BirdLife International 2011) is derived from an age at first breeding of 2.0 years, an annual survival of adults of 79% and a maximum longevity of 14.8 years, all extrapolated from congeners.

Threats

Threatened by wetland degradation in East Asia where it stages on migration (Bamford *et al.* 2008). Specifically threatened at Bohai Bay where up to 30% of the population stages on migration along 20 km of mudflats which is being reclaimed and developed (Rogers *et al.* 2010). General threats at migratory staging sites include environmental pollution, reduced river flows, human disturbance and reclamation for tidal power plants and barrages, industrial use and urban expansion (Barter 2002; Kelin and Qiang 2006; Moores 2006). However, also reported that a large proportion migrate across inland East Asia (Barter 2002; Bamford *et al.* 2008). Threats in Australia, especially eastern and southern Australia, include ongoing human disturbance, habitat loss and degradation from pollution, changes to the water regime and invasive plants (Rogers *et al.* 2006; Australian Government 2009).

Conservation objectives

1. Stable Australian non-breeding populations
2. Disturbance at key roosting and feeding sites reduced

Information required

1. Numbers and trends of Australian non-breeding population
2. Improved knowledge of dependence on key migratory staging sites
3. Improved knowledge of impacts of disturbance

Management actions required

1. Work with governments along the flyway at all levels, and local communities, to prevent destruction of key migratory staging sites
2. Negotiate with the Chinese government in particular to protect remaining tidal flats in Bohai Bay
3. Advocate for restoration of alienated wetland sites
4. Maintain and improve protection of roosting and feeding sites in Australia

Bibliography

Amano T, Székely T, Koyama K, Amano H, Sutherland WJ (2010) A framework for monitoring the status of populations: an example from wader populations in the East Asian–Australasian flyway. *Biological Conservation* **143**, 2238–2247.

Australian Government (2009) 'Draft significant impact guidelines for 36 migratory shorebirds'. Draft EPBC Act Policy Statement 3.21, Canberra. Retrieved 25 February, 2011 from <http://www.environment.gov.au/epbc/publications/migratory-shorebirds.html>.

Bamford M, Watkins D, Bancroft W, Tischler G, Wahl J (2008) 'Migratory shorebirds of the East Asian – Australasian flyway: population estimates and internationally important sites'. Wetlands International – Oceania, Canberra.

Barter MA (2002) 'Shorebirds of the Yellow Sea: importance, threats and conservation status'. Canberra, Australia.

BirdLife International (2011) 'Species factsheet: *Calidris ferruginea*'. Retrieved 17 January, 2011 from <http://www.birdlife.org/>.

Close DH (2008) Changes in wader numbers in the Gulf St Vincent, South Australia, 1979–2008. *Stilt* **54**, 24–27.

Department of Sustainability, Environment, Water, Population and Communities (2011) *Calidris ferruginea*. In 'Species Profile and Threats Database'. Department of Sustainability, Environment, Water, Population and Communities, Canberra. Retrieved 24 February, 2011 from <http://www.environment.gov.au/sprat>.

Fuller RA, Wilson HB, Kendall BE, Possingham HP (2009) 'Monitoring shorebirds using counts by the Queensland Wader Study Group'. Report to the Queensland Wader Study Group and the Department of Environment and Resource Management, Brisbane.

Higgins PJ, Davies SJJF (Eds) (1996) *Handbook of Australian, New Zealand and Antarctic Birds. Volume 3: Snipe to Pigeons*. Oxford University Press, Melbourne.

Kelin C, Qiang X (2006) Conserving migratory shorebirds in the Yellow Sea region. In *Waterbirds Around the World*. (Eds

G Boere, C Galbraith and D Stroud) p. 319. The Stationery Office, Edinburgh, UK.

Moores N (2006) South Korea's shorebirds: a review of abundance, distribution, threats and conservation status. *Stilt* **50**, 62–72.

Rogers D, Hassell C, Oldland J, Clemens R, Boyle A, Rogers K (2009) 'Monitoring Yellow Sea migrants in Australia (MYSMA): north-western Australian shorebird surveys and workshops, December 2008'.

Rogers DI, Piersma T, Hassell CJ (2006) Roost availability may constrain shorebird distribution: exploring the energetic costs of roosting and disturbance around a tropical bay. *Biological Conservation* **133**, 225–235.

Rogers DI, Yang H-Y, Hassell CJ, Boyle AN, Rogers KG, Chen B, Zhang Z-W, Piersma T (2010) Red Knots (*Calidris canutus piersmai* and *C. c. rogersi*) depend on a small threatened staging area in Bohai Bay, China. *Emu* **110**, 307–315.

Rogers KG, Gosbell K (2006) Demographic models for Red-necked Stint and Curlew Sandpiper in Victoria. *Stilt* **50**, 203–214.

Wainwright P, Christie M (2008) Wader surveys at the Coorong and S.E. Coastal Lakes, South Australia. *Stilt* **54**, 31–47.

Wetlands International (2006) 'Waterbird population estimates'. Wetlands International, Wageningen, The Netherlands.

Comments received from

Rob Clemens, Nicola Crockford, Ken Gosbell, Roger Jaensch, David Milton, Clive Minton, Danny Rogers

Species Conservation Summary

Black-breasted Button-quail

Turnix melanogaster (Gould, 1837)

Turnicidae

Conservation status

Near Threatened B2ab(i,ii,iii,iv,v)

Reasons for listing

Extent of occurrence may be <20 000 km^2 and area occupied <2000 km^2 with a continuing decline in both as well as area of habitat, number of locations and number of mature individuals but does not fluctuate and uncertainty about whether fragmented - birds probably move between locations.

Status 2000

Near Threatened B2ab(i,ii,iii,iv,v)

Status 1990

Near Threatened B2ab(i,ii,iii,iv,v)

Taxonomy

No infraspecific taxon described

Taxonomic uniqueness: high (2 genera/family, 15 species/genus, 1 subspecies/species)

Range

South-eastern Queensland from near Byfield in the north to the Border Ranges rainforests in the south, generally east of the Great Dividing Range, although records up to 300 km inland at locations at Palmgrove National Park and Barakula State Forest in Qld (Mathieson and Smith 2009; Department of Sustainability, Environment, Water, Population and Communities 2011). A survey in 1992/3 in south-east Qld found evidence of birds at 75 sites, with incidental observations having been made at another 50, clustered into 14 discrete areas (Hamley *et al.* 1997), with most being in the semi-evergreen vine thickets of the Yarraman-Jimna region and the coastal dune scrubs of the Great Sandy region (Mathieson and Smith 2009). The largest subpopulation is believed to occur in a 3 km × 100 km strip along the eastern edge of Fraser Island National Park (R. Hobson *in litt.*) but in habitat that it is extremely difficult to survey. Density in Qld has been reduced in many areas, particularly in the Dawson and Fitzroy River catchments (Bennett 1985; Hamley *et al.* 1997). There were 10 reliable yet unconfirmed records from NSW from 1990–2000 from 6 areas in the far north-east of the state (Milledge and McKinley 1998), but none since 2000 (M. Mathieson and G. Smith *in litt.*). A record of the species in open eucalypt forest (Smyth *et al.* 2001) distant from rainforest suggests that, like other button-quail, there can be movement between fragments. Extreme fragmentation is also unlikely since it is estimated that at least 40% of the population occurs in the Fraser I. subpopulation.

Abundance

The population size is effectively unknown. The number of mature individuals was estimated by an expert com-

Current eligibility against IUCN Red List Criteria

IUCN category	Criteria eligibility
A	Not applicable: past, current or future population declines are thought unlikely to exceed 20% in any 3-generation period
B	Near Threatened: AOO <2000 km^2, severe fragmentation uncertain, suspected continuing declines in EOO, AOO, habitat quality, no. locations and no. mature individuals
C	Not applicable: population 2500–10 000 mature individuals, continuing decline not observed or estimated, largest subpopulation >1000 mature individuals, >1 subpopulation
D	Not applicable: population >1000 mature individuals; >5 locations
E	Not applicable: no population viability analysis undertaken

IUCN Red List assessment data

	Estimate	Reliability
Extent of occurrence trend	83 000 km^2 stable	medium medium
Area of occupancy trend	750 km^2 decreasing	low low
No. of mature individuals trend	5000 decreasing	low low
No. subpopulations	9	low
No. locations	>10	high
Largest subpopulation	2000	low
Generation time	5.0 years	low
Global population share	100%	high

mittee as 5000±1700 but with very little baseline information. This figure is the same as that of Garnett and Crowley (2000).

Ecology

In south-east Qld most frequently reported in vine thicket rainforest that receives 800–1200 mm annual average rainfall, and have a largely closed canopy and deep litter layer. They also occur in softwood scrubs in the brigalow belt, vine scrub regrowth, mature Hoop Pine *Araucaria cunninghamii* plantations, especially where there is *Lantana camara*, dry sclerophyll forest adjacent to rainforest and *Acacia* and *Austromyrtus* scrubs on sandy coastal soils (Marchant and Higgins 1993; Hamley *et al.* 1997; Smyth 1997; Smith *et al.* 1998; Lees and Smith 2000; Mathieson and Smith 2009). Their principal food is invertebrates from the litter on the forest floor (Marchant and Higgins 1993; McConnell and Hobson 1995) and possibly seeds (Smyth 1997). The species is polyandrous and is commonly seen in pairs or occasionally in small groups (Hughes and Hughes 1991; Marchant and Higgins 1993). A generation time of 5.0 years is estimated, based on age at first breeding of 1.0 years, maximum longevity of 9.0 years (values elicited from a Birds Australia expert committee).

Threats

Currently the main threats are habitat degradation as a result of domestic stock and feral pigs *Sus scrofa* utilising Black-breasted Button-quail habitat, habitat loss or degradation from fire and predation by feral animals such as feral cats *Felis catus* and foxes *Vulpes vulpes* (Mathieson and Smith 2009). While at least 90% of Black-breasted Button-quail habitat has been cleared for agriculture or plantations of Hoop Pine, the area of bottletree scrubs being reduced from several hundred thousand to only a few thousand hectare by the 1960s (Hamley *et al.* 1997), almost all lands currently occupied are under state control and most are conserved.

Conservation objectives

1. All remaining subpopulations retained

Information required

1. A standard monitoring technique to determine presence and trends in abundance
2. Impact of cat and fox predation
3. Extent of movement between habitat patches
4. Size and distribution of population on Fraser I.
5. Definition of habitat requirements

Management actions required

1. Define assessment and monitoring strategies
2. Create a clear definition of occupied habitats
3. Develop a predictive model of the species' distribution
4. Confirm size and distribution of subpopulation on Fraser I.
5. Protect key ecosystems/habitat that support button-quail populations
6. Maintain or improve extent, condition and connectivity of habitat
7. If necessary, reduce the impacts of introduced predators and competitors

Bibliography

Bennett SC (1985) The distribution and status of the Black-breasted Button-quail *Turnix melanogaster. Emu* **85**, 157–162.

Department of Sustainability, Environment, Water, Population and Communities (2011) *Turnix melanogaster.* In 'Species Profile and Threats Database'. Department of Sustainability, Environment, Water, Population and Communities, Canberra. Retrieved 24 February, 2011 from <http://www.environment.gov.au/sprat>.

Hamley T, Flower P, Smith GC (1997) Present and past distribution of the Black-breasted Button-quail *Turnix melanogaster* (Gould) in Queensland. *Sunbird* **27**, 1–21.

Hughes P, Hughes B (1991) Notes on the Black-breasted Button-quail at Widgee, Queensland. *Australian Bird Watcher* **14**, 113–118.

Lees N, Smith GC (2000) Use of mature hoop pine plantation by the vulnerable Black-breasted Button-quail (*Turnix melanogaster*). *Australian Forestry* **62**, 330–335.

Marchant S, Higgins PJ (Eds) (1993) *Handbook of Australian, New Zealand and Antarctic Birds. Volume 2: Raptors to Lapwings.* Oxford University Press, Melbourne.

Mathieson MT, Smith GC (2009) 'National recovery plan for the buff-breasted button-quail *Turnix olivii*'. Report to Department of the Environment, Water, Heritage and the

Arts, Canberra. Department of Environment and Resource Management, Brisbane.

McConnell P, Hobson RG (1995) The diet and behaviour of the Black-breasted Button-quail *Turnix melanogaster. Sunbird* **25**, 18–23.

Milledge DR, McKinley AL (1998) 'The distribution, status and habitat of the Black-breasted Button-quail *Turnix melanogaster* in north-eastern NSW. Stage 1. Collation of habitat data from surveys of known and potential habitat in Queensland and records in New South Wales'. Report to NSW National Parks and Wildlife Service, Coffs Harbour.

Smith GC, Aridis J, Lees N (1998) Radio-tracking revealed home ranges of Black-breasted Button-quail *Turnix melanogaster* in remnant vine scrub between hoop pine plantation and agriculture. *Emu* **98**, 171–177.

Smyth A (1997) 'Draft recovery plan for Black-breasted Button-quail *Turnix melanogaster*'. Unpublished report to Queensland Department of Environment and Heritage., Brisbane.

Smyth AK, Noble D, Wiley C (2001) Black-breasted button-quail in open eucalypt forest in south-eastern Queensland. *Australian Bird Watcher* **19**, 45–47.

Comments received from

Geoff Smith, Mike Mathieson, Rod Hobson, Dez Wells

Species Conservation Summary

Buff-breasted Button-quail

Turnix olivii Robinson, 1900

Turnicidae

Conservation status

Endangered B2ac(ii)

Reasons for listing

Very small AOO and probably at ≤5 locations on average; this probably fluctuates greatly depending on annual variation in the pattern of burning. Ongoing decline possible as burning patterns have changed but uncertain so criteria do not include B2b

Status 2000

Endangered B2ac(ii)

Status 1990

Endangered B2ac(ii)

Taxonomy

No infraspecific taxa described

Taxonomic uniqueness: high (2 genera/family, 15 species/genus, 1 subspecies/species)

Range

Endemic to north-eastern Queensland. Most records from inland of Cairns including Mareeba, Mount Molloy, Hann Tableland National Park, and also further west at Rookwood Station near Chillagoe, the Karumba district and from around Lakefield National Park, Musgrave and Coen, with unconfirmed records near Ingham and Iron Range (Mathieson and Smith 2009). The first specimen was collected in Cooktown by a local resident but has not been recorded there since (M. Mathieson *in litt.*). Records near Musgrave and Chillagoe in 1984 coincided with peak in abundance at Mt Molloy (L. Nielsen pers. comm.). AOO probably fluctuates greatly depending on annual variations in the pattern of burning.

Abundance

Buff-breasted Button-quail is one of the least-known birds in Australia (Department of Sustainability, Environment, Water, Population and Communities 2011); it is very rarely recorded, e.g. no records in the 1977–1981 *Atlas of Australian Birds* (Blakers *et al.* 1984) and 4 records in 1998–2002 *The New Atlas of Australian Birds* (Barrett *et al.* 2003). In the absence of any new information, the population size is estimated as 500 mature individuals, following Garnett and Crowley (2000). Progressively, there were fewer birds each wet season near Mt Molloy over the 1990s (L. Nielsen pers. comm.) but no subsequent evidence of decline in number of birds or in quality of habitat. Habitat near Coen, described as sparsely wooded gravelly slopes in 1922 and where they were found breeding (White 1922b, 1922a), is now overgrown with woody weeds (S. Garnett pers. obs.).

Ecology

The few records of the species are mostly from stony and/or grassy woodlands and forests on plains and slopes, often with *Melaleuca viridiflora* and *M. minutifolia*

Current eligibility against IUCN Red List Criteria

IUCN category	Criteria eligibility
A	Not applicable: past, current or future population declines are thought unlikely to exceed 20% in any 3-generation period
B	Endangered: AOO <500 km^2, in some years at ≤5 locations with plausible future threat that there will be none, suspected that AOO subject to extreme fluctuations
C	Not applicable: no continuing population decline or extreme fluctuations
D	Vulnerable: population 250–1000 mature individuals, ≤5 locations with plausible future threat in some years
E	Not applicable: no population viability analysis undertaken

IUCN Red List assessment data

	Estimate	Reliability
Extent of occurrence trend	35 000 km^2 stable	low low
Area of occupancy trend	350 km^2 fluctuating	low low
No. of mature individuals trend	500 stable	low low
No. subpopulations	1	medium
No. locations	5	low
Generation time	5.0 years	low
Global population share	100%	high

(Mathieson and Smith 2009). The sparsely wooded, well-drained, slight-sloping bases of hills that support this vegetation appear critical during the breeding season (L. Nielsen pers. comm.). They have also been noted from grassy clearings within rainforest patches (Squire 1990; Marchant and Higgins 1993). Most records in the Mareeba area are in the late dry season and persist until the grass becomes very dense. Some remain and breed in weak or late wet seasons (Mathieson and Smith 2009). Fires after the first rains of the wet season, 'storm burns', can maintain the open structure of the grassland through the dry season: at least 2 sightings in the late wet season have been in storm burnt areas (R. Hobson pers. comm., S. Garnett pers. obs., L. Nielsen pers. comm.). Such fires are often lit by lightning and are characteristic of the fire regime that would have prevailed before Aboriginal people arrived. This pattern was maintained by Aboriginal people but storm burns became less frequent after Aboriginal fire regimes were disruped (Garnett and Crowley 2000; Crowley *et al.* 2009). The birds are assumed to be polyandrous and appear to feed on seeds and insects (Marchant and Higgins 1993). They build dome nests on the ground and lay 4 eggs that are tended by the male (White 1922b). A generation time of 5.0 years (BirdLife International 2011) is derived from an average age at first breeding of 1.0 years and a maximum longevity in the wild of 9.0 years, all extrapolated from expert opinion for *T. melanogaster.*

Threats

May have been eliminated from much of its range by inappropriate burning regimes and grazing that tends to promote woody weed invasion and results in few areas being storm-burnt in the early wet season (Crowley and Garnett 1998, 2000; Crowley *et al.* 2009). As a result there may be few places that retain an open structure that the button-quail appears to prefer for breeding. Predation by feral cat *Felis catus* or feral pig *Sus scrofa* may also affect abundance. The potential scarcity of suitable breeding habitat may mean that there are ≤5 locations in some years.

Conservation objectives

1. Stable numbers at monitored locations
2. Conservation burning for the species at least at 1 site

Information required

1. Reliable means of detecting birds
2. A photograph and a sound recording
3. Better knowledge of distribution, life history and ecology
4. Impacts of fire, cattle and pigs on habitat

Management actions required

1. Follow up all sighting to record ecological details and attempt to improve knowledge of habitat management needs
2. Undertake fire management at known locations, including storm burning

Bibliography

Barrett G, Silcocks A, Barry S, Cunningham R, Poulter R (2003) *The New Atlas of Australian Birds.* Royal Australian Ornithologists Union, Hawthorn East.

BirdLife International (2011) 'Species factsheet: *Turnix olivii*'. Retrieved 17 January, 2011 from <http://www.birdlife.org/>.

Blakers M, S., Davies JJF, Reilly PN (1984) *The Atlas of Australian Birds.* Melbourne University Press, Melbourne.

Crowley GM, Garnett ST (1998) Vegetation change in the grasslands and grassy woodlands of east-central Cape York Peninsula, Australia. *Pacific Conservation Biology* **4**, 132–148.

Crowley GM, Garnett ST (2000) Changing fire management in the pastoral lands of Cape York Peninsula of northeast Australia, 1623–1996. *Australian Geographical Studies* **38**, 10–26.

Crowley GM, Garnett ST, Shephard S (2009) Impact of storm-burns on *Melaleuca viridiflora* invasion of grasslands and grassy woodlands on Cape York Peninsula, Australia. *Austral Ecology* **34**, 196–209.

Department of Sustainability, Environment, Water, Population and Communities (2011) *Turnix olivii.* In 'Species Profile and Threats Database'. Department of Sustainability, Environment, Water, Population and Communities, Canberra. Retrieved 24 February, 2011 from <http://www.environment.gov.au/sprat>.

Garnett S, Crowley GM (2000) *The Action Plan for Australian Birds 2000.* Environment Australia, Canberra.

Marchant S, Higgins PJ (Eds) (1993) *Handbook of Australian, New Zealand and Antarctic Birds. Volume 2: Raptors to Lapwings*. Oxford University Press, Melbourne.

Mathieson MT, Smith GC (2009) 'National recovery plan for the buff-breasted button-quail *Turnix olivii*'. Report to Department of the Environment, Water, Heritage and the Arts, Canberra. Department of Environment and Resource Management, Brisbane.

Squire JE (1990) Some southern records and other observations of the Buff-breasted Button-quail *Turnix olivii*. *Australian Bird Watcher* **13**, 149–152.

White HL (1922a) A collecting trip to Cape York Peninsula. *Emu* **22**, 99–116.

White HL (1922b) Description of nest and eggs of *Turnix olivii* (Robinson). *Emu* **22**, 2–3.

Comments received from

Rod Hobson, Mike Mathieson, Lloyd Nielsen, Geoff Smith

Subspecies Conservation Summary

Painted Button-quail (Houtman Abrolhos)

Turnix varius scintillans (Gould, 1845)
Turnicidae

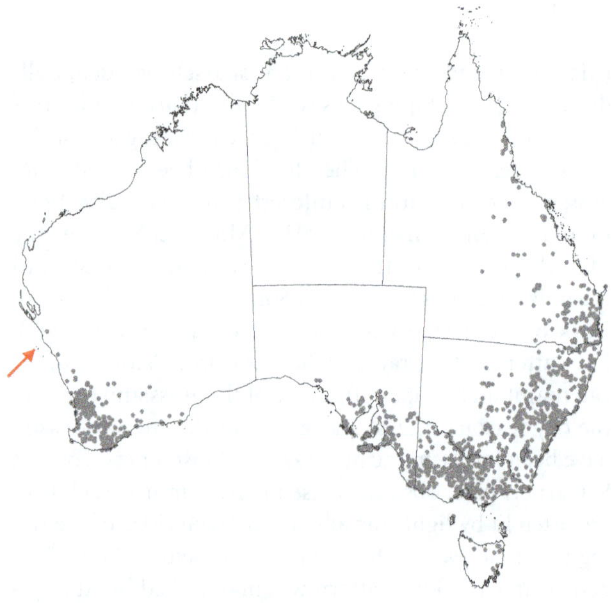

Conservation status

Endangered B1ab(v)+2ab(v), C2a(ii)

Reasons for listing

Occurs at only 2 locations and population on 1 island severely reduced with potential for local extinction

Status 2000

Vulnerable D2

Reason for change in 2010: evidence for scarcity at 1 location

Status 1990

Vulnerable D2

Taxonomy

T. v. varia (south-west Australia, Tasmania and eastern Australia) is Least Concern and *T. v. novaecaledoniae* (New Caledonia) is possibly extinct. The species is Least Concern.

Taxonomic uniqueness: medium (2 genera/family, 15 species/genus, 3 subspecies/species)

Range

Endemic to Houtman Abrolhos, c.60–80 km off central Western Australia. Resident on North I. (180 ha), East Wallabi I. (321 ha), West Wallabi I. (c.600 ha) (Storr and Johnstone 1984; Storr *et al.* 1986). Also recorded on Seagull I. (10 ha), which is on the same reef flat as East and West Wallabi I. and Pigeon I. (10 ha, and heavily settled) just offshore, but likely to be resident on neither. The presence of the birds on the Houtman Abrolhos, and the mobility of other subspecies, and other *Turnix* species, suggests that there is 1 subpopulation (contra Australian Government 2008) and 2 locations.

Abundance

It was reportedly common on East and West Wallabi Is. up until 1960 (Serventy 1943; Storr 1965, 1966) and also on North I. in the 1970s and 1980s (Storr *et al.* 1986; Western Australian Museum database). During 2 day-long targeted surveys in 2006 there was only 1 definite record and little evidence of foraging activity on North I. and 2 definite records and much evidence of foraging activity on East Wallabi I. (Blyth *et al.* 2006). The population size had been estimated as low thousands (Garnett 1992; Garnett and Crowley 2000) but is here estimated as 1000 (although the single published population density for the species is 1 bird/km^2 (Ford and Bell 1982)). On North I. the population may be declining following introduction of wallabies.

Ecology

This button-quail occupies all available habitats on the islands, excluding bare limestone pavements. It particularly favours *Spinifex longifolius* grasslands and flats of coral grit that support Grey Saltbush *Atriplex cinerea* and *Halosarcia halocnemoides* shrubland. It feeds on seeds, small insects and scraps from around fishing camps (Storr *et al.* 1986), and lays 3 eggs in scrapes under vegetation (Johnstone and Storr 1998). A generation time of 5.0 years (BirdLife International 2011) is derived from an age at first breeding of 1.0 years and a maximum longevity of 9.0 years, all extrapolated from expert opinion for *T. melanogaster*.

Current eligibility against IUCN Red List Criteria

IUCN category	Criteria eligibility
A	Not applicable: past, current or future population declines are thought unlikely to exceed 20% in any 3-generation period
B	Endangered: EOO <5000 km^2, AOO <500 km^2, at ≤5 locations with plausible future threat, suspected continuing decline no. mature individuals
C	Endangered: population 250–2500 mature individuals, inferred ongoing decline, 100% in 1 subpopulation
D	Vulnerable: population 250–1000 mature individuals, ≤5 locations with plausible future threat
E	Not applicable: no population viability analysis undertaken

IUCN Red List assessment data

	Estimate	Reliability
Extent of occurrence trend	30 km^2 stable	high high
Area of occupancy trend	11 km^2 decreasing	high medium
No. of mature individuals trend	<1000 decreasing	medium low
No. subpopulations	1	high
No. locations	2	medium
Generation time	5.0 years	low
Global population share	100%	high

Threats

The population on North I. has probably been severely reduced by grazing and trampling of vegetation by the recently introduced Tammar Wallaby *Macropus eugenii*, which are being removed by the Western Australian Government (Blyth *et al.* 2006; Department of Sustainability, Environment, Water, Population and Communities 2011), even though wallabies do live on the Wallabi Is. Other threats across all islands include fire, introduced predators and introduced competitors. However, the button-quail survived severe fires on North I. in 1935 and 1946, though its numbers may have been temporarily reduced (Storr *et al.* 1986). House Mice *Mus musculus* have been introduced to North I. and possibly East Wallabi I. and may compete with the button-quail for food or prey on their eggs (Garnett and Crowley 2000). None of the islands have other introduced mammals such as rats *Rattus* spp. or cats *Felis catus*, and their accidental introduction could be devastating.

Conservation objectives

1. A viable and secure population maintained on all 3 main islands

Information required

1. Population size
2. Relative impacts of threatening processes

Management actions required

1. Design and implement a monitoring program
2. Monitor islands biennially for predators and presence of button-quail in association with other monitoring for conservation purposes
3. Manage threats to areas of vegetation that contain button-quail populations
4. Develop and implement a management plan for the control and eradication of House Mouse
5. Develop and implement a management plan for Tammar Wallaby on North I.
6. Develop and implement a suitable fire management strategy for the Houtman Abrolhos archipelago

Bibliography

Australian Government (2008) 'Approved conservation advice for *Turnix varius scintillans* (Painted Button-quail (Houtman Abrolhos)'. Retrieved 5 May, 2011 from <http://www.environment.gov.au/biodiversity/threatened/species/pubs/26047-conservation-advice.pdf>.

BirdLife International (2011) 'Species factsheet: *Turnix varius*'. Retrieved 17 January, 2011 from <http://www.birdlife.org/>.

Blyth JD, Blyth JA, Agar G, Agar P (2006) 'Search for Painted Button-quail on North and East Wallabi Islands'. Unpublished report to Department of Environment and Conservation, Geraldton, Western Australia.

Department of Sustainability, Environment, Water, Population and Communities (2011) *Turnix varius scintillans.* In 'Species Profile and Threats Database'. Department of Sustainability, Environment, Water, Population and Communities, Canberra. Retrieved 24 February, 2011 from <http://www.environment.gov.au/sprat>.

Ford HA, Bell H (1982) Density of birds in eucalypt woodland affected to varying degrees by dieback. *Emu* **81**, 202–208.

Garnett S, Crowley GM (2000) *The Action Plan for Australian Birds 2000.* Environment Australia, Canberra.

Garnett ST (1992) 'Threatened and Extinct birds of Australia'. RAOU Report 82. RAOU, Melbourne & ANPWS, Canberra.

Johnstone RE, Storr GM (1998) *Handbook of Western Australian Birds. Volume 1: Non-passerines (Emu to Dollarbird).* Western Australian Museum, Perth.

Serventy VN (1943) Notes on nesting birds of the Abrolhos Islands. *Emu* **42**, 235–241.

Storr GM (1965) The physiography, vegetation and vertebrate fauna of the Wallabi group, Houtman Abrolhos. *Journal of the Royal Society of Western Australia* **48**, 1–14.

Storr GM (1966) Birds of the northern islands of the Houtman Abrolhos. *Emu* **65**, 209–221.

Storr GM, Johnstone RE (1984) The subspecific status of the Painted Button-quail of the Houtman Abrolhos, Western Australia. *Records of the Western Australian Museum* **11**, 315.

Storr GM, Johnstone RE, Griffin P (1986) Birds of the Houtman Abrolhos, Western Australia. *Records of the Western Australian Museum Supplement* **24**, 29–30.

Comments received from

John Blyth, Judy Blyth, Andrew Burbidge

Subspecies Conservation Summary

Lesser Noddy (Houtman Abrolhos)

Anous tenuirostris melanops (Gould, 1845)
Laridae

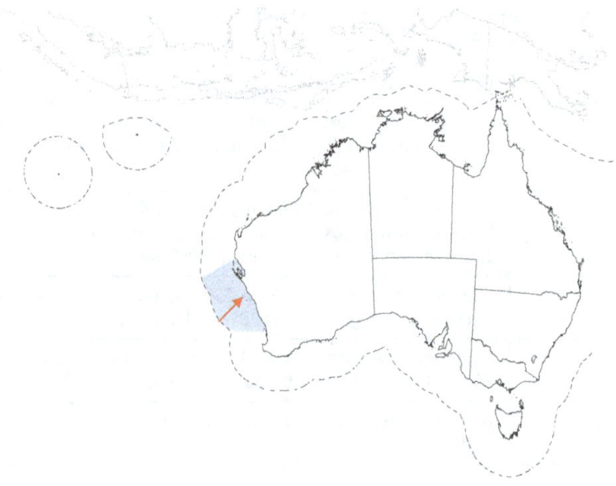

Conservation status

Endangered B2ab(iii,v)

Reasons for listing

Total area occupied is <1 km^2 across 3 locations; despite inter-annual variation, ongoing declines in number of mature individuals are probable, declines also possible in AOO, number of locations and, particularly, quality of habitat

Status 2000

Vulnerable D2
Reason for change in 2010: some evidence for decline in breeding habitat and population

Status 1990

Vulnerable D2

Taxonomy

A. t. tenuirostris (western Indian Ocean) is extralimital, although the subspecies of the birds at Ashmore Reef is unknown (Stokes and Hinchey 1990) and may be colonising Cocos (Keeling) Is. The species is Least Concern.

Taxonomic uniqueness: medium (14 genera/family, 3 species/genus, 2 subspecies/species)

Range

Breeding only in Houtman Abrolhos chain of islands, colonies are found on Pelsaert (4 colonies, total area 3.0 ha), Wooded (0.7 ha) and Morley (0.8 ha) Is. The area of currently occupied habitat on Wooded I. is contracting as a Pied Cormorant *Phalacrocorax varius* breeding colony expands but there are also areas of mangroves that are recovering following over-use by the noddies that will soon be available (Surman and Nicholson 2009b). Population appears to remain near these islands all year, although roosting flocks have been recorded at Shark Bay in 2009 and 2010 (C. Sims *in litt.*) and beach-washed birds have been recovered >250 km south of colonies.

Abundance

Numbers vary considerably between years and colonies. Total nest numbers fell from 77 360 in 1989 to 48 860 in 1993, returned gradually to 67 985 in 1999 but there were only 34 103 on the last survey in 2006. On each occasion Pelsaert I. holds about two-thirds of the nests with the remainder divided about equally between the other 2 islands. The low numbers in 2006 may have been because of a poor season but this has not been confirmed (Surman and Nicholson 2009b). Total numbers early in 20th century were possibly as low as 20 000 pairs (Burbidge and Fuller 1989).

Ecology

Nests in White Mangroves *Avicennia marina*, which occur in small scattered patches on a few of the islands of the Houtman Abrolhos (Surman and Wooller 2003). Nesting colonies move periodically because nesting birds retard tree growth and sometimes kill trees; trees take up to 13 years to recover (Surman and Nicholson 2009b). Feeds mostly on small fish and to a lesser degree squid in nearby waters (Surman and Wooller 2003). Fluctuations in size of the fish population may explain the variation in breeding numbers in any year (Surman and Nicholson 2009a) although the main determinant of breeding success is site selection: sites protected from strong westerly-north-westerly winds being more likely to succeed than exposed nests (Surman and Wooller 1995). A generation time of 11.5 years (BirdLife International 2011) is derived from an age of first breeding of 3.0 years, an annual survival of adults of 81.5% and a maximum longevity of 26.0 years, all extrapolated from congeners.

Threats

Currently 1 colony is contracting because an expanding population of Pied Cormorants is killing the mangroves. However, there is unused habitat elsewhere on the 3 islands and the birds have shown a capacity to move when they need to (Surman and Nicholson 2009b). Otherwise the main potential threat to breeding colonies is catastrophic destruction of habitat by cyclones, as storms destroy eggs (Surman and Wooller 1995) and could damage nesting habitat, pollution from oil spills that could damage birds and mangroves or over-fishing nearby. In the past, Lesser Noddies were probably driven off Pelsaert I. by guano extraction.

Conservation objectives

1. Stable population

Information required

1. Population trends

Current eligibility against IUCN Red List Criteria

IUCN category	Criteria eligibility
A	Near Threatened: population decline of 20–29% over 3 generations (35 years) estimated and projected from monitoring and inferred from deteriorating habitat quality
B	Endangered: AOO <500 km^2, at ≤5 locations with plausible future threat, estimated continuing decline in no. mature individuals
C	Not applicable: >10 000 mature individuals
D	Vulnerable: ≤5 locations with plausible future threat
E	Not applicable: no population viability analysis undertaken

IUCN Red List assessment data

	Estimate	Reliability
Extent of occurrence trend	300 000 km^2 stable	high high
Area of occupancy trend	0.5 km^2 stable	high high
No. of mature individuals trend	100 000 decreasing	medium medium
No. subpopulations	1	high
No. locations	3	high
Generation time	11.5 years	medium
Global population share	100%	high

2. Relationship between Lesser Noddy, its prey and commercial fishing

Management actions required

1. Maintain monitoring and current management of breeding islands.

Bibliography

BirdLife International (2011) 'Species factsheet: *Anous tenuirostris*'. Retrieved 17 January, 2011 from <http://www.birdlife.org/>.

Burbidge AA, Fuller PJ (1989) Numbers of breeding seabirds on Pelsaert Island, Houtman Abrolhos, Western Australia. *Corella* **13**, 57–61.

Stokes T, Hinchey M (1990) Which small Noddies breed at Ashmore Reef in Eastern Indian Ocean? *Emu* **90**, 269–271.

Surman CA, Nicholson LW (2009a) The good, the bad and the ugly: ENSO driven oceanographic variability and its influence on seabird diet and reproductive performance at the Houtman Abrolhos, Eastern Indian Ocean. *Marine Ornithology* **37**, 129–138.

Surman CA, Nicholson LW (2009b) A survey of the breeding seabirds and migratory shorebirds of the Houtman Abrolhos, Western Australia. *Corella* **33**, 81–98.

Surman CA, Wooller RD (1995) The breeding biology of the Lesser Noddy on Pelsaert Island, Western Australia. *Emu* **95**, 47–53.

Surman CA, Wooller RD (2003) Comparative foraging Ecology of five sympatric terns at a sub-tropical island in the eastern Indian Ocean. *Journal of The Zoological Society of London* **259**, 219–230.

Comments received from

Andrew Burbidge, Nic Dunlop, Colleen Sims

Species Conservation Summary

Fairy Tern

Sternula nereis (Gould, 1843)

Laridae

Conservation status

Australian breeding population: Vulnerable C1

Reasons for listing

Population <10 000 mature individuals and decline of 10–20% likely within 3 generations (33 years). Though it may be stable in Western Australia, numbers elsewhere in Australia have declined rapidly during the last 30 years; there are minimal numbers in New Zealand, New Caledonia and the Coral Sea

Status 2000

Least Concern

Reason for change in 2010: evidence of rapid decline

Status 1990

Least Concern

Taxonomy

S. n. nereis (southern Australia) is Vulnerable, *S.n.exsul* (New Caledonia and Coral Sea) is Endangered and extralimital *S. n. davisae* (New Zealand) is Critically Endangered

Taxonomic uniqueness: medium (14 genera/family, 7 species/genus, 3 subspecies/species)

Range

S. n. nereis occurs in coastal Southern Australia from the Pilbara, Western Australia east to Botany Bay, New South Wales as well as Tasmania but absent from the Great Australian Bight (Higgins and Davies 1996). *S. n. exsul* occurs primarily in coastal New Caledonia but also breeds at East Diamond Islet, near Lihou Reef National Nature Reserve in the Coral Sea (Carter and Mustoe 2007). *S. n. davisae* is confined to New Zealand (BirdLife International 2011).

Current eligibility against IUCN Red List Criteria

IUCN category	Criteria eligibility
Australian breeding population	
A	Not applicable: past, current or future population declines are thought unlikely to exceed 20% in any 3-generation period
B	Not applicable: EOO >20 000 km^2; AOO >2000 km^2
C	Vulnerable: population 2500–10 000 mature individuals, estimated continuing decline between 10% in 3 generations and 20% in 2 generations
D	Not applicable: population >1000 mature individuals, >5 locations
E	Not applicable: no population viability analysis undertaken

IUCN Red List assessment data

	Estimate	Reliability
Australian breeding population		
Extent of occurrence trend	290 000 km^2 stable	high medium
Area of occupancy trend	3300 km^2 decreasing	high medium
No. of mature individuals trend	9000 decreasing	medium medium
No. subpopulations	3	high
No. locations	>10	high
Largest subpopulation	8000	medium
Generation time	11.0 years	medium
Global population share	99%	high
Level of genetic exchange	low	high

Abundance

The total population of *S. n. nereis* is about 9000 mature individuals, of which 8000 occur in WA, the remainder in scattered breeding locations in eastern Australia, principally in SA and Tas (E. Woehler *in litt.*). There are only 100–200 pairs of *S. n. exsul* in New Caledonia, with <10 in the Coral Sea, and <50 pairs of *S. n. davisae* (BirdLife International 2011). Monitored populations of *S. n. nereis* have declined sharply in the last 2 decades.

Ecology

Frequents sheltered coasts, also fresh and saline wetlands, but *S. n. exsul* at least can traverse open ocean. Nests on sandy beaches, in exposed areas, laying 1–2 eggs, with low nesting success. It feeds primarily on small fish (BirdLife International 2011). A generation time of 11.0 years (BirdLife International 2011) is derived from an average age at first breeding of 3.2 years, an annual survival of adults of 82.4%, and a maximum longevity in the wild of 23.1 years, all extrapolated from congeners.

Threats

In south-eastern Australia nesting often fails because of interference by people and introduced predators while a collapse in fish stocks in the Coorong, SA, has led to a major decline there (Paton and Rogers 2009). There is some interbreeding with Little Tern *S. albifrons*, particularly in NSW. WA populations, which mostly nest on predator-free islands, and the tiny, possibly ephemeral, population of *S. n. exsul* in the Coral Sea, are largely free of imminent threats.

Conservation objectives

1. Populations retained and protected, particularly in south-eastern Australia

Information required

1. Population size and demographic trends
2. Prioritisation of south-eastern colonies for active management

Management actions required

1. Monitor demographic trends at important sites
2. Monitor sites for hybrids with Little Tern
3. Control introduced mammals and other nest predators at important breeding sites
4. Restrict human and vehicular access to important breeding colonies
5. Manage water quality in the lower Coorong to benefit remaining terns
6. Liaise with New Zealand and New Caledonia researchers and managers
7. Engage the public in research, monitoring, management and advocacy

Bibliography

BirdLife International (2011) 'Species factsheet: *Sterna nereis*'. Retrieved 17 January, 2011 from <http://www.birdlife.org/>.

Carter M, Mustoe S (2007) Another form of Fairy Tern *Sterna nereis* breeding in Australian Territory. *Australian Field Ornithology* **24**, 167–179.

Higgins PJ, Davies SJJF (Eds) (1996) *Handbook of Australian, New Zealand and Antarctic Birds. Volume 3: Snipe to Pigeons.* Oxford University Press, Melbourne.

Paton DC, Rogers DJ (2009) 'Ecology of breeding Fairy Terns *Sternula nereis* in the Coorong'. Final report for the Wildlife Conservation Fund.

Comments received from

Barry Baker, Stuart Butchart, Andrew Burbidge, Les Christidis, Kerryn Herman, Tim Holmes, Hugh Ford, Geoff Lacey, Richard Loyn, Peter Menkhorst, Simon Mustoe, David Paton, Don Saunders, Eric Woehler

Subspecies Conservation Summary

Fairy Tern (New Caledonian)

Sternula nereis exsul Mathews, 1912
Laridae

Conservation status
Australian breeding population: Endangered D

Reasons for listing
Australian population <50 mature individuals. Probably not isolated from population in New Caledonia, so downgraded, even though that population too is Endangered

Status 2000
Endangered D

Status 1990
Endangered D

Taxonomy
S. n. nereis (southern Australia) is Vulnerable,
S. n. davisae (New Zealand) is extralimital. The species is Vulnerable.
Taxonomic uniqueness: medium (14 genera/family, 7 species/genus, 3 subspecies/species)

Range
In 2006 a small breeding population was discovered on East Diamond Islet, near Lihou Reef National Nature Reserve in the Coral Sea, following 2 records in 1950 on Heron I. and unconfirmed sightings in late 1970s and 1980s (Carter and Mustoe 2007). Extralimitally this subspecies breeds around New Caledonia (Baling *et al.* 2009) and was last recorded breeding on the Chesterfield Is. in 1974 (Borsa *et al.* 2010).

Abundance
Five individuals, including a begging juvenile, were seen on East Diamond Islet in 2006. There are only 100–200 pairs of this subspecies in New Caledonia (BirdLife International 2011).

Ecology
The nominate subspecies frequents sheltered coasts, also fresh and saline wetlands, but this population must have travelled a long distance over open ocean to reach East Diamond Islet. Nests on sandy beaches, in exposed areas, feeds mostly on fish, and lays 1 or 2 eggs, with low nesting success in New Caledonia (BirdLife International 2011). A generation time of 11.0 years (BirdLife International 2011) is derived from an average age at first breeding of 3.2 years, an annual survival of adults of 82.4%, and a maximum longevity in the wild of 23.1 years, all extrapolated from congeners.

Threats
Disturbance by humans is a possible threat but visits to the islets are very few.

Conservation objectives
1. Population retained and protected

Information required
1. Population size and demographic trends
2. Analysis of threats

Management actions required
1. Assess colony size on occasional visits

Current eligibility against IUCN Red List Criteria

IUCN category	Criteria eligibility
Australian breeding population	
A	Not applicable: past, current or future population declines are thought unlikely to exceed 20% in any 3-generation period
B	Not applicable: EOO, AOO, habitat, locations or population not declining, population fluctuations not extreme
C	Not applicable: no continuing population decline or extreme fluctuations
D	Critically Endangered: population <50 mature individuals (downgraded to Endangered)
E	Not applicable: no population viability analysis undertaken

IUCN Red List assessment data

	Estimate	Reliability
Australian breeding population		
Extent of occurrence trend	280 000 km^2 stable	low medium
Area of occupancy trend	1 km^2 stable	high medium
No. of mature individuals trend	4 stable	medium low
No. subpopulations	1	high
No. locations	1	high
Generation time	11.0 years	medium
Global population share	1–10%	low
Level of genetic exchange	high	medium

2. Include islet within the Lihou Reef National Nature Reserve

Bibliography

Baling M, Jeffries D, Barré N, Brunton DH (2009) A survey of Fairy Tern (*Sterna nereis*) breeding colonies in the Southern Lagoon, New Caledonia. *Emu* **109**, 57–61.

BirdLife International (2011) 'Species factsheet: *Sterna nereis*'. Retrieved 17 January, 2011 from <http://www.birdlife.org/>.

Borsa P, Pandolfi M, Andréfouët S, Bretagnolle V (2010) Breeding avifauna of the Chesterfield Islands, Coral Sea: current population sizes, trends, and threats. *Pacific Science* **64**, 297–314.

Carter M, Mustoe S (2007) Another form of Fairy Tern *Sterna nereis* breeding in Australian Territory. *Australian Field Ornithology* **24**, 167–179.

Comments received from

Simon Mustoe

Subspecies Conservation Summary

Fairy Tern (Australian)

Sternula nereis nereis (Gould 1843)

Laridae

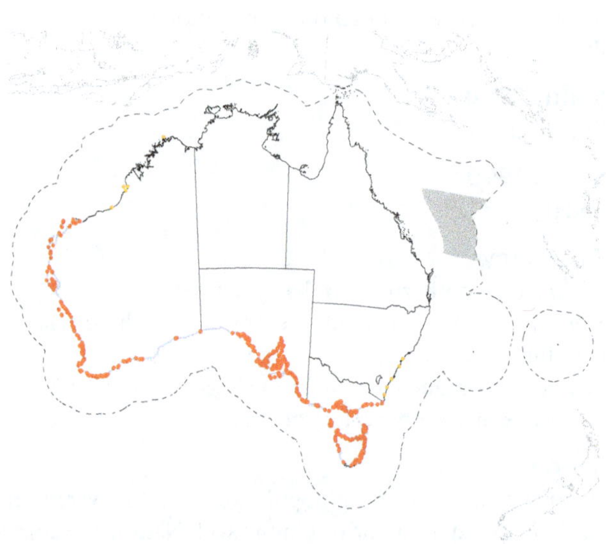

Conservation status

Vulnerable C1

Reasons for listing

Population <10 000 mature individuals and decline of 10–20% likely within 3 generations (33 years). Though it may be stable in Western Australia, numbers elsewhere in Australia have declined rapidly during the last 30 years

Status 2000

Least Concern

Reason for change in 2010: evidence of rapid decline

Status 1990

Least Concern

Taxonomy

S. n. exsul (New Caledonia) has a small Australian population in the Coral Sea and is Endangered in Australia, *S. n. davisae* is extralimital, breeding in New Zealand. The species is classified as Vulnerable.

Taxonomic uniqueness: medium (14 genera/family, 7 species/genus, 3 subspecies/species)

Range

Coastal southern Australia from the Montebello Is. off the Pilbara, Western Australia east to Botany Bay, New South Wales including Tasmania. The Great Australian Bight forms a gap in distribution between western and eastern subpopulations (Higgins and Davies 1996), and there are now substantial gaps between breeding colonies in eastern Australia, especially around the Victorian coastline.

Abundance

The total population in Australia estimated at 3300–5700 pairs (E. J. Woehler *in litt.*), the average is used here. The largest population of 3000–5000 pairs in WA is believed to be stable (A.A. Burbidge *in litt.*). In eastern Australia there have been rapid declines: there are currently about 180–300 pairs in SA in 2009–2010, 200–250 pairs in Tas (E. J. Woehler *in litt.*), 6–70 pairs in Vic (G. Lacey *in litt.*) and 12–40 pairs in NSW in 2009–10 (E. J. Woehler *in litt.*), where numbers are thought to have increased (NSW National Parks and Wildlife Service 2003). However, in SA, numbers in the South Lagoon of the Coorong declined from 1330 individuals in 1985 to 150–300 in 2000–2007, an 82% decline (Paton and Rogers 2009) and disappeared off the Fleurieu Peninsula in the late 1970s and around Kangaroo I. in the early 1990s. Breeding populations in Gulf St Vincent, Spencer Gulf and near Beachport are very small with low breeding success (D. Paton *in litt.*). In Vic, the number of breeding pairs at French I. fell from a peak of 108 in 1993 to an average of 21 for the decade 1996–2005 (D. Saunders *in litt.*). While the populations in WA are counted rarely, the declining populations in eastern Australia have been monitored for many years, justifying the C1 criterion.

Ecology

Fairy Terns feed almost entirely on fish in near-shore waters adjacent to nesting colonies and around island archipelagos (Higgins and Davies 1996). The subspecies nests in small colonies on coral shingle on continental islands or coral cays, on sandy islands and beaches inside estuaries, and on open sandy beaches, laying 1–2 eggs in a

Current eligibility against IUCN Red List Criteria

IUCN category	Criteria eligibility
A	Not applicable: past, current or future population declines are thought unlikely to exceed 20% in any 3-generation period
B	Not applicable: EOO >20 000 km^2; AOO >2000 km^2
C	Vulnerable: population 2500–10 000 mature individuals, estimated continuing decline between 10% in 3 generations and 20% in 2 generations
D	Not applicable: population >1000 mature individuals, >5 locations
E	Not applicable: no population viability analysis undertaken

IUCN Red List assessment data

	Estimate	Reliability
Extent of occurrence trend	13 200 km^2 stable	medium high
Area of occupancy trend	3300 km^2 decreasing	low medium
No. of mature individuals trend	9000 decreasing	medium medium
No. subpopulations	2	high
No. locations	>10	high
Largest subpopulation	8000	medium
Generation time	11.0 years	medium
Global population share	100%	high

sand scrape (Higgins and Davies 1996). There is a high natural rate of breeding failure as breeding sites can become unsuitable from overgrown vegetation, flood or be smothered by wind-blown sand. Extreme weather events, such as high tides, storm surges and extreme events can make nesting and efficient foraging very difficult, preventing successful fledging of chicks. Colonies tend to occupy areas rather than fixed sites, and nest sites are often abandoned after 1 year, even if they have been successful. A generation time of 11.0 years (BirdLife International 2011) is derived from an age at first breeding of 3.2 years, an annual survival of adults of 82.4% and a maximum longevity in the wild of 23.1 years, all extrapolated from congeners.

Threats

Failure of nests, which is naturally frequent, has increased as a result of human disturbance or predation by dogs, Black Rats *Rattus rattus*, Silver Gulls *Larus novaehollandiae* or ravens *Corvus* spp. Foxes *Vulpes vulpes* are a significant predator on the mainland, especially in Vic and NSW. In the Coorong inappropriate water management has lead to a collapse in the numbers of prey fish, and a subsequent abandonment of traditional breeding sites by Fairy Terns. Unfortunately there are no secure islands free from predators and close to areas that still support reasonable abundances of small fish (Paton and Rogers 2009). Interbreeding between this subspecies and Little Tern *S. albifrons sinensis* may constitute a threat to the genetic integrity of the 2 taxa in the future (Ross *et al.* 1999) with mixed species pairs and hybrids having been recorded at multiple sites (Cox and Close 1977; Norman *et al.* 1996).

Conservation objectives

1. Western Australian populations maintained
2. Populations in eastern states stable or increasing

Information required

1. Demographic trends including population size and breeding success at key locations
2. Top priority breeding populations in eastern Australia for conservation management

Management actions required

1. Monitor demographic trends at important sites
2. Monitor sites for hybrids with Little Tern
3. Control introduced mammals and other nest predators at important breeding sites
4. Restrict human and vehicular access to important breeding colonies
5. Manage water quality in the lower Coorong to benefit remaining terns
6. Liaise with New Zealand and New Caledonia researchers and managers
7. Engage the public in research, monitoring, management and advocacy

Bibliography

BirdLife International (2011) 'Species factsheet: *Sterna nereis*'. Retrieved 17 January, 2011 from <http://www.birdlife.org/>.

Cox JB, Close DH (1977) Interbreeding of Little and Fairy Terns. *Emu* **77**, 28–32.

Higgins PJ, Davies SJJF (Eds) (1996) *Handbook of Australian, New Zealand and Antarctic Birds. Volume 3: Snipe to Pigeons.* Oxford University Press, Melbourne.

Norman FI, Dann P, Menkhorst PW (1996) The status of seabirds in Victoria. In *The Status of Australia's Seabirds: Proceedings of the National Seabird Workshop*, Canberra, 1–2 November 1993. (Eds GJB Ross, K Weaver and JC Greig) pp. 185–200. Biodiversity Group, Environment Australia, Canberra.

NSW National Parks and Wildlife Service (2003) 'Little Tern (*Sterna albifrons*) recovery plan'. NSW NPWS, Hurstville.

Paton DC, Rogers DJ (2009) 'Ecology of breeding Fairy Terns *Sternula nereis* in the Coorong'. Final report for the Wildlife Conservation Fund.

Ross GE, Egan K, Priddel D (1999) Hybridization between Little tern *Sterna albifrons* and Fairy Tern *S. nereis* in Botany Bay, New South Wales. *Corella* **23**, 33–36.

Comments received from

Barry Baker, Stuart Butchart, Andrew Burbidge, Les Christidis, Kerryn Herman, Tim Holmes, Hugh Ford, Geoff Lacey, Richard Loyn, Peter Menkhorst, David Paton, Don Saunders, Eric Woehler

Species Conservation Summary

White-fronted Tern

Sterna striata Gmelin, 1789

Laridae

Conservation status

Australian breeding population: Near Threatened D1

Population visiting Australia: Least Concern

Reasons for listing

Population <1000 mature individuals; at few locations but currently no plausible future threats to nest sites. Downgraded 2 levels because of the high probability of genetic exchange with large Least Concern global population

Status 2000

Australian breeding population: Near Threatened D1

Population visiting Australia: Least Concern

Status 1990

Australian breeding population: Near Threatened D1

Population visiting Australia: Least Concern

Taxonomy

No infraspecific taxa described

Taxonomic uniqueness: medium (14 genera/family, 15 species/genus, 1 subspecies/species)

Range

Breeds in low numbers on Flinders and Cape Barren Is. off north-east Tasmania. There have been no surveys since 1986 when 13 locations were known: Mile I., Oyster Rocks, Billy Goat Reef, Briggs Islet, Penguin I., Little Dog I., Fisher I. Reef, Low Islets, Seal Point, Battery I., Key I., Neds Reef, Mid Woody I. (Brothers 1987). Also visits in large numbers from the primary breeding sites on and around the major islands of New Zealand, occurring from south Queensland to Tasmania and west to South Australia (Higgins and Davies 1996).

Abundance

Breeding population was estimated at 30 pairs in 1982, 53 in 1986 and 45 on 6 islands in 1996 (Brothers *et al.* 1996; Higgins and Davies 1996) and summarised as 53–63 pairs (Bryant 2002). There appear to have been no more recent surveys but it is possible that the numbers breeding in Australia are increasing: while breeding sites may be ephemeral and not used every season, terns have bred regularly on Fisher I. since the 1980s, but were never recorded breeding there earlier despite regular visits by researchers in the previous 40 years (Bryant 2002).

Ecology

Nest in colonies or solitarily on rocky or sandy beaches, laying 1–2 eggs in shallow depressions on rocky stacks or islets, often among succulent vegetation (Whinray 1980, 1982; Higgins and Davies 1996). They feed along the shore and in bays, and over oceanic waters in winter. They often feed in flocks, almost exclusively catching fish but will also take shrimp, feeding in the surf zone or several kilometres out to sea. A generation time of 10.1 years (BirdLife International 2011) is derived from an average age at first breeding of 2.0 years (Schreiber and Burger 2002), an annual survival of adults of 82.4%, and a maximum longevity in the wild of 23.1 years, both extrapolated from congeners.

Threats

No threats are likely in the foreseeable future, although introduced predators could affect some colonies. Feral cats *Felis catus* on islands and the increasing prevalence of coastal weed Sea Spurge *Euphorbia paralias* may reduce habitat suitability for nesting birds at some sites.

Conservation objectives

1. Persistence of Australian breeding population

Information required

1. Annual locations and sizes of breeding colonies
2. Potential threats to nesting birds at colonies

Management actions required

1. Map, count and monitor breeding colonies
2. Manage access to breeding sites as required

Bibliography

BirdLife International (2011) 'Species factsheet: *Sterna striata*'. Retrieved 17 January, 2011 from <http://www.birdlife.org/>.

Brothers N, Pemberton D, Gales R, Skira I (1996) The status of seabirds in Tasmania. In *The Status of Australia's Seabirds: Proceedings of the National Seabird Workshop*, Canberra, 1–2 November 1993. (Eds GJB Ross, K Weaver and JC Greig) pp. 181–184. Biodiversity Group, Environment Australia, Canberra.

Brothers NP (1987) White-fronted tern *Sterna striata*: Australian breeding distribution and population census. *Tasmanian Bird Report* **16**, 26–30.

Bryant S (2002) 'Conservation assessment of beach nesting and migratory shorebirds in Tasmania. Report to Birds Australia

Current eligibility against IUCN Red List Criteria

IUCN category	Criteria eligibility
Australian breeding population	
A	Not applicable: past, current or future population declines are thought unlikely to exceed 20% in any 3-generation period
B	Not applicable: EOO, AOO, habitat, locations or population not declining, population fluctuations not extreme
C	Not applicable: no continuing population decline or extreme fluctuations
D	Endangered: population 50–250 mature individuals (but downgraded to Near Threatened as global status Least Concern)
E	Not applicable: no population viability analysis undertaken

IUCN Red List assessment data

	Estimate	Reliability
Australian breeding population		
Extent of occurrence trend	1 300 000 km^2 stable	high high
Area of occupancy trend	13 km^2 stable	low medium
No. of mature individuals trend	120 stable	medium medium
No. subpopulations	1	high
No. locations	13	high
Generation time	10.1 years	medium
Global population share	<1%	medium
Level of genetic exchange	medium	low

and Natural Heritage Trust'. Department of Primary Industries, Water and Environment, Hobart.

Higgins PJ, Davies SJJF (Eds) (1996) *Handbook of Australian, New Zealand and Antarctic Birds. Volume 3: Snipe to Pigeons.* Oxford University Press, Melbourne.

Schreiber EA, Burger J (2002) *Biology of Marine Birds.* CRC Press, Boca Raton, FL.

Whinray JS (1980) The first Australian breeding record of the White-fronted Tern. *Australian Bird Watcher* **8**, 137–146.

Whinray JS (1982) More Australian breeding records of the White-fronted Tern. *Australian Bird Watcher* **9**, 160–164.

Comments received from

Eric Woehler

Subspecies Conservation Summary

Antarctic Tern (New Zealand)

Sterna vittata bethunei Buller, 1896

Laridae

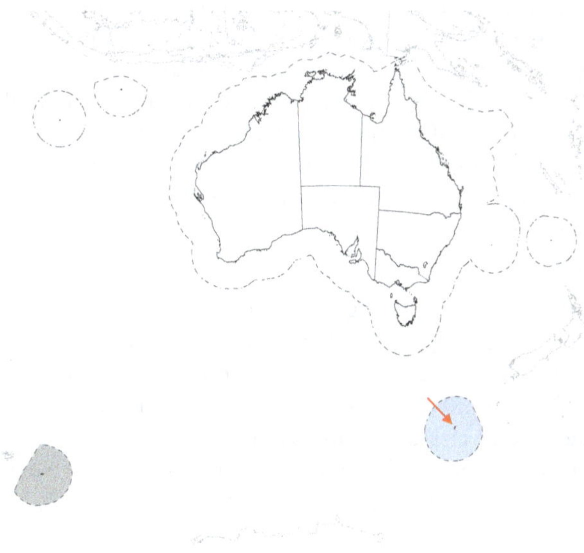

Conservation status

Australian breeding population: Endangered D

Population visiting Australia: Least Concern

Reasons for listing

Population <250 mature individuals

Status 2000

Australian breeding population: Endangered D

Population visiting Australia: Least Concern

Status 1990

Australian breeding population: Endangered D

Population visiting Australia: Least Concern

Taxonomy

S. v. vittata (Heard I.) is Endangered in Australia, 3 extralimital subspecies occur in the South Atlantic. The species is Least Concern.

Taxonomic uniqueness: medium (14 genera/family, 15 species/genus, 5 subspecies/species)

Range

In Australian territory, breeds on offshore stacks off Macquarie I. Extralimitally, breeds on Stewart I. and subantarctic islands of New Zealand. Some individuals from the Macquarie I. population remain in the area throughout the year, while others may disperse as far north as Australia and New Zealand. Otherwise, little is known about its movements outside the breeding season (Higgins and Davies 1996).

Current eligibility against IUCN Red List Criteria

IUCN category	Criteria eligibility
Australian breeding population	
A	Not applicable: past, current or future population declines are thought unlikely to exceed 20% in any 3-generation period
B	Not applicable: EOO, AOO, habitat, locations or population not declining, population fluctuations not extreme
C	Not applicable: no continuing population decline or extreme fluctuations
D	Endangered: population 50–250 mature individuals
E	Not applicable: no population viability analysis undertaken

IUCN Red List assessment data

	Estimate	Reliability
Australian breeding population		
Extent of occurrence trend	610 000 km^2 stable	high high
Area of occupancy trend	3 km^2 stable	high high
No. of mature individuals trend	100 stable	medium high
No. subpopulations	1	high
No. locations	1	high
Generation time	11.0 years	medium
Global population share	<1%	high
Level of genetic exchange	low	low

Abundance

There are no estimates of population size before exotic predators were introduced, but the population was apparently small in the early 20th century (Hamilton 1894). The most recent population estimate is from 1979, when 40 pairs were recorded on Macquarie I. (Rounsevell and Brothers 1984). Up to 1000 birds breed on New Zealand's subantarctic islands (Higgins and Davies 1996).

Ecology

Nests usually in live vegetation, though occasionally in unvegetated crevices. During summer, mostly feeds on small fish, but also crustaceans and other marine invertebrates in kelp beds near the island. During winter, they feed at the edge of ice and in patches of unfrozen inshore water (Higgins and Davies 1996). A generation length of 11.0 years (BirdLife International 2011) is derived from an age of first breeding of 3.2 years, a maximum age in the wild of 23.1 years and an adult survival of 82.4%, all extrapolated from congeners.

Threats

The absence of nesting terns from the main island of Macquarie I. suggests that predation by Black Rats *Rattus rattus*, and formerly feral cats *Felis catus* and Wekas *Gallirallus australis*, has historically reduced the population size. The species is also vulnerable to human disturbance, but is probably stable while the offshore stacks are free of rats. Erosion and landslips caused by over-grazing by rabbits have the potential to damage or completely destroy nesting habitat on Macquarie I., but the severity of this threat is unknown.

Conservation objectives

1. Persistence of population

Information required

1. Population trends

Management actions required

1. Eradicate rats from Macquarie I.
2. Maintain quarantine for offshore stacks

Bibliography

BirdLife International (2011) 'Species factsheet: *Sterna vittata*'. Retrieved 17 January, 2011 from <http://www.birdlife.org/>.

Hamilton A (1894) Notes on a visit to Macquarie Island. *Transactions and Proceedings of the New Zealand Institute* **27**, 559–579.

Higgins PJ, Davies SJJF (Eds) (1996) *Handbook of Australian, New Zealand and Antarctic Birds. Volume 3. Snipe to Pigeons.* Oxford University Press, Melbourne.

Rounsevell DE, Brothers NP (1984) The status of seabirds on Macquarie Island. In 'Status and Conservation of the World's Seabirds. ICBP Tech. Publ. 2'. (Eds JP Croxall, PGH Evans and RW Schreiber) pp. 587–592. ICBP, Cambridge, UK.

Subspecies Conservation Summary

Antarctic Tern (Indian Ocean)

Sterna vittata vittata Gmelin, 1789
Laridae

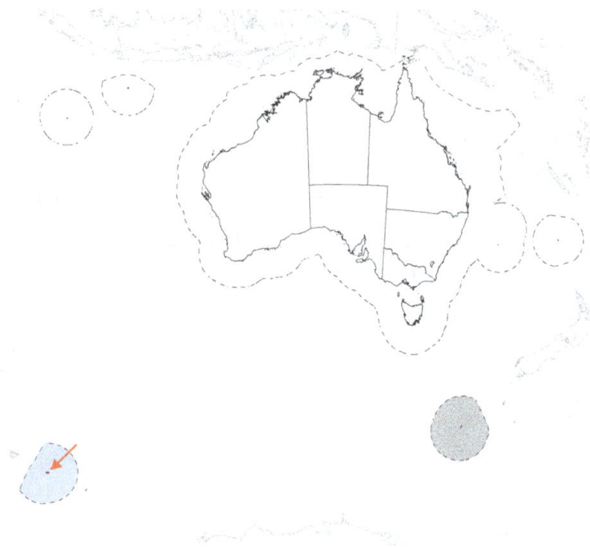

Conservation status

Australian breeding population: Endangered D
Population visiting Australia: Least Concern

Reasons for listing

Population <250 mature individuals

Status 2000

Australian breeding population: Endangered D
Population visiting Australia: Least Concern

Status 1990

Australian breeding population: Endangered D
Population visiting Australia: Least Concern

Taxonomy

S. v. bethunei (Macquarie I.) is Vulnerable in Australia, 3 extralimital subspecies occur in the South Atlantic. The species is Least Concern.
Taxonomic uniqueness: medium (14 genera/family, 15 species/genus, 5 subspecies/species)

Range

In Australian territory, breeds only on Heard I., occurring extralimitally on Crozet and Kerguelen I. (Higgins and Davies 1996). During the non-breeding season at least some visit Bird I., South Africa (Woehler *et al.* 2001). Birds wintering in South Africa were colour-banded and subsequently resighted on Heard I. during research expeditions (Woehler 2006). It is likely that birds from other subantarctic islands in the Indian Ocean (e.g. Kerguelen, Marion, Crozet) may also over-winter in southern Africa.

Abundance

The population is currently estimated at 100–200 pairs (Woehler 2006, 2010), based on recent surveys, the availability and distribution of suitable habitat and observations from around Heard I. However, actual breeding records are 48 in 1983/4, 13 in 1987/8, 33 in 2000/01 and 15 in 2003/4 (Woehler 2006).

Ecology

They usually nest on unvegetated scree slopes and river banks. During summer, they mostly feed on small fish, but also crustaceans and other marine invertebrates in kelp beds near the shore. A generation length of 11.0 years (BirdLife International 2011) is derived from an age of first breeding of 3.2 years, a maximum age of 23.1 years and an adult survival of 82.4%, all extrapolated from congeners.

Threats

Currently, there are no threats, but the population could be severely affected if Black Rats *Rattus rattus* or cats *Felis catus* were introduced.

Conservation objectives

1. Persistence of population

Information required

1. Population trends

Current eligibility against IUCN Red List Criteria

IUCN category	Criteria eligibility
Australian breeding population	
A	Not applicable: past, current or future population declines are thought unlikely to exceed 20% in any 3-generation period
B	Not applicable: EOO, AOO, habitat, locations or population not declining, population fluctuations not extreme
C	Not applicable: no continuing population decline or extreme fluctuations
D	Endangered: population 50–250 mature individuals
E	Not applicable: no population viability analysis undertaken

IUCN Red List assessment data

	Estimate	Reliability
Australian breeding population		
Extent of occurrence trend	520 000 km^2 stable	high high
Area of occupancy trend	2 km^2 stable	high high
No. of mature individuals trend	150 stable	medium high
No. subpopulations	1	high
No. locations	1	high
Generation time	11.0 years	medium
Global population share	26–50%	high
Level of genetic exchange	low	low

Management actions required

1. Enhance quarantine provisions, remove alien vertebrate predators

Bibliography

BirdLife International (2011) 'Species factsheet: *Sterna vittata*'. Retrieved 17 January, 2011 from <http://www.birdlife.org/>.

Higgins PJ, Davies SJJF (Eds) (1996) *Handbook of Australian, New Zealand and Antarctic Birds. Volume 3: Snipe to Pigeons.* Oxford University Press, Melbourne.

Woehler EJ (2006) Status and conservation of the seabirds of Heard Island and the McDonald Islands. In *Heard Island: Southern Ocean Sentinel*. (Eds K Green and EJ Woehler) pp. 128–165. Surrey Beatty & Sons, Chipping Norton, NSW.

Woehler EJ (2010) Status and conservation of the seabirds of Heard Island. *The State of Australia's Birds 2010: Islands and Birds. Wingspan* Supplement **20**, 8–9.

Comments received from

Eric Woehler

Species Conservation Summary

Norfolk Island Kaka

Nestor productus (Gould, 1836)
Psittacidae

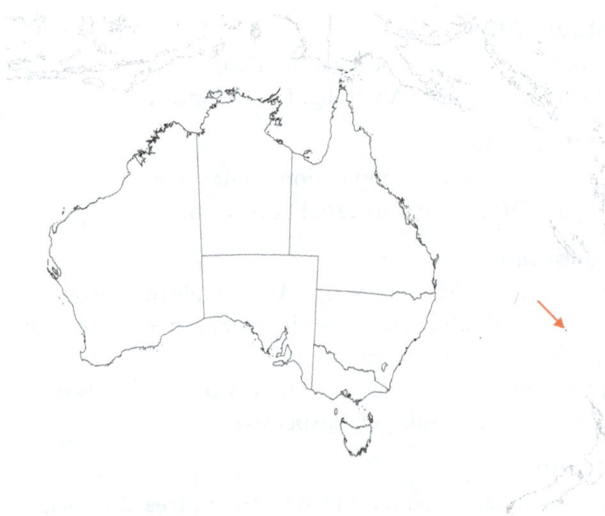

Conservation status

Extinct

Reasons for listing

The last individual of this species died in 1851

Status 2000

Extinct

Status 1990

Extinct

Taxonomy

No infraspecific taxa described

Taxonomic uniqueness: high (1 genus/family, 3 species/genus, 1 subspecies/species)

Former range

Endemic to Norfolk I. and adjacent Phillip I. (Forshaw 1981).

Former abundance

No historical information appears to be available regarding its numbers. The last known living bird is reported to have died in captivity in London in 1851 (Gould 1865).

Ecology

The Kaka appears to have had a broad niche, foraging among rocks, as well as up in the tallest trees (BirdLife International 2011). They laid 4 white eggs in a tree hollow. The birds were very tame, and fed on the blossoms of the White-wood tree or White Hibiscus (Gould 1865), presumably *Lagunaria patersonia*, although the size and strength of their bill suggests they must have eaten other foods as well (M. Christian pers. comm.).

Reasons for extinction

The most likely reason for extinction is clearance of habitat for agriculture, particularly of the large trees on which it probably relied for nesting hollows. The species had earlier survived being hunted for food by Polynesians who occupied the island until the 17th century (Holdaway and Anderson 2001). The fact that the species was still seen in the 1850s meant that they survived being hunted for food by convicts and settlers in the first wave of settlement as well, thought this would have depleted the population, besides being trapped to be kept as pets (BirdLife International 2011). The species may have survived on Phillip I. after its extinction on Norfolk I. (Forshaw 1981), before that habitat was also destroyed there by the introduction of rabbits *Oryctolagus cuniculus*, feral pigs *Sus scrofa* and feral goats *Hircus capra* (Coyne 2010).

Bibliography

BirdLife International (2011) 'Species factsheet: *Nestor productus*'. Retrieved 17 January, 2011 from <http://www.birdlife.org/>.

Coyne P (2010) Ecological rebound on Phillip Island, South Pacific. *Ecological Management and Restoration* **11**, 4–15.

Forshaw JM (1981) *Australian Parrots*. Lansdowne Press, Melbourne.

Gould J (1865) *Handbook to the Birds of Australia*. Vol. 2. Gould, London.

Holdaway RN, Anderson A (2001) Avifauna from the Emily Bay Settlement Site, Norfolk Island: a preliminary account. In *The Prehistoric Archaeology of Norfolk Island, Southwest Pacific*. (Eds A Anderson and P White) pp. 85–100. *Records of the Australian Museum*, Supplement **27**. Australian Museum, Sydney.

Comments received from

Margaret Christian, Joe Forshaw

Subspecies Conservation Summary

Palm Cockatoo (Australian)

Probosciger aterrimus macgillivrayi (Mathews, 1912)
Cacatuidae

Conservation status

Vulnerable C2a(ii), E

Reasons for listing

Population <10 000 mature individuals in a single subpopulation, undergoing a continuing decline, as suggested by modelling

Status 2000

Vulnerable C2a(ii)

Status 1990

Vulnerable C2a(ii)

Taxonomy

Three extralimital subspecies; the species is Least Concern

Taxonomic uniqueness: high (7 genera/family, 1 species/genus, 4 subspecies/species)

Range

Northern Cape York Peninsula, Queensland, from north of Pormpuraaw on the west coast to Saltwater Creek, Princess Charlotte Bay on the east (Storch 1996; Higgins 1999). No evidence of any change in range. If anything, the range has extended south by about 100 km in the last 20 years: it was not known from Saltwater Creek until the late 1990s. The AOO was estimated by putting a 2.5 km buffer around all closed and riparian forest within the EOO (S. Murphy *in litt.*).

Abundance

A tentative population estimate of 3000 mature individuals (Garnett and Crowley 2000) has not been confirmed but nor has it been superseded. Monitoring at Iron Range suggests a continuing decline (S. Murphy *in litt.*) and population viability analysis based on data from Iron Range suggests that population stability over 100 years would need extremely low mortality rates of juveniles (<8.8%) and adults (3.3–4.9%; Heinsohn *et al.* 2009). Otherwise, the modelling suggests that Iron Range would currently act as a sink from other populations (R. Heinsohn *in litt.*). Long-term observations at Weipa also suggest local declines (S. Gould *in litt.*). While more information from across the population's range would be desirable, a continuing decline, which is very hard to detect in such a long-lived bird, is assumed here on a precautionary basis.

Ecology

Occurs in closed forest and riparian systems but attains highest densities in the open woodlands adjacent to the closed forest dominated by *Corymbia* spp. and *Eucalytpus* spp. and has significantly lower densities in the closed forests themselves (Murphy 2006). Pairs are thought to occupy permanent breeding territories all year that are visited most mornings and afternoons, especially by males who defend nest trees aggressively (Murphy *et al.* 2003). Individuals and small flocks feed in closed forests and littoral systems during the middle of the day, mostly on the hard seeds of fibrous and woody fruits of woodland, littoral and closed forest species taken from the canopy and the ground (Wood 1988; Storch 1996). Nests are in large hollow trees, primarily large eucalypts in woodland, with an average distance of 320 m to rainforest (Murphy *et al.* 2003). Pairs also claim numerous other hollow trees as display sites (Wood 1988; Murphy *et al.* 2003). Low breeding success resulting from laying a single egg, high nesting failure and breeding on average every 2.2 years (Murphy *et al.* 2003; Heinsohn *et al.* 2009). A generation time of 17.7 years (BirdLife International 2011) is derived from an age at first breeding of 4.0 years (extrapolated from mean values for other genera in Psittacidae) and a longevity of 31.4 years, extrapolated from maximum longevity recorded in captivity.

Current eligibility against IUCN Red List Criteria

IUCN category	Criteria eligibility
A	Not applicable: past, current or future population declines are thought unlikely to exceed 20% in any 3-generation period
B	Not applicable: EOO >20 000 km², AOO >2000 km²
C	Vulnerable: population 2500–10 000 mature individuals, inferred continuing decline, 100% in 1 subpopulation
D	Not applicable: population >1000 mature individuals, >5 locations
E	Vulnerable: population viability analysis indicates >10% probability of extinction within 100 years

IUCN Red List assessment data

	Estimate	Reliability
Extent of occurrence trend	57 000 km² stable[e]	high high
Area of occupancy trend	27 000 km² stable	high high
No. of mature individuals trend	3000 decreasing	low medium
No. subpopulations	1	high
No. locations	>10	high
Generation time	17.7 years	medium
Global population share	100%	high

Threats

Some woodland, in which most nest-trees occur, is being invaded by rainforest (Russell-Smith *et al.* 2004) that has lower densities of nest and display hollows. Fire is used to maintain the stability of the rainforest/woodland ecotone and is needed to maintain recruitment of large trees, but can also destroy nest-trees, which are probably a limiting resource at Iron Range (Murphy *et al.* 2003; Murphy and Legge 2007). Cyclones both destroy old hollows and make available hollows by breaking newly hollowed trees (Murphy and Legge 2007; Heinsohn *et al.* 2009). Land clearing near Weipa for bauxite mining has also reduced hollow density, and some small patches of rainforest feeding habitat (S. Gould *in litt.*). Sulphur-crested Cockatoos *Cacatua galerita* compete for nest-trees (Heinsohn *et al.* 2003) and have increased over the last 2 decades around Weipa (S. Gould *in litt.*). While loss of nest contents to predators is high, this is thought to be a natural process (Murphy *et al.* 2003).

Conservation objectives

1. Stable population and demographic parameters
2. Persistence at multiple monitored sites

Information required

1. Techniques to understand age structure of populations
2. Longevity, survival and other demographic parameters
3. Demography of suitable nest-trees away from Iron Range
4. Optimal fire regime to ensure ongoing nest hollow availability

Management actions required

1. Monitor residence at key sites at Iron Range and along westward flowing rivers
2. Active fire management around rainforest blocks

Bibliography

BirdLife International (2011) 'Species factsheet: *Probosciger aterrimus*'. Retrieved 17 January, 2011 from <http://www.birdlife.org/>.

Garnett S, Crowley GM (2000) *The Action Plan for Australian Birds 2000.* Environment Australia, Canberra.

Heinsohn R, Murphy S, Legge S (2003) Overlap and competition for nest holes among eclectus parrots, palm cockatoos and sulphur-crested cockatoos. *Australian Journal of Zoology* **51**, 81–94.

Heinsohn R, Zeriga T, Murphy SA, Igag P, Legge S, Mack AL (2009) Do Palm Cockatoos (*Probosciger aterrimus*) have long enough lifespans to support their low reproductive success? *Emu* **109**, 183–191.

Higgins PJ (Ed.) (1999) *Handbook of Australian, New Zealand and Antarctic Birds. Volume 4: Parrots to Dollarbird.* Oxford University Press, Melbourne.

Murphy S, Legge S, Heinsohn R (2003) The breeding biology of palm cockatoos (*Probosciger aterrimus*): a case of a slow life history. *Journal of The Zoological Society of London* **261**, 327–339.

Murphy SA (2006) The ecology and conservation biology of palm cockatoos *Probosciger aterrimus*. PhD thesis, Australian National University, Canberra.

Murphy SA, Legge SM (2007) The gradual loss and episodic creation of Palm Cockatoo (*Probosciger aterrimus*) nest-trees in a fire- and cyclone-prone habitat. *Emu* **107**, 1–6.

Russell-Smith J, Stanton PJ, Whitehead PJ, Edwards A (2004) Rain forest invasion of eucalypt-dominated woodland savanna, Iron Range, north-eastern Australia: I. Successional processes. *Journal of Biogeography* **31**, 1293–1303.

Storch D (1996) 'The Palm Cockatoo: a survey of the nesting requirements and habitat use'. Report to Australian Nature Conservation Agency, Canberra.

Wood GA (1988) Further field observations of the Palm Cockatoo *Probosciger aterrimus* in the Cape York Peninsula, Queensland. *Corella* **12**, 48–52.

Comments received from

Sue Gould, Rob Heinsohn, Sarah Legge, Keith McDonald, Steve Murphy, Justin Perry

Subspecies Conservation Summary

Red-tailed Black-Cockatoo (south-eastern)

Calyptorhynchus banksii graptogyne Schodde, Saunders & Homberger, 1989

Cacatuidae

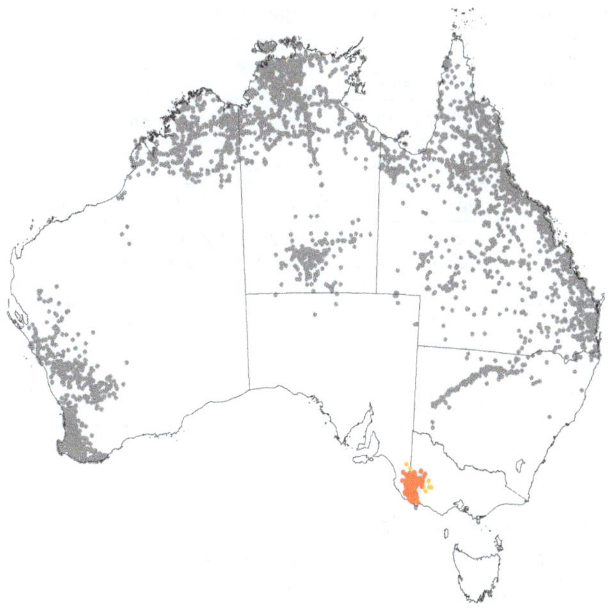

Conservation status
Endangered C2a(ii)

Reasons for listing
The single subpopulation contains <2500 mature individuals and a continuing decline probable given ongoing loss of habitat and low recruitment

Status 2000
Endangered C2a(ii)

Status 1990
Endangered C2a(ii)

Taxonomy
C. b. naso (south-west Western Australia) is Vulnerable, 3 other subspecies from northern and central Australia are Least Concern, as is the species

Taxonomic uniqueness: medium (7 genera/family, 5 species/genus, 5 subspecies/species)

Range
Occurs in a small area of south-eastern Australia delimited by Keith, Lucindale and Mt Gambier in South Australia and Portland, Casterton, Toolondo, Natimuk, Dimboola, Nhill and Kaniva in Victoria (Commonwealth of Australia 2007).

Abundance
The highest count of during annual autumn surveys is about 1460 birds, most other counts being around 600 (M. Marron *in litt.*). No survey can count all individuals of this forest bird, the true population size is thought to be around 1200 mature individuals. The proportion of mature males in flocks increased by 9% to 45% from 1999–2004, which may be indicative of reduced breeding success over this period (Commonwealth of Australia 2007) and appears to be linked to food availability in the preceding years (M. Marron *in litt.*).

Ecology
Occurs primarily in woodlands dominated by its 2 main food species, Desert Stringybark *Eucalyptus arenacea* and Brown Stringybark *E. baxteri*, as well as in adjacent woodlands of River Red Gum *E. camaldulensis*, Yellow Gum *E. leucoxylon* and Buloke *Allocasuarina luehmannii*, the third major food tree species. The birds feed both in blocks of forest and in scattered paddock trees. Desert and Brown Stringybarks have 3 and 2–4 year seeding cycles respectively, with the cockatoos feeding almost entirely on the species that has fruited most recently. They deplete the seed supply so effectively that recruitment is reduced in years when little seed is available (Commonwealth of Australia 2007). Buloke seed, which is only available from January to March in some years, is mostly taken from trees over 200 years old (Maron and Lill 2004). The cockatoos also prefer old trees with large hollows for breeding, fledging a maximum of 1 young per year (Joseph *et al.* 1991). A generation time of 19.2 years is estimated (BirdLife International 2011), based on an age at first breeding of 4.0 years and maximum longevity of 34.4 years, both extrapolated from Carnaby's Black-Cockatoo *C. latirostris* (Australian Bird and Bat Banding Scheme).

Threats
The principal constraint on population size appears to be food availability. Strongly cyclic variation in seed availability has been exacerbated by the loss and degradation of feeding habitat: 87% of the original stringybark habitat in SA has been cleared with 42% lost in Vic. While clearing of large forest blocks has largely ceased, the loss of isolated trees in paddocks continues, particularly of the very old Buloke trees favoured by the cockatoos (Maron 2005). Tree removal to make way for irrigation pivots continues to reduce the number of feed trees. Wild fire is the principal threat in forest blocks, fires leading to crown scorch reducing stringybark seed productivity for an average of 11 years (Commonwealth of Australia 2007). Fire management regimes that minimize crown scorch have been successfully implemented over the last decade so that controlled fire will only become a threat if fire management guidelines are relaxed. Eventually loss of nest hollows, which are largely in dead trees, may limit breeding but currently use of artificial nest-boxes is low, suggesting this is not a major constraint on population size.

Conservation objectives
1. Stable population
2. Effective stringybark fire management and retention, retention and recruitment of Buloke, and protection of nest trees

Current eligibility against IUCN Red List Criteria

IUCN category	Criteria eligibility
A	Not applicable: past, current or future population declines are thought unlikely to exceed 20% in any 3-generation period
B	Not applicable: EOO >20 000 km^2, AOO >2000 km^2
C	Endangered: population 250–2500 mature individuals, suspected continuing decline, >95% mature individuals in 1 subpopulation
D	Near Threatened: population near 1000 mature individuals
E	Not applicable: no population viability analysis undertaken

IUCN Red List assessment data

	Estimate	Reliability
Extent of occurrence trend	23 000 km^2 stable	high medium
Area of occupancy trend	1300 km^2 decreasing	medium high
No. of mature individuals trend	1100 decreasing	medium medium
No. subpopulations	1	medium
No. locations	>10	high
Generation time	19 years	medium
Global population share	100%	high

Information required

1. Trends in population size and demography
2. Relative value of paddock trees
3. Annual patterns of burning and fire intensity on private land

Management actions required

1. Continue annual counts
2. Incorporate habitat protection and management recommendations into local land planning.
3. Protect all Buloke and stringybark, particularly Desert Stringybark, from clearing
4. Continue to manage fire in stringybark to minimise canopy scorch
5. Continue to offer support for private landholders protecting and managing cockatoo habitat
6. If appropriate, maintain and expand availability of artificial nest sites
7. Maintain support for the Red-tailed Black-Cockatoo Recovery Team

Bibliography

BirdLife International (2011) 'Species factsheet: *Calyptorhynchus banksii*'. Retrieved 17 January, 2011 from <http://www.birdlife.org/>.

Commonwealth of Australia (2007) 'National recovery plan for the South-eastern Red-tailed Black-Cockatoo *Calyptorhynchus banksii graptogyne*'. Department of the Environment and Water Resources, Canberra.

Joseph L, Emison WB, Bren WM (1991) Critical assessment of the conservation status of the Red-tailed Black-Cockatoo in south-eastern Australia with special reference to nesting requirements. *Emu* **91**, 46–50.

Maron M (2005) Agricultural change and paddock tree loss: implications for an endangered subspecies of Redtailed Black-Cockatoo. *Ecological Management and Restoration* **6**, 206–211.

Maron M, Lill A (2004) Discrimination among potential buloke (*Allocasuarina leuhmannii*) feeding trees by the endangered south-eastern red-tailed black-cockatoo (*Calyptorhynchus banksii graptogyne*). *Wildlife Research* **31**, 311–317.

Comments received from

David Baker-Gabb, Richard Hill, Martine Maron, Penny Olsen

Subspecies Conservation Summary

Red-tailed Black-Cockatoo (south-western)

Calyptorhynchus banksii naso Gould, 1837
Cacatuidae

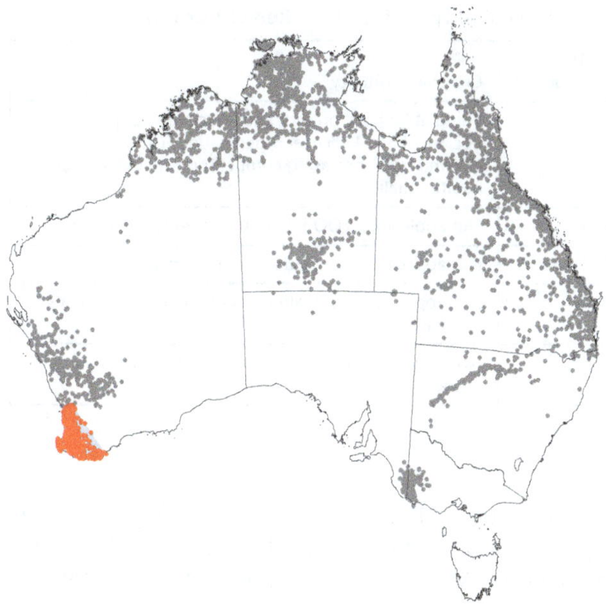

Conservation status
Vulnerable A2c+3c+4c

Reasons for listing
An inferred decline in the population of 30–49% in the last 3 generations (58 years) as a result of habitat loss which is probably continuing as a result of loss of nesting hollows from competitors (bees, other cockatoos) and fire

Status 2000
Vulnerable A2c+3c+4c

Status 1990
Vulnerable A2c+3c+4c

Taxonomy
C. b. graptogyne (south-east South Australia, south-west Victoria) is Endangered, 3 other subspecies from northern and central Australia are Least Concern, as is the species
Taxonomic uniqueness: medium (7 genera/family, 5 species/genus, 5 subspecies/species)

Range
Endemic to south-west Western Australia, between Albany, Augusta and Perth. Historically it occurred north along the Swan coastal plain as far north as Dandaragan, a historical range reduction of 25–36% of its range (Schodde and Mason 1997; Abbott 1998; Johnstone and Storr 1998). Because both EOO and AOO encompass occupied forest, the AOO is estimated to be at least a third of the EOO.

Abundance
Population is estimated at about 15 000 individuals (Abbott 1998), but only about 10% is thought to breed each year and the productivity is often low. In 2008 no breeding occurred in over 80 monitored nests and no juveniles were detected in flocks; 2006 was also a poor breeding year (R. Johnstone *in litt.*). It is most common in the northern Darling Range from about Collie north to Mundaring and is very local throughout the lower south-west and tiny remnant populations on the Swan Coastal Plain (Baldivis, Stakehill, Lake Mclarty and Capel) are not thought to be sustainable in the long term (R. Johnstone *in litt.*). A marked decline in numbers over the past 60 years has been postulated (Johnstone *et al.* 2010) though projected or suspected declines of ≥30% within the next 3 generations by Chapman (2007) lacked supporting data. The availability of nesting hollows, which may limit population size, has declined because of clearing, timber harvesting and competition from both introduced and native species. Given the lack of any recent quantitative data, the population is estimated here as 15 000 birds, declining at a rate inferred from changes in habitat and competition to be >30% in 3 generations (58 years).

Ecology
Lives in forests of Jarrah *Eucalyptus marginata*, Karri *E. diversicolor* and Marri *Corymbia calophylla*. It feeds primarily on seeds of Marri, which fruit heavily at intervals of 3 or more years, although it takes seed from other eucalypt spp. (Saunders *et al.* 1985; Johnstone and Kirkby 1999). However, since 1995, family groups and flocks have moved well out onto the Swan Coastal Plain in search of food, mainly the exotic tree White Cedar *Melia azedarach* (Johnstone *et al.* 2010). It nests in large tree hollows, particularly those in Marri, Jarrah and Karri (Johnstone and Storr 1998). It lays 2 eggs, but never raises more than 1 fledgling. Nesting may take place only in those years when Marri is fruiting heavily (Johnstone and Kirkby 1999). A generation time of 19.2 years (BirdLife International 2011) is derived from an age at first breeding of 4.0 years and maximum longevity of 34.4 years, both extrapolated from Carnaby's Black-Cockatoo *C. latirostris* (Australian Bird and Bat Banding Scheme).

Threats
Nest hollow shortage is considered the principal threat as suitable hollows are considered scarce, only forming in trees at least 130 to 220 years of age, many of which have been preferentially felled (Abbott and Whitford 2002; Chapman 2007). Nest hollows are likely to continue to be lost to mining (Chapman 2007) and fire. Competition for hollows with other cockatoos, including Carnaby's Black-Cockatoo and corellas *Cacatua* spp., Wood Ducks *Chenonetta jubatta* and feral European Honey Bees *Apis mellifera* (Johnstone and Cassarchis 2004; Chapman 2007). Fires caused by lightning or arson, and sometimes planned fires, cause older trees containing hollows to fall, and temporarily reduce food availability (Department of Sustainability, Environment, Water, Population and

Current eligibility against IUCN Red List Criteria

IUCN category	Criteria eligibility
A	Vulnerable: past, present and anticipated population decline of 30–49% over 3 generations (58 years) inferred from increasing competitors and deteriorating habitat quality
B	Not applicable: EOO >20 000 km^2, AOO >2000 km^2
C	Not applicable: population >10 000 mature individuals
D	Not applicable: population >1000 mature individuals, >5 locations
E	Not applicable: no population viability analysis undertaken

IUCN Red List assessment data

	Estimate	Reliability
Extent of occurrence trend	61 000 km^2 stable	high medium
Area of occupancy trend	20 000 km^2 declining	low low
No. of mature individuals trend	15 000 declining	medium low
No. subpopulations	1	high
No. locations	>10	high
Generation time	19 years	low
Global population share	100%	high

Communities 2011). The subspecies no longer occupies up to 36% of former habitat that has been cleared for agriculture (Abbott 1998; Johnstone and Kirkby 1999).

Conservation objectives

1. Extent of occurrence stable or increasing
2. Number of breeding pairs stable or increasing
3. Numbers in each roosting flock stable or increasing
4. Proportion of juveniles in each roosting flock stable or increasing

Information required

1. Factors affecting the number of breeding attempts and breeding success
2. Trends in numbers and recruitment
3. Relative impacts of threatening processes
4. Location of feeding and breeding habitat critical to survival

Management actions required

1. Monitor the progress of recovery, including the effectiveness of management actions and the need to adapt them if necessary
2. Determine and implement ways to minimise the effects of mining on habitat loss
3. Determine and implement ways to manage forests for the conservation of the subspecies
4. Develop and implement a management plan for the control and reduction of non-managed European Honey Bees in the region
5. Develop and implement a communication strategy to stop illegal shooting and improve compliance with the relevant legislation
6. Continue to support coordination of management by the Forest Cockatoo Recovery Group

Bibliography

Abbott I (1998) Conservation of the forest Red-tailed Black-Cockatoo, a hollow-dependent species, in the eucalypt forests of Western Australia. *Forest Ecology and Management* **109**, 175–185.

Abbott I, Whitford K (2002) Conservation of vertebrate fauna using hollows in forests of south-west Western Australia: strategic risk assessment in relation to ecology, policy, planning, and operations management. *Pacific Conservation Biology* 7, 240–255.

BirdLife International (2011) 'Species factsheet: *Calyptorhynchus banksii*'. Retrieved 17 January, 2011 from <http://www.birdlife.org/>.

Chapman TF (2007) 'Forest Black Cockatoo (Baudin's Cockatoo *Calyptorhynchus baudinii*) and Forest Red-tailed Black Cockatoo (*Calyptorhynchus banksii naso*) Recovery Plan 2007–2016'. Department of Conservation and Land Management, Perth.

Department of Sustainability, Environment, Water, Population and Communities (2011) *Calyptorhynchus banksii naso*. In 'Species Profile and Threats Database'. Department of Sustainability, Environment, Water, Population and Communities, Canberra. Retrieved 24 February, 2011 from <http://www.environment.gov.au/sprat>.

Johnstone RE, Cassarchis C (2004) 'Review of cockatoo research project and Cockatoo Care 2004'. Water Corporation and the Western Australian Museum, Perth.

Johnstone RE, Johnstone C, Kirkby T (2010) 'Carnaby's Cockatoo (*Calyptorhynchus latirostris*), Baudin's Cockatoo (*Calyptorhynchus baudinii*) and the Forest Red-tailed Black Cockatoo (*Calyptorhynchus banksii naso*) on the Swan Coastal Plain (Lancelin–Dunsborough), Western Australia. Studies on distribution, status, breeding, food, movements and historical changes'. Report to the Department of Planning, Perth.

Johnstone RE, Kirkby T (1999) Food of the Red-tailed Black-Cockatoo *Calyptorhynchus banksii naso* in south-west Western Australia. *Western Australian Naturalist* **22**, 167–178.

Johnstone RE, Storr GM (1998) *Handbook of Western Australian Birds. Volume 1: Non-passerines (Emu to Dollarbird)*. Western Australian Museum, Perth.

Saunders DA, Rowley I, Smith GT (1985) The effects of clearing for agriculture on the distribution of cockatoos in the southwest of Western Australia. In *Birds of Eucalypt Forests and Woodlands: Ecology, Conservation, Management*. (Eds A Keast, HF Recher, HA Ford and DA Saunders). RAOU and Surrey Beatty and Sons, Melbourne and Chipping Norton, NSW.

Schodde R, Mason IJ (Eds) (1997) *Zoological Catalogue of Australia 37.2: Aves (Columbidae to Coraciidae)*. CSIRO Publishing, Melbourne.

Comments received from

Allan Burbidge, Cheryl Gole, Ron Johnstone, Peter Mawson

Subspecies Conservation Summary

Glossy Black-Cockatoo (Kangaroo Island)

Calyptorhynchus lathami halmaturinus Mathews, 1912
Cacatuidae

Conservation status
Endangered D

Reasons for listing
Population <250 mature individuals

Status 2000
Endangered D

Status 1990
Critically Endangered C2a(ii)
Reason for change in 2000: effective conservation action resulting in sustained population increase

Taxonomy
C. l. lathami (south-eastern Australia) is Near Threatened and *C. l. erebus* (central Queensland, Dawson-Mackenzie basin to Paluma) is Least Concern, as is the species
Taxonomic uniqueness: medium (7 genera/family, 5 species/genus, 3 subspecies/species)

Range
Occurs only on Kangaroo I., South Australia. The birds aggregate in the non-breeding season in 6 main flocks, 5 along the north coast and 1 across the south with few birds on the Dudley Peninsula, in Flinders Chase National Park and in the north-east of the island, primarily in the north and west. In the 19th century it also occurred on the Fleurieu Peninsula and southern Mount Lofty Ranges, and possibly western Victoria (Mooney and Pedler 2005). There are about 50 km^2 of feeding habitat on Kangaroo I. (Crowley *et al.* 1998), most of which is used by the cockatoos, and 3300 km^2 of potential breeding habitat, though at least half this is not used as nests are within 12 km of feeding habitat, mostly within 1 km. The EOO and AOO were calculated by Harris *et al.* (2009).

Abundance
The population is currently estimated as 340–360 birds, based on the census of 2008. Counts in 2009 (minimum 236) and 2010 (minimum 262) omitted some major flocks and were much lower (Pedler and Sobey 2008, Pedler and Barth 2009, Barth and Mooney 2010). Given the male-biased sex ratio of 1:1.44 (average 2003–2008; Pedler and Sobey 2008) and the length of time before breeding, about 240 of these are thought likely to be mature individuals (L. Pedler *in litt.*): the total population needs to reach 400 before it can be considered Vulnerable (Burbidge and Raines 2003). After protection of nests successful juvenile recruitment has led to population increases from a minimum of 158 birds in 1995 (Department of Sustainability, Environment, Water, Population and Communities 2011). A population viability analysis estimated the carrying capacity of Kangaroo Island at 653 individuals (Harris *et al.* 2009).

Ecology
Usually occurs in pairs, or in groups of 3, which comprise a breeding pair and their offspring. Solitary males and small groups of juvenile males are also observed, as are loose flocks, which typically consist of 25 birds or less, but may comprise 50 or more birds in the post-breeding period (Department of Sustainability Environment Water Population and Communities 2011). They feed almost exclusively on the seeds of Drooping Sheoak *Allocasuarina verticillata*, particularly those growing on rocky hills and valleys that have kernels in more than half their seeds (Crowley and Garnett 2001). The cockatoos lay a single egg in hollows in tall trees, particularly Sugar Gum *Eucalyptus cladocalyx*. Trees that develop nest hollows suitable for this subspecies are thought to be at least 100 years old and possibly much older (Mooney and Pedler 2005). The cockatoos use both intact habitat and isolated Drooping Sheoak and tall gums within the semi-cleared agricultural land. A generation time of 14.1 years (BirdLife International 2011) is derived from an average age at first breeding of 3.0 years and an average adult survival of 91% (Harris *et al.* 2009).

Threats
The principal threat to the cockatoos is relaxation of current successful active management. The primary reason for decline on Kangaroo I. was the invasion of nesting hollows, and often predation of cockatoo eggs or young, by Common Brushtail Possums *Trichosurus vulpecula*, and interference at nests by Little Corellas *Cacatua sanguinea*, Galahs *Eolophus roseicapillus* and feral European Honey Bees *Apis mellifera*. Numbers of possums and open country cockatoos are thought to have increased as a result of partial clearing of Kangaroo I. (Garnett *et al.* 1999). Ongoing nest protection has doubled nest productivity. Nest sites are now being protected to prohibit entry by possums, Little Corella numbers are being controlled, and bees poisoned. Some nests and/or nest contents are lost due to flooding or collapse of the tree hollow, but artificial nest boxes are being erected and used by the

Current eligibility against IUCN Red List Criteria

IUCN category	Criteria eligibility
A	Not applicable: past, current or future population declines are thought unlikely to exceed 20% in any 3-generation period
B	Not applicable: EOO, AOO, habitat, locations or population not declining, population fluctuations not extreme
C	Not applicable: no continuing population decline or extreme fluctuations
D	Endangered: population 50–250 mature individuals
E	Not applicable: population viability analysis undertaken but needs new parameterisation

IUCN Red List assessment data

	Estimate	Reliability
Extent of occurrence trend	3500 km^2 stable	high high
Area of occupancy trend	150 km^2 stable	medium high
No. of mature individuals trend	240 increasing	medium high
No. subpopulations	1	high
No. locations	>10	high
Generation time	14.1 years	medium
Global population share	100%	high

cockatoos (Mooney and Pedler 2005). Fire is a threat to the food tree, as Drooping Sheoak takes at least 10 years to produce seed (Mooney and Pedler 2005): clearance of Drooping Sheoak, as well as loss of nest hollows, may have caused loss of the species from the mainland. Grazing by domestic, native or feral animals can also reduce regeneration of sheoaks as can land development for residential areas and rural subdivisions (Mooney and Pedler 2005). However, food is not currently limiting the population. Current potential threats include the spread of the infectious and fatal Psittacine Beak and Feather Disease, the loss of genetic diversity due to inbreeding and the loss of community support for the recovery program now that it is succeeding (Department of Sustainability, Environment, Water, Population and Communities 2011). So far no Psittacine Beak and Feather Disease has been found but levels of infertility seem high although population viability analysis suggests that the population is most sensitive to reduced adult survival, only moderately sensitive to severe fires and somewhat sensitive to disease outbreak (Harris *et al.* 2009). Community support remains strong.

Conservation objectives

1. A viable wild breeding population
2. Presence of the species on the mainland in South Australia

Information required

1. Reasons for eggs failure
2. Disease status of the population, particularly Psittacine Beak and Feather Disease

Management actions required

1. Continue to locate, protect and monitor nests in natural and artificial nest-sites.
2. Continue annual census to monitor demographic trends
3. Continue to partner with the community in research and management of the cockatoo
4. Continue revegetation of priority sites on both the island and the mainland

Bibliography

Barth M, Mooney T (2010) 'Draft Annual Census Report – October 2010'. Report to Department for Environment and Heritage, Kingscote.

BirdLife International (2011) 'Species factsheet: *Calyptorhynchus lathami*'. Retrieved 17 January, 2011 from <http://www.birdlife.org/>.

Burbidge A, Raines J (2003) 'South Australian Glossy Black-Cockatoo recovery program review 2003'. Department for Environment and Heritage, Kingscote, South Australia.

Crowley GM, Garnett ST (2001) Food Value and tree selection by Glossy Black-Cockatoos *Calyptorhynchus lathami*. *Austral Ecology* **26**, 116–126.

Crowley GM, Garnett ST, Carruthers S (1998) 'Mapping and spatial analysis of existing and potential Glossy Black-Cockatoo habitat on Kangaroo Island'. Report to Glossy Black-Cockatoo Recovery Team, Kingscote.

Department of Sustainability, Environment, Water, Population and Communities (2011) *Calyptorhynchus lathami halmaturinus*. In 'Species Profile and Threats Database'. Department of Sustainability, Environment, Water, Population and Communities, Canberra. Retrieved 24 February, 2011 from <http://www.environment.gov.au/sprat>.

Garnett ST, Pedler LP, Crowley GM (1999) The breeding biology of the Glossy Black-Cockatoo *Calyptorhynchus lathami* on Kangaroo Island, South Australia. *Emu* **99**, 262–279.

Harris JBC, Damien A, Fordham DA, Pedler LP, Mooney PA, Stead MG, Paton DC, Brook BW (2009) 'Spatially explicit population viability analysis of the South Australian subspecies of the Glossy Black-Cockatoo (*Calyptorhynchus lathami halmaturinus*) under climate change'. Report to the Glossy Black-Cockatoo Recovery Program, Kingscote.

Mooney PA, Pedler LP (2005) 'Recovery plan for the South Australian subspecies of the Glossy Black-Cockatoo (*Calyptorhynchus lathami halmaturinus*): 2005–2010'. Department for the Environment and Heritage, South Australia.

Pedler L, Sobey E (2008) 'Annual Census Report – October 2008'. Report for the Glossy Black-Cockatoo Recovery Team to Department for Environment and Heritage, Kingscote.

Pedler LP and Barth M (2009) 'Annual census report. Glossy Black-Cockatoo Recovery Program'. Report to Department for Environment and Heritage, Kingscote.

Comments received from

Mike Barth, Lynn Pedler

Subspecies Conservation Summary

Glossy Black-Cockatoo (south-eastern)

Calyptorhynchus lathami lathami (Temminck, 1807)
Cacatuidae

Conservation status

Near Threatened C2a(ii)

Reasons for listing

Population possibly as few as 10 000 mature individuals and declining; nearly 100% in largest subpopulation, but at least 1 small, potentially isolated population is known from the NSW Riverina.

Status 2000

Near Threatened C2a(ii)

Status 1990

Near Threatened C2a(ii)

Taxonomy

C. l. halmaturinus (Kangaroo I.) is Endangered and *C. l. erebus* (central Queensland, Dawson-Mackenzie basin to Paluma) is Least Concern, as is the species Taxonomic uniqueness: medium (7 genera/family, 5 species/genus, 3 subspecies/species)

Range

Patchy distribution from Mitchell, Queensland, through eastern New South Wales to Mallacoota, Victoria. A continuous distribution through the forested parts of the Great Dividing Range but a more scattered distribution inland, to as far west as the Riverina and Pilliga Scrub, NSW (Schodde and Mason 1997; Higgins 1999). While the cockatoos can cross substantial gaps between areas of forest and woodland, the Riverina population appears to be isolated. There is no measure of the area occupied within this range. The Birds Australia Atlas database has records from 5000 km^2. Given its often cryptic nature and the inaccessibility of much of its habitat, recording will be irregular. However, as the feeding and breeding habitat are themselves patchy, a minimum AOO of 10 000 km^2 is nominated here.

Abundance

Thinly distributed throughout range, with highest densities east of the Great Dividing Range (Higgins 1999). However, there is no estimate of the total population so the number of mature individuals was estimated by an expert Birds Australia committee as 10 000±1600 using a formal elicitation technique (Burgman 2005), using all available baseline information. The Riverina subpopulation probably numbers no more than 40 mature individuals (M. Cameron *in litt.*).

Ecology

Usually occur in pairs, or in groups of 3, which comprise a breeding pair and their offspring. They feed almost exclusively upon the seeds of *Allocasuarina* and *Casuarina* species, usually relying on only 1 or 2 species within a region (Higgins 1999). While the strength of their bill gives them access to this nutritious, abundant and relatively stable food resource, intake rates are low, and cannot be accelerated if food supply is short (Cameron 2005). The cockatoos lay a single egg in hollows in tall trees, particularly favouring dead trees (snags). Trees that develop nest hollows suitable for this subspecies are thought to be at least 100 years old and possibly much older (Cameron 2006b). A generation time of 19.2 years (BirdLife International 2011) is based on an average age at first breeding of 4.0 years and maximum longevity of 34.4 years, both extrapolated from Carnaby's Black-Cockatoo *C. latirostris* (Australian Bird and Bat Banding Scheme).

Threats

Historically clearance of habitat has reduced the range of the subspecies, particularly south and west of the Great Dividing Range. Fragmentation of habitat, especially when associated with agriculture, leads to penetration of competitors from more open habitats and/or ecotones (notably Common Brushtail Possum *Trichosurus vulpecula*, Little Corella *Cacatua sanguinea* and Galah *Eolophus roseicapillus*) and increasing competition for hollows (Downes et al., 1997, Crowley et al., 1998). Without nests being protected against such competitors, recruitment to the adult population may be minimal (Garnett *et al.* 1999). In the long term, however, the principal threat to the cockatoos is loss of nest hollows and a reduction in the extent and quality of foraging habitat. Large hollows required by the cockatoos are becoming less common and not being replaced (Cameron 2006a). Wildfire following drought can eliminate nest hollows over large areas (Cameron 2008). Food supply can also be limiting, especially during drought (Cameron 2005, 2006b) or when fire renders feeding habitat unsuitable for several years (Joseph 1982; Crowley 1986; Pepper 1997). Regeneration may be impeded by grazing, either by stock or feral herbivores. In

Current eligibility against IUCN Red List Criteria

IUCN category	Criteria eligibility
A	Not applicable: past, current or future population declines are thought unlikely to exceed 20% in any 3-generation period
B	Not applicable: EOO >20 000 km^2, AOO >2000 km^2
C	Near Threatened: population 2500–10 000 mature individuals, suspected continuing decline, near 100% in 1 subpopulation
D	Not applicable: population >1000 mature individuals, >5 locations
E	Not applicable: no population viability analysis undertaken

IUCN Red List assessment data

	Estimate	Reliability
Extent of occurrence trend	730 000 km^2 stable	high high
Area of occupancy trend	10 000 km^2 decreasing	low low
No. of mature individuals trend	10 000 decreasing	low low
No. subpopulations	3	medium
No. locations	>10	high
Largest subpopulation	9500	low
Generation time	19.2 years	medium
Global population share	100%	high

coastal areas, residential development is an increasing threat. Illegal trapping for the bird trade may be a localised problem, but is probably not a major threat.

Conservation objectives

1. Stable population
2. Retention of subpopulations in western NSW

Information required

1. Trends in numbers at key sites
2. Trends in habitat variables such as old trees with hollows and sheoak seed availability at a landscape scale

Management actions required

1. Determine age structure and population trends at selected sites throughout distribution
2. Identify and protect areas of high food value throughout distribution
3. Implement clearance controls and appropriate forest management practices to ensure long-term supply of trees suitable for nesting
4. Establish buffers of native forests and/or woodland around important nesting areas to limit incursions of competitors
5. Maintain links between populations

Bibliography

BirdLife International (2011) 'Species factsheet: *Calyptorhynchus lathami*'. Retrieved 17 January, 2011 from <http://www.birdlife.org/>.

Burgman MA (2005) *Risks and Decisions for Conservation and Environmental Management*. Cambridge University Press, Cambridge, New York.

Cameron M (2005) Group size and feeding rates of Glossy Black-Cockatoos in central New South Wales. *Emu* **105**, 299–304.

Cameron M (2006a) Distribution and cone production in *Allocasuarina diminuta* and *A. gymnanthera* (Casuarinaceae) in central New South Wales. *The Rangeland Journal* **28**, 153–161.

Cameron M (2006b) Nesting habitat of the glossy black-cockatoo in central New South Wales. *Biological Conservation* **127**, 402–410.

Crowley GM (1986) Ecology of *Allocasuarina littoralis* (Salisb.) L. Johnson and *A. torulosa* (Ait.) L. Johnson in North Queensland. PhD thesis, James Cook University of North Queensland, Townsville.

Crowley GM, Garnett ST, Carruthers S (1998) 'Mapping and spatial analysis of existing and potential Glossy Black-Cockatoo habitat on Kangaroo Island'. Report to Glossy Black-Cockatoo Recovery Team, Kingscote.

Downes SJ, Handasyde KA, Elgar MA (1997) The use of corridors by mammals in fragmented Australian eucalypt forests. *Conservation Biology* **11**, 718–725.

Garnett ST, Pedler LP, Crowley GM (1999) The breeding biology of the Glossy Black-Cockatoo *Calyptorhynchus lathami* on Kangaroo Island, South Australia. *Emu* **99**, 262–279.

Higgins PJ (Ed.) (1999) *Handbook of Australian, New Zealand and Antarctic Birds. Volume 4: Parrots to Dollarbird*. Oxford University Press, Melbourne.

Pepper JW (1997) A survey of the South Australian Glossy Black-Cockatoo (*Calyptorhynchus lathami halmaturinus*) and its habitat. *Wildlife Research* **24**, 209–224.

Schodde R, Mason IJ (Eds) (1997) *Zoological Catalogue of Australia 37.2: Aves (Columbidae to Coraciidae)*. CSIRO Publishing, Melbourne.

Comments received from

Allan Burbidge, Andrew Burbidge, Matt Cameron, Les Christidis, Hugh Ford, Tim Holmes, Rod Kavanagh, Richard Loyn, Peter Menkhorst, Penny Olsen, David Watson

Species Conservation Summary

Carnaby's Black-Cockatoo

Calyptorhynchus latirostris Carnaby, 1948

Cacatuidae

Conservation status

Endangered A2bcde+3bcde+4bcde

Reasons for listing

Rapid decline >50% in last 3 generations (58 years) likely to continue, based on sample counts, area of occupancy and quality of habitat, nest-robbing, disease and competition

Status 2000

Endangered A2bcde+3bcde+4bcde

Status 1990

Endangered A2bcde+3bcde+4bcde

Taxonomy

No infraspecific taxa described

Taxonomic uniqueness: high (7 genera/family, 5 species/genus, 1 subspecies/species)

Range

Endemic to south-western Australia (Schodde and Mason 1997), it currently ranges from Kalbarri south-east to Esperance (Department of Environment and Conservation (WA) 2007). Most breeding occurs between Three Springs and Stirling Range with a recent shift to use breeding sites in the west and south. Flocks of non-breeding birds then move west to coastal areas (Saunders *et al.* 1985; Johnstone *et al.* 2008, 2011). Disappeared from over one-third of its EOO between 1970s and 1990s (Saunders and Ingram 1995) and its AOO is continuing to contract, e.g. losses of habitat on the Swan Coastal Plain to urban development (Saunders *et al.* 1985; Department of Sustainability Environment Water Population and Communities 2011). The AOO has not been estimated but, based on breeding habitat, is unlikely to exceed 10 000 km^2.

Abundance

The total population is estimated at about 40 000 individuals, but possibly <10 000–60 000, in a single subpopulation in 4 main areas: northern wheatbelt, upper southern wheatbelt, south-eastern wheatbelt and Esperance coast (Mawson and Johnstone 1997; Department of Environment and Conservation (WA) 2007). There are few counts, including an estimated 8000–15 000 non-breeding birds on the northern Swan Coastal Plain in 2010 and a maximum flock size of 15 000 in Yanchep National Park in 2008 (Johnstone *et al.* 2008; Valentine and Stock 2008; Burnham *et al.* 2010). Initial trends based on roost counts suggest that declines are continuing: they are inferred here to be declining at a similar rate to historical range contractions, i.e. >50% in 3 generations (58 years).

Ecology

Forages in woodland and kwongan heath dominated by proteaceous species (Saunders and Ingram 1995). Mainly feeds on seeds of hakeas, grevilleas, banksias and eucalypts, also seeds of weeds and exotic pines *Pinus radiata*, and insect larvae (Higgins 1999). Sometimes forages in Canola *Brassica rapa* with negligible impact on the crop (Jackson 2009). Nests in hollows in large eucalypts, primarily Salmon Gum *Eucalyptus salmonophloia* and Wandoo *E. wandoo* (Saunders 1979). Suitable hollows can take c.120–230 years to develop but birds will breed in restored or artificial hollows (Pittman *et al.* 2007; Johnstone *et al.* 2008). Two eggs are laid, from which only 1 chick usually survives to fledging (Saunders *et al.* 1985). Breeding success depends on the presence of heathland feeding areas within 12 km of nesting habitat (Department of Environment and Conservation (WA) 2007). A generation time of 19.2 years (BirdLife International 2011) is derived from an age at first breeding of 4.0 years (Saunders 1986) and maximum known longevity in the wild of 34.4 years (Saunders and Dawson 2009).

Threats

Since the 1950s, most feeding habitat of breeding birds has been cleared for agriculture. The remnants are fragmented, threatened by rising soil salinity, weed invasion, *Phytophthora cinnamomi* dieback and fire, and are often too far from nesting areas for successful breeding (Saunders 1990; Saunders and Ingram 1998; Stojanovic 2008). Breeding habitat has also been extensively cleared, and there has been little regeneration of nest trees for >70 years because of grazing by sheep and rabbits *Oryctolagus cuniculus*. In 1 study, nest-hollows were lost at 2.2–4.8% per year (Saunders 1979). There are increasing observations of competition for hollows with other birds, especially Galah *Eolophus roseicapilla* and Western Corella *Cacatua pastinator* (Cale 2003; Johnstone and Cassarchis 2004; Johnstone *et al.* 2010). Competition for nest hollows and injury or death by European Honey Bees *Apis mellifera* is another current and future threat (Johnstone and

Current eligibility against IUCN Red List Criteria

IUCN category	Criteria eligibility
A	Endangered: past, current and anticipated population declines of 50–79% over periods of 3 generations (58 years) inferred from monitoring, and suspected from contracting AOO, deteriorating habitat quality, levels of exploitation and effects of competitors
B	Not applicable: EOO >20 000 km^2, AOO >2000 km^2
C	Not applicable: population >10 000 mature individuals
D	Not applicable: population >1000 mature individuals, >5 locations
E	Not applicable: no population viability analysis undertaken

IUCN Red List assessment data

	Estimate	Reliability
Extent of occurrence trend	180 000 km^2 declining	high low
Area of occupancy trend	10 000 km^2 declining	low medium
No. of mature individuals trend	40 000 declining	medium high
No. subpopulations	1	high
No. locations	>10	high
Generation time	19 years	medium
Global population share	100%	high

Kirkby 1999). Nest-robbing for the bird trade, although declining, also results in damage to nest hollows (Saunders *et al.* 1985; Mawson and Johnstone 1997). Approximately half of the Swan Coastal plain feeding habitat of non-breeding birds has also been cleared, and non-breeding birds on the Swan Coastal Plain and cockatoos in the northern wheatbelt now rely heavily on Monterey Pines *Pinus radiata*. These pines are also important as night roost sites and are being harvested but not replanted, at least on the Swan Coastal Plain (Valentine and Stock 2008). Birds are killed on roads or shot by landholders for damaging pines and nut and fruit crops (Pittman *et al.* 2007). Stochastic events have also killed significant numbers in recent years: in October 2009 an unknown disease killed up to 23 females in a single breeding colony at Koobabbie, a heatwave in January 2010 is known to have killed flocks of 63 and 145 birds near Hopetoun and 57 birds were killed and 24 injured by a hailstorm in March 2010 (Saunders *et al.* in press).

Conservation objectives

1. Area of occupancy does not fall below 18 000 km^2
2. Number of breeding pairs at priority areas remains stable or increases
3. Rate of population decline slows within 10 years

Information required

1. Priority areas for breeding, for feeding in breeding and non-breeding season and for revegetation
2. Population size and trends in numbers and productivity

Management actions required

1. Monitor population at key sites and times
2. Maintain and increase number of nesting hollows in priority areas
3. Maintain and increase night roost sites in priority areas
4. Regenerate breeding and feeding trees and banksia woodland in priority areas
5. Retain and plant feeding habitat in priority areas
6. Manage feeding habitat in priority non-breeding areas
7. Continue to support coordination of management by the Carnaby's Black-Cockatoo Recovery Team

Bibliography

BirdLife International (2011) 'Species factsheet: *Calyptorhynchus latirostris*'. Retrieved 17 January, 2011 from <http://www.birdlife.org/>.

Burnham Q, Barrett GW, Blythman M, Scott R (2010) 'Carnaby's Cockatoo (*Calyptorhynchus latirostris*) identification of nocturnal roost sites and the 2010 Great Cocky Count'. Unpublished report for WA Department of Environment and Conservation. Retrieved 9 March, 2011 from <http://www.birdsaustralia.com.au/images/stories/current-projects/cbc/CBC-GCC-report-Aug2010.pdf>.

Cale B (2003) 'Carnaby's Black-Cockatoo (*Calptorhynchus latirostris*) Recovery Plan 2002–2012. Western Australian Wildlife Management Program No. 36'. Department of Conservation and Land Management Western Australia, Perth.

Department of Environment and Conservation (WA) (2007) 'Fauna Note No. 5. Carnaby's Cockatoo'. Retrieved 5 May, 2011 from <http://www.agric.wa.gov.au/objtwr/imported_assets/content/pw/vp/bird/5_carnabys_cockatoo.pdf>.

Department of Sustainability, Environment, Water, Population and Communities (2011) *Calyptorhynchus latirostris*. In 'Species Profile and Threats Database'. Department of Sustainability, Environment, Water, Population and Communities, Canberra. Retrieved 24 February, 2011 from <http://www.environment.gov.au/sprat>.

Higgins PJ (Ed.) (1999) *Handbook of Australian, New Zealand and Antarctic Birds. Volume 4: Parrots to Dollarbird*. Oxford University Press, Melbourne.

Jackson C (2009) Assessing and quantifying canola crop damage by Carnaby's Black-Cockatoo *Calyptorhynchus latirostris* in the south-west of Western Australia. MSc thesis, The University of Sydney, Sydney.

Johnstone RE, Cassarchis C (2004) 'Review of cockatoo research project and Cockatoo Care 2004'. Water Corporation and the Western Australian Museum, Perth.

Johnstone RE, Johnstone C, Kirkby T (2008) 'Carnaby's Cockatoo (*Calyptorhynchus latirostris*) on the northern Swan Coastal Plain (Lancelin – Perth), Western Australia'. Western Australian Museum, Perth.

Johnstone RE, Johnstone C, Kirkby T (2010) 'Carnaby's Cockatoo (*Calyptorhynchus latirostris*), Baudin's Cockatoo (*Calyptorhynchus baudinii*) and the Forest Red-tailed Black Cockatoo (*Calyptorhynchus banksii naso*) on the Swan

Coastal Plain (Lancelin–Dunsborough), Western Australia. Studies on distribution, status, breeding, food, movements and historical changes'. Report to the Department of Planning, Perth.

Johnstone RE, Kirkby T (1999) Food of the Red-tailed Black-Cockatoo *Calyptorhynchus banksii naso* in south-west Western Australia. *Western Australian Naturalist* **22**, 167–178.

Mawson PR, Johnstone RE (1997) Conservation status of parrots and cockatoos in Western Australia. *Eclectus* **2**, 4–9.

Pittman H, Scott R, Stojanovic D, McLellan L (2007) Caring for Carnaby's. *Wingspan* **17**.

Saunders DA (1979) The availability of tree hollows for use as nest sites by White-tailed Black-Cockatoos. *Australian Wildlife Research* **6**, 205–216.

Saunders DA (1986) Breeding season, nesting success and nestling growth in Carnaby's Cockatoo, *Calyptorhynchus funereus latirostris*, over 16 years at Coomallo Creek, and a method for assessing the viability of populations in other areas. *Australian Wildlife Research* **13**, 261–273.

Saunders DA (1990) Problems of survival in an extensively cultivated landscape: the case of Carnaby's Cockatoo *Calyptorhynchus funereus latirostris*. *Biological Conservation* **54**, 277–290.

Saunders DA, Dawson R (2009) Update on longevity and movement of Carnaby's Black-Cockatoo. *Pacific Conservation Biology* **15**, 72–74.

Saunders DA, Ingram J (1995) 'Birds of Southwestern Australia. An atlas of changes in distribution and abundance of the wheatbelt fauna'. Surrey Beatty & Sons, Chipping Norton, NSW.

Saunders DA, Ingram JA (1998) Twenty-eight years of monitoring a breeding population of Carnaby's Cockatoo. *Pacific Conservation Biology* **4**, 261–270.

Saunders DA, Mawson P, Dawson R (in press) The impact of two extreme weather events and other causes of death on Carnaby's Black Cockatoo: a promise of things to come for a threatened species? *Pacific Conservation Biology.*

Saunders DA, Rowley I, Smith GT (1985) The effects of clearing for agriculture on the distribution of cockatoos in the southwest of Western Australia. In *Birds of Eucalypt Forests and Woodlands: Ecology, Conservation, Management*. (Eds A Keast, HF Recher, HA Ford and DA Saunders). RAOU and Surrey Beatty and Sons, Melbourne and Chipping Norton, NSW.

Schodde R, Mason IJ (Eds) (1997) *Zoological Catalogue of Australia 37.2: Aves (Columbidae to Coraciidae)*. CSIRO Publishing, Melbourne.

Stojanovic D (2008) Not so black and white: the complex conservation of Carnaby's black-cockatoo. *Wildlife Australia Magazine* **45**, 18–23.

Valentine LE, Stock W (2008) 'Food Resources of Carnaby's Black-Cockatoo (*Calyptorhynchus latirostris*) in the Gnangara Sustainability Strategy study area'. Unpublished report. Edith Cowan University and Department of Environment and Conservation. Retrieved 9 March, 2011 from <http://ro.ecu.edu.au/cgi/viewcontent.cgi?article=7147&context=ecuworks>.

Comments received from

Geoff Barrett, Allan Burbidge, Cheryl Gole, Ron Johnstone, Peter Mawson, Denis Saunders

---

Species Conservation Summary

Baudin's Black-Cockatoo

Calyptorhynchus baudinii Lear, 1832
Cacatuidae

Conservation status

Endangered A3cde+4cde

Reasons for listing

Ongoing population decline estimated >50% over 3 generations (58 years) inferred from shooting by orchardists and loss of nest hollows to bees and other cockatoos

Status 2000

Endangered A3cde+4cde

Status 1990

Endangered A3cde+4cde

Taxonomy

No infraspecific taxa described
Taxonomic uniqueness: high (7 genera/family, 5 species/genus, 1 subspecies/species)

Range

Endemic to higher rainfall parts of the south-west of Western Australia (Saunders 1979; Saunders *et al.* 1985). It ranges north to Gidgegannup and Hoddy Well and east to Clackline, Wundowie, the lower Darkin River, Boyagin

Current eligibility against IUCN Red List Criteria

IUCN category	Criteria eligibility
A	Endangered: current and projected future declines of >50% in 3 generations (58 years) inferred from decline in AOO, habitat quality, levels of exploitation and effects of competitors
B	Not applicable: EOO >20 000 km^2, AOO >2000 km^2
C	Near Threatened: population may approach 10 000 mature individuals with a continuing decline and all in 1 subpopulation
D	Not applicable: population >1000 mature individuals, >5 locations
E	Not applicable: no population viability analysis undertaken

IUCN Red List assessment data

	Estimate	Reliability
Extent of occurrence trend	70 000 km^2 stable	high medium
Area of occupancy trend	25 000 km^2 declining	low low
No. of mature individuals trend	12 500 declining	medium high
No. subpopulations	1	high
No. locations	>10	high
Generation time	19 years	low
Global population share	100%	high

Rock, Wandering, Williams, Kojonup and the King River also west to the eastern strip of the Swan Coastal Plain including West Midland, Armadale, Byford, Mundijong, Serpentine, North Dandalup and further west to the coast at Lake Clifton, Australind, Bunbury, Busselton, Dunsborough, Leeuwin-Naturaliste National Park and Augusta, also the Stirling and Porongurup Ranges and east along the south coast to Waychinicup National Park (Johnstone *et al.* 2007). The EOO probably stable since settlement (Saunders *et al.* 1985; Saunders and Ingram 1995), but lost from about 25% of its range since the 1950s (Johnstone 1997). Breeding occurs in the south, from Nornalup north to near Bridgetown, sometimes further north to Lowden and Harvey, following which most birds move to the Darling scarp south-west of Perth (Higgins 1999). The AOO is assessed here as about 10 000 km^2 based on the area most birds occupy during winter.

Abundance

Surveys in 1995–2004 suggest that the population is probably 10 000–15 000 individuals but only about 10% breed in any year (P. Mawson *in litt.*). The population density is believed to have declined (Johnstone 1997), a trend that is continuing at an unquantified rate (Johnstone *et al.* 2007). The reporting rate probably declined between the 1977–1981 and 1998–2001 (Olsen *et al.* 2003) but taxonomic changes in the first Birds Australia Atlas and confusion with other Black-Cockatoos since make comparisons difficult. Given the lack of any recent quantitative data, the rate of decline is inferred from changes in habitat and competition to be >50% in 3 generations (58 years).

Ecology

Occurs in temperate forest and woodland dominated by the eucalypts Jarrah *Eucalyptus marginata*, Karri *E. diversicolor* and Marri *Corymbia calophylla*. They mainly feed on the seeds and flowers of Marri, with their long beak making them particularly efficient at extracting the seed (Cooper *et al.* 2002), but also take seeds of Jarrah, cultivated apples and pears, *Banksia* and *Hakea* spp, *Erodium botrys* and insect larvae (Long 1985; Halse 1986; Department of Environment and Conservation (WA) 2007; Johnstone *et al.* 2010). They nest in hollows in mature eucalypts, particularly Marri, Karri and Wandoo *E. wandoo* and Bullich *E. megacarpa* (Johnstone *et al.* 2010), usually laying 2 eggs, although breeding success is only estimated at 0.6 young per pair (Johnstone and Storr 1998). Nests are thinly dispersed through the available feeding habitat, with greater densities perhaps being prevented by intense competition between nesting females (Saunders *et al.* 1985). A generation time of 19.2 years (BirdLife International 2011) is derived from an age at first breeding of 4.0 years and maximum longevity of 34.4 years, both extrapolated from Carnaby's Black-Cockatoo *C. latirostris*.

Threats

Nest hollow shortage is considered the principal threat, as suitable hollows are considered scarce, only forming in trees at least 130 to 220 years of age, many of which have been preferentially felled (Abbott and Whitford 2002; Chapman 2007). Nest hollows are likely to continue to be lost to mining (Chapman 2007) and fire. Fires caused by lightning or arson, and sometimes planned fires, cause older trees containing hollows to fall, and temporarily reduce food availability (Department of Sustainability, Environment, Water, Population and Communities 2011). Competition for hollows is also severe with other cockatoos, including Carnaby's Black-Cockatoo and corellas *Cacatua* spp., Wood Ducks *Chenonetta jubatta* and feral European Honey Bees *Apis mellifera* all successfully displacing Baudin's Black-Cockatoos (Johnstone and Cassarchis 2004; Chapman 2007; Johnstone and Kirkby 2007). Although the species has been fully protected since 1996 (Mawson and Johnstone 1997), illegal shooting by orchardists still occurs when they feed on apples and pears; in 1 study, 9.2% of 67 377 apples were damaged in 2 months (Halse 1986). The species no longer occupies up to 25% of former habitat that has been cleared for agriculture (Mawson and Johnstone 1997).

Conservation objectives

1. Area of occupancy stable or increasing
2. Number of breeding pairs stable or increasing
3. Numbers in each roosting flock stable or increasing
4. Proportion of juveniles in each roosting flock increasing

Information required

1. Demographic trends and relative impacts of threatening processes

2. Factors affecting the number of breeding attempts and breeding success
3. Spatial distribution of feeding and breeding habitat

Management actions required

1. Prevent illegal shooting in and around commercial orchard areas by enforcing anti-shooting legislation more vigorously
2. Assist orchardists in developing a non-lethal damage mitigation strategies
3. Monitor demographic indicators (population size, trends, recruitment), including the effectiveness of management actions and the need to adapt them if necessary
4. Map feeding and breeding habitat
5. Determine and implement ways to minimise the effects of mining on habitat loss
6. Determine and implement ways to manage forests for the conservation of the species
7. Develop and implement a management plan for the control and reduction of non-managed feral European Honey Bees in the region
8. Develop and implement a communication strategy to stop illegal shooting and improve compliance with the relevant legislation
9. Continue to support coordination of management by the Forest Cockatoo Recovery Group

Bibliography

Abbott I, Whitford K (2002) Conservation of vertebrate fauna using hollows in forests of south-west Western Australia: strategic risk assessment in relation to ecology, policy, planning, and operations management. *Pacific Conservation Biology* **7**, 240–255.

BirdLife International (2011) 'Species factsheet: *Calyptorhynchus baudinii*'. Retrieved 17 January, 2011 from <http://www.birdlife.org/>.

Chapman TF (2007) 'Forest Black Cockatoo (Baudin's Cockatoo *Calyptorhynchus baudinii*) and Forest Red-tailed Black Cockatoo (*Calyptorhynchus banksii naso*) Recovery Plan 2007–2016'. Department of Conservation and Land Management, Perth.

Cooper CE, Withers PC, Mawson PR, Bradshaw SD, Prince J, Roberston H (2002) Metabolic ecology of cockatoos in the south-west of Western Australia. *Australian Journal of Zoology* **50**, 67–76.

Department of Environment and Conservation (WA) (2007) 'Fauna Note No. 1. Reducing Fruit Damage by Baudin's Cockatoo'. Retrieved 9 March, 2011 from <http://www.dec.wa.gov.au/component/option,com_docman/Itemid,1427/gid,361/task,doc_details/>.

Department of Sustainability, Environment, Water, Population and Communities (2011) *Calyptorhynchus baudinii*. In 'Species Profile and Threats Database'. Department of Sustainability, Environment, Water, Population and Communities, Canberra. Retrieved 24 February, 2011 from <http://www.environment.gov.au/sprat>.

Halse SA (1986) 'Parrot damage in apple orchards in south-western Australia – a review'. Department of Conservation and Land Management Tech. Rep. 8.

Higgins PJ (Ed.) (1999) *Handbook of Australian, New Zealand and Antarctic Birds. Volume 4: Parrots to Dollarbird*. Oxford University Press, Melbourne.

Johnstone RE (1997) Current studies on three endemic Western Australian Cockatoos. *Eclectus* **3**, 34–35.

Johnstone RE, Cassarchis C (2004) 'Review of cockatoo research project and Cockatoo Care 2004'. Water Corporation and the Western Australian Museum, Perth.

Johnstone RE, Johnstone C, Kirkby T (2007) 'White-tailed Black Cockatoos (Baudin's Cockatoo *Calyptorhynchus baudinii* and Carnaby's Cockatoo *Calyptorhynchus latirostris*) on the southern Swan Coastal Plain (Bunbury - Dunsborough) Western Australia'. Western Australian Museum, Perth.

Johnstone RE, Johnstone C, Kirkby T (2010) 'Carnaby's Cockatoo (*Calyptorhynchus latirostris*), Baudin's Cockatoo (*Calyptorhynchus baudinii*) and the Forest Red-tailed Black Cockatoo (*Calyptorhynchus banksii naso*) on the Swan Coastal Plain (Lancelin–Dunsborough), Western Australia. Studies on distribution, status, breeding, food, movements and historical changes'. Report to the Department of Planning, Perth.

Johnstone RE, Kirkby T (2007) Feral European honey bees: a major threat to cockatoos and other tree hollow users. *Western Australian Naturalist* **25**, 252–254.

Johnstone RE, Storr GM (1998) *Handbook of Western Australian Birds. Volume 1: Non-passerines (Emu to Dollarbird)*. Western Australian Museum, Perth.

Long J (1985) Damage to cultivated fruit by parrots in the south of Western Australia. *Australian Wildlife Research* **12**, 75–80.

Mawson PR, Johnstone RE (1997) Conservation status of parrots and cockatoos in Western Australia. *Eclectus* **2**, 4–9.

Olsen P, Weston M, Silcocks A (2003) The state of Australia's birds 2003. *Wingspan (Supplement)* **13**.

Saunders DA (1979) Distribution and taxonomy of the White-tailed and Yellow-tailed Black-Cockatoos *Calyptorhynchus* spp. *Emu* **79**, 215–227.

Saunders DA, Ingram J (1995) *Birds of South-western Australia. An Atlas of Changes in Distribution and Abundance of the Wheatbelt Fauna*. Surrey Beatty & Sons, Chipping Norton, NSW.

Saunders DA, Rowley I, Smith GT (1985) The effects of clearing for agriculture on the distribution of cockatoos in the southwest of Western Australia. In *Birds of Eucalypt Forests and Woodlands: Ecology, Conservation, Management*. (Eds A Keast, HF Recher, HA Ford and DA Saunders). RAOU and Surrey Beatty and Sons, Melbourne and Chipping Norton, NSW.

Comments received from

Allan Burbidge, Cheryl Gole, Ron Johnstone, Peter Mawson, Denis Saunders

Subspecies Conservation Summary

Major Mitchell's Cockatoo (eastern)

Lophochroa leadbeateri leadbeateri (Vigors, 1831)
Cacatuidae

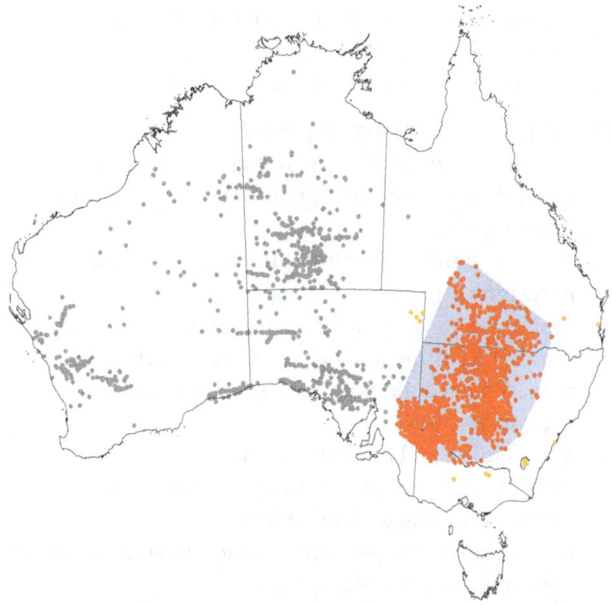

Conservation status
Near Threatened A3c

Reasons for listing
Projected declines in nest hollow availability are likely to increase existing declines to a rate approaching 30% within the next 3 generations (71 years)

Status 2000
Near Threatened A3c

Status 1990
Near Threatened A3c

Taxonomy
L. l. mollis (central and western arid zone and Nullarbor, west of the Eyre basin and Port Augusta) is Least Concern, as is the species
Taxonomic uniqueness: high (7 genera/family, 1 species/genus, 2 subspecies/species)

Range
Murray-Darling, Eyre and Bulloo River basins as far north as Isisford and Roma in Queensland, through western New South Wales to north-west Victoria and west to Adelaide and Mt Mary Plains (Higgins 1999). It had disappeared from the Adelaide and Mt Mary Plains by the 1950s (Boehm 1961), but is otherwise still extant across most of its historical range. As much of its range remains uncleared rangeland, it is assumed to be at least 25% is occupied.

Abundance
Population estimated, with low reliability, as 15 000 mature individuals based on the AOO and a density of 1 pair/30 km^2. The density of cockatoos has been greatly reduced in north-west Vic (Emison *et al.* 1987) and western NSW (Cooper and McAllan 1995) with reporting rates falling by >20% across most of its range between the 2 bird Atlases (Barrett *et al.* 2002). In the largest known nesting area the number of nests fell from 1992 to 2007, surveys of nest the number of occupied nests fell from 64 to a low of 11 in 2004 with 17 nesting pairs in 2006. A trial removal of Galahs *Eolophus roseicapillus* that competed for nest hollows in 2007 and 2008 at this site resulted in a 70.6% increase in the number of Major Mitchell's Cockatoos breeding (Hurley 2008). All of these new breeding pairs were recorded in tree hollows previously cleared of Galahs. This increase in the number of breeding pairs of Major Mitchell's Cockatoos was maintained in 2009 but not in 2010 (Hurley 2011). There has also been a substantial loss of nesting trees, for instance, only 47.6% of the 64 nest trees monitored in 1995 were standing in 2006 (Department of Sustainability and Environment 2008). This is likely to be representative across its range in extensive areas with a historic and continued use for stock grazing (Gibson *et al.* 2008; Cheal *et al.* 2010), so a population decline is estimated here to be 20–29% in 3 generations (71 years).

Ecology
Inhabits arid and semi-arid woodlands dominated by Mulga *Acacia aneura*, mallee and box eucalypts, Slender Cypress Pine *Callitris gracilis* or Belah *Casuarina cristata*. Often nests in hollows in Cypress Pines >80 and preferably 130–140 years old (Gibson *et al.* 2008). The main requirements of the species are trees with suitable nesting hollows and fresh surface water (Higgins 1999). In contrast to other cockatoos, pairs of Major Mitchell Cockatoos will generally not nest close to one another: breeding pairs of the western subspecies *L. l. mollis* occupy nests that are at least 1 km apart, and have densities of about 1 pair per 30 km^2 (Saunders *et al.* 1985; Rowley and Chapman 1991), though, in Vic, inter-nest distances are shorter (mean 444 m, minimum 50 m, with nests possibly more clumped because nest hollows are so restricted (Department of Sustainability and Environment 2008). They are scarce in fragmented landscapes, tending to travel along vegetated corridors to feeding sites (Department of Sustainability and Environment 2008). They feed on seeds of a wide variety of native shrubs and trees, especially cypress pines, and also some insect larvae and seeds of some agricultural crops and weeds (Higgins 1999). A generation time of 23.7 years (BirdLife International 2011) is derived from an age at first breeding of 4.3 years (extrapolated from *Cacatua* spp.) and a maximum longevity in the wild of 43.1 years, extrapolated from known oldest bird in captivity; 77 years.

Threats
The continued loss of hollow-bearing trees and the increasing competition for nest hollows with Galahs, which are increasing in semi-cleared agricultural landscapes, are serious immediate threats. The cockatoos also face competition for hollows with feral European Honey Bees *Apis mellifera*. Nest box trials conducted in Victoria (Hurley 2009) have so far attracted Galahs and Barn Owls *Tyto alba*

Current eligibility against IUCN Red List Criteria

IUCN category	Criteria eligibility
A	Near Threatened: anticipated population declines of 20–29% in 3 generations (71 years) suspected from deteriorating habitat quality
B	Not applicable: EOO >20 000 km^2, AOO >2000 km^2
C	Not applicable: population >10 000 mature individuals
D	Not applicable: population >1000 mature individuals, >5 locations
E	Not applicable: no population viability analysis undertaken

IUCN Red List assessment data

	Estimate	Reliability
Extent of occurrence trend	900 000 km^2 stable	high medium
Area of occupancy trend	225 000 km^2 decreasing	low low
No. of mature individuals trend	15 000 decreasing	low medium
No. subpopulations	1	high
No. locations	>10	high
Generation time	23.7 years	medium
Global population share	100%	high

but not Major Mitchell's Cockatoos (Hurley 2011). New hollows, however, are not being formed at a rate necessary to replace old ones because regeneration of the favoured hollow-bearing tree, Cypress Pine, is being prevented by grazing by stock, feral herbivores, such as European Rabbit *Oryctolagus cuniculus* and feral goats *Capra hircus*, and inflated populations of kangaroos *Macropus* spp. (Department of Sustainability and Environment 2008).

Conservation objectives

1. A population known to be stable
2. A successful recruitment event of *Callitris* pines within the species' range

Information required

1. Trends in population size
2. Definitive evidence that nest hollow availability is limiting population size
3. A nest box design that is used in the wild by Major Mitchell Cockatoos
4. Factors influencing *Callitris* pine regeneration and survival

Management actions required

1. Test effectiveness of Galah control
2. Design and trial an appropriate nest box
3. Where appropriate control nesting Galahs
4. Manage total grazing pressure in rangelands and protected areas by controlling stock, rabbits, feral goats and kangaroos
5. Manage fires to minimise loss of hollow-bearing trees, particularly known nest sites

Bibliography

Barrett GW, Silcocks A, Cunningham RB (2002) 'Australian Bird Atlas (1998–2001). Supplementary Report No. 1 – Comparison of Atlas 1 (1977–1981) and Atlas 2 (1998–2001)'. Birds Australia, Melbourne.

BirdLife International (2011) 'Species factsheet: *Cacatua leadbeateri*'. Retrieved 17 January, 2011 from <http://www.birdlife.org/>.

Boehm EF (1961) Pink Cockatoo in southern South Australia. *Emu* **61**, 138–139.

Cheal DC, Lucas A, Macaulay L (2010) 'Draft national recovery plan for Buloke woodlands of the Riverina and Murray Darling Depression Bioregions'. Department of Sustainability and Environment, Melbourne.

Cooper RP, McAllan IAW (1995) *The Birds of Western New South Wales. A Preliminary Atlas.* N.S.W. Bird Atlassers, Albury.

Department of Sustainability and Environment (2008) 'Draft Flora and Fauna Guarantee Action Statement Major Mitchell's Cockatoo *Lophocroa leadbeateri*'. Retrieved 13 April, 2011 from <http://www.dse.vic.gov.au/CA256F310024B628/0/62C9AA28BF539B8ECA257489000C0FF9/$File/Major+Mitchell+Draft+Action+Statement+.pdf>.

Emison WB, Beardsell CM, Norman FI, Loyn RH, Bennett SC (1987) *Atlas of Victorian Birds.* Department of Conservation, Forests and Lands and RAOU, Melbourne.

Gibson M, Singarayer F, Hurley VG (2008) 'Age distribution of Slender Cypress-pine (*Callitris gracilis*) within Pine Plains, Wyperfeld National Park'. Report to Department of Sustainability and Environment, Melbourne.

Higgins PJ (Ed.) (1999) *Handbook of Australian, New Zealand and Antarctic Birds. Volume 4: Parrots to Dollarbird.* Oxford University Press, Melbourne.

Hurley VG (2008) 'Results of pre-breeding removal of Galahs (*Eolophus roseicapillus*) from hollow bearing trees, Pine Plains, Wyperfeld NP'. Report to Department of Sustainability and Environment, Melbourne.

Hurley VG (2009) 'A report on installing nest boxes and repair of degraded nest hollows in Callitris Pine for use by Major Mitchell's Cockatoo (*Lophochroa l. leadbeateri*) Pine Plains, Wyperfeld NP'. Report to Mallee Catchment Management Authority, Mildura.

Hurley VG (2011) 'Draft report: results from the 2010 breeding survey of Major Mitchell's cockatoo (*Lophochroa l. leadbeateri*) Pine Plains, Wyperfeld NP'. Report to Department of Sustainability and Environment, Mildura.

Rowley I, Chapman G (1991) The breeding biology, food, social organisation, demography and conservation of the Major Mitchell or Pink Cockatoo, *Cacatua leadbeateri*, on the margin of the Western Australian wheatbelt. *Australian Journal of Zoology* **39**, 211–261.

Saunders DA, Rowley I, Smith GT (1985) The effects of clearing for agriculture on the distribution of cockatoos in the southwest of Western Australia. In *Birds of Eucalypt Forests and Woodlands: Ecology, Conservation, Management.* (Eds A Keast, HF Recher, HA Ford and DA Saunders). RAOU and Surrey Beatty and Sons, Melbourne and Chipping Norton, NSW.

Comments received from

Victor Hurley, Damon Oliver

Subspecies Conservation Summary

Double-eyed Fig-Parrot (southern, Coxen's)

Cyclopsitta diophthalma coxeni Gould, 1867

Psittacidae

Conservation status

Endangered D

Reasons for listing

Very small population which has been so small for so long that it is unlikely still to be declining

Status 2000

Endangered D

Status 1990

Endangered D

Taxonomy

C. d. marshalli (Marshall's Fig-Parrot; Cape York Peninsula) and *C. d. macleayana* (Macleay's Fig Parrot; north-east Queensland) are Least Concern, as is the species. There are 5 extralimital subspecies.

Taxonomic uniqueness: low (76 genera/family, 2 species/genus, 8 subspecies/species)

Range

Recorded occasionally between Rockhampton in central Queensland to the Richmond River in north-eastern New South Wales, and west to the Bunya Mountains, Main Range, Richmond Range and Koreelah Range (Department of Sustainability, Environment, Water, Population and Communities 2011). It has become very rare since at least the 1920s (Olsen 2007).

Abundance

The population size has been estimated, with low reliability, at 100 breeding birds and suspected to be declining in size (Garnett and Crowley 2000; Coxen's Fig-Parrot Recovery Team 2001). With even lower reliability, this was estimated to be divided into 4 subpopulations of maximum 50 individuals each. There were c.30 sightings in NSW and c.90 in Qld between 1970–2000 (Coxen's Fig-Parrot Recovery Team 2001). Although recent targeted surveys have been largely unsuccessful, incidental sightings continue to be reported sporadically by members of the public including a series of reports of up to 4 individuals in 2001–2004 from south-east Qld at Montville, Main Range NP, Imbil State Forest, Lamington NP, Beerwah Forest Reserve and Maleny (I. Gynther *in litt.*). Given a similar rate of sightings over previous decades, it is inferred to have had a very small population for at least several decades and can no longer to be declining or it would be extinct.

Ecology

The primary habitat is lowland subtropical rainforest, dry rainforest, littoral and developing littoral rainforest, sub-littoral mixed scrub, riparian corridors in woodland, open woodland and otherwise cleared land, and urbanised and agricultural areas with fig trees *Ficus* spp. It feeds on the seeds of figs, but also on fruit of other native and exotic trees, as well as nectar, lichen and insect larvae (Higgins 1999; Coxen's Fig-Parrot Recovery Team 2001). A generation time of 4.8 years (BirdLife International 2011) is derived from an age at first breeding of 1.0 years and a maximum longevity in the wild of 7.6 years, extrapolated from mean survival in captivity of Large Fig-Parrot *Psittaculirostris desmarestii* and Edwards's Fig-Parrot *P. edwardsii* (Brouwer *et al.* 2000)).

Threats

Widespread clearing of lowland rainforest and logging of rainforest trees that occurred around the turn of the 20th century have left just fragments of the parrot's former habitat (Coxen's Fig-Parrot Recovery Team 2001). Habitat that remains is vulnerable to weed invasion and there is little natural recruitment of large isolated figs that may constitute important lowland habitat (Department of Sustainability, Environment, Water, Population and Communities 2011). As no Coxen's Fig-Parrot are held in captivity, any nests found may be vulnerable to illegal egg or bird collectors (Holmes 1990).

Conservation objectives

1. Wild birds located
2. A captive population established
3. Extent, quality and connectivity of habitat extended

Information required

1. A reliable technique for finding Coxen's Fig-Parrots
2. Ecological requirements

Current eligibility against IUCN Red List Criteria

IUCN category	Criteria eligibility
A	Not applicable: past, current or future population declines are thought unlikely to exceed 20% in any 3-generation period
B	Not applicable: EOO, AOO, habitat, locations or population not declining, population fluctuations not extreme
C	Not applicable: no continuing population decline or extreme fluctuations
D	Endangered: population 50–250 mature individuals
E	Not applicable: no population viability analysis undertaken

IUCN Red List assessment data

	Estimate	Reliability
Extent of occurrence trend	65 000 km^2 stable	low low
Area of occupancy trend	330 km^2 stable	low low
No. of mature individuals trend	100 stable	low low
No. subpopulations	4	low
No. locations	>10	low
Largest subpopulation	50	low
Generation time	4.8 years	low
Global population share	100%	high

Management actions required

1. Encourage reporting of sightings by the community
2. Thoroughly investigate all sightings
3. Undertake extension to private land-holders and other land managers with potential habitat
4. Ensure currency of a contingency plan in the event of reliable sightings or the discovery of a nest

Bibliography

BirdLife International (2011) 'Species factsheet: *Cyclopsitta diophthalma*'. Retrieved 17 January, 2011 from <http://www.birdlife.org/>.

Brouwer K, Jones ML, King CE, Schifter H (2000) Longevity records for Psittaciformes in captivity. *International Zoo Yearbook* **37**, 299–216.

Coxen's Fig-Parrot Recovery Team (2001) 'Coxen's fig-parrot *Cyclopsitta diophthalma coxeni* recovery plan 2001–2005. Report to Environment Australia, Canberra'. Queensland Parks and Wildlife Service, Brisbane.

Department of Sustainability, Environment, Water, Population and Communities (2011) *Cyclopsitta diophthalma coxeni*. In 'Species Profile and Threats Database'. Department of Sustainability, Environment, Water, Population and Communities, Canberra. Retrieved 24 February, 2011 from <http://www.environment.gov.au/sprat>.

Garnett S, Crowley GM (2000) *The Action Plan for Australian Birds 2000*. Environment Australia, Canberra.

Higgins PJ (Ed.) (1999) *Handbook of Australian, New Zealand and Antarctic Birds. Volume 4: Parrots to Dollarbird*. Oxford University Press, Melbourne.

Holmes G (1990) 'The biology and ecology of Coxen's Fig-Parrot'. RAOU Report 65. RAOU, Melbourne.

Olsen P (2007) *Glimpses of Paradise: The Quest for the Beautiful Parakeet*. National Library of Australia, Canberra.

Comments received from

Ian Gynther

Subspecies Conservation Summary

Eclectus Parrot (Cape York Peninsula)

Eclectus roratus macgillivrayi Mathews, 1913

Psittacidae

Conservation status

Near Threatened D1

Reasons for listing

Population near 1000 mature individuals given male-biased sex ratio

Status 2000

Near Threatened D1

Status 1990

Near Threatened D1

Current eligibility against IUCN Red List Criteria

IUCN category	Criteria eligibility
A	Not applicable: past, current or future population declines are thought unlikely to exceed 20% in any 3-generation period
B	Not applicable: EOO, AOO, habitat, locations or population not declining, population fluctuations not extreme
C	Not applicable: no continuing population decline or extreme fluctuations
D	Near Threatened: population near 1000 mature individuals
E	Not applicable: population viability analysis predicts 0–5% chance of extinction over 100 years

IUCN Red List assessment data

	Estimate	Reliability
Extent of occurrence trend	5000 km² stable	high high
Area of occupancy trend	1000 km² stable	medium high
No. of mature individuals trend	2000 stable	medium medium
No. subpopulations	1	high
No. locations	>10	high
Generation time	19.9 years	medium
Global population share	100%	high

Taxonomy

E. r. polychloros (lowland New Guinea and offshore islands, including northern Torres Strait Is.) is Least Concern, as is the species. There are 8 extralimital subspecies.

Taxonomic uniqueness: medium (76 genera/family, 1 species/genus, 9 subspecies/species)

Range

Endemic to rainforests of Iron Range and McIlwraith Range of Cape York Peninsula, Queensland, between the Pascoe R. and Rocky R. (Forshaw and Cooper 1989; Higgins 1999). No evidence of any change in range.

Abundance

Population size is estimated as 2000 mature birds based on an estimate of 538–596 breeding females, and 1059–1173 males in 500 km^2 of rainforest around Iron Range National Park (Legge *et al.* 2004), which is c.50% of the subspecies' range, and compensating for the sex bias. No evidence of any decline.

Ecology

On Cape York Peninsula, Eclectus Parrots inhabit rainforest. They nest in hollows in emergent rainforest trees, often cooperatively with more than 1 male attending the female at the nest. The usual clutch size is 2 and about 18% of eggs result in fledglings (Heinsohn and Legge 2003), although more recent analyses suggest this is lower when there are extreme climatic events, such as El Niño events and cyclones (R. Heinsohn *in litt.*). The breeding season is sometimes curtailed because of nest hollow flooding (Heinsohn and Legge 2003) or can fail altogether after cyclones (R. Heinsohn *in litt.*). Eclectus Parrots feed on a variety of fruits, mostly from within the rainforest (Forshaw and Cooper 1989; Higgins 1999). A generation time of 19.9 years (BirdLife International 2011) is derived from an age at first breeding of 4.0 years (extrapolated from other Psittacidae) and an annual survival of adults of 93.7% (Heinsohn and Legge 2003).

Threats

There is less than 1 useable hollow per km^2 of rainforest, which limits population size, and at least 1 of their 3 major nest tree types, Black Bean *Castanospermum australe* has commercial value: if 20% of mature Black Bean trees are removed by logging, the probability of extinction as modelled by the Population Viability Analysis rises to 10–20% over 100 years (R. Heinsohn *in litt.*). Other pressures do not constitute threats: an average of 4% of nest-trees were lost each year to natural causes, a similar rate to which they are generated (Heinsohn and Legge 2003; Heinsohn *et al.* 2003). Sulphur-crested Cockatoos *Cacatua galerita* compete for nest-trees and permanently or temporarily take over 10–26% of nest-holes each year (Heinsohn *et al.* 2003). Amethystine Pythons *Morelia amethistina* and monitors *Varanus* spp. sometimes take adults and nestlings, and cause c.34% of nest failures (Heinsohn and Legge 2003). Adults are occasionally killed by Rufous Owl *Ninox rufa* and Peregrine Falcon *Falco peregrines* (Legge *et al.* 2003).

Conservation objectives

1. Stable population
2. Stable number of suitable nest-trees

Information required

1. Long-term trends in nest-tree turnover and occupancy

Management actions required

1. Effectively conserve rainforest in Iron and McIlwraith Ranges

Bibliography

BirdLife International (2011) 'Species factsheet: *Eclectus roratus*'. Retrieved 17 January, 2011 from <http://www.birdlife.org/>.

Forshaw JM, Cooper WD (1989) *Parrots of the World.* Lansdowne Press, Willoughby.

Heinsohn R, Legge S (2003) Breeding biology of the reverse-dichromatic, co-operative parrot *Eclectus roratus. Journal of The Zoological Society of London* **259**, 197–208.

Heinsohn R, Murphy S, Legge S (2003) Overlap and competition for nest holes among eclectus parrots, palm cockatoos and sulphur-crested cockatoos. *Australian Journal of Zoology* **51**, 81–94.

Higgins PJ (Ed.) (1999) *Handbook of Australian, New Zealand and Antarctic Birds. Volume 4: Parrots to Dollarbird.* Oxford University Press, Melbourne.

Legge S, Heinsohn R, Blackman C, Murphy S (2003) Predation by Rufous Owls on Eclectus Parrots and other animals at Iron Range National Park, Cape York. *Corella* **27**, 45–46.

Legge S, Heinsohn R, Garnett ST (2004) Availability of nest hollows and breeding population size of eclectus parrots, *Eclectus roratus*, on Cape York Peninsula, Australia. *Wildlife Research* **31**, 149–161.

Comments received from

Rob Heinsohn, Sarah Legge, Keith McDonald, Steve Murphy

Subspecies Conservation Summary

Regent Parrot (eastern)

Polytelis anthopeplus monarchoides Schodde, 1993
Psittacidae

Conservation status

Endangered B2ab(iii,v)

Reasons for listing

Probably occupies <500 km^2 scattered through 3 locations, the quality of which are declining as old trees die and habitat is overgrazed

Status 2000

Endangered B2ab(iii,v)

Status 1990

Endangered B2ab(iii,v)

Taxonomy

P. a. anthopeplus (south-west Western Australia) is Least Concern, as is the species
Taxonomic uniqueness: medium (76 genera/family, 3 species/genus, 2 subspecies/species)

Range

Occurs in the semi-arid interior of southeastern mainland Australia, breeding in 3 distinct and widely separated areas: southern mallee subpopulation along the Wimmera River/Outlet Creek system in western Victoria, predominantly in Wyperfeld National Park, Lake Albacutya and Lake Hindmarsh; the lower Murray River subpopulation in South Australia, upstream from Swan Reach to Lindsay I., far northwest Vic; the mid Murray River subpopulation in Vic and NSW, between Red Cliffs, south-east of Mildura and Piangil, including the lower Murrumbidgee and Wakool Rivers in NSW (Baker-Gabb and Hurley 2011; Department of Sustainability, Environment, Water, Population and Communities 2011). Non-breeding birds occur throughout the eastern mallee region (Higgins 1999) suggesting that there is mixing of the southern mallee and mid-Murray populations. Its range appears to have changed little in recent decades but has contracted since the late 19th century, when it bred in mallee Eucalypts in the Murray Sunset National Park, further south along the Wimmera River as far as Dimboola, much further north along the Darling River at Pooncarie and along the Murray River from upstream of Kerang and along the Avoca River to the lower reaches of the Murray River in SA (Baker-Gabb and Hurley 2011), although the birds may always have been sparse between the 2 extant Murray River breeding areas (Burbidge 1985).

Abundance

There are estimated to be 1500 mature individuals including 600 in NSW, 500 in Vic (Sluiter 2007), and 400 in SA (Smith 2011), although flock sizes of up to 900 individuals suggest that there is also a substantial number of immature birds (Baker-Gabb and Hurley 2011). The number of birds breeding along the Murray River in SA declined by about 30% between 1991 and 2003 (Schultz 2006), with a continued decline upstream of Lock 3 and loss of 60% of colonies and 66% pair from 1991 to 2010 SMITH 2011 . The number in Mallee Cliffs State Forest and Kemendoc Nature Reserve, NSW, declined from 400 individuals in 1993 to 150 in 1997 (Webster and Leslie 1997).

Ecology

Nests in hollows in loose colonies of up to 76 pairs (Smith 2011) over areas of up to 90 ha in littoral River Red Gums *Eucalyptus camaldulensis* and Black Box *E. largiflorens* woodlands, though most colonies are smaller (Baker-Gabb and Hurley 2011). An unusual outlier population was recently discovered breeding in Slender Cypress-Pine *Callitris gracilis* in northern Wyperfeld NP (Hurley 2011).

Current eligibility against IUCN Red List Criteria

IUCN category	Criteria eligibility
A	Not applicable: past, current or future population declines are thought unlikely to exceed 20% in any 3-generation period
B	Endangered: AOO <500 km^2, at ≤5 locations with plausible future threat, inferred continuing decline in habitat quality, no. individuals
C	Vulnerable: population 2500–10 000 mature individuals, suspected continuing decline, 100% in 1 subpopulation
D	Vulnerable: ≤5 locations with plausible future threat
E	Not applicable: no population viability analysis undertaken

IUCN Red List assessment data

	Estimate	Reliability
Extent of occurrence trend	61 000 km^2 stable	high high
Area of occupancy trend	500 km^2 decreasing	low medium
No. of mature individuals trend	3000 decreasing	medium medium
No. subpopulations	2	medium
No. locations	3	medium
Largest subpopulation	1500	low
Generation time	7.5 years	low
Global population share	100%	low

They lay 5–6 eggs but usually fledge only 2–4 young (Hurley 2009). They rely on vegetated landscapes and/or corridors to undertake foraging flights in both the breeding season (Sluiter 2007) and non-breeding seasons (Webster and Belcher 2008; Sluiter 2009). The parrots forage for seeds of grasses and herbaceous plants but also eat fruits, buds, flowers and occasionally insect larvae, psyllids and lerps. They usually feed in pairs and small parties, but sometimes in large flocks. They often forage in mallee within 20 km of the nesting site, particularly where Christmas Mallee *E. socialis* or Yellow Mallee *E. incrassata* dominate (Higgins 1999). They also forage on cereal crops in farmland and along roadsides and in open ground near remnant vegetation (Schultz 2006) although they are generally reluctant to cross open country (Webster and Leslie 1997; Sluiter 2007; Webster and Belcher 2008; Sluiter 2009). A generation time of 7.5 years (BirdLife International 2011) is derived from an average age at first breeding of 1.0 years and a maximum longevity in the wild of 14.0 years, both extrapolated from Superb Parrot *Polytelis swainsonii*.

Threats

The main threat has been the clearing, fragmentation and degradation of riparian woodlands that are used for breeding, and of mallee areas used for foraging within 20 km of nesting colonies and of wooded flight corridors between the 2 (Burbidge 1985; Webster and Leslie 1997). Such habitat deterioration was historically caused by deliberate clearing and timber extraction. This loss has largely ceased but degradation continues through waterlogging of riparian trees, resulting in their death and eventual collapse, and grazing by both domestic stock and feral herbivores such as goats *Capra hircus* and rabbits *Oryctolagus cuniculus*, which prevents regeneration of trees or may reduce food availability, particularly away from the breeding colony (Baker-Gabb and Hurley 2011). Other causes of death, though not necessarily threats to the subspecies, include trapping for the cage bird trade, deliberate killing to reduce agricultural damage, accidental killing through collisions with vehicles, ingestion of poison grain (Burbidge 1985) and competition for nest hollows from other birds advantaged by agricultural development such as Common Starling *Sturnus vulgaris*, Little Corella *Cacatua sanguinea* and Sulphur-crested Cockatoo *Cacatua galerita* (Robertson and Hurley 2010; Baker-Gabb and Hurley 2011). Competition with feral European Honey Bees *Apis mellifera* was not found to be a significant problem in dry years (Oldroyd *et al.* 1994). However, a subsequent survey in 2009 found increases in the number feral bee hives occupying River Red Gums within Regent Parrot breeding colonies during that breeding season (Robertson and Hurley 2010).

Conservation objectives

1. Existing subpopulations stable or increasing
2. Major nesting areas, including breeding trees, foraging habitat and intervening corridors, secure

Information required

1. Population and demographic trends in key populations
2. Non-lethal, economically viable, damage mitigation techniques for fruit-growers affected by the parrots
3. Significance of feral bee hives in breeding colonies during years of above average rainfall
4. Impacts of grazing on foraging and breeding success
5. Extent of movement between subpopulations

Management actions required

1. Locate and protect actual and potential nesting colonies, flyways to foraging areas and traditional watering points
2. Improve foraging habitat quality by reducing grazing pressure
3. Secure and supply environmental water to core breeding areas to simulate natural wetting and drying cycles
4. Control recreational impacts near nesting colonies
5. Reduce numbers of birds killed by shooting, poisoning and vehicles
6. Increase community involvement in many aspects of the recovery program

Bibliography

Baker-Gabb D, Hurley VG (2011) 'Draft National Recovery Plan for the Regent Parrot (eastern subspecies) *Polytelis anthopeplus monarchoides*'. Department of Sustainability and Environment, Melbourne.

BirdLife International (2011) 'Species factsheet: *Polytelis anthopeplus*'. Retrieved 17 January, 2011 from <http://www.birdlife.org/>.

Burbidge AH (1985) 'The Regent Parrot. A report on the breeding distribution and habitat requirements along the Murray River in south-eastern Australia'. Australian National Parks and Wildlife Service, Report Series No. 4.

Department of Sustainability, Environment, Water, Population and Communities (2011) *Polytelis anthopeplus monarchoides*. In 'Species Profile and Threats Database'. Department of Sustainability, Environment, Water, Population and Communities, Canberra. Retrieved 24 February, 2011 from <http://www.environment.gov.au/sprat>.

Higgins PJ (Ed.) (1999) *Handbook of Australian, New Zealand and Antarctic Birds. Volume 4: Parrots to Dollarbird*. Oxford University Press, Melbourne.

Hurley VG (2009) 'Results of nest site monitoring of active Regent Parrot (*Polytelis anthopeplus monarchoides*) nests, in the Outlet Creek system, Wyperfeld NP, during spring 2008'. Report to the Department of Sustainability and Environment, Mildura.

Hurley VG (2011) 'Results from the 2010 breeding survey of Major Mitchell's Cockatoo (*Lophochroa l. leadbeateri*) Pine Plains, Wyperfeld NP'. Report to the Department of Sustainability and Environment, Mildura.

Oldroyd BP, Lawler SH, Crozier RH (1994) Do feral honey bees (*Apis mellifera*) and regent parrots (*Polytelis anthopeplus*) compete for nest sites? *Australian Journal of Ecology* **19**, 444–450.

Robertson P, Hurley VG (2010) 'Nest site activity patterns of Regent Parrots (*Polytelis anthopeplus*) at Outlet Creek, Wyperfeld National Park'. Report to the Mallee Catchment Management Authority, Mildura.

Schultz MA (2006) 'Recovery Plan for the Regent Parrot (eastern subspecies) *Anthopeplus polytelis monarchoides* in the South Australian Murray Darling Basin'. Department for Environment and Heritage Adelaide, South Australia.

Sluiter IRK (2007) 'Observations on Spring Regent Parrot movements at Wemen between 1998 and 2007'. Ogyris P/L, Mildura.

Sluiter IRK (2009) 'Observations on the Regent Parrot in the Ouyen area: results and implications of non-breeding season survey in late March 2009'. Ogyris Ecological report no. 2009/02, Mildura.

Smith K (2011) 'Results of the 2010 Regent Parrot nest surveys in the SA Murray Darling Basin'. Report to the South Australian Department for Environment and Heritage, Murraylands, Berri.

Webster R, Belcher C (2008) 'Survey for the foraging habitat of the Regent Parrot (*Polytelis anthopeplus monarchoides*) in the breeding season in the Lower Murray Darling CMA area'. Ecosurveys P/L, Deniliquin.

Webster R, Leslie D (1997) 'Assessment of Regent Parrot *Polytelis anthopeplus* breeding habitat in south-western New South Wales'. Report to the NSW National Parks and Wildlife Service and State Forests of New South Wales, Sydney.

Comments received from

David Baker-Gabb, Victor Hurley, Rick Webster

Species Conservation Summary

Princess Parrot

Polytelis alexandrae Gould, 1863

Psittacidae

Conservation status

Near Threatened D1

Reasons for listing

Population possibly approaching 1000 individuals in poor years

Status 2000

Near Threatened D1

Status 1990

Near Threatened D1

Taxonomy

No infraspecific taxa described

Taxonomic uniqueness: medium (76 genera/family, 3 species/genus, 1 subspecies/species)

Range

Scattered and irregular across the arid regions from near Oodnadatta in South Australia, west to near Coolgardie and the east Murchison River in Western Australia, and north to near the Fitzroy River in Western Australia and to Howell Ponds in the Northern Territory (Higgins 1999; Baxter and Henderson 2000). It may be concentrated in the Great Sandy, Gibson, Tanami and Great Victoria Deserts, and in the central ranges (Blyth and Burbidge 1997; Higgins 1999). Irregularity confounds

Current eligibility against IUCN Red List Criteria

IUCN category	Criteria eligibility
A	Not applicable: past, current or future population declines are thought unlikely to exceed 20% in any 3-generation period
B	Not applicable: EOO >20 000 km^2, AOO >2000 km^2
C	Not applicable: no continuing population decline or extreme fluctuations
D	Near Threatened: population near 1000 mature individuals in poor years
E	Not applicable: no population viability analysis undertaken

IUCN Red List assessment data

	Estimate	Reliability
Extent of occurrence trend	960 000 km^2 stable	medium medium
Area of occupancy trend	10 000 km^2 stable	low low
No. of mature individuals trend	1200 stable	low low
No. subpopulations	1	high
No. locations	>10	medium
Generation time	7.7 years	medium
Global population share	100%	high

determination of any change in distribution but records from periphery of range, in south-west and north-west WA, northern NT and northern SA have become less frequent since the 1950s (Blyth and Burbidge 1997).

Abundance

Historical records include reports of large flocks and large breeding colonies (Department of Sustainability, Environment, Water, Population and Communities 2011). Contemporary records are mostly of very small numbers, with significant numbers only recorded twice: up to 300 on Canning Stock Route in early 1990s (Carter 1993) and several hundred west of Alice Springs in 2010 (A. Silcocks *in litt.*), extending to the eastern Gibson Desert where the population may have been in the low thousands (I. May *in litt.*). It is suspected that the total population probably varies with season and may be as low as 1000 mature individuals in poor years. Irregularity compounds determination of population trends; earlier suggestions that it is declining (Reid and Fleming 1992) were made before the recent breeding events in central Australia. The population is assessed here as being stable, although fluctuating in both time and space.

Ecology

Usually recorded from shrublands and savanna woodlands in swales between sand dunes, with occupied sites typically consisting of spinifex *Triodia* spp. and a variety of shrubs and scattered emergent trees. Some observers have noted birds feeding in areas recovering from recent fire (J. Reid *in litt.*), others noted that the occupied sites had not been burnt for a long time (I. May *in litt.*). Also sometimes in vegetated riverine and littoral areas. Feeds on seeds of grasses including spinifex, and on nectar from flowering trees (Carter 1993; Higgins 1999). Nests have been recorded in hollows in River Red Gum *Eucalyptus camaldulensis*, Marble Gum *E. gongylocarpa* and Desert Oak *Allocasuarina decaisneana*, with clutch sizes of 3 to 6 (Higgins 1999; Baxter and Henderson 2000); in a recent irruption near Mereenie the birds favoured Marble Gum for nesting (I. May *in litt.*). They may breed in core areas in wet years then irrupt or disperse more widely (Higgins 1999). A generation time of 7.7 years (BirdLife International 2011) is derived from an age at first breeding of 2.0 years and a maximum longevity of 13.4 years, both values extrapolated from captive birds where maximum recorded longevity 24.0 years.

Threats

No known threats exist, but Princess Parrots may be adversely affected by the same habitat changes that led to the extinction of many central Australian mammals over the 20th century (Burbidge and McKenzie 1989). Throughout their range, habitat may have been degraded and food availability reduced by altered fire regimes with a coarser mosaic of fire history, and the introduction of sheep and feral herbivores such as rabbits *Oryctolagus cuniculus* and camels *Camelus dromedarius*. Loss of tree-hollows from intense fire may affect breeding success but is unlikely to be limiting in the short term. Increased availability of water in areas grazed by domestic stock may have allowed other more water-dependent parrots to expand into the arid zone and compete with Princess Parrots. Psittacine Beak and Feather disease and poaching of eggs or young from nests are potential and minor threats, respectively (Carter 1993; Higgins 1999; Department of Sustainability, Environment, Water, Population and Communities 2011).

Conservation objectives

1. Populations known from more than 1 site at any 1 time
2. Successful breeding at monitored nests

Information required

1. Distribution (particularly core source areas) and population size
2. Ecology and breeding success at known nesting locations
3. Availability of suitable nesting habitat, specifically the extent of tall Marble Gum open forest groves in the North Mereenie area
4. Fire history, rainfall and grazing pressure correlated with records to determine drivers for dispersal

Management actions required

1. Protect any known breeding locations, including management of fire and feral herbivores

Bibliography

Baxter C, Henderson R (2000) A literature summary of the Princess Parrot *Polytelis alexandrae* and a suspected recent breeding event in South Australia. *South Australian Ornithologist* **33**, 93–108.

BirdLife International (2011) 'Species factsheet: *Polytelis alexandrae*'. Retrieved 17 January, 2011 from <http://www.birdlife.org/>.

Blyth J, Burbidge AH (1997) What do we know about the Princess Parrot *Polytelis alexandrae*? *Eclectus* **3**, 26–29.

Burbidge AA, McKenzie NL (1989) Patterns of modern decline of Western Australia's vertebrate fauna: causes and conservation implications. *Biological Conservation* **50**, 143–198.

Carter M (1993) Alexandra's or Princess Parrot – status and normal range. *Wingspan* **12**, 32–35.

Department of Sustainability, Environment, Water, Population and Communities (2011) *Polytelis alexandrae*. In 'Species Profile and Threats Database'. Department of Sustainability, Environment, Water, Population and Communities, Canberra. Retrieved 24 February, 2011 from <http://www.environment.gov.au/sprat>.

Higgins PJ (Ed.) (1999) *Handbook of Australian, New Zealand and Antarctic Birds. Volume 4: Parrots to Dollarbird*. Oxford University Press, Melbourne.

Reid J, Fleming M (1992) The conservation status of birds in arid Australia. *Rangeland Journal* **14**, 65–91.

Comments received from

Allan Burbidge, Ian May, Chris Pavey, Julian Reid, Andrew Silcocks

Subspecies Conservation Summary

Green Rosella (King Island)

Platycercus caledonicus brownii (Kuhl, 1820)
Psittacidae

Conservation status

Vulnerable D1

Reasons for listing

Population <1000 mature individuals

Status 2000

Vulnerable D1

Status 1990

Vulnerable D1

Taxonomy

P. c. caledonicus (Tasmania and nearby islands) is Least Concern, as is the species
Taxonomic uniqueness: medium (76 genera/family, 6 species/genus, 2 subspecies/species)

Range

Endemic to King I., Tasmania where it used to be common but is now largely confined to the Pegarah forest after large-scale clearance of vegetation (Green and McGarvie 1971).

Abundance

There have been no counts. The population estimate and trend is based on local knowledge (R. Dongahey pers. comm.).

Ecology

Green Rosellas on King I. are largely confined to eucalypt forests where, like the nominate subspecies in Tasmania, they probably feed on a wide range of seeds, including those of grasses, eucalypts and other trees as well as fruit, blossom and insects. They nest annually in hollows, in which they lay 4–5 eggs (Higgins 1999; Donaghey 2003). A generation time of 5.9 years (BirdLife International 2011) is derived from an age at first breeding of 1.5 years and a maximum longevity in the wild of 10.2 years, both extrapolated from congeners.

Threats

Approximately 70% of the native vegetation on King I. has been cleared, resulting in a highly fragmented and developed landscape (Green and McGarvie 1971). Fire frequency is the main problem in remaining vegetation. More than 95% of the vegetation burnt in 2001 is thought to have been burnt again in 2007, and the relatively short period between fires is likely to have disadvantaged the habitat regeneration, which will affect hollow availability in the long term (Threatened Species Section 2010). Scarcity of nesting sites may limit population size (McGarvie and Templeton 1974). Although some nest boxes are available (Threatened Species Section 2010), rosellas may face competition from Common Starlings *Sturnus vulgaris* for nest hollows, though this is

Current eligibility against IUCN Red List Criteria

IUCN category	Criteria eligibility
A	Not applicable: past, current or future population declines are thought unlikely to exceed 20% in any 3-generation period
B	Not applicable: EOO, AOO, habitat, locations or population not declining, population fluctuations not extreme
C	Not applicable: no continuing population decline or extreme fluctuations
D	Vulnerable: population 250–1000 mature individuals
E	Not applicable: no population viability analysis undertaken

IUCN Red List assessment data

	Estimate	Reliability
Extent of occurrence trend	800 km^2 stable	high high
Area of occupancy trend	50 km^2 stable	low high
No. of mature individuals trend	500 stable	medium low
No. subpopulations	1	high
No. locations	1	high
Generation time	5.9 years	medium
Global population share	100%	high

unconfirmed (Higgins 1999). Weeds and feral cats *Felis catus* are also listed as secondary threats (Threatened Species Section 2010), but without evidence.

Conservation objectives

1. Demonstrably stable population

Information required

1. Population size and trends
2. Hollow supplementation requirements
3. Extent of competition with Common Starlings and appropriate counter-measures
4. Effect of fire on habitat suitability

Management actions required

1. Monitor population size and trends
2. Investigate hollow availability and, if necessary, protect nest hollows against Common Starlings or erect extra, starling-proof nest-boxes for rosellas
3. Investigate effect of fire on food and hollow availability and, if necessary, reduce fire frequency
4. Encourage retention of trees with hollows for nesting

Bibliography

BirdLife International (2011) 'Species factsheet: *Platycercus caledonicus*'. Retrieved 17 January, 2011 from <http://www.birdlife.org/>.

Donaghey R (Ed.) (2003) *The Fauna of King Island. A Guide to Identification and Conservation Management.* King Island Natural Resource Management Group, Currie, King Island.

Green R, McGarvie AM (1971) The birds of King Island. *Records of the Queen Victoria Museum* **40**, 1–42.

Higgins PJ (Ed.) (1999) *Handbook of Australian, New Zealand and Antarctic Birds. Volume 4: Parrots to Dollarbird.* Oxford University Press, Melbourne.

McGarvie AM, Templeton MT (1974) Additions to the birds of King Island, Bass Strait. *Emu* **74**, 91–96.

Threatened Species Section (2010) 'King Island Biodiversity Management Plan'. Department of Primary Industries, Parks, Water and Environment, Hobart.

Comments received from

Richard Donaghey, Mark Holdsworth

Species Conservation Summary

Swift Parrot

Lathamus discolor (Shaw, 1790)

Psittacidae

Conservation status

Endangered C2a(ii)

Reasons for listing

Fewer than 2500 individuals, >95% in a single population and still declining

Status 2000

Endangered C2a(ii)

Status 1990

Endangered C2a(ii)

Current eligibility against IUCN Red List Criteria

IUCN category	Criteria eligibility
A	Not applicable: past, current or future population declines are thought unlikely to exceed 20% in any 3-generation period
B	Not applicable: >10 locations, not severely fragmented, population fluctuations not extreme
C	Endangered: population 250–2500 mature individuals, suspected continuing decline, >95% mature individuals in 1 subpopulation
D	Not applicable: population >1000 mature individuals, >5 locations
E	Not applicable: no population viability analysis undertaken

IUCN Red List assessment data

	Estimate	Reliability
Extent of occurrence trend	57 000 km^2 stable	medium medium
Area of occupancy trend	1400 km^2 stable	low low
No. of mature individuals trend	2000 decreasing	medium medium
No. subpopulations	1	high
No. locations	>10	high
Generation time	5.4 years	low
Global population share	100%	high

Taxonomy

No infraspecific taxa described

Taxonomic uniqueness: medium (76 genera/family, 1 species/genus, 1 subspecies/species)

Range

Breeding restricted to the region around Hobart in southern Tasmania, including Wielangta, Meehan Range, Wellington Range, Southern Forests, Tasman Peninsula, Bruny I. and Maria I., but also small numbers in central northern Tas. Non-breeding birds occur across Tas but particularly along the east coast south of Saint Helens. In non-breeding season, migrates to mainland Australia, principally Victoria and eastern New South Wales, but also irregularly in south-eastern Queensland, south-eastern South Australia and the Australian Capital Territory (Department of Sustainability, Environment, Water, Population and Communities 2011). The area of occupancy has declined significantly since European settlement, as can be inferred from the extent of habitat loss: 83% of box-ironbark woodland lost from Vic, 70% from NSW, and c.70% of grassy Tasmanian Blue Gum *Eucalyptus globulus* forest from Tas (Saunders and Tzaros 2010). The EOO includes all summer records from Tas. The AOO of all summer records is 1400 km^2. Recorded sites will not encompass all sites occupied, but this figure may also exaggerate the area actually used for breeding. AOO is particularly difficult to assess for this species, as different areas are suitable for breeding in each year.

Abundance

Population size and trends are poorly known as the species' habitat use within its broad range varies irregularly. Breeding population was estimated at 1320 breeding pairs in 1987–1988 and 940 breeding pairs in 1995–1996 (Brereton 1996). Annual surveys across non-breeding range fluctuate depending on the distribution of the birds across the landscape and flock sizes. A total of 2158 birds, including immatures, in 2010 was the highest count for many years (C. Tzaros and D. Ingwersen *in litt.*). There is little monitoring evidence of decreased abundance but much anecdotal evidence of localised declines and demonstrable loss of breeding habitat. The total population is estimated here as 2000 mature birds and declining.

Ecology

Breeds in tree-hollows in old-growth or other forest with suitable holes, in relatively close proximity to the main food source, flowering Tasmanian Blue Gum. Several pairs often nest in close proximity, in the same or neighbouring trees (Webb *et al.* 2007). Breeding success is correlated with the intensity and extent of flowering, which is highly variable between years. In poor years, Swamp (Black) Gum *E. ovata* is used as food source (Brereton *et al.* 2004). The species exhibits high site fidelity, returning to locations on an irregular cyclic basis (Saunders 2008). It disperses across eastern Tas after breeding and migrates to overwinter on the mainland in flowering woodlands and forests. They feed preferentially in largest trees (Kennedy and Overs 2001; Kennedy and Tzaros 2005). Their distribution fluctuates with food availability as they feed on psyllid lerps, seeds and fruit (Kennedy and Tzaros 2005). Non-breeding birds preferentially feed in inland box-ironbark and grassy woodlands, and coastal Swamp Mahogany *E. robusta* and Spotted Gum *Corymbia maculata* woodland when in flower; otherwise often in coastal forests from east Vic. to Central Coast of NSW. A generation time of 5.4 years (BirdLife International 2011) is derived from an age at first breeding of 2.0 years and maximum longevity of 8.8 years, both values estimated by an expert committee.

Threats

The main threat remains habitat loss and alteration within breeding and drought refuge habitats. In the breeding habitat the principal causes of loss are from forestry operation, including firewood collection. In the absence of adequate management prescriptions, foraging and nesting habitat in wet forests have been particularly prone to loss and alteration by forestry activities, either from conversion to plantation or from intensive native forest silviculture. On the mainland, besides intensive silviculture, the loss of old trees, including paddock and roadside trees, can also reduce resource availability (Saunders and Tzaros 2010). Residential and industrial development, agricultural tree senescence and dieback, regeneration suppression and frequent fire all contribute to the decline in habitat area and quality. Other less well understood potential threats include competition for food with large, aggressive nectarivorous honeyeaters and the introduced Bumblebee *Bombus terrestris*, and for nest-hollows with Common Starlings *Sturnus vulgaris*, which may be exacerbated by forest fragmentation (Hingston *et al.* 2004; Saunders and Tzaros 2010). High mortality also occurs through collisions with windows, wire netting, fences and vehicles (Pfennigwerth 2008). Wind energy turbines in south-eastern Australia are probably not a serious threat (Smales 2005). Another potential threat is Psittacine Beack and Feather Disease. Historically, both breeding and wintering habitat was extensively cleared for agriculture. Now, most box-ironbark woodlands are protected, especially in Vic.

Conservation objectives

1. Demonstrably stable or increasing population
2. Quantified improvements in the condition of both breeding and wintering habitats

Information required

1. Population trends across its range
2. Identity and priority of habitats and sites used across its range, on all land tenures
3. Breeding biology and reproductive success
4. Spatial and temporal habitat requirements
5. Migration movements and pathways
6. Disease prevalence and virulence
7. Population viability
8. Rate and type of collisions

Management actions required

1. Monitor key sites, particularly where there are management interventions
2. Monitor and manage the impact of collisions, competitions and disease
3. Identify the extent and quality of habitat
4. Promote ecological management of woodland remnants on private land
5. Place all areas of public land that support Swift Parrots under secure conservation management, particularly those in timber reserves, transport corridors and local government land
6. Develop more stringent guidelines for the management of production forests that may support breeding or feeding Swift Parrots
7. Using appropriate incentives, undertake extension with landholders who have suitable woodland habitat to promote sound management of remnants
8. Design conservation and regeneration plans at a landscape level
9. Promote revegetation and land reclamation that recreates woodland habitat with a full complement of biodiversity, including the parrot
10. Control and reduce firewood collection from areas occupied by the parrots, promoting wood-lot development close to markets, and reduce grazing densities where necessary
11. Continue to support conservation using a recovery team
12. Engage the public in research, monitoring, management and advocacy

Bibliography

BirdLife International (2011) 'Species factsheet: *Lathamus discolor*'. Retrieved 17 January, 2011 from <http://www.birdlife.org/>.

Brereton R (1996) 'The Swift Parrot Recovery Plan 1997–1999'. Tasmanian Parks and Wildlife Service, Hobart.

Brereton R, Mallick SA, Kennedy SJ (2004) Foraging preferences of Swift Parrots on Tasmanian Blue-gum: tree size, flowering frequency and flowering intensity. *Emu* **104**, 377–383.

Department of Sustainability, Environment, Water, Population and Communities (2011) *Lathamus discolor*. In 'Species Profile and Threats Database'. Department of Sustainability, Environment, Water, Population and Communities, Canberra. Retrieved 24 February, 2011 from <http://www.environment.gov.au/sprat>.

Hingston AB, Gartrell BD, Pinchbeck G (2004) How specialized is the plant-pollinator association between *Eucalyptus globulus* ssp. *globulus* and the swift parrot *Lathamus discolor*? *Austral Ecology* **29**, 624–630.

Kennedy SJ, Overs A (2001) Foraging ecology and habitat use of the Swift Parrot on the western slopes of New South Wales. *Corella* **25**, 68–74.

Kennedy SJ, Tzaros CL (2005) Foraging ecology of the Swift Parrot *Lathamus discolor* in the box-ironbark forests and woodlands of Victoria. *Pacific Conservation Biology* **11**, 158–173.

Pfennigwerth S (2008) 'Minimising the swift parrot collision threat. Guidelines and recommendations for parrot-safe building design'. WWF Australia, Sydney.

Saunders D (2008) Ecology and conservation of the Swift Parrot – an endangered austral migrant. PhD thesis, Fenner School of Environment and Society, Australian National University, Canberra.

Saunders DL, Tzaros CL (2010) 'National Recovery Plan for the Swift Parrot *Lathamus discolor*'. New South Wales Department of Environment and Climate Change, Queanbeyan, and Birds Australia, Melbourne.

Smales I (2005) 'Modelled cumulative impacts on the Swift Parrot of wind farms across the species' range in south-eastern Australia'. Biosis Research. Report for the Department of Environment and Heritage.

Webb M, Holdsworth M, Voogdt J (2007) 'Nesting requirements of the swift parrot *Lathamus discolor*'. Unpublished report to the Department of Primary Industries and Water, Hobart.

Comments received from

Barry Baker, Margaret Blakers, Allan Burbidge, Andrew Burbidge, Stuart Butchart, Les Christidis, Hugh Ford, Tim Holmes, Dean Ingwersen, Richard Loyn, Peter Menkhorst, Charles Meredith, Penny Olsen, Debbie Saunders, Andy Symes, Chris Tzaros, David Watson

Species Conservation Summary

Golden-shouldered Parrot

Psephotus chrysopterygius Gould, 1857
Psittacidae

Conservation status

Endangered B1ab(iii,v)

Reasons for listing

Extent of occurrence is <5000 km^2 at 5 locations with a continuing decline in habitat quality as grassland is invaded by ti-trees associated with a decline in nest density in parts of range

Status 2000

Endangered B1ab(iii,v)

Status 1990

Endangered B1ab(iii,v)

Taxonomy

No infraspecific taxa described
Taxonomic uniqueness: medium (76 genera/family, 5 species/genus, 1 subspecies/species)

Range

Endemic to southern and central Cape York Peninsula, Qld, where it has probably always been patchily distributed. A northern subpopulation comprising 4 locations occurs in the upper catchment of the Morehead R. and adjacent headwaters of the Alice River (1380 km^2), including parts of Alwal National Park, and Artemis, Killarney and Dixie Stations with a few pairs on Mary River and Imooya Stations,; a southern subpopulation breeds in the south-east corner of Staaten River National Park and adjacent section of Bulimba Station (400 km^2) for which fire management is combined, making it a single location (Crowley *et al.* 2004). Both subpopulations were more extensive until the early 1970s. Breeding subpopulations near Coen and Port Stewart area were last seen in the 1950s and 1 at Bulleringa persisted into the 1960s (J. Borgat pers. comm.). There have been some sightings outside this region since 2000 but sustained breeding has never been confirmed; the species has not been seen near the type locality at Normanton since 1855 nor from near Aurukun since 1913 (Crowley *et al.* 2004).

Abundance

The northern population is estimated at 1500 mature individuals based on surveys in 2009, and 1000 individuals are assumed to be in the southern population, based on partial surveys in 1999 and 2004. Continuing the trend since at least the 1920s, nest densities and presumably population numbers, have continued to decline between 2000 and 2009, as ti-tree *Melaleuca* spp. has invaded grassland (Crowley *et al.* 2004; Preece *et al.* 2009).

Ecology

Nests are excavated in mounds of termites *Amitermes* spp. on grassy areas within ti-tree or Shiny-leafed Box *Eucalyptus chlorophylla* savannas. After breeding, the birds disperse through open Darwin Stringybark *E. tetrodonta* woodland, concentrating their feeding activity on the super-abundant seeds of fire grasses *Schizachyrium* spp., on which they need to forage for less than 2 hours a day (Crowley and Garnett 1999; Crowley *et al.* 2004). After the first wet-season rains, they form flocks in association with breeding Black-faced Woodswallow *Artamus cinereus*, so predation by Pied Butcherbird *Cracticus nigrogularis* is less likely (Crowley *et al.* 2004). Through the wet season the parrots switch foods continually, feeding through most of the daylight hours. Nesting begins in the late wet season, as seeds of perennial grass, particularly *Alloteropsis semialata* (Crowley and Garnett 2001), and green legumes become abundant. A generation time of 4.1 years (BirdLife International 2011) is derived from an age at first breeding of 1.0 years and a maximum longevity in the wild of 8.0 years (Crowley *et al.* 2004).

Threats

The contraction in range of the Golden-shouldered Parrot has been linked to a change in fire regime, with an ongoing avoidance of intentionally hot burns in areas under pastoral management (Crowley and Garnett 2000; Crowley *et al.* 2009), and lower fuel loads as a result of cattle grazing. These 2 factors are thought to have caused an invasion of grassland by woodland (Stanton 1992; Neldner *et al.* 1997; Crowley and Garnett 1998; Crowley *et al.* 2009), which in turn appears to increase the success rate of the parrot's predators. Pied Butcherbirds, which can take advantage of the increased vegetation density to ambush prey, are thought to be the most significant predators. The parrots are particularly vulnerable to predation when feeding on the ground. The availability of seed from perennial grasses used by the parrot in the mid-wet season has been reduced by cattle grazing (Crowley and Garnett 2001), and, during the breeding season, most food is obtained where vegetation thickening is most marked (Crowley *et al.* 2004). In the longer term termite

Current eligibility against IUCN Red List Criteria

IUCN category	Criteria eligibility
A	Not applicable: past, current or future population declines are thought unlikely to exceed 20% in any 3-generation period
B	Endangered: EOO <5000 km^2, at ≤5 locations with plausible future threat, inferred continuing decline in habitat quality
C	Not applicable: population 500–2500 mature individuals, continuing decline not observed or estimated, largest subpopulation >1000 mature individuals but <95% of total population
D	Vulnerable: ≤5 locations with plausible future threat
E	Not applicable: no population viability analysis undertaken

IUCN Red List assessment data

	Estimate	Reliability
Extent of occurrence trend	3000 km^2 stable	high medium
Area of occupancy trend	1800 km^2 decreasing	medium medium
No. of mature individuals trend	2500 decreasing	medium medium
No. subpopulations	2	medium
No. locations	5	medium
Largest subpopulation	1500	medium
Generation time	4.1 years	medium
Global population share	100%	high

mound availability may be limiting: monitoring suggests that mounds large enough for the parrots are being lost faster than they are being replaced, largely from damage by cattle and feral pigs *Sus scrofa* (Crowley *et al.* 2004).

Conservation objectives

1. Persistence within current range
2. Improvement in habitat condition

Information required

1. Population trends
2. Trends in habitat condition
3. Location of early wet season flocks in the southern population
4. Western and southern breeding limits of southern population

Management actions required

1. Surveys of wet season and breeding distributions of southern population
2. Monitor both populations at 5-year intervals
3. Reduced grazing and effective fire management for northern population
4. Ongoing fire management for southern population
5. Habitat monitoring for northern population
6. Protection of nesting mounds in selected parts of range of northern population

Bibliography

BirdLife International (2011) 'Species factsheet: *Psephotus chrysopterygius*'. Retrieved 17 January, 2011 from <http://www.birdlife.org/>.

Crowley G, Garnett S (1999) Seeds of the annual grasses *Schizachyrium* spp. as a food resource for tropical granivorous birds. *Australian Journal of Ecology* **24**, 208–220.

Crowley GM, Garnett ST (1998) Vegetation change in the grasslands and grassy woodlands of east-central Cape York Peninsula, Australia. *Pacific Conservation Biology* **4**, 132–148.

Crowley GM, Garnett ST (2000) Changing fire management in the pastoral lands of Cape York Peninsula of northeast Australia, 1623–1996. *Australian Geographical Studies* **38**, 10–26.

Crowley GM, Garnett ST (2001) Growth, seed production and effect of defoliation in an early flowering perennial grass, *Alloteropsis semialata* (Poaceae), on Cape York Peninsula, Australia. *Australian Journal of Botany* **49**, 735–743.

Crowley GM, Garnett ST, Shephard S (2004) 'Management guidelines for golden-shouldered parrot conservation'. Queensland Parks and Wildlife Service, Brisbane.

Crowley GM, Garnett ST, Shephard S (2009) Impact of stormburns on *Melaleuca viridiflora* invasion of grasslands and grassy woodlands on Cape York Peninsula, Australia. *Austral Ecology* **34**, 196–209.

Neldner VJ, Fensham RJ, Clarkson JR, Stanton JP (1997) The natural grasslands of Cape York Peninsula, Australia: description, distribution and conservation status. *Biological Conservation* **81**, 121–136.

Preece N, Shephard S, Shephard T, Garnett ST (2009) 'Reassessment of the status of the Golden-shouldered Parrot *Psephotus chrysopterygius*'. Report to the Queensland Department of Environment and Resource Management Cape York Peninsula Biodiversity Technical Advisory Group.

Stanton JP (1992) J. P. Thomson oration: The neglected lands: recent changes in the ecosystem of Cape York Peninsula and the challenge of their management. *Queensland Geographical Journal* **7**, 1–18.

Comments received from

Gabriel Crowley, Lana Little, Steve Murphy, Noel Preece, Susan Shephard, Tom Shephard

Species Conservation Summary

Paradise Parrot

Psephotus pulcherrimus (Gould, 1845)
Psittacidae

Conservation status
Extinct

Reasons for listing
There have been no authenticated sightings since 1927

Status 2000
Extinct

Status 1990
Extinct

Taxonomy
No infraspecific taxa described
Taxonomic uniqueness: medium (76 genera/family, 5 species/genus, 1 subspecies/species)

Former range
Formerly occurred in central and southern Queensland, north near Brisbane to sites near Duaringa and on the Comet and Nogoa Rivers, and west to St George (Department of Sustainability, Environment, Water, Population and Communities 2011). Reports from northern Queensland are thought to be Golden-shouldered Parrots *P. chrysopterygius*. There are no authenticated records from New South Wales (Olsen 2007).

Former abundance
From the 1840s to the 1880s the parrots were fairly abundant (Campbell 1915), but by 1900 they had disappeared from most of their range. A few pairs were located in the 1920s, the last confirmed report dating from 1927 (Chisholm 1936). Though many Paradise Parrots were taken into captivity in the 19th century, there are no current captive populations of this species and none have been reintroduced into the wild. There was a proposal to begin captive breeding of the species for release in the 1920s, but action occurred too late to save the species (Olsen 2007).

Ecology
Pairs and family parties of the Paradise Parrot occurred along river valleys in open woodland and forest with a grassy understorey. They fed on native grass seeds and excavated nest hollows in termite mounds (Chisholm 1922). Nothing is known about the species' home ranges or territories other than adults are thought to have returned to the same nesting locality year after year (Chisholm 1922; Olsen 2007). Disappearance during periods of drought was interpreted as reflecting local extinctions rather than movements (Ashby 1924; Higgins 1999).

Reasons for extinction
The decline of the species to extinction was probably accelerated by the reduction in the availability of native grass seeds caused by overgrazing, particularly during drought, altered fire regimes, prickly pear *Opuntia* spp. invasion, removal of termite mounds or a combination of these factors (Joseph 1988). They were predated on by native (e.g. goannas *Varanus* spp.) as well as introduced predators, such as cats *Felis catus* and foxes *Vulpes vulpes* (Jerrard 2008). Local populations were also under heavy pressure for trapping for aviaries for Australian and European markets (Chisholm 1922). Once the populations were reduced to critical levels, inbreeding could also have been a problem, resulting in infertile eggs (Jerrard 2008).

Bibliography

Ashby E (1924) Notes on extinct or rare Australian birds, with suggestions as to some of the causes of their disappearance. *Emu* **23**, 294–298.

Campbell AJ (1915) Missing birds. *Emu* **14**(3), 167–168.

Chisholm AH (1922) The 'lost' Paradise Parrot. *Emu* **22**, 4–17.

Chisholm AH (1936) Various bird problems. *Emu* **35**, 317–323.

Department of Sustainability, Environment, Water, Population and Communities (2011) *Psephotus pulcherrimus.* In 'Species Profile and Threats Database'. Department of Sustainability, Environment, Water, Population and Communities, Canberra.

Higgins PJ (Ed.) (1999) *Handbook of Australian, New Zealand and Antarctic Birds.* Oxford University Press, Melbourne.

Jerrard CHH (2008) Historical article – Paradise Parrot. *Australian Field Ornithology* **25**, 59–68.

Joseph L (1988) A review of the conservation status of Australian parrots in 1987. *Biological Conservation* **46**, 261–280.

Olsen P (2007) *Glimpses of Paradise: The Quest for the Beautiful Parakeet.* National Library of Australia, Canberra.

Comments received from
Penny Olsen

Species Conservation Summary

Tasman Parakeet

Cyanoramphus cookii (G. R. Gray, 1859)
Psittacidae

Conservation status

Critically Endangered B1ab(iii,v)+2ab(iii,v), C2a(ii)

Reasons for listing

Population <250 mature individuals and possibly declining. One subspecies Extinct.

Status 2000

Endangered D

Reason for change in 2010: lack of monitoring and suggestions of decline

Status 1990

Critically Endangered B1ab(iii,v)+2ab(iii,v), C2a(ii)

Reasons for change in 2000: effective conservation action leading to population increase

Taxonomy

C. c. cookii (Norfolk I.) is Critically Endangered, *C. c. subflavescens* (Lord Howe I.) is Extinct

Taxonomic uniqueness: low (76 genera/family, 9 species/genus, 2 subspecies/species)

Range

C. c. cookii is endemic to Norfolk I. where historically widespread. It is now largely confined to the Mt Pitt region of Norfolk I. National Park (Hill 2002) and nearby forested areas, though small numbers occur regularly across island (Director of National Parks 2010). *C. c. subflavescens* was Endemic to Lord Howe I.

Abundance

On Norfolk I. an extrapolation based on banding and likely survivorship suggested that in 2008 *C. c. cookii* had a population of 150–200 birds (P. Olsen *in litt.*) and a population of >240 was estimated in 2009 based on 13 pairs recorded at point counts (G. Dutson pers. obs.). Numbers increased from a low of 32 bids in 1988 after nest protection, control of nest hollow competitors and erection of nest boxes. However, recent observations of disappearance from places where once common and evidence of cat predation (M. Christian pers. comm.; R. Ward pers. comm.; J. Forshaw pers. comm.) suggest the population may be declining again. A captive breeding program ended in 2009 when remaining birds were released. On Lord Howe I. *C. c. subflavescens* was common when the island was first settled but was already scarce when last seen in 1869 (McAllan *et al.* 2004).

Ecology

Both subspecies are, or were, forest birds, though they rapidly learnt to eat soft fruits in orchards. They feed on seeds, fruits, flowers and leaves of both native and introduced trees and shrubs in the canopy and on the ground. They nest in hollows in trees or treeferns, and will use nest boxes. A generation time of 4.6 years (BirdLife International 2011) is derived from an average age at first breeding of 2.0 years, extrapolated from mean values for other genera in Psittacidae, and maximum longevity of 7.3 years, extrapolated from maximum longevity recorded in captivity.

Threats

C. c. subflavescens is thought to have been hunted to extinction by early settlers wishing to prevent damage to fruit crops (McAllan *et al.* 2004). Birds survived early settlement of Norfolk I. but most nest trees were cut down. Most trees with hollows are now in Norfolk I. National Park or local government reserves. There they suffer competition from introduced Crimson Rosellas *Platycercus elegans* and predation by introduced Black Rats *Rattus rattus*. Feral cats *Felis catus* are also thought to be significant predators (Hill 2002).

Conservation objectives

1. Population continues to grow
2. If appropriate some *C. c. cookii* established on Lord Howe I.

Information required

1. Trends in population size and breeding success
2. Causes of breeding failure
3. Effects of competition for hollows from Crimson Rosellas

Current eligibility against IUCN Red List Criteria

IUCN category	Criteria eligibility
A	Not applicable: past, current or future population declines are thought unlikely to exceed 20% in any 3-generation period
B	Critically Endangered: EOO <100 km^2, AOO <10 km^2, 1 location with plausible future threat, suspected continuing decline in habitat quality, no. mature individuals
C	Critically Endangered: population <250 mature individuals, suspected continuing decline, >90% in 1 subpopulation
D	Endangered: population <250 mature individuals
E	Not applicable: no population viability analysis undertaken

IUCN Red List assessment data

	Estimate	Reliability
Extent of occurrence trend	35 km^2 stable	high high
Area of occupancy trend	8 km^2 stable	medium high
No. of mature individuals trend	250 decreasing	low low
No. subpopulations	1	high
No. locations	1	high
Generation time	4.6 years	low
Global population share	100%	high

4. Effective ways of removing Crimson Rosellas given constraints on firearm use

Management actions required

1. Continue to identify, monitor and protect known and potential parrot nest sites, and destroy introduced competitors using these sites (particularly Crimson Rosellas) and/or install and maintain rat-proof nesting hollows in Norfolk I. NP and on adjacent private land
2. Continue rat baiting and cat trapping across Norfolk I. NP
3. Advocate for rat eradication across Norfolk I.
4. Enhance control programs for Crimson Rosellas
5. Investigate reintroduction to Phillip I. when habitat suitable
6. Investigate reintroduction to Lord Howe I.

Bibliography

BirdLife International (2011) 'Species factsheet: *Cyanoramphus cookii*'. Retrieved 17 January, 2011 from <http://www.bird-life.org/>.

Director of National Parks (2010) 'Draft Norfolk Island Region Threatened Species Recovery Plan'. Department of the Environment, Water, Heritage and the Arts, Canberra.

Hill R (2002) 'Recovery Plan for the Norfolk Island Green Parrot'. Environment Australia, Canberra.

McAllan IAW, Curtis BR, Hutton I, Cooper RM (2004) The birds of the Lord Howe Island Group: a review of records. *Australian Field Ornithology* **21**, 1–82.

Comments received from

Margaret Christian, Joe Forshaw, Coral Rowston, Ron Ward

Subspecies Conservation Summary

Tasman Parakeet (Norfolk Island/Norfolk Island Green Parakeet)

Cyanoramphus cookii cookii (G. R. Gray, 1859)
Psittacidae

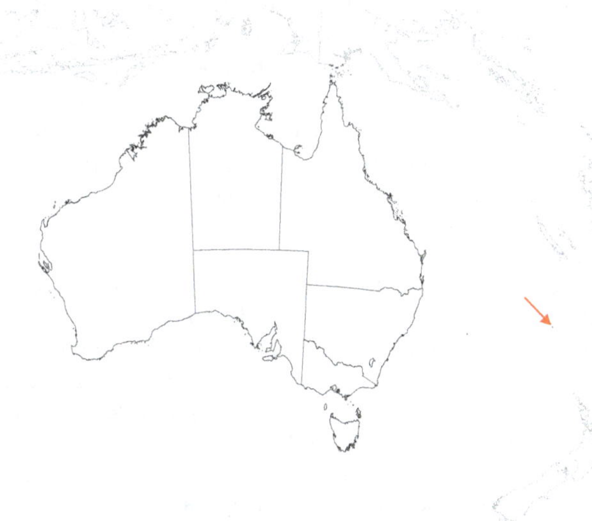

Conservation status

Critically Endangered B1ab(iii,v)+2ab(iii,v), C2a(ii)

Reasons for listing

There may be <250 mature individuals and not monitored. Population possibly declining so listed as Critically Endangered as a precautionary measure

Status 2000

Endangered D1
Reason for change in 2010: lack of monitoring and suggestions of decline

Status 1990

Critically Endangered B1ab(iii,v)+2ab(iii,v), C2a(ii)
Reasons for change in 2000: effective conservation action leading to population increase

Taxonomy

C. c. subflavescens (Lord Howe I.) is Extinct
Taxonomic uniqueness: low (76 genera/family, 9 species/genus, 2 subspecies/species)

Range

Endemic to Norfolk I. Historically widespread on Norfolk and Phillip Is. before European settlement, then contracted to remaining forest which is now Mt Pitt region of Norfolk I. National Park (Schodde *et al.* 1983; Hill 2002). The species now occurs throughout the park and botanic garden, is regularly seen in other forested areas outside the park with small numbers now occurring regularly across the island (Director of National Parks 2010). Breeding is largely confined to Norfolk I. National Park, as few suitable tree hollows elsewhere (Hill 2002), but there are also reports of birds breeding on private property nearby (C. Rowston pers. comm.).

Abundance

Population estimates include 190 pairs pre-settlement, a low of 4 breeding pairs and 32 individuals in 1988, increasing to at least 14 breeding pairs and 160 individuals in 2001 and at least 6 breeding pairs and 200–400 individuals in 2006–2007 (Hill 2002; Department of Sustainability, Environment, Water, Population and Communities 2011). The number of breeding pairs and their success is correlated with rainfall (Department of Sustainability, Environment, Water, Population and Communities 2011). An extrapolation based on numbers of chicks banded to 1985–2008 (588, 25/year) and likely survivorship suggested 150–200 birds (P. Olsen *in litt.*) and a population of >240 was estimated in 2009 based on 13 pairs recorded at point counts (G. Dutson pers. obs.). Observations of dis-

Current eligibility against IUCN Red List Criteria

IUCN category	Criteria eligibility
A	Not applicable: past, current or future population declines are thought unlikely to exceed 20% in any 3-generation period
B	Critically Endangered: EOO <100 km^2, AOO <10 km^2, 1 location with plausible future threat, suspected continuing decline in habitat quality, no. mature individuals
C	Critically Endangered: population <250 mature individuals, suspected continuing decline, >90% in 1 subpopulation
D	Endangered: population <250 mature individuals
E	Not applicable: no population viability analysis undertaken

IUCN Red List assessment data

	Estimate	Reliability
Extent of occurrence trend	35 km^2 stable	high high
Area of occupancy trend	8 km^2 stable	medium high
No. of mature individuals trend	250 decreasing	low low
No. subpopulations	1	high
No. locations	1	high
Generation time	4.6 years	low
Global population share	100%	high

appearance from places where once common and evidence of predation by cats *Felis catus* (M. Christian pers. comm., R. Ward pers. comm., J. Forshaw pers. comm.) suggest that the population could be declining. Even though its range has expanded, any suggestion of decline cannot be ignored for such a rare bird unless quantitative contrary evidence is available. Numbers banded declined since 2000, partly because of reduced effort, but also because fewer nesting sites were being maintained and perhaps because there were fewer breeding birds. A captive breeding program implemented on the island from the 1980s ended in 2009 when remaining birds were released.

Ecology

Largely restricted to forest, though they visit other areas, especially orchards to feed on soft fruits. They feed on seeds, fruits, flowers and leaves of both native and introduced trees and shrubs in the canopy and on the ground, including the seeds of the exotic Red Guava *Psidium cattleianum*, African Olive *Olea africana* and Lantana *Lantana camara* (Hicks and Preece 1991; Davidson 1997). They usually nest in hollows in native trees, mostly in Ironwood *Nestegis apetala*, and usually less than 2 m from the ground (Hicks and Greenwood, 1990) but also now using tree-ferns, *Cyathea brownii* and *C. australis* (C. Rowston pers. comm.). A generation time of 4.6 years (BirdLife International 2011) is derived from an average age at first breeding of 2.0 years, extrapolated from mean values for other genera in Psittacidae, and maximum longevity of 7.3 years, extrapolated from maximum longevity recorded in captivity.

Threats

Before 1950, forest was cleared for timber and agriculture, and most parrot nest trees, particularly Ironwood, were lost (Hicks and Greenwood 1989). Most remaining forest is now protected within Norfolk I. National Park or local government reserves but remnant large trees on private land are mostly unprotected (Hill 2002). There is an ongoing program to control the exotic shrubs that infest most native forests without providing nesting hollows. The limited number of nesting hollows is further reduced by competition from introduced Crimson Rosellas *Platycercus elegans*, but probably not by introduced Common Starlings *Sturnus vulgaris* and European Honey Bees *Apis mellifera* (Hermes *et al.* 1986; Hicks and Preece 1991; Hill 2002). This has been partly addressed in the past by erecting artificial nest hollows and by some culling of rosellas, which had a high population in 2009 (G. Dutson pers. obs.). Most known nest failures have resulted from predation by Black Rats *Rattus rattus*, which were introduced in the 1940s, and have probably also caused an imbalanced sex ratio observed in the 1990s (Robinson 1988; Hicks and Preece 1991; Hill 2002). Previous programs for erecting and maintaining nest hollows, banding nestlings and culling rosellas have ceased, although nest boxes retain rat-proofing collars and baiting is increased when chicks are detected. Feral cats are also thought to be significant predators, and have been implicated in the extinction of *Cyanoramphus* subspecies on other islands, but, while cat trapping effort has increased, since about 2000 there has been a gradual decline in the number of cats caught. Psittacine Circoviral Disease is present and may be fatal in stressed individuals (Hicks and Preece 1991). A small number of individuals die most years from collisions with windows (Hill 2002).

Conservation objectives

1. Population continues to grow
2. If appropriate subpopulations established on Phillip and Lord Howe Is.

Information required

1. Trends in population size and breeding success
2. Causes of breeding failure
3. Effects of competition for hollows from Crimson Rosellas
4. Effective ways of removing Crimson Rosellas given constraints on firearm use

Management actions required

1. Continue to identify, monitor and protect known and potential Tasman Parakeet nest sites, and destroy introduced competitors using these sites (particularly Crimson Rosellas) and/or install and maintain rat-proof nesting hollows in Norfolk I. NP and on adjacent private land
2. Continue rat baiting and cat trapping across Norfolk I. NP
3. Advocate for rat eradication across Norfolk I.
4. Enhance control programs for Crimson Rosellas

5. Investigate reintroduction to Phillip I. when habitat suitable
6. Investigate reintroduction of the species to Lord Howe I.

Bibliography

BirdLife International (2011) 'Species factsheet: *Cyanoramphus cookii*'. Retrieved 17 January, 2011 from <http://www.birdlife.org/>.

Davidson PM (1997) 'Movements and behaviour of juvenile Green Parrots determined by radio-tracking'. Unpublished report. Environment Australia, Canberra.

Department of Sustainability, Environment, Water, Population and Communities (2011) *Cyanoramphus cookii*. In 'Species Profile and Threats Database'. Department of Sustainability, Environment, Water, Population and Communities, Canberra. Retrieved 24 February, 2011 from <http://www.environment.gov.au/sprat>.

Director of National Parks (2010) 'Draft Norfolk Island Region Threatened Species Recovery Plan'. Department of the Environment, Water, Heritage and the Arts, Canberra.

Hermes N, Evans O, Evans B (1986) Norfolk Island birds: a review. *Notornis* **33**, 141–149.

Hicks J, Greenwood D (1989) Rescuing Norfolk Island's parrots. *Birds International* **1**, 35–47.

Hicks J, Preece M (1991) 'Green Parrot. 1991 Recovery Plan'. Unpublished report. Australian National Parks and Wildlife Service, Canberra.

Hill R (2002) 'Recovery Plan for the Norfolk Island Green Parrot'. Environment Australia, Canberra.

Robinson D (1988) 'Ecology and management of the Scarlet Robin, White-breasted White-eye and Long-billed White-eye on Norfolk Island'. Report to the Australian National Parks and Wildlife Service, Canberra.

Schodde R, Fullagar P, Hermes N (1983) 'A review of the status of Norfolk Island birds: past and present'. Australian National Parks and Wildlife Service, Canberra.

Comments received from

Margaret Christian, Joe Forshaw, Vanessa Lipianin, Coral Rowston, Ron Ward, Brooke Watson

Subspecies Conservation Summary

Tasman Parakeet (Lord Howe Island)

Cyanoramphus cookii subflavescens Salvadori, 1891
Psittacidae

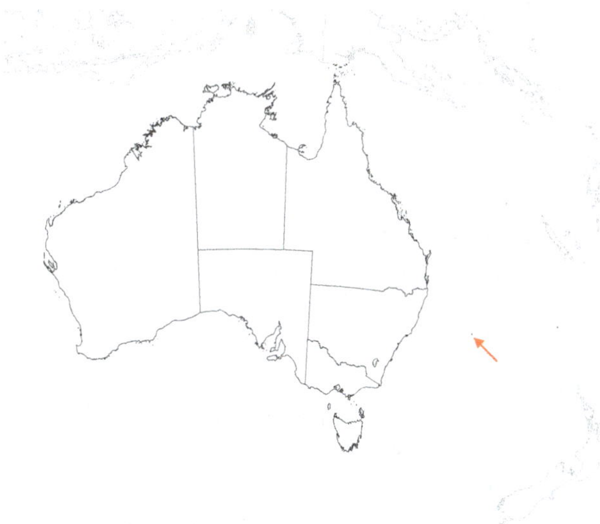

Conservation status
Extinct

Reasons for listing
The last record of this subspecies was in 1869

Status 2000
Extinct

Status 1990
Extinct

Taxonomy
C. c. cookii (Norfolk I.) is Critically Endangered
Taxonomic uniqueness: low (76 genera/family, 9 species/genus, 2 subspecies/species)

Former range
Endemic to Lord Howe I.

Former abundance
Common in the early days of settlement, and noted by most visitors (McAllan *et al.* 2004), but only 1 pair was recorded in 1869 and none thereafter (Hindwood 1940).

Ecology
The Tasman Parakeet must have lived in the forest and visited cultivated areas, but virtually nothing is known of its ecology. Their food probably consisted of plant material, such as seeds, fruits, buds and leaves of native trees and shrubs (Hutton 1991). They were considered a pest by the early settlers because they attacked and damaged crops (Hindwood 1940; Department of Sustainability, Environment, Water, Population and Communities 2011).

Reasons for extinction
The Tasman Parakeet was considered a pest around the crops and gardens of the first settlement on Lord Howe Island. Hunting and trapping to prevent the parrots feeding on crops are the only reasons that have been suggested to explain its extinction (Hindwood 1940; Hutton 1991).

Bibliography

Department of Sustainability, Environment, Water, Population and Communities (2011) *Cyanoramphus novaezelandiae subflavescens.* In 'Species Profile and Threats Database'. Department of Sustainability, Environment, Water, Population and Communities, Canberra. Retrieved 24 February, 2011 from <http://www.environment.gov.au/sprat>.

Hindwood KA (1940) The birds of Lord Howe Island. *Emu* **40**, 1–86.

Hutton I (1991) *Birds of Lord Howe Island, Past and Present.* The Author, Coffs Harbour.

McAllan IAW, Curtis BR, Hutton I, Cooper RM (2004) The birds of the Lord Howe Island Group: a review of records. *Australian Field Ornithology* **21**, 1–82.

Comments received from

David Priddel

Subspecies Conservation Summary

Red-fronted Parakeet (Macquarie Island)

Cyanoramphus novaezelandiae erythrotis (Wagler, 1832)
Psittacidae

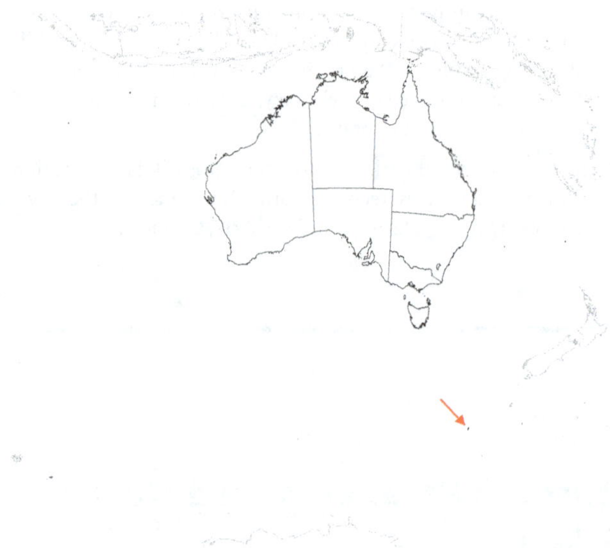

Conservation status

Extinct

Reasons for listing

The last recorded in 1890

Status 2000

Extinct

Status 1990

Extinct

Taxonomy

Four extralimital subspecies; the species is Vulnerable
Taxonomic uniqueness: low (76 genera/family, 9 species/genus, 6 subspecies/species)

Former range

The Macquarie island subspecies of Red-fronted Parakeet was endemic to Macquarie I.

Former abundance

The bird was common until 1879. Over the next decade, numbers fell quickly and it was not seen after 1890 (Taylor 1979).

Ecology

The parakeet lived in tussock grassland. Little is known about its diet, but it may have foraged on the seashore on crustaceans and other small invertebrates (Taylor 1979).

Reasons for extinction

The main reason for the decline of the Macquarie Island Red-fronted Parakeet was predation by feral cats *Felis catus* (Jones 1977), whose numbers increased dramatically following the introduction of rabbits *Oryctolagus cuniculus* (Taylor 1979). Large numbers of Parakeets were also killed for food by sealers in the 19th century (Department of Sustainability, Environment, Water, Population and Communities 2011). Predation by Black Rats *Rattus rattus* and Wekas *Gallirallus australis* was also likely to have been substantial (Taylor 1979).

Bibliography

Department of Sustainability, Environment, Water, Population and Communities (2011). *Cyanoramphus novaezelandiae erythrotis.* In 'Species Profile and Threats Database'. Department of Sustainability, Environment, Water, Population and Communities, Canberra. Retrieved 24 February, 2011 from: <http://www.environment.gov.au/sprat>.

Jones E (1977) Ecology of the Feral Cat *Felis catus* (L.), (Carnivora: Felidae) on Macquarie Island. *Australian Wildlife Research* **4**, 249–262.

Taylor RH (1979) How the Macquarie Island Parakeet became extinct. *New Zealand Journal of Ecology* **2**, 42–45.

Species Conservation Summary

Orange-bellied Parrot

Neophema chrysogaster (Latham, 1790)
Psittacidae

Conservation status
Critically Endangered A2abc+3bc+4abc, B1ab(i,ii,iii,iv,v), C1+2a(i,ii), D, E

Reasons for listing
Very small, rapidly declining population in 1 small area; models suggest extinction within 10 years; decline in the wild is being accelerated by capturing some juveniles to ensure genetic diversity of captive population

Status 2000
Critically Endangered C2a(ii)

Status 1990
Critically Endangered C2a(ii)

Taxonomy
No infraspecific taxa described
Taxonomic uniqueness: medium (76 genera/family, 6 species/genus, 1 subspecies/species)

Range
Breeds only at Melaleuca, south-western Tasmania, over-wintering in coastal mainland Australia, primarily in coastal Victoria and south eastern South Australia (Orange-bellied Parrot Recovery Team 2010). Formerly, the species bred along the south west coast of Tas from Birchs Inlet in the north to Louisa Plains in the south. The non-breeding range extended from the Yorke Peninsula, southeast through the Coorong, Robe, Beachport and Port MacDonnell in SA, through southwestern coastal Vic, Port Phillip Bay to South Gippsland, with isolated, outlying records from near Nowra and Sydney, New South Wales (Orange-bellied Parrot Recovery Team 2006).

Abundance
Population in June 2010 thought to be about 50 mature individuals (Orange-bellied Parrot Recovery Team 2010). Estimates based on marked individuals at Melaleuca for 1994–2004 showed an average minimum population of 71–116 birds (Orange-bellied Parrot Recovery Team 2006). Only 12 and 13 young are known to have fledged at Melaleuca in 2008–2009 and 2009–2010 with 21 of 27 young fledged in 2010–2011 taken to secure the captive population. Breeding females produce an average of over 3 fledglings per year but many females have not bred in recent years. Given the short generation time, 2 consecutive years of low recruitment resulted in a sharp decline (Orange-bellied Parrot Recovery Team 2010). The rate of decline in the wild is likely to culminate in extinction by 2015. During the late 1800s, and in the 1920s, the species was reported widely as being locally abundant, but first reported declining in 1917. There is a captive population of around 160 birds, but given the low genetic variation, a healthy captive population for the long term is unlikely (Orange-bellied Parrot Recovery Team 2010). Concern over this led to the collection of 21 new founders in Autumn 2011.

Ecology
Breeds in a mosaic of eucalypt forest and extensive moorland plains, nesting in hollows, mostly in live eucalypts. During the breeding season they feed on seeds of low shrubs and sedges from the surrounding moorlands, where food availability is partly a function of fire history: there is some evidence that early in the breeding season birds prefer to forage in areas burnt 7 to 15 years earlier, while during the middle of the breeding season they prefer areas burnt 3 to 5 years earlier. On migration they occur in dunes, heathland, coastal grasslands, saltmarsh and pasture. Non-breeding birds disperse to saltmarshes, dunes, beaches, pastures and shrublands, particularly on small islands and peninsulas, usually within 5 km of the coast, where there they feed on the ground or in low vegetation, almost exclusively on seeds and fruits, mainly of sedges, and salt-tolerant coastal and salt marsh plants. They occasionally feed on flowers and growing tips (Department of Sustainability Environment Water Population and Communities 2011). Generation length of 3.9 years based on age of first breeding of 1 year and annual adult survival of 65.0% (Holdsworth *et al.* in press).

Threats
Reasons for the species' decline are thought to be fragmentation and degradation of over-wintering habitat by drainage, grazing, agriculture or urban development and inappropriate fire regimes in the breeding range (Orange-bellied Parrot Recovery Team 2006; Ehmke *et al.* 2009). Given the low population size, stochastic factors, such as disease, loss of genetic variation, storms during migration, and destruction of nest sites, eggs and chicks by wildfire can affect the long-term survival of the species (Orange-bellied Parrot Recovery Team 2006). Other

Current eligibility against IUCN Red List Criteria

IUCN category	Criteria eligibility
A	Critically Endangered: population decline of >80% is currently being observed in AOO
B	Critically Endangered: EOO <100 km^2 at 1 location with a continuing decline in EOO, AOO, number of locations and subpopulations and number of mature individuals
C	Critically Endangered: population size <250 mature individuals, estimated continuing decline of >25% in 1 generation, <50 mature individuals in 1 subpopulation
D	Critically Endangered: population <50 mature individuals
E	Critically Endangered: >50% chance of extinction in the wild in the next 3 generations (12 years)

IUCN Red List assessment data

	Estimate	Reliability
Extent of occurrence trend	52 km^2 decreasing	high high
Area of occupancy trend	30 km^2 decreasing	medium high
No. of mature individuals trend	50 decreasing	high medium
No. subpopulations	1	high
No. locations	1	high
Generation time	3.9 years	high
Global population share	100%	high

pressures are also important, including predation by the introduced fox *Vulpes vulpes* and cat *Felis catus*, competition with the introduced seed-eaters European Goldfinch *Carduelis carduelis*, European Greenfinch *C. chloris* and House Mouse *Mus musculus* during winter, competition for nest hollows (including nest boxes) with introduced Common Starling *Sturnus vulgaris*, European Honey Bee *Apis mellifera* and sugar glider *Petaurus breviceps* and reduced fitness resulting from infection with Psittacine Circoviral Disease (Department of Sustainability, Environment, Water, Population and Communities 2011). Windfarm developments in Tas, western Vic and south-eastern SA may also kill about 1 bird per year (Smales *et al.* 2005).

Conservation objectives

1. A viable captive population
2. An increasing wild population
3. An understanding of trends in the wild population

Information required

1. Reason for low breeding participation
2. Nutritional requirements of breeding parrots, especially compared to the wild food available during breeding
3. Population size and demographic trends
4. Location of all nest sites
5. Influence of fire on feeding ecology

Management actions required

1. Capture enough wild parrots to maximise genetic variability in captive population
2. Maintain a healthy, genetically viable, captive population, increasing it from 160 to 400 individuals by 2014
3. Continue to manage wild breeding population including supplementary feeding, an active fire management program and protection of nests from competitors and predators
4. Secure known winter habitat from predators and competitors
5. Develop a re-introduction strategy and release healthy birds from the captive population into the wild population, preferably while the wild population is extant and displaying natural behaviours
6. Monitor the population size, productivity and survival
7. Maintain an active recovery team

Bibliography

Department of Sustainability, Environment, Water, Population and Communities (2011) *Neophema chrysogaster.* In 'Species Profile and Threats Database'. Department of Sustainability, Environment, Water, Population and Communities, Canberra. Retrieved 24 February, 2011 from <http://www.environment.gov.au/sprat>.

Ehmke G, Herrod A, Green R, Tzaros C (2009) 'Habitat protection and restoration plan for the Orange-bellied Parrot *Neophema chrysogaster* in the south east of South Australia'. Birds Australia, Melbourne.

Holdsworth M, Dettmann B, Baker GB (in press) Survival in the Orange-bellied Parrot *Neophema chrysogaster. Emu.*

Orange-bellied Parrot Recovery Team (2006) 'National Recovery Plan for the Orange-bellied Parrot (*Neophema chrysogaster*)'. Department of Primary Industries and Water (DPIW), Hobart.

Orange-bellied Parrot Recovery Team (2010) 'On the Brink. Current status of the Orange-bellied Parrot. The Trumped-Up Corella 4, June 2010'. Retrieved 10 April, 2011 from <http://www.birdsaustralia.com.au/images/stories/current-projects/obp/TUC-Jun10.pdf>.

Smales I, Muir S, Meredith C (2005) 'Modelled cumulative impacts on the Orange-bellied Parrot of wind farms across the species' range in south-eastern Australia'. Biosis Research Pty Ltd.

Comments received from

Barry Baker, Glenn Ehmke, Mark Holdsworth, Peter Menkhorst

Species Conservation Summary

Western Ground Parrot

Pezoporus flaviventris North, 1911
Psittacidae

Conservation status

Critically Endangered A2bc, C1+2a(ii)

Reasons for listing

Population estimated at <250 individuals, has declined at >80% in 3 generations (29 years) and >90% in 1 subpopulation

Status 2000

Endangered B2ab(i,ii,iii,iv,v), C2a(ii)
Reason for change in 2010: rate of decline has accelerated and population declined

Status 1990

Endangered B2ab(i,ii,iii,iv,v), C2a(ii)

Taxonomy

No infraspecific taxa described
Taxonomic uniqueness: medium (76 genera/family, 3 species/genus, 1 subspecies/species)

Range

In 2008, there were 2 subpopulations at 4 locations in Fitzgerald River and Cape Arid National Parks, south-west Western Australia (Gilfillan *et al.* 2009). Since 1990 has also been recorded in Nuytsland Nature Reserve (last recorded 2006) and Waychinicup-Many Peaks area (Burbidge *et al.* 1997; Comer *et al.* 2010; Gilfillan *et al.* 2009). Historically occurred on the coastal plains from Perth to Dongara, and from Israelite Bay to Augusta but last recorded on the coastal plain north of Perth at the end of the 19th century and west of Albany in 1983 (Watkins 1985; Gilfillan *et al.* 2009).

Abundance

Reliable estimates of numbers are difficult to obtain due to the parrots' cryptic nature and low population densities (Cale and Burbidge 1993). The average number of calls heard per sampling session in Fitzgerald National Park increased to about 60 in 1998–2000 after commencement of broadscale baiting for foxes *Vulpes vulpes*, then declined to 4 in 2004–2006 and 0–2 in 2008–2010 (Comer *et al.* 2010).

Ecology

Lives in floristically diverse heathlands, where it feeds on fruits, seeds and leaves, mostly taken green from the plants (Burbidge *et al.* 1989). Occurs in long unburnt (5 to >40 years), floristically diverse, near-coastal dry heath, especially areas relatively high in altitude, distant from rivers, gently sloping to level, with an intermediate cover of vegetation and a mosaic of vegetation ages (Gibson *et al.* 2007). Although Western Ground Parrots are usually found in long unburnt vegetation, they also feed in habitats 2–3 years post-fire provided there is older vegetation nearby, and a mosaic of burning that allows movement between patches of different ages may be important (Burbidge *et al.* 1997; Burbidge 1999; Burbidge *et al.* 2007; Gilfillan *et al.* 2009). A generation time of 9.7 years (BirdLife International 2011) is derived from an average age at first breeding of 2.0 years, and a maximum longevity of 17.8 years, both values extrapolated from mean values for other genera in Psittacidae.

Threats

Currently predation by introduced mammals, particularly foxes and feral cats *Felis catus*, is thought to be a major threat (Comer *et al.* 2010), and could lead rapidly to extinction of the species unless the current fox and cat baiting program is continued and extended. Frequent burning may have eliminated the species from many vegetation remnants and continues to be the main threat; e.g. habitat supporting >30% of the total population was burnt in 2004 (Burbidge 2003). Other potential threats include dieback caused by *Phytophthora cinnamomi*, disturbance by introduced animals (particularly hard-hoofed animals), weed invasion and changes in hydrological regimes (Gilfillan *et al.* 2009). Eight immature birds have been taken into captivity to trial captive-breeding techniques (Comer *et al.* 2010). Historical clearance for grazing and agriculture is considered responsible for contraction of its distribution, but no longer a threat as all parrots now occur in conservation reserves (Burbidge *et al.* 1997).

Conservation objectives

1. Viable captive population established
2. Population numbers and area of occupation increased
3. Knowledge of current distribution improved
4. Management options increased through improved ecological knowledge

Information required

1. Captive breeding techniques

Current eligibility against IUCN Red List Criteria

IUCN category	Criteria eligibility
A	Critically Endangered: past declines of >80% in 3 generations (29 years) estimated from decline in sample counts and EOO, AOO, habitat quality
B	Endangered: AOO <500 km^2; at ≤5 locations with continuing decline in EOO, AOO, habitat quality, number of locations and number of individuals
C	Critically Endangered: population estimated <250 individuals and ongoing decline estimated to continue at >25% in 1 generation (19 years); >90% in 1 subpopulation
D	Endangered: population <250 individuals
E	Not applicable: no population viability analysis undertaken

IUCN Red List assessment data

	Estimate	Reliability
Extent of occurrence trend	1400 km^2 decreasing	medium high
Area of occupancy trend	100 km^2 decreasing	low medium
No. of mature individuals trend	110 decreasing	medium high
No. subpopulations	1	high
No. locations	4	medium
Generation time	9.7 years	low
Global population share	100%	high

2. Trends in wild population
3. Potential habitat for new subpopulations
4. Constraints on wild population
5. Dietary requirements

Management actions required

1. Establish a viable captive breeding program
2. Continue habitat management and threat abatement of all occupied areas
3. Ensure that conservation of critical habitat is considered in fire management
4. Continue monitoring locations where wild populations have been recorded recently
5. Continue to support coordination of management by the South Coast Threatened Bird Recovery Team

Bibliography

BirdLife International (2011) 'Species factsheet: *Pezoporus wallicus*'. Retrieved 17 January, 2011 from <http://www.birdlife.org/>.

Burbidge AH (1999) Western Ground Parrot Interim Recovery Plan. *Eclectus* **6**, 23–26.

Burbidge AH (2003) Birds and fire in the Mediterranean climate of southwest Western Australia. In *Fire in Ecosystems of Southwest Western Australia: Impacts and Management*. (Eds I Abbott and N Burrows) pp. 321–348. Backhuys Publishers, Leiden.

Burbidge AH, Blyth J, Danks A, Gillen K, Newbey B (Eds) (1997) 'Western Ground Parrot Interim Recovery Plan, 1996–1999. Interim Recovery Plans 4–16 for Western Australian Critically Endangered Plants and Animals'. Department of Conservation and Land Management, Perth.

Burbidge AH, Rolfe J, McNee S, Newbey B, Williams M (2007) Monitoring population change in the cryptic and threatened Western Ground Parrot in relation to fire. *Emu* **107**, 79–88.

Burbidge AH, Watkins D, McNee S (1989) 'Conservation of the Ground Parrot in Western Australia'. Report to World Wildlife Fund (Australia) Project 118.

Cale PG, Burbidge AH (1993) 'Research Plan for the Western Ground Parrot, Western Whipbird and Western Bristlebird'. Australian National Parks & Wildlife Service.

Comer S, Burbidge AH, Tiller C, Berryman A, Utber D (2010) Heeding Kyloring's warning: south coast species under threat. *Landscope* **26**, 48–53.

Gibson L, Barrett B, Burbidge AH (2007) Dealing with uncertain absences in habitat modelling: a case study of a rare ground-dwelling parrot. *Diversity and Distributions* **13**, 704–713.

Gilfillan S, Comer S, Burbidge AH, Blyth J, Danks A (2009) 'South Coast Threatened Birds Recovery Plan. Western Australian Wildlife Management Program No. 44. WA Department of Environment and Conservation, Perth.

Watkins D (1985) 'Report of the RAOU Ground Parrot survey in Western Australia'. RAOU Report 15.

Comments received from

John Blyth, Allan Burbidge, Sarah Comer, Brenda Newby

Subspecies Conservation Summary

Eastern Ground Parrot (mainland)

Pezoporus wallicus wallicus (Kerr, 1792)
Psittacidae

Conservation status
Near Threatened C2a(i)

Reasons for listing
Population <10 000 mature individuals, declining with largest subpopulation approaching 1000 mature individuals

Status 2000
Near Threatened C2a(i)

Status 1990
Near Threatened C2a(i)

Taxonomy
P. w. leachi (Tasmania) is Least Concern, as is the species
Taxonomic uniqueness: medium (76 genera/family, 3 species/genus, 2 subspecies/species)

Range
This subspecies has a patchy distribution within 25 km of coastal eastern Australia from Wide Bay and Fraser I., Queensland, to Portland eastern Victoria (Higgins 1999). Current major population centres are, in Qld, Fraser Island and the adjacent coastline south to Noosa, in New South Wales on the north coast (Broadwater, Bundjalung, Yuraygir National Parks), central coast (Myall Lakes) and south coast (Barren Grounds Nature Reserve, Budderoo National Park, the Jervis Bay area, Nadgee Nature Reserve, Morton and Ben Boyd National Parks (Department of Environment and Conservation (NSW) 2005). In Vic this subspecies occurs in Croajingalong National Park, Bemm River-Marlo region and Wilsons Promontory National Park in the east and at Long Swamp, Discovery Bay in the far west of the state (Meredith and Isles 1980). The presence of vagrant birds at least 200 km from the nearest known subpopulation suggests that there is probably some movement between subpopulations. However, there are major breaks in distribution in central Vic, central NSW and south-east Qld and the subpopulations may have been small and isolated for some time given the low rate of genetic variability in Queensland (Chan *et al.* 2008). Once occurred on the Adelaide Plains in South Australia and there are also no recent records from the Carlisle River area in the Otway Ranges (Department of Sustainability, Environment, Water, Population and Communities 2011). The AOO is calculated on the basis of a density of about 15 birds/km^2 (based on Baker *et al.* 2010).

Abundance
The total population of 5500 is based on very little recent information. In Queensland the population was estimated at 2900–4900 birds (McFarland 1991c) but could be as low as 1600 if much of the habitat has been recently burnt (based on D. McFarland *in litt.*). In NSW, the population has been estimated at about 2000 birds, including 1000 birds at Barren Ground Nature Reserve and Budderoo National Park, less than 550 at Jervis Bay, 300 at Nadgee Nature Reserve and 100 at Ben Boyd National Park, with smaller numbers at other locations (Baker 1997). In Vic the population was estimated at 600–700 birds in 1979–80, including about 400 in Croajingalong National Park, <100 in the Bemm River-Marlo region, <100 in Wilsons Promontory National Park and about 60 at Long Swamp (Meredith and Isles 1980). At Cooloola, south-east Qld, a theoretical maximum population of 2400 can be as low as 800 if, as in 2010, the heath is either <1 year or >14 years post-fire (D. McFarland *in litt.*). However, Baker *et al.* (2010) concluded that densities up to 20 birds/km^2 can occur in habitat of any age post-fire. There is little evidence of decline given their capacity to cope with fire (Baker *et al.* 2010) but the loss from the Otways has probably occurred within the last 3 generations (29 years) and monitoring is at best patchy and has not been compiled and analysed across all sites.

Ecology
Lives in low heathland and sedgeland (Meredith 1984; Higgins 1999), eating seeds from a wide range of herbs, graminoids and heath plants (McFarland 1991a). Heathland becomes unsuitable immediately after fire (Meredith *et al.* 1984), in some cases, for a further 4 years (Jordan 1987; Baker and Whelan 1994). In Queensland habitat suitability may decline if left unburnt for more than 15 years (McFarland, 1991a) but the subspecies may not require fire over most of its range (Meredith *et al.* 1984; Baker and Whelan 1994; Baker *et al.* 2010). Nests are made on the ground, beneath dense vegetation, the clutch size averaging 3–4. Hatching failure is high relative to other parrots (McFarland 1991b), though whether this is related to the low genetic variability (Chan *et al.* 2008) is unknown, as are its demographic effects. A generation time of 9.7 years (BirdLife International 2011) is derived

Current eligibility against IUCN Red List Criteria

IUCN category	Criteria eligibility
A	Not applicable: past, current or future population declines are thought unlikely to exceed 20% in any 3-generation period
B	Not applicable: >10 locations, not severely fragmented, population fluctuations not extreme
C	Near Threatened: population 2500–10 000 mature individuals, possible continuing decline, largest subpopulation near 1000 mature individuals
D	Not applicable: population >1000 mature individuals, >5 locations
E	Not applicable: no population viability analysis undertaken

IUCN Red List assessment data

	Estimate	Reliability
Extent of occurrence trend	66 000 km^2 stable	high medium
Area of occupancy trend	370 km^2 decreasing	low low
No. of mature individuals trend	5500 decreasing	low low
No. subpopulations	6	medium
No. locations	>10	high
Largest subpopulation	1600	low
Generation time	9.7 years	low
Global population share	100%	high

from an age at first breeding of 2.0 years and a maximum longevity in the wild of 17.4 years, both extrapolated from other Psittacidae.

Threats

The most important current threat is extensive and intense fires that temporarily remove habitat and can extirpate isolated subpopulations. As a ground nester, the eastern Ground Parrot may be susceptible to predation by feral cats *Felis catus* and foxes *Vulpes vulpes* (Department of Sustainability, Environment, Water, Population and Communities 2011), particularly immediately after a fire. Historically much habitat has been lost through clearing for agriculture and residential development but this is no longer a direct threat and almost all remaining birds are on protected areas. The relatively high rate of hatching failure in the subpopulation in south-east Qld (McFarland 1991b) may be related to a particularly low rate of genetic variability (Chan *et al.* 2008).

Conservation objectives

1. All 6 major subpopulations known to be at least stable

Information required

1. Trends in all subpopulations: there is no trend data available for Vic or Myall Lakes, NSW and data from Qld is sporadic
2. Impact of feral predators on parrot population trends

Management actions required

1. Initiate or continue monitoring at key sites
2. Protect habitat from intense and extensive fire events
3. Undertake fox and feral cat control programs, particularly after fire events

Bibliography

Baker J (1997) The decline, response to fire, status and management of the Eastern Bristlebird. *Pacific Conservation Biology* **3**, 235–243.

Baker J, Whelan R (1994) Ground Parrots and fire at Barren Grounds, New South Wales: a long term study and an assessment of management implications. *Emu* **94**, 300–304.

Baker JR, Whelan RJ, Evans L, Moore S, Norton M (2010) Managing the Ground Parrot in its fiery habitat in south-eastern Australia. *Emu* **110**, 279–284.

BirdLife International (2011) 'Species factsheet: *Pezoporus wallicus*'. Retrieved 17 January, 2011 from <http://www.birdlife.org/>.

Chan K, Glover DR, Ramamge CM, Harrison DK (2008) Low genetic diversity in the Ground Parrot (*Pezoporus wallicus*) revealed by randomly amplified DNA fingerprinting. *Annales Zoologici Fennici* **45**, 211–216.

Department of Environment and Conservation (NSW) (2005) 'Eastern Ground Parrot – profile.' Retrieved 11 May, 2011 from <http://www.threatenedspecies.environment.nsw.gov.au/tsprofile/profile.aspx?id=10608>.

Department of Sustainability, Environment, Water, Population and Communities (2011) *Pezoporus wallicus wallicus*. In 'Species Profile and Threats Database'. Department of Sustainability, Environment, Water, Population and Communities, Canberra. Retrieved 24 February, 2011 from <http://www.environment.gov.au/sprat>.

Higgins PJ (Ed.) (1999) *Handbook of Australian, New Zealand and Antarctic Birds. Volume 4: Parrots to Dollarbird*. Oxford University Press, Melbourne.

Jordan R (1987) The Ground Parrot in Barren Grounds Nature Reserve. *RAOU Report* **27**, 19–23.

McFarland DC (1991a) The biology of the Ground Parrot, *Pezoporus wallicus*, in Queensland. I. Microhabitat use, activity, cycle and diet. *Wildlife Research* **18**, 169–184.

McFarland DC (1991b) The biology of the Ground Parrot, *Pezoporus wallicus*, in Queensland. II. Spacing, calling and breeding behaviour. *Wildlife Research* **18**, 185–197.

McFarland DC (1991c) The biology of the Ground Parrot, *Pezoporus wallicus*, in Queensland. III. Distribution and abundance. *Wildlife Research* **18**, 199–213.

Meredith CW (1984) 'The Ground Parrot *Pezoporus wallicus* Kerr'. RAOU Conservation Statement No. 1. RAOU, Melbourne.

Meredith CW, Gilmore AM, Isles AC (1984) The Ground Parrot in south-eastern Australia: a fire adapted species? *Australian Journal of Ecology* **9**, 367–380.

Meredith CW, Isles AC (1980) 'A study of the Ground Parrot (*Pezoporus wallicus*) in Victoria'. Ministry of Conservation Environmental Studies Division, Victoria.

Comments received from

Jack Baker, David McFarland

Species Conservation Summary

Night Parrot

Pezoporus occidentalis (Gould, 1861)
Psittacidae

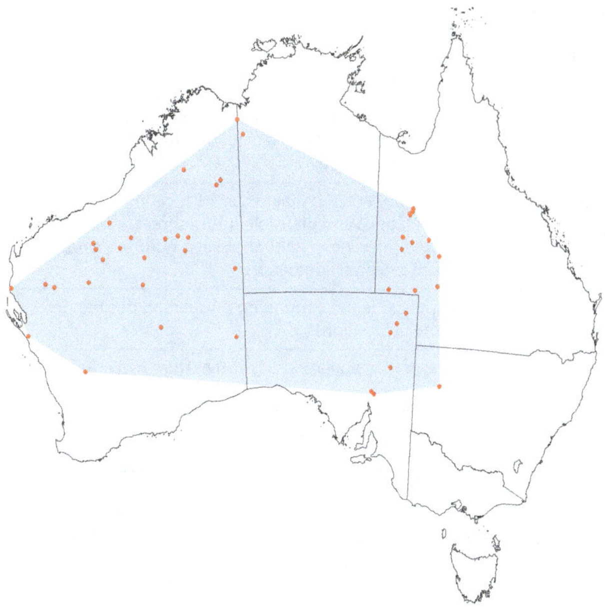

Conservation status

Endangered B2ab(iii)c(ii,iii,iv), D

Reasons for listing

Area occupied probably very small, scattered among few locations and probably fluctuates greatly with overall condition of habitat declining

Status 2000

Endangered B2ab(iii)c(ii,iii,iv), D

Status 1990

Endangered B2ab(iii)c(ii,iii,iv), D

Taxonomy

No infraspecific taxa described

Taxonomic uniqueness: medium (76 genera/family, 3 species/genus, 1 subspecies/species)

Range

Historically recorded from throughout arid and semi-arid Australia, mostly before the 1880s (Blyth 1997; Higgins 1999). There are few widely accepted records since 1935: sightings in the Pilbara, Western Australia, in 1980 and 2005, central WA in 1979, north-eastern South Australia in 1979, and western Queensland, where sightings in 1980 and 1993 and carcasses in 1990 and 2006 (Boles *et al.* 1994; Higgins 1999; Davis and Metcalf 2008; McDougall *et al.* 2009). Possibly still occurs at low density throughout its former range, with many unverified sight records, but several dedicated searches and publicity campaigns have failed to find any birds (Davies *et al.* 1988; Maher 1995; Jordan 1996; Blyth 1997; Blyth *et al.* 1997; Blyth and Boles 1999; Ramsay 2010). May be more scarce south of the Dog Fence in SA, NSW and Vic, based on habitat quality (J. Reid *in litt.*).

Abundance

Although population has been estimated as <50 mature individuals, an expert committee re-assessed this as untenable, given recent records from WA and Qld, and estimated that there might be 50–250 birds in ≤5 locations.

Ecology

Most records are from *Triodia* hummock grassland or chenopod shrublands (Higgins 1999). May persist in chenopod shrublands during dry years, and move into *Triodia* after there has been sufficient rain to cause grasses to set seed (Parker 1980). The 1990 specimen and associated reports were in *Astrebla* Mitchell grassland with scattered chenopods (Garnett *et al.* 1993; Boles *et al.* 1994), the 2006 specimen in sparse shrubland of Gidgee *Acacia cambagei*, Crimson Turkey Bush *Eremophila latrobei* and Blunt-leaf Cassia *Senna artemisioides* var. *helmsii* (McDougall *et al.* 2009) and there have been unconfirmed reports from mallee shrubland and in open *Eucalyptus* woodland with an understorey of grasses (Menkhorst and Isles 1981; Garnett *et al.* 1993). The 2 nests recorded have both been at the end of tunnels into dense vegetation and contained 3–6 eggs or young (Higgins 1999). A generation time of 9.7 years (BirdLife International 2011) is derived from an age at first breeding of 2.0 years and a maximum longevity of 17.4 years, both median values from an expert committee.

Threats

The sharp decline in reporting after the 1880s suggests an historical decline in abundance but all estimates of contemporary population parameters are essentially guesswork. One early account suggests the decline at Innamincka and Alice Springs coincided with the 'arrival' of feral cats *Felis catus* (Whitlock 1924). Similarly 'many were brought in by cats at Alice Springs Telegraph Station' in 1892 (Ashby 1924). Factors that have adversely affected medium-sized arid-zone mammals (Burbidge and McKenzie 1989) may have been responsible for the decline of the Night Parrot (Blyth *et al.* 1997). These include predation by feral cats and foxes *Vulpes vulpes*, altered fire regimes, competition for food and degradation of habitat near water by stock or rabbits *Oryctolagus cuniculus* and reduced availability of water as a result of over-use by feral camels *Camelus dromedarius*. Given the slow worsening of many of these threats, an expert committee assessed the species as suffering a continuing decline.

Conservation objectives

1. Discovered and stable at >1 location

Information required

1. Alternative detection techniques e.g. dogs trained to locate parrots by scent
2. Complete GIS analysis of records against habitat to identify priority areas to search

Current eligibility against IUCN Red List Criteria

IUCN category	Criteria eligibility
A	Not applicable: past, current or future population declines are thought unlikely to exceed 20% in any 3-generation period
B	Endangered: AOO <500 km², at ≤5 locations with ≤5 subpopulations and plausible future threat, inferred continuing decline in habitat quality, numbers probably undergo extreme fluctuations
C	Not applicable: no continuing population decline or extreme fluctuations
D	Endangered: population 50–250 mature individuals in poorest years
E	Not applicable: no population viability analysis undertaken

IUCN Red List assessment data

	Estimate	Reliability
Extent of occurrence trend	3 500 000 km² stable	low low
Area of occupancy trend	200 km² fluctuating	low low
No. of mature individuals trend	250 fluctuating	low low
No. subpopulations	2	low
No. locations	5	low
Largest subpopulation	250	low
Generation time	5.3 years	low
Global population share	100%	high

3 Captive-breeding and release techniques developed based on ground parrots *Pezoporus* spp.

Management actions required

1. Continue appeals for information
2. Using GIS, test early suggestions that Night Parrots move into higher elevation rangelands following above average rainfall
3. Investigate all plausible reports in a timely manner for independent verification
4. Develop a contingency plan for any Night Parrots found

Bibliography

Ashby E (1924) Notes on extinct or rare Australian birds, with suggestions as to some of the causes of their disappearance. *Emu* **23**, 294–298.

BirdLife International (2011) 'Species factsheet: *Pezoporus occidentalis*'. Retrieved 17 January, 2011 from <http://www.birdlife.org/>.

Blyth J (1997) Night Parrot (*Pezoporus occidentalis*) interim recovery plan for Western Australia 1996–1998. In 'Interim Recovery Plans 4–16 for Western Australian Critically Endangered Plants and Animals'. (Eds J Pryde, A Brown and AA Burbidge). Department of Conservation and Land Management, Perth.

Blyth J, Boles W (1999) An expedition to the Murchison, Gascoyne and East Pilbara areas to search for Night Parrots, November 1997. *Eclectus* **6**, 12–16.

Blyth J, Burbidge AH, Boles W (1997) Report on an expedition to the Western Desert and East Pilbara areas in search of the Night Parrot *Pezoporus occidentalis*. *Eclectus* **2**, 25–30.

Boles WE, Longmore NW, Thompson MC (1994) A recent specimen of the Night Parrot *Geopsittacus occidentalis*. *Emu* **94**, 37–40.

Burbidge AA, McKenzie NL (1989) Patterns of modern decline of Western Australia's vertebrate fauna: causes and conservation implications. *Biological Conservation* **50**, 143–198.

Davies SJJF, Bamford M, Bamford M (1988) 'The Night Parrot: a search in the Lake Disappointment Area, September, 1987'. RAOU Report 49. RAOU, Melbourne.

Davis RA, Metcalf BM (2008) The Night Parrot (*Pezoporus occidentalis*) in northern Western Australia: a recent sighting from the Pilbara region. *Emu* **108**, 233–236.

Garnett ST, Crowley GM, Duncan R, Baker N, Doherty P (1993) Notes on live Night Parrot sightings in north-western Queensland. *Emu* **93**, 292–297.

Higgins PJ (Ed.) (1999) *Handbook of Australian, New Zealand and Antarctic Birds. Volume 4: Parrots to Dollarbird.* Oxford University Press, Melbourne.

Jordan R (1996) 'A search for the Night Parrot in the Devoncourt area near Cloncurry'. Report to Queensland Department of Environment and Heritage, Brisbane.

Maher PN (1995) 'Night Parrot survey, techniques and habitat requirements with notes on the fauna of the Winton and Cloncurry districts'. Report to Queensland Department of Environment and Heritage, Brisbane.

McDougall A, Porter G, Mostert M, Cupitt R, Cupitt S, Joseph L, Murphy SA, Janetzki H, Gallagher A, Burbidge A (2009) Another piece in an Australian ornithological puzzle – a second Night Parrot is found dead in Queensland. *Emu* **109**, 198–203.

Menkhorst PW, Isles AC (1981) The Night Parrot *Geopsittacus occidentalis*: evidence of its occurrence in north-western Victoria during the 1950s. *Emu* **81**, 239–240.

Parker S (1980) Birds and conservation parks in the north-east of South Australia. *South Australian Parks and Conservation* **3**, 11–18.

Ramsay B (2010) Possible Night Parrots in WA. *The Bird Observer* **866**, 24.

Whitlock FL (1924) Journey to central Australia in search of the Night Parrot. *Emu* **23**, 248–281.

Comments received from

John Blyth, Allan Burbidge, Andrew Burbidge, Les Christidis, Hugh Ford, Tim Holmes, Leo Joseph, Richard Loyn, Peter Menkhorst, Steve Murphy, Penny Olsen, Mick Roderick, David Watson

Subspecies Conservation Summary

Barking Owl (Papuan)

Ninox connivens assimilis Salvadori & D'Albertis, 1875
Strigidae

Conservation status

Near Threatened D1

Reasons for listing

Australian population probably small enough for EN D but not isolated from Least Concern main population in New Guinea, so downgraded 2 categories to NT (following IUCN 2003)

Status 2000

Near Threatened D1

Status 1990

Near Threatened D1

Taxonomy

N. c. connivens is Near Threatened and *N. c. peninsularis* Least Concern, as is the species. There is 1 extralimital subspecies.
Taxonomic uniqueness: low (27 genera/family, 22 species/genus, 4 subspecies/species)

Range

There have been a recent series of records of this subspecies from Saibai (96 km^2) in the north Torres Strait islands (Clarke 2004) and it is also likely to occur on neighbouring Boigu (72 km^2). Records from the southern Torres Strait islands (e.g. Draffan *et al.* 1983) are probably the subspecies *N. c. peninsularis.*

Abundance

Away from the extensive mangroves, the population density is likely to be similar to the highest densities in Australia, where breeding pairs may be spaced at <1 km to 10 km (NPWS 2003). Assuming pairs are 1 km apart across the non-mangrove areas, the total Australian population of this subspecies may be 100 birds. Outside Australia, this subspecies is relatively common and widespread in eastern New Guinea, and the species is not threatened.

Ecology

In Papua New Guinea they inhabit savanna, gardens, open woodland, groves of trees and plantations, usually in the vicinity of swamps, lakes and streams in the lowlands, where they feed on insects, birds and mammals (Coates 1985). Nesting is presumed to be in large hollows of live eucalypts, with a clutch size of 1 to 3 (Higgins 1999). Given the lack of subspecific variation between relatively distant islands in its range (e.g. on Manam I and Karkar I), it is considered to be a dispersive species with regular exchange between the Torres Strait islands and New Guinea. A generation time of 8.5 years (BirdLife International 2011) is derived from an age at first breeding of 1.0 years (extrapolated from other Strigidae) and a maximum longevity in the wild of 15.9 years, extrapolated from Southern Boobook *N. novaehollandiae* (Australian Bird and Bat Banding Scheme).

Threats

The subspecies is listed as threatened because it has a very small population. There are no plausible threats; though

Current eligibility against IUCN Red List Criteria

IUCN category	Criteria eligibility
A	Not applicable: past, current or future population declines are thought unlikely to exceed 20% in any 3-generation period
B	Not applicable: EOO, AOO, habitat, locations or population not declining, population fluctuations not extreme
C	Not applicable: no continuing population decline or extreme fluctuations
D	Endangered: population 50–250 mature individuals (downgraded to Near Threatened as global status is Least Concern)
E	Not applicable: no population viability analysis undertaken

IUCN Red List assessment data

	Estimate	Reliability
Extent of occurrence trend	168 km^2 stable	high high
Area of occupancy trend	100 km^2 stable	medium high
No. of mature individuals trend	100 stable	low low
No. subpopulations	2	high
No. locations	2	high
Largest subpopulation	60	low
Generation time	8.5 years	low
Global population share	<1%	high
Level of genetic exchange	high	medium

Saibai and Boigu are low-lying, they are too far north for catastrophic cyclones.

Conservation objectives

1. Population size maintained

Information required

1. Population density and size estimated

Management actions required

1. Initiate community-based ecotourism business promoting the islands' unique avifauna. Monitoring of subpopulations and habitat could then be undertaken by visiting bird-watchers and local rangers

Bibliography

BirdLife International (2011) 'Species factsheet: *Ninox connivens*'. Retrieved 17 January, 2011 from <http://www.birdlife.org/>.

Clarke RH (2004) The avifauna of Northern Torres Strait: notes on a wet-season visit. *Australian Field Ornithology* **21**, 49–66.

Coates BJ (1985) *The Birds of Papua New Guinea, Vol I.* Dove Publications, Alderley, Australia,

Draffan RDW, Garnett ST, Malone GJ (1983) Birds of the Torres Strait: an annotated list and biogeographical analysis. *Emu* **83**, 207–234.

Higgins PJ (Ed.) (1999) *Handbook of Australian, New Zealand and Antarctic Birds. Volume 4: Parrots to Dollarbird.* Oxford University Press, Melbourne.

NPWS (2003) 'Draft Recovery Plan for the Barking Owl'. New South Wales National Parks and Wildlife Service, Hurstville, NSW.

Comments received from

Rohan Clarke, Stephen Debus

Subspecies Conservation Summary

Barking Owl (southern)

Ninox connivens connivens (Latham, 1802)

Strigidae

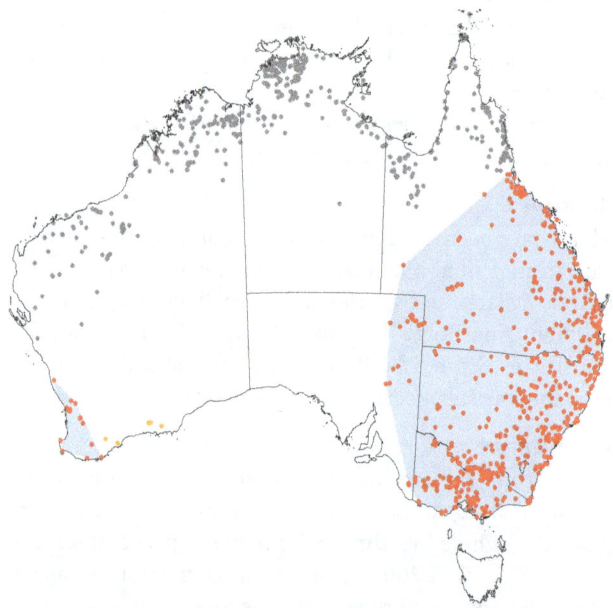

Conservation status

Near Threatened A2c+3c+4c

Reasons for listing

Past, current and expected declines in habitat quality across the owl's range approach 30% over 3 generations (26 years)

Status 2000

Near Threatened A2c+3c+4c

Status 1990

Near Threatened A2c+3c+4c

Taxonomy

N. c. peninsularis (Pilbara, Western Australia, to Burdekin Lynd Divide, Queensland) is Least Concern, *N. c. assimilis* (Papuan) is Near Threatened in Australia but Least Concern elsewhere, as is the species. There is 1 extralimital subspecies.

Taxonomic uniqueness: low (27 genera/family, 22 species/genus, 4 subspecies/species)

Range

Temperate and semi-arid regions from south of Townsville, Queensland to Flinders Ranges, South Australia, extending inland to the Lake Eyre, Bulloo and Murray Darling Basins. A second subpopulation inhabits coastal and subcoastal Western Australia from Esperance to Greenough R. (Schodde and Mason 1997; Johnstone and Storr 1998; Higgins 1999). In New South Wales, widespread on the coastal plain and foothills and the inland slopes and plains, but sparse at higher altitudes and in the arid zone, and usually absent from wet forest east of the Great Dividing Range (NPWS 2003).

Abundance

Its total population estimated as 1000 mature individuals across NSW, Vic, SA and WA (R. Kavanagh *in litt.*). Up to 200 in the Pilliga, its NSW stronghold (Kavanagh and Stanton 2009). They have declined across most of NSW in recent decades (Debus 1997). Recorded at 4 of 110 survey sites in north-west slopes of NSW 1995–1998 (Debus 2001). Recorded at a handful of sites in south-western NSW in 2006, and only 1 of 175 sites in 2009 (Parker *et al.* 2007). Twenty pairs in 170 km^2 of box-ironbark woodland in north-east Vic, but at only 8% of 75 sites in central Vic where previously reported, suggest Vic population is

Current eligibility against IUCN Red List Criteria

IUCN category	Criteria eligibility
A	Near Threatened: past, current and anticipated population declines of 20–29% over periods of 3 generations (26 years) suspected from declines in habitat quality
B	Not applicable: EOO >20 000 km^2, AOO >2000 km^2
C	Not applicable: population 2500–10 000 mature individuals, continuing decline not observed or estimated, largest subpopulation >1000 mature individuals, >1 subpopulation
D	Not applicable: population >1000 mature individuals, >5 locations
E	Not applicable: no population viability analysis undertaken

IUCN Red List assessment data

	Estimate	Reliability
Extent of occurrence trend	2 400 000 km^2 stable	high medium
Area of occupancy trend	8500 km^2 decreasing	medium high
No. of mature individuals trend	10 100 decreasing	low medium
No. subpopulations	2	high
No. locations	>10	high
Largest subpopulation	10 000	low
Generation time	8.5 years	low
Global population share	100%	high

<50 pairs (Silveira *et al.* 1997; Taylor *et al.* 2002). Small numbers monitored around Chiltern, Vic., were stable from 1980–1996 (Traill *et al.* 1996). WA subpopulation is very poorly-known but none recorded at 100 sites in 1999–2000 (Liddelow *et al.* 2002), c.10 records in 2000–2009 and recent trends are unknown (P. Mawson *in litt.*). In Qld much of the core range in the Brigalow Belt has been cleared and it was recorded in the second Atlas from about the same number of degree squares as in NSW (39; Barrett *et al.* 2003) and at a similar reporting rate. While the population in Qld is likely to be larger given the larger area of available habitat and higher population densities of *N. c. peninsularis* to the north, rates of land clearing were high in the early 2000s and log effects are still likely to be occurring in fragments. An overall rate of decline is inferred here from ongoing habitat loss and degradation to be 20–30% in 3 generations (12 years). A nominated total population size is provided here with AOO calculated based on 8.5 km^2/pair (Taylor *et al.* 2002).

Ecology

Occurs primarily in dry sclerophyll woodland, particularly in association with watercourses, wetlands and forest edges (Kavanagh *et al.* 1995a; Kavanagh *et al.* 1995b; Taylor *et al.* 2002). Nests in large hollows in live eucalypts, often near open country, where breeding pairs may be spaced at <1 km to 10 km (Higgins 1999; NPWS 2003). Hollows suitable for owl nesting do not form in eucalypts until they are at least 150–200 years old. In spite of feeding on a variety of birds, mammals and large insects, the owls prefer mammalian prey, especially rabbits *Oryctolagus cuniculus* and sugar gliders *Petaurus breviceps* (Debus *et al.* 2005). A generation time of 8.5 years (BirdLife International 2011) is derived from an age at first breeding of 1.0 years (extrapolated from other Strigidae), and a maximum longevity in the wild of 15.9 years, extrapolated from Southern Boobook *N. novaehollandiae* (Australian Bird and Bat Banding Scheme).

Threats

The main threat has been habitat clearance, fragmentation and degradation, and the subspecies may still be declining from historical habitat losses (NPWS 2003). Forestry practices, particularly those that include the felling of old-growth forests or over-mature trees, reduce the availability of nesting and roosting hollows for owls and some of their prey (Kavanagh *et al.* 1995b). On private land, much of the remaining habitat is fragmented and subject to further clearing, firewood collection and grazing, and there has been little regeneration (Barrett *et al.* 1994; Robinson and Traill 1996; Debus 1997; NPWS 2003). Other potential threats include feral European Honey Bees *Apis mellifera* competing for nesting hollows, predation by foxes *Vulpes vulpes*, feral cats *Felis catus* and goannas *Varanus* spp. and secondary poisoning by brodifacoum-based rodenticides (NPWS 2003).

Conservation objectives

1. Stable populations maintained in key regions

Information required

1. Population size and trends, especially in WA and Qld
2. Minimum habitat patch size to support successful breeding
3. Diet, particularly when breeding
4. Impact of competition and predation, e.g. by feral Honey Bees, goannas and foxes

Management actions required

1. Monitor key sites, particularly where there are management interventions
2. Develop guidelines for the management of owl populations in fragments
3. Promote ecological management of woodland remnants on private land as well as the protection of old, hollow-bearing trees in paddocks
4. Design conservation and regeneration plans at a landscape level
5. Protect from clearing all woodland in which owls are known to be resident
6. Place all areas of public land that contain the owls under secure conservation management, particularly those in timber reserves, transport corridors and local government land
7. Using appropriate incentives, undertake extension with landholders who have suitable woodland habitat to promote sound management of remnants and encourage greater connectivity between habitat patches

8. Promote revegetation and land reclamation that recreate woodland habitat with a full complement of biodiversity, including the owls
9. Control and reduce firewood collection from areas occupied by the owls, promoting wood-lot development close to markets, and reduce grazing densities where necessary

Bibliography

Barrett G, Ford HA, Recher HF (1994) Conservation of woodland birds in a fragmented rural landscape. *Pacific Conservation Biology* **1**, 245–256.

Barrett G, Silcocks A, Barry S, Cunningham R, Poulter R (2003) *The New Atlas of Australian Birds.* Royal Australian Ornithologists Union, Hawthorn East.

BirdLife International (2011) 'Species factsheet: *Ninox connivens*'. Retrieved 17 January, 2011 from <http://www.birdlife.org/>.

Debus SJS (1997) The Barking Owl in New South Wales. *Australian Birds* **30**, 53–80.

Debus SJS (2001) Surveys of the Barking Owl and Masked Owl on the north-west slopes of New South Wales. *Corella* **25**, 5–11.

Debus SJS, Ford JA, Rose AB (2005) Breeding-season diet of a pair of Barking Owls near Armidale, New South Wales. *Corella* **29**, 15–16.

Higgins PJ (Ed.) (1999) *Handbook of Australian, New Zealand and Antarctic Birds. Volume 4: Parrots to Dollarbird.* Oxford University Press, Melbourne.

Johnstone RE, Storr GM (1998) *Handbook of Western Australian Birds. Volume 1: Non-passerines (Emu to Dollarbird).* Western Australian Museum, Perth.

Kavanagh RP, Debus SJS, Rose AB, Turner RJ (1995a) Notes on the diet and habitat of the Barking Owl *Ninox connivens* in New South Wales. *Australian Bird Watcher* **16**, 137–144.

Kavanagh RP, Debus SJS, Tweedie T, Webster R (1995b) Distribution of nocturnal forest birds and mammals in north-eastern New South Wales: relationships with environmental variables and management history. *Wildlife Research* **22**, 359–377.

Kavanagh RP, Stanton MA (2009) Conserving Barking Owls in the Pilliga Forests. *Wingspan* **19**, 28–30.

Liddelow GL, Wheeler IB, Kavanagh RP (2002) Owls in the South-West Forests of Western Australia. In *Ecology and Conservation of Owls.* (Eds I Newton, RP Kavanagh, J Olsen and I Taylor) pp. 233–241. CSIRO Publishing, Melbourne.

NPWS (2003) 'Draft Recovery Plan for the Barking Owl'. New South Wales National Parks and Wildlife Service, Hurstville, NSW.

Parker DG, Webster R, Belcher CA, Leslie DJ (2007) A survey of large forest owls in State Forests of south-western New South Wales. *Australian Zoologist* **34**, 78–84.

Robinson D, Traill BJ (1996) 'Conserving Woodland Birds in the Wheat and Sheep Belts of Southern Australia'. RAOU (Birds Australia), Melbourne.

Schodde R, Mason IJ (Eds) (1997) *Zoological Catalogue of Australia 37.2: Aves (Columbidae to Coraciidae).* CSIRO Publishing, Melbourne.

Silveira CE, Yen AL, Bennett AF, Brown GW, Hinkley SD, Loyn RH, Lumsden LF, Smith W (1997) 'Fauna of the Box-Ironbark study area'. Report to the Land Conservation Council, Melbourne.

Taylor IR, Kirsten I, Peake P (2002) Habitat, breeding and conservation of the Barking Owls (*Ninox connivens*) in north-eastern Victoria, Australia. In *Ecology and Conservation of Owls.* (Eds I Newton, RP Kavanagh, J Olsen and I Taylor) pp. 116–124. CSIRO Publishing, Melbourne.

Traill BJ, Collins E, Peake P, Jessup S (1996) Current and past status of the birds of Chiltern – a box-ironbark forest in north-eastern Victoria. *Australian Bird Watcher* **16**, 309–326.

Comments received from

Graham Carpenter, Ron Johnstone, Rod Kavanagh, Peter Mawson, Todd Soderquist, David Watson

Subspecies Conservation Summary

Southern Boobook (Lord Howe Island)

Ninox novaeseelandiae albaria Ramsay, 1888
Strigidae

Conservation status
Extinct

Reasons for listing
Last recorded in the 1950s

Status 2000
Extinct

Status 1990

Extinct

Taxonomy

N. n. undulata/*novaeseelandiae* hybrids (Norfolk I.) are Critically Endangered. Five other subspecies from mainland Australia are Least Concern, as is the species. There are 7 extralimital subspecies
Taxonomic uniqueness: low (27 genera/family, 22 species/genus, 14 subspecies/species)

Former range

This subspecies of Boobook was endemic to Lord Howe I. (Department of Sustainability Environment Water Population and Communities 2011).

Former abundance

They were probably never numerous on Lord Howe I. Boobooks were last heard calling in the 1950s but the subspecies could not be determined (Hutton 1991).

Ecology

The Lord Howe Island Boobook inhabited native forests and also occurred around settlements. It presumably fed on birds and nocturnal insects, as well as on introduced House Mice *Mus musculus* after 1868 and nested in tree hollows (Department of Sustainability, Environment, Water, Population and Communities 2011).

Reasons for extinction

Between December 1922 and October 1930, Masked Owls from Tasmania *Tyto novaehollandiae castanops*, Barn Owls *T. alba* from Australia and North America and Southern Boobooks *N. n. boobook* from Australia were all shipped to Lord Howe I. in an attempt to control the Black Rats *Rattus rattus* that spread after a boat was wrecked in 1918 (Hindwood 1940). The Masked Owl has persisted since its introduction and may sometimes have competed for nesting hollows (McKean and Hindwood 1965), although Boobooks would usually have used smaller hollows than Masked Owls. The Masked Owl may also have preyed on the smaller native owl. Rats may also have reduced nesting success by preying on eggs and young, although this did not cause the extinction of the Norfolk Island Boobook. The local subspecies may also have interbred with the introduced subspecies, though neither form persisted.

Bibliography

Department of Sustainability, Environment, Water, Population and Communities (2011) *Ninox novaeseelandiae albaria*. In 'Species Profile and Threats Database'. Department of Sustainability, Environment, Water, Population and Communities, Canberra. Retrieved 24 February, 2011 from <http://www.environment.gov.au/sprat>.

Hindwood KA (1940) The birds of Lord Howe Island. *Emu* **40**, 1–86.

Hutton I (1991) *Birds of Lord Howe Island, Past and Present*. The Author, Coffs Harbour.

McKean JL, Hindwood KA (1965) Additional notes on the birds of Lord Howe Island. *Emu* **64**, 79–97.

Comments received from

Stephen Debus, David Priddel

Subspecies Conservation Summary

Southern Boobook (Norfolk Island × New Zealand)

Ninox novaeseelandiae undulata (Latham, 1802) × *N. n. novaeseelandiae* (Gmelin, 1788)
Strigidae

Conservation status

Critically Endangered D

Reasons for listing

Population <50 mature individuals

Status 2000

Critically Endangered D

Status 1990

Critically Endangered D

Taxonomy

N. n. albaria of Lord Howe I. is Extinct. Five other subspecies from mainland Australia are Least Concern, as is the species. There are 7 extralimital subspecies. The population on Norfolk I. retains some genes from the island endemic *N. n. undulata* and, because of the limited nature of the intentional hybridization, it is argued that the taxon does not meet the criteria for

Current eligibility against IUCN Red List Criteria

IUCN category	Criteria eligibility
A	Not applicable: past, current or future population declines are thought unlikely to exceed 20% in any 3-generation period
B	Not applicable: EOO, AOO, habitat, locations or population not declining, population fluctuations not extreme
C	Not applicable: no continuing population decline or extreme fluctuations
D	Critically Endangered: population <50 mature individuals
E	Not applicable: no population viability analysis undertaken

IUCN Red List assessment data

	Estimate	Reliability
Extent of occurrence trend	12 km^2 stable	medium medium
Area of occupancy trend	8 km^2 stable	medium medium
No. of mature individuals trend	40 stable	medium medium
No. subpopulations	1	high
No. locations	1	high
Generation time	8.5 years	medium
Global population share	100%	high

Extinct (Garnett *et al.* (in press) contra Garnett and Crowley (2000).
Taxonomic uniqueness: low (27 genera/family, 22 species/genus, 14 subspecies/species)

Range

Endemic to Norfolk I. Since at least 1909, however, it was largely confined to near Mt Pitt (Olsen *et al.* 1989).

Abundance

The subspecies was first recognised as Endangered in the 1960s, when visiting biologists neither heard nor saw owls. In 1973, 1978 and 1985, reports of calls were published, and local residents continued to hear at least 1 bird. A thorough survey in October 1986 found the sole surviving bird, a female, in the area in which many of the calls had been heard (Olsen *et al.* 1989). In 1988, 2 male individuals of *N. n. novaeseelandiae* from New Zealand, which was the closest relative of *N. n. undulata*, were introduced to the island. One of these birds survived, bred with the female and produced 2 broods of 2 offspring. Of these, a son and daughter produced at least 17 and probably 21 offspring. The pure-bred female was last seen late in 1996. The male survived at least a further 2 years and bred with his granddaughter, producing 4 offspring. Between December 1989 and December 1999, 28 *N. n. undulata/novaeseelandiae* hybrid nestlings were banded (Olsen 1996; Norman *et al.* 1998) and 24 have been banded since. Clutch size is smaller, possibly through inbreeding depression or an increased genetic load, but the population has persisted to 2010 and young, unbanded birds are seen, which is evidence of breeding outside the nestboxes provided (C. Rowston pers. comm.). Owls are heard calling at several locations across the island, although the exact number of individuals is unknown.

Ecology

All recent records have come from native forest or exotic plantations with a relatively clear understorey, where owls feed primarily on insects. Weed-infested native forest with a dense understorey is apparently avoided (Olsen *et al.* 1989). A generation time of 8.5 years (BirdLife International 2011) is derived from an age at first breeding of 1.0 years (extrapolated from congeners) and a maximum longevity of 15.9 years (mainland subspecies; Australian Bird and Bat Banding Scheme).

Threats

Extensive clearance for agriculture and felling of large trees destroyed most suitable nest holes on Norfolk I. (Olsen and Hicks 1989; Olsen 1996). Introduced Crimson Rosellas *Platycercus elegans*, Common Starlings *Sturnus vulgaris*, Pacific Rats *Rattus exulans* and Black Rats *R. rattus* and occasionally European Honey Bees *Apis mellifera* compete for the remaining hollows (Olsen 1996). Philip I., totally stripped of vegetation in the 19th century, is slowly recovering since the removal of pigs *Sus scrofa* and rabbits *Oryctolagus cuniculus*, but remains unsuitable for owls. As an added stress, in the 1910s many specimens of the owl were collected for museums (Olsen 1996). A change in forest structure, as a result of weed invasion, may also have reduced the area of habitat available (Olsen *et al.* 1989). The success of introductions and the associated nest-box program (which includes rat, Starling and Crimson Rosella control and exclusion) suggests that a shortage of hollows was the ultimate cause of population decrease, and remains the most potent threat (P. Olsen pers. obs.).

Conservation objectives

1. Population continues to grow
2. If appropriate, subpopulation established on Lord Howe I.

Information required

1. Trends in population size and breeding success
2. Causes of breeding failure
3. Effects of competition for hollows from Crimson Rosellas
4. Updated information on diet to minimise risk from rodent control programs
5. Understanding of genetic make-up and level of retention of original Norfolk I. owl genes

Management actions required

1. Continue to maintain rat-proof nesting hollows in Norfolk I. NP and on adjacent private land
2. Continue rat baiting and cat trapping across Norfolk I. NP
3. Advocate for rat eradication across Norfolk I.

4 Investigate reintroduction of the species to Lord Howe I.

Bibliography

BirdLife International (2011) 'Species factsheet: *Ninox novaeseelandiae*'. Retrieved 17 January, 2011 from <http://www.birdlife.org/>.

Garnett S, Crowley GM (2000) *The Action Plan for Australian Birds 2000.* Environment Australia, Canberra.

Garnett ST, Olsen P, Butchart SHM, Hoffmann AA (in press) Did hybridization save the Norfolk Island Boobook Owl? *Oryx.*

Norman JA, Olsen P, Christidis L (1998) Molecular genetics confirms taxonomic affinities of the endangered Norfolk Island Boobook Owl *Ninox novaeseelandiae undulata. Biological Conservation* **86**, 33–36.

Olsen P (1996) Re-establishment of an endangered subspecies: the Norfolk Island Boobook Owl. *Bird Conservation International* **6**, 63–80.

Olsen P, Hicks J (1989) The very last of the Norfolk Island Boobook. *Geo* **11**, 70–77.

Olsen PD, Mooney NJ, Olsen J (1989) Status and conservation of the Norfolk Island Boobook *Ninox novaeseelandiae undulata*. In *Raptors in the Modern World.* (Eds BU Meyburg and RD Chancellor) pp. 123–129. WWGBP, Berlin.

Comments received from

Stuart Butchart, Margaret Christian, Stephen Debus, Ary Hoffman, Penny Olsen, David Priddel, Coral Rowston, Ron Ward

Species Conservation Summary

Christmas Island Hawk-Owl

Ninox natalis (Lister, 1889)
Strigidae

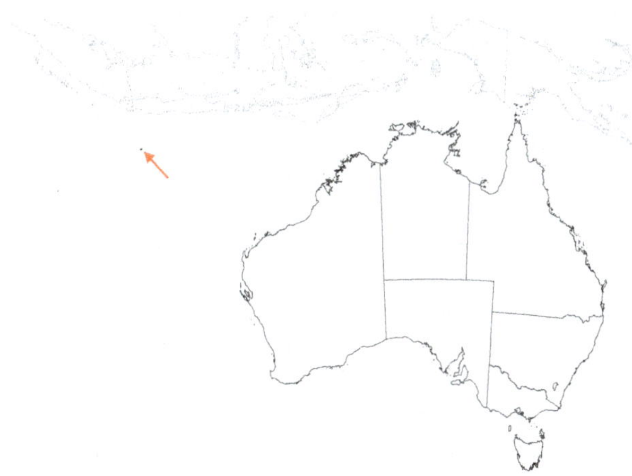

Conservation status

Vulnerable D1+2

Reasons for listing

No more than 1000 mature individuals at a single location with a plausible future threat from invasive species

Status 2000

Critically Endangered A3ce
Reason for change in 2010: ant populations temporarily reduced

Status 1990

Vulnerable D1+2
Reason for change in 2000: proliferation of invasive ants

Taxonomy

No infraspecific taxa described
Taxonomic uniqueness: low (27 genera/family, 22 species/genus, 1 subspecies/species)

Range

Endemic to Christmas I., Indian Ocean. As about a quarter of the suitable habitat has been cleared for mining (Department of Sustainability, Environment, Water, Population and Communities 2011), the Christmas Island Hawk-Owl population has probably decreased by at least 25% since settlement.

Abundance

Population estimates of 820–1200 birds (Hill and Lill 1998) are based on data collected in the 1990s. About 54 calling birds in a survey of 54 sites in 2005 and at 7 out of 9 sites sampled in 2009 are set against anecdotal evidence that the species may have declined abruptly over the last few years (Beeton *et al.* 2010). Initial results from a survey in 2010 suggest that the birds remain common throughout the forest (M. Smith pers. comm.). The population cited here, the average of the 1990s estimates, needs confirmation.

Ecology

This species occupies permanent territories in all forest types on the island, with highest densities in primary forest and lowest in post-mining regrowth (Department of Sustainability, Environment, Water, Population and Communities 2011). The owls only occasionally venture into unregenerated mine sites (Hill and Lill 1998). They nest in tree hollows in closed forest 30–40 m high with emergent trees up to 45 m tall, predominantly with *Syzygium nervosum*, *Planchonella nitida,* and *Hernandia ovigera* as canopy trees. They mainly feed on large insects, especially Orthoptera, Lepidoptera, and Coleoptera, but also small vertebrates including Black Rats *Rattus rattus,* geckos and Christmas Island White-eyes *Zosterops natalis* (Hill 2004). A generation time of 8.5 years (BirdLife International 2011) is derived from an age at first breeding of 1.0 years (extrapolated from other Strigidae), and a maximum longevity in the wild of 15.9 years (extrapo-

Current eligibility against IUCN Red List Criteria

IUCN category	Criteria eligibility
A	Not applicable: past, current or future population declines are thought unlikely to exceed 20% in any 3-generation period
B	Not applicable: EOO, AOO, habitat, locations and population not declining, population fluctuations not extreme
C	Not applicable: no continuing population decline or extreme fluctuations
D	Vulnerable: population 250–1000 mature individuals, ≤5 locations with plausible future threat
E	Not applicable: no population viability analysis undertaken

IUCN Red List assessment data

	Estimate	Reliability
Extent of occurrence trend	134 km^2 stable	high high
Area of occupancy trend	134 km^2 stable	medium medium
No. of mature individuals trend	1000 stable	low low
No. subpopulations	1	high
No. locations	1	high
Generation time	8.5 years	low
Global population share	100%	high

lated from Southern Boobook *N. novaehollandiae*; Australian Bird and Bat Banding Scheme).

Threats

Like other Christmas I. species, a proliferation of scale insects and a number of introduced ants that tend them, particularly the Yellow Crazy Ant *Anoplolepis gracilipes*, have the potential to change vegetation structure profoundly (O'Dowd *et al.* 2003; Abbott and Green 2007). Thus any faltering of ant control and infestation of all rainforest could eventually affect the owl and its prey, particularly nocturnal insects (A. A. Burbidge *in litt.*). However, the ants have been controlled and there is no evidence of decline in the owl even where ants are present. The species may also be affected by the pesticide Fipronil used for ant control, but this remains unproven (Beeton *et al.* 2010). Black Rats *Rattus rattus*, which are affecting some other bird species on Christmas I., are also likely to kill birds and reduce nesting success: rats have been implicated in the depletion or loss of owl populations on Lord Howe and Norfolk Is. and New Zealand. In the past, forest clearance has destroyed about 25% of the habitat. Although used for feeding, these areas of the island have not recovered sufficiently to contain nest hollows. Quarantine measures on the island are not currently sufficient enough to prevent invasion by additional unspecified predators.

Conservation objectives

1. Existing population maintained
2. Yellow Crazy Ant populations constrained to levels that do not threaten species
3. Black Rats eradicated

Information required

1. Secondary impacts of Fipronil and alternatives for controlling Yellow Crazy Ants
2. Population trends
3. Levels of ant abundance that will threaten species

Management actions required

1. Increase frequency and standardization of monitoring
2. Upgrade quarantine standards for Christmas I.
3. Continue control of Yellow Crazy Ants
4. Eradicate Black Rats
5. Create a multi-stakeholder team with responsibility for ensuring threatened species are conserved on Christmas I.

Bibliography

Abbott KL, Green PT (2007) Collapse of an ant-scale mutualism in a rainforest on Christmas Island. *Oikos* **116**, 1238–1246.

Beeton B, Burbidge AA, Grigg G, Harrison P, How RA, Humphries B, McKenzie N, Woinarski J (2010) 'Final report, Christmas Island Expert Working Group to Minister for the Environment, Heritage and the Arts'. Retrieved 1 April, 2010 from <http://www.environment.gov.au/parks/publications/christmas/pubs/final-report.pdf>.

BirdLife International (2011) 'Species factsheet: *Ninox natalis*'. Retrieved 17 January, 2011 from <http://www.birdlife.org/>.

Department of Sustainability, Environment, Water, Population and Communities (2011) *Ninox natalis*. In 'Species Profile and Threats Database'. Department of Sustainability, Environment, Water, Population and Communities, Canberra. Retrieved 24 February, 2011 from <http://www.environment.gov.au/sprat>.

Hill R (2004) 'National recovery plan for the Christmas Island Hawk-Owl *Ninox natalis*'. Commonwealth of Australia, Canberra.

Hill R, Lill A (1998) Density and total population estimates for the threatened Christmas Island Hawk-Owl *Ninox natalis*. *Emu* **98**, 209–220.

O'Dowd DJ, Green PT, Lake PS (2003) Invasional 'meltdown' on an oceanic island. *Ecology Letters* **6**, 812–817.

Comments received from

Stuart Butchart, Stephen Debus, Mark Holdsworth, Victor Hurley, David James, Penny Olsen, Mike Smith, John Woinarski

Subspecies Conservation Summary

Masked Owl (Tasmanian)

Tyto novaehollandiae castanops (Gould, 1837)
Tytonidae

Conservation status
Endangered C2a(ii)

Reasons for listing
Population <2500 mature individuals in a single subpopulation, in which a continuing decline is likely

Status 2000
Endangered C2a(ii)

Status 1990
Endangered C2a(ii)

Taxonomy
T. n. melvillensis (Tiwi Is) is Endangered, *T. n. kimberli* (northern mainland Australia) is Vulnerable and *T. n. novaehollandiae* (southern mainland Australia) is Least Concern, as is the species. There are 4 extralimital subspecies.
Taxonomic uniqueness: medium (2 genera/family, 14 species/genus, 8 subspecies/species)

Range
Endemic to mainland Tasmania, where recorded from all parts, except for the south-west. Highest densities are in the drier, lowland regions of the central valleys and the north coast (Bell *et al.* 1997).

Abundance
The population has been estimated to contain 1330 breeding birds or 615 breeding pairs, and probably occurs at a higher density than either of the mainland subspecies (Bell *et al.* 1997). Many cliff roosts are no longer used, which may indicate that the population was larger in the past (N. Mooney pers. comm.). It has been successfully introduced to Lord Howe I., where it is considered a threat to the Lord Howe Woodhen *Tricholimnas sylvestris* and culled when possible (Higgins 1999).

Ecology
The Masked Owl is secretive, relatively silent and strictly nocturnal. In Tas, most records are from lowland, dry sclerophyll forest, although it is also found occasionally in wet eucalypt forest, non-eucalypt dominated forest, scrub and suburban environments. Eucalypt forests and woodland containing old growth (150+ years old) trees or isolated old-growth trees containing large hollows are essential for breeding (Mooney 1997). Roost sites are usually in trees (among dense foliage or in tree hollows), cliffs (overhangs, potholes and caves) and occasionally human-made structures, such as farm sheds and open buildings (Bell *et al.* 1997). The main foods taken in agricultural areas are introduced rabbits *Oryctolagus cuniculus*, Black Rats *Rattus rattus* and House Mice *Mus musculus* (Fleay 1949; Mooney 1993). In less disturbed environments Tasmanian Masked Owls hunt marsupials and birds, and also beetles and moths (Hill 1955). A generation length of 6.1 years (BirdLife International 2011) is derived from an age of first breeding of 1.0 years, a maximum longevity in the wild of 17.9 years and an annual survival of adults of 42.1%, all values extrapolated from congeners.

Threats
Loss of old growth eucalypt forest from commercial timber harvesting, land clearance for rural residential development, tree felling for firewood and natural attrition of old growth trees have been ongoing threats for several decades, as is the loss of paddock trees (Koch *et al.* 2008). Habitat fragmentation has reduced hollow availability to the point that some pairs have problems breeding. There may be competition for remaining hollows with feral European Honey Bees *Apis mellifera*, introduced Laughing Kookaburras *Dacelo novaeguineae* and the increasing population of Common Brushtail Possums *Trichosurus vulpecula*. Death from collisions with vehicles, fences or power poles and electrocution on powerlines may also be a significant source of mortality (Bell *et al.* 1997).

Conservation objectives
1. Maintain the existing population
2. Secure habitat for nesting for now and in the future

Information required
1. Threshold hollow and prey densities

Management actions required
1. Undertake surveys throughout preferred habitat, including production forest on private and public land, to identify nest sites
2. Protect known nesting, roosting and priority foraging habitat from clearing, particularly old growth forest
3. Secure all known nesting sites on public land
4. Control firewood collection from areas occupied by Masked Owls

Current eligibility against IUCN Red List Criteria

IUCN category	Criteria eligibility
A	Not applicable: past, current or future population declines are thought unlikely to exceed 20% in any 3-generation period
B	Not applicable: EOO >20 000 km^2, AOO >2000 km^2
C	Endangered: population 250–2500 mature individuals, suspected continuing decline, >95% mature individuals in 1 subpopulation
D	Not applicable: >1000 mature individuals, >5 locations
E	Not applicable: no population viability analysis undertaken

IUCN Red List assessment data

	Estimate	Reliability
Extent of occurrence trend	50 000 km^2 stable	high high
Area of occupancy trend	7300 km^2 decreasing	medium medium
No. of mature individuals trend	1300 decreasing	medium medium
No. subpopulations	1	high
No. locations	>10	high
Generation time	6.1 years	medium
Global population share	100%	high

5. Investigate changes in prey abundance resulting from habitat fragmentation and forestry operations in the core range of the species
6. Maintain nesting habitat in parts of production forest considered important for owls
7. Undertake standardised surveys for Masked Owls and suitable nesting habitat before logging old-growth lowland dry sclerophyll forest and woodlands
8. Improve public knowledge and appreciation of the species

Bibliography

Bell P, Mooney N, Wiersma J (1997) 'Predicting the essential habitat for forest owls in Tasmania'. Australasian Raptor Association Report to the RFA Environment and Heritage Technical Committee, Hobart.

BirdLife International (2011) 'Species factsheet: *Tyto novaehollandiae*'. Retrieved 17 January, 2011 from <http://www.birdlife.org/>.

Fleay D (1949) The Tasmanian Masked Owl. *Emu* **48**, 169–176.

Higgins PJ (Ed.) (1999) *Handbook of Australian, New Zealand and Antarctic Birds. Volume 4: Parrots to Dollarbird.* Oxford University Press, Melbourne.

Hill LH (1955) Notes on the habits and breeding of the Tasmanian Masked Owl. *Emu* **55**, 203–210.

Koch AJ, Munks SA, Woehler EJ (2008) Hollow-using vertebrate fauna of Tasmania: distribution, hollow requirements and conservation status. *Australian Journal of Zoology* **56**, 323–349.

Mooney N (1993) Diet of the Masked Owl in Tasmania: past and present. In *Australian Raptor Studies.* (Ed. P Olsen) pp. 160–174. Australasian Raptor Association, RAOU, Melbourne.

Mooney N (1997) Habitat and seasonality of nesting Masked Owls in Tasmania. Australasian Raptor Studies. In *Australian Raptor Studies II.* (Eds GV Czechura and SJS Debus) pp. 34–39. Birds Australia Monograph 3. Birds Australia, Melbourne.

Comments received from

Stephen Debus, Rod Kavanagh, Nick Mooney, Penny Olsen, Mick Todd

Subspecies Conservation Summary

Masked Owl (northern)

Tyto novaehollandiae kimberli Mathews, 1912

Tytonidae

Conservation status

Vulnerable C2a(i)

Reasons for listing

Population probably <10 000 mature individuals in 3 subpopulations, none >1000 mature individuals, with a slow but continuing decline inferred from habitat loss and degradation

Status 2000

Vulnerable C2a(i)

Current eligibility against IUCN Red List Criteria

IUCN category	Criteria eligibility
A	Not applicable: past, current or future population declines are thought unlikely to exceed 20% in any 3-generation period
B	Not applicable: EOO >20 000 km², AOO >2000 km²
C	Vulnerable: population 2500–10 000 mature individuals, suspected continuing decline, possibly <1000 mature individuals in largest subpopulation
D	Not applicable: population >1000 mature individuals, >5 locations
E	Not applicable: no population viability analysis undertaken

IUCN Red List assessment data

	Estimate	Reliability
Extent of occurrence trend	1 800 000 km² decreasing	high medium
Area of occupancy trend	18 000 km² decreasing	low low
No. of mature individuals trend	3000 decreasing	low low
No. subpopulations	3	low
No. locations	>10	high
Largest subpopulation	1000	low
Generation time	6.1 years	low
Global population share	100%	high

Status 1990

Vulnerable C2a(i)

Taxonomy

T. n. castanops (Tasmania) and *T. n. melvillensis* (Tiwi Is) are Endangered, *T. n. novaehollandiae* (southern mainland Australia) is Least Concern, as is the species. There are 4 extralimital subspecies. Birds on Cape York Peninsula, which are sometimes separated as *T. n. galei* (Schodde and Mason 1997), are here included in *T. n. kimberli* (following Higgins 1999)

Taxonomic uniqueness: medium (2 genera/family, 14 species/genus, 8 subspecies/species)

Range

Found in the Kimberley Region, Western Australia to north-east Queensland (Higgins 1999). Recent records are from north and north-west coastal Kimberley, from Cambridge Gulf south-west to Yampi Sound, including Augustus Island (Johnstone and Storr 1998); the Top End of the Northern Territory, including Cobourg Peninsula, extending south to around Katherine, with a handful of isolated records from further south, including Jasper Gorge, McArthur River station and Avon Downs in the Barkly Tablelands (Woinarski 2004); north-eastern Qld, including a few early records from north-eastern Cape York Peninsula (Archer-Watson Rivers) as far south as the Atherton Tablelands and the Einasleigh-Burdekin divide (Schodde and Mason 1997; Higgins 1999).

Abundance

The population size and trends are very poorly known, partly because the species is unobtrusive and can be difficult to identify. There are no recent records from Cape York Peninsula (Higgins 1999; Woinarski 2004) and it appears to have declined in the Queensland Wet Tropics (Nielsen 1996) and may be declining in the NT at >10% per 15 years (Woinarski *et al.* 2007), although definitive data are lacking in both areas. The population is estimated here as 3000 mature individuals, spread evenly between the subpopulations – north Qld, northern part of the NT and the Kimberley, on the basis of the discontinuous known distribution, discontinuous suitable habitat and morphological variation close to subspecific status. If any subpopulation is found to be >1000 mature individuals the subspecies would qualify as Near Threatened, if all have <500 then it would be Endangered. A slow but continuing decline is inferred from habitat loss and degradation.

Ecology

In northern Australia, the Masked Owl has been recorded from riverside forests, rainforest, open forest, paperbark *Melaleuca* swamps, and the edge of mangroves, as well as along the margins of sugar cane fields (Storr 1980; Nielsen 1996; Higgins 1999). In the NT, they occur mainly in eucalypt tall open forests, but also roost in monsoon rainforests, and forage in more open vegetation types, including grasslands (Woinarski *et al.* 2007). They usually nest in tree hollows within forest patches. They feed mostly on small to medium-sized mammals, but also on insects and birds (Higgins 1999). Occurs at lower overall population densities than other subspecies which have home range sizes of 5–10 km² (Kavanagh and Murray 1996). They usually lay 2 or 3 eggs (Higgins 1999). A generation time of 6.1 years (BirdLife International 2011) is derived from an average age at first breeding of 1.0 years, an annual survival of adults of 42.0%, and a maximum longevity in the wild of 17.9 years, all values extrapolated from congeners.

Threats

The reason for the decline and low density of Masked Owls in northern Australia is a mystery. The most probable cause is a shortage of food, as small and medium-sized native mammals are becoming increasingly uncommon (Winter and Allison 1980; Woinarski *et al.* 2010). The current regime of more intense, frequent and extensive fires may reduce the availability of large trees and hollows (Williams *et al.* 1999) required for nesting, with some limited evidence from studies of hollow usurpation, competition between hollow-dependent mammal species and correlations between abundance of hollow-dependent mammals and hollow availability (J. C. Z. Woinarski *in litt.*).

Conservation objectives

1. Population stable as evidenced by monitoring, where feasible

2. Threatening processes are understood and ameliorated
3. All relevant stakeholders are aware of management guidelines, and where feasible, implement these

Information required

1. Population size, distribution and habitat requirements, including preferred fire regime
2. Trends in response to management interventions
3 Causes for decline of main prey species, i.e. small mammals

Management actions required

1. Maintain and enhance habitat suitability, through fire management

Bibliography

BirdLife International (2011) 'Species factsheet: *Tyto novaehollandiae*'. Retrieved 17 January, 2011 from <http://www.birdlife.org/>.

Higgins PJ (Ed.) (1999) *Handbook of Australian, New Zealand and Antarctic Birds. Volume 4: Parrots to Dollarbird.* Oxford University Press, Melbourne.

Johnstone RE, Storr GM (1998) *Handbook of Western Australian Birds. Vol. 1. Non-passerines (Emu to Dollarbird).* Western Australian Museum, Perth.

Kavanagh RP, Murray M (1996) Home range, habitat and behaviour of the masked owl *Tyto novaehollandiae* near Newcastle, New South Wales. *Emu* **96**, 250–257.

Nielsen L (1996) *Birds of Queensland's Wet Tropics and Great Barrier Reef.* Gerard Industries, Adelaide.

Schodde R, Mason IJ (Eds) (1997) *Zoological Catalogue of Australia 37.2: Aves (Columbidae to Coraciidae).* CSIRO Publishing, Melbourne.

Storr GM (1980) *Birds of the Kimberley Division, Western Australia.* Western Australian Museum, Perth, Western Australia.

Williams RJ, Cook GD, Gill AM, Moore PHR (1999) Fire regimes, fire intensity and tree survival in a tropical savanna in northern Australia. *Australian Journal of Ecology* **24**, 50–59.

Winter JW, Allison FR (1980) The native mammals of Cape York Peninsula – changes in status since the 1948 Archbold Expedition. In *Contemporary Cape York Peninsula.* (Eds NC Stevens and A Bailey) pp. 31–44. Royal Society of Queensland, St Lucia.

Woinarski J, Pavey C, Kerrigan R, Cowie I, Ward S (Eds) (2007) *Lost from Our Landscape: Threatened Species of the Northern Territory.* Northern Territory Government, Darwin.

Woinarski JCZ (2004) 'National multi-species recovery plan for the Partridge Pigeon [eastern subspecies] *Geophaps smithii smithii*, Crested Shrike-tit [northern (sub)species] *Falcunculus (frontatus) whitei*, Masked Owl [north Australian mainland subspecies] *Tyto novaehollandiae kimberli*; and Masked Owl [Tiwi Islands subspecies] *Tyto novaehollandiae melvillensis*, 2004– 2009'. Northern Territory Department of Infrastructure Planning and Environment, Darwin.

Woinarski JCZ, Armstrong M, Brennan K, Fisher A, Griffiths AD, Hill BM, Milne DJ, Palmer C, Ward S, Watson M, Winderlich S, Young S (2010) Monitoring indicates rapid and severe decline of native small mammals in Kakadu National Park, northern Australia. *Wildlife Research* **37**, 116–126.

Comments received from

Stephen Debus, Don Franklin, Keith McDonald, Lloyd Nielsen, Del Richards, James Smith, Mick Todd, Simon Ward, John Woinarski

Subspecies Conservation Summary

Masked Owl (Tiwi Islands)

Tyto novaehollandiae melvillensis Mathews, 1912

Tytonidae

Conservation status

Endangered C2a(ii)

Reasons for listing

Population <2500 mature individuals in a single subpopulation, with a continuing decline inferred from habitat loss and degradation

Status 2000

Near Threatened D2

Current eligibility against IUCN Red List Criteria

IUCN category	Criteria eligibility
A	Not applicable: past, current or future population declines are thought unlikely to exceed 20% in any 3-generation period
B	Not applicable: EOO >20 000 km^2, AOO >2000 km^2
C	Endangered: population 250–2500 mature individuals, suspected continuing decline, >95% mature individuals in 1 subpopulation
D	Near Threatened: population >1000 mature individuals, near 5 locations
E	Not applicable: no population viability analysis undertaken

IUCN Red List assessment data

	Estimate	Reliability
Extent of occurrence trend	7900 km^2 stable	high high
Area of occupancy trend	7000 km^2 stable	medium high
No. of mature individuals trend	2500 decreasing	low medium
No. subpopulations	1	high
No. locations	8	low
Generation time	6.1 years	low
Global population share	100%	high

Reason for change in 2010: substantial area of the highest quality habitat lost

Status 1990

Near Threatened D2

Taxonomy

T. n. castanops (Tasmania) is Endangered, *T. n. kimberli* (northern mainland Australia) is Vulnerable and *T. n. novaehollandiae* (southern mainland Australia) is Least Concern, as is the species. There are 4 extralimital subspecies.

Taxonomic uniqueness: medium (2 genera/family, 14 species/genus, 8 subspecies/species)

Range

Endemic to the Tiwi Islands (Bathurst and Melville I.), in the Northern Territory, where reasonably widespread on both islands (Woinarski *et al.* 2003).

Abundance

Reported to be common relative to mainland subspecies (Fensham and Woinarski 1992; Schodde and Mason 1997). Recorded at 44 out of 351 random survey sites in 2003 (Woinarski *et al.* 2003) and 217 from 837 survey sites in 2007–2009 and at 74% of 143 standard transect sites (J. Smith *in litt.*). Based on home range sizes of 5–10 km^2 for other subspecies (Kavanagh and Murray 1996), the population size is estimated at about 2500 mature individuals with a slow but continuing decline inferred from habitat degradation as well as loss through clearing for plantation development of some the highest-quality habitat between 2000 and 2009.

Ecology

The species is reasonably widespread on both islands, occurring mainly in tall open eucalypt forests, especially those dominated by Darwin Woollybutt *Eucalyptus miniata*, Darwin Stringybark *E. tetrodonta* and Melville Island Bloodwood *Corymbia nesophila*, and often roosts in monsoon rainforests and forages in more open vegetation (Woinarski *et al.* 2003). Commonest in the higher rainfall areas of north-west Melville Island, where eucalypt forests are tallest and there are many small patches of monsoon rainforest (Woinarski *et al.* 2003). Breeds in large tree hollows which are largely absent from plantations and mangroves (Higgins 1999). Feeds only on small to medium-sized mammals (J. Smith *in litt.*). A generation time of 6.1 years (BirdLife International 2011) is derived from an average age at first breeding of 1.0 years , an annual survival of adults of 42.0%, and a maximum longevity in the wild of 17.9 years (all values extrapolated from congeners).

Threats

The subspecies lost all nesting sites and probably much feeding habitat from 26 000 ha of tall eucalypt forest habitat, which was converted to plantation forestry in 2002–2009 (Department of Sustainability, Environment, Water, Population and Communities 2011). Additional probable ongoing threats include changed fire regime, which in turn has led to changes in the vegetation structure and may have reduced abundance of prey species (Fensham and Cowie 1998; Woinarski 2004; Woinarski *et al.* 2007). Fires and cyclones also reduce the number of suitable nesting trees (Department of Sustainability, Environment, Water, Population and Communities 2011), which, on the basis of frequent re-use, may be a limiting resource (J. Smith *in litt.*).

Conservation objectives

1. Stable population
2. Threatening processes ameliorated
3. All relevant stakeholders aware of management needs

Information required

1. Population trends
2. Effects of current fire regime on prey and nest hollow availability
3. Relationship to Masked Owls on Cobourg Peninsula

Management actions required

1. Monitor trends at selected sites, building on existing work
2. Manage fire to maintain prey numbers and hollow availability

Bibliography

BirdLife International (2011) 'Species factsheet: *Tyto novaehollandiae*'. Retrieved 17 January, 2011 from <http://www.birdlife.org/>.

Department of Sustainability, Environment, Water, Population and Communities (2011) *Tyto novaehollandiae melvillensis.*

In 'Species Profile and Threats Database'. Department of Sustainability, Environment, Water, Population and Communities, Canberra. Retrieved 24 February, 2011 from <http://www.environment.gov.au/sprat>.

Fensham RJ, Cowie ID (1998) Alien plant invasions on the Tiwi Islands: extent, implications and priorities for control. *Biological Conservation* **83**, 55–68.

Fensham RJ, Woinarski JCZ (1992) 'Yawalama: the ecology and conservation of monsoon forest on Tiwi Islands, Northern Territory'. Report to DASET. Conservation Commission of the Northern Territory.

Higgins PJ (Ed.) (1999) *Handbook of Australian, New Zealand and Antarctic Birds. Volume 4: Parrots to Dollarbird.* Oxford University Press, Melbourne.

Kavanagh RP, Murray M (1996) Home range, habitat and behaviour of the masked owl *Tyto novaehollandiae* near Newcastle, New South Wales. *Emu* **96**, 250–257.

Schodde R, Mason IJ (Eds) (1997) *Zoological Catalogue of Australia 37.2: Aves (Columbidae to Coraciidae)*. CSIRO Publishing, Melbourne.

Woinarski J, Brennan K, Hempel C, Armstrong M, Milne D, Chatto R (2003) 'Biodiversity conservation on the Tiwi islands, Northern Territory. Part 2. Fauna'. Department of Infrastructure Planning and Environment, Darwin.

Woinarski J, Pavey C, Kerrigan R, Cowie I, Ward S (Eds) (2007) *Lost from Our Landscape: Threatened Species of the Northern Territory.* Northern Territory Government, Darwin.

Woinarski JCZ (2004) 'National multi-species recovery plan for the Partridge Pigeon [eastern subspecies] *Geophaps smithii smithii*, Crested Shrike-tit [northern (sub)species] *Falcunculus (frontatus) whitei*, Masked Owl [north Australian mainland subspecies] *Tyto novaehollandiae kimberli*; and Masked Owl [Tiwi Islands subspecies] *Tyto novaehollandiae melvillensis*, 2004–2009'. Northern Territory Department of Infrastructure Planning and Environment, Darwin.

Comments received from

Stephen Debus, James Smith, Simon Ward, John Woinarski

Subspecies Conservation Summary

Azure Kingfisher (Tasmanian)

Ceyx azureus diemenensis (Gould, 1846)
Alcedinidae

Conservation status
Vulnerable D1

Reasons for listing
Population probably 250–1000 mature individuals

Status 2000
Vulnerable D1

Status 1990
Vulnerable D1

Taxonomy
Two mainland Australian subspecies are Least Concern, as is the species. There are 5 extralimital subspecies. Taxonomic uniqueness: medium (3 genera/family, 7 species/genus, 8 subspecies/species)

Range
This subspecies is endemic to Tasmania. It has been recorded in 30–33 major catchments and numerous additional sub-catchments in the south, west and northwest coast with outlying occurrences in the northeast, east, centre and Bass Strait islands, mirroring the distribution of suitable habitat along river systems (Wapstra *et al.* 2010).

Abundance
Although widely distributed, few birds are seen. Densities of 1–3 pairs along 10–20 km of Arthur River suggest very small numbers, but recent reports of birds every few kilometres along the Salisbury and Jane Rivers suggest a higher density in some rivers (M. Wapstra *in litt.*). The population is assumed to be very small and probably does not exceed 1000 mature individuals. Suggestions that it disappeared from around Hobart around 1903, and from 2 areas bordering Macquarie Harbour in about 1985 (Higgins 1999) are probably wrong.

Ecology
Occurs mainly along the forested margins of the major river systems of the west and northwest coast of Tas, potentially within a relatively wide range of forest types but mainly wet sclerophyll eucalypt forest with broad-leaved and rainforest shrub understoreys. Feeds on small

Current eligibility against IUCN Red List Criteria

IUCN category	Criteria eligibility
A	Not applicable: past, current or future population declines are thought unlikely to exceed 20% in any 3-generation period
B	Not applicable: EOO, AOO, habitat, locations or population not declining, population fluctuations not extreme
C	Not applicable: no continuing population decline or extreme fluctuations
D	Vulnerable: population 250–1000 mature individuals
E	Not applicable: no population viability analysis undertaken

IUCN Red List assessment data

	Estimate	Reliability
Extent of occurrence trend	2000 km^2 stable	medium low
Area of occupancy trend	500 km^2 stable	low medium
No. of mature individuals trend	1000 stable	low medium
No. subpopulations	1	high
No. locations	60	high
Generation time	4 years	medium

fish, freshwater crayfish, aquatic insects and their larvae, occasionally amphibians, and occasionally forages on the ground taking beetles and other terrestrial insects and it may catch dragonflies in the air. Nests are made in burrows excavated in stream banks; the usual clutch size is 5–6 (Higgins 1999). Generation length of 4.4 years (BirdLife International 2011) is derived from an age at first breeding of 1.0 years, an annual survival of adults of 32.9% and a maximum longevity in the wild of 11.7 years, all values extrapolated from congeners.

Threats

No obvious threats. Many major river systems that may once have supported Azure Kingfishers have had extensive clearing of the riparian vegetation, and this may have contributed to the localized extinctions of the species from several reported locations (Wapstra *et al.* 2010). Competition of prey species with introduced Brown Trout *Trutta trutta*, once thought to be a threat (Garnett and Crowley 2000), but is now thought not to be (Wapstra *et al.* 2010). Alteration to fish movement and flow regimes by hydro-electric dams may reduce habitat suitability and acid mine drainage from tailings dams may pose a localised threat in a limited number of catchments (Higgins 1999).

Conservation objectives

1. Stable or increasing population

Information required

1. Population size and a definitive understanding of distribution
2. Characteristics of occupied and unoccupied streams in terms of water quality, prey availability and likely threatening processes.

Management actions required

1. Survey streams in north-west, west and central Tasmania.

Bibliography

BirdLife International (2011) 'Species factsheet: *Alcedo azurea*'. Retrieved 17 January, 2011 from <http://www.birdlife.org/>.

Garnett S, Crowley GM (2000) *The Action Plan for Australian Birds 2000*. Environment Australia, Canberra.

Higgins PJ (Ed.) (1999) *Handbook of Australian, New Zealand and Antarctic Birds. Volume 4: Parrots to Dollarbird*. Oxford University Press, Melbourne.

Wapstra M, Bryant S, Bell P (2010) Conservation overview of the azure kingfisher *Ceyx azureus* subsp. *diemenensis* in Tasmania. *Tasmanian Bird Report* **34**, 8–23.

Comments received from

Sally Bryant, Mark Holdsworth, Nick Mooney, Mark Wapstra

Subspecies Conservation Summary

Sacred Kingfisher (Norfolk Island)

Todiramphus sanctus norfolkiensis (Tristram, 1885)
Alcedinidae

Conservation status

Near Threatened D1+2

Reasons for listing

Population probably near 1000 mature individuals in 1 location with a plausible but low probability threat

Status 2000

Near Threatened D1+2

Status 1990

Near Threatened D1+2

Taxonomy

T. s. vagans (New Zealand and other Tasman Sea Is., including Lord Howe I.) and *T. s. sanctus* (mainland Australia, Tasmania and off-shore islands) are Least Concern, as is the species. There are 2 extralimital subspecies.
Taxonomic uniqueness: low (11 genera/family, 21 species/genus, 5 subspecies/species)

Range

Endemic to Norfolk and Phillip Is. (Schodde and Mason 1997).

Abundance

Common in open and coastal parts of Norfolk I., and a few pairs around remnant and regenerating trees on Phillip I. (G. Dutson pers. obs.). Taking as a possible density the average of 0.2–1.7 birds/ha (Bell 1980) and 0.13 birds/ha (Skegg 1964), the total population would be around 750 pairs.

Ecology

Occurs in farmland, along rocky shores and the edges of rainforest, but eschews the forest centre (Schodde *et al.* 1983). It nests in banks or trees, including tree ferns and Norfolk Island Pines *Araucaria heterophylla* (Hermes *et al.* 1986). A generation time of 4.8 years (BirdLife International 2011) is derived from an age at first breeding of 1.0 years, extrapolated from other Alcedinidae, and a maximum longevity in the wild of 8.5 years, extrapolated from Collared Kingfisher *T. chloris*.

Threats

Any endemic confined to a single island as small as Norfolk I. might be considered Vulnerable D2 but the kingfishers have evidently survived the major disruptions of hunting and the introduction of cats *Felis catus* and rats *Rattus* spp., and have probably benefited from the clearing of forest. The only possible threat is the introduction of a new predator or disease, or some meteorological catastrophe.

Conservation objectives

1. Maintain current breeding population

Information required

1. Population trends on Norfolk I.

Management actions required

1. Maintain quarantine for Norfolk I.
2. Manage threats on Norfolk I. as required

Current eligibility against IUCN Red List Criteria

IUCN category	Criteria eligibility
A	Not applicable: past, current or future population declines are thought unlikely to exceed 20% in any 3-generation period
B	Not applicable: EOO, AOO, habitat, locations or population not declining, population fluctuations not extreme
C	Not applicable: no continuing population decline or extreme fluctuations
D	Near Threatened: population near 1000 mature individuals; at 1 location with plausible but low probability threat
E	Not applicable: no population viability analysis undertaken

IUCN Red List assessment data

	Estimate	Reliability
Extent of occurrence trend	36 km^2 stable	high high
Area of occupancy trend	28 km^2 stable	medium high
No. of mature individuals trend	1500 stable	low high
No. subpopulations	1	high
No. locations	2	high
Generation time	4.8 years	low
Global population share	100%	high

Bibliography

Bell HL (1980) Effects of power-line clearing on birds. *Corella* **4**, 8–19.

BirdLife International (2011) 'Species factsheet: *Todiramphus sanctus*'. Retrieved 17 January, 2011 from <http://www.birdlife.org/>.

Hermes N, Evans O, Evans B (1986) Norfolk Island birds: a review 1985. *Notornis* **33**, 141–149.

Higgins PJ (Ed.) (1999) *Handbook of Australian, New Zealand and Antarctic Birds. Volume 4: Parrots to Dollarbird.* Oxford University Press, Melbourne.

Schodde R, Mason IJ (Eds) (1997) *Zoological Catalogue of Australia 37.2: Aves (Columbidae to Coraciidae).* CSIRO Publishing, Melbourne.

Schodde R, Fullagar P, Hermes N (1983) 'A review of Norfolk Island birds: past and present'. Australian National Parks and Wildlife Service, Canberra.

Skegg PDG (1964) Birds of the Hen and Chicken Islands. *Notornis* **11**, 159–182.

Comments received from

Margaret Christian, Ron Ward

Species Conservation Summary

Albert's Lyrebird

Menura alberti Bonaparte, 1850

Menuridae

Conservation status

Near Threatened B1ab(i,ii,iv)+2ab(i,ii,iv)

Reasons for listing

Extent of occurrence <20 000 km^2 and occupies <2000 km^2, possibly near 10 locations, possible loss of outlying locations/subpopulations and thus extent of occurrence and area occupied

Status 2000

Vulnerable B1ab(iii,v)+2ab(iii,v)

Reason for change in 2010: protection of critical habitat from logging reduced probability of declines

Status 1990

Vulnerable B1ab(iii,v)+2ab(iii,v)

Taxonomy

No infraspecific taxa described

Taxonomic uniqueness: high (1 genus/family, 2 species/genus, 1 subspecies/species)

Range

Endemic to rainforests in far south-east Queensland and far north-east New South Wales. In Qld occurs on Tamborine Mountain, Lamington National Park, along the McPherson Range west to Mount Barney National Park and along the Great Dividing Range from Wilsons Peak north to Cunningham's Gap and Mount Mistake. In NSW it occurs along the McPherson and Tweed Ranges south to Koonyum and the Nightcap Ranges in the east and west to Boundary Creek, south-east of Kyogle, and the Richmond Range. There is an isolated population to the south at Uralba Nature Reserve in the Blackwall Range (Higgins *et al.* 2001). Much of the lyrebird's habitat was cleared during the 19th century (Garnett and Crowley 2000) but most lyrebirds now live in areas managed for conservation. The AOO of 600 km^2 is derived from records in the Birds Australia Atlas, rounding up from 532 km^2, given that some occupied areas will not have been recorded. There are assumed to be about 5 subpopulations, but the extent of genetic mixing is unknown.

Abundance

Based on a territory size of about 9 ha and an AOO of 600 km^2, the population size is calculated at about 13 000 mature individuals, which is much higher than estimated by Garnett and Crowley (2000). The population appears stable. Even what are now small isolated subpopulations at Blackwall Range (possibly 10 birds; Bower 1997) and Mt Tamborine (23 pairs; Nielsen 2010) have persisted for many decades. The largest population, based on area occupied, is probably along the Great Dividing Range and may contain about 5000 mature individuals.

Ecology

Mostly confined to rainforests and wet sclerophyll forests with mesic understorey, usually at over 300 m above sea level. Territories are typically located in gullies, along watercourses, and on the slopes and ridges of steep mountain ranges (Higgins *et al.* 2001). They feed on the ground,

Current eligibility against IUCN Red List Criteria

IUCN category	Criteria eligibility
A	Not applicable: past, current or future population declines are thought unlikely to exceed 20% in any 3-generation period
B	Near Threatened: EOO <20 000 km², AOO <2000 km², near 10 locations, suspected continuing decline in EOO, AOO, no. locations, no. subpopulations
C	Not applicable: population >10 000 mature individuals
D	Not applicable: population >1000 mature individuals, >5 locations
E	Not applicable: no population viability analysis undertaken

IUCN Red List assessment data

	Estimate	Reliability
Extent of occurrence trend	10 000 km² stable	high medium
Area of occupancy trend	600 km² stable	medium medium
No. of mature individuals trend	13 000 stable	low medium
No. subpopulations	5	low
No. locations	12	medium
Largest subpopulation	5000	low
Generation time	14.3 years	medium
Global population share	100%	high

mainly on insects (including beetles) and their larvae, and other soil-dwelling invertebrates. They have a clutch of 1, laid in a large domed nest built in trees, on rock escarpments or on the steep sides of gullies (Higgins *et al.* 2001). A generation time of 14.3 years (BirdLife International 2011) is derived from an age at first breeding of 6.5 years (extrapolated from Superb Lyrebird *M. superba*) and a maximum longevity in the wild of 22 years (Higgins *et al.* 2001).

Threats

Fire in exceptionally dry years is probably the main current threat, especially to outlying subpopulations (Higgins *et al.* 2001) and while unlikely to affect the entire populations, there may be as few as 12 locations during years with extensive fire. Historically much of the bird's habitat was cleared in the 19th century. It is also 1 of the fauna species most sensitive to logging (Kavanagh and Stanton 2005). However, almost all clearing has ceased and threat from intense forest management and the establishment of eucalypt plantations has abated in recent years as much of the Lyrebird's habitat is now reserved. Whian Whian Conservation Area, which was created in 2003, supports the highest density of the species (Department of Sustainability Environment Water Population and Communities 2011). Some populations, such as the isolated groups at Blackwall Range and Tamborine Mountain, are threatened simply because they are so small, and because population densities are lower than in optimal habitats (Gilmore 2000) but have nevertheless persisted with little change for decades (Nielsen 2010).

Conservation objectives

1. Outlying populations persist
2. Populations monitored at key sites are stable

Information required

1. Population trends at key sites

Management actions required

1. Ensure adequate fire protection is in place, particularly in dry years
2. Minimise development near remnant populations

Bibliography

BirdLife International (2011) 'Species factsheet: *Menura alberti*'. Retrieved 17 January, 2011 from <http://www.birdlife.org/>.

Bower H (1997) 'The suitability of isolated Big Scrub remnants as habitat for sooty owls, marbled frogmouth and Albert's lyrebird'. Integrated project. Southern Cross University, Lismore.

Department of Sustainability Environment Water Population and Communities (2011) *Menura alberti*. In 'Species Profile and Threats Database'. Department of Sustainability, Environment, Water, Population and Communities, Canberra. Retrieved 24 February, 2011 from <http://www.environment.gov.au/sprat>.

Garnett S, Crowley GM (2000) *The Action Plan for Australian Birds 2000*. Environment Australia, Canberra.

Gilmore A (2000) Distributional ecology of the Albert's Lyrebird, *Menura alberti*, in north-east New South Wales. In 'The 2nd Southern Hemisphere Ornithological Congress. Birds Australia Report Series 9'. (Eds LB Astheimer and MF Clarke) p. 65.

Higgins PJ, Peter JM, Steele WK (Eds) (2001) *Handbook of Australian, New Zealand and Antarctic Birds. Volume 5: Tyrant-Flycatchers to Chats*. Oxford University Press, Melbourne.

Kavanagh RP, Stanton MA (2005) Vertebrate species assemblages and species sensitivity to logging in the forests of north-eastern New South Wales. *Forest Ecology and Management* **209**, 309–341.

Nielsen L (2010) 'Albert's Lyrebirds on Tamborine Mountain'. Retrieved 8 May, 2011 from <http://birding-aus.org/?p=4216>.

Comments received from

Barry Baker, Andrew Burbidge, Stuart Butchart, Ian Gynther, Kerryn Herman, Tim Holmes, Lloyd Nielsen, John Woinarski

Species Conservation Summary

Rufous Scrub-bird

Atrichornis rufescens (Ramsay, 1867)
Atrichornithidae

Conservation status

Endangered B2ab(ii,iv,v)

Reasons for listing

Occupies <500 km^2, possibly severely fragmented, evidence of reduced area occupied, loss of some locations and thus of mature individuals

Status 2000

Endangered B2ab(ii, iv,v)

Status 1990

Least Concern
Reason for change in 2000: some evidence of decline

Taxonomy

Both *A. r. rufescens* (north-east New South Wales and south-east Queensland) and *A. r. ferrieri* (Dorrigo Plateau-Barrington Tops, NSW) are Endangered
Taxonomic uniqueness: high (1 genus/family, 2 species/genus, 2 subspecies/species)

Range

The species is occurs in high-rainfall areas above 600 m above sea level in south-eastern Queensland and north-eastern New South Wales. *A. r. rufescens* occurs in the Gibraltar Ranges, Border Ranges, the northern part of the McPherson Range and in parts of the Main Range, but formerly occurred in lowland habitats of the Richmond and Tweed River basins. *A. r. ferrieri* occurs on Barrington Tops, Hastings Range and in the Dorrigo/Ebor area. Some subpopulations of *A. r. rufescens* are thought to have disappeared within the last 2 decades including those at Mt Warning and Spicers Gap.

Abundance

In the early 19th century, the population size was probably about 12 000 pairs, but surveys from 1979–1983 estimated it at about 2500 pairs, *A. r. rufescens* numbering 730 pairs and *A. r. ferrieri* 1720 pairs at a density of about 6 pairs/km^2 (Ferrier 1984). Declines are suspected in both subspecies (Ekert 2005); *A. r. rufescens* because of a loss of suspected subpopulations (T. Holmes *in litt.*), *A. r. ferrieri* because of a reduction in area occupied by calling males in New England National Park (Garnett and Crowley 2000).

Ecology

Rainforests above 600 m elevation, including subtropical, warm temperate and cool temperate rainforests, and nearby moist and wet eucalypt forests. They require dense ground cover which is usually restricted to ecotones, forested watercourses and wetlands, and areas regenerating from fires, storms or along roadsides. They forage on small invertebrates, including snails and insects on the ground and over fallen logs, amongst leaf-litter and on other ground vegetation and debris, within the dense understorey (Higgins *et al.* 2001). A generation time of 4.9 years (BirdLife International 2011) is derived from an age at first breeding of 2.0 years and a maximum longevity in the wild of 7.8 years, both values extrapolated from Noisy Scrub-bird *A. clamosus*.

Threats

Eucalypt forest habitat, which supports the majority of the population and buffers wetter refuges, is potentially threatened by wild fires in dry periods (Ferrier 1984). Declines in density may also occur naturally as vegetation matures and ground cover provides less shelter, so some fire or other disturbance such as storms is helpful as the recovering habitat is readily used by the birds. In the longer term, however, the viability of the small remaining populations may be questionable. Most of the bird's lowland habitat was cleared in the 19th century, and, while clearance itself is not a continuing threat with almost all birds being in protected areas, the subdivision of a small population into even smaller fragments makes each subpopulation more susceptible to random events.

Conservation objectives

1. Each subpopulation known to be stable
2. Fire plans in place for each subpopulation

Information required

1. Change in distribution and abundance since first survey

Current eligibility against IUCN Red List Criteria

IUCN category	Criteria eligibility
A	Not applicable: population declines <30% in 3 generations (15 years)
B	Endangered: AOO <500 km^2, assumed to be severely fragmented, suspected continuing decline in AOO, no. locations, no. mature individuals
C	Not applicable: population <10 000 mature individuals, continuing decline not observed or estimated, largest subpopulation >1000 mature individuals, >1 subpopulation
D	Not applicable: population >1000 mature individuals, >5 locations
E	Not applicable: no population viability analysis undertaken

IUCN Red List assessment data

	Estimate	Reliability
Extent of occurrence trend	11 400 km^2 stable	high high
Area of occupancy trend	410 km^2 decreasing	medium low
No. of mature individuals trend	4900 decreasing	medium low
No. subpopulations	6	medium
No. locations	>10	high
Largest subpopulation	2000	low
Generation time	4.9 years	low
Global population share	100%	medium

2. Appropriate fire regime to maintain habitat suitability

Management actions required

1. Repeat first survey using same methods
2. Develop monitoring protocol for birds and habitat
3. Develop and implement fire management plans for all subpopulations
4. Assess whether experimental habitat manipulation is justified
5. Identify fire refugia at local and landscape levels

Bibliography

BirdLife International (2011) 'Species factsheet: *Atrichornis rufescens*'. Retrieved 17 January, 2011 from <http://www.birdlife.org/>.

Ekert PA (2005) 'Monitoring Rufous Scrub-birds (*Atrichornis rufescens*) in Central Eastern Rainforest Reserves of Australia (CERRA): Final Report 2005'. Report prepared by Ekerlogic Consulting Services (ECS) for the NSW National Parks and Wildlife Service, Coffs Harbour.

Ferrier S (1984) The status of the Rufous Scrub-bird *Atrichornis rufescens*: habitat, geographical variation and abundance. PhD thesis, University of New England, Armidale.

Garnett S, Crowley GM (2000) *The Action Plan for Australian Birds 2000*. Environment Australia, Canberra.

Higgins PJ, Peter JM, Steele WK (Eds) (2001) *Handbook of Australian, New Zealand and Antarctic Birds. Volume 5: Tyrant-Flycatchers to Chats*. Oxford University Press, Melbourne.

Comments received from

Barry Baker, Stuart Butchart, Peter Ekert, Hugh Ford, Tim Holmes, Richard Loyn, Mike Newman, David Paton, David Stewart, John Woinarski, David Watson

Subspecies Conservation Summary

Rufous Scrub-bird (southern)

Atrichornis rufescens ferrieri Schodde & Mason, 1999
Atrichornithidae

Conservation status

Endangered B2ab(ii,v)

Reasons for listing

Occupies <500 km^2, possibly severely fragmented and some evidence that the area occupied, and hence the number of mature individuals, is declining

Status 2000

Endangered B2ab(ii,v)

Status 1990

Least Concern

Reason for change in 2000: some evidence for population decline

Taxonomy

A. r. rufescens (north-east New South Wales and south-east Queensland) is Endangered, as is the species

Taxonomic uniqueness: high (1 genus/family, 2 species/genus, 2 subspecies/species)

Range

Found in high rainfall areas along the Great Dividing Range, north-central New South Wales (Schodde and Mason 1999). Known subpopulations are centred on Barrington Tops (including Gloucester Tops), Hastings Range (including Werrikimbie National Park) and Dorrigo/Ebor (New England National Park), mostly above 600 m in altitude (Ferrier 1984). All sites were still occupied in 2010 (M. Newman *in litt.*).A lowland record from near Wauchope (Boles and Tynan 1995) suggests birds may once have moved between what are now disjunct upland habitat patches.

Abundance

Relatively comprehensive surveys in 1979–1983 estimated c.1720 territories (Ferrier 1984). There have been no subsequent comprehensive counts. Although surveys 1999–2004 confirmed reasonable numbers persist in the Werrikimbie and Carrai National Parks with smaller numbers at other sites (Ekert 2005) and a survey above 1100 m on Barrington and Gloucester Tops in September 2010 recorded similar or higher densities than in 1981 (6 pairs/km^2; Ferrier 1984; Newman 2010). Given the small AOO, the population is assessed here as 3500 mature individuals fragmented into at least 3 subpopulations, the largest about 2000. Population size in the 1800s is estimated to have been about 14 200 pairs (Ferrier 1984). Anecdotal evidence of a decline in parts

Current eligibility against IUCN Red List Criteria

IUCN category	Criteria eligibility
A	Not applicable: past, current or future population declines are thought unlikely to exceed 20% in any 3-generation period
B	Endangered: AOO <500 km^2, possibly severely fragmented, declines in AOO, no. locations, no. mature individuals
C	Not applicable: population <10 000 mature individuals, continuing decline not observed or estimated, largest subpopulation >1000 mature individuals, >1 subpopulation
D	Not applicable: population >1000 mature individuals, >5 locations
E	Not applicable: no population viability analysis undertaken

IUCN Red List assessment data

	Estimate	Reliability
Extent of occurrence trend	9400 km^2 stable	high high
Area of occupancy trend	290 km^2 decreasing	medium low
No. of mature individuals trend	3400 decreasing	medium low
No. subpopulations	3	high
No. locations	>10	high
Largest subpopulation	2000	low
Generation time	4.9 years	medium
Global population share	100%	medium

of New England National Park close to agricultural land (Garnett and Crowley 2000) has not been corroborated, but any ongoing decline requires a precautionary approach given the small AOO and population size. At Gloucester Tops there has also been a suggestion of a decline below 600 m but this has not been confirmed (M. Newman *in litt.*).

Ecology

Now confined to high-altitude (above 600 m elevation) rainforests, including subtropical, warm temperate and cool temperate rainforests, and nearby moist and wet eucalypt forests. They require dense ground cover which is usually restricted to ecotones, forested watercourses and wetlands, and areas regenerating from fires or storms. They forage on small invertebrates, including snails and insects on the ground and over fallen logs, amongst leaf-litter and on other ground vegetation and debris, within the dense understorey (Higgins *et al.* 2001). Very little is known of their breeding behaviour because they are so shy and cryptic, living in dense vegetation. A generation time of 4.9 years is estimated (BirdLife International 2011), based on age at first breeding of 2.0 years and maximum longevity of 7.8 years, both values extrapolated from Noisy Scrub-bird *A. clamosus.*

Threats

Eucalypt forest habitat, which supports the majority of the population, is potentially threatened by wild fires in dry periods (Ferrier 1984). Declines in density may also occur naturally as vegetation matures and ground cover provides less shelter, so some fire or other disturbance such as storms is helpful as the recovering habitat is readily used by the birds. In the longer term, however, the viability of the small remaining populations may be questionable. Most of the bird's lowland habitat was cleared in the 19th century, and, while clearance itself is not a continuing threat with almost all birds being in protected areas, the subdivision of a small population into even smaller fragments makes each subpopulation more susceptible to random events.

Conservation objectives

1. Each subpopulation known to be stable
2. Fire plans in place for each subpopulation

Information required

1. Change in distribution and abundance since first survey
2. Appropriate fire regime to maintain habitat suitability

Management actions required

1. Maintain monitoring protocols established in 1981
2. At key sites monitor understorey vegetation density, ground level humidity, distance from rainforest, leaf litter volume and inter-territory separation
3. Develop and implement fire management plans for all subpopulations
4. Identify fire refugia at local and landscape levels

Bibliography

BirdLife International (2011) 'Species factsheet: *Atrichornis rufescens*'. Retrieved 17 January, 2011 from <http://www.birdlife.org/>.

Boles WE, Tynan B (1995) Low altitude record of Rufous Scrub-bird *Atrichornis rufescens. Corella* **19**, 91.

Ekert PA (2005) 'Monitoring Rufous Scrub-birds (*Atrichornis rufescens*) in Central Eastern Rainforest Reserves of Australia (CERRA): Final Report 2005'. Report to the Foundation for National Parks and Wildlife, Sydney.

Ferrier S (1984) The status of the Rufous Scrub-bird *Atrichornis rufescens*: habitat, geographical variation and abundance. PhD thesis, University of New England, Armidale.

Garnett S, Crowley GM (2000) *The Action Plan for Australian Birds 2000.* Environment Australia, Canberra.

Higgins PJ, Peter JM, Steele WK (Eds) (2001) *Handbook of Australian, New Zealand and Antarctic Birds. Volume 5: Tyrant-Flycatchers to Chats.* Oxford University Press, Melbourne.

Newman N (2010) 'Rufous Scrub-bird Surveys – Barrington/Gloucester Tops IBA – 15 to 17 September 2010'. Report to Hunter River Bird Observers' Club. Retrieved 8 May, 2011 from <http://www.hboc.org.au/resources/documents/RSB%20Sep%202010%20interim%20report.pdf>.

Schodde R, Mason IJ (1999) *The Directory of Australian Birds: Passerines.* CSIRO Publishing, Melbourne.

Comments received from

Barry Baker, Stuart Butchart, Peter Ekert, Hugh Ford, Tim Holmes, Richard Loyn, Mike Newman, David Paton, David Stewart, John Woinarski, David Watson

Subspecies Conservation Summary

Rufous Scrub-bird (northern)

Atrichornis rufescens rufescens (Ramsay, 1866)
Atrichornithidae

Conservation status

Endangered B1ab(ii,iv,v)+2ab(ii,iv,v)

Reasons for listing

Extent of occurrence <5000 km^2, occupies <500 km^2 and possibly severely fragmented with a continuing decline suspected in area occupied, number of locations and thus of mature individuals because fire has eliminated 1 small population and at least 1 more has not been recorded recently

Status 2000

Endangered B1ab(ii,iv,v)+2ab(ii,iv,v)

Status 1990

Near Threatened D1
Reason for change in 2000: some evidence for population decline

Taxonomy

A. r. ferrieri (Dorrigo Plateau-Barrington Tops, New South Wales) is Endangered, as is the species
Taxonomic uniqueness: high (1 genus/family, 2 species/genus, 2 subspecies/species)

Range

Found in high rainfall areas in north-east New South Wales and south-east Queensland (Schodde and Mason 1999), now mostly in areas above 600 m in altitude as a result of clearing of lowland habitat (Ferrier, 1984). There are 4 subpopulations: Border Ranges (Border Ranges National Park), Gibraltar Ranges (Gibraltar Ranges and Barool National Parks), northern part of the McPherson Range (Lamington National Park west to Mount Barney National Park) and in the Main Range (upper reaches of Dalrymple Creek, Mount Cordeaux, Mt Mitchell and Cunninghams Gap). There have been no recent records from Mt Warning NP, Richmond Range NP, Whian Whian State Forest, Mt Jerusalem NP, Spicers Gap (T. Holmes *in litt.*) and Mt Mathieson, although some of these sites have not been thoroughly searched recently (D. Stewart *in litt.*). The AOO is estimated from the density and the estimate of population size of Ferrier (1984).

Abundance

Relatively comprehensive surveys in 1979–1983 estimated c.730 territories (Ferrier 1984). There appear to have been no subsequent comprehensive counts. In 1999–2004, reasonable numbers were recorded in the Gibraltar Ranges, although numbers were lower after a fire in 2002 (Ekert 2005). Seven were heard from 58 survey points in Main Range and Lamington National Park in 2004, and 11 heard in 2006 (Stewart 2006). In surveys of Border Ranges in 2010 35 birds were heard calling on 8/12 transects, 7 permanent territories were confirmed, another 11 were considered possible but no scrub-birds were heard calling below 700 m altitude (Hill 2010). Even in suitable habitat, territories are widely spaced with a maximum density of 6 pairs/km^2 (Ferrier 1984). At Wiangarie territories sometimes occurred in clusters (Ferrier 1984), possibly because the understorey vegetation in the rainforest supporting the northern subspecies was less homogeneous than for the southern subspecies (M. Newman *in litt.*). The population appears to be fragmented into at least 4 subpopulations, none >1000. Population size in the 1800s is estimated to have been about 4800 pairs (Ferrier 1984). The apparent loss of birds from parts of its range suggests that declines of the smallest subpopulations are continuing. There is a danger that even the larger remaining subpopulations, only recently separated by the lowland habitat loss, are also not viable.

Ecology

Now confined to high-altitude (above 600 m elevation) rainforests, including subtropical, warm temperate and cool temperate rainforests, and nearby moist and wet eucalypt forests. They require dense ground cover which is usually restricted to ecotones, forested watercourses and wetlands, and areas regenerating from fires or storms. They forage on small invertebrates, including snails and insects on the ground and over fallen logs, amongst leaf-litter and on other ground vegetation and debris, within the dense understorey (Higgins *et al.* 2001). Very little is known of their breeding behaviour because they are so shy and cryptic, living in dense vegetation. A generation time of 4.9 years (BirdLife International 2011) is derived from an age at first breeding of 2.0 years and maximum longevity of 7.8 years, both values extrapolated from Noisy Scrub-bird *A. clamosus.*

Threats

Eucalypt forest habitat, which supports the majority of the population, is potentially threatened by wild fires in dry periods (Ferrier 1984). Declines in density may also occur naturally as habitat matures and ground cover provides less shelter, so some fire or other disturbance such as storms is

Current eligibility against IUCN Red List Criteria

IUCN category	Criteria eligibility
A	Not applicable: past, current or future population declines are thought unlikely to exceed 20% in any 3-generation period
B	Endangered: EOO <5000 km^2, AOO <500 km^2, assumed to be severely fragmented, suspected continuing decline in AOO, no. locations, no. mature individuals
C	Vulnerable: population <10 000 mature individuals, suspected continuing decline, largest subpopulation <1000 mature individuals
D	Near Threatened: population near 1000 mature individuals
E	Not applicable: no population viability analysis undertaken

IUCN Red List assessment data

	Estimate	Reliability
Extent of occurrence trend	2000 km^2 stable	high high
Area of occupancy trend	120 km^2 decreasing	medium low
No. of mature individuals trend	1500 decreasing	medium low
No. subpopulations	4	medium
No. locations	>10	high
Largest subpopulation	800	medium
Generation time	4.9 years	medium
Global population share	100%	medium

helpful as the recovering habitat is readily used by the birds (D. Stewart *in litt.*). In the Border Ranges disturbed roadside habitat is used by the birds (P. Ekert *in litt.*). In the longer term, however, the viability of the small remaining populations may be questionable. Most of the bird's lowland habitat was cleared in the 19th century, and, while clearance itself is not a continuing threat with almost all birds being in protected areas, the subdivision of a small population into even smaller fragments makes each subpopulation more susceptible to random events, such as appears to have happened at Mt Warning and Spicers Gap.

Conservation objectives

1. Each subpopulation known to be stable
2. Fire plans in place for each subpopulation

Information required

1. Change in distribution and abundance since first surveys
2. Appropriate fire regime to maintain habitat suitability
3. Potential to manipulate habitat to advantage scrub-birds

Management actions required

1. Repeat first survey using same methods
2. Refine monitoring protocol for birds and habitat
3. Develop and implement fire management plans for all subpopulations
3. Assess whether experimental habitat manipulation is justified
4. Identify fire refugia at local and landscape levels

Bibliography

BirdLife International (2011) 'Species factsheet: *Atrichornis rufescens*'. Retrieved 17 January, 2011 from <http://www.birdlife.org/>.

Ekert PA (2005) 'Monitoring Rufous Scrub-birds (*Atrichornis rufescens*) in Central Eastern Rainforest Reserves of Australia (CERRA): Final Report 2005'. Report prepared by Ekerlogic Consulting Services (ECS) for the NSW National Parks and Wildlife Service, Coffs Harbour.

Ferrier S (1984) The status of the Rufous Scrub-bird *Atrichornis rufescens*: habitat, geographical variation and abundance. PhD thesis, University of New England, Armidale.

Higgins PJ, Peter JM, Steele WK (Eds) (2001) *Handbook of Australian, New Zealand and Antarctic Birds. Volume 5: Tyrant-Flycatchers to Chats*. Oxford University Press, Melbourne.

Hill F (2010) 'Rufous Scrub-bird Monitoring Border Ranges Section of the Scenic Rim IBA'. Report to Tweed Bird Observers Club.

Schodde R, Mason IJ (1999) *The Directory of Australian Birds: Passerines*. CSIRO Publishing, Melbourne.

Stewart DA (2006) 'Rufous Scrub-bird surveys in the Queensland CERRA; November 2006'. Unpublished report. QPWS, Moggill, Qld.

Comments received from

Barry Baker, Stuart Butchart, Peter Ekert, Hugh Ford, Tim Holmes, Richard Loyn, Mike Newman, David Paton, David Stewart, John Woinarski, David Watson

Species Conservation Summary

Noisy Scrub-bird

Atrichornis clamosus (Gould,1844)
Atrichornithidae

Conservation status

Endangered B1ab(iii,v)+2ab(iii,v)

Reasons for listing

The population is restricted to few locations resulting in a very small extent of occurrence and area occupied. The recent series of large wildfires and some unexplained smaller-scale declines suggests a probable continuing decline in habitat quality and number of mature individuals

Status 2000

Vulnerable D2
Reason for change in 2010: multiple large fires and some smaller local losses suggest a continuing decline

Status 1990

Endangered B1ab(ii,v)+2ab(ii,v), C2a(ii)
Reason for change in 2000: decline reversed and a second viable subpopulation established

Taxonomy

No infraspecific taxa described
Taxonomic uniqueness: high (1 genus/family, 2 species/genus, 1 subspecies/species)

Range

Endemic to south-west Western Australia. In the 19th century, recorded from 3 main areas: near Waroona, Margaret River-Augusta area and Albany area (Whittell 1943). No confirmed records from 1889 to 1961, when <100 individuals rediscovered at Two Peoples Bay, east of Albany (Webster 1962; Danks *et al.* 1996). Natural colonists and translocations have created 1 extensive subpopulation at a handful of locations from Two Peoples Bay to Cheyne Beach (Danks *et al.* 1996; Gilfillan *et al.* 2009; Department of Sustainability, Environment, Water, Population and Communities 2011). There is a second introduced subpopulation at Bald Island; but birds reintroduced to the Darling Ranges, Porongurup National Park and Mt Taylor have not been recorded breeding (Tiller 2009a; Comer *et al.* 2010; Department of Sustainability, Environment, Water, Population and Communities 2011).

Abundance

Fires in 2000–2004 reduced the mainland subpopulation from 569 singing males in 1999 and 733 in 2001 to 278 in 2005; numbers are expected to recover, although it may take 10 years before the habitat is suitable (Comer 2002; Comer *et al.* 2005; Danks and Comer 2006). In 2005, the mainland subpopulation included 4 singing males at Moates Lake-Gardner Lake, 126 at Mount Gardner, 37 at Angove River-Normans Inlet, 32 at Mount Manypeaks, 53 at Waychinicup and 26 at Mermaid (Department of Sustainability, Environment, Water, Population and Communities 2011). Based on a ratio of 2.5 individuals for each territorial male (Danks *et al.* 1996), the mainland subpopulation was estimated at 695 mature birds. Numbers counted during surveys at Mt Gardner were 132 in 2007, 121 in 2008 and 119 in 2009 (A. Burbidge and S. Comer *in litt.*). The second subpopulation on Bald I. totalled 99 territorial males (c. 248 mature birds) in 2010 and 95 in 2007, an increase from 2 territorial males in 1993, 10 in 1997 and 37 in 2001 (Danks *et al.* 1996; Cowen 2009; Department of Sustainability, Environment, Water, Population and Communities 2011).

Ecology

Good Noisy Scrub-bird habitat comprises dense vegetation in particular understorey and shrub layer, with sedges or piles of debris for nesting interspersed with small open areas with a thick accumulation of leaf-litter and a well developed litter fauna for feeding (Danks 1991; Danks *et al.* 1996). The female lays a single egg in a domed nest, which is usually built in a clump of sedge. The main food is terrestrial arthropods (Danks and Calver 1993). Most occupied sites have not been burnt for >10 years, the sites on Mt Gardner having a post-fire age of >50 years and on Bald I. of >120 years. All habitat on Manypeaks was burnt in a wildfire in 2004, and some territories recolonised by 2006, but only in areas where the fire was intense and vegetation regeneration had been rapid (S. Comer *in litt.*). Historically, the species may have occupied the ecotone between swamp vegetation and forest dominated by Jarrah *Eucalyptus marginata* and Marri *E. calophylla* (Smith 1985), although such habitat is rare within the species' current range. Dispersal appears to be along corridors of closed vegetation, the birds readily crossing roads but not cleared land (Danks 1991). A generation time of 6.0 years (BirdLife International 2011) is derived from an average age at first breeding of 2.0 years and a maximum longevity of banded birds of 10.0 years (Australian Bird and Bat Banding Scheme).

Current eligibility against IUCN Red List Criteria

IUCN category	Criteria eligibility
A	Not applicable: past, current or future population declines are thought unlikely to exceed 20% in any 3-generation period
B	Endangered: EOO <5000 km², AOO <500 km², at ≤5 locations with plausible future threat, inferred continuing decline in area and quality of habitat, and no. birds
C	Vulnerable: population <10 000 mature individuals, inferred continuing decline, 250–1000 mature individuals in largest subpopulation
D	Vulnerable: population 250–1000 mature individuals, ≤5 locations with plausible future threat
E	Not applicable: no population viability analysis undertaken

IUCN Red List assessment data

	Estimate	Reliability
Extent of occurrence trend	250 km² stable	high high
Area of occupancy trend	140 km² stable	high high
No. of mature individuals trend	950 decreasing	medium medium
No. subpopulations	2	high
No. locations	5	medium
Largest subpopulation	700	medium
Generation time	6.0 years	medium
Global population share	100%	high

Threats

Extensive high-intensity bushfire is the major threat, as the species depends on long unburnt habitat and has limited dispersal abilities. In 1994, a fire at Mt Taylor destroyed most of a recently translocated population, with remaining birds disappearing during the subsequent year (A. Danks *in litt.*). A fire burnt 4500 ha at Mt Manypeaks in 2004–2005, a loss of one-third of suitable habitat and 56% of singing males in the Albany area (Comer and Burbidge 2006; Danks and Comer 2006). Although a small amount of breeding was recorded within 4 years after the fire, it could be up to 10 years before the habitat is suitable again (Danks and Comer 2006; Tiller 2009b). Except for a handful, all birds are on protected land, but habitat clearance on private land could cut corridors, preventing dispersal and fragmenting subpopulations. Breeding success is relatively good, but various native and introduced mammals and reptiles may raid nests and kill adults (Smith and Robinson 1976; Danks *et al.* 1996). The reasons for the decline in the subpopulation at Mt Gardner, and the disappearance of the self-established subpopulation around Lake Gardner, are unknown but could be related to the length of time since fire, increased predation by feral cats *Felis catus* or the removal of birds for translocation (Danks 1997; Comer and Danks 2000).

Conservation objectives

1. Population numbers and area of occupation increased or maintained
2. Management options increased through improved ecological knowledge

Information required

1. Trends in all known subpopulations
2. Effect of fire age on vegetation structure and food availability

Management actions required

1. Active fire management around all scrub-bird locations
2. Continue habitat management and threat abatement of all occupied areas
3. Continue translocation program
4. Continue to support coordination of management by the South Coast Threatened Bird Recovery Team

Bibliography

BirdLife International (2011) 'Species factsheet: *Atrichornis clamosus*'. Retrieved 17 January, 2011 from <http://www.birdlife.org/>.

Comer S (2002) Noisy Scrub-birds in the Albany Management Area. *South Coast Threatened Birds News* **6**, 3.

Comer S, Burbidge A (2006) Manypeaks rising from the ashes. *Landscope* **22**, 51–55.

Comer S, Danks A (2000) 'Noisy Scrub-bird Recovery Plan Annual Report 1999'. South Coast Threatened Birds Recovery Team, Albany.

Comer S, Danks A, Burbidge AH (2005) Noisy Scrub-birds, Western Whipbirds and wildfire at Mt Manypeaks. *Western Australian Bird Notes* **113**, 16–17.

Comer S, Danks A, Burbidge AH, Tiller C (2010) The history and success of Noisy Scrub-bird re-introductions in Western Australia: 1983–2005. In *Global Re-Introduction Perspectives: Additional Case-studies from Around the Globe.* (Ed. PS Soorae) pp. 187–192. IUCN/SSC Re-Introduction Specialist Group, Abu Dhabi, UAE.

Cowen S (2009) DNA of a 'phoenix': a study of genetic variability in the Noisy Scrub-bird. *South Coast Threatened Birds News* **13**, 5.

Danks A (1991) The role of corridors in the management of an endangered passerine. In *Nature Conservation 2: The Role of Corridors.* (Eds DA Saunders and RJ Hobbs) pp. 291–296. Surrey Beatty, Sydney.

Danks A (1997) Conservation of the Noisy Scrub-bird: a review of 35 years of research and management. *Pacific Conservation Biology* **3**, 341–349.

Danks A, Burbidge AA, Burbidge AH, Smith GT (1996) 'Noisy Scrub-bird Recovery Plan'. WA Dept Conservation & Land Management, Perth.

Danks A, Calver MC (1993) Diet of the Noisy Scrub-bird *Atrichornis clamosus* at Two Peoples Bay, south-western Western Australia. *Emu* **93**, 203–205.

Danks M, Comer S (2006) Mt Manypeaks update: Project Phoenix rises! *South Coast Threatened Birds News* **10**, 10–11.

Department of Sustainability Environment Water Population and Communities (2011) *Atrichornis clamosus.* In 'Species Profile and Threats Database'. Department of Sustainability, Environment, Water, Population and Communities, Canberra. Retrieved 24 February, 2011 from <http://www.environment.gov.au/sprat>.

Gilfillan S, Comer S, Burbidge AH, Blyth J, Danks A, Newell J (2009) 'South Coast Threatened Birds Recovery Plan. Western Australian Wildlife Management Program No. 44'. WA Department of Environment and Conservation, Perth.

Smith GT (1985) Population and habitat selection of the Noisy Scrub bird, *Atrichornis clamosus*. *Australian Wildlife Research* **12**, 479–485.

Smith GT, Robinson FN (1976) The Noisy Scrub-bird: an interim report. *Emu* **76**, 37–42.

Tiller C (2009a) Noisy Scrub-bird translocation news. *South Coast Threatened Birds News* **13**, 10.

Tiller C (2009b) Noisy Scrub-birds return to Mt Manypeaks. *South Coast Threatened Birds News* **13**, 2.

Webster HO (1962) Rediscovery of the Noisy Scrubbird *Atrichornis clamosus. Western Australian Naturalist* **8**, 57–59.

Whittell HM (1943) The Noisy Scrub-bird (*Atrichornis clamosus*). *Emu* **42**, 217–234.

Comments received from
Allan Burbidge, Sarah Comer

Subspecies Conservation Summary

Brown Treecreeper (south-eastern)

Climacteris picumnus victoriae Mathews, 1912
Climacteridae

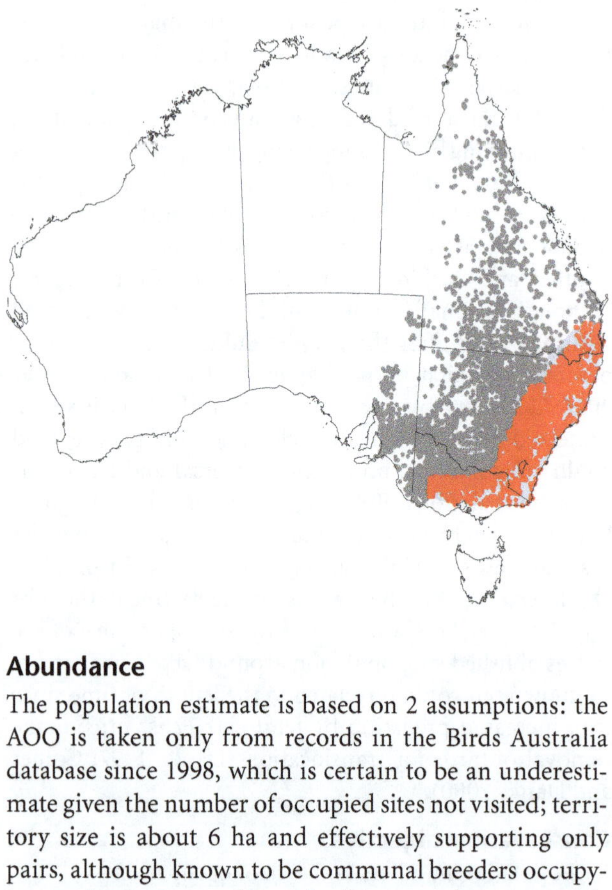

Conservation status
Near Threatened A2bc+3c+4bc

Reasons for listing
Declining across a substantial proportion of its range based on surveys and local extirpations, probably at 20–29% within 3 generations (22 years), because of the many ongoing effects of habitat loss and fragmentation

Status 2000
NT A2bc+3c+4bc

Status 1990
NT A2bc+3c+4bc

Taxonomy
C. p. melanotus (Cape York Peninsula) and *C. p. picumnus* (inland Eastern Australia, eastern Queensland) are Least Concern, as is the species

Taxonomic uniqueness: high (2 genera/family, 5 species/genus, 3 subspecies/species)

Range
Occurs in south-eastern Australia from the Grampians in western Victoria, through central New South Wales to the Bunya Mountains in the Queensland (Schodde and Mason 1999), from coast to the inland slopes of Great Diving Range. However, it is replaced inland by *C. p. picumnus.* The EOO has not declined but treecreepers have disappeared from many habitat fragments (Cooper and Walters 2002b; Ford *et al.* 2009), inferring a continuing decline in AOO.

Abundance
The population estimate is based on 2 assumptions: the AOO is taken only from records in the Birds Australia database since 1998, which is certain to be an underestimate given the number of occupied sites not visited; territory size is about 6 ha and effectively supporting only pairs, although known to be communal breeders occupying territories as small as 3 ha (Doerr and Doerr 2006). The largest subpopulation is probably along the inland fringes of the Great Dividing Range, although the subpopulation in the Pilliga Scrub is also likely to be large: densities of 16.5 birds/km^2 in 5000 km^2 gives a population of 80 000 (H. Ford *in litt.*). While there was no consistent decline between the 1977–1981 and 1998–2002 Atlases in most of the bioregions occupied by this subspecies (Barrett *et al.* 2002), and some populations are stable (e.g. Traill *et al.* 1996; Cunningham and Olsen 2009) or increasing (Lindenmayer and Cunningham 2011), there are many reports of local declines (e.g. Taylor and COG

Current eligibility against IUCN Red List Criteria

IUCN category	Criteria eligibility
A	Near Threatened: past, current and anticipated population declines of 20–29% over periods of 3 generations (22 years) suspected from monitoring and deteriorating habitat quality
B	Not applicable: EOO >20 000 km², AOO >2000 km²
C	Not applicable: population >10 000 mature individuals
D	Not applicable: population >1000 mature individuals, >5 locations
E	Not applicable: no population viability analysis undertaken

IUCN Red List assessment data

	Estimate	Reliability
Extent of occurrence trend	600 000 km² stable	high medium
Area of occupancy trend	12 000 km² decreasing	low high
No. of mature individuals trend	500 000 decreasing	low medium
No. subpopulations	>10	low
No. locations	>10	low
Largest subpopulation	300 000	low
Generation time	7.2 years	high
Global population share	100%	high

1992; Barrett *et al.* 1994; Paton *et al.* 1994; Reid 1999, 2000; Traill and Duncan 2000; Olsen 2008; Ford *et al.* 2009). Overall it is inferred here that, given the losses from fragmented habitats, there is a substantial decline likely to be occurring.

Ecology

Inhabits dry eucalypt woodland and adjoining vegetation, though absent from degraded woodlands and steep rocky hills (Noske 1982) and has higher breeding success in territories with lower densities of shrubs, moderate levels of ground cover, greater amounts of foraging substrate and greater invertebrate biomass (Doerr *et al.* 2006). It feeds on ants, beetles and larvae, which it takes from the trunks of live and dead trees as well as from fallen branches and on ground and lays 2–3 eggs, usually in tree-hollows (Higgins and Peter 2002). Fragmentation of habitat disrupts dispersal and recruitment, especially of immature females (Cooper and Walters 2002a). Connectivity with scattered trees is more important than fragment size for reducing extirpation risk; it is not found in patches >700 m from the nearest patch >10 ha unless there are scattered trees <100 m apart (Cooper and Walters 2002b; Doerr and Doerr 2007). A generation time of 7.2 years (BirdLife International 2011) is derived from an age at first breeding of 2.0 years (Higgins and Peter 2002), mean adult annual survival of 77.0% (extrapolated from congeners) and a maximum longevity in the wild of 13.9 years (Australian Bird and Bat Banding Scheme).

Threats

One of a suite of declining temperate woodland species (Reid 1999; Montague-Drake *et al.* 2009). Historical clearance destroyed much habitat and most of the remainder is fragmented leading to ongoing declines as 'extinction debt' which is likely to result in local extirpations for at least 50 years after fragmentation has occurred (Ford 2011). Remnant treecreeper subpopulations are particularly at risk because of their limited capacity to cross gaps of >80 m and so recolonise isolated habitat patches (Cooper and Walters 2002b, 2002a). With >90% of the remaining vegetation in some regions restricted to patches <500 ha, local extinctions likely to continue for many decades to come (Szabo *et al.* 2011). These pressures on the treecreepers may be compounded by a reliance on what is becoming a depauperate terrestrial fauna (Watson 2011). Other ongoing threats include inappropriate fire regimes, residential and commercial development, firewood collection and tidying of farmland, grazing by stock, rabbits *Oryctolagus cuniculus* and kangaroos *Macropus* spp., increased predation from introduced mammals, invasive weeds and exclusion by Noisy Miners *Manorina melanocephala* (Willson and Bignall 2009). Most of these threats are greater at habitat edges so exacerbated by fragmentation.

Conservation objectives

1. Stable populations established in key sites across its range

Information required

1. Population densities and trends in arid parts of the range
2. Criteria for sites where management investment to save treecreeper populations is warranted
3. Appropriate management interventions for isolated subpopulations including predator and competitor control, fire regime, ground cover manipulation and assisted recolonisation of vacated patches as part of metapopulation management

Management actions required

1. Monitor key sites, particularly where there are management interventions
2. Develop guidelines for the management of treecreeper populations in fragments
3. Promote ecological management of woodland remnants on private land
4. Design conservation and regeneration plans at a landscape level, restore and reconnect habitat with corridors and large trees spaced at <100 m
5. Protect all woodland in which treecreepers are known to be resident from clearing
6. Place all areas of public land that contain the treecreepers under secure conservation management,

particularly those in timber reserves, transport corridors and local government land

7. Using appropriate incentives, undertake extension with landholders who have suitable woodland habitat to promote sound management of remnants and encourage greater connectivity between habitat patches
8. Promote revegetation and land reclamation that recreate woodland habitat with a full complement of biodiversity, including the treecreeper
9. Control and reduce firewood collection from areas occupied by the treecreepers, promoting wood-lot development close to markets, and reduce grazing densities where necessary

Bibliography

Barrett G, Ford HA, Recher HF (1994) Conservation of woodland birds in a fragmented rural landscape. *Pacific Conservation Biology* **1**, 245–256.

Barrett GW, Silcocks A, Cunningham RB (2002) 'Australian Bird Atlas (1998–2001). Supplementary Report No. 1 – Comparison of Atlas 1 (1977–1981) and Atlas 2 (1998–2001)'. Birds Australia, Melbourne.

BirdLife International (2011) 'Species factsheet: *Climacteris picumnus*'. Retrieved 17 January, 2011 from <http://www.birdlife.org/>.

Cooper CB, Walters JR (2002a) Experimental evidence of disrupted dispersal causing decline of an Australian passerine in fragmented habitat. *Conservation Biology* **16**, 471–478.

Cooper CB, Walters JR (2002b) Independent effects of woodland loss and fragmentation on Brown Treecreeper distribution. *Biological Conservation* **105**, 1–10.

Cunningham R, Olsen P (2009) A statistical methodology for tracking long-term change in reporting rates of birds from volunteer-collected presence–absence data. *Biodiversity Conservation* **18**, 1305–1327.

Doerr ED, Doerr VAJ (2006) Comparative demography of treecreepers: evaluating hypotheses for the evolution and maintenance of cooperative breeding. *Animal Behaviour* **72**, 147–159.

Doerr ED, Doerr VAJ (2007) Gene flow in fragmented landscapes: final report. In *Better Knowledge Better Bush Project*. (Eds PG Ryan, D Freudenberger and S Briggs). CSIRO Sustainable Ecosystems, Canberra.

Doerr VAJ, Doerr ED, Jenkins SH (2006) Habitat selection in two Australasian treecreepers: what cues should they use? *Emu* **106**, 93–103.

Ford HA (2011) The causes of decline of birds of eucalypt woodlands: advances in our knowledge over the last 10 years. *Emu* **111**, 1–9.

Ford HA, Walters JR, Cooper CB, Debus SJS, Doerr VAJ (2009) Extinction debt or habitat change? Ongoing losses of woodland birds in north-eastern New South Wales, Australia. *Biological Conservation* **142**, 3182–3190.

Higgins PJ, Peter JM (Eds) (2002) *Handbook of Australian, New Zealand and Antarctic Birds. Volume 6: Pardalotes to Spangled Drongo*. Oxford University Press, Melbourne.

Lindenmayer DB, Cunningham RB (2011) Longitudinal patterns in bird reporting rates in a threatened ecosystem: Is change regionally consistent? *Biological Conservation* **144**, 430–440.

Montague-Drake RM, Lindenmayer DB, Cunningham RB (2009) Factors affecting site occupancy by woodland bird species of conservation concern. *Biological Conservation* **142**, 2896–2903.

Noske RA (1982) Comparative behaviour and ecology of some Australian bark-foraging birds. PhD thesis, University of New England, Armidale, NSW.

Olsen P (2008) The state of Australia's birds. *Supplement to Wingspan* **18**.

Paton DC, Carpenter G, Sinclair RG (1994) A second bird atlas of the Adelaide region. Part 1: Changes in the distribution of birds: 1974–75 vs 1984–85. *South Australian Ornithologist* **31**, 151–193.

Reid JRW (1999) 'Threatened and declining birds in the New South Wales Sheep-Wheat Belt: I. Diagnosis, characteristics and management'. Consultancy report to NSW National Parks and Wildlife Service. CSIRO Sustainable Ecosystems, Canberra.

Reid JRW (2000) 'Threatened and declining birds in the New South Wales Sheep-Wheat Belt: II. Landscape relationships – modelling bird atlas data against vegetation cover'. Consultancy report to NSW National Parks and Wildlife Service. CSIRO Sustainable Ecosystems, Canberra.

Schodde R, Mason IJ (1999) *The Directory of Australian Birds: Passerines*. CSIRO Publishing, Melbourne.

Szabo JK, Baxter PWJ, Vesk PA, Possingham HP (2011) Paying the extinction debt: declining woodland birds in the Mount Lofty Ranges, South Australia. *Emu* **111**, 59–70.

Taylor M, COG (1992) *Birds of the Australian Capital Territory: An Atlas*. Canberra Ornithologists Club and National Capital Planning Authority, Canberra.

Traill BJ, Collins E, Peake P, Jessup S (1996) Current and past status of the birds of Chiltern – a box-ironbark forest in north-eastern Victoria. *Australian Bird Watcher* **16**, 309–326.

Traill BJ, Duncan S (2000) 'Status of birds in the New South Wales temperate woodlands region'. Report to New South Wales National Parks and Wildlife Service, Dubbo.

Watson DM (2011) A productivity-based explanation for woodland bird declines: poorer soils yield less food. *Emu* **111**, 10–18.

Willson A, Bignall J (2009) 'Regional recovery plan for threatened species and ecological communities of Adelaide and the Mount Lofty Ranges, South Australia'. Department for Environment and Heritage, South Australia.

Comments received from

Hugh Ford, David Watson

Species Conservation Summary

Fawn-breasted Bowerbird

Ptilonorhynchus cerviniventris (Gould, 1850)
Ptilonorhynchidae

Conservation status

Australian breeding population: Near Threatened C2a(ii)

Reasons for listing

Population may be <10 000 individuals, in a single subpopulation and may be declining based on recent anecdotal reports

Status 2000

Least Concern
Reason for change in 2010: multiple anecdotal reports that fewer seen than previously and missing from previously known locations

Status 1990

Least Concern

Taxonomy

No infraspecific taxa described. The global status of the species is Least Concern
Taxonomic uniqueness: high (8 genera/family, 5 species/genus, 1 subspecies/species)

Range

New Guinea and Australia, where confined to eastern Cape York Peninsula, from near Cape York south to Massey Creek, east of Coen (Higgins *et al.* 2006). Mapped distribution and contiguous habitat suggests that all birds are in a single subpopulation in Australia but Torres Strait appears to act as a barrier to movement (Garnett 1992). The AOO is assumed to be about 10% of the EOO.

Abundance

The bowerbirds are nowhere common in Australia and few records have been submitted to the Birds Australia Atlas (18, 3, 8, 6, 1, 4, 3, 1, 4, 2 and 8 for 2000–2010). No recent records from several sites where previously known for a long time (A. Kutt *in litt.*; D. Chaplin *in litt.*) and there have been <20 reports of the species in the last 2 years from travelling birdwatchers (D. Chaplin *in litt.*). Given the paucity of records, and the patchy distribution even in areas occupied, a population size of 9000 mature individuals at a density of 3/km^2 of AOO, is given here. While the ease of finding this species does vary over time (B. Venables *in litt.*), the bowerbird would be Vulnerable if this population size and a decline were to be corroborated.

Ecology

In Australia, recorded mainly in eucalypt and ti-tree woodlands and adjacent mangroves; occasionally in or at the edge of rainforest. Often in dense thickets of shrubs within woodland and also dense coastal and riparian shrubland (Higgins *et al.* 2006). Feeds mainly on fruit, also some seeds and insects (Higgins *et al.* 2006). A generation time of 13.5 years (BirdLife International 2011) is derived from an average age at first breeding of 4.4 years, extrapolated from *Ptilonorhynchus* and *Sericulus*, and an annual survival of adults of 89.0%, extrapolated from Great Bowerbird *P. nuchalis* (Doerr 2010).

Threats

Reasons for any decline are entirely speculative. Altered fire regimes are affecting the vegetation structure in some parts of the species' range (Russell-Smith *et al.* 2004) but the area is otherwise little changed. Evidence for a decline is speculative and listing here is precautionary.

Conservation objectives

1. Stable numbers at monitoring sites

Information required

1. Population size
2. Appropriate monitoring sites
3. Trends at monitoring sites
4. Impact of contemporary fire regimes

Management actions required

1. Identify sites for monitoring and monitor both birds and potential environmental drivers of population change until trend established
2. Support return to traditional fire management

Bibliography

BirdLife International (2011) 'Species factsheet: *Chlamydera cerviniventris*'. Retrieved 17 January, 2011 from <http://www.birdlife.org/>.

Doerr NR (2010) Annual survival rates of bower-owning male Great Bowerbirds *Ptilonorhynchus nuchalis* in Townsville, Queensland. *Corella* **34**, 14–16.

Garnett ST (1992) The Torres Strait as a barrier to rainforest bird movement. In 'Australian National Rainforest Study Report Vol. 1. in Proceedings of a workshop on the past, present and future of Australian rainforests, Griffith University, Dec. 1983'. (Eds GL Werren and AP Kershaw). Australian Heritage Commission, Canberra.

Current eligibility against IUCN Red List Criteria

IUCN category	Criteria eligibility
A	Not applicable: past, current or future population declines are thought unlikely to exceed 20% in any 3-generation period
B	Not applicable: >10 locations, not severely fragmented, population fluctuations not extreme
C	Near Threatened: population possibly 2500–10 000 mature individuals, possible continuing decline, 100% in 1 subpopulation
D	Not applicable: population >1000 mature individuals, >5 locations
E	Not applicable: no population viability analysis undertaken

IUCN Red List assessment data

	Estimate	Reliability
Extent of occurrence trend	31 500 km^2 stable	high medium
Area of occupancy trend	3000 km^2 stable	low low
No. of mature individuals trend	9000 decreasing	low low
No. subpopulations	1	high
No. locations	>10	high
Generation time	13.5 years	medium
Global population share	1–10%	medium

Higgins PJ, Peter JM, Cowling SJ (Eds) (2006) *Handbook of Australian, New Zealand and Antarctic Birds. Volume 7: Boatbill to Starlings. Part A. Boatbill to Larks*. Oxford University Press, Melbourne.

Russell-Smith J, Stanton PJ, Whitehead PJ, Edwards A (2004) Rain forest invasion of eucalypt-dominated woodland savanna, Iron Range, north-eastern Australia: I. Successional processes. *Journal of Biogeography* **31**, 1293–1303.

Comments received from

Dominic Chaplin, Cliff Frith, Dawn Frith, Alex Kutt, Keith McDonald, Brian Venables

Subspecies Conservation Summary

Purple-crowned Fairy-wren (western)

Malurus coronatus coronatus Gould, 1858
Maluridae

Conservation status

Endangered B2ab(ii,iii,iv,v)

Reasons for listing

The linear riparian habitat occupied is contracting and deteriorating in quality, causing declines in AOO, no. of locations, subpopulations and mature individuals

Status 2000

Endangered B2ab(ii,iii,iv,v)

Status 1990

Endangered B2ab(ii,iii,iv,v)

Taxonomy

M. c. macgillivrayi (eastern) is Near Threatened. The species is Least Concern.

Taxonomic uniqueness: medium (5 genera/family, 13 species/genus, 2 subspecies/species)

Range

Occurs along the Fitzroy, Isdell, Drysdale, Durack and Pentecost River systems in the Kimberley region, Western Australia, and patchily along the Victoria R. system in north-west Northern Territory (Rowley 1993). Extirpated from historical range on Ord R., upper reaches of Pentecost R. and Chamberlin R., lower reaches of Fitzroy R. and perhaps from other sites with unconfirmed historical records (Skroblin and Legge 2010). Because the habitat is linear, the AOO is calculated on the basis of the length of river occupied: 330 km in the Kimberley (A. Skroblin *in litt.*) and 170 km in the Victoria River District, although only 12.2 km^2 of habitat (van Doorn 2007). A genetic analysis using mitochondrial markers revealed no historical population structure between river catchments, suggesting dispersal between rivers over recent evolutionary time (A. Skroblin *in litt.*). However, extirpation of the subpopulation from the Kimberley rivers closest to the Victoria River is likely to prevent effective migration.

Current eligibility against IUCN Red List Criteria

IUCN category	Criteria eligibility
A	Not applicable: past, current or future population declines are thought unlikely to exceed 20% in any 3-generation period
B	Endangered: AOO <500 km^2, population severely fragmented, inferred continuing decline in AOO, quality of habitat, no. locations, no. subpopulations, no. mature individuals
C	Near Threatened: population 2500–10 000 mature individuals, inferred continuing decline, largest subpopulation near 1000 mature individuals
D	Not applicable: population >1000 mature individuals, >5 locations
E	Not applicable: no population viability analysis undertaken

IUCN Red List assessment data

	Estimate	Reliability
Extent of occurrence trend	169 000 km^2 stable	high high
Area of occupancy trend	500 km (linear) decreasing	high high
No. of mature individuals trend	7300 decreasing	medium high
No. subpopulations	5	medium
No. locations	>10	high
Largest subpopulation	1500	high
Generation time	8.3 years	low
Global population share	100%	high

Given limitations to the species' dispersal, increasing fragmentation of its already patchy habitat and the small size and isolation of remnant subpopulations means that it qualifies as severely fragmented.

Abundance

Population estimated at <5000 individuals in 3 subpopulations the in NT (van Doorn 2007; Woinarski *et al.* 2007). In the Kimberley, populations projected as <1000 on the upper Fitzroy R., c.1500 on Drysdale R., c.1500 on Durack R., <500 on King George R., and unknown at Isdell R. and Pentecost R. based on 10 birds/km (A. Skrobolin *in litt.*, Skroblin and Legge 2010). Recent counts totalling 500 birds in at least 5 subpopulations at 31/92 survey locations across its Kimberley range in 2007 suggest that the total population may be smaller, as much habitat is degraded (Skroblin and Legge 2010). The total population size is estimated here as 5000–9500. Declines are confirmed on Fitzroy (c. 1920s), Ord (c. 1970s after dam building and recently extirpated from upper reaches; Skroblin and Legge 2010) and Victoria R. (recent; van Doorn 2007; Department of Environment and Conservation 2009) and probably also elsewhere because of habitat degradation (Department of Environment and Conservation 2009). Has not declined at Drysdale River Crossing where habitat remains unchanged (Rowley 1993; Skroblin and Legge 2010). Northern Territory population is expected to decline >10% over the next 3 generations (Woinarski *et al.* 2007), which may also be valid across the Kimberley.

Ecology

Its preferred habitat comprises a well-developed mid-storey under a dense canopy of emergent *Eucalyptus* and *Melaleuca* spp. In the Kimberley region, usually occurs where mid-storey is dominated by *Pandanus aquaticus* (Boekel 1979) or *Barringtonia acutangula*, a freshwater mangrove (A. Skroblin, S. Legge *in litt.*), accompanied by a variety of shrubs. Along the Victoria R., and formerly on the lower Fitzroy R., usually associated with areas of dense river grass dominated by *Chionachne cyanthopoda* (Rowley 1993; van Doorn and Low Choy 2009). Pairs or family parties forage in the shade, on the ground or in the undergrowth, and take a range of insects and also small quantities of seeds (Rowley and Russell 1993; Woinarski *et al.* 2007). A generation time of 8.3 years (BirdLife International 2011) is derived from an average age at first breeding of 2.3 years, an annual survival of adults of 78.0%, and a maximum longevity in the wild of 17.0 years, all values extrapolated from other *Malurus* spp.

Threats

The greatest threat is degradation or loss of habitat. Livestock and feral herbivores seeking water eat and trample riparian vegetation, and more frequent and/or more intense fires have also been detrimental in some places (Smith and Johnstone 1977; NRETA 2006). Populations decreased 50% over a 2-year period at 2 sites, where grazing and trampling was allowed around habitat patches (van Doorn 2007) and the historical loss of fairy-wrens on the lower Fitzroy was caused by the replacement of native riparian vegetation by weeds. Birds may be exposed to predation by feral cats *Felis catus* after cattle destroyed the habitat during bad droughts, and have failed to recolonise after habitat had apparently recovered (Boekel 1979; van Doorn 2007). Density of birds increased from 8.7–13.5 birds/km over 3 years after stock removal and active fire management at Mornington Station (Legge and Fleming 2008). Increased flood energy due to rangeland degradation has also destroyed some areas of riparian vegetation. Heavy weed invasion may also degrade habitat and support more fires. At 1 site the breeding success was very low, largely due to nest predation, perhaps from Black Rat *Rattus rattus* (van Doorn 2007). Declined on Ord R. (c. 1970) following dam construction and flooding (Smith and Johnstone 1977; Rowley 1993), and extirpated after 2003 due to continued degradation of the remaining habitat by grazing and weed invasion (Skroblin and Legge 2010).

Conservation objectives

1. Stable or improved trends at key sites
2. Viable numbers maintained at all subpopulations

Information required

1. Trends at key accessible subpopulations
2. Trends on population persistence of rarely visited, less-accessible subpopulations

3. Impact of predation by Black Rat and feral cats
4. Dispersal ability

Management actions required

1. Exclude livestock from riparian vegetation at key sites
2. Provide watering points away from riparian areas where stock have access to rivers
3. Reduce the abundance of feral herbivores near key sites
4. Reduce the frequency of intense fires that affect riparian vegetation
5. Identify and remove weeds
6. Restore habitat at key sites

Bibliography

BirdLife International (2011) 'Species factsheet: *Malurus coronatus*'. Retrieved 17 January, 2011 from <http://www.birdlife.org/>.

Boekel C (1979) Notes on the status and behaviour of the Purple-crowned Fairy-wren *Malurus coronatus* in the Victoria River Downs area, Northern Territory. *Australian Bird Watcher* **8**, 91–97.

Department of Environment and Conservation (2009) 'A synthesis of scientific knowledge to support conservation management in the Kimberley region of Western Australia'. Department of Environment and Conservation (WA), Perth.

Legge S, Fleming A (2008) Monitoring to evaluate conservation management: fire and grazing. *The State of Australia's Birds. Supplement to Wingspan* **18**.

NRETA (2006) 'Threatened species information sheet: Purple-crowned Fairy-wren (Western subspecies) *Malurus coronatus coronatus*'. Retrieved 18 March, 2011 from <http://www.nt.gov.au/nreta/wildlife/animals/threatened/pdf/birds/purplecrowned_fairywren_vu.pdf>.

Rowley I (1993) The Purple-crowned Fairy-wren *Malurus coronatus*. I History, distribution and present status. *Emu* **93**, 220–234.

Rowley I, Russell E (1993) The Purple-crowned Fairy-wren *Malurus coronatus*. II Breeding biology, social organisation, demography and management. *Emu* **93**, 235–250.

Skroblin A, Legge S (2010) The distribution and status of the western subspecies of the Purple-crowned Fairy-wren (*Malurus coronatus coronatus*). *Emu* **110**, 339–347.

Smith LA, Johnstone RE (1977) Status of the Purple-crowned Wren (*Malurus coronatus*) and Buff-sided robin (*Poecilodryas superciliosa*) in Western Australia. *Western Australian Naturalist* **13**, 185–188.

van Doorn A (2007) Ecology and conservation of the purple-crowned fairy-wren *Malurus coronatus coronatus* in the Northern Territory, Australia. PhD thesis, University of Florida, Gainesville, Florida, USA.

van Doorn A, Low Choy J (2009) A description of the primary habitat of the Purple-crowned Fairy-wren *Malurus coronatus coronatus* in the Victoria River District, N.T. *Northern Territory Naturalist* **21**, 24–33.

Woinarski J, Pavey C, Kerrigan R, Cowie I, Ward S (Eds) (2007) *Lost from Our Landscape: Threatened Species of the Northern Territory*. Northern Territory Government, Darwin.

Comments received from

Sarah Legge, Anja Skroblin, Anne-Marie van Doorn, John Woinarski

Subspecies Conservation Summary

Purple-crowned Fairy-wren (eastern)

Malurus coronatus macgillivrayi Mathews, 1913
Maluridae

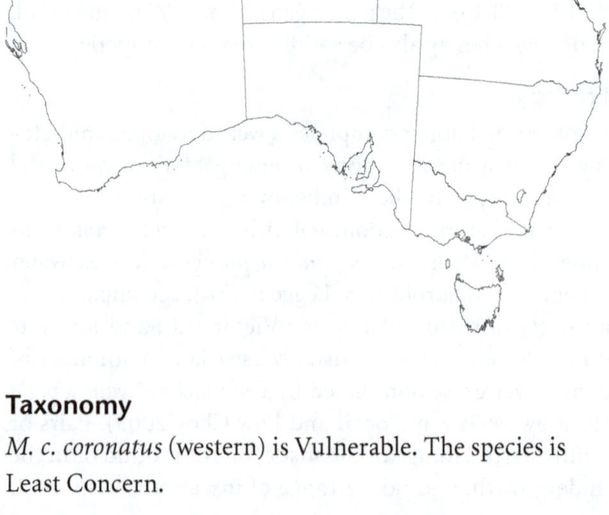

Conservation status

Near Threatened C2a(ii)

Reasons for listing

Population may be close to 10 000 mature individuals in a single subpopulation, with an ongoing deterioration in habitat quality

Status 2000

Least Concern
Reason for change in 2010: surveys in western Queensland found that habitat quality had declined in many sites outside protected areas

Status 1990

Least Concern

Taxonomy

M. c. coronatus (western) is Vulnerable. The species is Least Concern.

Current eligibility against IUCN Red List Criteria

IUCN category	Criteria eligibility
A	Not applicable: past, current or future population declines are thought unlikely to exceed 20% in any 3-generation period
B	Not applicable: >10 locations, not severely fragmented, population fluctuations not extreme
C	Near Threatened: population near 10 000 mature individuals, inferred continuing decline, 100% in a single subpopulation
D	Not applicable: population >1000 mature individuals, >5 locations
E	Not applicable: no population viability analysis undertaken

IUCN Red List assessment data

	Estimate	Reliability
Extent of occurrence trend	119 000 km^2 stable	high high
Area of occupancy trend	1800 km^2 decreasing	medium medium
No. of mature individuals trend	15 000 decreasing	medium medium
No. subpopulations	1	medium
No. locations	>10	medium
Generation time	8.3 years	low
Global population share	100%	high

Taxonomic uniqueness: medium (5 genera/family, 13 species/genus, 2 subspecies/species)

Range

Occurs along most rivers draining into south-west and south Gulf of Carpentaria from Roper River in Northern Territory to Leichhardt River in Queensland (Rowley 1993). Mapped as occurring on >15 river systems (Rowley 1993).

Abundance

Population size was estimated as 18 000 with up to 2500 in Nicholson River subpopulation based on 10 birds/km (Rowley 1993). Recent surveys of *M. c. coronatus* suggest lower populations as much habitat is degraded (Skroblin and Legge 2010). The overall population size is estimated here as possibly as low as 15 000. Declined along the Leichardt River (Rowley 1993). More general declines are inferred from anecdotal evidence of habitat degradation mirroring Victoria River subpopulations of *M. c. coronatus*.

Ecology

No specific information for this subspecies, but assumed to be similar to *M. c. coronatus*. Preferred habitat comprises a well-developed mid-storey under a dense canopy of emergent *Eucalyptus* and *Melaleuca* spp. usually where the mid-storey is dominated by *Pandanus aquaticus* accompanied by a variety of shrubs, or with areas of dense river grass dominated by *Chionachne cyanthopoda* (Rowley 1993). Pairs or family parties forage in the shade, on the ground or in undergrowth, and take a range of insects and also small quantities of seeds (Rowley and Russell 1993; NRETA 2006). A generation time of 8.3 years (BirdLife International 2011) is derived from an average age at first breeding of 2.3 years, an annual survival of adults of 78.0%, and a maximum longevity in the wild of 17.0 years, all values extrapolated from other *Malurus* spp.

Threats

As with *M. c. coronatus*, the greatest threat is degradation or loss of habitat from cattle grazing, weeds, fire, flooding and mining (A. Skroblin *in litt.*). Livestock and feral herbivores seeking water eat and trample riparian vegetation, and more frequent and/or more intense fires have also been detrimental in some places (Smith and Johnstone 1977; NRETA 2006). Populations of western subspecies decreased 50% over a 2-year period at 2 sites where grazing and trampling was allowed around habitat patches (van Doorn 2007). Birds may be exposed to predation by feral cats *Felis catus* after cattle have damaged habitat (Boekel 1979; van Doorn 2007). Heavy weed invasion may also degrade habitat and support more fires. Very low breeding success from nest predation was attributed to Black Rat *Rattus rattus* at 1 site (van Doorn 2007).

Conservation objectives

1. Stable or improved trends at key sites
2. Viable numbers maintained at accessible subpopulations

Information required

1. Trends at key accessible subpopulations
2. Site occupancy of rarely visited, less-accessible sites
3. Impact of predation by Black Rat and feral cat
4. Dispersal ability

Management actions required

1. Exclude livestock from riparian vegetation at key sites
2. Provide watering points away from riparian areas where stock have access to rivers
3. Reduce the abundance of feral herbivores near key sites
4. Reduce the frequency of intense fires that affect riparian vegetation
5. Identify and remove weeds
6. Restore habitat at key sites

Bibliography

BirdLife International (2011) 'Species factsheet: *Malurus coronatus*'. Retrieved 17 January, 2011 from <http://www.birdlife.org/>.

Boekel C (1979) Notes on the status and behaviour of the Purple-crowned Fairy-wren *Malurus coronatus* in the Victoria River Downs area, Northern Territory. *Australian Bird Watcher* **8**, 91–97.

NRETA (2006) 'Threatened species information sheet: Purple-crowned Fairy-wren (Western subspecies) *Malurus coronatus coronatus*'. Retrieved 18 March, 2011 from <http://www.nt.gov.au/nreta/wildlife/animals/threatened/pdf/birds/purplecrowned_fairywren_vu.pdf>.

Rowley I (1993) The Purple-crowned Fairy-wren *Malurus coronatus*. I History, distribution and present status. *Emu* **93**, 220–234.

Rowley I, Russell E (1993) The Purple-crowned Fairy-wren *Malurus coronatus*. II Breeding biology, social organisation, demography and management. *Emu* **93**, 235–250.

Skroblin A, Legge S (2010) The distribution and status of the western subspecies of the Purple-crowned Fairy-wren (*Malurus coronatus coronatus*). *Emu* **110**, 339–347.

Smith LA, Johnstone RE (1977) Status of the Purple-crowned Wren (*Malurus coronatus*) and Buff-sided robin (*Poecilodryas superciliosa*) in Western Australia. *Western Australian Naturalist* **13**, 185–188.

van Doorn A (2007) Ecology and conservation of the purple-crowned fairy-wren *Malurus coronatus coronatus* in the Northern Territory, Australia. PhD thesis, University of Florida, Gainesville, Florida.

Comments received from

Alastair Freeman, Sarah Legge, Anja Skroblin, Anne-Marie van Doorn, John Woinarski

Subspecies Conservation Summary

White-winged Fairy-wren (Barrow Island)

Malurus leucopterus edouardi A. J. Campbell, 1901
Maluridae

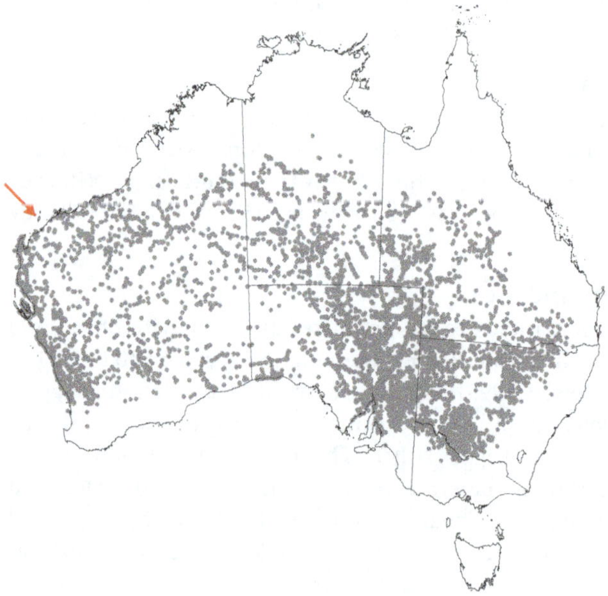

Conservation status

Vulnerable D2

Reasons for listing

Occurs at single location where large-scale fire or introduction of exotic predators a plausible future threat

Status 2000

Vulnerable D2

Status 1990

Vulnerable D2

Taxonomy

M. l. leucopterus (Dirk Hartog I.) is also Vulnerable;
M. l. leuconotus (semi-arid Australia) is Least Concern
Taxonomic uniqueness: medium (5 genera/family, 13 species/genus, 3 subspecies/species)

Range

Restricted to Barrow I. off Western Australia (Schodde and Mason 1999), and probably historically on Trimouille I. and Hermite I., in the Montebello Is. (Burbidge *et al.* 2000). In 2010, 27 birds were translocated to Hermite I.; their survival is being monitored and a further translocation will occur in 2011 (A. H. Burbidge *in litt.*).

Abundance

Three population estimates: 8150 birds in 1976 (Sedgwick 1978), 7519 in 2001 (Pruett-Jones and O'Donnell 2004) and 9336 in 2004 (Bamford and Wilcox 2005). Given the similarity of these estimates despite differences in methods, season and climate, the population is assessed as having been stable at about 8000 individuals. The proposed developments are expected to cause a small decline in population, probably <10% in 3 generations (29 years) and not ongoing.

Ecology

Occurs across Barrow I. in grasslands and low shrublands with a dense ground cover of *Triodia* spp., patches of bare ground and often scattered clumps of shrubs that are used for foraging, nesting and roosting (Pruett-Jones and Tarvin 2001; Bamford and Bamford 2005; Bamford and Wilcox 2005). They eat a wide range of insects, as well as seeds and fruits and are a cooperative breeder (Pruett-Jones and Tarvin 2001). A generation time of 8.3 years (BirdLife International 2011) is derived from an age at first breeding of 2.3 years, an annual survival of adults of 78.0%, and a maximum longevity in the wild of 17.0 years, all extrapolated from congeners.

Threats

Large industrial facilities for processing oil and gas are planned for Barrow I. Operations up to 2004 appear to have had little effect on the species' abundance (Pruett-Jones and O'Donnell 2004; Bamford and Wilcox 2005). A proposed gas processing facility would result in the loss of habitat for c.315 birds or c.4% of the total population (Bamford and Wilcox 2005; Chevron Australia 2009b). It should benefit from the strict quarantine and effective fire control offered by the industry for the conservation of a

Current eligibility against IUCN Red List Criteria

IUCN category	Criteria eligibility
A	Not applicable: past, current or future population declines are thought unlikely to exceed 20% in any 3-generation period
B	Not applicable: EOO, AOO, habitat, locations or population not declining, population fluctuations not extreme
C	Not applicable: population >10 000 mature individuals
D	Vulnerable: ≤5 locations with plausible future threat
E	Not applicable: no population viability analysis undertaken

IUCN Red List assessment data

	Estimate	Reliability
Extent of occurrence trend	236 km^2 stable	high high
Area of occupancy trend	236 km^2 stable	high high
No. of mature individuals trend	8000 stable	high high
No. subpopulations	1	high
No. locations	1	high
Generation time	8.3 years	low
Global population share	100%	high

suite of species, notably for 5 Vulnerable marsupials (Maxwell *et al.* 1996; Chevron Australia 2009a). There is a very small chance that almost all of the habitat on Barrow I. could be destroyed by an extensive wildfire (Department of Sustainability, Environment, Water, Population and Communities 2011). Given that threats of fire and colonisation by invasive species could impact the whole island, the subspecies is assessed as having a single location.

Conservation objectives

1. Population maintained on Barrow I.
2. Viable subpopulations on Montebello Is.

Information required

1. Population trends

Management actions required

1. Monitor subspecies during the routine monitoring of the 5 Vulnerable marsupial taxa
2. Ensure industrial activities are sited and managed in a manner that does not adversely affect the subspecies
3. Minimise the extent of vegetation clearing
4. Develop conservation agreements and/or covenants that aid the conservation of this subspecies
5. Maintain strict quarantine procedures to prevent the introduction of invasive alien species
6. Develop and implement a suitable fire management strategy for this subspecies
7. Liaise with existing and proposed industrial facilities to minimize and monitor impacts of industrial operations on this subspecies
8. Continue translocations to previously occupied islands

Bibliography

Bamford MJ, Bamford AR (2005) 'Gorgon Development on Barrow Island Technical Report: Avifauna'. Report to ChevronTexaco Australia, Perth.

Bamford MJ, Wilcox JA (2005) 'Gorgon Development on Barrow Island Technical Report: White-winged Fairy-wren (*Malurus leucopterus edouardi*)'. Attachment to Avifauna technical report. Report to ChevronTexaco Australia, Perth.

BirdLife International (2011) 'Species factsheet: *Malurus leucopterus*'. Retrieved 17 January, 2011 from <http://www.birdlife.org/>.

Burbidge AA, Blyth JD, Fuller PJ, Kendrick PG, Stanley FJ, Smith LA (2000) The terrestrial vertebrate fauna of the Montebello Islands, Western Australia. *CALMScience* **3**, 95–107.

Chevron Australia (2009a) 'Gorgon Gas Development and Jansz Feed Gas Pipeline Fire Management Plan'. Unpublished report. Chevron Australia, Perth.

Chevron Australia (2009b) 'Terrestrial and Subterranean Baseline State and Environmental Impact Report'. Unpublished report. Chevron Australia, Perth.

Department of Sustainability, Environment, Water, Population and Communities (2011) *Malurus leucopterus edouardi*. In 'Species Profile and Threats Database'. Department of Sustainability, Environment, Water, Population and Communities, Canberra. Retrieved 24 February, 2011 from <http://www.environment.gov.au/sprat>.

Maxwell S, Burbidge AA, Morris K (Eds) (1996) *The 1996 Action Plan for Australian Marsupials and Monotremes.* Environment Australia, Canberra.

Pruett-Jones S, O'Donnell E (2004) Land birds on Barrow Island: status, population estimates, and responses to an oilfield development. *Journal of the Royal Society of Western Australia* **87**, 101–108.

Pruett-Jones S, Tarvin KA (2001) Aspects of the ecology and behaviour of White-winged Fairy-wrens on Barrow Island. *Emu* **101**, 73–78.

Schodde R, Mason IJ (1999) *The Directory of Australian Birds: Passerines.* CSIRO Publishing, Melbourne.

Sedgwick EH (1978) A population study of Barrow Island avifauna. *West Australian Naturalist* **14**, 85–108.

Comments received from

Andrew Burbidge

Subspecies Conservation Summary

White-winged Fairy-wren (Dirk Hartog Island)

Malurus leucopterus leucopterus Dumont, 1824
Maluridae

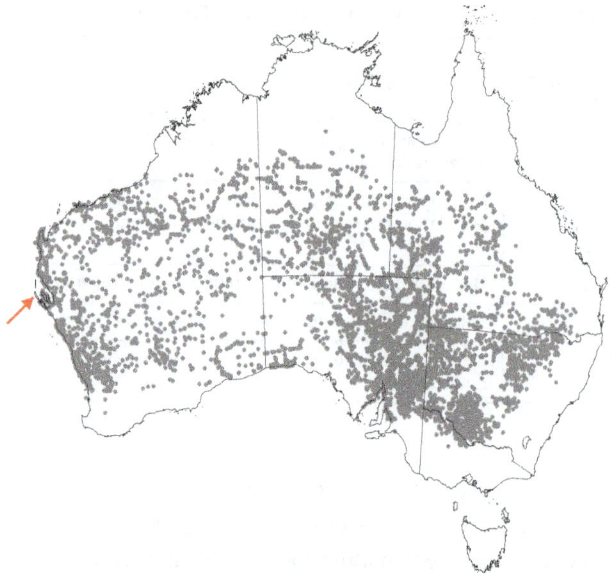

Conservation status

Vulnerable D2

Reasons for listing

Occurs at single location where large-scale fire or introduction of exotic predators a plausible threat

Status 2000

Vulnerable D2

Status 1990

Vulnerable D2

Taxonomy

M. l. edouardi (Barrow I.) is also Vulnerable; *M. l. leuconotus* (semi-arid Australia) is Least Concern
Taxonomic uniqueness: medium (5 genera/family, 13 species/genus, 3 subspecies/species)

Range

Dirk Hartog I. off Western Australia, where common and widespread (Whitlock 1921; Johnstone and Storr 2004). There have also been several reports of White-winged Fairy-wrens with dark plumage, and potentially of this subspecies, from mainland sites adjacent to Dirk Hartog I. (Schodde and Mason 1999; Johnstone and Storr 2004).

Abundance

Given territory sizes of 1.5–6 ha for the mainland subspecies *M. l. leuconotus* (Rowley and Russell 1997; Higgins *et al.* 2001) and reliable estimates of c.8000 birds on the 250 km^2 Barrow I. (Pruett-Jones and O'Donnell 2004; Bamford and Wilcox 2005), which suffers no grazing by livestock and lacks House Mice *Mus musculus* and feral cats *Felis catus*, the population has been estimated at 7000 adult birds (Department of Sustainability, Environment, Water, Population and Communities 2011). There is no evidence for a historical decline in population size but it should benefit from current conservation management including removal of all introduced herbivores (Department of Environment and Conservation (WA) 2008).

Ecology

The subspecies inhabits low heathlands and shrublands that support a variety of shrubs, including *Acacia*, *Atriplex* and *Scaevola*, and, in many areas, hummocks of Spinifex *Triodia* ssp. (Rathburn and Montgomerie 2003; Johnstone and Storr 2004). It eats a wide range of insects, as well as seeds and fruits (Schodde 1982; Johnstone and Storr 2004). In the single study of breeding on Dirk Hartog I., c.40% of nests were depredated (mostly by snakes, lizards, Pied Butcherbirds *Cracticus nigrogularis*, Nankeen Kestrels *Falco cenchroides* and feral cats) and c.20% were parasitised by Horsfield's Bronze-Cuckoo *Chrysococcyx basalis* (Rathburn and Montgomerie 2003). A generation time of 8.3 years (BirdLife International 2011) is derived from an age at first breeding of 2.3 years, an annual survival of adults of 78% and a maximum longevity in the wild of 17.0 years, all extrapolated from congeners.

Threats

The main threats to terrestrial birds on Dirk Hartog I. are the introduction of harmful non-endemic fauna, especially Black Rats *Rattus rattus*, invasive alien plants or pathogens (Department of Environment and Conservation (WA) 2008). The subspecies has persisted despite the presence of sheep, feral goats *Hircus capra*, House Mice and feral cats, and occasional extensive fires, but the island has lost the Western Grasswren *Amytornis textilis textilis*. The island has recently become a National Park; sheep and goats are being eradicated, and feral cats are being controlled (Department of Environment and Conservation (WA) 2008). This small narrow island with uniform habitat is also potentially at risk from large or frequent fires that could affect the whole island (Department of Environment and Conservation (WA) 2008), so the subspecies is assessed as having a single location.

Conservation objectives

1. Population maintained

Information required

2. Population trends, particularly in relation to management actions

Management actions required

1. Implement management plans for the control and eradication of goats, cats and House Mice
2. Develop protocols to prevent or minimise the introduction of invasive alien species
3. Monitor presence of non-native species and undertake remedial action if needed
4. Develop and implement a suitable fire management strategy

Bibliography

Bamford MJ, Wilcox JA (2005) 'Gorgon Development on Barrow Island Technical Report: White-winged Fairy-wren

Current eligibility against IUCN Red List Criteria

IUCN category	Criteria eligibility
A	Not applicable: past, current or future population declines are thought unlikely to exceed 20% in any 3-generation period
B	Not applicable: EOO, AOO, habitat, locations or population not declining, population fluctuations not extreme
C	Not applicable: population not declining
D	Vulnerable: ≤5 locations with plausible future threat
E	Not applicable: no population viability analysis undertaken

IUCN Red List assessment data

	Estimate	Reliability
Extent of occurrence trend	620 km^2 stable	high high
Area of occupancy trend	620 km^2 stable	high high
No. of mature individuals trend	7000 stable	low high
No. subpopulations	1	high
No. locations	1	high
Generation time	5.3 years	low
Global population share	100%	high

(*Malurus leucopterus edouardi*)'. Attachment to Avifauna technical report. Report to ChevronTexaco Australia, Perth.

BirdLife International (2011) 'Species factsheet: *Malurus leucopterus*'. Retrieved 17 January, 2011 from <http://www.birdlife.org/>.

Department of Environment and Conservation (WA) (2008) 'Shark Bay World Heritage Property Strategic Plan 2008–2020'. Department of Environment and Conservation, Perth.

Department of Sustainability Environment Water Population and Communities (2011) *Malurus leucopterus leucopterus*. In 'Species Profile and Threats Database'. Department of Sustainability, Environment, Water, Population and Communities, Canberra. Retrieved 24 February, 2011 from <http://www.environment.gov.au/sprat>.

Higgins PJ, Peter JM, Steele WK (Eds) (2001) *Handbook of Australian, New Zealand and Antarctic Birds. Volume 5: Tyrant-Flycatchers to Chats*. Oxford University Press, Melbourne.

Johnstone RE, Storr GM (2004) *Handbook of Western Australian Birds. Volume 2: Passerines (Blue-winged Pitta to Goldfinch)*. Western Australian Museum, Perth.

Pruett-Jones S, O'Donnell E (2004) Land birds on Barrow Island: status, population estimates, and responses to an oilfield development. *Journal of the Royal Society of Western Australia* **87**, 101–108.

Rathburn MK, Montgomerie R (2003) Breeding biology and social structure of White-winged Fairy-wrens (*Malurus leucopterus*): comparison between island and mainland subspecies having different plumage phenotypes. *Emu* **103**, 295–306.

Rowley I, Russell E (1997) *Fairy-wrens and Grasswrens: Maluridae, 102*. Oxford University Press, Oxford.

Schodde R (1982) *The Fairy-Wrens. A Monograph of the Maluridae*. Lansdowne Editions, Melbourne.

Schodde R, Mason IJ (1999) *The Directory of Australian Birds: Passerines*. CSIRO Publishing, Mlebourne.

Whitlock FL (1921) Notes on Dirk Hartog Island and Peron Peninsula, Shark Bay, Western Australia. *Emu* **20**, 168–186.

Comments received from

Andrew Burbidge

Subspecies Conservation Summary

Variegated Fairy-wren (Shark Bay)

Malurus lamberti bernieri Ogilvie-Grant, 1909

Maluridae

Conservation status

Vulnerable D2

Reasons for listing

Occurs at 1 or 2 locations, where large-scale fire or introduction of exotic predators a plausible future threat

Status 2000

Vulnerable D2

Status 1990

Vulnerable D2

Current eligibility against IUCN Red List Criteria

IUCN category	Criteria eligibility
A	Not applicable: past, current or future population declines are thought unlikely to exceed 20% in any 3-generation period
B	Not applicable: EOO, AOO, habitat, locations or population not declining, population fluctuations not extreme
C	Not applicable: not declining
D	Vulnerable: ≤5 locations with plausible future threat
E	Not applicable: no population viability analysis undertaken

IUCN Red List assessment data

	Estimate	Reliability
Extent of occurrence trend	95 km^2 stable	high high
Area of occupancy trend	95 km^2 stable	high high
No. of mature individuals trend	5400 stable	low high
No. subpopulations	2	high
No. locations	2	high
Largest subpopulation	3000	low
Generation time	5.3 years	low
Global population share	100%	high

Taxonomy

Four mainland subspecies are Least Concern, as is the species

Taxonomic uniqueness: medium (5 genera/family, 13 species/genus, 5 subspecies/species)

Range

Dorre I. and Bernier I. on the edge of Shark Bay, Western Australia (Schodde and Mason 1999). These islands are separated by a channel c.500 m wide, but other similar birds, e.g. Rufous Fieldwren *Calamanthus campestris dorrie*, only occur on 1 of the islands, so the fairy-wrens probably have 2 subpopulations.

Abundance

Given recorded territory size of the subspecies in western New South Wales (3.75 ha; Tidemann 1990), the population is here estimated as possibly 5400 adult birds. There is no evidence for any decline in population size, but it should be stable given that the management of the island as a nature reserve.

Ecology

Variegated Fairy-wrens live in dense heath and low shrubs, taking ants, beetles, bugs, wasps and other insects, and, occasionally, spiders, seeds and vegetation from amongst the foliage (Schodde 1982; Rowley and Russell 1997). A generation time of 8.3 years (BirdLife International 2011) is derived from an age at first breeding of 2.3 years, an annual survival of adults of 78% and a maximum longevity of 17.0 years, all extrapolated from congeners.

Threats

The main threats to terrestrial birds on Dorre and Bernier Is. are the introduction of harmful non-endemic fauna, especially Black Rats (*Rattus rattus*), invasive alien plants or pathogens (Department of Environment and Conservation (WA) 2008, 2009). These small islands with uniform habitat are also potentially at risk from large or frequent fires that could affect a whole island (Department of Environment and Conservation (WA) 2008), so the subspecies is assessed as having just 2 locations.

Conservation objectives

1. Population maintained

Information required

1. Population trends

Management actions required

1. Develop protocols to prevent or minimise the introduction of invasive species
2. Monitor presence of non-native species and undertake remedial action if needed
3. Develop and implement a suitable fire management strategy

Bibliography

BirdLife International (2011) 'Species factsheet: *Malurus lamberti*'. Retrieved 17 January, 2011 from <http://www.birdlife.org/>.

Department of Environment and Conservation (WA) (2008) 'Shark Bay World Heritage Property Strategic Plan 2008–2020'. Department of Environment and Conservation, Perth.

Department of Environment and Conservation (WA) (2009) 'Shark Bay Variegated Fairy-wren *(Malurus lamberti bernieri)*'. Retrieved 10 March, 2011 from <http://www.sharkbay.org/Variegatedfairy-wrenfactsheet.aspx>.

Rowley I, Russell E (1997) *Fairy-wrens and Grasswrens: Maluridae, 102.* Oxford University Press, Oxford.

Schodde R (1982) *The Fairy-Wrens. A Monograph of the Maluridae.* Lansdowne Editions, Melbourne.

Schodde R, Mason IJ (1999) *The Directory of Australian Birds: Passerines.* CSIRO Publishing, Melbourne.

Tidemann SC (1990) Factors affecting territory establishment, size, and use by three co-existing species of fairy-wrens (*Malurus*). *Emu* **90**, 7–14.

Comments received from

Andrew Burbidge

Subspecies Conservation Summary

Southern Emu-wren (Dirk Hartog Island)

Stipiturus malachurus hartogi Carter, 1916
Maluridae

Conservation status
Vulnerable D2

Reasons for listing
Occurs at single location where large-scale fire or introduction of exotic predators a plausible future threat

Status 2000
Vulnerable D2

Status 1990
Vulnerable D2

Taxonomy
S. m. parimeda (Eyre Peninsula) and *S. m. intermedius* (Adelaide region) are Endangered; 4 other subspecies are Least Concern, as is the species
Taxonomic uniqueness: medium (5 genera/family, 3 species/genus, 8 subspecies/species)

Range
Confined to Dirk Hartog I. off Western Australia (Schodde and Mason 1999), where probably it is widespread, though scarce in the southern sheep-grazed part of the island (Garnett and Crowley 2000).

Abundance
Population densities for other subspecies are 0.04–4.5 birds/ha (Higgins *et al.* 2001), giving a mean population of about 15 000. However, it is less common than the White-winged Fairy-wren *Malurus l. leucopterus* on Dirk Hartog I. (Carter 1917; Whitlock 1921; Garnett and Crowley 2000), the population of which was estimated at c.7000 (L. Brooker *in litt.*). Therefore the population of Southern Emu-wren (Dirk Hartog I.) is here estimated as possibly 5000 adult birds. A lower density than other locations might be expected given the aridity of the site relative to other Southern Emu-wren habitat. There is no evidence for any decline in population size.

Ecology
On Dirk Hartog I., lives in low *Thryptomene/Triodia* heathlands and *Acacia/Diplolaena/Triodia* thickets, foraging amongst the foliage and taking insects, spiders, seeds and leaves (Johnstone and Storr 2004). A generation time of 9.7 years (BirdLife International 2011) is derived from an age at first breeding of 2.3 years and a maximum longevity of 17.0 years, both values extrapolated from fairy-wrens *Malurus* spp.

Threats
The main threats to terrestrial birds on Dirk Hartog I. are the introduction of harmful non-endemic fauna, especially Black Rats *Rattus rattus*, invasive alien plants or pathogens (DEC 2007; 2009). The subspecies has persisted despite the presence of sheep, feral goats *Hircus capra*, House Mice *Mus musculus* and feral cats *Felis catus*, and occasional extensive fires, but the island has lost the Western Grasswren *Amytornis textilis textilis*. The island has recently become a National Park; sheep and goats are being eradicated, and feral cats are being controlled (Department of Environment and Conservation (WA) 2008). This small narrow island with uniform habitat is also potentially at risk from large or frequent fires that could affect the whole island (Department of Environment and Conservation (WA) 2008), so the subspecies is assessed as having a single location.

Conservation objectives
1. Population maintained

Information required
1. Population trends, particularly in relation to management actions

Management actions required
1. Implement management plans for the control and eradication of goats, cats and House Mice
2. Develop protocols to prevent or minimise the introduction of invasive alien species
3. Monitor presence of non-native species and undertake remedial action if needed
4. Develop and implement a suitable fire management strategy

Bibliography
BirdLife International (2011) 'Species factsheet: *Stipiturus malachurus*'. Retrieved 17 January, 2011 from <http://www.birdlife.org/>.

Carter T (1917) The birds of Dirk Hartog Island and Peron Peninsula, Shark Bay, Western Australia, 1916-17. *Ibis* **59**, 564–611.

Department of Environment and Conservation (WA) (2008) 'Shark Bay World Heritage Property Strategic Plan 2008–2020'. Department of Environment and Conservation, Perth.

Current eligibility against IUCN Red List Criteria

IUCN category	Criteria eligibility
A	Not applicable: past, current or future population declines are thought unlikely to exceed 20% in any 3-generation period
B	Not applicable: EOO, AOO, habitat, locations or population not declining, population fluctuations not extreme
C	Not applicable: no continuing population decline or extreme fluctuations
D	Vulnerable: ≤5 locations with plausible future threat
E	Not applicable: no population viability analysis undertaken

IUCN Red List assessment data

	Estimate	Reliability
Extent of occurrence trend	620 km^2 stable	high high
Area of occupancy trend	620 km^2 stable	high high
No. of mature individuals trend	5000 stable	low high
No. subpopulations	1	high
No. locations	1	high
Generation time	9.7 years	low
Global population share	100%	high

Garnett S, Crowley GM (2000) *The Action Plan for Australian Birds 2000.* Environment Australia, Canberra.

Higgins PJ, Peter JM, Steele WK (Eds) (2001) *Handbook of Australian, New Zealand and Antarctic Birds. Volume 5: Tyrant-Flycatchers to Chats.* Oxford University Press, Melbourne.

Johnstone RE, Storr GM (2004) *Handbook of Western Australian Birds. Volume 2: Passerines (Blue-winged Pitta to Goldfinch).* Western Australian Museum, Perth.

Schodde R, Mason IJ (1999) *The Directory of Australian Birds: Passerines.* CSIRO Publishing, Melbourne.

Whitlock FL (1921) Notes on Dirk Hartog Island and Peron Peninsula, Shark Bay, Western Australia. *Emu* **20**, 168–186.

Comments received from

Andrew Burbidge

Subspecies Conservation Summary

Southern Emu-wren (Fleurieu Peninsula)

Stipiturus malachurus intermedius Ashby, 1920

Maluridae

Conservation status

Endangered B1ab(i,ii,iii,iv,v)+2ab(i,ii,iii,iv,v), C2a(i)

Reasons for listing

Extent of occurrence <5000 km^2 and area occupied <500 km^2, with the population of <2500 mature individuals, none in a subpopulation of >250 mature individuals, continuing to decline, as does the EOO, AOO, habitat quality and no. locations and subpopulations, as a result of fire and rural land use intensification

Status 2000

Endangered B1ab(i,ii,iii,iv,v)+2ab(i,ii,iii,iv,v), C2a(i)

Status 1990

Endangered B1ab(i,ii,iii,iv,v)+2ab(i,ii,iii,iv,v), C2a(i)

Taxonomy

S. m. parimeda (Eyre Peninsula) is Endangered, *S. m. hartogi* (Dirk Hartog I.) is Vulnerable, 5 other subspecies are Least Concern, as is the species

Taxonomic uniqueness: medium (5 genera/family, 3 species/genus, 8 subspecies/species)

Range

Fleurieu Peninsula, South Australia, between Kuitpo in the north and Deep Creek in the south (Schodde and Mason 1999). At present, confined to a few scattered subpopulations on Fleurieu Peninsula around Deep Creek, Parawa, Myponga, Mt Compass, Nangkita and Finniss. Locally extinct from several sites: Yundi, Mt Compass, Cox's Scrub, Ashbourne and possibly Kuitpo, with little likelihood of recolonisation. There was an estimated 24% decline in area of occupancy from 25 km^2 in 1993 to 19 km^2 in 2010, which would have been 16.25 km^2 in 2010

Current eligibility against IUCN Red List Criteria

IUCN category	Criteria eligibility
A	Not applicable: past, current or future population declines are thought unlikely to exceed 20% in any 3-generation period
B	Endangered: EOO <5000 km^2, AOO <500 km^2, population severely fragmented, estimated continuing decline in EOO, AOO, area, extent and quality of habitat, no. locations, no. subpopulations, no mature individuals
C	Endangered: population 250–1000 mature individuals
D	Vulnerable: population 250–1000 mature individuals
E	Not applicable: no population viability analysis undertaken

IUCN Red List assessment data

	Estimate	Reliability
Extent of occurrence trend	683 km^2 decreasing	high medium
Area of occupancy trend	19 km^2 decreasing	medium high
No. of mature individuals trend	300 decreasing	medium high
No. subpopulations	>10	medium
No. locations	>10	medium
Largest subpopulation	180	medium
Generation time	9.7 years	low
Global population share	100%	high

had it not been for translocated populations (Pickett 2010).

Abundance

In 2010 the population was estimated at about 210–410 mature individuals with most individuals (41–74%) based in dry-heath, but most occupied sites (63%) in swamp habitats. The Deep Creek Conservation Park subpopulation was the largest (115–240 mature individuals). In 2001–2002, 46 individuals were translocated from Deep Creek Conservation Park to Cox Scrub CP, a site from which they were extirpated by fire in 1983. These birds survived, settled and reproduced, but this did not offset substantial losses elsewhere and the population has continued to decline (M. Pickett *in litt.*).

Ecology

Occurs in 2 general habitat types, both characterised by high density of vegetation up to 1 m: most populations occur in dense swamp (which includes wet-heath, sedgeland, reedland and occasionally lignum), although 1 of the 2 largest populations inhabits dry-heath (MLR Southern Emu-wren & Fleurieu Peninsula Swamps Recovery Team 2007). They feed on small invertebrates, chiefly insects and build well-concealed domed nests in the spring and summer (Higgins *et al.* 2001). A generation time of 9.7 years (BirdLife International 2011) is derived from an age at first breeding of 2.3 years and a maximum longevity of 17.0 years, both values extrapolated from fairy-wrens *Malurus* spp.

Threats

The largest heathland populations are on protected areas where the major threat is bushfires. Swamp habitat, however, occurs almost entirely on private property, is fragmented, rare and subject to incremental alteration or clearance. Most habitat patches have been drained and cleared for agriculture, dried out after feeder streams have been dammed or are heavily grazed. During drought, these patches are also vulnerable to fire, especially extensive wildfires such as occurred in 1983, but also deliberate burning to create pasture. In the past, habitat patches could be replenished after local extinctions by birds dispersing from refuges, but as the birds' dispersal is limited because of their poor flying abilities, the increasing separation of patches means that this is now much more difficult or impossible and subpopulations are prone to local extinction (MLR Southern Emu-wren & Fleurieu Peninsula Swamps Recovery Team 2007) Existing and potential major threats to Emu-wren populations and habitats are bushfire, drought, climate change, livestock grazing, mining, water extraction, swamp (surface-water) drainage, plantation forestry, demographic stochasticity, vegetation succession, fragmented habitat and isolated habitat, and lack of habitat. Other threats are flood, storm damage, vegetation clearance, deliberate burning, residential development, changes in land-use affecting alienated but potential habitat, weed invasion, dieback due to *Phytophthora*, genetic stochasticity, introduced predators, native predators and parasites (MLR Southern Emu-wren & Fleurieu Peninsula Swamps Recovery Team 2007; Willson and Bignall 2009).

Conservation objectives

1. Existing subpopulations securely managed as a metapopulation
2. Population larger, habitat quality improved and extent of habitat expanded and fragmentation reduced

Information required

1. Trends in population parameters, especially where management is occurring

Management actions required

1. Monitor key indicators of performance, including population size, breeding success and vegetation condition, especially in association with management
2. Rehabilitate swamp vegetation to improve its quality and reduce fragmentation
3. Undertake translocations to depleted sites as necessary
4. Protect and enhance known habitat, including the active management of emu-wren populations in existing conservation reserves, and close liaison with private owners of known locations
5. Minimise risk from fire
6. Coordinate conservation through the Mount Lofty Ranges Recovery Team
7. Coordinate conservation with that of the Chestnut-rumped Heathwren *Hylacola pyrrhopygia parkeri*

8. Liaise with land managers to encourage their involvement in conservation

Bibliography

BirdLife International (2011) 'Species factsheet: *Stipiturus malachurus*'. Retrieved 17 January, 2011 from <http://www.birdlife.org/>.

Higgins PJ, Peter JM, Steele WK (Eds) (2001) *Handbook of Australian, New Zealand and Antarctic Birds. Volume 5: Tyrant-Flycatchers to Chats*. Oxford University Press, Melbourne.

MLR Southern Emu-wren & Fleurieu Peninsula Swamps Recovery Team (2007) 'Recovery Plan for the Mount Lofty Ranges Southern Emu-wren *Stipiturus malachurus intermedius*: 2006–2011'. Conservation Council of South Australia, Adelaide.

Pickett M (2010) 'MLRSEW IUCN Parameters Based on 2010 Census Results'. Unpublished report to Mount Lofty Ranges Southern Emu-wren and Fleurieu Peninsula Swamps Recovery Program.

Schodde R, Mason IJ (1999) *The Directory of Australian Birds: Passerines*. CSIRO Publishing, Melbourne.

Willson A, Bignall J (2009) 'Regional recovery plan for threatened species and ecological communities of Adelaide and the Mount Lofty Ranges, South Australia'. Department for Environment and Heritage, South Australia.

Comments received from

Marcus Pickett

Subspecies Conservation Summary

Southern Emu-wren (Eyre Peninsula)

Stipiturus malachurus parimeda Schodde & Weatherly, 1981

Maluridae

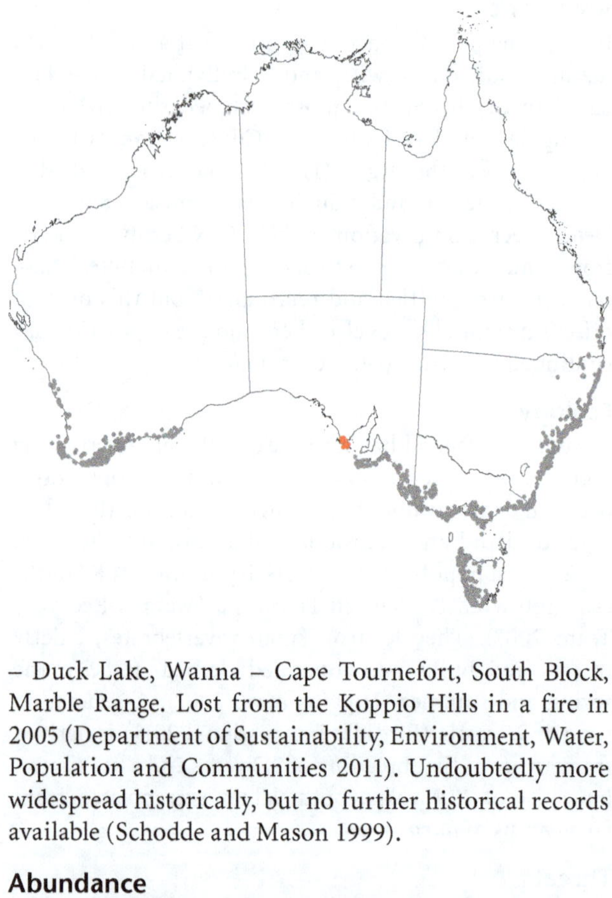

Conservation status

Endangered B1ab(ii,iii,iv,v)+2ab(ii,iii,iv,v), C2a(i)

Reasons for listing

Extent of occurrence <5000 km^2 and area occupied <500 km^2, with the population of <2500 mature individuals, none in a subpopulation of >250 mature individuals, continuing to decline, as does the EOO, AOO, habitat quality and no. locations and subpopulations, largely as a result of fire

Status 2000

Endangered B1ab(ii,iii,iv,v)+2ab(ii,iii,iv,v), C2a(i)

Status 1990

Endangered B1ab(ii,iii,iv,v)+2ab(ii,iii,iv,v), C2a(i)

Taxonomy

S. m. intermedius (Fleurieu Peninsula) is Endangered, *S. m. hartogi* (Dirk Hartog I.) is Vulnerable; 5 other subspecies are Least Concern, as is the species

Taxonomic uniqueness: medium (5 genera/family, 3 species/genus, 8 subspecies/species)

Range

Southern tip of Eyre Peninsula, South Australia. Currently known from 11 locations, all likely to be isolated (Department of Sustainability, Environment, Water, Population and Communities 2011): Shoal Point – D'Anville Bay – Whalers Way – Fishery Bay, Sleaford Bay West – Tulka, West Point – Jussieu Bay, Carcase Point – MacLaren Point – Point Haselgrove – Taylors Landing, Point Avoid, Yangie Bay, Kellidie Bay – Wanilla , Edillilie – Salt Creek – Duck Lake, Wanna – Cape Tournefort, South Block, Marble Range. Lost from the Koppio Hills in a fire in 2005 (Department of Sustainability, Environment, Water, Population and Communities 2011). Undoubtedly more widespread historically, but no further historical records available (Schodde and Mason 1999).

Abundance

The total population is provisionally estimated to be less than 1000 mature individuals, which are severely fragmented with none of the subpopulations likely to be >100 pairs (Pickett 2002). The only relatively large subpopulations evident were in the MacLaren Point, West Point, Whalers Way and Kellidie Bay areas, all of which are declining (Pickett 2008). Surveys in 2008 found birds at only around one-third of previously occupied sites, although the surveys were rapid and are only indicative of decline (Pickett 2008).

Ecology

Preferred habitat types include shrubland, mallee and sedgeland, these being characterised by having at least 1 or 2

Current eligibility against IUCN Red List Criteria

IUCN category	Criteria eligibility
A	Not applicable: past, current or future population declines are thought unlikely to exceed 20% in any 3-generation period
B	Endangered: EOO <5000 km^2, AOO <500 km^2, population severely fragmented, inferred continuing decline in EOO, AOO, area, extent and quality of habitat, no. locations, no. subpopulations, no. mature individuals
C	Endangered: population 250–2500 mature individuals, inferred continuing decline, largest subpopulation <250 mature individuals
D	Vulnerable: population 250–1000 mature individuals
E	Not applicable: no population viability analysis undertaken

IUCN Red List assessment data

	Estimate	Reliability
Extent of occurrence trend	2500 km^2 decreasing	medium medium
Area of occupancy trend	54 km^2 decreasing	low high
No. of mature individuals trend	1000 decreasing	low high
No. subpopulations	11	high
No. locations	11	high
Largest subpopulation	200	medium
Generation time	9.7 years	low
Global population share	100%	high

low dense layers. Southern Emu-wrens feed on small invertebrates, chiefly insects, and generally form pairs that produce 1 or 2 broods of 1–3 young in well-concealed domed nests during the spring–summer months (Higgins *et al.* 2001). A generation time of 9.7 years (BirdLife International 2011) is derived from an age at first breeding of 2.3 years and a maximum longevity of 17.0 years, both values extrapolated from fairy-wrens *Malurus* spp.

Threats

The major threat is fire. Two major bushfires in 2005 burnt 17 out of 20 former sites surveyed. All habitat at 5 or 6 sites, comprising 3 or 4 small populations (from 25 recognised small populations) were burnt including all birds in the Koppio Hills, thus extinguishing this subpopulation (Australian Government 2009). In the past, patches of suitable habitat were probably continuous, but they have since been fragmented by clearance for agriculture and altered fire frequency outside conservation reserves. Emu-wrens are particularly poor dispersers that depend on having suitable connective corridors between patches. As a result, when isolated subpopulations are extirpated, recolonisation is prevented. Though clearance has now largely ceased, extensive fire remains a threat at all occupied sites except 3 in samphire (Pickett 2002). Moderate grazing by high densities of kangaroos *Macropus* spp. and Emus *Dromaius novaehollandiae* at 1 minor site and development at an unconfirmed cliff-top site constitute minor threats (Pickett 2002).

Conservation objectives

1. Existing subpopulations secure
2. Extirpated subpopulations re-established where suitable habitat exists

Information required

1. Population trends at key sites
2. Potential for habitat and corridor restoration

Management actions required

1. Monitor trends in key populations, especially where management occurring
2. Institute appropriate fire management to protect remaining subpopulations
3. Link sites and subpopulations through revegetation
4. Translocate to unoccupied habitat if appropriate, managing the birds as a metapopulation
5. Fence habitat where appropriate
6. Raise the profile with landholders and local government
7. Create a multi-stakeholder team with responsibility for ensuring threatened species are conserved on Eyre Peninsula

Bibliography

BirdLife International (2011) '*Species Factsheet: Stipiturus malachurus*'. Retrieved 17 January, 2011 from <http://www.birdlife.org/>.

Department of Sustainability, Environment, Water, Population and Communities (2011) *Stipiturus malachurus parimeda.* In 'Species Profile and Threats Database'. Department of Sustainability, Environment, Water, Population and Communities, Canberra. Retrieved 24 February, 2011 from <http://www.environment.gov.au/sprat>.

Higgins PJ, Peter JM, Steele WK (Eds) (2001) *Handbook of Australian, New Zealand and Antarctic Birds. Volume 5: Tyrant-Flycatchers to Chats.* Oxford University Press, Melbourne.

Pickett M (2002) 'Status review and action plan for the Eyre Peninsula Southern Emu-wren *Stipiturus malachurus parimeda*'. Prepared for Southern Eyre Birds Inc. and National Parks and Wildlife SA, Department for Environment and Heritage.

Pickett M (2008) 'Eyre Peninsula Southern Emu-wren 2008 survey'. Report prepared for the Department for Environment & Heritage.

Schodde R, Mason IJ (1999) *The Directory of Australian Birds: Passerines.* CSIRO Publishing, Melbourne.

Comments received from

Marcus Pickett

Species Conservation Summary

Mallee Emu-wren

Stipiturus mallee A. J. Campbell, 1908
Maluridae

Conservation status
Endangered A2c

Reasons for listing
Population declines of 50–79% in the last 3 generations (29 years) inferred from the loss of habitat to fires

Status 2000
Vulnerable B2ab(i,ii,iii,iv,v)
Reason for change in 2010: extensive fires through habitat since 2000

Status 1990
Vulnerable B2ab(i,ii,iii,iv,v)

Taxonomy
No infraspecific taxa described
Taxonomic uniqueness: high (5 genera/family, 3 species/genus, 1 subspecies/species)

Range
Victorian and South Australian mallee regions, south and east of the Murray R. Thirty years ago it occurred between Nadda, Peebinga, Pinnaroo, Comet Bore, Coonalpyn and Coombe in SA (Eckert 1977; Carpenter and Matthew 1992; Rowley and Russell 1997), east across the Big Desert in Vic. to Bronzewing and Wathe Flora and Fauna Reserves as well as the Murray-Sunset reserve complex, Hattah-Kulkyne National Park and Annuello Fauna and Flora Reserve (Emison *et al.* 1987; Rowley and Russell 1997). It is currently widespread in Hattah-Kulkyne National Park, patchily distributed across the Murray-Sunset reserve complex and extremely rare in a small area in the eastern part of Wyperfeld National Park (Clarke and Brown 2007). In SA it occurs at a few small sites in Ngarkat National Park and, though last recorded from Billiatt Conservation Park in 1987, a few birds persist at the park's edge (S. Brown *in litt.*). It was last recorded in Bronzewing Flora and Fauna Reserve in 1997 and is considered extinct from Annuello and Wathe Fauna and Flora Reserves (Silveira 1993; Clarke and Brown 2007; Watson 2011). The capacity of large fires to remove entire populations means there are 4 locations: Ngarkat, Wyperfeld, Murray-Sunset and Hattah-Kulkyne. The area of occupancy has been estimated at 460 km^2 in Murray-Sunset region with 275 km^2 in Hattah Kulkyne, about 1 km^2 in Wyperfeld (Brown *et al.* 2009) and may be about 50 km^2 in Ngarkat adding up to about 786 km^2.

Abundance
Based on surveys and habitat modelling population has been estimated at about 14 900 (7700–35 600) mature individuals (taking into account a male-skewed sex ratio), 14 300 of which are in the Murray-Sunset region with 500 in Hattah Kulkyne, <50 in Wyperfeld and 100 in Ngarkat (Brown *et al.* 2009), with the last 2 sites at least unlikely to be viable. The population in Ngarkat Conservation Park was 400–7000 birds in 2000, where it had been recovering from fires in the 1980s (Baker-Gabb 2005). Given its potential range and population in Ngarkat, Billiatt, the Big Desert and Wyperfeld as well as the 2 smaller reserves, it is considered likely that >50% of the population has been lost since 1981.

Ecology
Occupies habitats containing Spinifex *Triodia* spp., in which it builds well-hidden domed nests containing 2–3 eggs, usually within low woodland dominated by mallee eucalypts and Slender Cypress Pine *Callitris gracilis*. It also occurs in heath containing banksias, casuarinas and grass-trees *Xanthorrhoea* (Emison *et al.* 1987; Rowley and Russell 1997; Clarke 2005). Highest densities, which vary from 0.8–2.1 birds/km^2, occur in vegetation that has not been burnt for at least 15 years (Brown *et al.* 2009). Fire eliminates birds but recolonisation can occur within 3–4 years in some circumstances, but be very slow in others (L. Pedler pers. comm.). A generation time of 9.7 years (BirdLife International 2011), is derived from an age at first breeding of 2.3 years and a maximum longevity of 17.0 years, both values extrapolated from fairy-wrens *Malurus* spp.

Threats
Clearance of habitat has eliminated the species from the better soils in its former range. Though clearing has nearly stopped, the population has been fragmented (Emison *et al.* 1987) into isolated subpopulations that have proved vulnerable to extensive fires and other accidents that happen to small populations. The threat from fire is acute in times of drought, such as in 2006 when fire almost eliminated the species from SA and during the 2002 Big Desert fire. Remaining populations are then susceptible to other threats, such as predation by feral cats *Felis catus* and foxes *Vulpes vulpes* (Baker-Gabb

Current eligibility against IUCN Red List Criteria

IUCN category	Criteria eligibility
A	Endangered: population decline of 50–89% over the last 3 generations (29 years) suspected from deteriorating habitat quality
B	Vulnerable: AOO <2000 km^2, population at ≤10 locations, inferred continuing decline in EOO, AOO, area, extent and quality of habitat, no. locations, no. subpopulations, no. mature individuals
C	Not applicable: population >10 000 mature individuals
D	Vulnerable: 5 locations
E	Not applicable: no population viability analysis undertaken

IUCN Red List assessment data

	Estimate	Reliability
Extent of occurrence trend	20 000 km^2 decreasing	high medium
Area of occupancy trend	800 km^2 decreasing	medium high
No. of mature individuals trend	14 900 decreasing	low high
No. subpopulations	4	medium
No. locations	4	medium
Largest subpopulation	14 300	low
Generation time	9.7 years	low
Global population share	100%	high

2005), as the ebb and flow of local extinction and recolonisation can no longer occur.

Conservation objectives

1. All subpopulations retained
2. Connectivity between subpopulations enhanced
3. Populations at sites burnt 2000–2010 recovering

Information required

1. Trends in major subpopulations, particularly rates of recovery from fire
2. Interactions between grazing intensity, underlying geomorphology, climactic conditions, time since fire and habitat quality
3. Extent of genetic isolation of geographically separated populations and the potential need for active translocations
4. Efficacy of feral cat and fox control

Management actions required

1. Monitor populations at key sites, especially those where management occurring
2. Determine the need for population translocations, and develop a re-introduction program if necessary
3. Minimise the frequency and scale of fire at occupied sites

Bibliography

Baker-Gabb D (2005) 'Recovery Plan for the Mallee Emu-wren *Stipiturus mallee*, Red-lored Whistler *Pachycephala rufogularis*, Western Whipbird, *Psophodes nigrogularis leucogaster*, and other threatened mallee birds. Second Draft'.

BirdLife International (2011) 'Species factsheet: *Stipiturus mallee*'. Retrieved 17 January, 2011 from <http://www.birdlife.org/>.

Brown S, Clarke MF, Clarke R (2009) Fire is a key element in the landscape-scale habitat requirements and global population status of a threatened bird: the Mallee Emu-wren (*Stipiturus mallee*). *Biological Conservation* **142**, 432–445.

Carpenter GA, Matthew J (1992) Western records of the Mallee Emu-wren *Stipiturus mallee*. *South Australian Ornithologist* **31**, 125.

Clarke R (2005) 'Ecological requirements of birds specialising in mallee habitats'. Unpublished La Trobe University report to Department for Environment and Heritage, Adelaide.

Clarke R, Brown S (2007) 'The Mallee Emu-wren in Victoria: recent surveys and an assessment of habitat preferences and population densities'. A report to the Mallee Catchment Authority by Deakin University, Burwood.

Eckert J (1977) The distribution of the emu-wrens *Stipiturus malachurus* and *S. ruficeps mallee* in South Australia. *South Australian Ornithologist* **27**, 186–187.

Emison WB, Beardsell CM, Norman FI, Loyn RH, Bennett SC (1987) *Atlas of Victorian Birds*. Department of Conservation, Forests and Lands and RAOU, Melbourne.

Rowley I, Russell E (1997) *Fairy-wrens and Grasswrens: Maluridae, 102*. Oxford University Press, Oxford.

Silveira CE (1993) 'The recovery plan for Australia's threatened mallee birds – addressing fire as a threatening process: research phase'. Report to the Australian National Parks and Wildlife Service, Royal Australasian Ornithologists Union, Melbourne.

Watson S (2011) 'Research survey of Mallee Emu-wren *Stipiturus mallee* at Murray Sunset National Park and Annuelllo Flora and Fauna Reserve'. Report for the mallee Catchment Management Authority by Deakin University, Burwood.

Comments received from

Sarah Brown, Stuart Butchart, Rohan Clarke, Hugh Ford, Victor Hurley, Richard Loyn, Simon Mustoe, David Paton, Don Saunders

Subspecies Conservation Summary

Grey Grasswren (Bulloo)

Amytornis barbatus barbatus Favaloro & McEvey, 1968
Maluridae

Conservation status
Endangered B1ab(i,ii,iii,v)+2ab(i,ii,iii,v)

Reasons for listing
Restricted to <5 locations across a very small EOO and AOO and with a continuing decline in habitat quality and number of individuals

Status 2000
Endangered B1ab(i,ii,iii,v)+2ab(i,ii,iii,v)

Status 1990
Endangered B1ab(i,ii,iii,v)+2ab(i,ii,iii,v)

Taxonomy
A. b. diamantina (Lake Eyre basin) is Least Concern. Birds from the Cooper Creek system and Lake Cudappan are assumed to be *A. b. diamantina* but their taxonomic status is uncertain (Schodde and Christidis 1987; Carpenter 2002). The species is Least Concern. Taxonomic uniqueness: medium (5 genera/family, 11 species/genus, 2 subspecies/species)

Range
Floodplain of Bulloo River on New South Wales/Queensland border including Bulloo River overflow, Caryapundy and Jerrira Swamps (Favaloro and McEvey 1968; McAllan and Cooper 1995; Rowley and Russell 1997). In NSW, restricted to c.80 km² of suitable habitat on Narriearra and Delalah Downs Stations in 2000–2009 compared to c.910 km² of suitable habitat in 1967 (Hardy 2010). No recent information from Qld but not found at Caryapundy Swamp in 2006 (Department of Sustainability, Environment, Water, Population and Communities 2011).

Abundance
Population density was estimated to be less than 0.2 birds/ha in 'prime' lignum habitat in dry years on Narriearra Station, suggesting a NSW population of 1600 birds (Hardy 2010). In another study on Pyampa Station, from zero up to 82 birds were trapped in 11 ha of optimal habitat each year from 1984–1996 (Hardy 2002). Population probably fluctuates greatly according to the level of flooding and the intensity of cattle grazing, with birds retreating to refuges of dense habitat during dry periods and dispersing widely in wet periods (Hardy 2002, 2010). As a conservative estimate, population during droughts is putatively estimated at 20 birds/km^2 over 500 km^2 of refuge habitat.

Ecology
Live in swamps dominated by Lignum *Muehlenbeckia cunninghamii* and Swamp Canegrass *Eragrostis australasica*, where these plants form well-separated clumps at least 1 m high (Favaloro and McEvey 1968). They usually nest in Lignum in the most flood-prone areas and sometimes in Swamp Canegrass in irregularly-flooded areas when Lignum habitat is inundated. They will also feed in Old Man Saltbush *Atrioplex nummularia* and samphire *Halosarcia* spp. where flooding is infrequent (Hardy 2010). They eat seeds and insects, build semi-domed nests and lay 2–3 eggs (Rowley and Russell 1997). A generation time of 9.7 years (BirdLife International 2011) is derived from an age at first breeding of 2.3 years and a maximum longevity of 17.0 years, both extrapolated from fairy-wrens *Malurus* spp.

Threats
Cattle grazing is the main threat, particularly in dry years (McAllan and Cooper 1995; Hardy 2010). A specific example is the type-locality of the Caryapundy outflow channel, where cattle grazing young and regenerating Lignum and Swamp Canegrass rendered all habitat unsuitable within >10 km (Hardy 2010). Much of the recent historical habitat in NSW is now unsuitable, degraded by over-grazing and drought (Hardy 2010). Rabbits *Oryctolagus cuniculus* and feral pigs *Sus scrofa* are suspected to degrade habitat at some sites (Hardy 2002; Hardy 2010). Fire and predation by feral cats *Felis catus* and foxes *Vulpes vulpes* are potential threats (McAllan and Cooper 1995; Australian Government 2005). Water extraction from the Bulloo River system for irrigation or invasion by noxious weeds would be significant threats (Australian Government 2005). As cattle grazing is the main threat, the Bulloo Grey Grasswren is assessed as having <5 locations but if hydrology becomes a more potent threat, the subspecies would have only 1 location.

Conservation objectives
1. Stable numbers at surveyed sites
2. Adequate area of suitable habitat

Information required
1. Population size and distribution in Queensland
2. Impact of feral cats and foxes
3. Hydrological requirements of habitat

Current eligibility against IUCN Red List Criteria

IUCN category	Criteria eligibility
A	Not applicable: past, current or future population declines are thought unlikely to exceed 20% in any 3-generation period
B	Endangered: EOO <5000 km^2, AOO <500 km^2, at ≤5 locations with plausible future threat, inferred continuing decline in EOO, AOO, habitat quality and no. of individuals
C	Vulnerable: population 2500–10 000 mature individuals, suspected continuing decline, 100% in 1 subpopulation
D	Vulnerable: ≤5 locations with plausible future threat
E	Not applicable: no population viability analysis undertaken

IUCN Red List assessment data

	Estimate	Reliability
Extent of occurrence trend	1700 km^2 decreasing	high medium
Area of occupancy trend	500 km^2 decreasing	low medium
No. of mature individuals trend	10 000 decreasing	low medium
No. subpopulations	1	high
No. locations	≤5	medium
Generation time	9.7 years	low
Global population share	100%	high

Management actions required

1. Erect and maintain effective fencing to exclude livestock from key areas
2. Consider options for de-stocking some or all of Narriearra Station
3. Manage feral herbivores
4. Develop fire management plan
5. Manage any disruptions to water flow

Bibliography

Australian Government (2005) 'Commonwealth listing advice on Grey Grasswren (Bulloo) (*Amytornis barbatus barbatus*).' Retrieved 5 April, 2011 from <http://www.environment.gov.au/biodiversity/threatened/species/amytornis-barbatus-barbatus.html>.

BirdLife International (2011) 'Species factsheet: *Amytornis barbatus*'. Retrieved 17 January, 2011 from <http://www.birdlife.org/>.

Carpenter GA (2002) The Grey Grasswren on Cooper Creek, south-west Queensland. *Sunbird* **32**, 52–54.

Department of Sustainability, Environment, Water, Population and Communities (2011) *Amytornis barbatus barbatus.* In 'Species Profile and Threats Database'. Department of Sustainability, Environment, Water, Population and Communities, Canberra. Retrieved 24 February, 2011 from <http://www.environment.gov.au/sprat>.

Favaloro NJ, McEvey A (1968) A new species of Australian Grass-wren. *Memoirs of the National Museum of Victoria* **28**, 1–9.

Hardy JW (2002) A banding study of the Grey Grasswren *Amytornis brabatus barbatus* in the Caryapundy Swamp of south-western Queensland. *Corella* **26**, 106–109.

Hardy JW (2010) Distribution, status and options for the future management of the Grey Grasswren *Amytornis barbatus barbatus* in New South Wales. *Corella* **34**, 25–35.

McAllan IAW, Cooper RM (1995) Distribution of the Grey Grasswren in New South Wales. *Australian Birds* **28**, 65–70.

Rowley I, Russell E (1997) *Fairy-wrens and Grasswrens: Maluridae, 102*. Oxford University Press, Oxford.

Schodde R, Christidis L (1987) Genetic differentiation and subspeciation in the Grey Grasswren *Amytornis barbatus* (Maluridae). *Emu* **87**, 188–192.

Comments received from

Jeff Hardy, Danny Rogers

Subspecies Conservation Summary

Striated Grasswren (sandplain)

Amytornis striatus striatus (Gould, 1840)
Maluridae

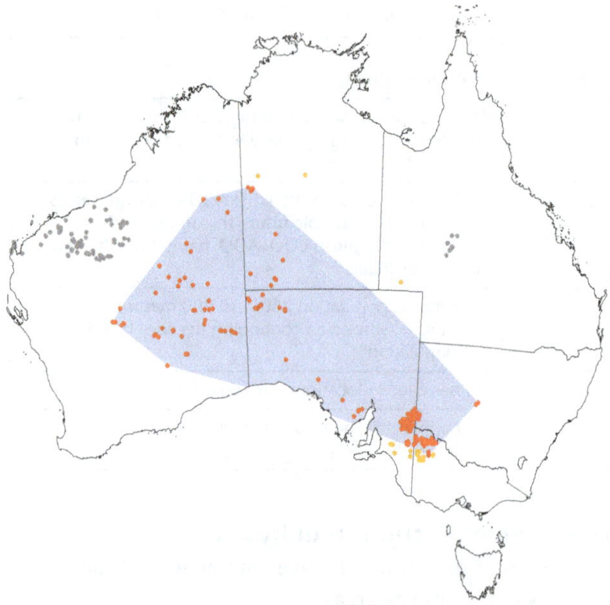

Conservation status

Near Threatened A2bc+3c+4bc

Reasons for listing

Losses and declines in the central and south-eastern parts of the range suggest past, ongoing and probable future declines of 20–39% in the last 3 generations (29 years), primarily from more extensive and frequent fires

Status 2000

Least Concern

Status 1990

Least Concern

Taxonomy

A. s. rowleyi (Opalton) and *A. s. whitei* (Pilbara) are both Least Concern, as is the species
Taxonomic uniqueness: medium (5 genera/family, 11 species/genus, 3 subspecies/species)

Range

Scattered across much of arid Australia, perhaps continuous distribution in Great Victoria, Gibson and Tanami deserts, with subpopulations in northern Eyre Peninsula, Murray Mallee and central New South Wales around Yathong Nature Reserve (Schodde and Mason 1999). Other isolated subpopulations include near Queen Victoria Spring, Western Australia, far south-east Northern Territory, far west Queensland and west of Lake Frome, South Australia (Higgins *et al.* 2001; Barrett *et al.* 2003; Johnstone and Storr 2004). Historically more widespread in NSW (Higgins *et al.* 2001).

Abundance

Population size unknown, but estimated nominally as 200 000 mature individuals, based on an AOO of 20 000 km^2 and a conservative average density in occupied areas of 10 birds/km^2. Population trends poorly known, but evidence of loss and declines include extirpation of subpopulations from the Wiluna district, WA (Garnett and Crowley 2000) and Bronzewing reserve, Vic (Clarke 2005), and absence from 9/10 historic sites at Uluru (J. Augusteyn *in litt.*). Given extent of fires in south-eastern parts of its range in the last decade and recent scarcity in central Australia, an overall rate of decline is estimated here as 20–30% in 3 generations (29 years) and continuing.

Ecology

Lives on sandplains dominated by mature spinifex *Triodia* hummock grassland with an overstorey of shrubs, usually mallee eucalypts (Schodde 1982). They occur less frequently in heathland dominated by banksias or hakeas or in mallee shrublands with dense *Beyeria opaca* from which spinifex is absent (Eckert 1982; Clarke 2005). Mean territory size was found to be 3 ha in 1 study (Karubian 2001). Striated Grasswrens may recolonise habitat within 3 years after fires (Carpenter and Matthew 1986), but show no clear preferences for age of habitat after burning (Clarke 2005). Their abundance increased at Uluru soon after mosaic burning was reinstated (Garnett and Crowley 2000). They forage mostly on the ground, eating seeds, fruits, insects and other invertebrates (Higgins *et al.* 2001). A generation time of 9.7 years (BirdLife International 2011) is derived from an age at first breeding of 2.3 years and a maximum longevity of 17.0 years, both extrapolated from fairy-wrens *Malurus* spp.

Threats

Primarily threatened by extensive fires that burn mature hummock grassland. Large fires have occurred over much of the species range in south-eastern Australia in the last decade. Such fires can have long-term effects where refuges from which it can recolonise are eliminated (Rowley and Russell 1997); e.g. the subspecies' continuing absence from western NSW has been attributed to wildfires in 1974–1975 (Clarke 2005). Grazing by livestock and feral herbivores, such as feral goats *Hircus capra* and rabbit *Oryctolagus cuniculus*, particularly in mallee habitats, may also affect habitat structure (Garnett and Crowley 2000). The effect of predation by feral cats *Felis catus* and foxes *Vulpes vulpes* is uncertain, but widespread fox baiting may be the reason for recent increases of other grasswren species in the Flinders Ranges and north Olary Plains (Pedler *et al.* 2007). Habitat in the south-east of the subspecies' range has been historically reduced and fragmented, and such losses continue at a slow rate in NSW.

Conservation objectives

1. Stable Area of Occupancy, especially in areas which have suffered losses
2. Stable numbers at surveyed sites

Information required

1. Optimal fire regimes

Current eligibility against IUCN Red List Criteria

IUCN category	Criteria eligibility
A	Near Threatened: past, current and anticipated population declines of 20–29% over periods 3 generations (29 years) suspected from monitoring and deteriorating habitat quality
B	Not applicable: EOO >20 000 km^2, AOO >2000 km^2
C	Not applicable: population >10 000 mature individuals
D	Not applicable: population >1000 mature individuals, >5 locations
E	Not applicable: no population viability analysis undertaken

IUCN Red List assessment data

	Estimate	Reliability
Extent of occurrence trend	2 000 000 km^2 stable	high medium
Area of occupancy trend	20 000 km^2 decreasing	low medium
No. of mature individuals trend	200 000 decreasing	low medium
No. subpopulations	>10	low
No. locations	>10	low
Largest subpopulation	130 000	low
Generation time	9.7 years	low
Global population share	100%	high

2. Impact of feral predators
3. Impact of grazing by livestock and feral mammals

Management actions required

1. Work with pastoralists and fire management authorities to minimise burning frequency in mallee
2. Work with traditional owners in arid zone to reinstate mosaic burning of sandplains

Bibliography

Barrett G, Silcocks A, Barry S, Cunningham R, Poulter R (2003) *The New Atlas of Australian Birds.* Royal Australian Ornithologists Union, Hawthorn East.

BirdLife International (2011) 'Species factsheet: *Amytornis striatus*'. Retrieved 17 January, 2011 from <http://www.birdlife.org/>.

Carpenter GA, Matthew JS (1986) The birds of Billiatt Conservation Park. *South Australian Ornithologist* **30**, 29–37.

Clarke R (2005) 'Ecological requirements of birds specialising in mallee habitats'. Unpublished La Trobe University report to Department for Environment and Heritage, Adelaide.

Eckert J (1982) Striated Grasswrens in atypical habitat. *South Australian Ornithologist* **29**, 25.

Garnett S, Crowley GM (2000) *The Action Plan for Australian Birds 2000.* Environment Australia, Canberra.

Higgins PJ, Peter JM, Steele WK (Eds) (2001) *Handbook of Australian, New Zealand and Antarctic Birds. Volume 5: Tyrant-Flycatchers to Chats.* Oxford University Press, Melbourne.

Johnstone RE, Storr GM (2004) *Handbook of Western Australian Birds. Volume 2: Passerines (Blue-winged Pitta to Goldfinch).* Western Australian Museum, Perth.

Karubian J (2001) The social organization and mating system of the Striated Grasswren. *Condor* **103**, 412–417.

Pedler LP, Watson M, Langdon P, Pedler R (2007) 'Rare bird surveys, Mt Lyndhurst Station March 2007'. Report prepared for SA Arid Lands Natural Resources Management Board.

Rowley I, Russell E (1997) *Fairy-wrens and Grasswrens: Maluridae, 102.* Oxford University Press, Oxford.

Schodde R (1982) *The Fairy-Wrens. A Monograph of the Maluridae.* Lansdowne Editions, Melbourne.

Schodde R, Mason IJ (1999) The Directory of Australian Birds: Passerines. CSIRO Publishing, Melbourne.

Comments received from

John Augusteyn, Graham Carpenter, Rohan Clarke, Lynn Pedler

Species Conservation Summary

Short-tailed Grasswren

Amytornis merrotsyi Mellor, 1913

Maluridae

Conservation status

Near Threatened B2ab(ii,iii,v)

Reasons for listing

Small area occupied, near 10 locations and a possible continuing decline as a result of high fire frequencies

Status 2000

Least Concern

Reason for change in 2010: fires believed to have become more extensive and more frequent

Status 1990

Least Concern

Taxonomy

Both *A. m. merrotsyi* in the Flinders Ranges and *A. m. pedleri* in the Gawler Ranges are Vulnerable

Taxonomic uniqueness: medium (5 genera/family, 11 species/genus, 2 subspecies/species)

Range

Southern South Australia; *A. m. merrotsyi* in 4 subpopulations between Mt Neil in the north and Nelshaby in the south and *A. m. pedleri* in at least 3 subpopulations in the central and western Gawler Ranges (Parker 1982; Baxter and Paton 1998; Higgins *et al.* 2001; Christidis *et al.*

Current eligibility against IUCN Red List Criteria

IUCN category	Criteria eligibility
A	Not applicable: past, current or future population declines are thought unlikely to exceed 20% in any 3-generation period
B	Near Threatened: AOO <2000 km², near 10 locations, inferred continuing decline in AOO, habitat quality, no. mature individuals
C	Not applicable: population 2500–10 000 mature individuals, continuing decline not observed or estimated, largest subpopulation >1000 mature individuals, >1 subpopulation
D	Not applicable: population >1000 mature individuals, >5 locations
E	Not applicable: no population viability analysis undertaken

IUCN Red List assessment data

	Estimate	Reliability
Extent of occurrence trend	35 000 km² stable	high medium
Area of occupancy trend	1100 km² declining	medium low
No. of mature individuals trend	6000 declining	medium low
No. subpopulations	7	low
No. locations	17	low
Largest subpopulation	2000	medium
Generation time	9.7 years	low
Global population share	100%	high

2008). *A. m. merrotsyi* may no longer occur at the southern extremity of the range (Carpenter *et al.* 2003).

Abundance

Total population is probably <10 000 mature individuals in at least 7 subpopulations, the largest of which, in the south-east of Flinders Ranges National Park, is around 1000 pairs (Carpenter 2004). The population of *A. m. pedleri* is about 900 mature individuals and stable, but, given the frequency of extensive fires, it is assumed that the AOO, habitat and the population of *A. m. merrotsyi* are all likely to be declining.

Ecology

Inhabits rocky hilltops, ridges and hillsides with spinifex *Triodia* tussock grassland and usually with scattered shrubs (Christidis *et al.* 2008). *A. m. merrotsyi* recolonises 5–7 years after a fire and reaches peak densities 10–30 years after fire (G. Carpenter *in litt.*). It forages mostly on the ground, eating seeds, fruits, insects and other invertebrates (Higgins *et al.* 2001). A generation time of 9.7 years (BirdLife International 2011) is derived from an age at first breeding of 2.3 years and a maximum longevity of 17.0 years, both values extrapolated from fairy-wrens *Malurus* spp.

Threats

Excessive frequencies of fires, both natural and human mediated, along with grazing, have been identified as the most immediate threats to the species (Higgins *et al.* 2001). A single fire event has the potential to eliminate a whole subpopulation (G. Carpenter *in litt*), so were there fewer locations, the species would be at least Vulnerable. The habitat of the southern subpopulation near Quorn is burned regularly to make it suitable for sheep grazing and much of the region is burned on average every 10 years (Whisson 1999). Predation by foxes *Vulpes vulpes* may be significant as intensive fox baiting over 20 years has resulted in comparatively high population densities in south-east of Flinders Ranges National Park (Carpenter 2004).

Conservation objectives

1. Stable population size
2. Increased area of suitable habitat

Information required

1. Population trends at regularly monitored sites
2. Optimal fire regimes
3. Impacts of predation by foxes

Management actions required

1. Actively manage fires to reduce overall extent and frequency
2. If appropriate continue and extend fox baiting

Bibliography

Baxter C, Paton PA (1998) Further notes on the birds of the Gawler Ranges. *South Australian Ornithologist* **33**, 1–15.

BirdLife International (2011) 'Species factsheet: *Amytornis merrotsyi*'. Retrieved 17 January, 2011 from <http://www.birdlife.org/>.

Carpenter GA (2004) 'The Short-tailed Grasswren *Amytornis merrotsyi* in the Flinders Ranges, South Australia 2000–2003'. Report for Flinders Ranges Bounceback, SA Dept Environment & Heritage, Adelaide.

Carpenter GA, Black AB, Harper D, Horton P (2003) Bird report, 1982–1999. *South Australian Ornithologist* **34**, 93–151.

Christidis L, Horton P, Norman JA (2008) Subspeciation in the Short-tailed Grasswren (*Amytornis merrotsyi*, Maluridae). *Emu* **108**, 275–282.

Higgins PJ, Peter JM, Steele WK (Eds) (2001) *Handbook of Australian, New Zealand and Antarctic Birds. Volume 5: Tyrant-Flycatchers to Chats.* Oxford University Press, Melbourne.

Parker SA (1982) Notes on *Amytornis striatus merrotsyi* Mellor, a subspecies of the Striated Grasswren inhabiting the Flinders Ra. *South Australian Ornithologist* **29**, 13–15.

Whisson C (1999) The impact of burning and grazing on *Triodia* hummock grassland on the ranges north of Quorn. In *Proceedings of the Flinders Ranges National Park Vegetation Management Workshop.* (Eds P Alexander, B Arkell and L Edmunds). SA Dept Environment, Heritage and Aboriginal Affairs, Adelaide.

Comments received from

Andrew Black, Graham Carpenter, Lynn Pedler

Subspecies Conservation Summary

Short-tailed Grasswren (Flinders Ranges)

Amytornis merrotsyi merrotsyi Mellor, 1913

Maluridae

Conservation status

Vulnerable B1ab(ii,iii,v)+2ab(ii,iii,v)

Reasons for listing

Small area occupied, <10 locations and a possible continuing decline as a result of high fire frequencies

Status 2000

Least Concern

Reason for change in 2010: fires believed to have become more extensive and more frequent

Status 1990

Least Concern

Taxonomy

A. m. pedleri (Gawler Ranges) is Vulnerable. The species is Near Threatened.

Taxonomic uniqueness: medium (5 genera/family, 11 species/genus, 2 subspecies/species)

Range

Flinders Ranges, South Australia, between Mt Neil in the north and Nelshaby in the south. Occurs in 4 main subpopulations: north of Quorn between Dutchman's Stern and Buckaringa and formerly extending to the Ragless Range; in and adjacent to the south-east corner of Flinders Ranges National Park; between Black Range and Belton; and north of Arkaroola (Christidis *et al.* 2008). May no longer occur at southern extremity of range at Telowie Gorge near Nelshaby, where last recorded in 1981 (Parker 1982; Higgins *et al.* 2001; Carpenter *et al.* 2003). The documented AOO of 93 km^2 (Birds Australia Atlas database) is likely to be an order of magnitude too low, given the extent of habitat and intensity of surveying.

Abundance

The largest subpopulation, around Stokes Hill in the south-east of Flinders Ranges National Park, is around 1000 pairs (Carpenter 2004), with another 3000 mature individuals assumed to be spread among the other 3 subpopulations. Given the frequency of extensive fires, it is assumed that the AOO, habitat and the population are all likely to be declining.

Ecology

They inhabit rocky (quartzitic) hillsides and hilltops, steep-sided gullies, stony rises and ridge-crests and, less often, foothills. The vegetation is spinifex *Triodia* tussock grassland, usually with scattered low shrubs, particularly *Acacia*, *Daviesia genistifolia*, *Xanthorrhoea quadrangulata* and *Cassia*, and occasionally an open overstorey of Cypress Pine *Callitris* or mallee *Eucalyptus* (Christidis *et al.* 2008). Short-tailed Grasswrens recolonise patches 5–7 years after a fire and reach peak densities 10–30 years after fire (G. Carpenter *in litt.*). They forage mostly on the ground, eating seeds, fruits, insects and other invertebrates (Higgins *et al.* 2001). A generation time of 9.7 years (BirdLife International 2011) is derived from an age at first breeding of 2.3 years and a maximum longevity of 17.0 years, both extrapolated from fairy-wrens *Malurus* spp.

Threats

Excessive frequencies of fires, both natural and human mediated, along with grazing, have been identified as the most immediate threats to the species (Higgins *et al.* 2001). Given that single fire events have the potential to eliminate entire subpopulations (G. Carpenter *in litt.*), the species is assessed here as having no more than 10 locations. Much of the region is burned on average every 10 years (Whisson 1999) and the grasswren's habitat near Quorn is burned regularly for sheep grazing. Predation by foxes *Vulpes vulpes*, possibly in association with fire, may also be significant, given the comparatively high population densities of *A. m. merrotsyi* in the heavily baited south-east of Flinders Ranges National Park (Carpenter 2004).

Conservation objectives

1. Stable population size
2. Increased area of suitable habitat

Information required

1. Population trends at regularly monitored sites
2. Optimal fire regimes
3. Impacts of predation by foxes

Management actions required

1. Actively manage fires to reduce overall extent and frequency
2. If appropriate continue and extend fox baiting

Bibliography

BirdLife International (2011) 'Species factsheet: *Amytornis merrotsyi*'. Retrieved 17 January, 2011 from <http://www.birdlife.org/>.

Current eligibility against IUCN Red List Criteria

IUCN category	Criteria eligibility
A	Not applicable: past, current or future population declines are thought unlikely to exceed 20% in any 3-generation period
B	Vulnerable: EOO <20 000 km², AOO <2000 km², ≤10 locations with plausible future threat, inferred continuing decline likely in AOO, habitat quality, no. mature individuals
C	Not applicable: population 2500–10 000 mature individuals, continuing decline not observed or estimated, largest subpopulation >1000 mature individuals, >1 subpopulation
D	Not applicable: population >1000 mature individuals, >5 locations
E	Not applicable: no population viability analysis undertaken

IUCN Red List assessment data

	Estimate	Reliability
Extent of occurrence trend	9300 km² stable	high medium
Area of occupancy trend	1000 km² decreasing	medium low
No. of mature individuals trend	5000 decreasing	medium low
No. subpopulations	4	low
No. locations	10	low
Largest subpopulation	2000	medium
Generation time	9.7 years	low
Global population share	100%	high

Carpenter GA (2004) 'The Short-tailed Grasswren *Amytornis merrotsyi* in the Flinders Ranges, South Australia 2000–2003'. Report for Flinders Ranges Bounceback, SA Dept Environment & Heritage, Adelaide.

Carpenter GA, Black AB, Harper D, Horton P (2003) Bird report, 1982–1999. *South Australian Ornithologist* **34**, 93–151.

Christidis L, Horton P, Norman JA (2008) Subspeciation in the Short-tailed Grasswren (*Amytornis merrotsyi*, Maluridae). *Emu* **108**, 275–282.

Higgins PJ, Peter JM, Steele WK (Eds) (2001) *Handbook of Australian, New Zealand and Antarctic Birds. Volume 5: Tyrant-Flycatchers to Chats*. Oxford University Press, Melbourne.

Parker SA (1982) Notes on *Amytornis striatus merrotsyi* Mellor, a subspecies of the Striated Grasswren inhabiting the Flinders Ra. *South Australian Ornithologist* **29**, 13–15.

Whisson C (1999) The impact of burning and grazing on *Triodia* hummock grassland on the ranges north of Quorn. In *Proceedings of the Flinders Ranges National Park Vegetation Management Workshop*. (Eds P Alexander, B Arkell and L Edmunds). SA Dept Environment, Heritage and Aboriginal Affairs, Adelaide.

Comments received from

Andrew Black, Graham Carpenter, Lynn Pedler

Subspecies Conservation Summary

Short-tailed Grasswren (Gawler Ranges)

Amytornis merrotsyi pedleri Christidis, Horton & Norman 2008

Maluridae

Conservation status

Vulnerable D1

Reasons for listing

Population of about 900 mature individuals

Status 2000

Vulnerable D1

Status 1990

Vulnerable D1

Taxonomy

A. m. merrotsyi (Flinders Ranges) is Vulnerable. The species is Near Threatened.

Taxonomic uniqueness: medium (5 genera/family, 11 species/genus, 2 subspecies/species)

Range

Central and western Gawler Ranges, South Australia (Baxter and Paton 1998; Christidis *et al.* 2008), where there are 3 main subpopulations: the eastern side of Mt Ive Station, the eastern side of Paney Station in Gawler

Current eligibility against IUCN Red List Criteria

IUCN category	Criteria eligibility
A	Not applicable: past, current or future population declines are thought unlikely to exceed 20% in any 3-generation period
B	Not applicable: EOO, AOO, habitat, locations or population not declining, population fluctuations not extreme
C	Not applicable: no continuing population decline or extreme fluctuations
D	Vulnerable: population 250–1000 mature individuals
E	Not applicable: no population viability analysis undertaken

IUCN Red List assessment data

	Estimate	Reliability
Extent of occurrence trend	1100 km^2 stable	medium medium
Area of occupancy trend	90 km^2 stable	low medium
No. of mature individuals trend	900 stable	low medium
No. subpopulations	3	medium
No. locations	7	medium
Largest subpopulation	300	low
Generation time	9.7 years	low
Global population share	100%	high

Ranges National Park and south-east of Hiltaba Homestead. There are outlying or intermediate records from 4 other locations (G. Carpenter *in litt.*).

Abundance

The largest subpopulation, in Gawler Ranges National Park, is estimated at a few hundred individuals (G. Carpenter *in litt.*). Assuming 300 birds at Paney Station, 200 at Mt Ive and Hiltaba and 50 at each of the 4 less important sites, the total population would be 900 mature individuals, with no evidence of decline (A. Black *in litt.*).

Ecology

They inhabit rocky (granitic) hilltops, ridges and hillsides with spinifex *Triodia* tussock grassland and usually with scattered spiny shrubs, particularly *Acacia* and *Grevillea* (Christidis *et al.* 2008). *A. m. merrotsyi* recolonises sites 5–7 years after a fire and reaches peak densities 10–30 years after fire (G. Carpenter *in litt.*). They forage mostly on the ground, eating seeds, fruits, insects and other invertebrates (Higgins *et al.* 2001). A generation time of 9.7 years (BirdLife International 2011) is derived from an age at first breeding of 2.3 years and a maximum longevity of 17.0 years, both values extrapolated from fairy-wrens *Malurus* spp.

Threats

Excessive frequencies of fires, both natural and human mediated, along with grazing have been identified as the most immediate threats to the species (Higgins *et al.* 2001). Given that single fire events have the potential to eliminate whole subpopulations (G. Carpenter *in litt*), the species would be listed as Endangered were there any evidence of decline. Predation by foxes *Vulpes vulpes*, possibly in association with fire, may also be significant given the comparatively high population densities of *A. m. merrotsyi* in the heavily baited south-east of Flinders Ranges National Park (Carpenter 2004).

Conservation objectives

1. Stable population size
2. Increased area of suitable habitat

Information required

1. Population trends at regularly monitored sites
2. Optimal fire regimes
3. Impacts of predation by foxes

Management actions required

1. Actively manage fires to reduce overall extent and frequency
2. If appropriate continue and extend fox baiting

Bibliography

Baxter C, Paton PA (1998) Further notes on the birds of the Gawler Ranges. *South Australian Ornithologist* **33**, 1–15.

BirdLife International (2011) 'Species factsheet: *Amytornis merrotsyi*'. Retrieved 17 January, 2011 from <http://www.birdlife.org/>.

Carpenter GA (2004) 'The Short-tailed Grasswren *Amytornis merrotsyi* in the Flinders Ranges, South Australia 2000–2003'. Report for Flinders Ranges Bounceback, SA Dept Environment & Heritage, Adelaide.

Christidis L, Horton P, Norman JA (2008) Subspeciation in the Short-tailed Grasswren (*Amytornis merrotsyi*, Maluridae). *Emu* **108**, 275–282.

Higgins PJ, Peter JM, Steele WK (Eds) (2001) *Handbook of Australian, New Zealand and Antarctic Birds. Volume 5: Tyrant-Flycatchers to Chats.* Oxford University Press, Melbourne.

Comments received from

Andrew Black, Graham Carpenter, Lynn Pedler

Species Conservation Summary

White-throated Grasswren

Amytornis woodwardi Hartert, 1905
Maluridae

Conservation status

Vulnerable B1ab(ii,iii,iv,v)+2ab(ii,iii,iv,v), C2a(i)

Reasons for listing

Population <10 000 with <1000 in any subpopulation, small AOO, severely fragmented and continuing decline in AOO, habitat quality and number of locations

Status 2000

Vulnerable B1ab(ii,iii,iv,v)+2ab(ii,iii,iv,v), C2a(i)

Status 1990

Vulnerable B1ab(ii,iii,iv,v)+2ab(ii,iii,iv,v), C2a(i)
Reason for change in 2000: Fire frequency and extent increasing

Taxonomy

No infraspecific taxa described
Taxonomic uniqueness: medium (5 genera/family, 11 species/genus, 1 subspecies/species)

Range

Western Arnhem Land, Northern Territory, on and around the Arnhem Land escarpment, between Katherine and Maningrida (Noske 1992). There have been few surveys in this inaccessible range since Noske (1992), but anecdotal information from birdwatchers at the periphery of the range suggest that the White-throated Grasswren is increasingly hard to locate at sites where once seen regularly (R. Noske pers. comm.) and is certainly absent at some sites mentioned by Noske (1992), such as Plum Tree Creek.

Abundance

Based on population density estimates at 8 sites, Noske (1992) estimated the total population at c.50 000 (14 000–182 000) individuals, but Woinarski (1992) suggested a total population of c.5000–10 000 individuals. Given the fire history of the habitat, numbers have almost certainly declined and the population is estimated here as no more than 10 000 mature birds. NRETA (2006) assessed it as severely fragmented into subpopulations of <1000 mature individuals each, but noted that the subpopulation structure is not well known, and the fragmentation or continuity of populations is uncertain. Based on the fire history, a precautionary assessment is made here that >50% of the species' AOO could be in habitat patches too small to support a viable population, and separated by distances several times its long-term average dispersal distance. Woinarski (2009) reported a significant decline in Kakadu National Park, from a mean of 0.08 individuals/quadrat in 2001–2004 (recorded in 6/142 quadrats) to zero in 2010. The reporting rate from the best-known site at Gunlom has declined through the 2000s. Given the size of wildfires in the species' rocky habitat, it is assessed here as having >10 locations.

Ecology

Occurs on bare flat plateaux and stepped or terraced hillsides along broad valleys, with or without narrow rocky gullies (Schodde and Mason 1975; Noske 1992). The habitat is characterised by mature spinifex *Triodia* spp., which is used for nesting, and bare rock, with most occupied sites containing bare pavement and/or boulders, sometimes with open shrubland or woodland overstorey. The diet comprises invertebrates, seeds and other vegetable matter (Noske 1992). Breeding occurs from December to June, and territory size is around 10 ha (Noske 1992). A generation time of 9.7 years (BirdLife International 2011) is derived from an age at first breeding of 2.3 years and a maximum longevity of 17.0 years, both values extrapolated from fairy-wrens *Malurus* spp.

Threats

Probably threatened by increased extent and frequency of fire (Woinarski *et al.* 2007; Woinarski *et al.* 2009). In 1980–1995, in heath habitats of Kakadu National Park, 69% of plots had fires recurring within 3 years and only 11% burnt only once or remained unburnt (Russell-Smith *et al.* 2002) and fire regimes across western Arnhem land are characterised by high annual frequencies of large late dry season fires (Edwards and Russell-Smith 2009) that cover large areas (Yates *et al.* 2008). Despite the rocky habitat offering protection from fires (Noske 1992), many fires are too hot and extensive to leave functional unburned refugia. The ecologically similar Carpentarian Grasswren *A. dorotheae* may not re-occupy areas until at least 3 years after a fire (Harrington *et al.* 2009). The progressive vegetational change, notably replacement of spinifex by annual sorghum *Sarga* spp., may reduce habitat suitability as well as supporting further hot fires (Woinarski 1992).

Conservation objectives

1. Stable numbers at monitored sites
2. Patchy fire across its range

Current eligibility against IUCN Red List Criteria

IUCN category	Criteria eligibility
A	Not applicable: declines <30% in 3 generations (29 years)
B	Vulnerable EOO <20 000 km^2, AOO <2000 km^2, severely fragmented and inferred continuing decline in AOO, habitat quality and number of locations
C	Vulnerable: population <10 000 mature individuals with a continuing decline and <1000 mature individuals in any subpopulation
D	Not applicable: population >1000 mature individuals, >5 locations
E	Not applicable: no population viability analysis undertaken

IUCN Red List assessment data

	Estimate	Reliability
Extent of occurrence trend	10 000 km^2 stable	high medium
Area of occupancy trend	1000 km^2 decreasing	low medium
No. of mature individuals trend	10 000 decreasing	low medium
No. subpopulations	>10	low
No. locations	>10	low
Largest subpopulation	1000	low
Generation time	9.7 years	low
Global population share	100%	high

Information required

1. Population trends at nominated monitoring sites
2. Relationship between grasswrens, habitat suitability and fire regimes
3. Fire histories of occupied and unoccupied habitats
4. Metapopulation structure

Management actions required

1. Survey all sites from which recorded, and pockets of habitat known to have had few fires in the last 2 decades
2. Initiate regular monitoring at a selection of known sites
3. Continue intensive fire management of Kakadu and western Arnhem Land

Bibliography

BirdLife International (2011) 'Species factsheet: *Amytornis woodwardi*'. Retrieved 17 January, 2011 from <http://www.birdlife.org/>.

Edwards AC, Russell-Smith J (2009) Ecological thresholds and the status of fire-sensitive vegetation in western Arnhem Land, northern Australia: implications for management. *International Journal of Wildland Fire* **18**, 127–146.

Harrington G, Perry J, Forsyth R, Venables B (2009) A tale of two Grasswrens. *Wingspan* **19**, 23–25.

Noske RA (1992) The status and ecology of the White-throated Grasswren *Amytornis woodwardi*. *Emu* **92**, 39–51.

NRETA (2006) 'Threatened species information sheet: White-throated Grass-wren *Amytornis woodwardi*'. Retrieved 30 March, 2011 from <http://www.nt.gov.au/nreta/wildlife/animals/threatened/pdf/birds/whitethroated_grasswren_vu.pdf>.

Russell-Smith J, Ryan PG, Cheal DC (2002) Fire regimes and the conservation of sandstone heath in monsoonal northern Australia: frequency, interval, patchiness. *Biological Conservation* **104**, 91–106.

Schodde R, Mason IJ (1975) Occurrence, nesting and affinities of the White-throated Grass-wren *Amytornis woodwardi* and White-lined Honeyeater *Meliphaga albilineata*. *Emu* **75**, 12–18.

Woinarski J, Pavey C, Kerrigan R, Cowie I, Ward S (Eds) (2007) *Lost from Our Landscape: Threatened Species of the Northern Territory*. Northern Territory Government, Darwin.

Woinarski JCZ (1992) The conservation status of the white-throated grass-wren *Amytornis woodwardi*, an example of problems in status designation. *Northern Territory Naturalist* **13**, 1–5.

Woinarski JCZ (2009) 'Fauna monitoring at fire plots in Kakadu National Park, 2008–09'. Interim report to Parks Australia, NRETAS, Darwin.

Woinarski JCZ, Russell-Smith J, Andersen AN, Brennan K (2009) Fire management and biodiversity of the western Arnhem Land plateau. In *Culture, Ecology, and Economy of Fire Management in North Australian Savannas: Rekindling the Wurrk Tradition*. (Eds J Russell-Smith, P Whitehead and P Cooke) pp. 201–228. CSIRO Publishing, Melbourne.

Yates CP, Edwards AC, Russell-Smith J (2008) Big fires and their ecological impacts in Australian savannas: size and frequency matters. *International Journal of Wildland Fire* **17**, 768–781.

Comments received from

Don Franklin, Richard Noske, Simon Ward, John Woinarski

Species Conservation Summary

Carpentarian Grasswren

Amytornis dorotheae (Mathews, 1914)
Maluridae

Conservation status
Near Threatened A2c, C2a(ii)

Reasons for listing
Population has probably declined as a result of fire by 20–29% in the last 3 generations (29 years) and is now near 10 000 mature individuals, almost all in a single subpopulation

Status 2000
Near Threatened A2c

Status 1990
Least Concern
Reason for change in 1990: declines assumed to have started

Taxonomy
No infraspecific taxa described
Taxonomic uniqueness: medium (5 genera/family, 11 species/genus, 1 subspecies/species)

Range
Known to have occurred in 4 separate areas between the Tawallah Range/Limmen Bight River, Northern Territory, and Gunpowder, north-west Queensland (McKean and Martin 1989; Murphy *et al.* 2011), which may have been a single subpopulation within the last century. There have been no records from the subpopulation around Borroloola, including Limmen National Park, since 1986, despite several targeted surveys in the last decade. No suitable unburnt habitat could be found in 2009 (Harrington *et al.* 2009). The Wollogorang subpopulation survives in 2 small locations in the NT but there have been no records in the adjacent area of Qld since 1986, or from the China Wall area to the south since 1974. Recent visits have found severe burning and no suitable habitat (Harrington *et al.* 2009; Perry *et al.* in press). The Boodjamulla (Lawn Hill) subpopulation was exposed to very extensive fires in 2003 and 2006, but survives in greatly reduced numbers (Harrington *et al.* 2009). The Mount Isa subpopulation was recorded across c.29 000 km^2 in 1990–1995 and again in 2008–2009 (Harrington *et al.* 2009; Harris and Stewart 2009; Murphy *et al.* 2011).

Abundance
The Mount Isa subpopulation is estimated to number 13 000 mature individuals based on an extent of occurrence of c.29 000 km^2 of which a third is likely to be occupied, and an estimated breeding territory size of 1.5 km^2 based on survey data from 2400 1-km^2 survey sites (G. Harrington *in litt.*). The northern populations have all declined substantially in the last 29 years (3 generations). Were they as large as the Mount Isa population at that time, the status now would be Vulnerable or Endangered. However, given they were fairly scattered in the NT in the early 1980s (McKean and Martin 1989), the decline is assumed to have been <30% of the total population. There is no evidence for historical declines in the Mount Isa subpopulation, but the predicted spread of Buffel Grass *Cenchrus ciliaris* is likely to increase the frequency and extent of fires. The Boodjamulla and Wollogorang subpopulations are estimated here as about 500 mature individuals each, and both of them are probably declining.

Ecology
In the northern part of its range, confined to sandstone outcrops in mature spinifex *Triodia* hummock grassland. In the south occupies long-unburnt spinifex with stony areas between the hummocks on which grow a range of short grasses, forbs and patchy low trees and shrubs (McKean and Martin 1989; Rowley and Russell 1997). Grasswrens normally abandon burnt areas, unless substantial unburnt pockets of spinifex remain. Recolonisation can occurs after 4 years, or 3 years in the north where higher rainfall enables faster regeneration, if nearby source populations persist (Harrington *et al.* 2009). A generation time of 9.7 years (BirdLife International 2011) is derived from an age at first breeding of 2.3 years and a maximum longevity of 17.0 years, both values extrapolated from fairy-wrens *Malurus* spp.

Threats
Fire caused the extirpation of 1 historical subpopulation, has greatly reduced 2 others and was probably responsible for the separation of the subpopulations in the first place. Rainfall in the northern part of the historical range has increased over the last century (Woinarski *et al.* 2007) because the wet season has been starting earlier (Garnett and Williamson 2010). This allows spinifex to recover from fire more quickly but also enables fires to be more frequent and extensive. The extent of control burning may also have declined. An average of 28% of

Current eligibility against IUCN Red List Criteria

IUCN category	Criteria eligibility
A	Near Threatened: past population decline of 20–29% over 3 generations (29 years) suspected from deteriorating habitat quality
B	Not applicable: EOO >20 000 km^2, AOO >2000 km^2
C	Near Threatened: population near 10 000 mature individuals, inferred continuing decline, largest subpopulation approaching 100% of total population
D	Not applicable: population >1000 mature individuals, >5 locations
E	Not applicable: no population viability analysis undertaken

IUCN Red List assessment data

	Estimate	Reliability
Extent of occurrence trend	35 000 km^2 decreasing	high high
Area of occupancy trend	11 000 km^2 decreasing	low medium
No. of mature individuals trend	14 000 decreasing	low high
No. subpopulations	3	medium
No. locations	>10	high
Largest subpopulation	13 000	low
Generation time	9.7 years	low
Global population share	100%	high

the area of the Borroloola subpopulation burned annually 2004–2010 (41% in 2004), 24% across Wollogorang (48% in 2004), 11% across Boojamulla (51% in 2006) and 4% across Mount Isa (maximum of 11% in 2010; Murphy *et al.* 2011). However, the frequency and scale of fires is likely to increase in the south as the exotic invasive Buffel Grass fills the stony ground between spinifex hummocks, allowing fires to spread more frequently and to burn deeper into rocky refuges. There may be ongoing very localised threats from mining and associated development.

Conservation objectives

1. Specific fire regimes at key sites
2. Stable or improved trends in monitoring at key sites

Information required

1. Population trends at key monitoring sites
2. The effect of Buffel Grass on habitat occupancy and fires

Management actions required

1. Monitor fire patterns from satellite imagery and model impact on habitat
2. Map Buffel Grass and determine relationship to fire data
3. Monitor key sites across Range
4. Create a mosaic of spinifex ages by burning < 10% of the total area each year, burning any site at intervals of >10 years, each fire <20 km^2 and mostly late or very early in the wet season

Bibliography

BirdLife International (2011) 'Species factsheet: *Amytornis dorotheae*'. Retrieved 17 January, 2011 from <http://www.birdlife.org/>.

Garnett ST, Williamson G (2010) Spatial and temporal variation in precipitation at the start of the rainy season in tropical Australia. *The Rangeland Journal* **32**, 215–226.

Harrington G, Perry J, Forsyth R, Venables B (2009) A tale of two Grasswrens. *Wingspan* **19**, 23–25.

Harris PL, Stewart DA (2009) Grasswren *Amytornis dorotheae* surveys near Mt Isa (1990–1995). *Sunbird* **39**, 3–13.

McKean JL, Martin KC (1989) Distribution and status of the Carpentarian Grass-wren *Amytornis dorotheae. Northern Territory Naturalist* **11**, 12–19.

Murphy S, Harrington G, Felderhof L (2011) 'Preliminary spatial model using fire scar data to monitor Carpentarian Grasswrens'. Report by Map IT and Birds Australia North Queensland.

Perry J, Fisher A, Palmer C (in press) Status and habitat of the Carpentarian grasswren (*Amytornis dorotheae*) in the Northern Territory. *Emu.*

Rowley I, Russell E (1997) *Fairy-wrens and Grasswrens: Maluridae, 102.* Oxford University Press, Oxford.

Woinarski J, Pavey C, Kerrigan R, Cowie I, Ward S (Eds) (2007) *Lost from Our Landscape: Threatened Species of the Northern Territory.* Northern Territory Government, Darwin.

Comments received from

Don Franklin, Graham Harrington, Keith McDonald, Steve Murphy, Justin Perry

Subspecies Conservation Summary

Thick-billed Grasswren (MacDonnell Ranges)

Amytornis modestus modestus (North, 1902)
Maluridae

Conservation status
Extinct

Reasons for listing
Last recorded in 1936

Status 2000
Extinct

Status 1990
Extinct

Taxonomy

A. m. inexpectatus (central New South Wales) is Extinct, *A. m. obscurior* (NW NSW) is Critically Endangered, *A. m. raglessi* (Flinders Ranges) is Vulnerable, *A. m. curnamona* (Lake Frome Basin) is Near Threatened and *A. m. indulkanna* (western) is Least Concern, as is the species.
Taxonomic uniqueness: medium (5 genera/family, 11 species/genus, 6 subspecies/species)

Range

Historically recorded in the Finke River Drainage Basin near Alice Springs, Northern Territory, from Mereenie Bluff and the upper reaches of Hugh River, south to the lower reaches of the Finke River, and at Laurie's Creek and McMinn's Creek (Storr 1977; Higgins *et al.* 2001). The last record was a clutch of eggs taken in 1936 (Parker 1972).

Ecology

Uniquely within the species, it inhabited Sandhill Canegrass *Zygochloa paradoxa* shrubland in the sandy watercourses of the MacDonnell Ranges (Black *et al.* 2010; Black 2011) as well as 'much eaten saltbushes' (Whitlock 1924) and 'flood debris in the river bed' (White 1914).

Reasons for extinction

The most likely reason for extinction is overgrazing by cattle and rabbits *Oryctolagus cuniculus* in combination with drought, as it is thought to happened to *A. m. inexpectatus* (Schodde 1982; McAllan 1987). The extinction of this subspecies has much in common with the loss of mammals in central Australia – the birds were most abundant in areas near watercourses or on alluvial plains, where nutrient levels are slightly higher than in the surrounding landscape. Such areas were among the most heavily grazed by introduced stock and rabbits, and thus attracted the highest densities of introduced predators like foxes *Vulpes vulpes* and feral cats *Felis catus* (Morton 1990).

Bibliography

Black AB (2011) Subspecies of the Thick-billed Grasswren *Amytornis modestus* (Aves-Maluridae). *Transactions of the Royal Society of South Australia* **135**, 26–38.

Black AB, Joseph L, Pedler LP, Carpenter GA (2010) A taxonomic framework for interpreting evolution within the *Amytornis textilis–modestus* complex of grasswrens. *Emu* **110**, 358–363.

Higgins PJ, Peter JM, Steele WK (Eds) (2001) *Handbook of Australian, New Zealand and Antarctic Birds. Volume 5: Tyrant-Flycatchers to Chats.* Oxford University Press, Melbourne.

McAllan IAW (1987) Early records of the Thick-billed Grasswren *Amytornis textilis* and Striated Grasswren *Amytornis striatus* in New South Wales. *Australian Birds* **28**, 65–70.

Morton SR (1990) The impact of European settlement on the vertebrate animals of arid Australia: a conceptual model. *Proceedings of the Ecological Society of Australia* **16**, 210–213.

Parker SA (1972) Remarks on distribution and taxonomy of the grass wrens *Amytornis textilis, modestus* and *purnelli. Emu* **72**, 157–166.

Schodde R (1982) *The Fairy-Wrens. A Monograph of the Maluridae.* Lansdowne Editions, Melbourne.

Storr GM (1977) *Birds of the Northern Territory.* Western Australian Museum, Perth.

White SA (1914) Scientific notes on an expedition into the interior of Australia. *Transactions of the Royal Society of South Australia* **38**, 407–415.

Whitlock FL (1924) Journey to central Australia in search of the Night Parrot. *Emu* **23**, 248–281.

Comments received from

Andrew Black, Graham Carpenter, Lynn Pedler, Danny Rogers

Species Conservation Summary

Thick-billed Grasswren (north-west New South Wales)

Amytornis modestus obscurior (Mathews, 1923)
Maluridae

Conservation status
Critically Endangered D

Reasons for listing
The known population of this subspecies was only about 10 adults at the time of assessment

Status 2000
Critically Endangered D

Status 1990
Critically Endangered D

Taxonomy
A. m. modestus (MacDonnell Ranges) and *A. m. inexpectatus* (central NSW) are Extinct, *A. m. raglessi* (Flinders Ranges) is Vulnerable, *A. m. curnamona* (Lake Frome Basin) is Near Threatened, *A. m. indulkanna* (western) is Least Concern, as is the species.
Taxonomic uniqueness: medium (5 genera/family, 11 species/genus, 4 extant subspecies/species)

Range
Specimens collected in 1912 near Milparinka, north of Broken Hill (McAllan 1987) and a clutch of eggs collected north-east of Tibooburra near the Queensland border in 1936 (Black and Longmore 2009). A population was rediscovered in 2008 near Packsaddle, east of the Milparinka record (Parker *et al.* 2010).

Abundance
The population is very small. Up to 10 adult birds were seen in 2008–2010 (Black 2011). It is assessed here as having a stable population, since any decline would have caused its extinction. As it is, any evidence of a continuing decline would qualify it as Critically Endangered under the additional criteria B1ab+2ab, C2a(ii).

Ecology
The existing birds were recorded on a low ridgeline covered with gibber, with scattered Blackbush *Maireana pyramidata* and Thorny Saltbush *Rhagodia spinescens* the dominant vegetation as well as scattered trees and taller shrubs (Parker *et al.* 2010). Other Thick-billed Grasswren subspecies inhabit chenopod shrublands, particularly those dominated by saltbush *Atriplex* spp. and bluebush *Maireana* spp. They forage on the ground for berries, seeds and insects (Rowley and Russell 1997). They lay 2–3 eggs in domed or cup nests built in shrubs (Rowley and Russell 1997). A generation time of 9.7 years (BirdLife International 2011) is derived from an age at first breeding of 2.3 years and a maximum longevity of 17.0 years, both values extrapolated from fairy-wrens *Malurus* spp.

Threats
Overgrazing by sheep, cattle, feral goats *Hircus capra* and rabbits *Oryctolagus cuniculus*, particularly in combination with drought, is thought to be the main threat, and is already thought to have caused the extinction of *A. m. inexpectatus*. Although Blackbush and similar shrubs are resilient and long-lived, recruitment is infrequent and their seedlings are highly palatable to stock (Crisp 1978; Tiver and Andrew 1997). Predation by foxes *Vulpes vulpes* may be significant and intensive fox baiting may underlie recent increases in *A. m. curnamona* on the north Olary Plains (Pedler *et al.* 2007).

Conservation objectives
1. Viable population size
2. Expanded area of occupancy

Information required
1. All individuals located in intensive surveys of nearby potential habitat
2. Status assessed across historical and potential new locations
3. Optimal levels of grazing by livestock and feral mammals assessed
4. Impacts of predation by foxes

Management actions required
1. Ensure grazing by stock is kept at very low levels until the relationship between grazing and grasswren abundance is better understood

Bibliography

BirdLife International (2011) 'Species factsheet: *Amytornis textilis*'. Retrieved 17 January, 2011 from <http://www.birdlife.org/>.

Black AB (2011) Subspecies of the Thick-billed Grasswren *Amytornis modestus* (Aves-Maluridae). *Transactions of the Royal Society of South Australia* **135**, 26–38.

Black AB, Longmore NW (2009) Notes on Grasswren eggs in Len Harvey's collection, Museum Victoria. *Australian Field Ornithology* **26**, 132–141.

Current eligibility against IUCN Red List Criteria

IUCN category	Criteria eligibility
A	Not applicable: past, current or future population declines are thought unlikely to exceed 20% in any 3-generation period
B	Not applicable: EOO, AOO, habitat, locations or population not thought to be declining, population fluctuations not extreme
C	Not applicable: no continuing population decline or extreme fluctuations
D	Critically Endangered: population <50 mature individuals
E	Not applicable: no population viability analysis undertaken

IUCN Red List assessment data

	Estimate	Reliability
Extent of occurrence trend	<100 km² stable	medium medium
Area of occupancy trend	<10 km² stable	medium medium
No. of mature individuals trend	10 stable	medium low
No. subpopulations	1	high
No. locations	1	high
Generation time	9.7 years	low
Global population share	100%	high

Crisp MD (1978) Demography and survival under grazing of three Australian semi-desert shrubs. *Oikos* **30**, 520–528.

McAllan IAW (1987) Early records of the Thick-billed Grasswren *Amytornis textilis* and Striated Grasswren *Amytornis striatus* in New South Wales. *Australian Birds* **28**, 65–70.

Parker DG, Egan D, Ballestrin M (2010) Recent observations of the Thick-billed Grasswren *Amytornis textilis modestus* in New South Wales. *Australian Field Ornithology* **27**, 159–164.

Pedler LP, Watson M, Langdon P, Pedler R (2007) 'Rare bird surveys, Mt Lyndhurst Station March 2007'. Report prepared for SA Arid Lands Natural Resources Management Board.

Rowley I, Russell E (1997) *Fairy-wrens and Grasswrens: Maluridae, 102.* Oxford University Press, Oxford.

Tiver F, Andrew MH (1997) Relative effects of herbivory by sheep, rabbits, goats and kangaroos on recruitment and regeneration of shrubs and trees in eastern South Australia. *Journal of Applied Ecology* **34**, 903–914.

Comments received from

Andrew Black, Graham Carpenter, Lynn Pedler, Ian McAllan

Subspecies Conservation Summary

Thick-billed Grasswren (central New South Wales)

Amytornis modestus inexpectatus (Mathews, 1912)
Maluridae

Conservation status

Extinct

Reasons for listing

Last recorded in 1886

Status 2000

Extinct

Status 1990

Extinct

Taxonomy

A. m. modestus (MacDonnell Ranges) is also Extinct, *A. m. obscurior* (NW NSW) is Critically Endangered, *A. m. raglessi* (Flinders Ranges) is Vulnerable, *A. m. curnamona* (Lake Frome Basin) is Near Threatened, *A. m. indulkanna* (western) is Least Concern, as is the species
Taxonomic uniqueness: medium (5 genera/family, 11 species/genus, 6 subspecies/species)

Former range

Known historically from 2 areas in central New South Wales, but last recorded in about 1886 (Parker 1972; McAllan 1987; Black 2011). The northern population has been recorded just once, when Gould (1848) collected it on the lower Namoi River. The southern population was known from sightings and 3 specimens taken near Mossgiel on the Hay Plains around 1886 by K. H. Bennett (Parker 1972; McAllan 1987). It was recorded from the catchments of only 2 Murray-Darling tributaries and, while possibly more widespread than these 2 records suggest, both localities are well separated from other populations to the west of the Darling River drainage.

Ecology

Habitat along the lower Namoi River was scrubs and clumps of a shrub-like tree resembling the Barilla *Atriplex cinerea* of the coast (McAllan 1987). The lower Namoi and Mossgiel areas still have plenty of Nitrebush *Nitraria billardierei*, and chenopods also occur in the Mossgiel district (Parker 1972). Other Thick-billed Grasswren subspecies inhabit chenopod shrublands, particularly those dominated by saltbush *Atriplex* and bluebush *Maireana*.

Reasons for extinction

Its extinction has been attributed to destruction of habitat by livestock (Schodde 1982; McAllan 1987). Soon after the second population was last seen, the region was affected by an exceptionally severe drought ending a decade of extraordinarily high stocking rates throughout the region (McAllan 1987). However, other subspecies are known to survive in very sparse cover in areas degraded by sheep, cattle, feral goats *Hircus capra* and rabbits *Oryctolagus cuniculus* and drought. Predation by foxes *Vulpes vulpes* may have exacerbated other stresses.

Bibliography

Black AB (2011) Subspecies of the Thick-billed Grasswren *Amytornis modestus* (Aves-Maluridae). *Transactions of the Royal Society of South Australia* **135**, 26–38.

Gould J (1848) *The Birds of Australia*. Gould, London.

McAllan IAW (1987) Early records of the Thick-billed Grasswren *Amytornis textilis* and Striated Grasswren *Amytornis striatus* in New South Wales. *Australian Birds* **28**, 65–70.

Parker SA (1972) Remarks on distribution and taxonomy of the grass wrens *Amytornis textilis, modestus* and *purnelli*. *Emu* **72**, 157–166.

Schodde R (1982) *The Fairy-Wrens. A Monograph of the Maluridae*. Lansdowne Editions, Melbourne.

Comments received from

Andrew Black, Lynn Pedler

Subspecies Conservation Summary

Thick-billed Grasswren (Flinders Ranges)

Amytornis modestus raglessi Black 2011
Maluridae

Conservation status

Vulnerable B1ab(i,ii,iii,iv)+2ab(i,ii,iii,iv)

Reasons for listing

A small extent of occurrence and area occupied at <10 locations and a continuing decline in EOO, AOO, habitat quality, no. locations inferred from recent historical losses and ongoing grazing pressure

Status 2000

Vulnerable Blab(i,ii,iii,iv) + 2ab(i,ii,iii,iv)

Status 1990

Vulnerable Blab(i,ii,iii,iv) + 2ab(i,ii,iii,iv)

Taxonomy

A. m. modestus (MacDonnell Ranges) and
A. m. inexpectatus (central NSW) are Extinct,
A. m. obscurior (NW NSW) is Critically Endangered,
A. m. curnamona (Lake Frome Basin) is Near Threatened, *A. m. indulkanna* (western) is Least Concern, as is the species
Taxonomic uniqueness: medium (5 genera/family, 11 species/genus, 6 subspecies/species)

Range

Recently described as occurring across watercourses draining the northern North Flinders Ranges towards southern Lake Eyre, Lake Blanche and Lake Callabonna (Black 2011). Known from specimens at 8 locations and sightings at another 6 (Black *et al.* 2010) from Dulkaninna Station c.80 km north of Marree, to North Moolooloo Station 20 km south of Copley, and north-east to Murnpeowie Station and the western margin of Lake Callabonna (A.Black *in litt.*). The number of subpopulations is estimated here as about 4, based on gaps of >10 km between records.

Current eligibility against IUCN Red List Criteria

IUCN category	Criteria eligibility
A	Not applicable: past, current or future population declines are thought unlikely to exceed 20% in any 3-generation period
B	Vulnerable: EOO <20 000 km², AOO <2000 km², ≤10 locations with plausible future threat, suspected continuing decline in EOO, AOO, habitat quality, no. locations
C	Not applicable: population 2500–10 000 mature individuals, continuing decline not observed or estimated, largest subpopulation >1000 mature individuals, >1 subpopulation
D	Not applicable: population >1000 mature individuals, >5 locations
E	Not applicable: no population viability analysis undertaken

IUCN Red List assessment data

	Estimate	Reliability
Extent of occurrence trend	12 000 km² decreasing	medium low
Area of occupancy trend	600 km² decreasing	low low
No. of mature individuals trend	5000 decreasing	low low
No. subpopulations	4	low
No. locations	10	low
Largest subpopulation	2000	low
Generation time	9.7 years	low
Global population share	100%	high

Abundance

There are no measures of abundance, but an average density of 8/km², extrapolated from estimates for Western Grasswren *Amytornis textilis myall* (A. Black *in litt.*), within an AOO that is 5% of EOO gives an approximate population size of 5000 mature individuals. A largest subpopulation of 2000 is allocated to the south-eastern subpopulation with the largest range and number of records.

Ecology

Inhabits low chenopod shrublands, with 62% recent records in habitats dominated by Blackbush *Maireana pyramidata*, 14% by Low Bluebush *Maireana astrotricha* and 9% by Spiny Saltbush *Rhagodia spinescens* (A. Black *in litt.*). Other Thick-billed Grasswren subspecies forage on the ground for berries, seeds and insects (Rowley and Russell 1997). They lay 2–3 eggs in domed or cup nests built in shrubs (Rowley and Russell 1997). A generation time of 9.7 years (BirdLife International 2011) is derived from an age at first breeding of 2.3 years and a maximum longevity of 17.0 years, both values extrapolated from fairy-wrens *Malurus* spp.

Threats

Overgrazing by sheep, cattle, feral goat *Hircus capra* and rabbit *Oryctolagus cuniculus* in combination with drought is thought to be the main threat, and is already thought to have caused the extinction of *A. m. inexpectatus* (Schodde 1982; McAllan 1987). Although Blackbush and similar shrubs are resilient and long-lived, recruitment is infrequent and their seedlings are highly palatable to stock (Crisp 1978; Tiver and Andrew 1997). Predation by foxes *Vulpes vulpes* may be significant and intensive fox baiting may underlie recent increases in *A. m. curnamona* on the north Olary Plains (Pedler *et al.* 2007).

Conservation objectives

1. Stable population size
2. Stable area of occupancy

Information required

1. Status and trends at historical and potential new locations
2. Optimal levels of grazing by livestock and feral mammals
3. Impacts of predation by foxes

Management actions required

1. Ensure grazing by stock is kept at low levels at a selection of locations until the relationship between grazing and grasswren abundance is better understood

Bibliography

BirdLife International (2011) 'Species factsheet: *Amytornis textilis*'. Retrieved 17 January, 2011 from <http://www.birdlife.org/>.

Black AB (2011) Subspecies of the Thick-billed Grasswren *Amytornis modestus* (Aves-Maluridae). *Transactions of the Royal Society of South Australia* **135**, 26–38.

Black AB, Joseph L, Pedler LP, Carpenter GA (2010) A taxonomic framework for interpreting evolution within the *Amytornis textilis–modestus* complex of grasswrens. *Emu* **110**, 358–363.

Crisp MD (1978) Demography and survival under grazing of three Australian semi-desert shrubs. *Oikos* **30**, 520–528.

McAllan IAW (1987) Early records of the Thick-billed Grasswren *Amytornis textilis* and Striated Grasswren *Amytornis striatus* in New South Wales. *Australian Birds* **28**, 65–70.

Pedler LP, Watson M, Langdon P, Pedler R (2007) 'Rare bird surveys, Mt Lyndhurst Station March 2007'. Report prepared for SA Arid Lands Natural Resources Management Board, April 2007.

Rowley I, Russell E (1997) *Fairy-wrens and Grasswrens: Maluridae, 102*. Oxford University Press, Oxford.

Schodde R (1982) *The Fairy-Wrens. A Monograph of the Maluridae*. Lansdowne Editions, Melbourne.

Tiver F, Andrew MH (1997) Relative effects of herbivory by sheep, rabbits, goats and kangaroos on recruitment and regeneration of shrubs and trees in eastern South Australia. *Journal of Applied Ecology* **34**, 903–914.

Comments received from

Andrew Black, Lynn Pedler

Subspecies Conservation Summary

Thick-billed Grasswren (Lake Frome Basin)

Amytornis modestus curnamona Black 2011
Maluridae

Conservation status
Near Threatened D1

Reasons for listing
Population size about 2500 mature individuals

Status 2000
Near Threatened D1

Status 1990
Near Threatened D1

Taxonomy
A. m. modestus (MacDonnell Ranges) and *A. m. inexpectatus* (central New South Wales) are Extinct, *A. m. obscurior* (NW NSW) is Critically Endangered, *A. m. raglessi* (Flinders Ranges) is Vulnerable, *A. m. indulkanna* (western) is Least Concern, as is the species
Taxonomic uniqueness: medium (5 genera/family, 11 species/genus, 6 subspecies/species)

Range
Recently described as occurring in the south-western Lake Frome Basin (Black *et al.* 2010; Black 2011). Known from specimens from 3 locations and sightings from another 5 locations from Wirrealpa Station in the north to Bimbowrie Conservation Park in the east (Black and Baxter 2003; Bellchambers and Baker-Gabb 2006; Black *et al.* 2010), including Wirrealpa, Martin's Well, Erudina, Baratta, Curnamona, Koonamore, Mount Victor, Plumbago and Bimbowrie Stations (Black *in litt.*). No evidence of a wider historical distribution, despite absence from suitable habitat (Black *in litt.*). It is estimated here as having 5–10 subpopulations and locations, but is poorly known.

Abundance
Population size estimated at 2000–3000 mature individuals (A. Black *in litt.*). No evidence of any decline at Curnamona Station, where it was first discovered in 1975, but is poorly known (Reid 1976; Black and Baxter 2003).

Ecology
Inhabits low chenopod shrublands, with 92% recent records in habitats dominated by Blackbush *Maireana pyramidata* (Black and Baxter 2003; A. Black *in litt.*). Other Thick-billed Grasswren subspecies forage on the ground for berries, seeds and insects (Rowley and Russell 1997). They lay 2–3 eggs in domed or cup nests built in shrubs (Rowley and Russell 1997). A generation time of 9.7 years (BirdLife International 2011) is derived from an age at first breeding of 2.3 years and a maximum longevity of 17.0 years, both values extrapolated from fairy-wrens *Malurus* spp.

Threats
The greatest threat is overgrazing by sheep, cattle, feral goats *Hircus capra* and rabbits *Oryctolagus cuniculus* in combination with drought is thought to be the main threat, and is already thought to have caused the extinction of *A. m. inexpectatus* (Schodde 1982; McAllan 1987). Although Blackbush and similar shrubs are resilient and long-lived, recruitment is infrequent and their seedlings are highly palatable to stock (Crisp 1978; Tiver and Andrew 1997). Predation by foxes *Vulpes vulpes* may be significant and intensive fox baiting may underlie recent increases in the subspecies on the north Olary Plains (Pedler *et al.* 2007).

Conservation objectives
1. Stable population size
2. Stable area of occupancy

Information required
1. Status and trends at historical and potential new locations
2. Optimal levels of grazing by livestock and feral mammals
3. Impacts of predation by foxes

Management actions required
1. Ensure grazing by stock is kept at low levels at a selection of locations until the relationship between grazing and grasswren abundance is better understood

Bibliography

Bellchambers K, Baker-Gabb D (2006) 'A survey of Plains-wanderers and Thick-billed Grasswrens in the north-east pastoral zone of South Australia'. South Australian Arid Lands Natural Resources Management Board, Adelaide.

BirdLife International (2011) 'Species factsheet: *Amytornis textilis*'. Retrieved 17 January, 2011 from <http://www.birdlife.org/>.

Black AB (2011) Subspecies of the Thick-billed Grasswren *Amytornis modestus* (Aves-Maluridae). *Transactions of the Royal Society of South Australia* **135**, 26–38.

Current eligibility against IUCN Red List Criteria

IUCN category	Criteria eligibility
A	Not applicable: past, current or future population declines are thought unlikely to exceed 20% in any 3-generation period
B	Not applicable: EOO, AOO, habitat, locations or population not declining, population fluctuations not extreme
C	Not applicable: no continuing population decline or extreme fluctuations
D	Near Threatened: population near 1000 mature individuals
E	Not applicable: no population viability analysis undertaken

IUCN Red List assessment data

	Estimate	Reliability
Extent of occurrence trend	6000 km^2 stable	medium medium
Area of occupancy trend	300 km^2 stable	medium medium
No. of mature individuals trend	2500 stable	medium medium
No. subpopulations	8	low
No. locations	8	low
Largest subpopulation	1000	low
Generation time	9.7 years	low
Global population share	100%	high

Black AB, Baxter CI (2003) Observations of Thick-billed Grasswrens on the North Olary Plains. *South Australian Ornithologist* **34**, 70–74.

Black AB, Joseph L, Pedler LP, Carpenter GA (2010) A taxonomic framework for interpreting evolution within the *Amytornis textilis–modestus* complex of grasswrens. *Emu* **110**, 358–363.

Crisp MD (1978) Demography and survival under grazing of three Australian semi-desert shrubs. *Oikos* **30**, 520–528.

McAllan IAW (1987) Early records of the Thick-billed Grasswren *Amytornis textilis* and Striated Grasswren *Amytornis striatus* in New South Wales. *Australian Birds* **28**, 65–70.

Pedler LP, Watson M, Langdon P, Pedler R (2007) 'Rare bird surveys, Mt Lyndhurst Station March 2007'. Report prepared for SA Arid Lands Natural Resources Management Board.

Reid N (1976) Bird Report 1975. *South Australian Ornithologist* **27**, 147–158.

Rowley I, Russell E (1997) *Fairy-wrens and Grasswrens: Maluridae, 102.* Oxford University Press, Oxford.

Schodde R (1982) *The Fairy-Wrens. A Monograph of the Maluridae.* Lansdowne Editions, Melbourne.

Tiver F, Andrew MH (1997) Relative effects of herbivory by sheep, rabbits, goats and kangaroos on recruitment and regeneration of shrubs and trees in eastern South Australia. *Journal of Applied Ecology* **34**, 903–914.

Comments received from

Andrew Black, Lynn Pedler

Subspecies Conservation Summary

Western Grasswren (Gawler Ranges)

Amytornis textilis myall (Mathews, 1916)

Maluridae

Conservation status

Vulnerable C2a(ii)

Reasons for listing

Population probably 2500–10 000 mature individuals in a single subpopulation that is declining as a result of habitat degradation

Status 2000

Least Concern

Reason for change in 2010: change in land use making population decline more likely

Status 1990

Least Concern

Taxonomy

A. t. textilis (Shark Bay) is Least Concern, as is the species

Current eligibility against IUCN Red List Criteria

IUCN category	Criteria eligibility
A	Not applicable: past, current or future population declines are thought unlikely to exceed 20% in any 3-generation period
B	Near Threatened: EOO <20 000 km², AOO <2000 km², near 10 locations, suspected continuing decline in AOO, habitat quality
C	Vulnerable: population 2500–10 000 mature individuals, suspected continuing decline, 100% in 1 subpopulation
D	Not applicable: population >1000 mature individuals, >5 locations
E	Not applicable: no population viability analysis undertaken

IUCN Red List assessment data

	Estimate	Reliability
Extent of occurrence trend	12 000 km^2 stable	high medium
Area of occupancy trend	600 km^2 decreasing	medium low
No. of mature individuals trend	8400 decreasing	medium low
No. subpopulations	1	medium
No. locations	>10	medium
Generation time	9.7 years	low
Global population share	100%	high

Taxonomic uniqueness: medium (5 genera/family, 11 species/genus, 2 subspecies/species)

Range

Known from many locations between Munyaroo Conservation Park south of Whyalla and the eastern Gawler Ranges, concentrated in the Myall Creek and Pine Creek drainages of the north-eastern Eyre Peninsula (Matthew and Carpenter 1993; Black *et al.* 2009; Black *et al.* 2010). In 2006, found at 76% of 62 located sites of previous records (Black *et al.* 2009). Historically possibly much more widespread, as indicated by specimens from the Yellabinna area of western South Australia in 1909 (Black 2004; Black *et al.* 2010). The AOO from the survey by Black *et al.* (2009) is only 190 km^2 but is likely to be about 5% of the EOO on average: as low as 2% of the landscape in parts of the Gawler Ranges but could possibly be >10% at its core in the Whyalla/Iron Knob corridor (A. Black *in litt.*).

Abundance

Densities are estimated at 8–20 mature individuals/km^2 within the AOO giving a population of 4800–12 000 (A. Black *in litt.*); here we opt for the mid-point of these, 8400. A continuing decline is suspected based on apparent extirpation from some historical sites (Black *et al.* 2009) and ongoing threats.

Ecology

Of 62 sites surveyed by Black *et al.* (2009), 64% were low shrublands, chiefly Blackbush *Maireana pyramidata* (40%) and Australian Boxthorn *Lycium australe* (21%), 28% low woodlands, mostly Western Myall *Acacia papyrocarpa* (19%). Most sites were along drainage lines but some were on low rocky hills and in semi-arid low woodlands. The density of shrub cover in occupied sites was higher than in unoccupied sites (Black *et al.* 2009). The grasswrens forage mostly on the ground, eating seeds, fruits, insects and other invertebrates (Higgins *et al.* 2001). A generation time of 9.7 years (BirdLife International 2011) is derived from an age at first breeding of 2.3 years and a maximum longevity of 17.0 years, both extrapolated from fairy-wrens *Malurus* spp.

Threats

Threatened by over-grazing, land use change and/or introduced predators. The species has been extirpated from several locations due to habitat destruction by stock, sand quarrying and recreational activities. Large areas of Blackbush are believed to have succumbed from recent extended drought conditions, independent of grazing activity (Black *et al.* 2009). The massive decline of *A. t. textilis* in the early 20th century across southern Western Australia and the extinction of the related Thick-billed Grasswren *A. m. inexpectatus* from most of NSW have been attributed to destruction of habitat by livestock and feral herbivores at times of severe drought and possibly introduced predators (Schodde 1982; McAllan 1987; Higgins *et al.* 2001). The reason for the disappearance of this subspecies from the Yellabinna region is unknown, perhaps predation by foxes *Vulpes vulpes* or feral cats *Felis catus* or overgrazing by rabbits *Oryctolagus cuniculus*, as the region has not been grazed by livestock (Australian Government 2009). Widespread fox baiting may underlie recent increases in other grasswrens in the Flinders Ranges and north Olary Plains (Pedler *et al.* 2007). About 20% of the population is on lands nominated for military purposes, including tank training which poses potential risks of direct physical habitat damage, as well as fires (Black *et al.* 2009). Given the likely primary threat is over-grazing, it is assessed here that the subspecies occurs at as few as 10 locations.

Conservation objectives

1. Stable population size
2. Stable area of occupancy

Information required

1. Status across historical and potential new locations
2. Optimal levels of grazing by livestock and feral mammals
3. Impact of predation by foxes and feral cats

Management actions required

1. Ensure grazing by stock is kept at low levels at a selection of locations until the relationship between grazing and grasswren abundance is better understood

2. Ensure that military authorities monitor and conserve populations on land they control

Bibliography

Australian Government (2009) 'Commonwealth listing advice on *Amytornis textilis myall* (Thick-billed Grasswren (Gawler Ranges))'. Retrieved 5 April, 2011 from <http://www.environment.gov.au/biodiversity/threatened/species/pubs/64454-listing-advice.pdf>.

BirdLife International (2011) 'Species factsheet: *Amytornis textilis*'. Retrieved 17 January, 2011 from <http://www.birdlife.org/>.

Black AB (2004) The 'Immarna' grasswrens of R. C. Chandler: locality, habitat, identity and taxonomic implications. *South Australian Ornithologist* **34**, 199–211.

Black AB, Carpenter GA, Pedler LP (2009) Distribution and habitats of the Thick-billed Grasswren *Amytornis textilis* subspecies *myall*. *South Australian Ornithologist* **35**, 161–177.

Black AB, Joseph L, Pedler LP, Carpenter GA (2010) A taxonomic framework for interpreting evolution within the *Amytornis textilis–modestus* complex of grasswrens. *Emu* **110**, 358–363.

Higgins PJ, Peter JM, Steele WK (Eds) (2001) *Handbook of Australian, New Zealand and Antarctic Birds. Volume 5: Tyrant-Flycatchers to Chats*. Oxford University Press, Melbourne.

Matthew J, Carpenter G (1993) Recent bird records from east of Lake Gairdner, Gawler Ranges, S. A. *South Australian Ornithologist* **31**, 127–131.

McAllan IAW (1987) Early records of the Thick-billed Grasswren *Amytornis textilis* and Striated Grasswren *Amytornis striatus* in New South Wales. *Australian Birds* **28**, 65–70.

Pedler LP, Watson M, Langdon P, Pedler R (2007) 'Rare bird surveys, Mt Lyndhurst Station March 2007'. Report prepared for SA Arid Lands Natural Resources Management Board.

Schodde R (1982) *The Fairy-Wrens. A Monograph of the Maluridae*. Lansdowne Editions, Melbourne.

Comments received from

Andrew Black, Graham Carpenter, Lynn Pedler

Species Conservation Summary

Black Grasswren

Amytornis housei (Milligan, 1902)
Maluridae

Conservation status

Near Threatened B1ab(i,ii,iii,iv,v)+2ab(i,ii,iii,iv,v)

Reasons for listing

Small but poorly-known EOO and AOO, possibly <10 locations and declining EOO, AOO and habitat quality likely as a result of excessive fire

Status 2000

Least Concern

Reason for change in 2010: both fire frequency, and rainfall which helps drive it, have been increasing

Status 1990

Least Concern

Taxonomy

No infraspecific taxa described

Taxonomic uniqueness: medium (5 genera/family, 11 species/genus, 1 subspecies/species)

Range

Endemic to the north-west Kimberley region, Western Australia from the Admiralty Gulf (Crystal Head) through Mitchell Plateau south to the Isdell River (S. Legge pers. comm.) and Manning Creek (Johnstone and Storr 2004). Recorded near Manning Creek 1968–1990s, from which they were possibly extirpated by fire (Hadden 2007). They are generally under-recorded in this inaccessible region, and area of occupancy, population size and trends poorly known. For want of better information the AOO has been nominally put at 10% of the EOO.

Abundance

Reported to be moderately common to common e.g. 16 pairs counted in a 2 km transect along a creek in sandstone near Mertens Falls (Johnstone and Storr 2004). Based on a density of 14/km^2 across the landscape (extrapolated from Western Grasswren *A. textilis myall*), a population of 15 000 mature individuals is postulated. A continuing decline is inferred from recent and likely ongoing fire regimes.

Ecology

Lives in areas where the ground cover is dense spinfex *Triodia* spp. in low open woodland, tall shrubland or tall

Current eligibility against IUCN Red List Criteria

IUCN category	Criteria eligibility
A	Not applicable: past, current or future population declines are thought unlikely to exceed 20% in any 3-generation period
B	Near Threatened: EOO <20 000 km^2, AOO near 2000 km^2, near 10 locations, inferred continuing decline in EOO, AOO, habitat quality, no. locations
C	Not applicable: population >10 000 mature individuals
D	Not applicable: population >1000 mature individuals, >5 locations
E	Not applicable: no population viability analysis undertaken

IUCN Red List assessment data

	Estimate	Reliability
Extent of occurrence trend	11 000 km^2 decreasing	medium low
Area of occupancy trend	1100 km^2 decreasing	low low
No. of mature individuals trend	15 000 decreasing	low medium
No. subpopulations	1	high
No. locations	10	low
Generation time	9.7 years	low
Global population share	100%	high

open shrubland in the vicinity of large sandstone boulders and crevices. Mainly found in dissected sandstone areas with massive boulders (Rowley and Russell 1997; Johnstone and Storr 2004). Feeds on insects, other invertebrates and seeds of grasses and sedges (Johnstone and Storr 2004; Hadden 2007). Builds a domed nest in the spinifex, in which 1–2 eggs laid (Johnstone and Kolichis 1999). A generation time of 9.7 years (BirdLife International 2011) is derived from an age at first breeding of 2.3 years and a maximum longevity of 17.0 years, both extrapolated from fairy-wrens *Malurus* spp.

Threats

Probably threatened by increased extent and frequency of fire, 1 of the main threatening processes to biodiversity in the region (Carwardine *et al.* 2011). Fire frequency may have increased partly because of an increase in rainfall over the last 100 years (Woinarski *et al.* 2009) that is accelerating recovery of the spinifex following fire and increasing fuel loads. There have also been changes in how people in the region use and manage fire. From 1990–1999, an average of 31% of the North Kimberley region was burned each year, mostly in hot dry-season burns (Fisher *et al.* 2003). However, parts of its range are naturally protected by the extremely rugged terrain that breaks up fire and protects the old spinifex. The ecologically similar Carpentarian Grasswren *A. dorotheae* may not re-occupy areas until at least 3 years after a fire (Harrington *et al.* 2009). Grasswrens are relatively poor dispersers and recolonisation is dependent on a nearby source population and hence dependent on the size, intensity and pattern of recent fires. Given the primary threat of large wildfires, the species is assessed here as having about 10 locations.

Conservation objectives

1. Stable numbers at monitored sites
2. Patchy fire across range

Information required

1. Population trends at nominated monitoring sites
2. Fire histories of occupied and unoccupied habitat patches

Management actions required

1. Initiate regular monitoring at a selection of known sites
2. Maintain and enhance habitat suitability through fire management

Bibliography

BirdLife International (2011) 'Species factsheet: *Amytornis housei*'. Retrieved 17 January, 2011 from <http://www.birdlife.org/>.

Carwardine J, O'Connor T, Legge SM, Mackey B, Possingham HP, Martin TG (2011) 'Priority threat management to protect Kimberley wildlife'. CSIRO, Canberra.

Fisher R, Vigilante T, Yates C, Russell-Smith J (2003) Patterns of landscape fire and predicted vegetation response in the North Kimberley region of Western Australia. *International Journal of Wildland Fire* **12**, 369–379.

Hadden D (2007) Black Grasswrens feast on termites. *Wingspan* **17**, 29.

Harrington G, Perry J, Forsyth R, Venables B (2009) A tale of two Grasswrens. *Wingspan* **19**, 23–25.

Johnstone RE, Kolichis N (1999) First description of the nest and eggs of the Black Grasswren *Amytornis housei* (Milligan) with notes on breeding. *Records of the Western Australian Museum* **19**, 259–265.

Johnstone RE, Storr GM (2004) *Handbook of Western Australian Birds. Volume 2: Passerines (Blue-winged Pitta to Goldfinch)*. Western Australian Museum, Perth.

Rowley I, Russell E (1997) *Fairy-wrens and Grasswrens: Maluridae, 102*. Oxford University Press, Oxford.

Woinarski J, Mackey B, Nix H, Traill B (2009) *The Nature of Northern Australia*. ANU E-Press, Canberra.

Comments received from

Allan Burbidge, Ron Johnstone, Sarah Legge

Species Conservation Summary

Eastern Bristlebird

Dasyornis brachypterus (Latham, 1802)
Dasyornithidae

Conservation status

Endangered B1ab(i,ii,iii,iv)+2ab(i,ii,iii,iv)

Reasons for listing

Extent of occurrence <5000 km^2 and occupies <500 km^2, occurs at 5 locations and there is a continuing decline in extent of occurrence, area occupied, habitat quality and no. locations and subpopulations, particularly for the northern subspecies, though reintroductions probably balance out the population losses

Status 2000

Endangered B1ab(i,ii,iii,iv,v)+2ab(i,ii,iii,iv,v)

Status 1990

Endangered B1ab(i,ii,iii,iv,v)+2ab(i,ii,iii,iv,v)

Taxonomy

D. b. brachypterus (eastern Victoria, southern and central coastal New South Wales) is Endangered, *D. b. monoides* (north-eastern New South Wales and south-eastern Queensland) is Critically Endangered. Subspecies are not supported by genetic analysis (Roberts *et al.* 2011), but are nevertheless phenotypically distinct (Schodde and Mason 1999).

Taxonomic uniqueness: high (1 genus/family, 3 species/genus, 2 subspecies/species)

Range

There are 3 main areas where the species occurs. *D. b. monoides* occurs on the Queensland/New South Wales border including 12 locations across the Lamington, Border Ranges and Gibraltar Ranges National Parks. *D. b. brachypterus* occurs on and near the central NSW coast, including Barren Grounds Nature Reserve and Budderoo National Park and the Bherwerre and Beecroft Peninsulas around Jervis Bay. *D. b. brachypterus* also occurs on the NSW/Victoria border including Nadgee and Croajingalong National Parks (Baker 1997; Bramwell 2008). Small numbers occur elsewhere, either remnants, natural colonists or the results of planned translocations. A continuing decline in EOO and AOO is likely given the localised extirpations in at least the north.

Abundance

No more than 50 *D. b. monoides* are likely to persist (Stewart 2006; D. Stewart pers. comm.; D. Rohweder pers. comm.). Subpopulations of *D. b. brachypterus* include about 1250 mature individuals at Barren Grounds, 1200 at Jervis Bay and 300 at Nadgee/Howe Flat. There are newly established subpopulations at Cataract of 50 individuals and about 10 at Red Rocks (Baker 1997, 1998, 2009). Population trends are assessed as stable, with increasing translocated subpopulations in the south balancing declines in the north.

Ecology

D. b. monoides mostly occurs in tall, dense, grassy groundcover in open *Eucalyptus* forests or woodlands, often interspersed with, or within 750 m of, mature subtropical rainforest. *D. b. brachypterus* primarily inhabits heathland, sedgeland and swamp communities, but also open sclerophyll woodland or forest with an understorey of shrubs, grass or Bracken *Pteridium esculentum* and occasionally in temperate rainforest (Department of Sustainability, Environment, Water, Population and Communities 2011). Population density increases until 10–20 years after fire (Baker 2001; Bain and McPhee 2005). Feeds near the ground, mainly on small invertebrates with some seeds and small fruits (Gibson and Baker 2004). Breeds annually in spring, laying 2 eggs (Higgins and Peter 2002). They breed readily in captivity and appear to have high survival rates after translocation or captive breeding and release. A generation time of 5.2 years (BirdLife International 2011) is derived from an average age at first breeding of 1.6 years, an annual survival of adults of 77.0%, and a maximum longevity in the wild of 7.3 years, all values elicited from an expert committee.

Threats

Extensive fire is currently the main threat (Baker 2000). Both subspecies have been substantially affected by fire in the last few decades. Fire destroys cover, but too infrequent fire can cause the vegetation to become too dense for nesting (Bain *et al.* 2008). However, bristlebirds can avoid fire by moving to unburned areas and returning later when conditions are more suitable (Bain *et al.* 2008). Predation by foxes *Vulpes vulpes* and feral cats *Felis catus* is also significant (Lindenmayer *et al.* 2009). Other possible threats include overgrazing, invasion of habitat by exotic plants and disturbance by humans (Department of Sustainability, Environment, Water, Population and Communities 2011). Historically the range has been greatly reduced by clearing for residential development, agriculture and forestry, and degraded by fire and grazing mammals.

Conservation objectives

1. An increasing population
2. Viable populations at more than 5 locations

Information required

1. Impacts of introduced predators, particularly in combination with fire
2. Reasons for the failure of the Nadgee subpopulation to recover

Management actions required

1. Establish and maintain a viable captive population of *D. b. monoides*
2. Prepare and implement fire management plans for all subpopulations
3. Develop contingency plans in case of fire including emergency evacuation, assisted habitat recovery and translocation

Current eligibility against IUCN Red List Criteria

IUCN category	Criteria eligibility
A	Not applicable: past, current or future population declines are thought unlikely to exceed 20% in any 3-generation period
B	Endangered: EOO <5000 km^2, AOO <500 km^2, at ≤5 locations with plausible future threat, inferred continuing decline in EOO, AOO, habitat quality, no. locations, no. subpopulations
C	Not applicable: no continuing population decline or extreme fluctuations
D	Not applicable: population >1000 mature individuals, >5 locations
E	Not applicable: no population viability analysis undertaken

IUCN Red List assessment data

	Estimate	Reliability
Extent of occurrence trend	1270 km^2 decreasing	medium medium
Area of occupancy trend	214 km^2 decreasing	medium medium
No. of mature individuals trend	2550 stable	high medium
No. subpopulations	7	low
No. locations	17	low
Largest subpopulation	1250	medium
Generation time	5.2 years	low
Global population share	100%	high

4. Control foxes and cats where vulnerable to predation, particularly after large-scale fire
5. Control weeds that are incompatible with bristlebirds
6. Undertake further translocations and or reintroductions

Bibliography

Bain D, McPhee N (2005) Resurveys of the Eastern Bristlebird *Dasyornis brachypterus* in central-eastern New South Wales 1999–2001: their relationship with fire and observer competence. *Corella* **29.**

Bain DW, Baker JR, French KO, Whelan RJ (2008) Post-fire recovery of eastern bristlebirds (*Dasyornis brachypterus*) is context-dependent. *Wildlife Research* **35**, 44–49.

Baker J (1997) The decline, response to fire, status and management of the Eastern Bristlebird. *Pacific Conservation Biology* **3**, 235–243.

Baker J (1998) 'Eastern Bristlebird Recovery Plan 1997–2002'. NSW NPWS, Sydney.

Baker J (2000) The Eastern Bristlebird: cover-dependent and fire-sensitive. *Emu* **100**, 286–298.

Baker J (2001) Population density and home range estimates for the Eastern Bristlebird at Jervis Bay, south-eastern Australia. *Corella* **25**, 62–67.

Baker J (2009) Assessment of Eastern Bristlebird habitat: refining understanding of appropriate habitats for reintroductions. *Ecological Management and Restoration* **10**, S136–S139.

BirdLife International (2011) 'Species factsheet: *Dasyornis brachypterus*'. Retrieved 17 January, 2011 from <http://www.birdlife.org/>.

Bramwell MD (2008) The Eastern Bristlebird *Dasyornis brachypterus* in East Gippsland, Victoria, 1997–2002. *Australian Field Ornithology* **25**, 2–11.

Department of Sustainability, Environment, Water, Population and Communities (2011) *Dasyornis brachypterus*. In 'Species Profile and Threats Database'. Department of Sustainability, Environment, Water, Population and Communities, Canberra. Retrieved 24 February, 2011 from <http://www.environment.gov.au/sprat>.

Gibson L, Baker J (2004) Diet of the Eastern Bristlebird *Dasyornis brachypterus* in New South Wales. *Corella* **28**, 79–81.

Higgins PJ, Peter JM (Eds) (2002) *Handbook of Australian, New Zealand and Antarctic Birds. Volume 6: Pardalotes to Spangled Drongo.* Oxford University Press, Melbourne.

Lindenmayer DB, MacGregor CA, Wood JT, Cunningham RB, Crane M, Michael D, Montague-Drake R, Brown D, Fortescue M, Dexter N, Hudson M, Gill M (2009) What factors influence rapid post-fire site re-occupancy? A case study of the endangered Eastern Bristlebird in eastern Australia. *International Journal of Wildland Fire* **18**, 84–95.

Roberts DG, Baker J, Perrin C (2011) Population genetic structure of the endangered Eastern Bristlebird, *Dasyornis brachypterus*; implications for conservation. *Conservation Genetics.*

Schodde R, Mason IJ (1999) The Directory of Australian Birds: Passerines. CSIRO Publishing, Melbourne.

Stewart DA (2006) 'Recovery plan for the northern population of the Eastern Bristlebird *Dasyornis brachypterus monoides* 2007–2011'. Report to the Department of the Environment and Heritage, Canberra. Queensland Parks and Wildlife Service, Brisbane.

Comments received from

Jack Baker, Tim Holmes, David Rohweder, David Stewart

Subspecies Conservation Summary

Eastern Bristlebird (southern)

Dasyornis brachypterus brachypterus (Latham, 1802)
Dasyornithidae

Conservation status

Endangered B1ab(ii)+2ab(ii)

Reasons for listing

Extent of occurrence <5000 km^2 and occupies <500 km^2, occurs at 5 locations and there is a continuing decline in AOO

Status 2000

Endangered B1ab(ii)+2ab(ii)

Status 1990

Endangered B1ab(ii)+2ab(ii)

Taxonomy

D. b. monoides (north-eastern New South Wales and south-eastern Queensland) is Critically Endangered, and the species is Endangered
Taxonomic uniqueness: high (1 genus/family, 3 species/genus, 2 subspecies/species)

Range

Near-coastal distribution in New South Wales and east Gippsland, Victoria, historically from Myall Lakes to Gippsland Lakes (Schodde and Mason 1999). In central NSW, subpopulations persist around Barren Grounds Nature Reserve including Budderoo National Park, and Jervis Bay including Bherwerre and Beecroft Peninsulas, and remnants and/or recent colonists in some additional locations including Red Rock Nature Reserve (Baker 1997, 1998). Near NSW/Vic border, remains at Nadgee National Park and Howe Flat, Croajingalong National Park with small numbers elsewhere in this contiguous habitat (Baker 1997; Bramwell 2008). In 2003–2005, successfully reintroduced to Beecroft Peninsula (Bain and French 2009). In 2008, released at Cataract on the Woronora Plateau, where breeding (Baker 2009). However, AOO probably continuing to decline with relatively recent contraction in range in Vic.

Abundance

The total population is estimated as 2500 mature individuals. Potentially viable subpopulations at Barren Grounds (maximum c.1250 individuals), Jervis Bay (c.1100 individuals) and Nadgee/Howe Flat (300), newly established at Cataract (50) and very few individuals elsewhere including c.10 at Red Rocks (Bain and McPhee 2005; Bramwell 2008; J. Baker *in litt.*). Genetic analysis indicates 3 closed subpopulations (Roberts *et al.* 2011). The number of individuals is assessed here as stable, as increases in translocated subpopulations compensate for likely losses to fire.

Ecology

Inhabits dense, low vegetation, in a variety of habitat types. Often found in heath, including a variety of heathland, sedgeland and swamp communities, but also in open sclerophyll woodland or forest with an understorey of shrubs and a dense ground layer of grasses or Bracken *Pteridium esculentum*. Occasionally in temperate rainforest (Department of Sustainability, Environment, Water, Population and Communities 2011). Eliminated by intense fire but can recolonise recently burnt areas if there are adequate nearby unburned refuges (Bain *et al.* 2008). Population density increases until 10–20 years after fire, when 3–5 birds/10 ha, but individuals have overlapping home territories of c.10 ha (Baker 2001; Bain and McPhee 2005). Most remaining subpopulations are probably isolated as they are poor at long distance dispersal, however records 15 km west of Barren Grounds at Fitzroy Falls and Upper Kangaroo Valley indicate some dispersal through suitable corridor habitat. Feeds mainly on small invertebrates, also seeds and small fruits (Gibson and Baker 2004). Breeds annually in spring, laying 2 eggs, but rarely raises more than 1 fledgling (Higgins and Peter 2002). A generation time of 5.2 years (BirdLife International 2011) is derived from an average age at first breeding of 1.6 years, an annual survival of adults of 77.0%, and a maximum longevity in the wild of 7.3 years, all values elicited from an expert committee.

Threats

Extensive fire is the main current threat (Baker 2000). The species was lost from 10 of 11 known locations in Vic. from fire during 1978–1994 (Clarke and Bramwell 1998). In NSW, fires almost eliminated it from Barren Grounds in 1968 and from Nadgee in 1980 (Baker 1998). Single fires could be catastrophic across Howe Flat and Nadgee or Barren Grounds and Budderoo (J. Baker *in litt.*). Fire destroys cover, but too infrequent fire can cause the vegetation to become too dense for nesting (Bain *et al.* 2008). However, birds can avoid fire by moving to unburnt areas and returning later when conditions are more suitable, so patches of unburnt habitat and other refuges are of great importance (Bain *et al.* 2008). However, unlike the populations further north, the Nadgee population has not recovered from extensive fires in 1972 and 1980 (J. Baker *in*

Current eligibility against IUCN Red List Criteria

IUCN category	Criteria eligibility
A	Not applicable: past, current or future population declines are thought unlikely to exceed 20% in any 3-generation period
B	Endangered: EOO <5000 km^2, AOO <500 km^2, at ≤5 locations with plausible future threat, inferred continuing decline in AOO
C	Not applicable: no continuing population decline or extreme fluctuations
D	Vulnerable: ≤5 locations with plausible future threat
E	Not applicable: no population viability analysis undertaken

IUCN Red List assessment data

	Estimate	Reliability
Extent of occurrence trend	670 km^2 stable	high medium
Area of occupancy trend	205 km^2 decreasing	medium low
No. of mature individuals trend	2500 stable	high medium
No. subpopulations	3	high
No. locations	5	low
Largest subpopulation	1250	high
Generation time	5.2 years	low
Global population share	100%	medium

litt.) and the range has contracted substantially in Victoria. Predation by foxes *Vulpes vulpes* and feral cats *Felis catus* may be significant (Lindenmayer *et al.* 2009). Other possible threats include overgrazing, infestations of habitat by exotic plants and disturbance by humans (Department of Sustainability, Environment, Water, Population and Communities 2011). Historically the range has been greatly reduced by clearing and degraded by fire.

Conservation objectives

1. An increasing population
2. Viable populations at more than 5 locations

Information required

1. Impacts of introduced predators, particularly in combination with fire
2. Reasons for the failure of the Nadgee subpopulation to recover

Management actions required

1. Prepare and implement fire management plans for all subpopulations
2. Develop contingency plans in case of fire including emergency evacuation, assisted habitat recovery and translocation
3. Control foxes and cats where vulnerable to predation, particularly after large-scale fire
4. Control weeds that exclude bristlebirds
5. Investigate the need for further translocations

Bibliography

Bain D, French K (2009) Impacts on a threatened bird population of removals for translocation. *Wildlife Research* **36**, 516–521.

Bain D, McPhee N (2005) Resurveys of the Eastern Bristlebird *Dasyornis brachypterus* in central-eastern New South Wales 1999–2001: their relationship with fire and observer competence. *Corella* **29**.

Bain DW, Baker JR, French KO, Whelan RJ (2008) Post-fire recovery of eastern bristlebirds (*Dasyornis brachypterus*) is context-dependent. *Wildlife Research* **35**, 44–49.

Baker J (1997) The decline, response to fire, status and management of the Eastern Bristlebird. *Pacific Conservation Biology* **3**, 235–243.

Baker J (1998) 'Eastern Bristlebird Recovery Plan 1997–2002'. NSW NPWS, Sydney.

Baker J (2000) The Eastern Bristlebird: cover-dependent and fire-sensitive. *Emu* **100**, 286–298.

Baker J (2001) Population density and home range estimates for the Eastern Bristlebird at Jervis Bay, south-eastern Australia. *Corella* **25**, 62–67.

Baker J (2009) Assessment of Eastern Bristlebird habitat: refining understanding of appropriate habitats for reintroductions. *Ecological Management and Restoration* **10**, S136–S139.

BirdLife International (2011) 'Species factsheet: *Dasyornis brachypterus*'. Retrieved 17 January, 2011 from <http://www.birdlife.org/>.

Bramwell MD (2008) The Eastern Bristlebird *Dasyornis brachypterus* in East Gippsland, Victoria, 1997–2002. *Australian Field Ornithology* **25**, 2–11.

Clarke R, Bramwell MD (1998) The Eastern Bristlebird *Dasyornis brachypterus* in East Gippsland, Victoria. *Australian Bird Watcher* **17**, 245–253.

Department of Sustainability, Environment, Water, Population and Communities (2011) *Dasyornis brachypterus*. In 'Species Profile and Threats Database'. Department of Sustainability, Environment, Water, Population and Communities, Canberra. Retrieved 24 February, 2011 from <http://www.environment.gov.au/sprat>.

Gibson L, Baker J (2004) Diet of the Eastern Bristlebird *Dasyornis brachypterus* in New South Wales. *Corella* **28**, 79–81.

Higgins PJ, Peter JM (Eds) (2002) *Handbook of Australian, New Zealand and Antarctic Birds. Volume 6: Pardalotes to Spangled Drongo*. Oxford University Press, Melbourne.

Lindenmayer DB, MacGregor CA, Wood JT, Cunningham RB, Crane M, Michael D, Montague-Drake R, Brown D, Fortescue M, Dexter N, Hudson M, Gill M (2009) What factors influence rapid post-fire site re-occupancy? A case study of the endangered Eastern Bristlebird in eastern Australia. *International Journal of Wildland Fire* **18**, 84–95.

Roberts DG, Baker J, Perrin C (2011) Population genetic structure of the endangered Eastern Bristlebird, *Dasyornis brachypterus*; implications for conservation. *Conservation Genetics*.

Schodde R, Mason IJ (1999) The Directory of Australian Birds: Passerines. CSIRO Publishing, Melbourne.

Comments received from

Jack Baker, Allan Burbidge, Andrew Burbidge, Les Christidis, Hugh Ford, Tim Holmes, Richard Loyn, Peter Menkhorst, Charles Meredith, Damon Oliver, Penny Olsen, David Watson

Subspecies Conservation Summary

Eastern Bristlebird (northern)

Dasyornis brachypterus monoides Schodde & Mason, 1999

Dasyornithidae

Conservation status

Critically Endangered B2ab(ii,iii,iv,v), C2a(i), D

Reasons for listing

Very small fragmented population with declining numbers and a contracting range

Status 2000

Critically Endangered B2ab(ii,iii,iv,v), C2a(i)

Status 1990

Critically Endangered B2ab(ii,iii,iv,v), C2a(i)

Taxonomy

D. b. brachypterus (Sydney to east Victoria) is Endangered, as is the species

Taxonomic uniqueness: high (1 genus/family, 3 species/genus, 2 subspecies/species)

Range

Endemic to south-east Queensland and north-east New South Wales. Currently occurs at Conondale Range and in and around Lamington and Border Ranges National Parks, and possibly Mt Barney. A record from Gibraltar Range National Park needs confirmation (D. Rohweder pers. comm., D. Stewart pers. comm.). Historically it has been recorded from isolated upland areas between Conondale National Park, Qld, and Dorrigo Plateau, NSW, up to 1160 m elevation (Department of Sustainability Environment Water Population and Communities 2011). It is presumed locally extinct at Spicer's Gap, Stretcher Track, Razorback Mountains, Mt Burrell, Big Scrub, Mt Richmond and Dorrigo Plateau (D. Stewart pers. comm.).

Abundance

There are no more than 50 mature individuals. In Qld the most complete survey in 2006 found birds at 12 locations, mostly with 3 or fewer pairs, the largest group being 13 individuals (Stewart 2006). There have been no complete surveys subsequently, but birds were heard in the Conondale Ranges in 2010 (D. Stewart pers. comm.). Surveys in NSW up to 2010 have found birds at 10–15 sites with a total population of 25–30 individuals; detection at a site varies from year to year (D. Rohweder pers. comm.). Their numbers have declined rapidly: 154 individuals were located in 103 territories in 1988, 36 birds in 30 territories in 1996 and 26 birds in 16 territories in 1997–1998 (Department of Sustainability, Environment, Water, Population and Communities 2011). None of the tiny subpopulations will be viable in the long term. Two siblings were captured in 2004 and are breeding successfully at David Fleay Wildlife Park in Qld (Department of Sustainability, Environment, Water, Population and Communities 2011). One female produced 5 fertile clutches in a season (R. Booth *in litt.*) but the breeding program has been scaled back to about 2 clutches/year and has not been supplemented from the wild (D. Stewart pers. comm.). Eight birds have been released in 2008: 4 at Spicers Gap, Qld, where some persisted until at least 2009, but have not been heard since (D. Stewart pers. comm.), 4 at Grady's Creek, Border Ranges, where birds, possibly a banded bird, were still present when last checked in 2010 (D. Rohweder pers. comm.).

Ecology

Mostly occurs in tall, dense, grassy ground-cover in open *Eucalyptus* forests or woodlands, often interspersed with, or within 750 m of, mature subtropical rainforest (Holmes 1989; Rohweder 2000). Typical ground cover includes tussock-grasses such as *Sorghum leiocladum* (Higgins and Peter 2002). At Mt Barney it inhabited montane heath and on the Dorrigo Plateau sedges and mat-rushes along watercourses (Holmes 1989). Probably requires a mosaic of habitat types and fire histories. Eliminated by intense fire but can recolonise recently burnt areas if there are adequate nearby unburnt refuges (Bain *et al.* 2008). Population density increases until 10–20 years after fire (Baker 2001; Bain and McPhee 2005). Feeds near the ground, mainly on small invertebrates with some seeds and small fruits (Gibson and Baker 2004). Breeds annually in spring, laying 2 eggs, but rarely raises more than 1 fledgling (Higgins and Peter 2002). A generation time of 5.2 years (BirdLife International 2011) is derived from an average age at first breeding of 1.6 years, an annual survival of adults of 77.0%, and a maximum longevity in the wild of 7.3 years, all values elicited from an expert committee.

Threats

The main current threat is an inappropriate fire regime. Usually found in habitat >5 years since burning; if fires are too frequent they eliminate tussocks and enable the invasion by introduced woody weeds. However, if fires are too infrequent the vegetation becomes too dense for nesting. Recolonisation of recently burnt areas is hin-

Current eligibility against IUCN Red List Criteria

IUCN category	Criteria eligibility
A	Not applicable: past, current or future population declines are thought unlikely to exceed 20% in any 3-generation period
B	Critically Endangered: AOO <10 km², population severely fragmented, inferred continuing decline in AOO, extent and/or quality of habitat, no. locations, no. individuals
C	Critically Endangered: population <250 mature individuals, suspected continuing decline, <50 mature individuals in largest subpopulation
D	Critically Endangered: population <50 mature individuals
E	Not applicable: no population viability analysis undertaken

IUCN Red List assessment data

	Estimate	Reliability
Extent of occurrence trend	600 km² decreasing	medium high
Area of occupancy trend	9 km² decreasing	medium high
No. of mature individuals trend	50 decreasing	medium high
No. subpopulations	4	low
No. locations	12	medium
Largest subpopulation	13	medium
Generation time	5.2 years	low
Global population share	100%	high

dered if the population is too fragmented and refuges are too distant (Bain *et al.* 2008). However, birds can avoid fire by moving to unburnt areas and returning later when conditions are more suitable, if patches of unburnt habitat remain (Bain *et al.* 2008). Habitat is also degraded by feral pigs *Sus scrofa* and domestic livestock. Several studies suggest that predation by foxes *Vulpes vulpes* and feral cats *Felis catus* may be significant, especially where habitat has been burnt or trampled by livestock (Lindenmayer *et al.* 2009; Department of Sustainability, Environment, Water, Population and Communities 2011). Other possible threats include over-grazing, infestation of habitat by exotic plants and disturbance by humans. Natural nest predation, such as by Pied Currawong *Strepera graculina*, may be important for very small populations (Department of Sustainability Environment Water Population and Communities 2011). Bristlebirds occur in tiny fragmented locations and are now so rare that there is a risk of inbreeding depression and stochastic events causing extinction. Given the main threat of fire, they probably occur at no more than 10 locations, but many of these may not be viable.

Conservation objectives

1. At least 1 viable subpopulation

Information required

1. Fire management strategy to maintain suitable grassy habitat
2. Feasibility of captive breeding and translocations, including consolidation of fragmented subpopulations
3. Impacts of introduced predators

Management actions required

1. Survey for undiscovered populations
2. Establish and maintain a viable captive population
3. Undertake releases to recently occupied sites in combination with fox and cat baiting
4. Secure conservation agreements for occupied habitat on private land
5. Prepare and implement fire management plans for all locations
6. Exclude stock by fencing known habitat
7. Control foxes and cats where bristlebirds are vulnerable to predation, particularly after large-scale fire

Bibliography

Bain D, McPhee N (2005) Resurveys of the Eastern Bristlebird *Dasyornis brachypterus* in central-eastern New South Wales 1999–2001: their relationship with fire and observer competence. *Corella* **29**.

Bain DW, Baker JR, French KO, Whelan RJ (2008) Post-fire recovery of eastern bristlebirds (*Dasyornis brachypterus*) is context-dependent. *Wildlife Research* **35**, 44–49.

Baker J (2001) Population density and home range estimates for the Eastern Bristlebird at Jervis Bay, south-eastern Australia. *Corella* **25**, 62–67.

BirdLife International (2011) 'Species factsheet: *Dasyornis brachypterus*'. Retrieved 17 January, 2011 from <http://www.birdlife.org/>.

Department of Sustainability, Environment, Water, Population and Communities (2011) *Dasyornis brachypterus*. In 'Species Profile and Threats Database'. Department of Sustainability, Environment, Water, Population and Communities, Canberra. Retrieved 24 February, 2011 from <http://www.environment.gov.au/sprat>.

Gibson L, Baker J (2004) Diet of the Eastern Bristlebird *Dasyornis brachypterus* in New South Wales. *Corella* **28**, 79–81.

Higgins PJ, Peter JM (Eds) (2002) *Handbook of Australian, New Zealand and Antarctic Birds. Volume 6: Pardalotes to Spangled Drongo*. Oxford University Press, Melbourne.

Holmes G (1989) 'Eastern Bristlebird: species management plan for northern populations'. Queensland Parks and Wildlife Service & NSW National Parks and Wildlife Service, Brisbane & Sydney.

Lindenmayer DB, MacGregor CA, Wood JT, Cunningham RB, Crane M, Michael D, Montague-Drake R, Brown D, Fortescue M, Dexter N, Hudson M, Gill M (2009) What factors influence rapid post-fire site re-occupancy? A case study of the endangered Eastern Bristlebird in eastern Australia. *International Journal of Wildland Fire* **18**, 84–95.

Rohweder D (2000) 'Assessment of Eastern Bristlebird (*Dasyornis brachypterus*) habitat, northern New South Wales: vegetation structure and floristics'. Unpublished report to New South Wales National Parks and Wildlife Service, Kyogle.

Stewart DA (2006) 'Recovery plan for the northern population of the Eastern Bristlebird *Dasyornis brachypterus monoides* 2007–2011'. Report to the Department of the Environment and Heritage, Canberra. Queensland Parks & Wildlife Service, Brisbane.

Comments received from
Jack Baker, Tim Holmes, David Rohweder, David Stewart

Species Conservation Summary

Western Bristlebird

Dasyornis longirostris Gould, 1841
Dasyornithidae

Conservation status
Endangered B1ab(iii)

Reasons for listing
The very small population is restricted to a very small extent of occurrence at very few locations. The recent series of large wildfires suggests a probable continuing decline in habitat quality

Status 2000
Vulnerable D2
Reason for change in 2010: multiple large fires suggest a continuing decline in habitat quality

Status 1990
Vulnerable D2

Taxonomy
No infraspecific taxa described
Taxonomic uniqueness: high (1 genus/family, 3 species/genus, 1 subspecies/species)

Range
Currently occurs east of Albany at Two Peoples Bay Nature Reserve, Betty's Beach, Mt Manypeaks to Bluff Creek, and at 14 different sites in and near Fitzgerald River National Park (Gilfillan *et al.* 2009). Historic records of the Western Bristlebird suggest that it occurred in coastal areas from Perth to Augusta and from Albany to Fitzgerald River National Park, Western Australia (Cale and Burbidge 1993). Eighteen birds were translocated in 1999–2000 and 2007 from Two People's Bay to near Walpole, west of Albany, but there was no evidence of breeding (Burbidge 2003; Gilfillan *et al.* 2009; Burbidge *et al.* 2010).

Abundance
A population of c.620 pairs in 2001 was reduced by fires to c.320 pairs in 2005 (Burbidge *et al.* 2010); here estimated as 1000 mature birds. The density of birds is greater in the Manypeaks-Waychinicup areas than in the Fitzgerald River National Park, but reasons for this are unknown. The Albany to Mt Manypeaks area population declined from c.500 pairs in 2001 to 200–315 pairs in 2005 and 2006, largely as a result of wildfires, although the cause for the decline in some areas is unclear (Comer and McNee 2001; Tiller *et al.* 2006). The Fitzgerald River National Park subpopulation numbered c.125 pairs in 2005 (Burbidge *et al.* 2010). At the time of assessment, the total population was considered not to be declining over a timescale of 3 generations (16 years), but an increasing risk of wildfires was considered to be probably causing a continuing decline in habitat quality.

Ecology
At Two Peoples Bay Nature Reserve lives in dense closed heath 1–1.5 m high. Near Waychinicup R. and in the Fitzgerald R. National Park, the main habitat is closed heath 0.5–1 m high, sometimes with scattered patches of mallee eucalypts, though more open heaths may be used if there are enough patches of dense shrubs in the area (McNee 1986). Territory size is estimated to be 7 ha, and territories may remain unchanged for at least 30 years (Smith 1987; A. H. Burbidge *in litt.*). In Fitzgerald River National Park, subpopulations appear to have survived fire where patches of habitat remained unburnt. After fires, unburnt swampy vegetation, predominantly sedges and thickets, may be important as refugia, but severe or large scale fires can destroy all suitable habitat (Smith 1977; McNee 1986; Smith 1987). At Two People's Bay Nature Reserve, moist heaths were reoccupied 2–3 years after fire (Burbidge 2003). Heaths in drier areas may not be reoccupied until 11–14 years after fire (Smith 1987). The species was found in heaths 5–12 years after fire between Boulder Hill and east of Waychinicup R., and

Current eligibility against IUCN Red List Criteria

IUCN category	Criteria eligibility
A	Not applicable: past, current or future population declines are thought unlikely to exceed 20% in any 3-generation period
B	Endangered: EOO <5000 km^2, at ≤5 locations with plausible future threat, inferred continuing decline in habitat quality
C	Vulnerable: population 2500–10 000 mature individuals, inferred continuing decline, 250–1000 mature individuals in largest subpopulation
D	Vulnerable: population 250–1000 mature individuals, ≤5 locations with plausible future threat
E	Not applicable: no population viability analysis undertaken

IUCN Red List assessment data

	Estimate	Reliability
Extent of occurrence trend	5000 km^2 stable	medium high
Area of occupancy trend	700 km^2 decreasing	medium medium
No. of mature individuals trend	1000 decreasing	medium medium
No. subpopulations	2	high
No. locations	5	medium
Largest subpopulation	660	medium
Generation time	5.2 years	medium
Global population share	100%	high

14–28 years after fire in northern Fitzgerald River National Park (McNee 1986). A generation time of 5.2 years (BirdLife International 2011) is derived from an age at first breeding of 1.5 years, a maximum longevity of 7.3 years and annual adult longevity of 77%, all values extrapolated from expert estimates for Eastern Bristlebird *D. brachypterus*.

Threats

Fire is the main current threat, with fires at less than 5–10 year intervals leading to local extinction (Smith 1987). At the other end of the scale, coastal heath (at least at Two Peoples Bay) remains suitable habitat for at least 60 years after fire, though carrying capacity may be reduced with time (Smith 1987; A. H. Burbidge *in litt.*). The incidence and extent of wildfires have been increasing in recent years, despite increased skills, capacity and effort to stop them. Given that fire is the main threat, locations are defined as being independent during massive wildfires; here estimated to be <5. Predation by introduced mammals, particularly foxes *Vulpes vulpes* and feral cats *Felis catus*, may be significant (Gilfillan *et al.* 2009; Burbidge *et al.* 2010). Other potential threats include dieback caused by *Phytophthora cinnamomi*, disturbance by introduced animals (particularly hard-hoofed animals), weed invasion and changes in hydrological regimes (Gilfillan *et al.* 2009). Clearance for grazing and agriculture is considered responsible for contraction of its distribution, but is no longer a threat as almost all bristlebirds now occur in conservation reserves (Department of Sustainability, Environment, Water, Population and Communities 2011).

Conservation objectives

1. Population numbers and area of occupation increased or maintained
2. Knowledge of current distribution improved
3. Management options increased through improved ecological knowledge

Information required

1. Size of all known subpopulations
2. Habitat requirements, in particular in relation to fire age, vegetation structure and food availability
3. Relationship between *Phytophthora cinnamomi* and habitat requirements
4. Impacts of cat and fox predation

Management actions required

1. Active fire management around all bristlebird locations
2. Continue habitat management and threat abatement of all occupied areas within an adaptive management framework
3. Continue translocation program
4. Survey potential habitat for new subpopulations
5. Continue to investigate habitat requirements, in particular in relation to fire age, vegetation structure, food availability, *Phytophthora cinnamomi* and fox and cat density
6. Continue to support coordination of management by the South Coast Threatened Bird Recovery Team

Bibliography

BirdLife International (2011) 'Species factsheet: *Dasyornis longirostris*'. Retrieved 17 January, 2011 from <http://www.birdlife.org/>.

Burbidge AH (2003) Birds and fire in the Mediterranean climate of southwest Western Australia. In *Fire in Ecosystems of Southwest Western Australia: Impacts and Management.* (Eds I Abbott, N Burrows) pp. 321–348. Backhuys Publishers, Leiden.

Burbidge AH, Comer S, Danks A, Berryman A, Hamilton N (2010) Attempted reintroduction of the Western Bristlebird in south-western Australia. In *Global Re-Introduction Perspectives: Additional Case-studies from Around the Globe.* (Ed. PS Soorae) pp. 171–175. IUCN/SSC Re-Introduction Specialist Group, Abu Dhabi, UAE.

Cale PG, Burbidge AH (1993) 'Research plan for the Western Ground Parrot, Western Whipbird and Western Bristlebird'. Report to Australian National Parks and Wildlife Service.

Comer S, McNee S (2001) 'Surveys for the Westen Bristlebird and Western Whipbird. Albany 2001'. Unpublished report to the South Coast Threatened Bird Recovery Team.

Department of Sustainability, Environment, Water, Population and Communities (2011) *Dasyornis longirostris.* In 'Species Profile and Threats Database'. Department of Sustainability, Environment, Water, Population and Communities, Canberra. Retrieved 24 February, 2011 from <http://www.environment.gov.au/sprat>.

Gilfillan S, Comer S, Burbidge AH, Blyth J, Danks A, Newell J (2009) 'South Coast Threatened Birds Recovery Plan. Western Australian Wildlife Management Program No. 44'. WA Department of Environment and Conservation, Perth.

McNee S (1986) 'Surveys of the Western Whipbird and Western Bristlebird in Western Australia, 1985'. RAOU Report 18.

Smith GT (1977) The effect of environmental change on six rare birds. *Emu* **77**, 173–179.

Smith GT (1987) Observations on the biology of the Western Bristle-bird *Dasyornis longirostris*. *Emu* **87**, 111–118.

Tiller C, Comer S, Danks A (2006) 'Noisy Scrub-bird Recovery Program and Project Phoenix Annual Report 2006" Unpublished report to the South Coast Threatened Birds Recovery Team.

Comments received from

John Blyth, Allan Burbidge, Sarah Comer

Subspecies Conservation Summary

Rufous Bristlebird (western)

Dasyornis broadbenti litoralis (Milligan, 1902)
Dasyornithidae

Conservation status

Extinct

Reasons for listing

Last recorded in 1906

Status 2000

Extinct

Status 1990

Extinct

Taxonomy

D. b. broadbenti (Coorong, South Australia) and *D. b. caryochrous* (Otways, Victoria) are Least Concern, as is the species

Taxonomic uniqueness: high (1 genus/family, 3 species/genus, 3 subspecies/species)

Former range

Known only from the coast between Cape Mentelle and Cape Naturaliste at the south-western tip of Western Australia (Milligan 1901), where last recorded reliably in 1906 (Smith 1977). A search in 1922 between Cape Leeuwin and Cape Naturaliste failed to find any birds (Whitley 1971), there were unconfirmed reports in 1940 (Serventy and Whittell 1967) and more recently in 1977 (Garnett 1992), but subsequent attempts to find it have failed (Blakers *et al.* 1984; Garnett 1992).

Former abundance

Reported to have been 'moderately common' (Storr 1991).

Ecology

It was reported to have inhabited coastal thickets about 40 cm high (Milligan 1901). Western Whipbird *Psophodes nigrogularis* and Noisy Scrub-bird *Atrichornis clamosus*, both of which are highly fire-sensitive, were recorded in the same area, but have since become regionally Extinct.

Reasons for extinction

Almost certainly fire. By 1916, many parts of the subspecies' former habitat had become 'bare sand drifts' as a result of frequent burning to improve the land for cattle grazing (Carter 1924).

Bibliography

Blakers MS, Davies JJF, Reilly PN (1984) *The Atlas of Australian Birds*. Melbourne University Press, Melbourne.

Carter T (1924) Birds of the Broome Hill district. *Emu* **23**, 306–318.

Garnett ST (1992) 'Threatened and extinct birds of Australia'. RAOU Report 82. RAOU, Melbourne and ANPWS, Canberra.

Milligan AW (1901) Description of a new bristlebird (*Sphenura*). *Emu* **1**, 67–69.

Schodde R, Mason IJ (1999) *The Directory of Australian Birds: Passerines*. CSIRO Publishing, Melbourne.

Serventy DL, Whittell HM (1967) *Birds of Western Australia*. University of Western Australia Press, Perth.

Smith GT (1977) The effect of environmental change on six rare birds. *Emu* **77**, 173–179.

Storr GM (1991) Birds of the South-west Division of Western Australia. *Records of the Western Australian Museum* **Supplement 35**.

Whitley GP (1971) Field notes on birds by Thomas Carter. *Western Australian Naturalist* **12**, 41–44.

Comments received from

John Blyth, Allan Burbidge

Subspecies Conservation Summary

Scrubtit (King Island)

Acanthornis magnus greenianus Schodde and Mason, 1999

Acanthizidae

Conservation status

Critically Endangered B1ab(i,ii,iii,iv,v)+2ab(i,ii,iii,iv,v), C2a(i,ii), D

Reasons for listing

Population <50 mature individuals, probably declining, at a single small site that is deteriorating in condition following fires

Status 2000

Critically Endangered C2a(ii)

Status 1990

Critically Endangered C2a(ii)

Taxonomy

A. m. magnus (mainland Tasmania and south Bruny I.) is Least Concern

Taxonomic uniqueness: high (14 genera/family, 1 species/genus, 2 subspecies/species)

Range

Endemic to King I., Tasmania. Until the late 1960s it was present in habitat remnants at The Nook, Pass River, Yellow Rock, Yarra Creek and Pegarah Forestry Reserve (Green and McGarvie 1971). It disappeared from Pegarah State Forest in the 1970s and from the other 3 locations in the 1990s (Donaghey 2011). Since 2004 known only from the Nook Swamps in Lavinia State Reserve and a new locality at Collier's Swamp with a small number of birds in it. In February 2007 fire burnt about 60% of the core critical Scrubtit habitat in the Nook Swamps. This killed many Scrubtits and destroyed much habitat.

Abundance

The estimated population declined by more than half to fewer than 50 birds (Donaghey 2011).

Ecology

Inhabits remnant unburnt patches of mature Swamp Paperbark *Melaleuca ericifolia* forest that occur in flat, low-lying and poorly-drained swamps (Threatened Species Section 2010) that support a dense understorey of young paperbark and ground-layer plants (Department of Sustainability, Environment, Water, Population and Communities 2011). It lays 3–4 eggs in domed nests near the ground (Green and McGarvie 1971) and probably feeds on insects, spiders and snails, like the nominate subspecies (Department of Sustainability, Environment, Water, Population and Communities 2011). A generation length of 5.3 years (BirdLife International 2011) is derived from an age at first breeding of 1.0 years and a maximum longevity in the wild of 9.5 years, both extrapolated from other Acanthizids (Higgins and Peter 2002).

Threats

The principal threat is currently that fire burns remaining habitat patches (Threatened Species Section 2010; Donaghey 2011). The tiny remaining populations are then vulnerable to minor changes in mortality, such as from predation by feral cat *Felis catus*, Black Rat *Rattus rattus* or a range of native mammals, birds or reptiles (Donaghey 2011), ticks (Schodde and Mason 1999) or inbreeding. Historically land clearance has removed most of the habitat of the species and its scarcity is probably an extinction debt from that process; extirpated subpopulations can no longer be replaced by natural dispersal from other sites.

Conservation objectives

1. Secure existing populations
2. Increase the number of individuals and subpopulations

Information required

1. Feasibility of translocation from Nook Swamp to other locations with suitable habitat
2. Feasibility of cross fostering by Tasmanian Thornbills *Acanthiza ewingii rufifrons* or Tasmanian Scrubwren *Sericornis humilis tregallasi*
3. Demographic trends including population size, density, trends, survival, productivity and movements
4. Ecological resource bottlenecks such as habitat structure and food availability

Management actions required

1. Develop and implement fast-response fire protection plans for occupied sites
2. Work with neighbours to Lavinia State Reserve and Collier's Swamp to reduce the incidence of fire
3. Monitor number and productivity of Scrubtit, ensuring research intervention does not increase rates of nest predation
4. Conduct basic demographic and ecological research
5. Identify other sites with potential habitat

Current eligibility against IUCN Red List Criteria

IUCN category	Criteria eligibility (check)
A	Endangered: population decline of 50–79% over the last 3 generations (29 years) inferred from deteriorating habitat quality
B	Critically Endangered: EOO <100 km^2, AOO <10 km^2, severely fragmented, continuing decline in EOO, AOO, habitat quality, no. locations, no. mature individuals
C	Critically Endangered: population <250 mature individuals, estimated continuing decline, >90% in 1 subpopulation
D	Critically Endangered: population <50 mature individuals
E	Not applicable: no population viability analysis undertaken

IUCN Red List assessment data

	Estimate	Reliability
Extent of occurrence trend	10 km^2 decreasing	high high
Area of occupancy trend	5 km^2 decreasing	high low
No. of mature individuals trend	50 decreasing	medium low
No. subpopulations	1	low
No. locations	2	low
Generation time	5.3 years	low
Global population share	100%	high

6. If appropriate, undertake cross-fostering experiments, food supplementation, habitat restoration and translocation
7. Increase community awareness of and pride in the Scrubtit

Bibliography

BirdLife International (2011) 'Species factsheet: *Acanthornis magna*'. Retrieved 17 January, 2011 from <http://www.birdlife.org/>.

Department of Sustainability, Environment, Water, Population and Communities (2011) *Acanthornis magna greeniana*. In 'Species Profile and Threats Database'. Department of Sustainability, Environment, Water, Population and Communities, Canberra. Retrieved 24 February, 2011 from <http://www.environment.gov.au/sprat>.

Donaghey R (2011) 'Survival, distribution and recovery of the King Island Scrubtit in Lavinia State Reserve following the 2007 wildfire'. Unpublished report.

Green R, McGarvie AM (1971) The birds of King Island. *Records of the Queen Victoria Museum* **40**, 1–42.

Higgins PJ, Peter JM (Eds) (2002) *Handbook of Australian, New Zealand and Antarctic Birds. Volume 6: Pardalotes to Spangled Drongo*. Oxford University Press, Melbourne.

Schodde R, Mason IJ (1999) The Directory of Australian Birds: Passerines. CSIRO Publishing, Melbourne.

Threatened Species Section (2010) 'King Island Biodiversity Management Plan'. Department of Primary Industries, Parks, Water and Environment, Hobart.

Comments received from

Richard Donaghey, Mark Holdsworth, Danny Rogers

Subspecies Conservation Summary

Chestnut-rumped Heathwren (Mount Lofty Ranges)

Hylacola pyrrhopygia parkeri Schodde and Mason, 1999

Acanthizidae

Conservation status

Endangered B1ab(i,ii,iii,iv,v)+2ab(i,ii,iii,iv,v)

Reasons for listing

Extent of occurrence <5000 km^2 and area occupied <500 km^2; severely fragmented population; and continuing decline observed, inferred or projected in extent of occurrence, area of occupancy, area, extent and/or quality of habitat, number of locations or subpopulations and number of individuals

Current eligibility against IUCN Red List Criteria

IUCN category	Criteria eligibility
A	Not applicable: past, current or future population declines are thought unlikely to exceed 20% in any 3-generation period
B	Endangered: EOO <5000 km^2, AOO <500 km^2, severely fragmented, inferred continuing decline in EOO, AOO, habitat quality, no. locations, no. mature individuals
C	Vulnerable: population 2500–10 000 mature individuals, inferred continuing decline, 250–1000 mature individuals in largest subpopulation
D	Not applicable: population >1000 mature individuals, >5 locations
E	Not applicable: no population viability analysis undertaken

IUCN Red List assessment data

	Estimate	Reliability
Extent of occurrence trend	2200 km^2 decreasing	high medium
Area of occupancy trend	62 km^2 decreasing	low medium
No. of mature individuals trend	1000 decreasing	low low
No. subpopulations	7	low
No. locations	>10	low
Largest subpopulation	550	low
Generation time	5.0 years	low
Global population share	100%	high

Status 2000

Endangered B1ab(i,ii,iii,iv,v)+2ab(i,ii,iii,iv,v)

Status 1990

Endangered B1ab(i,ii,iii,iv,v)+2ab(i,ii,iii,iv,v)

Taxonomy

H. p. pedleri (Flinders Ranges) is Vulnerable and *H. p. pyrrhopygia* (south-eastern Australia) is Least Concern, as is the species

Taxonomic uniqueness: medium (14 genera/family, 2 species/genus, 3 subspecies/species)

Range

Endemic to the Fleurieu Peninsula and the southern Mount Lofty Ranges in South Australia (Schodde and Mason 1999). Its known former range extends from Tanunda in the north to Deep Creek in the south and east to the Mount Barker-Wistow area (Pickett 2007). The range contracted considerably between the 1970s and 1980s (Paton *et al.* 1994) and is continuing to decline (Szabo *et al.* 2011). At present there are 17–18 locations in 6–7 subpopulations: Deep Creek Conservation Park, Newland Head Conservation Park, Myponga – Mt Billy, Cuitpo-Cox Scrub Conservation Park, Mount Bold – Cherry Gardens-Cleland Conservation Park – Black Hill Conservation Park – Aristey Hill, Mt Gawler Native Forest Reserve and Warren Conservation Park. An AOO of 62 km^2 is derived from records for 2000–2007 with accuracy ≤5 km. This is getting smaller: there have been no records from Tanunda–Angaston, Mt Barker–Callington or at Scott Conservation Park since the 1970s, there was a decline in 1977–2004 based on Birds Australia regional atlas data, a decline in average recording rate from 26% to 2% of surveys for the pre-1980 to post-1995 periods, it could not be found during surveys at Deep Creek Conservation Park Apr–Jun 2007 surveys and there have been few recent records from the Kuitpo–Mt Compass–Myponga region (Pickett 2007).

Abundance

A population size of about 1000 mature individuals is estimated as the AOO multiplied by a density of about 16/km^2 derived from transects (Pickett 2007) with the largest population being about 540 mature individuals in the Mount Bold – Aristey Hill subpopulation. The decline in AOO represents a substantial decline in the population size, which is now highly fragmented.

Ecology

Heathwrens are recorded mostly from sandy to rocky hillcrest, ridges or hillslope low closed mallee shrubland to open forest, usually with multi-layered understoreys taller than 2-m tree with the canopy dominated by either mallee tree/shrubs or heath (Pickett 2007). They feed on insects and seeds of grasses and sedges, and build dome-shaped nests near or on the ground and lay 2–3 eggs (Higgins and Peter 2002). A generation time of 5.0 years (BirdLife International 2011) is derived from an age at first breeding of 1.0 years (extrapolated from *Acanthiza* spp.) and a maximum longevity of 9.1 years (Australian Bird and Bat Banding Scheme).

Threats

Most suitable habitat has been cleared and the remainder is fragmented and degraded. Broad-scale land clearance ceased in the mid-1980s (Paton *et al.* 1994), but habitat loss and degradation continue. At present the major threat is fire, followed by simplification of the habitat through vegetation dieback, caused by the plant fungal disease *Phytophthora*, and incremental residential expansion (Pickett 2007). Invasion by woody weeds, and possibly grazing by stock and rabbits *Oryctolagus cuniculus*, predation by feral cats *Felis catus* and foxes *Vulpes vulpes*, mining and quarrying, recreational activities and site disturbance (Willson and Bignall 2009; Department of Sustainability, Environment, Water, Population and Communities 2011). Fragmentation means that local extinction due to catastrophic events (such as wildfire or drought) or inbreeding depression, is not remedied through immigration. The cumulative effects of the many threats listed are expected to cause the extinction of the Chestnut-rumped Heathwren in the foreseeable future (Ford and Howe 1980; Possingham and Field 2001).

Conservation objectives

1. All viable existing subpopulations secured

2. Population size and the area and extent of habitat expanded, fragmentation reduced and degradation at key sites reversed

Information required

1. Trends in abundance at key sites
2. Criteria for identifying subpopulations and locations that warrant conservation investment
3. Refine understanding of effects of habitat management on suitability
4. Impact of introduced predators and other putative threats
5. Appropriate reintroduction strategies

Management actions required

1. Monitor key indicators of performance, including population size, breeding success and vegetation condition, especially in association with management
2. Identify locations/subpopulations as targets for conservation investment
3. Protect and enhance known habitat, including the active management of Heathwren populations in existing conservation reserves, and the acquisition of additional sites for conservation.
4. Determine feasibility of translocation
5. Coordinate conservation with that of the Mount Lofty Ranges Southern Emu-wren *Stipiturus malachurus intermedius*
6. Liaise with land managers to encourage their involvement in conservation
7. Engage the community in research, management, raising awareness and advocacy

Bibliography

BirdLife International (2011) 'Species factsheet: *Hylacola pyrrhopygia*'. Retrieved 17 January, 2011 from <http://www.birdlife.org/>.

Department of Sustainability Environment Water Population and Communities (2011) *Hylacola pyrrhopygia parkeri*. In 'Species Profile and Threats Database'. Department of Sustainability, Environment, Water, Population and Communities, Canberra. Retrieved 24 February, 2011 from <http://www.environment.gov.au/sprat>.

Ford HA, Howe RW (1980) The future of birds in the Mount Lofty Ranges. *South Australian Ornithologist* **28**, 85–89.

Higgins PJ, Peter JM (Eds) (2002) *Handbook of Australian, New Zealand and Antarctic Birds. Volume 6: Pardalotes to Spangled Drongo*. Oxford University Press, Melbourne.

Paton DC, Carpenter G, Sinclair RG (1994) A second bird atlas of the Adelaide region. Part 1: Changes in the distribution of birds: 1974–75 vs 1984–85. *South Australian Ornithologist* **31**, 151–193.

Pickett M (2007) 'Assessment of the distribution, habitat and conservation status of the Chestnut-rumped Heathwren *Hylacola pyrrhopygia parkeri* in the Mount Lofty Ranges'. Report prepared for the Department for Environment and Heritage.

Possingham HP, Field SA (2001) Regional bird extinctions and their implications for vegetation clearance policy. *Lifelines* **7**, 15–16.

Schodde R, Mason IJ (1999) The Directory of Australian Birds: Passerines. CSIRO Publishing, Melbourne.

Szabo JK, Baxter PWJ, Vesk PA, Possingham HP (2011) Paying the extinction debt: declining woodland birds in the Mount Lofty Ranges, South Australia. *Emu* **111**, 59–70.

Willson A, Bignall J (2009) 'Regional recovery plan for threatened species and ecological communities of Adelaide and the Mount Lofty Ranges, South Australia'. Department for Environment and Heritage, South Australia.

Comments received from

Marcus Pickett

Subspecies Conservation Summary

Chestnut-rumped Heathwren (Flinders Ranges)

Hylacola pyrrhopygia pedleri Schodde and Mason, 1999

Acanthizidae

Conservation status

Vulnerable D1

Reasons for listing

Probably <1000 mature individuals

Status 2000

Vulnerable D1

Status 1990

Vulnerable D1

Current eligibility against IUCN Red List Criteria

IUCN category	Criteria eligibility
A	Not applicable: past, current or future population declines are thought unlikely to exceed 20% in any 3-generation period
B	Not applicable: EOO, AOO, habitat, locations or population not declining, population fluctuations not extreme
C	Not applicable: no continuing population decline or extreme fluctuations
D	Vulnerable: population 250–1000 mature individuals
E	Not applicable: no population viability analysis undertaken

IUCN Red List assessment data

	Estimate	Reliability
Extent of occurrence trend	800 km^2 stable	high low
Area of occupancy trend	100 km^2 stable	medium low
No. of mature individuals trend	570 stable	low low
No. subpopulations	4	low
No. locations	>10	low
Largest subpopulation	240	low
Generation time	5.0 years	low
Global population share	100%	high

Taxonomy

H. p. parkeri (Mt Lofty Ranges, South Australia) is Endangered, *H. p. pyrrhopygia* (south-eastern Australia) is Least Concern, as is the species

Taxonomic uniqueness: medium (14 genera/family, 2 species/genus, 3 subspecies/species)

Range

Found only north-east of Port Pirie in the southern Flinders Ranges, in South Australia, from Beetaloo and Telowie Gorge in the south past Mt Remarkable National Park to Mt Brown and Dutchman's Stern in the north (Ford and Howe 1980; Schodde and Mason 1999; Higgins and Peter 2002). Searches of Devils Peak area between Mt Brown and Dutchman's Stern found no birds (L. Pedler *in litt.*). The habitat is now divided into 4 forest blocks in a matrix of drier, mostly grazed, land (L. Pedler *in litt.*). Records are few, but all the forest is protected and has not been burnt by large fires, so there is no suggestion of decline. The total area of forest in the 3 areas is about 350 km^2, but suitable habitat is unlikely to exceed 10%, giving an AOO of about 35 km^2. Densities are probably lower in the north, where birds are likely to be confined to the best pockets of habitat left on high ridges and gorges, and higher in the wetter southern blocks (L. Pedler *in litt.*).

Abundance

A population size of 570 mature individuals estimated as the AOO x a density of about 16/km^2 estimated for *H. p. parkeri* (based on Pickett 2007) with about 240 in the 150 km^2 Mt Remarkable National Park, 180 in the 110 km^2 encompassed by Telowie Gorge Conservation Park, Beetaloo Reservoir Reserve and the native forest remaining in Wirrabara Forest Reserve, 80 in the 50 km^2 Mt Brown Conservation Park and 60 in the 40 km^2 Dutchman's Stern Conservation Park. There have been no major perturbations to suggest that any of these populations are declining, but on the other hand there has been no monitoring to demonstrate that these small and isolated populations are persisting. If there were declines, the subspecies would be Endangered.

Ecology

Lives in the dense undergrowth in heath, eucalypt forests and woodlands, particularly where there are rocks or along gullies. They feed on insects and seeds and lay 2 eggs in domed nests built low in the vegetation (Higgins and Peter 2002). A generation time of 5.0 years (BirdLife International 2011) is derived from an age at first breeding of 1.0 years (extrapolated from *Acanthiza* spp.) and a maximum longevity of 9.1 years (Australian Bird and Bat Banding Scheme).

Threats

Clearing has probably reduced the area of habitat occupied by the subspecies, fragmenting it into 4 subpopulations. Currently, the major threat is probably large-scale bushfires, although no fire is likely to be large enough to remove an entire subpopulation unless refuge areas are entirely removed (L. Pedler *in litt.*), and fires occur at a lower enough frequency that recovery would be expected between fires. Feral cats *Felis catus* and foxes *Vulpes vulpes* probably take some birds but whether they constitute a threat to the taxon is unknown.

Conservation objectives

1. Persistence at all 4 locations verified through observations

Information required

1. Sites regularly occupied by the subspecies
2. Habitat requirements and population density/size
3. Trends in population
4. Effects of fire and introduced predators

Management actions required

1. Develop an appropriate monitoring protocol for each location
2. Determine size, density, demography and habitat requirements of subpopulations
3. Work with local fire authorities to develop an appropriate fire response plan

Bibliography

BirdLife International (2011) 'Species factsheet: *Hylacola pyrrhopygia*'. Retrieved 17 January, 2011 from <http://www.birdlife.org/>.

Ford HA, Howe RW (1980) The future of birds in the Mount Lofty Ranges. *South Australian Ornithologist* **28**, 85–89.

Higgins PJ, Peter JM (Eds) (2002) *Handbook of Australian, New Zealand and Antarctic Birds. Volume 6: Pardalotes to Spangled Drongo.* Oxford University Press, Melbourne.

Pickett M (2007) 'Assessment of the distribution, habitat and conservation status of the Chestnut-rumped Heathwren *Hylacola pyrrhopygia parkeri* in the Mount Lofty Ranges'. Report prepared for the Department for Environment & Heritage.

Schodde R, Mason IJ (1999) *The Directory of Australian Birds: Passerines*. CSIRO Publishing, Melbourne.

Comments received from

Lynn Pedler

Subspecies Conservation Summary

Rufous Fieldwren (Dorre Island)

Calamanthus campestris dorrie Mathews, 1912

Acanthizidae

Conservation status
Vulnerable D2

Reasons for listing
Occurs at single location where large-scale fire or introduction of exotic predators a plausible future threat

Status 2000
Vulnerable D2

Status 1990
Vulnerable D2

Taxonomy
C. c. hartogi (Dirk Hartog I.) is Vulnerable. Six other subspecies are Least Concern, as is the species

Taxonomic uniqueness: medium (14 genera/family, 2 species/genus, 8 subspecies/species)

Range
Confined to Dorre I. in Shark Bay, Western Australia (Schodde and Mason 1999).

Abundance
Recorded as moderately common to very common (Johnstone and Storr 2004). Given recorded population densities for Striated Fieldwren *C. fuliginosus* of 0.06–0.9 birds/ha (Higgins and Peter 2002), the population is here estimated as possibly 2500 adult birds. There is no evidence for any decline in population size.

Ecology
Fieldwrens live in low, sparse heath, saltmarsh or samphire, with or without emergent trees, and feed on insects and seeds (Higgins and Peter 2002). A generation time of 5.3 years (BirdLife International 2011) is derived from an estimated age at first breeding of 1.0 years and a maximum longevity of 9.5 years, both extrapolated from mean values for Acanthizidae.

Threats
The main threats to terrestrial birds on Dorre I. are the introduction of harmful non-endemic fauna, both predators and herbivores, invasive alien plants or pathogens (Department of Environment and Conservation (WA) 2007, 2009). This small island with uniform habitat is also potentially at risk from large or frequent fires that could affect the whole island (Department of Environment and Conservation (WA) 2007), so the subspecies is assessed as having a single location.

Conservation objectives
1. Population maintained

Information required
1. Monitor the population, and relate back to management actions

Management actions required
1. Develop protocols to prevent or minimise the introduction of invasive alien species
2. Monitor presence of non-native species and undertake remedial action if needed
3. Develop and implement a suitable fire management strategy

Bibliography
BirdLife International (2011) 'Species factsheet: *Calamanthus campestris*'. Retrieved 17 January, 2011 from <http://www.birdlife.org/>.

Department of Environment and Conservation (WA) (2007) 'Draft management plan for the Shark Bay Terrestrial Reserves and proposed reserve additions'. Retrieved 10 May, 2011 from <http://www.sharkbay.org/assets/documents/sbmp_draft.pdf>.

Department of Environment and Conservation (WA) (2009) 'Rufous Fieldwrens of Shark Bay'. Retrieved 10 March, 2011 from <http://www.sharkbay.org/assets/documents/fact%20sheets/rufous%20fieldwrens%202%20v2.pdf>.

Current eligibility against IUCN Red List Criteria

IUCN category	Criteria eligibility
A	Not applicable: past, current or future population declines are thought unlikely to exceed 20% in any 3-generation period
B	Not applicable: EOO, AOO, habitat, locations or population not declining, population fluctuations not extreme
C	Not applicable: no continuing population decline or extreme fluctuations
D	Vulnerable: ≤5 locations with plausible future threat
E	Not applicable: no population viability analysis undertaken

Higgins PJ, Peter JM (Eds) (2002) *Handbook of Australian, New Zealand and Antarctic Birds. Volume 6: Pardalotes to Spangled Drongo.* Oxford University Press, Melbourne.

Johnstone RE, Storr GM (2004) *Handbook of Western Australian Birds: Volume 2: Passerines (Blue-winged Pitta to Goldfinch).* Western Australian Museum, Perth.

IUCN Red List assessment data

	Estimate	Reliability
Extent of occurrence trend	53 km^2 stable	high high
Area of occupancy trend	53 km^2 stable	high high
No. of mature individuals trend	2500 stable	low high
No. subpopulations	1	high
No. locations	1	high
Generation time	5.3 years	low
Global population share	100%	high

Schodde R, Mason IJ (1999) *The Directory of Australian Birds: Passerines.* CSIRO Publishing, Melbourne.

Comments received from
Andrew Burbidge

Subspecies Conservation Summary

Rufous Fieldwren (Dirk Hartog Island)

Calamanthus campestris hartogi Carter, 1916
Acanthizidae

Conservation status
Vulnerable D2

Reasons for listing
Occurs at single location where large-scale fire or introduction of exotic predators a plausible threat

Status 2000
Vulnerable D2

Status 1990
Vulnerable D2

Taxonomy
C. c. dorrie (Dorre I.) is Vulnerable. Six other subspecies are Least Concern, as is the species.
Taxonomic uniqueness: medium (14 genera/family, 2 species/genus, 8 subspecies/species)

Range
Confined to Dirk Hartog I. off Western Australia (Schodde and Mason 1999), where widely distributed (Whitlock 1921).

Abundance
Recorded as common (Carter 1917; Whitlock 1921) or very common (Johnstone and Storr 2004). Given recorded population densities for Striated Fieldwren *C. fuliginosus* of 0.06–0.9 birds/ha (Higgins and Peter 2002), the population is here estimated as 30 000 adult birds. There is no evidence for any decline in population size.

Ecology
Fieldwrens live in low, sparse heath, saltmarsh or samphire, with or without emergent trees, and feed on insects and seeds (Higgins and Peter 2002). They are recorded as common on Dirk Hartog I. in open scrub but not in dense dark growths of large wattle bushes (Carter 1917). A generation time of 5.3 years (BirdLife International 2011) is derived from an estimated age at first breeding of 1.0 years and a maximum longevity of 9.5 years, both extrapolated from mean values for Acanthizidae.

Threats
The main threats to terrestrial birds on Dirk Hartog I. is the introduction of harmful non-endemic fauna, partic-

Current eligibility against IUCN Red List Criteria

IUCN category	Criteria eligibility
A	Not applicable: past, current or future population declines are thought unlikely to exceed 20% in any 3-generation period
B	Not applicable: EOO, AOO, habitat, locations or population not declining, population fluctuations not extreme
C	Not applicable: no continuing population decline or extreme fluctuations
D	Vulnerable: ≤5 locations with plausible future threat
E	Not applicable: no population viability analysis undertaken

IUCN Red List assessment data

	Estimate	Reliability
Extent of occurrence trend	620 km^2 stable	high high
Area of occupancy trend	620 km^2 stable	high high
No. of mature individuals trend	30 000 stable	low high
No. subpopulations	1	high
No. locations	1	high
Generation time	5.3 years	low
Global population share	100%	high

ularly rats *Rattus* spp., invasive alien plants or pathogens (Department of Environment and Conservation (WA) 2007, 2009). The subspecies has persisted despite the presence of sheep, goats *Hircus capra*, House Mice *Mus musculus* and feral cats *Felis catus*, and occasional extensive fires, but the island has lost the Thick-billed Grasswren *Amytornis textilis textilis*. The island has recently become a National Park; sheep and goats are being eradicated, and cats are being controlled (Department of Environment and Conservation (WA) 2007). This small narrow island with uniform habitat is also potentially at risk from large or frequent fires that could affect the whole island (Department of Environment and Conservation (WA) 2007), so the subspecies is assessed as having a single location.

Conservation objectives

1. Population maintained

Information required

1. Monitor the population, and relate back to management actions

Management actions required

1. Implement management plans for the control and eradication of goats, cats and House Mice
2. Develop protocols to prevent or minimise the introduction of invasive alien species
3. Monitor presence of non-native species and undertake remedial action if needed
4. Develop and implement a suitable fire management strategy

Bibliography

BirdLife International (2011) 'Species factsheet: *Calamanthus campestris*'. Retrieved 17 January, 2011 from <http://www.birdlife.org/>.

Carter T (1917) The birds of Dirk Hartog Island and Peron Peninsula, Shark Bay, Western Australia, 1916–17. *Ibis* **59**, 564–611.

Department of Environment and Conservation (WA) (2007) 'Draft management plan for the Shark Bay Terrestrial Reserves and proposed reserve additions'. Retrieved 10 May, 2011 from <http://www.sharkbay.org/assets/documents/sbmp_draft.pdf>.

Department of Environment and Conservation (WA) (2009) 'Rufous Fieldwrens of Shark Bay'. Retrieved 10 March, 2011 from <http://www.sharkbay.org/assets/documents/fact%20sheets/rufous%20fieldwrens%202%20v2.pdf>.

Higgins PJ, Peter JM (Eds) (2002) *Handbook of Australian, New Zealand and Antarctic Birds. Volume 6: Pardalotes to Spangled Drongo*. Oxford University Press, Melbourne.

Johnstone RE, Storr GM (2004) *Handbook of Western Australian Birds: Volume 2: Passerines (Blue-winged Pitta to Goldfinch)*. Western Australian Museum, Perth.

Schodde R, Mason IJ (1999) *The Directory of Australian Birds: Passerines*. CSIRO Publishing, Melbourne.

Whitlock FL (1921) Notes on Dirk Hartog Island and Peron Peninsula, Shark Bay, Western Australia. *Emu* **20**, 168–186.

Comments received from

Andrew Burbidge

Species Conservation Summary

Norfolk Island Gerygone

Gerygone modesta Pelzeln, 1860
Acanthizidae

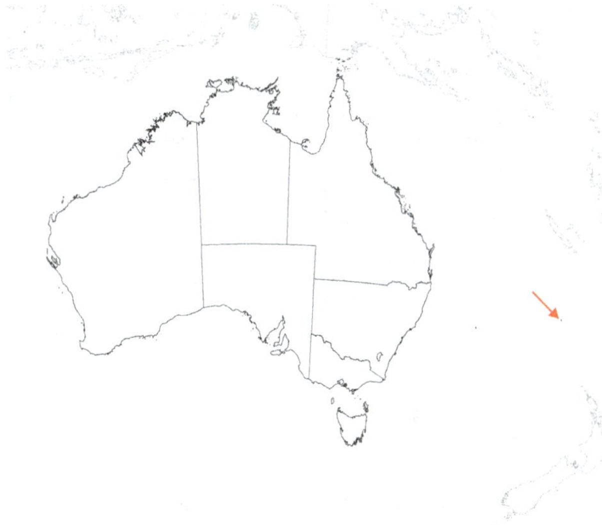

Conservation status
Near Threatened D2

Reasons for listing
Occurs at 1 location with a plausible but low probability threat

Status 2000
Near Threatened D2

Status 1990
Near Threatened D2

Taxonomy
No infraspecific taxa described. Sometimes included with Grey Gerygone *G. igata* of New Zealand (Schodde and Mason 1999).

Taxonomic uniqueness: medium (14 genera/family, 18 species/genus, 1 subspecies/species)

Range
Endemic to Norfolk I. Widespread on Norfolk I but does not occur on Phillip I. Most are in the areas of remaining forest, notably in and around Norfolk I. National Park (Schodde *et al.* 1983; M. Christian *in litt.*).

Abundance
The population Norfolk I. National Park estimated to be at least 3800 pairs (G. Dutson pers. obs.) and many more hundreds outside the park. A comparison with previous surveys suggests that the population is stable or slightly increasing (Robinson 1988; 1997, G. Dutson pers. obs.), supported by other regular observers (BirdLife International 2011).

Ecology
Occurs in remnant areas of tree or shrub growth, such as rainforest, thickets and gardens (Schodde *et al.* 1983; M. Christian *in litt.*). They are common in weedy forest dominated by the exotic Red Guava *Psidium cattleianum*, African Olive *Olea africana* but at a density about half that in native forest (G. Dutson pers. obs.). They are scattered and much less common in patchy forest and scrub away from Norfolk I. National Park (G. Dutson pers. obs.). Their behaviour and the location of their domed nests reduces the probability of predation by cats *Felis catus* or Black Rats *Rattus rattus.* (Garnett and Crowley 2000). A generation time of 5.5 years (BirdLife International 2011) is derived from an age at first breeding of 1.0 years, extrapolated from congeners, and an annual survival of 82.0%, extrapolated from Grey Gerygone *G. igata* (Gill 1982).

Threats
Most remaining forest is now protected within Norfolk Island National Park or local government reserves but remnant trees and shrubs on private land are mostly unprotected and may be cleared gradually for development (BirdLife International 2011). This may be balanced by an ongoing program to control the exotic shrubs which infest most native forest in Norfolk I. National Park. The species coexists with the introduced Black Rat and feral cat which are likely predators but not to the extent of causing the population to decline. However, the similar Lord Howe Gerygone *G. insularis* became extinct probably because of predation by Black Rats (Hindwood 1940) or possibly disease from introduced passerines (McAllan *et al.* 2004). There is a low probability of threat from introduction of a novel predator or disease (Director of National Parks 2010).

Conservation objectives
1. Maintain current breeding population

Information required
2. Population trends on Norfolk I.

Management actions required
1. Maintain quarantine for Norfolk I.
2. Manage threats on Norfolk I. as required
3. Investigate reintroduction to Phillip I. and introduction to Lord Howe I.

Bibliography

BirdLife International (2011) 'Species factsheet: *Gerygone modesta*'. Retrieved 17 January, 2011 from <http://www.birdlife.org/>.

Director of National Parks (2010) 'Draft Norfolk Island region threatened species recovery plan'. Department of the Environment, Water, Heritage and the Arts, Canberra.

Garnett S, Crowley GM (2000) *The Action Plan for Australian Birds 2000.* Environment Australia, Canberra.

Gill BJ (1982) Breeding the Grey Warbler *Geryone igata* at Kaioura, New Zeland. *Ibis* **124**, 123–147.

Hindwood KA (1940) The birds of Lord Howe Island. *Emu* **40**, 1–86.

McAllan IAW, Curtis BR, Hutton I, Cooper RM (2004) The birds of the Lord Howe Island Group: a review of records. *Australian Field Ornithology* **21**, 1–82.

Current eligibility against IUCN Red List Criteria

IUCN category	Criteria eligibility
A	Not applicable: past, current or future population declines are thought unlikely to exceed 20% in any 3-generation period
B	Not applicable: EOO, AOO, habitat, locations or population not declining, population fluctuations not extreme
C	Not applicable: no continuing population decline or extreme fluctuations
D	Near Threatened: at 1 location with plausible but low probability threat
E	Not applicable: no population viability analysis undertaken

IUCN Red List assessment data

	Estimate	Reliability
Extent of occurrence trend	36 km^2 stable	high high
Area of occupancy trend	36 km^2 stable	medium high
No. of mature individuals trend	10 000 stable	low high
No. subpopulations	1	high
No. locations	1	high
Generation time	4.8 years	low
Global population share	100%	high

Robinson D (1988) 'Ecology and management of the Scarlet Robin, White-breasted White-eye and Long-billed White-eye on Norfolk Island'. Report to the Australian National Parks and Wildlife Service, Canberra.

Robinson D (1997) 'An evaluation of the status of the Norfolk island Robin following rat-control and weed-control works in the Norfolk island National Park'. Report to Environment Australia, Canberra.

Schodde R, Fullagar P, Hermes N (1983) *A Review of Norfolk Island Birds: Past and Present*. Australian National Parks and Wildlife Service, Canberra.

Schodde R, Mason IJ (1999) *The Directory of Australian Birds: Passerines*. CSIRO Publishing, Melbourne.

Comments received from

Margaret Christian, Ron Ward

Species Conservation Summary

Lord Howe Gerygone

Gerygone insularis Ramsay, 1879

Acanthizidae

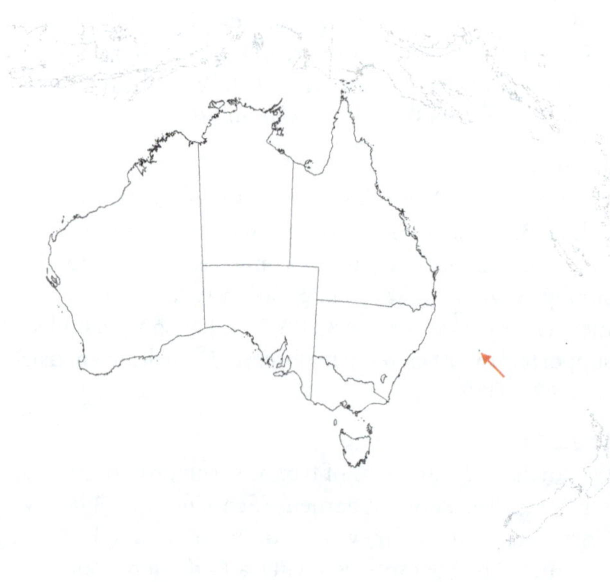

Conservation status

Extinct

Reasons for listing

Last recorded in 1928

Status 2000

Extinct

Status 1990

Extinct

Taxonomy

No infraspecific taxa described. Sometimes included with Grey Gerygone *G. igata* of New Zealand (Schodde and Mason 1999).

Taxonomic uniqueness: medium (14 genera/family, 18 species/genus, 1 subspecies/species)

Former range

Endemic to Lord Howe I. (Schodde and Mason 1999).

Former abundance

Until the early 20th century, it was considered abundant (Hull 1909). Even 10 years after the arrival of Black Rats *Rattus rattus* in 1918 its call was heard frequently (Sharland 1929). However, it was neither seen nor heard during searches in 1936 (Hindwood 1940), and there have been no confirmed records since (McAllan *et al.* 2004).

Ecology

Almost nothing is known of the biology of the Lord Howe Gerygone except that it was found in forests around the settlement, particularly in the canopy (Hull 1909), and fed on small insects and spiders (Sharland 1929; Department of Sustainability, Environment, Water, Population and Communities 2011).

Reasons for extinction

Black Rats, which colonised the island from a wrecked ship in 1918, are thought to be responsible for the Lords Howe Gerygone's extinction (Hindwood 1940). Although still heard frequently 10 years later (Sharland 1929), rats probably did not kill adults which could have persisted for a decade even as rates of nest failure increased. However, the similar Norfolk Island Gerygone *G. modesta* has remained common on Norfolk I. for many years in the presence of rats and it has been speculated that some other cause, possibly disease from introduced passerines, may have contributed to the species' extinction on Lord Howe I. (McAllan *et al.* 2004).

Bibliography

Department of Sustainability, Environment, Water, Population and Communities (2011) *Gerygone insularis*. In 'Species Profile and Threats Database'. Department of Sustainability, Environment, Water, Population and Communities, Canberra. Retrieved 24 February, 2011 from <http://www.environment.gov.au/sprat>.

Hindwood KA (1940) The birds of Lord Howe Island. *Emu* **40**, 1–86.

Hull AFB (1909) The birds of Lord Howe and Norfolk Islands. *Proceedings of the Linnean Society of New South Wales* **34**, 636–693.

McAllan IAW, Curtis BR, Hutton I, Cooper RM (2004) The birds of the Lord Howe Island Group: a review of records. *Australian Field Ornithology* **21**, 1–82.

Schodde R, Mason IJ (1999) *The Directory of Australian Birds: Passerines*. CSIRO Publishing, Melbourne.

Sharland MSR (1929) Land birds of Lord Howe Island. *Emu* **29**, 5–11.

Subspecies Conservation Summary

Slender-billed Thornbill (Gulf St Vincent)

Acanthiza iredalei rosinae Mathews, 1913
Acanthizidae

Conservation status

Vulnerable B1ab(i,ii,iii,iv,v)+2ab(i,ii,iii,iv,v)

Reasons for listing

Extent of occurrence is <5000 km^2 and birds occupy <500 km^2 but not severely fragmented and risk from coastal development, with a consequent decline in EOO, AOO, quality of habitat, and number of locations, subpopulations and mature individuals, is spread across at least 10 locations

Status 2000

Vulnerable B1ab(i,ii,iii,iv,v)+2ab(i,ii,iii,iv,v)

Status 1990

Vulnerable B1ab(i,ii,iii,iv,v)+2ab(i,ii,iii,iv,v)

Taxonomy

A. i. hedleyi (south-east South Australia and central western Victoria) and *A. i. iredalei* (southern arid zone of Western Australia and South Australia west to Spencer Gulf) are Least Concern, as is the species
Taxonomic uniqueness: medium (14 genera/family, 12 species/genus, 3 subspecies/species)

Range

This subspecies is patchily distributed around the northern shores of the Gulf of St Vincent, South Australia (Higgins and Peter 2002) from St Kilda to Ardrossan with 3 population foci: Price, Clinton Conservation Park to Sandy Point and Port Prime south to Torrens I. (Matthew 1994). The Price and Clinton Conservation Park regions are assumed to exchange birds occasionally along the narrow coastal strip as there is a record from between the 2 at Port Arthur (Matthew 1994). Specimens were collected at Port Broughton, Spencer Gulf, in 1917, but there have been no records since (Matthew 1994) despite searching. This subpopulation is assumed to have been lost, and may in case have been more closely related to *A. i. iredalei* than *A. i. rosinae* (L. Pedler *in litt.*). The bird has also probably disappeared from around Port Adelaide (Matthew 1994). It is assumed that some of the loss has occurred within the last 3 generations (37 years), constituting a small but continuing decline. The AOO is calculated from the approximate area of suitable habitat around Gulf St Vincent. Given the number of tenures within the AOO, there are at least 10 locations, though many of the smaller ones would not hold viable populations on their own.

Abundance

There are probably about 4300 mature individuals. Based on densities of about 5.6 birds/ha, and adjusting the esti-

Current eligibility against IUCN Red List Criteria

IUCN category	Criteria eligibility
A	Not applicable: past, current or future population declines are thought unlikely to exceed 20% in any 3-generation period
B	Vulnerable: EOO <20 000 km^2, AOO <2000 km^2, population at ≤10 locations, inferred continuing decline in EOO, AOO, area, extent and quality of habitat, no. locations, no. subpopulations, no. mature individuals
C	Not applicable: population 2500–10 000 mature individuals, continuing decline not observed or estimated, largest subpopulation >1000 mature individuals, >1 subpopulation
D	Not applicable: population >1000 mature individuals, >5 locations
E	Not applicable: no population viability analysis undertaken

IUCN Red List assessment data

	Estimate	Reliability
Extent of occurrence trend	1600 km^2 decreasing	high medium
Area of occupancy trend	60 km^2 decreasing	medium medium
No. of mature individuals trend	4300 decreasing	low medium
No. subpopulations	2	medium
No. locations	10	medium
Largest subpopulation	2200	low
Generation time	12.3 years	low
Global population share	100%	high

mates of maximum numbers from Matthew (1994) to his subsequent lower estimate of density (Matthew 2002), there are likely to be no more than 600 mature individuals at Price, 2200 from Clinton Conservation Park to Sandy Point, and about 1500 from Port Prime south to Torrens I. (assuming lower densities than at Clinton Conservation Park; Matthew 1994). Matthew (1994) thought there might still be 1000 birds on Spencer Gulf but, after 93 years, this assumption is not made here, especially given the taxonomic uncertainty. The subspecies has also disappeared from around Port Adelaide (Matthew 1994).

Ecology

Occupies chenopod shrublands, particularly samphire dominated by *Sclerostegia arbuscula*, on narrow coastal saline mudflats, usually within 20 m of a tidal channel or saline lakes. It also occasionally feeds in Grey Mangrove *Avicennia marina* and among other chenopod species (Mathew 1994, 2002). They build globular nests in low shrubs, laying 2–4 eggs (Higgins and Peter 2002). Densities of 2.9–8.4 birds/ha have been recorded in suitable habitat at Clinton Conservation Park, but are thought to be lower in the southern subpopulation (Matthew 1994). A generation time of 12.3 years (BirdLife International 2011) is derived from an age at first breeding of 1.0 years (extrapolated from congeners) and a maximum longevity of 23.5 years (extrapolated from Striated Thornbill *A. striatus*, Australian Bird and Bat Banding Scheme).

Threats

This subspecies has suffered from habitat loss, particularly cumulative minor developments in the southern part of its range (Grady and Brook 2002), though its disappearance from Port Broughton is unexplained. Though generally unattractive for rural or residential development, the fringes of some salt-flats have been reclaimed for recreational and industrial use, such as for salt works. Other coastal residential and commercial developments, such as marinas, are proposed in parts of the subspecies' range. Inappropriate water management and use also threaten their habitat. Other threats include recreational activities and site disturbance, grazing and disturbance by rabbits *Oryctolagus cuniculus* and predation by foxes *Vulpes vulpes* and feral and uncontrolled cats *Felis catus* (Willson and Bignall 2009).

Conservation objectives

1. Existing subpopulations retained and viable

Information required

1. Population trends, particularly near developed areas
2. Impacts of proposed threats
3. Extent of movement between subpopulations/regions

Management actions required

1. Establish annual monitoring sites, particularly at fringes of range
2. Determine genetic connectivity between subpopulations/regions
3. Work with local government and the Coast Protection Board to secure all remaining habitat

Bibliography

BirdLife International (2011) 'Species factsheet: *Acanthiza iredalei*'. Retrieved 17 January, 2011 from <http://www.birdlife.org/>.

Grady M, Brook J (2000) 'Submission by The Conservation Council of South Australia Inc. to the Senate Inquiry into Gulf St Vincent'. The Conservation Council of South Australia Inc., Adelaide. Retrieved 10 May, 2011 from <http://www.ccsa.asn.au/submissions/gulfstvincent.htm>.

Higgins PJ, Peter JM (Eds) (2002) *Handbook of Australian, New Zealand and Antarctic Birds. Volume 6: Pardalotes to Spangled Drongo.* Oxford University Press, Melbourne.

Matthew J (1994) The status, distribution and habitat of the Slender-billed Thornbill *Acanthiza iredalei* in South Australia. *South Australian Ornithologist* **32**, 1–23.

Matthew JS (2002) Notes on the ecology of the Slender-Billed Thornbill *Acanthiza iredalei rosinae. South Australian Ornithologist* **34**, 15–22.

Willson A, Bignall J (2009) 'Regional recovery plan for threatened species and ecological communities of Adelaide and the Mount Lofty Ranges, South Australia'. Department for Environment and Heritage, South Australia.

Comments received from

Graham Carpenter, Jamie Matthew, Lynn Pedler

Subspecies Conservation Summary

Brown Thornbill (King Island)

Acanthiza pusilla archibaldi Mathews, 1910
Acanthizidae

Conservation status
Critically Endangered D

Reasons for listing
Population probably <50 mature individuals

Status 2000
Critically Endangered D

Status 1990
Critically Endangered D

Taxonomy
The other 4 subspecies in eastern Australia, Tasmania and Kangaroo I. are all Least Concern, as is the species Taxonomic uniqueness: medium (14 genera/family, 12 species/genus, 5 subspecies/species)

Range
Endemic to King I., Tasmania (Boles 1983). Since the discovery of the subspecies, there are only about a dozen records (Department of Sustainability, Environment, Water, Population and Communities 2011). They were recorded in Pegarah Forestry Reserve in 1968 (Green and McGarvie 1971), at Loorana in 1971 (McGarvie and Templeton 1974) and not until 2002 in Pegarah again (S. Lloyd pers. comm.).

Abundance
Extraordinarily scarce. There have never been many records of this species on King I. so there are assumed to be no more than 50 mature individuals.

Ecology
The subspecies evidently persisted in dense mixed scrub in the 1 remaining forest fragment on the island until 1968 (Green and McGarvie 1971) and was captured in a garden adjacent to open farmland in 1971 (McGarvie and Templeton 1974). The record in March 2002 was in wet forest dominated by Tasmanian Blue Gum *Eucalyptus globulus*, with an understorey of Swamp Paperbark *Melaleuca ericifolia*, Rough Dogwood *Pomaderris apetala* and Goldenwood *Monotoca glauca* (Department of Sustainability Environment Water Population and Communities 2011). Its ecology is probably similar to that of the Tasmanian subspecies *A. p. diemenensis* (Higgins and Peter 2002). A generation length of 5.7 years (BirdLife International 2011) is derived from an age at first breeding of 1.0 years (Higgins and Peter 2002) and maximum longevity in the wild of 17.6 years (Australian Bird and Bat Banding Scheme).

Threats
The principal threat is currently loss of remaining habitat, primarily from fire but also from clearing. Any remaining population will also be vulnerable to minor changes in mortality, such as from predation by feral cats *Felis catus*, Black Rat *Rattus rattus* or a range of native mammals, birds or reptiles, tick infestation (Schodde and Mason 1999) or inbreeding. Historically land clearance has removed most of the habitat of the species and its scarcity is probably an extinction debt from that process; extirpated subpopulations can no longer be replaced by natural dispersal from other sites.

Conservation objectives
1. Secure existing populations
2. Increase the number of individuals and subpopulations

Information required
1. Location of birds
2. Demographic trends including population size, density, trends, survival, productivity and movements
3. Ecological resource bottlenecks such as habitat structure and food availability

Management actions required
1. Search for remaining birds
2. Develop and implement fast-response fire protection plans for any sites discovered
3. Work with neighbours to any sites found to reduce the incidence of fire
4. Monitor number and productivity of thornbill
5. Conduct basic demographic and ecological research
6. Identify other sites with potential habitat
7. If appropriate undertake cross-fostering experiments, food supplementation, habitat restoration and translocation
8. Increase community awareness of the thornbill

Bibliography
BirdLife International (2011) 'Species factsheet: *Acanthiza pusilla*'. Retrieved 17 January, 2011 from <http://www.birdlife.org/>.

Current eligibility against IUCN Red List Criteria

IUCN category	Criteria eligibility
A	Not applicable: past, current or future population declines are thought unlikely to exceed 20% in any 3-generation period
B	Not applicable: EOO, AOO, habitat, locations or population not declining, population fluctuations not extreme
C	Not applicable: no continuing population decline or extreme fluctuations
D	Critically Endangered: population <50 mature individuals
E	Not applicable: no population viability analysis undertaken

IUCN Red List assessment data

	Estimate	Reliability
Extent of occurrence trend	5 km^2 stable	low low
Area of occupancy trend	5 km^2 stable	low low
No. of mature individuals trend	50 stable	low low
No. subpopulations	1	low
No. locations	1	low
Generation time	5.7 years	high
Global population share	100%	high

Boles WE (1983) A taxonomic revision of the Brown Thornbill *Acanthiza pusilla* (Shaw) 1790 with description of a new subspecies. *Emu* **83**, 51–58.

Department of Sustainability Environment Water Population and Communities (2011) *Acanthiza pusilla archibaldi.* In 'Species Profile and Threats Database'. Department of Sustainability, Environment, Water, Population and Communities, Canberra. Retrieved 24 February, 2011 from <http://www.environment.gov.au/sprat>.

Green R, McGarvie AM (1971) The birds of King Island. *Records of the Queen Victoria Museum* **40**, 1–42.

Higgins PJ, Peter JM (Eds) (2002) *Handbook of Australian, New Zealand and Antarctic Birds. Volume 6: Pardalotes to Spangled Drongo.* Oxford University Press, Melbourne.

McGarvie AM, Templeton MT (1974) Additions to the birds of King Island, Bass Strait. *Emu* **74**, 91–96.

Schodde R, Mason IJ (1999) *The Directory of Australian Birds: Passerines.* CSIRO Publishing, Melbourne.

Comments received from

Nigel Burgess, Richard Donaghey, Mark Holdsworth, Sarah Lloyd, Eric Woehler

Species Conservation Summary

Chestnut-breasted Whiteface

Aphelocephala pectoralis (Gould, 1871)
Acanthizidae

Conservation status

Near Threatened C2a(i)

Reasons for listing

A single population <10 000 mature individuals but probably more than 1 subpopulation and >1000 mature individuals in smallest subpopulation, also uncertainty about whether continuing decline in habitat quality has been translated into long-term population declines

Status 2000

Near Threatened C2a(i)

Status 1990

Near Threatened C2a(i)

Taxonomy

No infraspecific taxa described
Taxonomic uniqueness: high (14 genera/family, 3 species/genus, 1 subspecies/species)

Range

Widely but patchily distributed in central and northern South Australia as far west as Coober Pedy and Tarcoola, through the Marla-Oodnadatta area and south-east to Marree, around the Birdsville Track and Leigh Creek districts (Pedler 1992; Higgins *et al.* 2001). Recorded at 14 locations in 1990, including only 7 out of 33 historical sites, with extirpation from several locations. However, found at some additional sites in 2007 (Pedler *et al.* 2007). Although documented AOO is only 172 km^2 (Birds Australia Atlas database), actual occupancy of 2000 km^2 (Garnett and Crowley 2000) is probably conservative.

Current eligibility against IUCN Red List Criteria

IUCN category	Criteria eligibility
A	Not applicable: past, current or future population declines are thought unlikely to exceed 20% in any 3-generation period
B	Not applicable: >10 locations, not severely fragmented, population fluctuations not extreme
C	Near Threatened: population 2500–10 000 mature individuals, possible continuing decline, largest subpopulation near 1000 mature individuals
D	Not applicable: population >1000 mature individuals, >5 locations
E	Not applicable: no population viability analysis undertaken

IUCN Red List assessment data

	Estimate	Reliability
Extent of occurrence trend	183 000 km^2 stable	high medium
Area of occupancy trend	2000 km^2 stable	low low
No. of mature individuals trend	6000 stable	low low
No. subpopulations	3	low
No. locations	>10	high
Largest subpopulation	2000	low
Generation time	5.3 years	low
Global population share	100%	high

Abundance

In 1990, the population was estimated at <1500 mature individuals between Lyndhurst and Birdsville Track, <2000 near Oodnadatta and <2500 between Coober Pedy and Port Augusta (Pedler 1991, 1992). No declines were noted on a repeat survey in 1999 or part survey in 2007, and some new locations discovered (Pedler *et al.* 2007). Monitoring in 8–11 km^2 of Mt Lyndhurst Station suggested stable numbers, higher after several wet years leading up to 1991 (c.70 in 1991, 27+ in 1999 and 30+ in 2007; Pedler *et al.* 2007). The meta-population structure is poorly known but is assessed here as having at least 3 subpopulations, some with >1000 mature individuals.

Ecology

There appear to be 2 important components to the species' preferred habitat – a stony, generally open terrain, and an often patchy cover of perennial chenopod shrubs, in particular Low Bluebush *Maireana astrotricha* (Pedler 1992). True stony gibber plains, characterised by the absence of perennial shrubs, are avoided. Most observations have been made in hilly or similar stony environments such as tablelands and breakaways, where the topographic relief gives rise to creek lines that frequently support denser shrubby vegetation. Feeds on the ground, taking seeds and arthropods, builds a domed nest in shrubs, usually bluebush *Maireana* spp., and lays 2 eggs (Pedler 1991, 1992). A generation time of 5.3 years (BirdLife International 2011) is derived from an age at first breeding of 1.5 years (value extrapolated from *Acanthiza*) and a maximum longevity in the wild of 9.5 years (extrapolated from congeners).

Threats

Overgrazing by livestock and rabbits *Oryctolagus cuniculus* is the primary long-term threat, as it reduces the recruitment of perennial chenopod shrubs. The species has disappeared from a number of locations from which it was recorded historically (Pedler 1991, 1992), apparently as a result of heavy grazing of the chenopod shrubland by sheep, cattle, feral goats *Hircus capra* and rabbit in combination with drought. Fire may also be a threat where there is sufficient vegetation to act as fuel. More recently there has been a large increase in mining activity, some of which has directly affected areas that have supported relatively high densities of the species (Pedler 2000; Pedler *et al.* 2007; BirdLife International 2011). The impact of predation by foxes *Vulpes vulpes* and feral cats *Felis catus* is unknown (Pedler *et al.* 2007).

Conservation objectives

1. Increasing population size

Information required

1. Population size and trends at accessible locations
2. Trends in abundance correlated with land use, notably stocking density
3. Quantitative characterisation of habitat requirements
4. Impacts of predation by foxes and cats

Management actions required

1. Maintain stocking densities suitable for healthy perennial shrubs
2. Manage fire at sites where a potential threat
3. Ensure appropriate management of mining at sites occupied by the whiteface

Bibliography

BirdLife International (2011) 'Species factsheet: *Aphelocephala pectoralis*'. Retrieved 17 January, 2011 from <http://www.birdlife.org/>.

Garnett S, Crowley GM (2000) *The Action Plan for Australian Birds 2000.* Environment Australia, Canberra.

Higgins PJ, Peter JM, Steele WK (Eds) (2001) *Handbook of Australian, New Zealand and Antarctic Birds. Volume 5: Tyrant-Flycatchers to Chats.* Oxford University Press, Melbourne.

Pedler LP (1991) 'The status and ecology of the Chestnut-breasted Whiteface *Aphelocephala pectoralis*'. Report to the Australian National Parks and Wildlife Service, Canberra.

Pedler LP (1992) Review of the status and distribution of the Chestnut-breasted Whiteface *Aphelocephala pectoralis. South Australian Ornithologist* **31**, 79–93.

Pedler LP (2000) 'Chestnut-Breasted Whiteface survey, July–August 1999'. Report to the Wildlife Advisory Committee, National Parks and Wildlife South Australia, South Australia.

Pedler LP, Watson M, Langdon P, Pedler R (2007) 'Rare bird surveys, Mt Lyndhurst Station March 2007'. Report prepared for SA Arid Lands Natural Resources Management Board.

Comments received from

Andrew Black, Lynn Pedler, Julian Reid

Species Conservation Summary

Forty-spotted Pardalote

Pardalotus quadragintus Gould, 1838
Pardalotidae

Conservation status

Endangered A2a, B2ab(ii,iii,iv,v)

Reasons for listing

Population has declined by >50% in the last 3 generations (12 years); AOO <500 km^2, severely fragmented and contracting with decline in habitat quality, decline in number of individuals in existing colonies and permanent loss of some colonies

Status 2000

Endangered B2ab(ii,iii,iv,v)

Status 1990

Endangered B2ab(ii,iii,iv,v)

Taxonomy

No infraspecific taxa described
Taxonomic uniqueness: high (1 genus/family, 4 species/genus, 1 subspecies/species)

Range

Endemic to Tasmania, where it occurs in 3 subpopulations at 6 locations: (1) Bruny I. (including Partridge and Snake Is.) and, on mainland Tasmania, Taroona and Tinderbox Peninsula (including Coffee Creek/Howden) and Conningham Peninsula; (2) Maria Island and (3) Flinders Island (Department of Sustainability, Environment, Water, Population and Communities 2011). It is not known whether the 4–6 birds seen in Bob Smiths Gully in August 2010 (Bryant, Hume and Webb *in litt.*), represent a seventh location. It appears to have gone from locations at Taroona and at Lime Bay State Reserve (Bryant 2010). Historically it is thought to have been widely distributed in lowland forests of White Gum *Eucalyptus viminalis* in eastern Tas.

Abundance

Most early reports note the species as being uncommon or rare (Woinarski and Bulman 1985). In 1986, 3520 individuals in 110 colonies were located on 3789 ha (Brown 1986, 1989); in 1994–1997, 3840 individuals in 121 colonies on 4107 ha (Bryant 1997) but in 2009 only 1486 birds found in 54 of 102 colonies surveyed, representing a loss of 47% of colonies, despite the same techniques being used. The largest population of 600–900 is on Maria I. (Bryant 2010).

Ecology

Mostly in forest or woodland containing White Gum with the tree itself providing most of the pardalote's food in the form of invertebrates, lerp secretions and manna (Woinarski and Bulman 1985), as well as hollows for nesting (Woinarski and Rounsevell 1983), although it will nest in other tree species and on the ground (M. Webb *in litt.*). On Maria and Bruny Is. areas with as little as 5% White Gum nonetheless support small numbers of pardalotes (Department of Sustainability, Environment, Water, Population and Communities 2011). A generation length of 3.9 years (BirdLife International 2011) is derived from an average age at first breeding of 1.5 years, extrapolated from similar-sized Passeriformes, and a maximum longevity in the wild of 6.3 years, extrapolated from congeners.

Threats

In most areas habitat loss through degradation and fragmentation are the major cause of decline (Bryant 2010). Extinction debt from fragmentation is continuing, especially on Bruny I. where White Gum degradation has mostly been caused by extended periods of low rainfall. The large numbers of White Gum seedlings planted to assist the pardalote are still too young to make up for the loss of mature trees that continue to be lost, mostly now through natural attrition (Department of Sustainability, Environment, Water, Population and Communities 2011) although some occupied sites have been logged on Bruny I. However, habitat loss does not explain declines on Maria I. with counts there possibly being lower because of extended drought. The Noisy Miner *Manorina melanocephala* and introduced Laughing Kookaburra *Dacelo novaeguineae* are potential competitors or predators across the range of the pardalote. The Noisy Miner is absent from all known pardalote locations but present through much of former range (M. Webb *in litt.*). Similarly the Striated Pardalote *P. striatus* may have a competitive advantage in altered or drought-affected habitat. Human activity, noise and other habitat disturbances are likely to be contributing to the decline of this sensitive bird in urban and public use areas (Threatened Species Section 2006; Bryant 2010).

Conservation objectives

1. All known colonies protected and managed
2. Population size larger
3. Area of occupancy expanded
4. Public awareness and community involvement in the recovery process expanded

Current eligibility against IUCN Red List Criteria

IUCN category	Criteria eligibility
A	Endangered: >50% decline in the last 3 generations (12 years) by direct observation
B	Endangered: AOO <500 km^2, severely fragmented with a continuing decline in AOO, area, extent and/or quality of habitat, no. locations, no. mature individuals
C	Vulnerable: population <10 000 mature individuals, continuing decline and largest population <1000 mature individuals
D	Vulnerable: number of location ≤5 and plausible future threat
E	Not applicable: no population viability analysis undertaken

IUCN Red List assessment data

	Estimate	Reliability
Extent of occurrence trend	330 km^2 stable	high high
Area of occupancy trend	50 km^2 decreasing	high high
No. of mature individuals trend	1500 decreasing	low medium
No. subpopulations	3	high
No. locations	6	high
Largest subpopulation	650	medium
Generation time	3.9 years	high
Global population share	100%	high

5. Recovery program integrated with other nature conservation programs

Information required

1. Threats and factors quantified that limit expansion of colonies and colonisation of potential habitat
2. Relationship between site variables and food productivity
3. Extent of juvenile dispersal, home range and colony dynamics
4. Extent to which nest hollow availability limits breeding

Management actions required

1. Protect and manage known colonies, including improving and linking habitat
2. Monitor the population and habitat quality more frequently
3. Identify threats and factors limiting expansion of colonies and colonisation of potential habitat
4. Increase the size of the population and the area of occupancy
5. Maintain community awareness of the species and involvement in recovery actions
6. Review management prescriptions for areas subject to production forestry
7. Revise management approaches and survey methods to better reflect the need for hollow-bearing trees for nesting

Bibliography

BirdLife International (2011) 'Species factsheet: *Pardalotus quadragintus*'. Retrieved 17 January, 2011 from <http://www.birdlife.org/>.

Bryant SL (2010) 'Conservation assessment of the endangered forty-spotted pardalote 2009–2010'. Report to Threatened Species Section, DPIPWE and NRM South, Hobart.

Department of Sustainability Environment Water Population and Communities (2011) *Pardalotus quadragintus*. In 'Species Profile and Threats Database'. Department of Sustainability, Environment, Water, Population and Communities, Canberra. Retrieved 24 February, 2011 from <http://www.environment.gov.au/sprat>.

Threatened Species Section (2006) 'Fauna recovery plan: Forty-Spotted Pardalote 2006–2010'. Department of Primary Industries and Water, Hobart.

Woinarski JCZ, Bulman C (1985) Ecology and breeding biology of the Forty-spotted Pardalote and other pardalotes on North Bruny Island. *Emu* **85**, 106–120.

Woinarski JCZ, Rounsevell DEB (1983) Comparative ecology of Pardalotes, including the Forty-Spotted Pardalote, *Pardalotus quadragintus* (Aves: Pardalotidae) in south-eastern Tasmania. *Australian Wildlife Research* **10**, 351–361.

Comments received from

Sally Bryant, Matt Webb

Subspecies Conservation Summary

Yellow-tufted Honeyeater (Helmeted)

Lichenostomus melanops cassidix (Gould, 1867)
Meliphagidae

Conservation status

Critically Endangered B1ab(iii,v)+2ab(iii,v), C2a(i), D

Reasons for listing

The extent of occurrence is <100 km^2 and area occupied <10 km^2 with a severely fragmented population and a net decline in the no. mature individuals. Neither of the benign introductions at other sites is sufficiently well-established to be considered self-sustaining. The population has <50 mature individuals.

Status 2000

Critically Endangered B1ab(iii)+2ab(iii)

Status 1990

Critically Endangered B1ab(iii, v)+2ab(iii, v), C2a(i), D

Taxonomy

L. m. melanops (south-eastern Australia, south or east of the Great Dividing Range) and *L. m. meltoni* (south-eastern Australia, inland slopes of the Great Dividing Range) are Least Concern, as is the species
Taxonomic uniqueness: low (43 genera/family, 21 species/genus, 3 subspecies/species)

Range

Endemic to southern-central Victoria. Currently an original population inhabits Yellingbo Nature Conservation Reserve with small reintroduced colonies 30 km south-east at Tonimbuk in Bunyip State Park (birds first released in 2001; Menkhorst 2008; Friends of the Helmeted Honeyeater 2011). Formerly occurred in suitable riparian habitats of the mid-Yarra River and adjacent Western Port catchments and nearby south Gippsland (Blackney and Menkhorst 1993).

Abundance

In 2011 there were 16 breeding pairs at Yellingbo and 5 breeding pairs at Bunyip State Park, making a total of 42 mature individuals with 15 pairs in captivity at Healesville Sanctuary and Taronga Park Zoo (Menkhorst 2008). The population size has declined throughout the 20th century with 2 of the last 3 natural colonies being eliminated by fire in 1983 (Smales *et al.* 1990). The population had declined to 15 breeding pairs and about 50 mature individuals by late 1989 and has been fluctuating since then.

Ecology

Inhabits streamside lowland swamp forest, particularly that dominated by Mountain Swamp Gum *Eucalyptus camphora*, rarely occurring far from water with all current birds living in closed riparian forest. They feed on invertebrates, lerps, honeydew, manna and sap, all gleaned from foliage, twigs and branches of eucalypts or tall shrubs (Menkhorst 2008).They lay 2 or 3 eggs and are multi-brooded. While survival of eggs and chicks is low (mean survivorship of nests from laying to fledging: 0.17), extended parental care of juveniles leads to high survivorship of juveniles (survivorship from 40 days old to 1 year of age: 0.63). The generation length is 3.2 years based on mean age at first breeding of 2.1 years and mean adult female survivorship of 0.7 (Smales *et al.* 2009).

Threats

The primary current threats to the Helmeted Honeyeater relate to the small population size and consequent demographic uncertainty (McCarthy *et al.* 1994) and their concentration in a tiny geographic area and isolated linear habitat patches. The remaining colonies at Yellingbo are threatened by maturation of the principal eucalypt without regeneration of the understory, proximity to colonies of Bell Miner *Manorina melanophrys*, which reduces breeding success, and predation, both by native species and probably by feral cats *Felis catus* and foxes *Vulpes vulpes*. Past decline in range and abundance of the subspecies was caused by extensive destruction of its habitat, largely through clearance for agriculture, with colonies in remnant habitat patches having been destroyed by fire, exclusively occupied by Bell Miners or simply been overcome by the accidents that can befall very small, isolated populations (Menkhorst 2008).

Conservation objectives

1. A stable population of at least 1000 individuals in at least 10 separate, but interconnected, colonies dispersed along several creek systems in the mid-Yarra and Western Port catchments
2. Captive colony maintained
3. Genetic diversity maintained
4. Habitat value of Yellingbo Nature Conservation Reserve improved

Information required

1. Ongoing demographic trends

Current eligibility against IUCN Red List Criteria

IUCN category	Criteria eligibility
A	Not applicable: past, current or future population declines are thought unlikely to exceed 20% in any 3-generation period
B	Critically Endangered: EOO <100 km^2, AOO <10 km^2, severely fragmented, observed continuing decline in no. mature individuals
C	Critically Endangered: population <250 mature individuals, observed continuing decline, <50 mature individuals in largest subpopulation
D	Critically Endangered: population <50 mature individuals in an established population
E	Not applicable: population viability analysis undertaken but may need recalibrating

IUCN Red List assessment data

	Estimate	Reliability
Extent of occurrence trend	75 km^2 stable	high high
Area of occupancy trend	8 km^2 decreasing	high high
No. of mature individuals trend	42 decreasing	high low
No. subpopulations	2	high
No. locations	4	high
Largest subpopulation	33	high
Generation time	3.2 years	high
Global population share	100%	high

2. Techniques to maximise reintroduction success
3. Refined habitat suitability index
4. Causes of eucalypt dieback throughout Yellingbo NCR
5. Techniques to accelerate revegetation, particularly of Mountain Swamp Gum, and reduce tree death
6. Causes of low breeding success in captivity

Management actions required

1. Monitor both established and new populations
2. Control and reduce the spread of Bell Miners as necessary
3. Reintroduce birds to new sites, reinforcing new sites as necessary and minimising disease transfer risks
4. Improve the management of stream flows, water quality and riparian environments throughout catchments of existing and potential sites
5. Manage the captive population of Helmeted Honeyeaters to provide insurance against the demise of the wild population and to meet the needs of the recovery program
6. Manage the populations to maintain genetic diversity
7. Protect and rehabilitate riparian vegetation, particularly sedge-rich Mountain Swamp Gum swamp
8. Ensure regional fire plan is adequate for protecting key habitats and develop a honeyeater-specific fire response plan, including as emergency evacuation of birds from Healesville Sanctuary
9. Continue to work closely with the Friends of the Helmeted Honeyeater, both within and outside the Yellingbo NCR
10. Continue to manage the conservation effort through the Helmeted Honeyeater Recovery Team

Bibliography

BirdLife International (2011) 'Species factsheet: *Lichenostomus melanops*'. Retrieved 17 January, 2011 from <http://www.birdlife.org/>.

Blackney JR, Menkhorst PW (1993) Distribution of subspecies of the Yellow-tufted Honeyeater in the Yarra Valley Region, Victoria. *Emu* **93**, 209–213.

Friends of the Helmeted Honeyeater (2011) 'Helmeted Honeyeater Fact Sheet'. Retrieved 8 May, 2011 from <http://www.helmetedhoneyeater.org.au/helmeted_honeyeater.htm>.

McCarthy MA, Franklin DC, Burgman MA (1994) The importance of demographic uncertainty: An example from the Helmeted honeyeater *Lichenostomus melanops cassidix*. *Biological Conservation* **67**, 135–142.

Menkhorst P (2008) 'National recovery plan for the Helmeted Honeyeater *Lichenostomus melanops cassidix*'. Department of Sustainability and Environment, Melbourne.

Smales IJ, Craig SA, Williams GA, Dunn RW (1990) The Helmeted Honeyeater: decline, conservation and recent initiatives for recovery. In *Management and Conservation of Small Populations*. (Eds TW Clark and JH Seebeck) pp. 225–238. Chicago Zoological Society, Chicago.

Smales IJ, Quin BR, Menkhorst PW, Franklin DC (2009) Demography of the Helmeted Honeyeater (*Lichenostomus melanops cassidix*). *Emu* **109**, 352–359.

Comments received from

Peter Menkhorst

Species Conservation Summary

Black-eared Miner

Manorina melanotis (Wilson, 1911)
Meliphagidae

Conservation status

Endangered A2ce, B2ab(ii,iii,iv,v), C2a(i)

Reasons for listing

Over half of the known population lost to fire in last 3 generations (21 years); integrity of taxon continues to be threatened by hybridisation with Yellow-throated Miner; AOO small and contracting; small declining population with <1000 mature individuals in largest subpopulation

Status 2000

Endangered B2ab(ii,iv,v), C2a(ii), D
Reason for change in 2010: fire-induced losses increased the number of criteria under which listed

Status 1990

Endangered B2ab(ii,iv,v), C2a(ii), D

Taxonomy

No infraspecific taxa described. Interbreeds freely with the Yellow-throated Miner *M. f. flavigula*, producing fertile hybrids (Clarke *et al.* 2001)
Taxonomic uniqueness: medium (43 genera/family, 4 species/genus, 1 subspecies/species)

Range

Endemic to Murray Mallee region of Victoria, South Australia and New South Wales where they currently occur in Riverland Biosphere Reserve, SA and the Murray-Sunset National Park, Vic. Small captive colonies of hybrid birds were established at 3 zoos (Baker-Gabb 2003). These have now all been released as 5 colonies (c.100 birds) translocated to Murray-Sunset National Park, Bronzewing Flora and Fauna Reserve, Scotia Sanctuary and Tarawi Nature Reserve in western NSW (Clarke *et al.* 2002; Baker-Gabb 2007). Two colonies persist in Murray-Sunset National Park but, as of 2010, the colonies in Scotia and Tarawi are being overwhelmed by Yellow-throated Miners and hybrids.

Abundance

At present, Riverland Biosphere Reserve in SA has the biggest population of about 200 colonies (Clarke *et al.* 2008). Two colonies persist in Murray-Sunset National Park. This represents a substantial decline. In 2006, the Riverland Biosphere Reserve supported an estimated 501 (270–927) colonies containing 3758 (2026–6954) Black-eared Miners, 2255 (1215–4170) hybrids and small numbers of Yellow-throated Miners; Murray-Sunset National Park contained about 53 (32–85) Black-eared Miner/hybrid colonies; Scotia Sanctuary and Tarawi Nature Reserve in western NSW held about 14 colonies; and Bronzewing Flora and Fauna Reserve in northwestern Vic had about 4 colonies (Baker-Gabb 2007). Only those in Riverland and Murray-Sunset are now considered sufficiently pure to be worth conserving (D. Baker-Gabb pers. comm.). Given the skewed adult sex ratio (1 female: 1.8 males) and the bird's complex social organization (colonial and cooperative), the number of mature Black-eared Miners is only about 10 per cent of its total population size, i.e. twice the number of colonies.

Ecology

Black-eared Miners live in the semiarid zone, and display a distinct preference for mallee habitats that have not been burnt for at least 45 years (Clarke *et al.* 2005), although they will forage in much younger mallee if available. They are obligate cooperative breeders, with up to 13 helpers (mostly male) assisting at the nest of a single breeding pair (Clarke *et al.* 2005). When breeding, adults typically forage short distances from the nest (up to 0.8 km). When not breeding, birds move as groups (either as an entire colony or in smaller aggregations) over greater distances to forage (Baker-Gabb 2003). Their diet consists of invertebrates and lerp and also nectar. A generation time of 7.8 years (BirdLife International 2011) is derived from an age at first breeding of 2.0 years and a maximum longevity in the wild of 12.1 years, both values extrapolated from congeners.

Threats

Past vegetation clearance and consequent mallee fragmentation have contributed to the decline of the Black-eared Miner and their contact and interbreeding with Yellow-throated Miner. This introgression (or genetic swamping) caused by landuse changes is now the major threat to the Black-eared Miner (Clarke *et al.* 2001). The purity of the Riverland Biosphere Reserve is only maintained through active removal of hybrids, while both the Murray-Sunset colonies have 1 Yellow-throated Miner among them (D. Baker-Gabb pers. comm.). The other major threat is large uncontrolled wildfires that destroy and fragment remaining old-growth mallee in large reserves. Much of the remaining habitat in Vic and NSW

Current eligibility against IUCN Red List Criteria

IUCN category	Criteria eligibility
A	Endangered: population decline of 50–79% over the last 3 generations (21 years) inferred from deteriorating habitat quality and effects of hybridisation
B	Endangered: AOO <500 km^2, at ≤5 locations with plausible future threat, inferred continuing decline in AOO, habitat quality, no. locations, no. mature individuals
C	Endangered: population 250–2500 mature individuals, inferred continuing decline, no subpopulation of phenotypically pure Black-eared Miners >250 mature individuals
D	Vulnerable: population 250–1000 mature individuals, ≤5 locations with plausible future threat
E	Not applicable: no population viability analysis undertaken

IUCN Red List assessment data

	Estimate	Reliability
Extent of occurrence trend	34 000 km^2 decreasing	high medium
Area of occupancy trend	600 km^2 decreasing	low high
No. of mature individuals trend	500 decreasing	medium high
No. subpopulations	4	medium
No. locations	4	medium
Largest subpopulation	400	medium
Generation time	7.8 years	low
Global population share	100%	medium

has been burnt within the last 25 years, although the miners can use habitats that have been burnt patchily if some long-unburnt vegetation remains. A particularly large fire in 2006 in Riverland Biosphere Reserve reduced the largest remaining area of long unburnt habitat by about a third. Habitat degradation by grazing is also a threat (Baker-Gabb 2007). Female dispersion might be compromised in fragmented populations resulting in uneven sex ratio (Ewen *et al.* 2001).

Conservation objectives

1. Existing colonies and habitat retained
2. Increases in the number of viable colonies in the wild

Information required

1. Alternative means for reducing genetic introgression
2. Comparison of genetic variability in captive and wild populations
3. Analysis of population viability
4. Responses to fire and validity of habitat predictive models

Management actions required

1. Monitor existing colonies, removing Yellow-throated Miners or hybrids as required
2. Train people to distinguish between pure and hybrid miners
3. Put in place fire prevention measures around colonies and ensure rapid fire response capability is in place
4. Reduce grazing pressure on regenerating habitat as required, including removal of artificial watering points
5. Develop landscape-scale fire planning tools to ensure adequate fire age classes exist with sufficient connectivity
6. Continue to manage the conservation program through the Black-eared Miner Recovery Team

Bibliography

Baker-Gabb D (2003) 'Recovery plan for the Black-eared Miner *Manorina melanotis* 2002–2006: conservation of old-growth dependent mallee fauna'. Department for Environment and Heritage, Adelaide.

Baker-Gabb D (2007) *The Black-eared Miner. A Decade of Recovery.* The Black-eared Miner Recovery Team, Melbourne.

BirdLife International (2011) 'Species factsheet: *Manorina melanotis*'. Retrieved 17 January, 2011 from <http://www.birdlife.org/>.

Clarke R, Stewart A, Lumb K (2008) 'Assessing the impact of a large bushfire on Black-eared Miner populations within the Bookmark Mallee, SA'. Report prepared for SA Department for Environment and Heritage and the National Black-eared Miner Recovery Team.

Clarke RH, Boulton RL, Clarke MF (2002) Translocation of the socially complex Black-eared Miner *Manorina melanotis*: a trial using hard and soft release techniques. *Pacific Conservation Biology* **8**, 223–234.

Clarke RH, Boulton RL, Clarke MF (2005) Estimating population size of the Black-eared Miner, with an assessment of landscape-scale habitat requirements. *Pacific Conservation Biology* **11**, 174–188.

Clarke RH, Gordon IR, Clarke MF (2001) Intraspecific phenotypic variability in the black-eared miner (*Manorina melanotis*); human facilitated introgression and the consequences for an endangered taxon. *Biological Conservation* **99**, 145–155.

Ewen JG, Clarke R, Moysey E, Boulton RL, Crozier RH, Clarke MF (2001) Primary sex ratio bias in an endangered cooperatively breeding bird, the black-eared miner, and its implications for conservation. *Biological Conservation* **101**, 137–145.

Comments received from

David Baker-Gabb, Rohan Clarke, Arthur Keates, Victor Hurley

Species Conservation Summary

Regent Honeyeater

Anthochaera phrygia (Shaw, 1794)
Meliphagidae

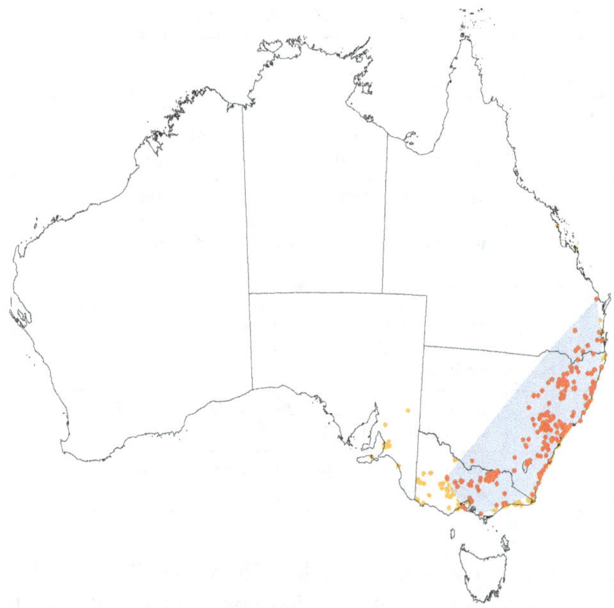

Conservation status
Critically Endangered A2b

Reasons for listing
Population thought to have declined by >80% over the last 3 generations (24 years) notwithstanding recent releases of captive reared birds

Status 2000
Endangered C2a(ii)
Reason for change: population decline has greatly accelerated

Status 1990
Endangered C2a(ii)

Taxonomy
No infraspecific taxa described
Taxonomic uniqueness: medium (43 genera/family, 5 species/genus, 1 subspecies/species)

Range
An extremely patchy distribution within a range stretching from south-east Queensland to central Victoria (Menkhorst *et al.* 1999). Key breeding areas are the Chiltern section of Chiltern–Mt Pilot National Park, in northeastern Victoria (Menkhorst 2003), Capertee Valley in central eastern New South Wales and Bundarra-Barraba region in northern NSW (Oliver and Lollback 2010) with a few birds breeding in other areas, such as the Wangaratta-Mansfield region in Victoria, Warrumbungle National Park, Pilliga forests, the Mudgee-Wollar region, and the Hunter and Clarence Valleys (NSW Scientific Committee 2010). Also breeds irregularly in Qld. While the species has regional variation in calls (Powys 2010), banded birds have been recorded moving between all main sites. Historically, they occurred from Adelaide to 100 km north of Brisbane within 300 km of the coast (Franklin *et al.* 1989). There have been no records from South Australia and western Vic since the 1970s. The AOO has been estimated at <200 km^2 in NSW (NSW Scientific Committee 2010) and would be much less than that in the other states.

Abundance
The population size is difficult to assess, as numbers fluctuate greatly between years at a given site. Expert opinion suggests that there are currently 350–400 mature individuals (D. Ingwersen and P. Menkhorst *in litt.*). The population in NSW was estimated at maximum of 1000 birds in 1997, but there have been many fewer seen subsequently, with a maximum count of just 40 in 2009. In north-eastern Vic there are now probably fewer than 10 pairs and there have been around 30 birds in the Bundarra-Barraba region, NSW, in the last 3 years (NSW Scientific Committee 2010). In the 19th century, recorded in 'great' or 'immense' numbers, occasionally in 'thousands' or 'very large flocks' during influxes (Higgins *et al.* 2001). Now usually seen singly, in twos, or in small groups, (Department of Sustainability, Environment, Water, Population and Communities 2011). A captive breeding population is now well established and managed to produce birds for release. Captive birds have been released into Capertee (9 birds in 2000) and Chiltern (27 birds in 2008 and 44 in 2010) some of which have been seen in the wild over a year later (Regent Honeyeater Recovery Team, unpub. data).

Ecology
Occurs in Box-Ironbark forests and woodlands in areas of low to moderate relief on the inland slopes of the Great Dividing Range, particularly the moister, more fertile patches. The birds also frequently occur in broad valleys that extend into the ranges (Menkhorst *et al.* 1999) and in NSW, riparian forests of River Oak *Casuarina cunninghamiana*, with Needle-leaf Mistletoe *Amyema cambagei*, are also important for feeding and breeding. The birds also use wet lowland coastal forest dominated by Swamp Mahogany *Eucalyptus robusta* or Spotted Gum *E. maculata*. They feed on nectar from the flowers of Mugga Ironbark *E. sideroxylon*, other eucalypts (White Box *E. albens*, Yellow Box *E. melliodora* and Yellow Gum *E. leucoxylon*) and mistletoe species, as well as lerp, honeydew and insects (Oliver 2000). Generally, they build nests in the crowns of tall trees, mostly eucalypts and sometimes among mistletoe. While they exhibit some fidelity to nesting areas, pairs have also been recorded breeding up to 85 km from sites used in the previous breeding season (Geering and French 1998). A generation time of 8.0 years (BirdLife International 2011) is derived from an age at first breeding of 1.5 years (extrapolated from *Manorina*) and a maximum longevity in the wild of 14.4 years (extrapolated from *A. chrysoptera*; Australian Bird and Bat Banding Scheme).

Current eligibility against IUCN Red List Criteria

IUCN category	Criteria eligibility
A	Critically Endangered: past population decline of >80% over 3 generations (24 years) inferred from monitoring
B	Not applicable: >10 locations, not severely fragmented, population fluctuations not extreme
C	Endangered: population 250–2500 mature individuals, inferred continuing decline, >95% in a single subpopulation
D	Vulnerable: population 250–1000 mature individuals
E	Not applicable: no population viability analysis undertaken

IUCN Red List assessment data

	Estimate	Reliability
Extent of occurrence trend	600 000 km^2 decreasing	high high
Area of occupancy trend	300 km^2 decreasing	low high
No. of mature individuals trend	350 decreasing	medium high
No. subpopulations	1	high
No. locations	>10	high
Generation time	8.0 years	low
Global population share	100%	high

Threats

The Regent Honeyeater is a nomadic species with a large range and relatively broad resource and habitat requirements. It has proved difficult to identify any 1 proximal cause of decline. The fundamental reason for the decline, however, is that at least 75% of the Regent Honeyeater's habitat, particularly the most-favoured vegetation communities, has been cleared. As the species requires a number of habitats of adequate quality and age structure to produce a range of different foods (Oliver 2000), it is vulnerable in years when flowering is poor in the fragments that remain, especially as some fragments continue to be degraded by removal of the larger trees for posts, sleepers and firewood, and by ongoing declines in tree health. The poor health of many remnants, the lack of regeneration due to stock grazing, and the very slow growth rate of replacement trees is also likely to be contributing to the decline in honeyeater numbers. The recent dramatic population decline coincides with a 12-year period of reduced rainfall in south-eastern Australia. Fragmentation has apparently advantaged more aggressive honeyeaters, particularly the Noisy Miner *Manorina melanocephala* and the Noisy Friarbird *Philemon corniculatus*, with competition for scarce resources possibly reducing the time available for gathering food (Oliver 2001). What was once a very large population has declined so quickly that a severe loss of genetic variability must now be a threat.

Conservation objectives

1. Persistence of the species in the wild
2. Breeding in the wild of the offspring of reintroduced birds
3. A viable captive population

Information required

1. Movement patterns, particularly after breeding, and environmental correlates of sightings and absences
2. Trends in the number and location of breeding birds
3. Impacts of Noisy Miners, and of their absence
4. Silvicultural techniques that accelerate maturity in key food species
5. Genetic variability, particularly the extent to which the captive population is representative of the wild variability

Management actions required

1. Continue to monitor wild birds at all recently used sites
2. Establish and maintain a reintroduced/translocated population
3. Continue to restore habitat at a landscape scale
4. Protect all regularly-used Regent Honeyeater breeding and feeding sites on public land including Travelling Stock Routes
5. Continue to support conservation management through the Regent Honeyeater Recovery Team and its operations groups
6. Continue to support community, particularly landholder, involvement in the recovery program

Bibliography

BirdLife International (2011) 'Species factsheet: *Xanthomyza phrygia*'. Retrieved 17 January, 2011 from <http://www.birdlife.org/>.

Department of Sustainability, Environment, Water, Population and Communities (2011) *Anthochaera phrygia*. In 'Species Profile and Threats Database'. Department of Sustainability, Environment, Water, Population and Communities, Canberra. Retrieved 24 February, 2011 from <http://www.environment.gov.au/sprat>.

Franklin DC, Menkhorst PW, Robinson J (1989) Ecology of the Regent Honeyeater *Xanthomyza phrygia*. *Emu* **89**, 140–154.

Geering D, French K (1998) Breeding biology of the Regent Honeyeater *Xanthomyza phrygia* in the Capertee Valley, New South Wales. *Emu* **98**, 104–116.

Higgins PJ, Peter JM, Steele WK (Eds) (2001) *Handbook of Australian, New Zealand and Antarctic Birds. Volume 5: Tyrant-Flycatchers to Chats*. Oxford University Press, Melbourne.

Menkhorst P, Schedvin N, Geering D (1999) 'Regent Honeyeater Recovery Plan 1999–2003'. Department of Natural Resources and Environment.

NSW Scientific Committee (2010) 'Regent Honeyeater *Anthochaera phrygia* – critically endangered species listing NSW Scientific Committee – final determination'. Retrieved 8 May, 2011 from <http://www.environment.nsw.gov.au/determinations/regenthoneyeaterFD.htm>.

Oliver DL (2000) Foraging behaviour and resource selection of the Regent Honeyeater *Xanthomyza phrygia* in northern New South Wales. *Emu* **100**, 12–30.

Oliver DL (2001) Activity budget of the regent honeyeater, *Xanthomyza phrygia*, in northern New South Wales. *Australian Journal of Zoology* **49**, 695–712.

Oliver DL, Lollback GW (2010) Breeding habitat selection by the endangered Regent Honeyeater *Anthochaera phrygia* (Meliphagidae) at the local and landscape scale. *Pacific Conservation Biology* **16**, 27–35.

Powys V (2010) Regent honeyeaters – mapping their movements through song. *Corella* **34**, 92–102.

Comments received from

David Geering, Ian Gynther, Dean Ingwersen, Peter Menkhorst

Subspecies Conservation Summary

Yellow Wattlebird (King Island)

Anthochaera paradoxa kingi (Mathews, 1925)
Meliphagidae

Conservation status

Near Threatened D2

Reasons for listing

Present only at only 1 location with a plausible threat but of low probability

Status 2000

Near Threatened D2

Status 1990

Near Threatened D2

Taxonomy

A. p. paradoxa (Tasmania) is Least Concern, as is the species

Taxonomic uniqueness: medium (43 genera/family, 5 species/genus, 2 subspecies/species)

Range

Endemic to King I., Tasmania (Schodde and Mason 1999). It is most common in the south but is patchily distributed in areas of forest such as Pegarah Forestry Reserve (Green and McGarvie 1971; Donaghey 2003; R. Donaghey *in litt.*). The AOO is estimated at 150 km^2. While the birds can move between habitat patches, only about a third of the island retains native vegetation and, of that, perhaps half is occupied by the wattlebirds (R. Donaghey pers. comm.).

Abundance

Based on a density of 0.46 birds/ha (Higgins *et al.* 2001), the population would be about 8000 mature individuals based on the AOO. Extensive wildfires and clearance of forests for agriculture, especially in the first half of the 20th century, reduced the population until about 1960 after which numbers probably increased because native vegetation was given greater protection and regrowing eucalypts matured (Donaghey 2003).

Ecology

Lives in Blue Gum *Eucalyptus globulus* forests and woodlands with White Gum *E. viminalis* and Brooker's Gum *E. brookeriana*, scrub and heath with Silver Banksia *Banksia marginata* as well as homestead and urban gardens, foraging at every level of the vegetation on nectar, fruit, insects, spiders, honeydew and lerp (Donaghey 2003). Lays 2–3 eggs in open cup nests built in the mid-canopy (Higgins *et al.* 2001). A generation time of 8.0 years (BirdLife International 2011) is derived from an age at first breeding of 1.5 years (extrapolated from *Manorina*) and a maximum longevity in the wild of 14.4 years (extrapolated from *A. chrysoptera*; Australian Bird and Bat Banding Scheme).

Threats

The area of available habitat has been greatly reduced by clearing but the population size is now probably stable. The most likely threat is unusual climatic conditions that unexpectedly reduce food availability on the island, although more needs to be known about population trends. Fire and the spread of exotic weeds are also considered a threat (Threatened Species Section 2010). The small area of remaining habitat leaves the subspecies susceptible to random events and exotic predators and disease are low-probability threats given the resilience of the nominate Tasmanian subspecies.

Conservation objectives

1. A stable population

Information required

1. Population size and trends

Current eligibility against IUCN Red List Criteria

IUCN category	Criteria eligibility
A	Not applicable: past, current or future population declines are thought unlikely to exceed 20% in any 3-generation period
B	Not applicable: EOO, AOO, habitat, locations or population not declining, population fluctuations not extreme
C	Not applicable: no continuing population decline or extreme fluctuations
D	Near Threatened: ≤5 locations with a plausible but low probability threat
E	Not applicable: no population viability analysis undertaken

IUCN Red List assessment data

	Estimate	Reliability
Extent of occurrence trend	800 km^2 stable	high high
Area of occupancy trend	150 km^2 stable	low medium
No. of mature individuals trend	8000 stable	low medium
No. subpopulations	1	high
No. locations	1	high
Generation time	8.0 years	low
Global population share	100%	high

Management actions required

1. Monitor population
2. If trends negative, determine necessary conservation management.

Bibliography

BirdLife International (2011) 'Species factsheet: *Anthochaera paradoxa*'. Retrieved 17 January, 2011 from <http://www.birdlife.org/>.

Donaghey R (Ed.) (2003) *The Fauna of King Island. A Guide to Identification and Conservation Management.* King Island Natural Resource Management Group, Currie, King Island.

Green R, McGarvie AM (1971) The birds of King Island. *Records of the Queen Victoria Museum* **40**, 1–42.

Higgins PJ, Peter JM, Steele WK (Eds) (2001) *Handbook of Australian, New Zealand and Antarctic Birds. Volume 5: Tyrant-Flycatchers to Chats.* Oxford University Press, Melbourne.

Schodde R, Mason IJ (1999) *The Directory of Australian Birds: Passerines.* CSIRO Publishing, Melbourne.

Threatened Species Section (2010) 'King Island biodiversity management plan'. Department of Primary Industries, Parks, Water and Environment, Hobart.

Comments received from

Richard Donaghey, Mark Holdsworth

Subspecies Conservation Summary

Yellow Chat (Capricorn)

Epthianura crocea macgregori Keast, 1958

Meliphagidae

Conservation status

Endangered B2ab(i,ii,iv,v), C2a(i), D

Reasons for listing

Population <250 mature individuals occupying <2000 km^2 occupying <2000 km^2 with the rapid decline of the Curtis I. population suggesting a continuing decline in extent of occurrence, area occupied, number of locations and the population size

Status 2000

Endangered D

Status 1990

Endangered D

Taxonomy

E. c. tunneyi (Alligator Rivers) is Endangered, *E. c. crocea* (inland northern Australia) is Least Concern, as is the species

Taxonomic uniqueness: medium (43 genera/family, 4 species/genus, 3 subspecies/species)

Range

Endemic to the central Queensland coast near the Fitzroy R. (Keast 1958; Higgins *et al.* 2001). Recorded at 15 sites (Houston *et al.* 2009), but only regularly at 5 loca-

Current eligibility against IUCN Red List Criteria

IUCN category	Criteria eligibility
A	Not applicable: past, current or future population declines are thought unlikely to exceed 20% in any 3-generation period
B	Endangered: AOO <2000 km^2, at ≤5 locations with plausible future threat, estimated continuing decline in EOO, AOO, no. locations, no. mature individuals
C	Endangered: population <2500 mature individuals, observed continuing decline, <250 mature individuals in largest subpopulation
D	Endangered: population 50–250 mature individuals
E	Not applicable: no population viability analysis undertaken

IUCN Red List assessment data

	Estimate	Reliability
Extent of occurrence trend	6900 km^2 stable	high medium
Area of occupancy trend	60 km^2 decreasing	high low
No. of mature individuals trend	250 decreasing	medium low
No. subpopulations	1	high
No. locations	5	medium
Generation time	5.8 years	low
Global population share	100%	high

tions: St Lawrence East, Herbert West, Torilla Plain, Northern Fitzroy Delta and Southern Fitzroy Delta (W. Houston *in litt.*). About 70% of the population occurs at Torilla Plains (Houston 2011). Numbers on Curtis Island Marine Plain suggest that the location is no longer used (Houston 2011).

Abundance

Average recent population size estimated at 250 mature individuals, but a maximum of 390 in 2008–2009, which were wetter years (Houston 2011). Populations fluctuate (e.g. at Torilla Plain, it halved in 2 years and then increased 4-fold in 2 years (Houston 2011)), but declining at Curtis I. (2001: 30 and 42, 2002: 40, 2005: 16, 2008: 10, 2009 and 2010: 0) (Houston *et al.* 2004b; Houston 2011). The pattern of disappearance resembles the loss of White-fronted Chat *E. albifrons* around Sydney, NSW (Jenner *et al.* 2011), suggesting immediate recovery is unlikely.

Ecology

Inhabits marine plain wetlands that are subject to extensive seasonal inundation and varying degrees of both fresh and saltwater (tidal) influence (Houston *et al.* 2004b; Jaensch *et al.* 2004). Typically inhabits areas with a network of shallow drainage channels and depressions supporting a mosaic of grassland, dense beds of rush or sedge, patches of low chenopods, areas of bare mud and/or shallow water and occasionally low mangroves (Houston *et al.* 2004a; Houston *et al.* 2004b; Jaensch *et al.* 2004). Small groups of 2–10 birds feed on insects and other invertebrates, mainly on the ground, in grassy or open areas including the muddy edges of wetlands, and amongst low vegetation (Houston *et al.* 2004a; Houston *et al.* 2004b; Jaensch *et al.* 2004). A generation time of 5.8 years (BirdLife International 2011) is derived from an age at first breeding of 1.5 years (extrapolated from *Manorina* and *Phylidonyris*) and a maximum longevity in the wild of 10.1 years (extrapolated from other Meliphagidae).

Threats

Potentially threatened by habitat degradation caused by grazing, trampling and digging by mammalian herbivores, invasion by exotic pasture grasses, alteration of natural water flows, expansion of industrial operations and wildfire (Houston *et al.* 2004b; Houston *et al.* 2006). Grazing and associated trampling and/or digging by feral pigs *Sus scrofa* has degraded habitat on Curtis I. (Houston *et al.* 2004b). Grazing and trampling by cattle may be degrading habitat on Torilla Plain but may also limit the spread of invasive grasses (Houston and McCabe 1996; Houston *et al.* 2004a). There is a risk of invasion by exotic pasture grasses such as *Brachiaria mutica* and aquatic weeds, such as *Hymenachne amplexicaulis*, although this may not affect the chats greatly: on the central Torilla Plain *B. mutica* has become dominant without obvious impact (Houston and Melzer 2008). Hydrological changes could affect the chat's habitat, e.g. tidal exclusion works have reduced vegetation on Torilla Plain (Jaensch *et al.* 2004). Drought and reduced freshwater inflow might impede successful breeding (Houston *et al.* 2004a; Houston *et al.* 2004b; Houston *et al.* 2006). Wildfire is considered a minor threat (Houston *et al.* 2004b). Industrial expansion may cause habitat loss in the Fitzroy Delta (Houston and Melzer 2008).

Conservation objectives

1. Increasing population size

Information required

1. Population and habitat trends
2. Movements and threats

Management actions required

1. Maintain unregulated surface flows and current drainage patterns
2. Maintain grazing at levels that do not cause habitat deterioration
3. Manage fire
4. Manage pests, particularly feral pigs
5. Consider fencing exclosures around key habitat
6. Establish habitat management strategies on Curtis I.

Bibliography

BirdLife International (2011) 'Species factsheet: *Epthianura crocea*'. Retrieved 17 January, 2011 from <http://www.birdlife.org/>.

Higgins PJ, Peter JM, Steele WK (Eds) (2001) *Handbook of Australian, New Zealand and Antarctic Birds. Volume 5: Tyrant-Flycatchers to Chats.* Oxford University Press, Melbourne.

Houston W, Elder R, Black R, McCabe J (2006) Conservation significance of coastal wetland habitats for birds at Twelve Mile Creek, Fitzroy River, central Queensland. *Sunbird* **36**, 20–36.

Houston W, Jaensch R, Black R, Elder R, Black L (2009) Further discoveries extend the range of Capricorn yellow chat in coastal central Queensland *Sunbird* **39**, 29–38.

Houston W, McCabe J (1996) 'Waterbirds and Freshwater Wetland Habitats of the Lower Fitzroy River, Queensland'. Report to Queensland Department of Environment and Australian Heritage Commission. Capricorn Conservation Council, Rockhampton, Queensland.

Houston W, Melzer A (2008) 'Yellow chat (Capricorn subspecies) *Epthianura crocea macgregori* recovery plan'. Report to Department of the Environment, Water, Heritage and the Arts, Canberra. Queensland Environmental Protection Agency, Brisbane.

Houston W, Porter G, Elder R, Black R, Sheaves M (2004a) Rediscovery of Yellow Chats (Capricorn subspecies) on the Fitzroy River Delta central Queensland. *Sunbird* **34**, 36–42.

Houston W, Porter G, O'Neill P, Elder R (2004b) The ecology of the critically endangered Yellow Chat *Epthianura crocea macgregori* on Curtis Island. *Sunbird* **34**, 10–23.

Houston WA (2011) Distribution, breeding ecology, population and habitat use of the Capricorn Yellow Chat, *Epthianura crocea macgregori* Keast (Aves: Melaphagidae), a critically endangered bird from central Queensland, Australia. Master's Degree, Faculty of Sciences, Engineering and Health, Central Queensland University, Rockhampton.

Jaensch R, Houston W, Black R, Campbell L, McCabe J, Elder R, Porter G (2004) Rediscovery of the Capricorn subspecies of Yellow Chat *Epthianura crocea macgregori* at Torilla Plain, on the mainland coast of central Queensland. *Sunbird* **34**, 24–35.

Jenner B, French KO, Oxenham K, Major RE (2011) Population decline of the White-fronted Chat (*Epthianura albifrons*) in New South Wales, Australia. *Emu* **111**, 84–91.

Keast A (1958) The relationship between seasonal movements and the development of geographic variation in the Australian chats *Epthianura* Gould and *Ashbyia* North (Passeres: Muscicapidae, Malurinae). *Australian Journal of Zoology* **6**, 53–68.

Comments received from

Allan Briggs, Tim Holmes, Wayne Houston

Subspecies Conservation Summary

Yellow Chat (Alligator Rivers)

Epthianura crocea tunneyi Mathews, 1912

Meliphagidae

Conservation status

Endangered B2ab(ii,iii,iv,v), C2a(ii)

Reasons for listing

Population of 250–2500 mature individuals occupying an area <500 km^2 in a single subpopulation at <5 locations with a continuing decline in area occupied, habitat quality, no. locations and no. mature individuals, probably from a combination of introduced weeds, feral animals and saline intrusion

Status 2000

Endangered B2ab(ii,iii,iv,v), C2a(ii)

Status 1990

Endangered B2ab(ii,iii,iv,v), C2a(ii)

Taxonomy

E. c. macgregori (Capricorn) is Endangered, *E. c. crocea* (inland northern Australia) is Least Concern, as is the species

Taxonomic uniqueness: medium (43 genera/family, 4 species/genus, 3 subspecies/species)

Range

Recorded from a small number of sites in Adelaide R., Mary R., Wildman R., South Alligator R. and East Alligator R. floodplains in the Northern Territory (Keast 1958; Armstrong 2004). Most records are from within Kakadu National Park but the relative importance of the different floodplains is unknown. Occasional records elsewhere, mostly from Darwin to Oenpelli, are probably of dispersive individuals which make up what is presumed to be a single subpopulation.

Abundance

Based on recording 96 individuals in 32 observations over 2 weeks, Armstrong (2004) estimated a population of <300 in Kakadu National Park. Anecdotally the subspecies was more widespread in the past and a continuing

Current eligibility against IUCN Red List Criteria

IUCN category	Criteria eligibility
A	Not applicable: past, current or future population declines are thought unlikely to exceed 20% in any 3-generation period
B	Endangered: AOO <500 km^2, at ≤5 locations with plausible future threat, inferred continuing decline in AOO, habitat quality, no. individuals
C	Endangered: population 250–2500 mature individuals, suspected continuing decline, >95% mature individuals in 1 subpopulation
D	Vulnerable: population <1000 mature individuals
E	Not applicable: no population viability analysis undertaken

IUCN Red List assessment data

	Estimate	Reliability
Extent of occurrence trend	1700 km^2 stable	medium low
Area of occupancy trend	170 km^2 decreasing	low low
No. of mature individuals trend	500 decreasing	medium medium
No. subpopulations	1	high
No. locations	5	medium
Generation time	5.8 years	low
Global population share	100%	high

decline is inferred from the nature of the threats and their expanding influence on chat habitat.

Ecology

Inhabits alluvial coastal and subcoastal grassy floodplains, primarily near floodplain depressions and channels, typically sparsely vegetated with a small number of grasses, sedges and forbs, and aggregates around refugial waterholes in the late dry season (Armstrong 2004). Also recorded from tall grasslands and samphire shrublands (Woinarski *et al.* 2007). The habitat is frequently burnt and is flooded every wet season. It feeds in small groups on insects, mainly on the ground, in grassy or open areas including the muddy edges of wetlands, and amongst low vegetation (Higgins *et al.* 2001; Armstrong 2004). A generation time of 5.8 years (BirdLife International 2011) is derived from an age at first breeding of 1.5 years (extrapolated from *Manorina* and *Phylidonyris*) and a maximum longevity in the wild of 10.1 years (extrapolated from other Meliphagidae).

Threats

Threatened by habitat degradation caused by exotic plant species, grazing, feral pigs *Sus scrofa*, altered fire regimes and saltwater intrusion. Invasion by the weeds *Mimosa pigra*, Para Grass *Urochloa mutica*, Gamba Grass *Andropogon gayanus* and perhaps Aleman Grass *Echinochloa polystachya* and Olive Hymenachne *Hymenachne amplexicaulis*, which are established across much of the chat's range may render the habitat unsuitable, although the relationship between these weeds and chat abundance has not been established and the largest population of *E. c. macgregori* occurs on grazed land dominated by *H. amplexicaulis* (Houston and Melzer 2008). However, while grazing by cattle does not seem to affect *E. c. macgregori*, wallowing by feral Water Buffalo *Bubalus bubalis* will degrade habitat of *E. c. tunneyi* if buffalo numbers are allowed to return to high densities. Armstrong (2004) did not record the species in otherwise suitable habitats where mud-banks had been significantly disturbed by pig-rooting. At the end of the dry season, both feral pigs and chats concentrate in relatively small areas of suitable habitat. Regular patch burning of parts of the floodplain may be necessary to maintain habitat diversity (Lucas *et al.* 1997). The plains are also at risk from saltwater incursion as a result of sea level rise, though this is not straightforward as mobilisation of sediments by rising sea levels can lead to re-emergence of freshwater wetlands at a higher level (Winn *et al.* 2006). Predation by feral cats *Felis catus* has been suggested as a threat (Woinarski *et al.* 2007). Given the threats of weeds and saltwater intrusion, the subspecies is assessed as having 5 locations: the floodplains of the Adelaide R., Mary R., Wildman R., South Alligator and East Alligator Rivers.

Conservation objectives

1. A viable population
2. Evidence-based management of chat habitat

Information required

1. Population trends
2. Movements, habitat and threats
3. Effect of management actions

Management actions required

1. Survey population and establish a monitoring protocol
2. On pastoral properties maintain grazing at levels that do not cause habitat deterioration
3. Manage pests, particularly feral pigs
4. Control or eradicate invasive weeds that are affecting taxon

Bibliography

Armstrong M (2004) 'The Yellow Chat *Epthianura crocea tunneyi* in Kakadu National Park'. Unpublished report. Parks Australia (North).

BirdLife International (2011) 'Species factsheet: *Epthianura crocea*'. Retrieved 17 January, 2011 from <http://www.birdlife.org/>.

Higgins PJ, Peter JM, Steele WK (Eds) (2001) *Handbook of Australian, New Zealand and Antarctic Birds. Volume 5: Tyrant-Flycatchers to Chats.* Oxford University Press, Melbourne.

Houston W and Melzer A (2008) 'Yellow chat (Capricorn subspecies) *Epthianura crocea macgregori* recovery plan'. Report to Department of the Environment, Water, Heritage and the Arts, Canberra. Queensland Environmental Protection Agency, Brisbane.

Keast A (1958) The relationship between seasonal movements and the development of geographic variation in the Australian chats *Epthianura* Gould and *Ashbyia* North (Passeres: Muscicapidae, Malurinae). *Australian Journal of Zoology* **6**, 53–68.

Lucas D, Gapindi M, Russell-Smith J (1997) Cultural perspectives of the South Alligator River floodplain: continuity and change. In *Tracking Knowledge in North Australian Landscapes: Studies in Indigenous and Settler Ecological Knowledge Systems.* (Eds DB Rose and A Clarke) pp. 120–140. North Australia Research Unit, Darwin.

Winn KO, Saynor MJ, Eliot MJ, Eliot (2006) Saltwater intrusion and morphological change at the mouth of the East Alligator River, Northern Territory. *Journal of Coastal Research* **22**, 137–150.

Woinarski J, Pavey C, Kerrigan R, Cowie I, Ward S (Eds) (2007) *Lost from Our Landscape: Threatened Species of the Northern Territory.* Northern Territory Government, Darwin.

Comments received from

Martin Armstrong, Peter Kyne, Simon Ward, John Woinarski

Subspecies Conservation Summary

Black-chinned Honeyeater (south-eastern)

Melithreptus gularis gularis (Gould, 1837)
Meliphagidae

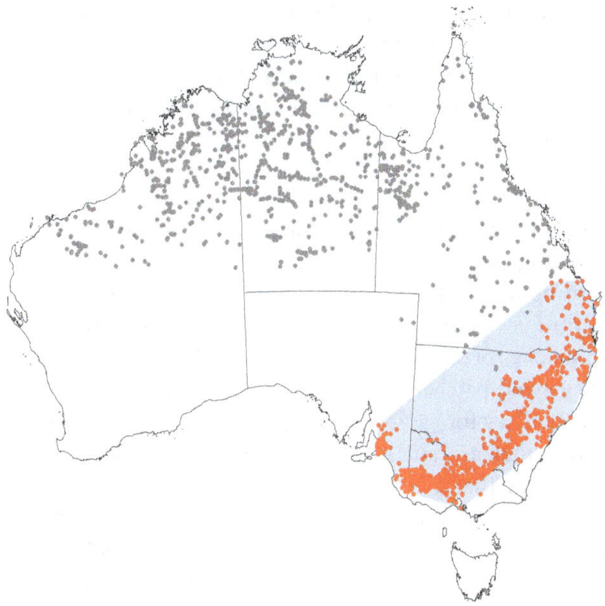

Conservation status

Near Threatened A2bc+3c+4bc

Reasons for listing

Declining across a substantial proportion of its range based on surveys and local extirpations, probably at 20–29% within 3 generations (21 years), because of the many ongoing effects of 200 years of habitat loss and fragmentation

Status 2000

Near Threatened A2bc+3c+4bc

Status 1990

Near Threatened A2bc+3c+4bc

Taxonomy

M. g. laetior (northern Australia) is Least Concern, as is the species
Taxonomic uniqueness: medium (43 genera/family, 7 species/genus, 2 subspecies/species)

Range

This subspecies is found in eastern Australia, along the inland slopes of the Great Dividing Range, extending to the coast between Sydney and Newcastle and between Brisbane and Rockhampton, as well as westward into south-eastern South Australia (Schodde and Mason 1999). From Dubbo, New South Wales to Rockhampton, Queensland, intergrades with *M. g. laetior* in a broad band stretching west to southern Cape York Peninsula, and again around Mt Isa. A subpopulation in the Mt Lofty Ranges may have been isolated historically (Schodde and Mason 1999). Its distribution has contracted in the Mt Lofty Ranges during the course of the 20th century to 11 sites by the mid 1990s (Paton *et al.* 1994). The AOO, derived solely from records in the Birds Australia database since 1998, is 5400 km^2. However, the EOO includes many areas that have been, at best, partially surveyed so is likely to be at least double this area, and probably much greater.

Abundance

Taking an average density of 0.21 birds/ha (0.02–0.26/ha in box-ironbark forests in Vic and 0.28/ha northern NSW; Oliver *et al.* 1999; NSW Scientific Committee 2001) and AOO of 11 000 km^2 gives a population estimate of >200 000 birds. The subspecies has been identified as 1 of a suite of declining species in western New South Wales and northern Victoria (Robinson 1993; Barrett *et al.* 1994; Robinson and Traill 1996; Reid 1999; Olsen *et al.* 2005; Mac Nally *et al.* 2009) where it is apparently absent from large areas that would once have contained suitable habitat (Mac Nally *et al.* 2009). Though still present in SA, they are very rare (Possingham and Possingham

Current eligibility against IUCN Red List Criteria

IUCN category	Criteria eligibility
A	Near Threatened: past, current and anticipated population declines of 20–29% over periods 3 generations (21 years) suspected from monitoring, contracting AOO and deteriorating habitat quality
B	Not applicable: EOO >20 000 km^2, AOO >2000 km^2
C	Not applicable: population >10 000 mature individuals
D	Not applicable: population >1000 mature individuals, >5 locations
E	Not applicable: no population viability analysis undertaken

IUCN Red List assessment data

	Estimate	Reliability
Extent of occurrence trend	1 200 000 km^2 stable	high high
Area of occupancy trend	11 000 km^2 decreasing	low high
No. of mature individuals trend	200 000 decreasing	low high
No. subpopulations	2	medium
No. locations	>10	high
Largest subpopulation	200 000	low
Generation time	6.9 years	high
Global population share	100%	high

2008). Although reporting rates were either stable or increased between Birds Australia atlases (Barrett *et al.* 2002), this may have been because observers sought them out, biasing the results. The overall rate of decline is difficult to estimate but is inferred here from ongoing habitat loss and degradation to be 20–29% in 3 generations (21 years).

Ecology

Occupies dry eucalypt woodland within an annual rainfall range of 400–700 mm, particularly associations containing ironbark and box eucalypts as well as River Red Gum *Eucalyptus camaldulensis*. While they do not occur in patches of <200 ha (NSW Scientific Committee 2001), they avoid patches embedded in plantations (Loyn *et al.* 2010). They forage in the canopy mostly by prising apart leaves that are stuck together by insects, a specialised technique that may help explain their rarity (Lollback *et al.* 2008). They build a suspended nest in which 2 eggs are usually laid (Higgins *et al.* 2001). As most Meliphagidae are highly mobile, it is assumed that the eastern birds have operated as 1 subpopulation, but it is suspected that the birds in the Mt Lofty Ranges have long been isolated (Schodde and Mason 1999) and it is possible that birds in fragments are also now separated from larger populations in more continuous habitat (NSW Scientific Committee 2001). A generation time of 6.9 years (BirdLife International 2011) is derived from an age at first breeding of 1.5 years, extrapolated from *Manorina*, and a maximum longevity in the wild of 12.3 years, extrapolated from Eastern White-naped Honeyeater *Melithreptus lunatus* (Australian Bird and Bat Banding Scheme).

Threats

One of a suite of declining temperate woodland species (Reid 1999; Montague-Drake 2009). Historical clearance destroyed much habitat so populations in the agricultural regions are now fragmented and suffering from extinction debt (Ford 2011). Birds in fragments can suffer such high nest predation that they produce too few recruits to replace adult mortality (Ford *et al.* 2009) and are subject to competition with Noisy Miners *Manorina melanocephala* (Maron and Lill 2005). These pressures may be compounded by a reliance on what is becoming a depauperate food base (Watson 2011). Most threats are greater at habitat edges so are exacerbated by fragmentation. However, the species also uses isolated paddock trees, themselves a diminishing resource (Gibbons and Boak 2002).

Conservation objectives

1. Stable populations in key sites across its range

Information required

1. Population densities and trends in arid parts of the range
2. Criteria for sites where management investment to keep populations is warranted
3. Appropriate management interventions for isolated subpopulations including predator and competitor control, fire regime and ground cover manipulation

Management actions required

1. Monitor key sites, particularly where there are management interventions
2. Develop guidelines for the management of populations in fragments
3. Promote ecological management of woodland remnants on private land
4. Design conservation and regeneration plans at a landscape level
5. Protect all woodland in which honeyeaters are known to be resident from clearing
6. Place all areas of public land that contain honeyeaters under secure conservation management, particularly those in timber reserves, transport corridors and local government land
7. Using appropriate incentives, undertake extension with landholders who have suitable woodland habitat to promote sound management of remnants and encourage greater connectivity between habitat patches
8. Promote revegetation and land reclamation that recreates woodland habitat with a full complement of biodiversity, including the honeyeater
9. Control and reduce firewood collection from areas occupied by honeyeaters, promoting wood-lot development close to markets, and reduce grazing densities where necessary

Bibliography

Barrett G, Ford HA, Recher HF (1994) Conservation of woodland birds in a fragmented rural landscape. *Pacific Conservation Biology* **1**, 245–256.

Barrett GW, Silcocks A, Cunningham RB (2002) 'Australian Bird Atlas (1998–2001). Supplementary Report No. 1 – Comparison of Atlas 1 (1977–1981) and Atlas 2 (1998–2001)'. Birds Australia, Melbourne.

BirdLife International (2011) 'Species factsheet: *Melithreptus gularis*'. Retrieved 17 January, 2011 from <http://www.birdlife.org/>.

Ford HA (2011) The causes of decline of birds of eucalypt woodlands: advances in our knowledge over the last 10 years. *Emu* **111**, 1–9.

Ford HA, Walters JR, Cooper CB, Debus SJS, Doerr VAJ (2009) Extinction debt or habitat change? Ongoing losses of woodland birds in north-eastern New South Wales, Australia. *Biological Conservation* **142**, 3182–3190.

Gibbons P, Boak M (2002) The value of paddock trees for regional conservation in an agricultural landscape. *Ecological Management and Restoration* **3**, 205–210.

Higgins PJ, Peter JM, Steele WK (Eds) (2001) *Handbook of Australian, New Zealand and Antarctic Birds. Volume 5: Tyrant-Flycatchers to Chats.* Oxford University Press, Melbourne.

Lollback GW, Ford HA, Cairns SC (2008) Is the uncommon Black-chinned Honeyeater a more specialised forager than the co-occurring and common Fuscous Honeyeater? *Emu* **108**, 125–132.

Loyn RH, McNabb EG, Cheers G (2010) Embedded remnant forest patches as habitat for birds in the Green Triangle of south-western Victoria and south-eastern South Australia. *Arthur Rylah Institute for Environmental Research Technical Report Series* **209**, 1–24.

Mac Nally R, Bennett AF, Thomson JR, Radford JQ, Unmack G, Horrocks G, Vesk PA (2009) Collapse of an avifauna: climate change appears to exacerbate habitat loss and degradation. *Diversity and Distributions* **15**, 720–730.

Maron M, Lill A (2005) The influence of livestock grazing and weed invasion on habitat use by birds in grassy woodland remnants. *Biological Conservation* **124**, 439–450.

Montague-Drake RM, Lindenmayer DB, Cunningham RB (2009) Factors affecting site occupancy by woodland bird species of conservation concern. *Biological Conservation* **142**, 2896–2903.

NSW Scientific Committee (2001) 'Black-chinned honeyeater (eastern subspecies) – vulnerable species listing'. Retrieved 24 February, 2011 from <http://www.environment.nsw.gov.au/>.

Oliver DL, Ley A, Ford HA, Williams B (1999) Habitat of the Regent Honeyeater, *Xanthomyza phrygia* and the value of the Bundarra-Barraba region for the conservation of avifauna. *Pacific Conservation Biology* **5**, 224–239.

Olsen P, Weston M, Tzaros C, Silcocks A (2005) The state of Australia's birds 2005. Woodlands and birds. *Supplement to Wingspan* **15**, 1–32.

Paton DC, Carpenter G, Sinclair RG (1994) A second bird atlas of the Adelaide region. Part 1: Changes in the distribution of birds: 1974–75 vs 1984–85. *South Australian Ornithologist* **31**, 151–193.

Possingham ML, Possingham HP (2008) Species richness and abundance of birds in Mt Lofty Ranges Gum woodland habitat: year 2000 survey. *South Australian Ornithologist* **35**, 69–85.

Reid JRW (1999) 'Threatened and declining birds in the New South Wales Sheep-Wheat Belt: I. Diagnosis, characteristics and management'. Consultancy report to NSW National Parks and Wildlife Service. CSIRO Sustainable Ecosystems, Canberra.

Robinson D (1993) Vale Toolern Vale: the loss of our woodland birds. *Wingspan* **March**, 1–4.

Robinson D, Traill BJ (1996) 'Conserving Woodland Birds in the Wheat and Sheep Belts of Southern Australia'. RAOU, Melbourne.

Schodde R, Mason IJ (1999) *The Directory of Australian Birds: Passerines.* CSIRO Publishing, Melbourne.

Watson DM (2011) A productivity-based explanation for woodland bird declines: poorer soils yield less food. *Emu* **111**, 10–18.

Comments received from

David Baker-Gabb, Hugh Ford, Damon Oliver, Steve Priday, David Watson

Species Conservation Summary

Painted Honeyeater

Grantiella picta (Gould, 1838)
Meliphagidae

Conservation status
Vulnerable C2a(ii)

Reasons for listing
Population 2500–10 000 in a single subpopulation with a continuing decline projected from habitat loss

Status 2000
Near Threatened C2a(ii)
Reason for change in 2010: population thought to have declined to <10 000

Status 1990
Near Threatened C2a(ii)

Taxonomy
No infraspecific taxa described
Taxonomic uniqueness: high (43 genera/family, 1 species/genus, 1 subspecies/species)

Range
Sparsely distributed from south-eastern Australia to north-western Queensland and eastern Northern Territory (Higgins *et al.* 2001). Greatest concentrations and almost all records of breeding come from south of 26°S, on inland slopes of Great Dividing Range between Grampians, Victoria, and Roma, Qld (Higgins *et al.* 2001; Barrett *et al.* 2003). Seasonal north–south movements are governed principally by the fruiting of mistletoe, with many birds moving after breeding to semi-arid regions, including north-eastern South Australia, central and western Qld and central NT (Higgins *et al.* 2001). The only records from SA in 1998–2008 are from Gluepot reserve (Birds Australia Atlas). Given its dispersive habits, it is assessed here as having a single subpopulation. The AOO, based on a density of about 30 ha/pair when breeding (D. Watson *in litt.*), is unlikely to exceed 1000 km^2 scattered across multiple sites.

Abundance
Total population is unlikely to exceed 10 000 mature individuals. Based on an estimated maximum 150 pairs in the 4300 ha Binya State Forest, up to 6 such concentrations across in NSW and similar or lower numbers elsewhere, the total NSW population is estimated to be <2000 pairs (D. Watson *in litt.*). Much smaller numbers nest in Vic, but numbers in Qld are poorly known and are assessed here as similar to those in NSW. The maximum documented count was 74 birds in 4 days of surveys in Culgoa floodplain, NSW (Oliver *et al.* 2003). The species has declined in abundance in western NSW (Maher 1988; Smith *et al.* 1995) and Vic (Brindley 1991; Department of Sustainability and Environment 2003). Although the number of records submitted to Birds Australia Atlas has increased as the species has become a target for birdwatchers, the distribution may have contracted (recorded from 164 grids in 1977–81; 142 grids in 1998–2003 and 90 grids in 2003–2008, despite greater search effort).

Ecology
The most specialised of Australia's honeyeaters, being largely dependent on mistletoe fruits (Watson 2011, in press). Breeding Painted Honeyeaters depend primarily on just 2 species: Needle-leaved Mistletoe *Amyema cambagei* and Grey Mistletoe *A. quandang*. These grow in acacia-dominated woodlands, particularly those dominated by Brigalow *Acacia harpophylla*, Yarran *A. homalophylla*, Boree or Weeping Myall *A. pendula* and Mulga *A. aneura*, as well as woodlands of Belah *Casuarina cristata* and Buloke *Allocasuarina luehmanni*, riparian woodlands of Black Box *Eucalyptus largiflorens* and River Red Gum *E. camaldulensis*, box-ironbark woodlands and White Cypress-pine *Callitris glaucophylla*. River She-oak *Casuarina cunninghamiana* gallery forests with abundant mistletoes may serve as a drought refuge (Higgins *et al.* 2001; Barea and Watson 2007; Barea 2008b; Watson in press). Prefers woodland with many mature trees, as these host more mistletoes (Oliver *et al.* 2003). Besides mistletoes, also feeds on nectar and insects especially in the non-breeding season (Oliver *et al.* 2003). Less common in strips of remnant box-ironbark woodlands, such as along roadsides and in windbreaks, than in wider blocks (Robinson 1994). Breeds during peak Grey Mistletoe fruit availability (Barea and Watson 2007). Nesting success is relatively low in trees (43%) and especially within mistletoe clumps (17%), with 83% of nest failures caused by predation (Barea 2008a). A generation time of 5.8 years (BirdLife International 2011) is derived from an age at first breeding of 1.5 years (extrapolated from mean values for *Manorina* and *Phylidonyris*) and a maximum longevity in the wild of 10.1 years (extrapolated from mean values for other genera in Meliphagidae).

Current eligibility against IUCN Red List Criteria

IUCN category	Criteria eligibility
A	Near Threatened: Near Threatened: population decline of 20–29% over the last 3 generations (17 years) suspected from monitoring, a reduced AOO and deteriorating habitat quality
B	Not applicable: >10 locations, not severely fragmented, population fluctuations not extreme
C	Vulnerable: population 2500–10 000 mature individuals, suspected continuing decline, 100% in 1 subpopulation
D	Not applicable: population >1000 mature individuals, >5 locations
E	Not applicable: no population viability analysis undertaken

IUCN Red List assessment data

	Estimate	Reliability
Extent of occurrence trend	2 800 000 km^2 stable	high high
Area of occupancy trend	1000 km^2 decreasing	low medium
No. of mature individuals trend	10 000 decreasing	medium medium
No. subpopulations	1	high
No. locations	>10	high
Generation time	5.8 years	low
Global population share	100%	high

Threats

Brigalow and other suitable habitats have been extensively cleared in the recent past, especially in Qld (Oliver *et al.* 2003). In Qld, much suitable habitat was cleared historically (only 11% of Brigalow woodland in the Brigalow Belt South remained in 1999), but annual rates of loss remained at c.5% in 1997–1999 (Wilson *et al.* 2002); the rate of clearance has subsequently declined but is unquantified. In NSW, 640 000 ha native habitat was approved for clearance in 1998–2005, and illegal clearing was estimated at c.30 000 ha in 2005 alone; much of this was suitable for Painted Honeyeaters. Small strips and patches of habitat are not protected from clearance. Most remaining habitat is on private land and continues to be degraded by over-grazing by livestock, native macropods and rabbits *Oryctolagus cuniculus*. Overgrazing also prohibits regeneration (Oliver *et al.* 2003). Within this habitat rates of nest predation are particularly high and may not be sustainable (Barea 2008a).

Conservation objectives

1. Stable population indices at key sites
2. No further clearance of suitable habitat

Information required

1. Population trends at key sites
2. Ecology and location during non-breeding season
3. Principal causes of nest failure and their management implications

Management actions required

1. Monitor key sites, particularly where there are management interventions
2. Investigate causes of nest failure and determine influence on population viability
3. Promote ecological management of woodland remnants on private land, particularly those heavily parasitised by mistletoe
4. Protect all woodland in which the honeyeaters are known to be resident from clearing
5. Place all areas of public land that contain the honeyeaters under secure conservation management, particularly those in timber reserves, transport corridors and local government land
6. Using appropriate incentives, undertake extension with landholders who have suitable woodland habitat to promote sound management of remnants
7. Promote revegetation and land reclamation that recreates woodland habitat with a full complement of biodiversity, including the honeyeater

Bibliography

Barea LP (2008a) Interactions between frugivores and their resources: case studies with the Painted Honeyeater *Grantiella picta*. PhD thesis, Charles Sturt Univeristy, NSW.

Barea LP (2008b) Nest-site selection by the Painted Honeyeater (*Grantiella picta*), a mistletoe specialist. *Emu* **108**, 213–220.

Barea LP, Watson DM (2007) Temporal variation in food resources determines onset of breeding in an Australian mistletoe specialist. *Emu* **107**, 203–209.

Barrett G, Silcocks A, Barry S, Cunningham R, Poulter R (2003) *The New Atlas of Australian Birds*. Royal Australian Ornithologists Union, Hawthorn East.

BirdLife International (2011) 'Species factsheet: *Grantiella picta*'. Retrieved 17 January, 2011 from <http://www.birdlife.org/>.

Brindley A (1991) *Birds of the Bendigo District*. Bendigo Field Naturalists Club, Bendigo.

Department of Sustainability and Environment (2003) 'Action Statement. Painted Honeyeater *Grantiella picta*'. Retrieved 12 May, 2011 from <http://www.dse.vic.gov.au/__data/assets/pdf_file/0017/103175/193_Painted_Honeyeater_2003.pdf>.

Higgins PJ, Peter JM, Steele WK (Eds) (2001) *Handbook of Australian, New Zealand and Antarctic Birds. Volume 5: Tyrant-Flycatchers to Chats*. Oxford University Press, Melbourne.

Maher PN (1988) Threatened avifauna of New South Wales. *National Parks Journal* **32**, 11–16.

Oliver DL, Chambers MA, Parker DG (2003) Habitat and resource selection of the Painted Honeyeater (*Grantiella picta*) on the northern floodplains region of New South Wales. *Emu* **103**, 171–176.

Robinson D (1994) 'Research plan for threatened woodland birds of south-eastern Australia'. Arthur Rylah Institute for Environmental Research Technical Report Series 133.

Smith PJ, Smith JE, Pressey RL, Whish GL (1995) 'Birds of particular conservation concern in the Western Division of New South Wales: distributions, habitats and threats'. Occasional Paper 20. National Parks and Wildlife Service (NSW), Hurstville.

Watson DM (2011) *Mistletoes of Southern Australia.* CSIRO Publishing, Melbourne.

Watson DM (in press) Painted Honeyeater. In *Queensland's Threatened Animals.* (Eds LK Curtis, AJ Dennis, KR McDonald, PM Kyne, SJS Debus). CSIRO Publishing, Melbourne.

Wilson BA, Neldner VJ, Accad A (2002) The extent and status of remnant vegetation in Queensland and its implications for state wide management and legislation. *The Rangeland Journal* **24**, 6–36.

Comments received from

Barry Baker, Laurence Barea, Jez Bird, Andrew Burbidge, Hugh Ford, Tim Holmes, Richard Johnson, Peter Menkhorst, James O'Connor, Andy Symes, David Watson, John Woinarski

Subspecies Conservation Summary

Spotted Quail-thrush (Mt Lofty Ranges)

Cinclosoma punctatum anachoreta Schodde & Mason, 1999

Psophodidae

Conservation status

Critically Endangered (Possibly Extinct) D

Reasons for listing

Possibly Extinct since 1992, based on its documented decline, survey effort, and degradation of habitat, balanced against the inconspicuous nature of the species and continued existence of some potentially suitable habitat

Status 2000

Critically Endangered (Possibly Extinct) A2b, D

Status 1990

Critically Endangered A2bc+3c+4bc, B1ab(i,ii,iii,iv,v)+B 2ab(i,ii,iii,iv,v), C2a(i,ii), D

Reason for change in 2000: last recorded in 1992

Taxonomy

C. p. punctatum (south-eastern Australia) and *C. p. dovei* (eastern Tasmania) are Least Concern, as is the species

Taxonomic uniqueness: medium (4 genera/family, 5 species/genus, 3 subspecies/species)

Former range

This subspecies was confined to the Mount Lofty Ranges and Fleurieu Peninsula in South Australia from Para Wira Conservation Park and Cape Jervis in the south to Angaston in the north (Higgins and Peter 2002).

Former abundance

It was regarded as 'numerous' in the early 20th century (McGilp 1922), became rare by the 1930s (Ashby 1932) and very rare by the 1950s (Terrill and Rix 1950). Subsequent records were sporadic: it was seen at Clarendon Ridge until 1963, near Happy Valley around 1963, at Deep Creek Conservation Park around 1964, in the Mount Crawford area of Warren Conservation Park in 1965 and around 1969, and at Onkaparinga Gorge, near Clarendon in 1972, near Ashbourne and at Waitpinga Conservation Park, Spring Mount Forest and Parawa in 1977 and near Spring Mount Conservation Park in 1981 (Department of Sustainability Environment Water Population and Communities 2011). The last 1 was seen at Mt Bold in 1992 (R. Bates *in litt.*).

Ecology

Spotted Quail-thrushes inhabit eucalypt forest and feed on seeds, insects and other small invertebrates. They lay 2 to 3 eggs in a well-disguised cup built on the ground (Higgins and Peter 2002). A generation time of 4.1 years (BirdLife International 2011) is derived from an age at first breeding of 1.5 years, extrapolated from passerines of similar size and ecology, and a maximum longevity in the wild of 8.1 years, extrapolated from Chestnut Quail-thrush *C. castanotus* (Australian Bird and Bat Banding Scheme).

Reason for possible extinction

The decline and eventual extinction of this subspecies was caused by the loss, fragmentation and degradation of its habitat in the Mt Lofty Ranges (Ford and Howe 1980). Severe fires in 1983 may have reduced the population to just a few individuals but fires are just 1 of many cumulative pressures such as predation by feral cats *Felis catus* and foxes *Vulpes vulpes*, disturbance by dogs and people and the invasion of its habitat by noxious weeds (Department of Sustainability, Environment, Water, Population and Communities 2011).

Current eligibility against IUCN Red List Criteria

IUCN category	Criteria eligibility
A	Not applicable: population declines are thought unlikely to have exceeded 20% in the last 3 generations (12 years)
B	Not applicable: if extant, EOO, AOO, habitat, locations or population not declining, population fluctuations not extreme
C	Not applicable: no continuing population decline or extreme fluctuations
D	Critically Endangered (Possibly Extinct): if extant, population <50 mature individuals
E	Not applicable: no population viability analysis undertaken

Conservation objectives

1. A viable population in the Mt Lofty Ranges

Information required

1. Evidence that it remains extant
2. If extant, habitat requirements, threats, population size and trends

Management actions required

1. Undertake surveys of habitat that might still hold populations, using call back and other techniques, particularly in Deep Creek National Park
2. Develop a contingency plan should the bird be rediscovered, including establishing a captive breeding population

Bibliography

Ashby E (1932) Bird notes. *South Australian Ornithologist* **11**, 232–233.

BirdLife International (2011) 'Species factsheet: *Cinclosoma punctatum*'. Retrieved 17 January, 2011 from <http://www.birdlife.org/>.

Department of Sustainability, Environment, Water, Population and Communities (2011) *Cinclosoma punctatum anachoreta*. In 'Species Profile and Threats Database'. Department of Sustainability, Environment, Water, Population and Communities, Canberra. Retrieved 24 February, 2011 from <http://www.environment.gov.au/sprat>.

Ford HA, Howe RW (1980) The future of birds in the Mount Lofty Ranges. *South Australian Ornithologist* **28**, 85–89.

Higgins PJ, Peter JM (Eds) (2002) *Handbook of Australian, New Zealand and Antarctic Birds. Volume 6: Pardalotes to Spangled Drongo.* Oxford University Press, Melbourne.

McGilp JN (1922) Notes by the late Dr Angove, Tea-tree gully, S. A. *South Australian Ornithologist* **6**, 114–118.

Terrill SE, Rix CE (1950) The birds of South Australia: their distribution and habitat. *South Australian Ornithologist* **19**, 53–100.

Comments received from

Graham Carpenter, Peter Copley, David Paton, Lynn Pedler, Hugh Possingham, Adrian Stokes

---

Subspecies Conservation Summary

Western Whipbird (Kangaroo Island)

Psophodes nigrogularis lashmari Schodde & Mason, 1991
Eupetidae

Taxonomic uniqueness: medium (4 genera/family, 5 species/genus, 4 subspecies/species)

Conservation status

Near Threatened D1

Reasons for listing

Population may be near 1000 mature individuals in the years immediately following large fires, but evolved in the presence of very large fires and no other immediate Threats

Status 2000

Near Threatened D1

Status 1990

Near Threatened D1

Taxonomy

P. n. nigrogularis (western heath, Western Australia) is Endangered, *P. n. leucogaster* (eastern) is Vulnerable, *P. n. oberon* (Stirling Ranges to Ravensthorpe, Western Australia) is Least Concern, as is the species

Range

Endemic to Kangaroo I., South Australia, inhabiting remnant mallee on the western and southern part of the island (Baxter 1995), found within the major forest blocks of Flinders Chase National Park, Cape Gantheaume

Current eligibility against IUCN Red List Criteria

IUCN category	Criteria eligibility
A	Not applicable: past, current or future population declines are thought unlikely to exceed 20% in any 3-generation period
B	Not applicable: >10 locations, not severely fragmented, population fluctuations not extreme
C	Not applicable: no continuing population decline or extreme fluctuations
D	Near Threatened: population near 1000 mature individuals
E	Not applicable: no population viability analysis undertaken

IUCN Red List assessment data

	Estimate	Reliability
Extent of occurrence trend	4400 km^2 stable	high high
Area of occupancy trend	200 km^2 stable	low low
No. of mature individuals trend	2000 stable	low low
No. subpopulations	1	medium
No. locations	>10	high
Generation time	6.4 years	low
Global population share	100%	high

Conservation Park and the southern half of Dudley Peninsula between Dudley and Cape Hart Conservation Parks. Although cover is sparse beside Pennington Bay, there is probably enough to connect populations in the eastern and western side of the island. Based on the area of habitat unburnt in the last 3 generations, particularly on Dudley Peninsula, but also along most of the south and north-west coasts, the AOO is currently at least 200 km^2. This is despite fires in 2007 burning 85% of Flinders Chase National Park, and 19% of the island, including the northern part of the second largest block of native vegetation on the island, Cape Gantheaume Conservation Park, which had burnt almost entirely in 1996. Sporadic observations at localities in the north of the island (e.g. Kangaroo gully, Middle River gorge, Kellys Heritage area, Ballast Head) are generally within months to a year after large fires, which suggest an ability to disperse considerable distances if needed. These birds could also represent dispersal of juveniles or post breeding dispersal (L. Pedler pers. comm.). Because there are so many unburnt sites, there are more than 10 locations.

Abundance

The population is probably at least 2000 mature individuals. Densities of *P. n. leucogaster* are conservatively estimated at 20 birds/km^2 (Higgins and Peter 2002) but may be lower on Kangaroo I. (L. Pedler pers. comm.) so taken here as 10/km^2. It was still common at multiple unburnt sites in Flinders Chase National Park in 2008 (Johnstone 2011). Fires are a major perturbation, but they do not constitute a major population decline since there is a long history of large fires from which whipbirds have recovered; indeed they evolved with lightning-lit fires that were probably more extensive than modern fires, since there are no natural fire breaks. However, no fire within recent history has ever been so large that there would have been a 90% drop in the population: i.e. fluctuations from fire are not extreme.

Ecology

Occurs in coastal and subcoastal mallee and heath on limestone and dunes (Baxter 1995) and forages on the ground for invertebrates. Its habitat is characterised more by vegetation structure than species composition and the frequency of fire has an important role in maintaining a suitable structure. Most records for *P. n. leucogaster* are from vegetation last burnt 10–25 years earlier (Department of Sustainability, Environment, Water, Population and Communities 2011). It can also survive fire in pockets of unburnt vegetation (Johnstone 2011). The subspecies forages on the ground primarily for insects, building a domed nest close to the ground (Higgins and Peter 2002). A generation time of 6.4 years (BirdLife International 2011) is derived from an age at first breeding of 0.5 years and a maximum longevity of 12.3 years, both values extrapolated from Eastern Whipbird *P. olivaceus*.

Threats

Apart from fire, with which whipbirds have evolved, whipbirds are probably killed by feral cats *Felis catus*, but it is not known whether this constitutes a threat to the subspecies. In the past, half of the native vegetation on Kangaroo Island was cleared for agriculture. Most of the remainder is protected.

Conservation objectives

1. Population persists at multiple sites across the island

Information required

1. Distribution and abundance at multiple sites across the island

Management actions required

1. Survey subpopulations when species is calling during spring to determine current distribution, particularly in areas burnt in the last decade (Cape Gantheaume Conservation Park, northern half of Flinders Chase National Park)
2. Maintain a fire management regime that limits the extent of burnt habitat within any 1 habitat block

Bibliography

Baxter C (1995) 'An annotated list of birds of Kangaroo Island'. Department of Environment and Natural Resources, Kingscote.

BirdLife International (2011) 'Species factsheet: *Psophodes nigrogularis*'. Retrieved 17 January, 2011 from <http://www.birdlife.org/>.

Department of Sustainability, Environment, Water, Population and Communities (2011) *Psophodes nigrogularis leucogaster.* In 'Species Profile and Threats Database'. Department of Sustainability, Environment, Water, Population and Communities, Canberra. Retrieved 24 February, 2011 from <http://www.environment.gov.au/sprat>.

Higgins PJ, Peter JM (Eds) (2002) *Handbook of Australian, New Zealand and Antarctic Birds. Volume 6: Pardalotes to Spangled Drongo*. Oxford University Press, Melbourne.

Johnstone G (2011) 'Distribution and conservation status of western whipbirds *(Psophodes nigrogularis)*'. Retrieved 14 April, 2011 from <http://www.zoossa.com.au/conservation-ark/research/past-projects?project=Distribution%20and%20conservation%20status%20of%20western%20whipbirds%20%28Psophodes%20nigrogularis%29>

Comments received from

Lynn Pedler

Subspecies Conservation Summary

Western Whipbird (eastern)

Psophodes nigrogularis leucogaster Howe & Ross, 1933

Eupetidae

Conservation status

Vulnerable B1ab(i,ii,iii,iv,v)+2ab(i,ii,iii,iv,v)

Reasons for listing

Extent of occurrence across Eyre and Yorke Peninsulas and the Murray Mallee is <5000 km^2 and birds occupy <500 km^2 but not severely fragmented, and at risk from fire, with a consequent decline in EOO, AOO, quality of habitat, and number of locations, subpopulations and mature individuals, spread across at least 10 locations

Status 2000

Vulnerable B1ab(i,ii,iii,iv,v)+2ab(i,ii,iii,iv,v)

Status 1990

Vulnerable B1ab(i,ii,iii,iv,v)+2ab(i,ii,iii,iv,v)

Taxonomy

P. n. nigrogularis (western heath, Western Australia) is Endangered, *P. n. lashmari* (Kangaroo I.) is Near Threatened, *P. n. oberon* (Stirling Ranges to Ravensthorpe, WA) is Least Concern, as is the species

Taxonomic uniqueness: medium (4 genera/family, 5 species/genus, 4 subspecies/species)

Range

Found in 3 widely separated regions: southern Eyre Peninsula, south-western Yorke Peninsula and the Murray Mallee of South Australia and Victoria (Schodde and Mason 1991). In Eyre Peninsula, occurs in Coffin Bay and Lincoln Bay National Parks (total area 590 km^2). In Yorke Peninsula, occurs in Innes National Park and Warrenben Conservation Park (130 km^2; Higgins and Peter 2002). In the Murray Mallee known from Billiatt Conservation Park south to Ninety-Mile Plain, Mount Rescue Conservation Park, Ngarkat Conservation Park, Comet Bore and at several sites further west. However, the population is so low (Department of Sustainability, Environment, Water, Population and Communities 2011) that the area occupied could be <10 km^2. Probably extirpated from Victoria, where previously recorded at Manya, Murrayville, Red Bluff and Moonlight Tank, but the last definite record was 1974 (Hunt 1976; Higgins and Peter 2002) though it may survive in unsurveyed areas of Big Desert (Department of Sustainability, Environment, Water, Population and Communities 2011). Assessed here as occurring in 4 subpopulations assuming that Eyre Peninsula birds can still move along the coast and that the 2 reserves on the Yorke Peninsula are connected. Birds of the Murray-Mallee region are separated into at least 2 groups (Billiatt and Ngarkat). It is unlikely to have survived in unsurveyed areas of Big Desert (Clarke 2007). Assessed as having at least 10 locations, with large fires posing plausible threats to several of the subpopulations, but unlikely ever to eliminate any of the western subpopulations in 1 event. Given that not all habitat is suitable in the western parks, the AOO is estimated to be around 400 km^2.

Abundance

With an AOO of 400 km^2 and a density of around 20 birds/km^2 (Higgins and Peter 2002), a population of 6000 mature individuals is estimated, with the largest subpopulation of about 5000 on Eyre Peninsula. However, there are thought to be <100 individuals in the Murray Mallee (Department of Sustainability, Environment, Water, Population and Communities 2011). Eyre and Yorke Peninsula populations are noted as small but stable in 2006–2007, where recorded from <5% survey sites (Johnstone 2011), but large fires in Ngarkat and Billiatt Conservation Park have almost certainly reduced the tiny population in the east still further, and probably also reduced the number of locations.

Current eligibility against IUCN Red List Criteria

IUCN category	Criteria eligibility
A	Not applicable: past, current or future population declines are thought unlikely to exceed 20% in any 3-generation period
B	Vulnerable: EOO <20 000 km², AOO <2000 km², population at ≤10 locations, inferred continuing decline in EOO, AOO, area, extent and quality of habitat, no. locations, no. subpopulations, no. mature individuals
C	Not applicable: population 2500–10 000 mature individuals, continuing decline not observed or estimated, largest subpopulation >1000 mature individuals, >1 subpopulation
D	Not applicable: population >1000 mature individuals, >5 locations
E	Not applicable: no population viability analysis undertaken

IUCN Red List assessment data

	Estimate	Reliability
Extent of occurrence trend	4900 km² decreasing	medium medium
Area of occupancy trend	300 km² decreasing	low medium
No. of mature individuals trend	6000 decreasing	low medium
No. subpopulations	4	medium
No. locations	10	medium
Largest subpopulation	5000	low
Generation time	6.4 years	low
Global population share	100%	high

Ecology

On Eyre and Yorke Peninsulas, the Western Whipbird occurs in dense vegetation of mallee eucalypts, heath or acacias. In the eastern mallee, it occupies thickets with a dense shrubby understorey below an open mallee eucalypt layer. Its habitat is characterised more by vegetation structure than species composition and the frequency of fire has an important role in maintaining a suitable structure. Most records are from vegetation last burnt 10–25 years earlier (Department of Sustainability, Environment, Water, Population and Communities 2011). The subspecies forages on the ground primarily for insects. They build a domed nest close to the ground (Higgins and Peter 2002). A generation time of 6.4 years (BirdLife International 2011) is derived from an age at first breeding of 0.5 years and a maximum longevity of 12.3 years, both values extrapolated from Eastern Whipbird *P. olivaceus*.

Threats

Past clearance of mallee has been responsible for contraction and fragmentation of the habitat of the Western Whipbird. Fire is currently the major threat, fragmenting subpopulations, even in continuous habitat by making recolonisation between fire events difficult and leading to local extinctions (Baker-Gabb 2005). Major fires in Billiatt, Ngarkat and Mt Rescue Conservation Parks and Lincoln National Park in the last decade have undoubtedly killed many birds and rendered much habitat unsuitable for extended periods. Given their poor flight and subsequent low dispersal ability, Western Whipbirds are not able to disperse easily across cleared land. Ground-foraging and low nesting makes them vulnerable to introduced predators such as feral cats *Felis catus* and foxes *Vulpes vulpes* (Baker-Gabb 2005). Given the main threat of fires, it is assessed here as occurring at 10 locations but the number of viable locations may be <10.

Conservation objectives

1. All subpopulations retained
2. Populations at sites burnt 2000–2010 recovering
3. Re-established and secure in Victoria

Information required

1. Location of birds in the Murray Mallee region
2. Extent of genetic isolation of geographically separated populations and the potential need for active translocations
3. Efficacy of herbivore and fox control
4. Relationships between whipbirds and fire

Management actions required

1. Monitor populations at key sites, especially where management is occurring
2. Determine the need for population translocations, and develop a re-introduction program if necessary
3. Manage fire at a landscape scale so that there is both adequate connectivity across the Murray Mallee reserve system and a variety of ages since fire
4. Manage total grazing pressure, not just from introduced herbivores
5. Create a multi-stakeholder team with responsibility for ensuring threatened species are conserved on Eyre Peninsula

Bibliography

Baker-Gabb D (2005) 'Recovery plan for the Mallee Emu-wren *Stipiturus mallee*, Red-lored Whistler *Pachycephala rufogularis*, Western Whipbird, *Psophodes nigrogularis leucogaster*, and other threatened mallee birds. Second Draft'.

BirdLife International (2011) 'Species factsheet: *Psophodes nigrogularis*'. Retrieved 17 January, 2011 from <http://www.birdlife.org/>.

Clarke R (2007) 'Surveys for threatened mallee birds within the Murray mallee reserve system, Victoria, spring 2006'. Report to Department of Sustainability and Environment, Mildura.

Department of Sustainability, Environment, Water, Population and Communities (2011) *Psophodes nigrogularis leucogaster.* In 'Species Profile and Threats Database'. Department of Sustainability, Environment, Water, Population and Communities, Canberra. Retrieved 24 February, 2011 from <http://www.environment.gov.au/sprat>.

Higgins PJ, Peter JM (Eds) (2002) *Handbook of Australian, New Zealand and Antarctic Birds. Volume 6: Pardalotes to Spangled Drongo*. Oxford University Press, Melbourne.

Hunt T (1976) Birds in the Big Desert region of Victoria and South Australia. *Victorian Ornithological Research Group Notes* **12**, 3–13.

Johnstone G (2011) 'Distribution and conservation status of western whipbirds *(Psophodes nigrogularis)*.' Retrieved 14 April, 2011 from <http://www.zoossa.com.au/conservation-ark/research/past-projects?project=Distribution%20and%20conservation%20status%20of%20western%20whipbirds%20%28Psophodes%20nigrogularis%29>.

Schodde R, Mason IJ (1991) Subspeciation in the Western Whipbird *Psophodes nigrogularis* and its zoogeographical significance, with descriptions of two new subspecies. *Emu* **91**, 133–144.

Comments received from

Graham Carpenter, Peter Copley, Victor Hurley, Peter Menkhorst, David Paton, Lynn Pedler, Danny Rogers, Adrian Stokes

Subspecies Conservation Summary

Western Whipbird (western heath)

Psophodes nigrogularis nigrogularis Gould, 1844
Eupetidae

Conservation status

Endangered B2ab(iii)

Reasons for listing

The very small population is restricted to a very small AOO at very few locations. The recent series of large wildfires suggests a probable continuing decline in habitat.

Status 2000

Vulnerable D2

Reason for change: multiple large fires suggest a continuing decline in habitat quality

Status 1990

Vulnerable D2

Taxonomy

P. n. leucogaster (east S. A. and west Vic.) is Vulnerable, *P. n. lashmari* (Kangaroo I., S. A.) is Near Threatened, *P. n. oberon* (Stirling Ranges to Ravensthorpe, W. A.) is Least Concern, as is the species

Taxonomic uniqueness: medium (4 genera/family, 5 species/genus, 4 subspecies/species)

Range

Restricted to a small area east of Albany in south-west Western Australia from Two Peoples Bay/Mt Gardiner in the south west to about Cape Riche Road in the north east, with the South Coast Highway as an approximate inland boundary (Gilfillan *et al.* 2009). Historically occupied south-west WA, along the west coast from Perth to Augusta and on the south coast from King Georges Sound east to at least Two People's Bay (Milligan 1904; Whittell 1951; Webster 1966; Smith 1977; Schodde and Mason 1991).

Abundance

Subpopulation at Two Peoples Bay increased from about 60 pairs in 1970 to about 100 pairs in 1982 (Smith 1991). Following a series of fires in the Two Peoples Bay-Manypeaks area between 2000 and 2004, the number of singing males dropped from over 400 in 2001 to c.200 in 2005 and c.175 in 2006 (Comer *et al.* 2005; A.H. Burbidge *in litt.*). Not recorded at Mt Taylor since 2004 (S. Comer *in litt.*). The population size for the whole subspecies is assessed here as 500 mature individuals. At the time of assessment, it was considered not to be declining over a timescale of 3 generations (16 years), but an increasing risk of wildfires was considered to be probably causing a continuing decline in habitat quality.

Ecology

At Two Peoples Bay occurs in dense shrubland with an open overstorey (Higgins and Peter 2002). On Mt Manypeaks the subspecies also occurs in dense low mallee and shrubland (Cale and Burbidge 1993). All of the domed nests found have been in dense bushes in heath adjacent to thickets. Clutch size is usually 2 (Smith 1991). The birds are insectivorous, feeding mostly on or near the ground (Smith 1991). They fly short distances, are poor dispersers and the population was assessed as fragmented into 3 subpopulations by Garnett and Crowley (2000). A generation time of 6.4 years (BirdLife International 2011) is derived from an age at first breeding of 0.5 years and a maximum longevity of 12.3 years, both values extrapolated from published figures for Eastern Whipbird *P. olivaceus*.

Current eligibility against IUCN Red List Criteria

IUCN category	Criteria eligibility
A	Not applicable: past, current or future population declines are thought unlikely to exceed 20% in any 3-generation period
B	Endangered: AOO <500 km^2, at ≤5 locations with plausible future threat, inferred continuing decline in area and quality of habitat
C	Vulnerable: population 2500–10 000 mature individuals, inferred continuing decline, 250–1000 mature individuals in largest subpopulation
D	Vulnerable: population 250–1000 mature individuals, ≤5 locations with plausible future threat
E	Not applicable: no population viability analysis undertaken

IUCN Red List assessment data

	Estimate	Reliability
Extent of occurrence trend	500 km^2 stable	high high
Area of occupancy trend	350 km^2 stable	high high
No. of mature individuals trend	500 decreasing	low high
No. subpopulations	1	high
No. locations	5	low
Generation time	5.2 years	low
Global population share	100%	high

Threats

Fire appears to be the main threat and is thought to have caused the species' disappearance from the Cape Mentelle – Margaret R. area (Whitley 1971). The population increase at Two Peoples Bay has been attributed to local exclusion of fire since 1970 (Smith 1985). The species has been observed to recolonise areas of burnt vegetation 7–10 years after fire in 1 area (Smith 1985), and 4–6 years after fire in another, the recolonisation time probably being related to the growth rate of the vegetation (Smith 1991). Given that fire is the main threat, locations are defined as those that are independent during massive wildfires; here estimated to be 5. Predation by introduced mammals, particularly foxes *Vulpes vulpes* and feral cats *Felis catus*, may be significant. Other potential threats include dieback caused by *Phytophthora cinnamomi*, disturbance by introduced animals, weed invasion and changes in hydrological regimes (Gilfillan *et al.* 2009). Historical clearance for grazing and agriculture is considered responsible for contraction of its distribution, but no longer a threat as almost all individuals now live in conservation reserves.

Conservation objectives

1. Population numbers and area of occupation increased or maintained
2. Knowledge of current distribution improved
3. Management options increased through improved ecological knowledge

Information required

1. Size of all known subpopulations
2. Habitat requirements, in particular in relation to fire age, vegetation structure and food availability
3. Relationship between *Phytophthora cinnamomi* and habitat requirements
4. Impacts of fox and cat predation

Management actions required

1. Active fire management around all locations
2. Continue habitat management and threat abatement of all occupied areas
3. Survey potential habitat for new subpopulations
4. Continue to investigate habitat requirements, in particular in relation to fire age, vegetation structure, food availability, *Phytophthora cinnamomi* and fox and cat density
5. Continue to support coordination of management by the South Coast Threatened Bird Recovery Team

Bibliography

BirdLife International (2011) 'Species factsheet: *Psophodes nigrogularis*'. Retrieved 17 January, 2011 from <http://www.birdlife.org/>.

Cale PG, Burbidge AH (1993) 'Research Plan for the Western Ground Parrot, Western Whipbird and Western Bristlebird'. Report to Australian National Parks and Wildlife Service.

Comer S, Danks A, Burbidge AH (2005) Noisy Scrub-birds, Western Whipbirds and wildfire at Mt Manypeaks. *Western Australian Bird Notes* **113**, 16–17.

Garnett S, Crowley GM (2000) *The Action Plan for Australian Birds 2000.* Environment Australia, Canberra.

Gilfillan S, Comer S, Burbidge AH, Blyth J, Danks A, Newell J (2009) 'South Coast Threatened Birds Recovery Plan. Western Australian Wildlife Management Program No. 44'. WA Department of Environment and Conservation, Perth.

Higgins PJ, Peter JM (Eds) (2002) *Handbook of Australian, New Zealand and Antarctic Birds. Volume 6: Pardalotes to Spangled Drongo.* Oxford University Press, Melbourne.

Milligan AW (1904) Notes on a trip to the Wongan Hills, Western Australia, with a description of a new *Ptilotis. Emu* **3**, 217–226.

Schodde R, Mason IJ (1991) Subspeciation in the Western Whipbird *Psophodes nigrogularis* and its zoogeographical significance, with descriptions of two new subspecies. *Emu* **91**, 133–144.

Smith GT (1977) The effect of environmental change on six rare birds. *Emu* **77**, 173–179.

Smith GT (1985) Fire effects on populations of the Noisy Scrub-bird (*Atrichornis clamosus*), Western Bristlebird (*Dasyornis longirostris*), and Western Whipbird (*Psophodes nigrogularis*). In *Fire Ecology and Management of Ecosystems in Western Australia.* (Ed. JR Ford) pp. 95–102. Western Australian Institute of Technology, Perth.

Smith GT (1991) Ecology of the Western Whipbird *Psophodes nigrogularis* in Western Australia. *Emu* **91**, 145–157.

Webster HO (1966) The Western Whipbird at Two People Bay. *Western Australian Naturalist* **10**, 25–28.

Whitley GP (1971) Field notes on birds by Thomas Carter. *Western Australian Naturalist* **12**, 41–44.

Whittell HM (1951) A review of the work of John Gilbert in Western Australia. *Emu* **51**, 17–29.

Comments received from

Allan Burbidge, Sarah Comer

Subspecies Conservation Summary

Long-tailed Triller (Norfolk Island)

Lalage leucopyga leucopyga (Gould, 1838)
Campephagidae

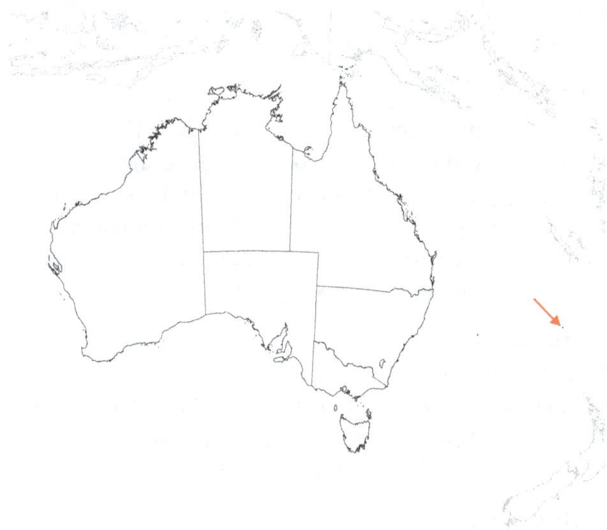

Conservation status

Extinct

Reasons for listing

No records since 1942

Status 2000

Extinct

Status 1990

Extinct

Taxonomy

Extant extralimital subspecies of the Long-tailed Triller occur on New Caledonia, Vanuatu and the Solomon Is.; the species is Least Concern
Taxonomic uniqueness: medium (6 genera/family, 13 species/genus, 6 subspecies/species)

Former range

Endemic to Norfolk I. (Schodde *et al.* 1983).

Former abundance

It was abundant in 1941, but has not been seen since 1942 (Schodde *et al.* 1983).

Ecology

Ecology is unknown but probably occupied all wooded habitats on the island.

Reasons for extinction

The subspecies almost certainly became extinct as result of predation by the Black Rat *Rattus rattus*. The rats arrived in the mid-1940s (Robinson 1988) and the triller has not been seen since 1942.

Bibliography

Robinson D (1988) 'Ecology and management of the Scarlet Robin, White-breasted White-eye and Long-billed White-eye on Norfolk Island'. Report to the Australian National Parks and Wildlife Service, Canberra.

Schodde R, Fullagar P, Hermes N (1983) *A Review of Norfolk Island Birds: Past and Present.* Australian National Parks and Wildlife Service, Canberra.

Species Conservation Summary

Red-lored Whistler

Pachycephala rufogularis Gould, 1841
Pachycephalidae

Conservation status

Vulnerable B2ab(i,ii,iii,iv,v), C2a(i)

Reasons for listing

Occupies a small area at ≤10 locations which fire threatens to reduce in extent of occurrence, area occupied, habitat quality, no. locations and no. mature individuals. The population size is poorly-known but estimated to be <10 000 mature individuals with <1000 mature individuals in each subpopulation

Status 2000

Near Threatened C2a(i)
Reason for change in 2010: it is assumed that large fires reduced the AOO to <2000 km^2 and the size of largest subpopulation < 1000 mature individuals

Status 1990

Near Threatened C2a(i)

Taxonomy

No infraspecific taxa described
Taxonomic uniqueness: medium (11 genera/family, 33 species/genus, 1 subspecies/species)

Range

Semi-arid regions of New South Wales, Victoria and South Australia. The largest occupied areas are the Murray-Sunset National Park and smaller contiguous reserves and the Big Desert in Vic., and Billiatt Conservation Park, Ninety-Mile Desert and Upper South-East region including Riverland Biosphere Reserve, in SA (Baker-Gabb 2005) as well as Scotia in NSW (Higgins and Peter 2002). This region is fragmented by historical clearance and contemporary fires into at least 4 subpopulations. There is also an isolated subpopulation at Noombinnie and Round Hill Nature Reserves in central NSW (Chapman 1990; Barrett *et al.* 2003) and possibly 1 as far east as Annuello Fauna and Flora Reserve (Watson 2011). A small population at Pinkawillinie Conservation Park on the northern Eyre Peninsula in SA (Matthew *et al.* 1995; Matthew *et al.* 1996) has not been found again despite searching (Way 2007). It is also thought to have become locally extirpated at Pulletop Nature Reserve in central NSW (last seen 1982), near Ennisvale, Springwood Stations and Tarawi Nature Reserves in south-west NSW (last seen 1996) and, during the 1980s, from a number of locations within its core range including Mount Boothby and Ferries-McDonald Conservation Parks, Hattah-Kulkyne National Park, Bronzewing and Wathe Flora and Fauna Reserves. Occasional records outside these areas are likely to be dispersing birds or misidentifications (Baker-Gabb 2005; Department of Sustainability, Environment, Water, Population and Communities 2011). Such records suggest that apparently isolated subpopulations are connected. However, it is also known that some have become extinct and not been recolonised, suggesting genetic isolation. Even in the extensive habitat areas where it occurs, the distribution is patchy (Woinarski *et al.* 1987). Given the small population size and measured densities, the AOO is estimated here to be no more than about 500 km^2 across all sites.

Abundance

The largest populations are probably in Riverland Biosphere Reserve but only part of this is occupied (Harrington 2002) since a large fire burnt a substantial proportion of Taylorville Station. If 20% of Riverland is occupied and the home range is about 50 ha with 30% overlap (Moise 2008), the total population might be about 1000 mature individuals. Recent surveys suggest that, following large fires during the last decade, there are now about 100–200 pairs in the Ngarkat/Big Desert/Wyperfeld complex, and 100+ pairs in Murray-Sunset/Hattah complex (R. Clarke *in litt.*). Given that there are only likely to be small numbers in Scotia and central NSW, the total population is thus likely to be no greater than 2000 mature individuals. Population declines have been recorded at Billiatt, Mount Rescue and Ngarkat Conservation Parks, probably since the 1980s, and only the largest tracts of mallee at these 3 locations remain occupied (Clarke 2004). However, 'there is no evidence to suggest that the documented regional declines are not continuing within the species remaining area of occupation' (Clarke 2007).

Ecology

Usually found in habitats with spinifex *Triodia* hummock grassland or tall open heath, especially mallee which has been unburned for 21–44 years or heath-dominated mallee unburned for 10–24 years. At such sites it can attain densities of 0.02 birds/ha (Woinarski 1987; Higgins and Peter 2002; Clarke 2005). In good habitat, pairs can occupy home ranges of >200 ha, but most had around 50 ha with some overlap between pairs (Moise 2008). Birds dispersing in the

Current eligibility against IUCN Red List Criteria

IUCN category	Criteria eligibility
A	Near Threatened: population decline of 20–29% over the last 3 generations (19 years) suspected from deteriorating habitat quality
B	Vulnerable: AOO <2000 km^2, at ≤10 locations with plausible future threat, estimated continuing decline in EOO, AOO, habitat quality, no. locations, no. mature individuals
C	Vulnerable: population <10 000 mature individuals, inferred continuing decline, 250–1000 mature individuals in largest subpopulation
D	Not applicable: population >1000 mature individuals, >5 locations
E	Not applicable: no population viability analysis undertaken

IUCN Red List assessment data

	Estimate	Reliability
Extent of occurrence trend	130 000 km^2 decreasing	medium medium
Area of occupancy trend	500 km^2 decreasing	low high
No. of mature individuals trend	2,000 decreasing	low high
No. subpopulations	6	medium
No. locations	10	high
Largest subpopulation	<1000	low
Generation time	9.8 years	low
Global population share	100%	high

non-breeding season can occupy a variety of other woodland habitats. They forage on and close to the ground, capturing insects and other invertebrates by gleaning or sallying, and build a cupped nest, in which they lay 2 to 3 eggs (Higgins and Peter 2002). A generation time of 9.8 years (BirdLife International 2011) is derived from an age at first breeding of 1.5 years (extrapolated from other passerines of similar size and ecology) and a maximum longevity in the wild of 19.0 years (extrapolated from congeners).

Threats

The current major threat is fire, given the preference for long unburnt habitat. Other threats include degradation of habitat by grazing mammalian herbivores including stock, feral goats *Hircus capra* and rabbits *Oryctolagus cuniculus*, which may explain why observations at Gluepot reserve were all >6 km from water (Harrington 2002), and predation by foxes *Vulpes vulpes* (Baker-Gabb 2005). Broombush *Melaleuca uncinata* harvesting continues in some areas used by the whistlers and may be locally deleterious. This species coexists at several sites with the similar Gilbert's Whistler *P. inornata* and competition is probably secondary to other threats. Historically, the major reason for decline was the extensive clearance of mallee vegetation in the 19th and 20th centuries.

Conservation objectives

1. All subpopulations retained
2. Populations at sites burnt in 2000–2010 are recovering

Information required

1. Trends in major subpopulations, particularly rates of recovery from fire
2. Extent of genetic isolation of geographically separated populations and the potential need for active translocations
3. Efficacy of herbivore and fox control
4. Impact of Broombush harvesting operations
5. Distribution outside protected areas
6. Recolonisation time after fire

Management actions required

1. Monitor populations at key sites, especially where management occurring
2. Determine the need for population translocations, and develop a re-introduction program if necessary
3. Minimise the frequency and scale of fire in occupied sites
4. Manage total grazing pressure especially from introduced herbivores
5. Amend Broombush harvesting protocols if necessary
6. Work with landholders outside protected areas to manage their land appropriately

Bibliography

Baker-Gabb D (2005) 'Recovery plan for the Mallee Emu-wren *Stipiturus mallee*, Red-lored Whistler *Pachycephala rufogularis*, Western Whipbird, *Psophodes nigrogularis leucogaster*, and other threatened mallee birds. Second Draft'.

Barrett G, Silcocks A, Barry S, Cunningham R, Poulter R (2003) *The New Atlas of Australian Birds*. Royal Australian Ornithologists Union, Hawthorn East.

BirdLife International (2011) 'Species factsheet: *Pachycephala rufogularis*'. Retrieved 17 January, 2011 from <http://www.birdlife.org/>.

Chapman G (1990) Profile: the Red-lored Whistler. *Birds International* **2**, 20–25.

Clarke R (2004) 'Threatened bird species recorded within the Billiatt and Ngarkat Conservation Park complexes, South Australia, Spring 2003'. Unpublished report to the Department for Environment and Heritage, Adelaide.

Clarke R (2005) 'Ecological requirements of birds specialising in mallee habitats'. Unpublished La Trobe University report to Department for Environment and Heritage, Adelaide.

Clarke, R (2007) 'Surveys for threatened mallee birds within the Murray mallee reserve system, Victoria, spring 2006'. Report to Department of Sustainability and Environment.

Department of Sustainability, Environment, Water, Population and Communities (2011) *Pachycephala rufogularis*. In 'Species Profile and Threats Database'. Department of Sustainability, Environment, Water, Population and Communities, Canberra. Retrieved 24 February, 2011 from <http://www.environment.gov.au/sprat>.

Harrington R (2002) The effects of artificial watering points on the distribution and abundance of avifauna in an arid and semi-arid mallee environment. PhD thesis, University of Melbourne, Melbourne.

Higgins PJ, Peter JM (Eds) (2002) *Handbook of Australian, New Zealand and Antarctic Birds. Volume 6: Pardalotes to Spangled Drongo*. Oxford University Press, Melbourne.

Matthew J, Carpenter G, Croft T (1996) Revision of the distribution of the Red-lored Whistler in South Australia. *South Australian Ornithologist* **32**, 103–107.

Matthew J, Croft T, Carpenter G (1995) A record of the Red-lored Whistler on Eyre Peninsula. *South Australian Ornithologist* **32**, 39–40.

Moise D (2008) Ecology of Red-lored and Gilbert's Whistler in South Australia. PhD thesis, University of Adelaide, Adelaide.

Watson S (2011) 'Research survey of Mallee Emu-wren *Stipiturus mallee* at Murray Sunset National Park and Annuello Flora and Fauna Reserve'. Report for the Mallee Catchment Management Authority by Deakin University, Burwood.

Way S (2007) 'Re-surveying for Red-lored Whistler *Pachycephala rufogularis* in Pinkawillinie Conservation Park, northern Eyre Peninsula'. Department for Environment and Heritage, South Australia.

Woinarski JCZ (1987) Notes on the status and ecology of the Red-lored Whistler *Pachycephala rufogularis*. *Emu* **87**, 224–231.

Comments received from

David Baker-Gabb, Sarah Brown, Rohan Clarke, Victor Hurley

Subspecies Conservation Summary

Golden Whistler (Lord Howe Island)

Pachycephala pectoralis contempta Hartert, 1898

Pachycephalidae

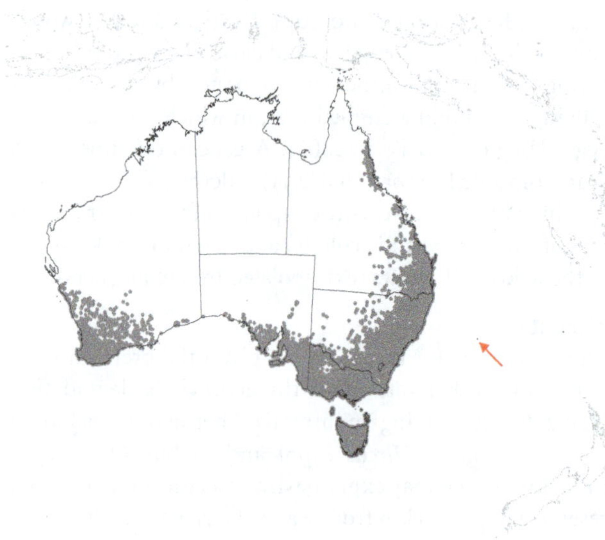

Conservation status
Near Threatened D2

Reasons for listing
Occurs at a single location where susceptible to introduction of predators, competitors or disease, although the risk is low

Status 2000
Near Threatened D2

Status 1990
Near Threatened D2

Taxonomy
P. p. xanthoprocta (Norfolk I.) is Near Threatened; 4 mainland Australian subspecies are Least Concern, as is the species. There are 58 extralimital subspecies. Taxonomic uniqueness: low (11 genera/family, 33 species/genus, 64 subspecies/species)

Range
Endemic to Lord Howe I. where it occurs throughout forested parts (McAllan *et al.* 2005).

Abundance
Described as an uncommon breeding resident (McAllan *et al.* 2004). There are no measures of abundance but, based on a density of about 5 birds/ha in Norfolk I. National Park (G. Dutson pers. obs.) and an occupied forest area on Lord Howe I. of about 1000 ha, there could be 10 000 mature individuals on the island. To be precautionary this estimate is halved as many of the forest types on Lord Howe I. are likely to be less productive than in Norfolk I. National Park and may not be as suitable.

Ecology
The subspecies lives in forests throughout Lord Howe I., feeding primarily on insects, and nests high in the trees away from most predators (Hutton 1991). A generation time of 9.8 years (BirdLife International 2011) is derived from a mean age at first breeding of 1.5 years, extrapolated from passerines of similar size and ecology, and a maximum longevity in the wild of 18.0 years (Australian Bird and Bat Banding Scheme).

Threats
Golden Whistlers on Lord Howe I. have survived the introduction of domestic and feral cats *Felis catus*, Black Rats *Rattus rattus*, feral pigs *Sus scrofa*, feral goats *Hircus capra* and Masked Owls *Tyto novaehollandiae*. The restricted area of occupancy, however, makes the subspecies susceptible to catastrophe, such as the introduction of another predator or disease, though the risk is low.

Conservation objectives
1. Maintain current breeding population

Information required
1. Population trends on Lord Howe I.

Current eligibility against IUCN Red List Criteria

IUCN category	Criteria eligibility
A	Not applicable: past, current or future population declines are thought unlikely to exceed 20% in any 3-generation period
B	Not applicable: EOO, AOO, habitat, locations or population not declining, population fluctuations not extreme
C	Not applicable: no continuing population decline or extreme fluctuations
D	Near Threatened: ≤5 locations with a plausible but low probability threat
E	Not applicable: no population viability analysis undertaken

IUCN Red List assessment data

	Estimate	Reliability
Extent of occurrence trend	18 km^2 stable	high high
Area of occupancy trend	12 km^2 stable	medium high
No. of mature individuals trend	5000 stable	low high
No. subpopulations	1	high
No. locations	1	high
Generation time	9.8 years	low
Global population share	100%	high

Management actions required

1. Maintain quarantine for Lord Howe I.
2. Manage threats on Lord Howe I. as required

Bibliography

BirdLife International (2011) 'Species factsheet: *Pachycephala pectoralis*'. Retrieved 17 January, 2011 from <http://www.birdlife.org/>.

Hutton I (1991) *Birds of Lord Howe Island, Past and Present*. The Author, Coffs Harbour.

McAllan IAW, Curtis BR, Hutton I, Cooper RM (2004) The birds of the Lord Howe Island Group: a review of records. *Australian Field Ornithology* **21**, 1–82.

Subspecies Conservation Summary

Golden Whistler (Norfolk Island)

Pachycephala pectoralis xanthoprocta Gould, 1838
Pachycephalidae

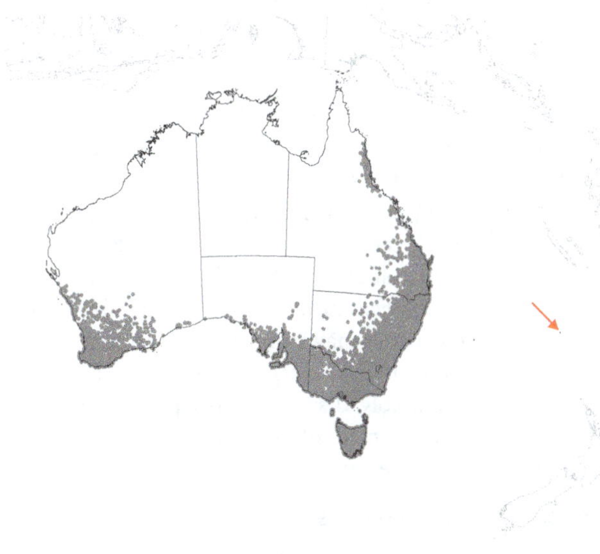

Conservation status

Near Threatened D2

Reasons for listing

Occurs at a single location where susceptible to introduction of predators, competitors or disease, although the risk is low

Status 2000

Near Threatened D2

Status 1990

Near Threatened D2

Taxonomy

P. p. contempta (Lord Howe I.) is Near Threatened; 4 mainland Australian subspecies are Least Concern, as is the species. There are 58 extralimital subspecies. Taxonomic uniqueness: low (11 genera/family, 33 species/genus, 64 subspecies/species)

Range

Endemic to Norfolk I., where it is largely confined to the Norfolk Island National Park and nearby forested areas (Robinson 1997; Commonwealth of Australia 2005; Director of National Parks 2010).

Abundance

A steady decline recorded through 1960s and 1970s, but the subspecies is still present over nearly half the island in 1978 (Schodde *et al.* 1983). By 1990, virtually confined to Norfolk Island National Park (Bell 1990) with a population estimated at 535 pairs (Robinson 1988). Recent estimates suggest the population has now stabilised (Robinson 1997) and a survey in 2010 estimated there were about 2200 birds in the park on the basis of 121 birds detected in point counts at a density of about 500/km^2 (G. Dutson pers. obs.).

Current eligibility against IUCN Red List Criteria

IUCN category	Criteria eligibility
A	Not applicable: past, current or future population declines are thought unlikely to exceed 20% in any 3-generation period
B	Not applicable: EOO, AOO, habitat, locations or population not declining, population fluctuations not extreme
C	Not applicable: no continuing population decline or extreme fluctuations
D	Near Threatened: ≤5 locations with a plausible but low probability threat
E	Not applicable: no population viability analysis undertaken

IUCN Red List assessment data

	Estimate	Reliability
Extent of occurrence trend	40 km^2 stable	high high
Area of occupancy trend	8 km^2 stable	high high
No. of mature individuals trend	3000 stable	low high
No. subpopulations	1	high
No. locations	1	high
Generation time	9.8 years	low
Global population share	100%	high

Ecology

Prefers the shrubby understorey of rainforest, palm forest and indigenous pine forest (Smithers and Disney 1969), but also uses plantations of exotic species. It has been recorded in, or at the edges of, pockets of suitable habitat throughout the island, but does not occur near gardens (Schodde *et al.* 1983). However, densities in weedy forest (2.5 singing males/ha) is very similar to non-weedy native forest (2.8 males/ha; G. Dutson pers. obs.). A generation time of 9.8 years (BirdLife International 2011) is derived from a mean age at first breeding of 1.5 years, extrapolated from passerines of similar size and ecology, and a maximum longevity in the wild of 18.0 years (Australian Bird and Bat Banding Scheme).

Threats

Much suitable habitat has been cleared or fragmented, and the subspecies appears to be confined to the largest tract of remaining forest. The reason for the decline in the 1960s and 1970s is not known given the island has been settled for 2 centuries. However, both Black Rats *Rattus rattus* (introduced in the mid 1940s; Robinson 1988) and feral cats *Felis catus* may take some birds (Bell 1990), and the whistler may be more vulnerable to Black Rats in disturbed environments. The addition of further predators and possibly disease is also possible but all threats are here deemed of low likelihood given the density of birds in the national park.

Conservation objectives

1. Maintain current breeding population

Information required

1. Population trends

Management actions required

1. Maintain quarantine
2. Manage additional threats as they emerge
3. Consider reintroduction to Phillip I. when vegetation there is sufficiently mature

Bibliography

Bell BD (1990) 'The status and management of the White-breasted White-eye and other birds on Norfolk Island'. Report to Australian National Parks and Wildlife Service, Canberra.

BirdLife International (2011) 'Species factsheet: *Pachycephala pectoralis*'. Retrieved 17 January, 2011 from <http://www.birdlife.org/>.

Commonwealth of Australia (2005) 'National recovery plan for the Norfolk Island Scarlet Robin *Petroica multicolor multicolor* and the Norfolk Island Golden Whistler *Pachycephala pectoralis xanthroprocta*'. Department of the Environment and Heritage, Canberra.

Director of National Parks (2010) 'Draft Norfolk Island region threatened species recovery plan'. Department of the Environment, Water, Heritage and the Arts, Canberra.

Robinson D (1988) 'Ecology and management of the Scarlet Robin, White-breasted White-eye and Long-billed White-eye on Norfolk Island'. Report to the Australian National Parks and Wildlife Service, Canberra.

Robinson D (1997) 'An evaluation of the status of the Norfolk island Robin following rat-control and weed-control works in the Norfolk island National Park'. Report to Environment Australia, Canberra.

Schodde R, Fullagar P, Hermes N (1983) *A Review of Norfolk Island Birds: Past and Present*. Australian National Parks and Wildlife Service, Canberra.

Smithers CN, Disney HJ (1969) The distribution of terrestrial and freshwater birds on Norfolk Island. *Australian Zoologist* **15**, 127–140.

Comments received from

Margaret Christian, Joe Forshaw, Penny Olsen

Subspecies Conservation Summary

Pied Currawong (Grampians)

Strepera graculina ashbyi Mathews, 1913

Conservation status
Extinct

Reasons for listing
Probably extinct before 1900; even when described only hybrids seem to have been present

Status 2000
Extinct

Status 1990
Extinct

Taxonomy
S. g. crissalis (Lord Howe Island) is Endangered; 4 other subspecies are Least Concern, as is the species
Taxonomic uniqueness: medium (4 genera/family, 3 species/genus, 6 subspecies/species)

Former range
Presumed to have been an isolated population west of the grassy western plains in Victoria and extending west to about the South Australian border. Before the 1950s, birds resembling *S. g. ashbyi* were recorded east to the upper Yarra valley (Amadon 1951); now such forms only occur west of a line between Ballarat and Cape Otway (Schodde and Mason 1999). However, blacker birds, with more extensive white than is thought to have been characteristic of pure *S. g. ashbyi*, were already present within its former range in 1925 (Ashby 1927).

Former abundance
Pied Currawongs were notably common in the Grampians in 1925 (Ashby 1927), as they are now in western Vic (Farnes 2007), but both forms were already present and would have been interbreeding.

Ecology
Pied Currawongs generally live in forest and woodland, where they take insects and small birds and animals. They also frequent settlements and agricultural land (Higgins *et al.* 2006). Nothing is known about the ecology of *S. g. ashbyi* in particular.

Reasons for extinction
This subspecies is believed to have been hybridised out of existence by the more widespread *S. g. nebulosa*. While uniformly slaty-grey individuals with vestigial white in the wings do still occur in western Vic, so do blacker birds with much more white as well as intermediate forms so that no distinct subpopulation can be identified. Schodde and Mason (1999) suggested that the sooty form could be either a morph under simple genetic control, or a regional form whose range has been invaded secondarily by *S. g. nebulosa*. Introgression may have escalated, or been enabled, as a result of increases in tree density across western Vic since the late the 18th century after Aboriginal burning regimes were abandoned, but before deliberate tree-clearing was universal (Barr and Cary 1992).

Bibliography

Amadon D (1951) Taxonomic notes on the Australian Butcherbirds (Family Cracticidae). *American Museum Novitates* **1504**, 1–33.

Ashby E (1927) The Grampian Range of Victoria and its bird life. *Emu* **26**, 285–292.

Barr N, Cary J (1992) *Greening a Brown Land*. Macmillan Education Australia, Melbourne.

Farnes RF (2007) *Birds of Portland District Victoria*. Portland Field Naturalists Club Inc., Portland.

Higgins PJ, Peter JM, Cowling SJ (Eds) (2006) *Handbook of Australian, New Zealand and Antarctic Birds. Volume 7: Boatbill to Starlings. Part A. Boatbill to Larks*. Oxford University Press, Melbourne.

Schodde R, Mason IJ (1999) *The Directory of Australian Birds: Passerines*. CSIRO Publishing, Melbourne.

Subspecies Conservation Summary

Pied Currawong (Lord Howe Island)

Strepera graculina crissalis Sharpe, 1877
Artamidae

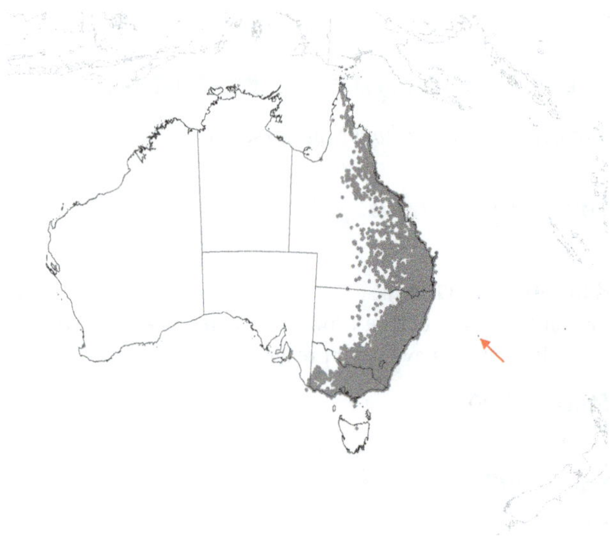

Conservation status
Endangered D

Reasons for listing
Population <250 mature individuals

Status 2000
Endangered D

Status 1990
Endangered D

Taxonomy
S. g. ashbyi (western Victoria) is Extinct, the other 4 mainland subspecies are Least Concern, as is the species
Taxonomic uniqueness: medium (4 genera/family, 3 species/genus, 6 subspecies/species)

Range
Endemic to islands of the Lord Howe I. group. The currawongs are widely distributed on the main island, mostly in forests, particularly in the south. Individuals are highly mobile and occasionally visit offshore islands (Department of Environment and Climate Change (NSW) 2007). In autumn and winter they form flocks and are more common around the settled areas (McAllan *et al.* 2004).

Abundance
The population, which is thought to be stable, has been estimated at 215±11 birds, including 42 juveniles, based on a banding study in 2006 (Carlile and Priddel 2007).

Ecology
Breeds in rainforest and palm forest types on Lord Howe I., particularly in undisturbed forests along streams. All territories contain a section of stream or gully that is lined by tall trees. They are omnivorous, foraging widely in all parts of the island and among seabird colonies offshore. Takes chicks of both land and sea birds, including those of Lord Howe Woodhen *Tricholimnas sylvestris* and White Tern *Gygis alba*, as well as live and dead Black Rats *Rattus rattus* and House Mice *Mus musculus*. They also feed on fruits, seeds and insects (Carlile and Priddel 2007).

Threats
The subspecies has successfully persisted alongside rats since 1918 (Hindwood 1940; McAllan *et al.* 2004). However, because currawongs scavenge dead rodents, they are likely to be in danger of secondary poisoning from brodifacoum, the bait favoured for planned eradication of introduced rodents from the Lord Howe I. group, which is why it is planned to take as many as possible into captivity while the corpses of poisoned rodents are toxic (Lord Howe Island Board 2009). Until the 1980s local residents persecuted currawongs for attacking poultry, woodhens or terns, though now do so rarely (Department of Environment and Climate Change (NSW) 2007). The small and restricted population is also susceptible to catastrophe, such as the introduction of another predator.

Conservation objectives
1. Existing population maintained

Current eligibility against IUCN Red List Criteria

IUCN category	Criteria eligibility
A	Not applicable: past, current or future population declines are thought unlikely to exceed 20% in any 3-generation period
B	Not applicable: EOO, AOO, habitat, locations or population not declining, population fluctuations not extreme
C	Not applicable: no continuing population decline or extreme fluctuations
D	Endangered: population 50–250 mature individuals
E	Not applicable: no population viability analysis undertaken

IUCN Red List assessment data

	Estimate	Reliability
Extent of occurrence trend	18 km^2 stable	high high
Area of occupancy trend	12 km^2 stable	high high
No. of mature individuals trend	200 stable	high medium
No. subpopulations	1	high
No. locations	1	high
Generation time	10 years	medium
Global population share	100%	high

Information required

1. Population trends

Management actions required

1. During the rat eradication period capture and retain in captivity on the island as many individuals as possible, releasing back to the wild as soon as possible after poisoned rodents have decayed
2. Continue annual population monitoring
3. Maintain high quarantine standards

Bibliography

Carlile N, Priddel D (2007) 'Population size and distribution of the Lord Howe Currawong *Strepera graculina crissalis*'. Report to Lord Howe Island Board, Sydney.

Department of Environment and Climate Change (NSW) (2007) 'Lord Howe Island Biodiversity Management Plan'. Department of Environment and Climate Change (NSW), Sydney.

Hindwood KA (1940) The birds of Lord Howe Island. *Emu* **40**, 1–86.

Lord Howe Island Board (2009) 'Draft Lord Howe Island rodent eradication plan'. Lord Howe Island Board, Lord Howe Island.

McAllan IAW, Curtis BR, Hutton I, Cooper RM (2004) The birds of the Lord Howe Island Group: a review of records. *Australian Field Ornithology* **21**, 1–82.

Comments received from

Nicholas Carlile, David Priddel

Subspecies Conservation Summary

Black Currawong (King Island)

Strepera fuliginosa colei Mathews, 1916

Artamidae

Conservation status

Endangered D

Reasons for listing

Population <250 mature individuals

Status 2000

Endangered D

Status 1990

Endangered D

Taxonomy

S. f. fuliginosa (Tasmania) and *S. f. parvior* (Flinders I.) are Least Concern, as is the species

Taxonomic uniqueness: high (4 genera/family, 3 species/genus, 3 subspecies/species)

Range

Endemic to King I., Tasmania, the species was described as sporadically present in all parts of the island in the early 1970s (Green 1971) but declining with the clearing of scrub for agriculture (McGarvie 1974).

Abundance

There have been no population counts; an estimate of no more than 250 individuals is based on expert knowledge (R. Donaghey pers. comm.). Black Currawongs were considered locally common on King I. in the 1960s with winter flocks of 150 (Donaghey 2003).

Ecology

On King Island, the currawongs live in wet sclerophyll forest, woodland and heath. They also feed on beaches amongst seaweed and in pasture (Green 1971; Donaghey 2003). They are omnivorous, eating insects, carrion, fruit and small vertebrates. They build their open stick nests in trees, laying 2-3 eggs, like the Tasmanian subspecies (Higgins *et al.* 2007). A generation time of 10.1 years (BirdLife International 2011) is derived from a mean age at first breeding of 1.5 years and a maximum longevity in the wild of 24.2 years, both extrapolated from Pied Currawong *S. graculina* (Australian Bird and Bat Banding Scheme).

Threats

The scarcity of the Black Currawong on King I. has been attributed to a lack of forested habitat in which to breed but no major threats have been identified (Threatened Species Section 2010).

Conservation objectives

1. A stable population

Information required

1. Population size and trends

Management actions required

1. Monitor population
2. If declining, determine necessary conservation management

Current eligibility against IUCN Red List Criteria

IUCN category	Criteria eligibility
A	Not applicable: past, current or future population declines are thought unlikely to exceed 20% in any 3-generation period
B	Not applicable: EOO, AOO, habitat, locations or population not declining, population fluctuations not extreme
C	Not applicable: no continuing population decline or extreme fluctuations
D	Endangered: population 50–250 mature individuals
E	Not applicable: no population viability analysis undertaken

IUCN Red List assessment data

	Estimate	Reliability
Extent of occurrence trend	800 km^2 stable	high high
Area of occupancy trend	500 km^2 stable	low medium
No. of mature individuals trend	250 stable	low low
No. subpopulations	1	high
No. locations	1	high
Generation time	10 years	high
Global population share	100%	high

Bibliography

BirdLife International (2011) 'Species factsheet: *Strepera fuliginosa*'. Retrieved 17 January, 2011 from <http://www.birdlife.org/>.

Donaghey R (Ed.) (2003) *The Fauna of King Island. A Guide to Identification and Conservation Management.* King Island Natural Resource Management Group, Currie, King Island.

Green R, McGarvie AM (1971) The birds of King Island. *Records of the Queen Victoria Museum* **40**, 1–42.

Higgins PJ, Peter JM, Cowling SJ (Eds) (2006) *Handbook of Australian, New Zealand and Antarctic Birds. Volume 7: Boatbill to Starlings. Part B. Dunnock to Starlings.* Oxford University Press, Melbourne.

McGarvie AM, Templeton MT (1974) Additions to the birds of King Island, Bass Strait. *Emu* **74**, 91–96.

Threatened Species Section (2010) 'King Island biodiversity management plan'. Department of Primary Industries, Parks, Water and Environment, Hobart.

Comments received from

Richard Donaghey, Mark Holdsworth, Eric Woehler

Subspecies Conservation Summary

Grey Fantail (Lord Howe Island)

Rhipidura fuliginosa cervina Ramsay, 1879
Dicruridae

Conservation status

Extinct

Reasons for listing

Last recorded in 1924

Status 2000

Extinct

Status 1990

Extinct

Taxonomy

R. f. pelzelni (Norfolk I.) is Near Threatened, 5 mainland subspecies are Least Concern, as is the species. There are 5 extralimital subspecies.

Taxonomic uniqueness: medium (1 genus/family, 42 species/genus, 13 subspecies/species)

Range

This subspecies of Grey Fantail was endemic to Lord Howe I., where it probably occupied all wooded habitats (Hindwood 1940).

Abundance

It was common in 1908 (Hull 1909), but has not been recorded since 1924, when already scarce (Hindwood 1940; McAllan *et al.* 2004), and could not be found in 1928 (Sharland 1929).

Ecology

On Lord Howe I., it was a tame bird that occurred in both the forest and around the settlement. It nested on horizontal branches and fed on flying insects (McAllan *et al.*

2004; Department of Sustainability Environment Water Population and Communities 2011).

Reasons for extinction

Black Rats *Rattus rattus*, which colonised the island from a shipwreck in 1918, almost certainly caused its extinction between 1924 and 1928 (Sharland 1928; Hindwood 1940).

Bibliography

Department of Sustainability Environment Water Population and Communities (2011) *Rhipidura fuliginosa cervina.* In 'Species Profile and Threats Database'. Department of Sustainability, Environment, Water, Population and Communities, Canberra. Retrieved 24 February, 2011 from <http://www.environment.gov.au/sprat>.

Hindwood KA (1940) The birds of Lord Howe Island. *Emu* **40**, 1–86.

Hull AFB (1909) The birds of Lord Howe and Norfolk Islands. *Proceedings of the Linnean Society of New South Wales* **34**, 636–693.

McAllan IAW, Curtis BR, Hutton I, Cooper RM (2004) The birds of the Lord Howe Island Group: a review of records. *Australian Field Ornithology* **21**, 1–82.

Sharland MSR (1929) Land birds of Lord Howe Island. *Emu* **29**, 5–11.

Subspecies Conservation Summary

Grey Fantail (Norfolk Island)

Rhipidura fuliginosa pelzelni G. R. Gray, 1862
Dicruridae

Conservation status

Near Threatened D2

Reasons for listing

Occurs at a single location where susceptible to introduction of predators, competitors or disease, although the risk is low

Status 2000

Near Threatened D2

Status 1990

Near Threatened D2

Taxonomy

R. f. cervina (Lord Howe I.) is Extinct, 5 mainland subspecies are Least Concern, as is the species. There are 5 extralimital subspecies.
Taxonomic uniqueness: medium (1 genus/family, 42 species/genus, 13 subspecies/species)

Range

Endemic to Norfolk I. (Schodde and Mason 1999). Present over almost entire island (Schodde *et al.* 1983) although their ability to co-exist with people and cleared environments, and their high visibility, may make them seem more numerous than they are (M. Christian pers. comm.).

Abundance

Based on a survey in 2010, the population in Norfolk I. National Park is estimated to be at least 1600 adults on the basis of 155 birds detected in point counts at a density of about 400/km^2 (G. Dutson pers. obs.), with many more outside the park. There is no suggestion of a decline in the population.

Ecology

Common in native and weedy forest, and less common in all other habitats with some trees and bushes. The fantails nest and feed on insects where trees or shrubs remain on the island, laying 2 eggs in cup nests (Schodde *et al.* 1983). A generation time of 5.1 years (BirdLife International 2011) is derived from a mean age at first breeding of 0.6 years (Higgins and Peter 2002), and a maximum longevity in the wild of 9.7 years (Australian Bird and Bat Banding Scheme).

Threats

Clearing has removed some habitat but the subspecies has persisted in that which remains. Unlike *R. f. cervina* on Lord Howe I., predation by Black Rats *Rattus rattus* seems not to have affected the population since the rats' introduction in the 1940s, possibly because the earlier introduction of the Pacific Rat *R. exulans* by Polynesians 800 years ago (Holdaway and Anderson 1998) had already exerted some selection pressure. Domestic and feral cats *Felis catus* probably kill some individuals. The principal threat is the introduction of a new predator or some other catastrophe, both of which are plausible, given the extinction of *R. f. cervina*, but low probability.

Conservation objectives

1. Maintain current breeding population

Information required

1. Population trends on Norfolk I.

Current eligibility against IUCN Red List Criteria

IUCN category	Criteria eligibility
A	Not applicable: past, current or future population declines are thought unlikely to exceed 20% in any 3-generation period
B	Not applicable: EOO, AOO, habitat, locations or population not declining, population fluctuations not extreme
C	Not applicable: no continuing population decline or extreme fluctuations
D	Near Threatened: ≤5 locations with a plausible but low probability threat
E	Not applicable: no population viability analysis undertaken

IUCN Red List assessment data

	Estimate	Reliability
Extent of occurrence trend	40 km^2 stable	high high
Area of occupancy trend	30 km^2 stable	medium high
No. of mature individuals trend	5000 stable	low high
No. subpopulations	1	high
No. locations	1	high
Generation time	5.1 years	low
Global population share	100%	high

Management actions required

1. Maintain quarantine for Norfolk I.
2. Manage threats on Norfolk I. as required
3. Reintroduce to Phillip I. when vegetation there is sufficiently mature

Bibliography

BirdLife International (2011) 'Species factsheet: *Rhipidura fuliginosa*'. Retrieved 17 January, 2011 from <http://www.birdlife.org/>.

Higgins PJ, Peter JM (Eds) (2002) *Handbook of Australian, New Zealand and Antarctic Birds. Volume 6: Pardalotes to Spangled Drongo.* Oxford University Press, Melbourne.

Holdaway RN, Anderson AJ (1998) 14C AMS dates on *Rattus exulans* bones from natural and archaeological contexts on Norfolk Island, South-west Pacific. *Archaeology in New Zealand* **41**, 195–198.

Schodde R, Fullagar P, Hermes N (1983) *A Review of Norfolk Island Birds: Past and Present.* Australian National Parks and Wildlife Service, Canberra.

Schodde R, Mason IJ (1999) *The Directory of Australian Birds: Passerines.* CSIRO Publishing, Melbourne.

Comments received from

Margaret Christian, Joe Forshaw, Penny Olsen

Subspecies Conservation Summary

Scarlet Robin (Norfolk Island)

Petroica multicolor multicolor (Gmelin, 1789)

Petroicidae

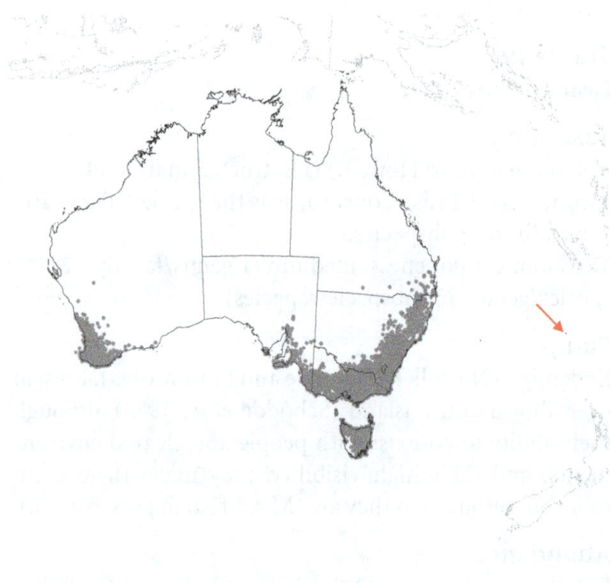

Conservation status

Endangered B1ab(ii,v)+B2ab(ii,v), C2a(ii)

Reasons for listing

Small EOO, AOO and population size, and a gradual decline in abundance that appears to be continuing

Status 2000

Endangered B1ab(ii,v)+B2ab(ii,v), C2a(ii)

Status 1990

Endangered B1ab(ii,v)+B2ab(ii,v), C2a(ii)

Taxonomy

Species follows BirdLife International (2011); sometimes separated from mainland Australian species (Schodde and Mason 1999; Christidis and Boles 2008). *P. m. boodang, P. m. campbelli* and *P. m. leggii* in Australia, all Least Concern, as is the species. There are 13 extralimital subspecies.

Taxonomic uniqueness: low (14 genera/family, 10 species/genus, 17 subspecies/species)

Range

Endemic to Norfolk I. where largely confined to the Norfolk I. National Park and nearby forested areas including the valleys between Prince Phillip Drive and Mt Pitt Road, between Douglas Drive and the park boundary and valleys near Duncombe Bay (Commonwealth of Australia

Current eligibility against IUCN Red List Criteria

IUCN category	Criteria eligibility
A	Not applicable: past, current or future population declines are thought unlikely to exceed 20% in any 3-generation period
B	Endangered: EOO <5000 km^2, AOO <500 km^2, at ≤5 locations with plausible future threat, inferred continuing decline in AOO and no. mature individuals
C	Endangered: population 250–2500 mature individuals, suspected continuing decline, >95% mature individuals in 1 subpopulation
D	Vulnerable: population 250–1000 mature individuals, ≤5 locations with plausible future threat
E	Not applicable: no population viability analysis undertaken

IUCN Red List assessment data

	Estimate	Reliability
Extent of occurrence trend	40 km^2 stable	high high
Area of occupancy trend	8 km^2 stable	high high
No. of mature individuals trend	1000 decreasing	medium low
No. subpopulations	1	high
No. locations	3	medium
Generation time	6.2 years	low
Global population share	100%	high

2005; Director of National Parks 2010). It disappeared from many sites elsewhere on the island during the 1980s and this decline is reported to have continued in the 1990s and 2000s (Schodde *et al.* 1983; Robinson 1988; Bell 1990; M. Christian pers. comm.). There are 3 locations: the National Park, government reserves and nearby private land, as each have the potentially different levels of threat and threat mitigation.

Abundance

The population estimated at 400–500 pairs in 1988 (Robinson 1988) is used as the basis for the estimate here. The population appeared little changed in 1997 (Robinson 1997). There have been no subsequent quantitative surveys but numbers seemed broadly similar in Norfolk I NP in 2009 (G. Dutson pers. obs.). Trends in numbers outside the park are unknown but birds are absent from sites where seen regularly until recently (M. Christian pers. comm.). A decline is assumed as a precautionary measure; taxon would otherwise be Vulnerable D1.

Ecology

The most favoured habitats are rainforest with or without dense thickets of exotic Red Guava *Psidium cattleianum*. Habitats dominated by Norfolk Island Palm *Rhopalostylis baueri* or exotic African Olive *Olea africana* hold lower densities and exotic eucalypt forests support very few robins. Most breeding appears to be in rainforest. The main requirements appear to be deep litter, a dense understorey and a fairly open ground layer to give the robins sufficient visibility to feed on the ground (Robinson 1988; Department of Sustainability Environment Water Population and Communities 2011). Nests are usually placed near the top of the subcanopy at a mean height of 5.5 m, most commonly in Norfolk Island Pine *Araucaria heterophylla*, Red Guava and various rainforest trees (Robinson 1988; Major 1989). A generation time of 6.2 years (BirdLife International 2011) is derived from an age at first breeding of 1.5 years, extrapolated from congeners, and a maximum adult annual survival in the wild of 10.9 years, extrapolated from Flame Robin *P. phoenicea* (Australian Bird and Bat Banding Scheme).

Threats

The decline of the Scarlet Robin was first noticed about 20 years after the introduction of Black Rats *Rattus rattus* in the 1940s. Most known nest losses were due to predation, mainly by rats (Major 1989). Feral cats *Felis catus* may also be a problem and have been recorded taking adult birds which often feed on the ground. The effects of predation and competition have been exacerbated by habitat clearance and fragmentation (Robinson 1988). Invasion of habitat by exotic shrubs and trees reduces habitat quality (Robinson 1988) but only until it develops an open understorey (Robinson 1997), so that Guava control in particular may be detrimental to the robin. Most remaining forest is now protected within the National Park or local government reserves and invasive shrubs are gradually being controlled (Director of National Parks 2008). Rats and feral cats are baited in the National Park and some other reserves. The species is not monitored.

Conservation objectives

1. Maintain stable and viable population on Norfolk I.
2. Introduce to Phillip I. following revegetation.

Information required

1. Population size and trends
2. Efficacy of rat and cat baiting
3. Effect of weed control

Management actions required

1. Monitor populations
2. Continue rat baiting and cat trapping across Norfolk I. National Park
3. Advocate for rat eradication across Norfolk I.
4. Continue to restore native habitat across Norfolk I.
5. Continue to provide incentives and information for habitat restoration on private land
6. Introduce to Phillip I. when habitat is suitable

Bibliography

Bell BD (1990) 'The status and management of the White-breasted White-eye and other birds on Norfolk Island'. Report to Australian National Parks and Wildlife Service, Canberra.

BirdLife International (2011) 'Species factsheet: *Petroica multicolor*'. Retrieved 17 January, 2011 from <http://www.birdlife.org/>.

Christidis L, Boles WE (2008) *Systematics and Taxonomy of Australian Birds*. CSIRO Publishing, Melbourne.

Commonwealth of Australia (2005) 'National recovery plan for the Norfolk Island Scarlet Robin *Petroica multicolor multicolor* and the Norfolk Island Golden Whistler *Pachycephala pectoralis xanthroprocta*'. Department of the Environment and Heritage, Canberra.

Department of Sustainability, Environment, Water, Population and Communities (2011) *Petroica multicolor multicolor.* In 'Species Profile and Threats Database'. Department of Sustainability, Environment, Water, Population and Communities, Canberra. Retrieved 24 February, 2011 from <http://www.environment.gov.au/sprat>.

Director of National Parks (2010) 'Draft Norfolk Island region threatened species recovery plan'. Department of the Environment, Water, Heritage and the Arts, Canberra.

Major RE (1989) 'Reproductive output and recruitment of the Norfolk Island Scarlet Robin *Petroica multicolor multicolor* – Phase II'. Report to the Australian National Parks and Wildlife Service, Canberra.

Robinson D (1988) 'Ecology and management of the Scarlet Robin, White-breasted White-eye and Long-billed White-eye on Norfolk Island'. Report to the Australian National Parks and Wildlife Service, Canberra.

Robinson D (1997) 'An evaluation of the status of the Norfolk island Robin following rat-control and weed-control works in the Norfolk island National Park'. Report to Environment Australia, Canberra.

Schodde R, Fullagar P, Hermes N (1983) *A Review of Norfolk Island Birds: Past and Present.* Australian National Parks and Wildlife Service, Canberra.

Schodde R, Mason IJ (1999) *The Directory of Australian Birds: Passerines.* CSIRO Publishing, Melbourne.

Comments received from

Margaret Christian, Penny Olsen, Coral Rowston, Ron Ward

Species Conservation Summary

Flame Robin

Petroica phoenicea Gould, 1837
Petroicidae

Conservation status

Near Threatened A2b

Reasons for listing

Rapid decline approaching 30% in 3 generations (19 years)

Status 2000

Near Threatened A2b

Status 1990

Near Threatened A2b

Taxonomy

No infraspecific taxa described
Taxonomic uniqueness: medium (14 genera/family, 10 species/genus, 1 subspecies/species)

Range

Breeds in temperate south-eastern Australia from central Victoria along the Great Dividing Range to south-east Queensland, including Tasmania and Bass Strait Is. In winter migrates to warmer lowland areas including the inland slopes of Great Dividing Range and many coastal sites. Details of the migration are poorly understood but many Tasmanian birds cross Bass Strait to winter on the mainland (Higgins and Peter 2002). Historically birds bred in South Australia and reached Yorke Peninsula and Kangaroo I. but now does so very rarely (Paton *et al.* 1994). The AOO is estimated here as about 10% of the EOO.

Abundance

The population is estimated on the basis of the AOO and a density of 0.5 birds/ha (Higgins and Peter 2002). A decline in reporting rate from 6.0% to 2.9% between the first Atlas (1977–1981) and the second Atlas (1998–2002), including a significant decline across its whole range except for some regions of Tas (Barrett *et al.* 2003, 2007), is corroborated by other observations. While it remains common in the high country of the Great Dividing Range, it has become rarer in other regions. In SA it has become particularly scarce (Paton *et al.* 1994) where Mt Lofty Ranges population appears to be unviable (Possingham and Field 2001). In Tas

Current eligibility against IUCN Red List Criteria

IUCN category	Criteria eligibility
A	Near Threatened: population decline of 20–29% over the last 3 generations (19 years) suspected from monitoring
B	Not applicable: EOO >20 000 km^2, AOO >2000 km^2
C	Not applicable: population >10 000 mature individuals
D	Not applicable: population >1000 mature individuals, >5 locations
E	Not applicable: no population viability analysis undertaken

IUCN Red List assessment data

	Estimate	Reliability
Extent of occurrence trend	570 000 km^2 stable	high medium
Area of occupancy trend	57 000 km^2 stable	low medium
No. of mature individuals trend	2 800 000 decreasing	low medium
No. subpopulations	1	high
No. locations	>10	high
Generation time	6.2 years	low
Global population share	100%	high

it is common but less so than in the past (S. Lloyd pers. comm., S. Bryant pers. comm.). It has declined in the ACT since 1998 (Bounds and Taws 2007). Some results are contradictory – there was a sharp decline in reporting rate across NSW in the early 1990s that has since stabilised at about half its previous level, but reporting rates in the Lachlan biome have been cyclic, with a peak in the early 1990s (Olsen 2008). Given the tendency of the species to favour recently burnt areas, the large forest fires of the last decade may have been to the species' advantage. Nevertheless, as a precautionary measure, it is assessed here as declining at 20–29% in 3 generations (20 years) recognizing that the data are poor and it may be declining at significantly higher or lower rates.

Ecology

Usually breeds in wet eucalypt forest in hilly or mountainous terrain. Generally prefers mountain tops and slopes with more clearings and less understorey (Robinson 1990a,b, 1992a,b). Often more abundant in areas recently cleared or burnt by bushfires, but moves away once the undergrowth regrows and becomes dense (Loyn 1980, 1985). In the autumn and winter, moves to low altitudes where they occur in open woodlands and grasslands with remnant native bushes and trees (Robinson 1992b; Higgins and Peter 2002). They are more common in regions retaining a high percentage of remnant native vegetation (Barrett *et al.* 2007). In the last 20 years birds have been returning later to the Australian Alps, even though temperatures have increased, and can be disadvantaged if they return early (Green 2006, 2010). A generation time of 6.2 years (BirdLife International 2011) is derived from an age at first breeding of 1.5 years, extrapolated from congeners, and a maximum adult annual survival in the wild of 10.9 years (Australian Bird and Bat Banding Scheme).

Threats

The most obvious change in the habitat of the Flame Robin is clearing, cultivation, overgrazing and other historical and ongoing degradation of non-breeding habitat but there have been no studies linking these changes to robin demography. On the breeding grounds, clearance for timber production and extensive wildfires are likely to increase the area of suitable habitat episodically while, in some areas, nest loss from native and exotic predators is high (Higgins and Peter 2002). The complex relationship with climate change (Green 2006, 2010) is barely understood.

Conservation objectives

1. Stable reporting rates across range

Information required

1. The effect of climate and habitat condition on breeding productivity and winter survival

Long-term trends in abundance

1. Reasons for potential decline

Management actions required

1. Regularly analyse reporting rate for each bioregion
2. Determine breeding productivity and survivorship in different contexts

Bibliography

Barrett G, Silcocks A, Barry S, Cunningham R, Poulter R (2003) *The New Atlas of Australian Birds.* RAOU, Melbourne.

Barrett GW, Silcocks AF, Cunningham R, Oliver DL, Weston MA, Baker J (2007) Comparison of atlas data to determine the conservation status of bird species in New South Wales, with an emphasis on woodland-dependent species. *Australian Zoologist* **34**, 37–77.

BirdLife International (2011) 'Species factsheet: *Petroica phoenicea*'. Retrieved 17 January, 2011 from <http://www.birdlife.org/>.

Bounds J, Taws N (2007) A statistical analysis of trends in occupancy rates of woodland birds in the ACT December 1998 to December 2005. *Canberra Bird Notes* **32**, 2, 73–98.

Green K (2006) The effect of variation in snowpack on timing of bird migration in the Snowy Mountains. *Emu* **106**, 187–192.

Green K (2010) Alpine taxa exhibit differing responses to climate warming in the Snowy Mountains of Australia. *Journal of Mountain Science* **7**, 167–175.

Higgins PJ, Peter JM (Eds) (2002) *Handbook of Australian, New Zealand and Antarctic Birds. Volume 6: Pardalotes to Spangled Drongo.* Oxford University Press, Melbourne.

Loyn RH (1980) Bird populations in a mixed eucalypt forest used for production of wood in Gippsland, Victoria. *Emu* **80**, 145–156.

Loyn RH (1985) Ecology, distribution and density of birds in Victorian forests. In *Birds of Eucalypt Forests and Woodlands: Ecology, Conservation, Management.* (Eds A Keast, HF Recher, H Ford, D Saunders) pp. 33–46. RAOU, Melbourne, and Surrey Beatty and Sons, Chipping Norton, NSW.

Olsen P (2008) The state of Australia's birds 2008. *Wingspan* **18**(suppl. 4), 1–32.

Paton DC, Carpenter G, Sinclair RG (1994) A second bird atlas of the Adelaide region. Part 1: Changes in the distribution of birds: 1974–75 vs 1984–85. *South Australian Ornithologist* **31**, 151–193.

Possingham HP, Field SA (2001) Regional bird extinctions and their implications for vegetation clearance policy. *Lifelines* **7**, 15–16.

Robinson D (1990a) The nesting ecology of sympatric Scarlet Robin *Petroica multicolor* and Flame Robin *Petroica phoenicea* populations in open eucalypt forest. *Emu* **90**, 40–52.

Robinson (1990b) The social organisation of the Scarlet Robin *Petroica multicolor* and the Flame Robin *P. phoenicea* in southeastern Australia: a comparison between sedentary and migratory flycatchers. *Ibis* **132**, 78–94.

Robinson D (1992a) Why do Flame Robins *Petroica phoenicea* migrate? A comparison between the social and feeding ecologies of the Flame Robin and Scarlet Robin *P. multicolor. Corella* **16**, 1–14.

Robinson D (1992b) Habitat use and foraging behaviour of the Scarlet Robin and Flame Robin at a site of breeding-season sympatry. *Wildlife Research* **19**, 377–395.

Comments received from

Sarah Lloyd, Richard Loyn, Damon Oliver, Doug Robinson

Subspecies Conservation Summary

Hooded Robin (south-eastern)

Melanodryas cucullata cucullata (Latham, 1802)
Petroicidae

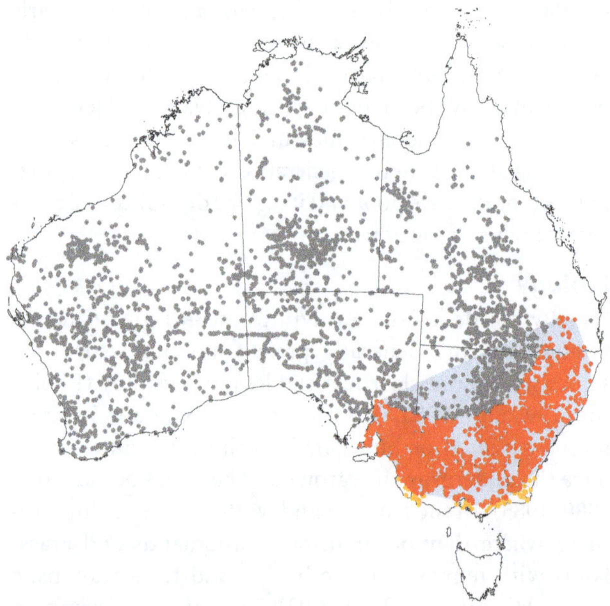

Conservation status

Near Threatened A2bc+3c+4bc

Reasons for listing

Declining across a substantial proportion of its range based on surveys and local extirpations, probably at 20–29% within 3 generations (16 years), because of the many ongoing effects of 2 centuries of habitat loss and fragmentation

Status 2000

Near Threatened A2bc+3c+4bc

Status 1990

Near Threatened A2bc+3c+4bc

Taxonomy

M. c. melvillensis (Tiwi Is., Northern Territory) is Critically Endangered (Possibly Extinct). Two other subspecies are Least Concern, as is the species

Taxonomic uniqueness: medium (14 genera/family, 2 species/genus, 4 subspecies/species)

Range

This subspecies occurs in south-eastern Australia from Yorke Peninsula, South Australia, to far south-east Queensland, intergrading with *M. c. westralensis* in the northern Murray-Darling basin (Schodde and Mason 1999). It has been extirpated from many habitat fragments and regions across its range, particularly in the wetter areas of the south and east (Robinson 1993; Barrett *et al.* 1994; Paton *et al.* 1994; Fitri and Ford 1996; Robinson and Traill 1996; Olsen *et al.* 2005; Priday 2010; Ford 2011).

Abundance

Taking the AOO only from records in the Birds Australia database since 1998, which is certain to be an underestimate given many occupied areas will not have been visited, and assuming a density of 5.5 birds/km^2 (average of creek and non-creek values in Pilliga Scrub; Date *et al.* 2002) gives a population of about 30 000 in the 5000 km^2 Pilliga Scrub, which may be optimal habitat, and a total of 165 000 altogether (H. Ford *in litt.*). This population has been declining in agricultural landscapes for many decades (Robinson 1993, Robinson and Traill 1996). The reporting rate declined by 41% in NSW between the 1977–1981 and 1998–2002 Atlases, with no variation between bioregions (Barrett *et al.* 2007). The greatest declines appear to have been in the eucalypt-dominated

Current eligibility against IUCN Red List Criteria

IUCN category	Criteria eligibility
A	Near Threatened: past, current and anticipated population declines of 20–29% over periods 3 generations (29 years) suspected from monitoring, contracting AOO and deteriorating habitat quality
B	Not applicable: EOO >20 000 km^2, AOO >2000 km^2
C	Not applicable: population >10 000 mature individuals
D	Not applicable: population >1000 mature individuals, >5 locations
E	Not applicable: no population viability analysis undertaken

IUCN Red List assessment data

	Estimate	Reliability
Extent of occurrence trend	1 100 000 km^2 stable	high high
Area of occupancy trend	10 000 km^2 decreasing	low high
No. of mature individuals trend	165 000 decreasing	low high
No. subpopulations	1	medium
No. locations	>10	high
Generation time	5.3 years	high
Global population share	100%	high

woodlands to the south and east of its range with populations in acacia and casuarina woodlands faring better (D. Watson *in litt.*), possibly because soils and litter-dwelling prey are richer in such communities (Watson *et al.* 2000; Watson 2011). Populations on the drier fringe of the range may also be less likely to decline, e.g. Port Wakefield, SA (Taylor 1987). The overall rate of decline is difficult to estimate but is inferred here from ongoing habitat loss and degradation to be 20–29% in 3 generations (16 years).

Ecology

Inhabits dry eucalypt and acacia woodlands and shrublands, with an open understorey, some grassy areas, and a complex ground layer. Occurrence in agricultural landscapes is correlated with habitat patch size, tree canopy cover, shrub cover, ground cover, logs, fallen branches and litter (Watson *et al.* 2003). Particularly favours the edges of open vegetation dominated by ungrazed or lightly grazed native perennial tussock-grassland next to timbered native vegetation cover on moderately deep to deep soils (Priday 2010). Feeds on insects and small lizards from the ground by pouncing from perches, preferring sites where there is small fallen timber (Antos *et al.* 2008). Builds a cup nest and usually lays 2 eggs. Breeding pairs or groups defend exclusive nesting territories of c.6 ha, but have larger winter home ranges of c.18 ha, sometimes up to 50 ha (Fitri and Ford 2003a,b; Higgins and Peter 2002), although they can be present in patches as small as 2.9 ha (Montague-Drake *et al.* 2009). As most Petroicidae are highly mobile, it is assumed that there is still 1 subpopulation. A generation time of 5.3 years (BirdLife International 2011) is derived from an age at first breeding of 1.0 years (Higgins and Peter 2002) and a maximum longevity in the wild of 9.6 years (Australian Bird and Bat Banding Scheme).

Threats

One of a suite of declining temperate woodland species (Reid 1999; Montague-Drake *et al.* 2009). Like many of these species (e.g. Jacky Winter *Microeca fascinans*, Crested Bellbird *Oreoica gutturalis*), Hooded Robins are an arid-adapted species and the declines are in the wetter portions of their distribution. Historical clearance destroyed much habitat with that in the agricultural regions now fragmented and suffering from extinction debt (Ford 2011). Birds in fragments can suffer such high nest predation that they produce too few recruits to compensate adult mortality (Ford *et al.* 2009) and are subject to competition with Noisy Miners *Manorina melanocephala* (Maron and Lill 2005). These pressures on the robins may be compounded by a reliance on what is becoming a depauperate terrestrial fauna (Watson 2011). Probable ongoing threats include inappropriate fire regimes, particularly those that lead to an increase in understorey density (H. Ford *in litt.*), firewood collection and tidying of farmland, grazing by stock, rabbits *Oryctolagus cuniculus* and kangaroos *Macropus* spp., increased predation from introduced mammals and invasive weeds (Willson and Bignall 2009). Most of these threats are greater at habitat edges so are exacerbated by fragmentation.

Conservation objectives

1. Stable populations maintained in key sites across its range

Information required

1. Population densities and trends in arid parts of the range
2. Criteria for sites where management investment to save robin populations is warranted
3. Appropriate management interventions for isolated subpopulations including predator and competitor control, fire regime and ground cover manipulation

Management actions required

1. Monitor key sites, particularly where there are management interventions
2. Develop guidelines for the management of robin populations in fragments
3. Promote ecological management of woodland remnants on private land
4. Design conservation and regeneration plans at a landscape level
5. Protect all woodland in which robins are known to be resident from clearing
6. Place all areas of public land that contain the robins under secure conservation management, particularly those in timber reserves, transport corridors and local government land

7. Using appropriate incentives, undertake extension with landholders who have suitable woodland habitat to promote sound management of remnants and encourage greater connectivity between habitat patches
8. Promote revegetation and land reclamation that recreates woodland habitat with a full complement of biodiversity, including the robin
9. Control and reduce firewood collection from areas occupied by the robins, promoting wood-lot development close to markets, and reduce grazing densities where necessary

Bibliography

Antos MJ, Bennett AF, White JG (2008) Where exactly do ground-foraging woodland birds forage? Foraging sites and microhabitat selection in temperate woodlands of southern Australia. *Emu* **108**, 201–211.

Barrett G, Ford HA, Recher HF (1994) Conservation of woodland birds in a fragmented rural landscape. *Pacific Conservation Biology* **1**, 245–256.

Barrett GW, Silcocks AF, Cunningham R, Oliver D, Weston MA, Baker J (2007) Comparison of atlas data to determine the conservation status of bird species in New South Wales, with an emphasis on woodland-dependent species. *Australian Zoologist* **34**, 37–77.

BirdLife International (2011) 'Species factsheet: *Melanodryas cucullata*'. Retrieved 17 January, 2011 from <http://www.birdlife.org/>.

Date EM, Ford HA, Recher HF (2002) Impacts of logging, fire and grazing regimes on bird species assemblages of the Pilliga Woodlands of New South Wales. *Pacific Conservation Biology* **8**, 177–195.

Fitri L, Ford HA (1997) Status, habitat and social organisation of the Hooded Robin *Melanodryas cucullata* in the New England region of New South Wales. *Australian Bird Watcher* **17**, 142–155.

Fitri L, Ford HA (2003a) Breeding biology of Hooded Robins *Melanodryas cucullata* in New England. *Corella* **27**, 68–74.

Fitri L, Ford HA (2003b) Foraging behaviour of Hooded Robins *Melanodryas cucullata* in the Northern Tablelands of New South Wales. *Corella* **27**, 61–67.

Ford HA (2011) The causes of decline of birds of eucalypt woodlands: advances in our knowledge over the last 10 years. *Emu* **111**, 1–9.

Ford HA, Walters JR, Cooper CB, Debus SJS, Doerr VAJ (2009) Extinction debt or habitat change? Ongoing losses of woodland birds in north-eastern New South Wales, Australia. *Biological Conservation* **142**, 3182–3190.

Higgins PJ, Peter JM (Eds) (2002) *Handbook of Australian, New Zealand and Antarctic Birds. Volume 6: Pardalotes to Spangled Drongo.* Oxford University Press, Melbourne.

Maron M, Lill A (2005) The influence of livestock grazing and weed invasion on habitat use by birds in grassy woodland remnants. *Biological Conservation* **124**, 439–450.

Montague-Drake RM, Lindenmayer DB, Cunningham RB (2009) Factors affecting site occupancy by woodland bird species of conservation concern. *Biological Conservation* **142**, 2896–2903.

Olsen P, Weston M, Tzaros C, Silcocks A (2005) The state of Australia's birds 2005. Woodlands and birds. *Supplement to Wingspan* **15**, 1–32.

Paton DC, Carpenter G, Sinclair RG (1994) A second bird atlas of the Adelaide region. Part 1: Changes in the distribution of birds: 1974–75 vs 1984–85. *South Australian Ornithologist* **31**, 151–193.

Priday SD (2010) Beyond the 'woody remnant' paradigm in conservation of woodland birds: habitat requirements of the Hooded Robin (*Melanodryas cucullata cucullata*). *Emu* **110**, 118–124.

Reid JRW (1999) 'Threatened and declining birds in the New South Wales Sheep-Wheat Belt: I. Diagnosis, characteristics and management'. Consultancy report to NSW National Parks and Wildlife Service. CSIRO Sustainable Ecosystems, Canberra.

Robinson D (1993) Vale Toolern Vale: the loss of our woodland birds. *Wingspan* **March**, 1–4.

Robinson D, Traill BJ (1996) 'Conserving Woodland Birds in the Wheat and Sheep Belts of Southern Australia'. RAOU (Birds Australia), Melbourne.

Schodde R, Mason IJ (1999) *The Directory of Australian Birds: Passerines.* CSIRO Publishing, Melbourne.

Watson DM (2011) A productivity-based explanation for woodland bird declines: poorer soils yield less food. *Emu* **111**, 10–18.

Watson DM, Mac Nally RC, Bennett AF (2000) The avifauna of remnant Buloke (*Allocasuarina luehmanni*) woodlands in western Victoria. *Pacific Conservation Biology* **6**, 46–60.

Willson A, Bignall J (2009) 'Regional recovery plan for threatened species and ecological communities of Adelaide and the Mount Lofty Ranges, South Australia'. Department for Environment and Heritage, South Australia.

Comments received from

David Baker-Gabb, Hugh Ford, Damon Oliver, Steve Priday, David Watson

Subspecies Conservation Summary

Hooded Robin (Tiwi Islands)

Melanodryas cucullata melvillensis (Zietz, 1914)
Petroicidae

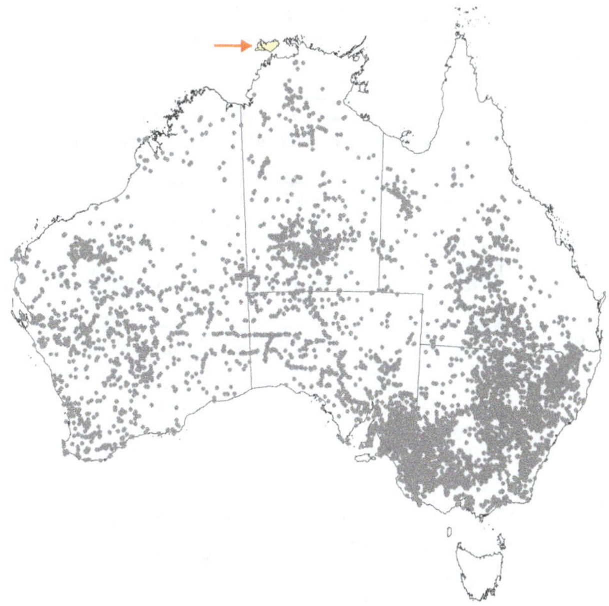

Conservation status
Critically Endangered (Possibly Extinct) D

Reasons for listing
Possibly Extinct since 1992, based on its documented decline, survey effort, conspicuous nature of the species and plausible threat of changed fire regimes, balanced against the lack of known attributes predisposing it to extinction and continued existence of potentially suitable habitat

Status 2000
Critically Endangered (Possibly Extinct) A2b, D

Status 1990
Critically Endangered A2bc+3c+4bc, B1ab(i,ii,iii,iv,v)+B 2ab(i,ii,iii,iv,v), C2a(i,ii), D

Taxonomy
M. c. cucullata (south-eastern Australia) is Near Threatened. Two other subspecies are Least Concern, as is the species
Taxonomic uniqueness: medium (14 genera/family, 2 species/genus, 4 subspecies/species)

Former range
Endemic to the Tiwi Is. (Bathurst and Melville Is.), in the Northern Territory (Schodde and Mason 1999).

Former abundance
Historically described as common on Melville I., but without any further details (Mathews 1914). There have been very few records, despite recent comprehensive and systematic general wildlife surveys within its range (including systematic surveys of 351 1-hectare quadrats and 400 person-days of opportunistic surveys elsewhere in 2000–2001), and the generally reasonably conspicuous nature of the species. The last known records were at 1 site on each island in 1991 and 1992 (Fensham and Woinarski 1992; Mason and Schodde 1997; Woinarski *et al.* 2000, 2003, 2007; Department of Sustainability, Environment, Water, Population and Communities 2011).

Ecology
This subspecies has been reported from eucalypt tall open forests and treeless plains or *Acacia* shrublands (Woinarski *et al.* 2003). Other subspecies forage a range of ground-dwelling invertebrates, preferentially choosing areas where the ground vegetation is sparse, such as in burnt areas or in thickets where a dense canopy limits the cover of grasses (Woinarski and Fisher 1995). Hooded Robins make cup nests in which they lay 2 eggs (Higgins and Peter 2002). Near Townsville the species has been recorded as more common in areas with frequent fires (Kutt and Woinarski 2007), but the frequency of fire on the Tiwi Is. is much greater than near Townsville (J. Woinarski *in litt.*). A generation time of 5.3 years (BirdLife International 2011) is derived from an age at first breeding of 1.0 years (Higgins and Peter 2002) and a maximum longevity in the wild of 9.6 years (Australian Bird and Bat Scheme).

Reasons for possible extinction
The most likely cause of extinction is change in fire regime, from an intricate fine-scaled mosaic imposed by Aboriginal management to a more polarised regime, now characterised by extensive areas burnt by larger hotter fires around more accessible areas, and a low frequency of fires in the more remote areas (Woinarski *et al.* 2000). Probably disadvantaged by a low frequency of fires because the resulting dense grass cover (Fensham and Cowie 1998; Rossiter *et al.* 2003; Woinarski *et al.* 2003) is likely to reduce foraging efficiency. A high frequency of extensive hot fires may also be disadvantageous (Woinarski *et al.* 2003). Over much of their range elsewhere, Hooded Robins have declined or become locally extinct, possibly due to vegetation loss and change, increased predation rates or lower abundance of some invertebrate prey. Habitat was lost when 26 000 ha of tall eucalypt forest was converted to plantation forestry in 2002–2007.

Conservation objectives
1. A viable population on the Tiwi Is.

Information required
1. Evidence that it remains extant
2. If extant, habitat requirements, threats, population size and trends

Management actions required
1. Undertake surveys of habitat that might still hold populations, especially at the drier, less frequently visited, eastern end of the island

Current eligibility against IUCN Red List Criteria

IUCN category	Criteria eligibility
A	Not applicable: population declines are thought unlikely to have exceeded 20% in the last 3 generations (16 years)
B	Not applicable: EOO, AOO, habitat, locations or population not declining, population fluctuations not extreme
C	Not applicable: no continuing population decline or extreme fluctuations
D	Critically Endangered (Possibly Extinct): if extant, population <50 mature individuals
E	Not applicable: no population viability analysis undertaken

2. Develop a contingency plan should the subspecies be rediscovered, including establishing a captive breeding population

Bibliography

BirdLife International (2011) 'Species factsheet: *Melanodryas cucullata*'. Retrieved 17 January, 2011 from <http://www.birdlife.org/>.

Department of Sustainability, Environment, Water, Population and Communities (2011) *Melanodryas cucullata melvillensis*. In 'Species Profile and Threats Database'. Department of Sustainability, Environment, Water, Population and Communities, Canberra. Retrieved 24 February, 2011 from <http://www.environment.gov.au/sprat>.

Fensham RJ, Cowie ID (1998) Alien plant invasions on the Tiwi Islands: extent, implications and priorities for control. *Biological Conservation* **83**, 55–68.

Fensham RJ, Woinarski JCZ (1992) 'Yawalama: the ecology and conservation of monsoon forest on Tiwi Islands, Northern Territory'. Report to DASET. Conservation Commission of the Northern Territory.

Higgins PJ, Peter JM (Eds) (2002) *Handbook of Australian, New Zealand and Antarctic Birds. Volume 6: Pardalotes to Spangled Drongo.* Oxford University Press, Melbourne.

Kutt AS, Woinarski JCZ (2007) The effects of grazing and fire on vegetation and the vertebrate assemblage in a tropical savanna woodland in north-eastern Australia. *Journal of Tropical Ecology* **23**, 95–106.

Mason IJ, Schodde R (1997) 'Bird survey of the Tiwi Islands, October 1996'. Report to Tiwi Land Council, Northern Territory.

Mathews GM (1914) A list of the birds of Melville Island, Northern Territory, Australia. *Ibis* **2**, 91–132.

Rossiter NA, Setterfield SA, Douglas MM, Hutley LB (2003) Testing the grass-fire cycle: alien grass invasion in the tropical savannas of northern Australia. *Diversity and Distributions* **9**, 169–176.

Schodde R, Mason IJ (1999) *The Directory of Australian Birds: Passerines.* CSIRO Publishing, Melbourne.

Woinarski J, Brennan K, Hempel C, Armstrong M, Milne D, Chatto R (2003) 'Biodiversity conservation on the Tiwi islands, Northern Territory. Part 2. Fauna'. Department of Infrastructure Planning and Environment, Darwin.

Woinarski J, Pavey C, Kerrigan R, Cowie I, Ward S (Eds) (2007) *Lost from Our Landscape: Threatened Species of the Northern Territory.* Northern Territory Government, Darwin.

Woinarski JCZ, Brennan K, Hempel C, Firth R, Watt F (2000) 'Biodiversity conservation on the Tiwi Islands: plants, vegetation types and terrestrial vertebrates on Melville Island'. Report to Tiwi Land Council, Darwin.

Woinarski JCZ, Fisher A (1995) Wildlife of lancewood (*Acacia shirleyi*) thickets and woodlands in northern Australia. 2. Comparisons with other environments of the region (*Acacia* woodlands, *Eucalyptus* savanna woodlands and monsoon rainforests). *Wildlife Research* **22**, 413–443.

Comments received from

Simon Ward, John Woinarski

Subspecies Conservation Summary

Horsfield's Bushlark (Tiwi Islands)

Mirafra javanica melvillensis Mathews, 1912
Alaudidae

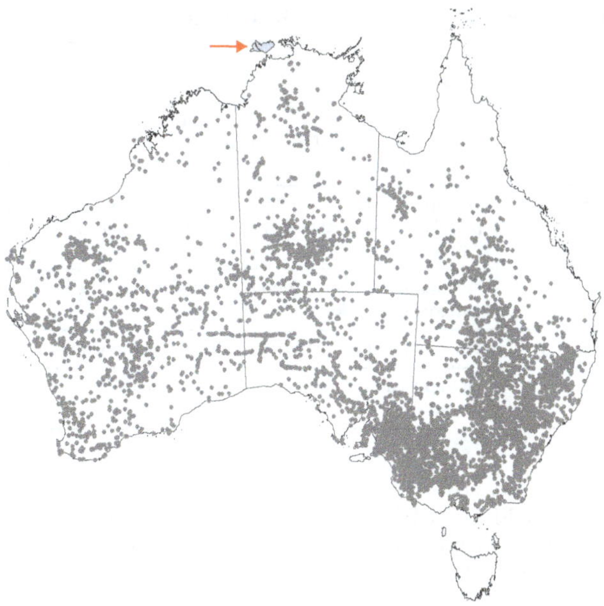

Conservation status
Vulnerable D1

Reasons for listing
Population probably <1000 mature individuals based on paucity of records and restricted habitat

Status 2000
Vulnerable D1

Status 1990
Vulnerable D1

Taxonomy
Eight other subspecies across Australia are Least Concern, as is the species. There are 7 extralimital subspecies
Taxonomic uniqueness: low (19 genera/family, 28 species/genus, 16 subspecies/species)

Range
Endemic to the Tiwi Islands (Bathurst and Melville Is.), in the Northern Territory (Schodde and Mason 1999). Few records: recorded historically (Zietz 1914; Mathews 1914), then at 2 sites in 1992 (Fensham and Woinarski 1992), not recorded in CSIRO survey (Mason and Schodde 1997), recorded at 2 sites in surveys of 351 1-hectare quadrats and 400 person-days opportunistic surveys in 2000–2001 (Woinarski *et al.* 2003b) and 2 other records in the Atlas of Australian Birds (Woinarski *et al.* 2003b). Treeless plains (*Acacia* open shrubland with *Sorghum* grassland understorey) cover about 191 km^2 on the Tiwi Is (Wilson *et al.* 1990; Woinarksi *et al.* 2003a) but not all may be suitable for this open-country species.

Abundance
Given the paucity of records despite its conspicuous song-flight, its restricted habitat and the lack of obvious threatening processes, the population is estimated here to be probably <1000 mature birds with no evidence of decline.

Ecology
Other subspecies occur in a range of native and introduced grasslands. In the NT, most common in perennial tussock grasslands; far less abundant as tree cover increases and only rarely seen in open woodlands. Conspicuous song-flight high above ground. Feeds on seeds, insects and other invertebrates. Forages entirely on the ground. Three or 4 eggs are laid in a nest on ground beneath a tussock of grass, sedge or small shrub (Higgins *et al.* 2001). A generation time of 3.8 years (BirdLife International 2011) is derived from an age at first breeding of 1.0 years and a maximum longevity in the wild of 6.6 years, both values extrapolated from mean values for other Alaudidae genera.

Threats
Threats to its treeless plains habitat include conversion to forestry, inappropriate fire regimes, altered water regimes and feral animals (Woinarksi *et al.* 2003a). Between 1975 and 1994, about 10% (2000 ha) of the plains of western Melville I. were replaced by plantations of the exotic Caribbean Pine *Pinus caribaea* but these have grown poorly and further plantations are not planned (Wilson and Fensham 1994; ForSci 1999). Fire regimes have changed significantly from an intricate fine-scaled mosaic imposed by Aboriginal management to a more polarised regime now characterised by extensive areas burnt by larger hotter fires around more accessible areas, and a low frequency in the more remote areas (Woinarski *et al.* 2000). The low frequency of fires, exacerbated by the spread of invasive weeds, is likely to result in denser grass cover, which may be unsuitable for this species (Fensham and Cowie 1998; Rossiter *et al.* 2003; Woinarski *et al.* 2003b). Locally high densities of feral Water Buffalo *Bubalus bubalis* and pigs *Sus scrofa* may degrade habitat (Woinarski *et al.* 2003c), though may also reduce grass density.

Conservation objectives
1. Stable population

Information required
1. Population size and trends
2. Extent of treeless plains occupied and likely threats

Management actions required
1. Assess population size and distribution

Bibliography

BirdLife International (2011) 'Species factsheet: *Mirafra javanica*'. Retrieved 17 January, 2011 from <http://www.birdlife.org/>.

Fensham RJ, Cowie ID (1998) Alien plant invasions on the Tiwi Islands: extent, implications and priorities for control. *Biological Conservation* **83**, 55–68.

Current eligibility against IUCN Red List Criteria

IUCN category	Criteria eligibility
A	Not applicable: past, current or future population declines are thought unlikely to exceed 20% in any 3-generation period
B	Not applicable: EOO, AOO, habitat, locations or population not declining, population fluctuations not extreme
C	Not applicable: no continuing population decline or extreme fluctuations
D	Vulnerable: population 250–1000 mature individuals
E	Not applicable: no population viability analysis undertaken

IUCN Red List assessment data

	Estimate	Reliability
Extent of occurrence trend	7900 km^2 stable	medium low
Area of occupancy trend	200 km^2 stable	low low
No. of mature individuals trend	1000 stable	low low
No. subpopulations	1	high
No. locations	<5	low
Generation time	3.8 years	low
Global population share	100%	high

Fensham RJ, Woinarski JCZ (1992) 'Yawalama: the ecology and conservation of monsoon forest on Tiwi Islands, Northern Territory'. Report to DASET. Conservation Commission of the Northern Territory.

ForSci Pty Ltd (1999) 'Environmental Impact Assessment of fast grown plantations on the Tiwi Islands'. Report to Tiwi Land Council, Darwin.

Higgins PJ, Peter JM, Steele WK (Eds) (2001) *Handbook of Australian, New Zealand and Antarctic Birds. Volume 5: Tyrant-Flycatchers to Chats.* Oxford University Press, Melbourne.

Mason IJ, Schodde R (1997) 'Bird survey of the Tiwi Islands, October 1996'. Report to Tiwi Land Council, Northern Territory.

Mathews GM (1914) A list of the birds of Melville Island, Northern Territory, Australia. *Ibis* **2**, 91–132.

Rossiter NA, Setterfield SA, Douglas MM, Hutley LB (2003) Testing the grass-fire cycle: alien grass invasion in the tropical savannas of northern Australia. *Diversity and Distributions* **9**, 169–176.

Schodde R, Mason IJ (1999) *The Directory of Australian Birds: Passerines.* CSIRO Publishing, Melbourbe.

Wilson BA, Fensham RJ (1994) A comparison of classification systems for the conservation of sparsely wooded plains on Melville Island, northern Australia. *Australian Geographer* **25**, 18–31.

Woinarski J, Brennan K, Hempel C, Armstrong M, Milne D, Chatto R (2003a) 'Biodiversity conservation on the Tiwi islands, Northern Territory. Part 2. Fauna'. Department of Infrastructure Planning and Environment, Darwin.

Woinarski JCZ, Brennan K, Cowie I, Kerrigan R, Hempel C (2003b) 'Biodiversity conservation on the Tiwi Islands, Northern Territory. Part 1. Environment and plants'. Report to National Heritage Trust, Tiwi Land Council and Department of Infrastructure Planning and Environment.

Woinarski JCZ, Brennan K, Hempel C, Firth R, Watt F (2000) 'Biodiversity conservation on the Tiwi Islands: plants, vegetation types and terrestrial vertebrates on Melville Island'. Report to Tiwi Land Council, Darwin.

Woinarski JCZ, Hadden K, Hicks J, McLeod D (2003c) 'Biodiversity conservation on the Tiwi islands, Northern Territory. Part 3. Management and planning for biodiversity conservation'. Report to Department of Infrastructure Planning and Environment, Darwin.

Zietz FR (1914) The avifauna of Melville Island, Northern Territory. *South Australian Ornithologist* **1**, 11–18.

Comments received from

Simon Ward, John Woinarski

Species Conservation Summary

Christmas Island White-eye

Zosterops natalis Lister, 1889
Timaliidae

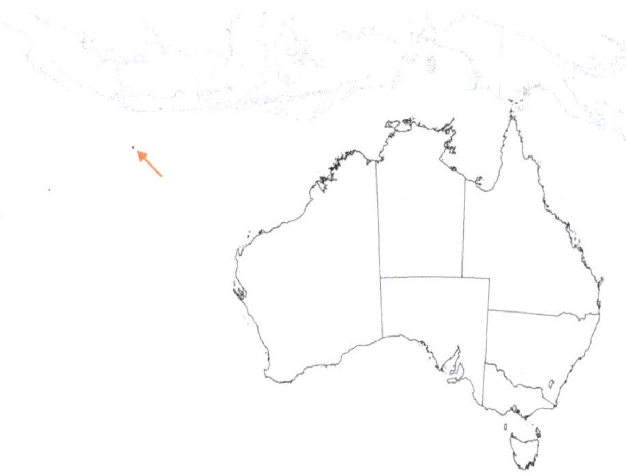

Conservation status
Near Threatened D2

Reasons for listing
Occurs at a single location where susceptible to introduction of predators, competitors or disease, although the risk is now considered low

Status 2000
Near Threatened D2

Status 1990
Near Threatened D2

Taxonomy
No infraspecific taxa described
Taxonomic uniqueness: low (13 genera/family, 73 species/genus, 1 subspecies/species)

Range
Endemic to Christmas I., Indian Ocean, where it occurs in all habitats (Stokes 1988). Also occurs on Cocos Keeling Is. where introduced between 1885 and 1900 and persist around the settlement there (Stokes *et al.* 1984).

Abundance
The most abundant bird on Christmas I. with estimates of 100 000–1 000 000 (van Tets 1975), 20 000 (Garnett and Crowley 2000) and 80 000–170 000 (Beeton *et al.* 2010). It occurred at 99% of 206 sites sampled for birds (James and Retallick 2007). The conservative figure of 20 000 is retained here pending more accurate estimates.

Ecology
The Christmas Island White-eye occupies all forested habitats, including suburban gardens and weed fields in abandoned mine sites, foraging in flocks in all levels of vegetation (Stokes 1988; James and Retallick 2007). A generation time of 9.8 years (BirdLife International 2011) is derived from a mean age at first breeding of 1.0 years, extrapolated from congeners, and a maximum longevity in the wild of 18.7 years, extrapolated from Silvereye *Z. lateralis* (Australian Bird and Bat Banding Scheme).

Threats
The disappearance of white-eyes from Norfolk and Lord Howe Is. suggests that island populations of *Zosterops* are particularly vulnerable to extinction. Fears that the species would be affected by the the Yellow Crazy Ant *Anoplolepis gracilipes* (Garnett and Crowley 2000) have proved wrong: both counts and foraging success of the white-eye, were higher in places where the ants were present because of an increase in scale insects (Davis *et al.* 2008), although this may be because there were fewer Red Crabs *Gecaroidea natalis*, resulting in a more complex vegetation structure (Beeton *et al.* 2010). While Black Rats *Rattus rattus*, which have been responsible for the extinction of white-eyes elsewhere, are present, there is no indication that Christmas Island White-eye numbers are currently being affected. Feral cats *Felis catus* may kill some birds, and exposure to the insecticide Fipronil may be harmful, but there is no primary evidence for this (Beeton *et al.* 2010). There is a plausible, but low probability, threat from new introductions of predators or disease.

Conservation objectives
1. Existing population maintained
2. Black Rats and feral cats eradicated

Information required
1. Population trends
2. Alternatives to Fipronil for controlling Yellow Crazy Ants
3. Effects of rodenticides

Management actions required
1. Continue population monitoring
2. Upgrade quarantine standards for Christmas I.
3. Eradicate Black Rats and feral cats
4. Create a multi-stakeholder team with responsibility for ensuring threatened species are conserved on Christmas I.

Bibliography

Beeton B, Burbidge AA, Grigg G, Harrison P, How RA, Humphries B, McKenzie N, Woinarski J (2010) 'Final report, Christmas Island Expert Working Group to Minister for the Environment, Heritage and the Arts'. Retrieved 1 April, 2010 from <http://www.environment.gov.au/parks/publications/christmas/pubs/final-report.pdf>.

BirdLife International (2011) 'Species factsheet: *Zosterops natalis*'. Retrieved 17 January, 2011 from <http://www.birdlife.org/>.

Davis NE, O'Dowd DJ, Green PT, Mac Nally R (2008) Effects of an alien ant invasion on abundance, behavior, and reproductive success of endemic island birds. *Conservation Biology* **22**, 1165–1176.

Current eligibility against IUCN Red List Criteria

IUCN category	Criteria eligibility
A	Not applicable: past, current or future population declines are thought unlikely to exceed 20% in any 3-generation period
B	Not applicable: EOO, AOO, habitat, locations or population not declining, population fluctuations not extreme
C	Not applicable: no continuing population decline or extreme fluctuations
D	Near Threatened: ≤5 locations with a plausible but low probability threat
E	Not applicable: no population viability analysis undertaken

IUCN Red List assessment data

	Estimate	Reliability
Extent of occurrence trend	134 km^2 stable	high high
Area of occupancy trend	134 km^2 stable	high high
No. of mature individuals trend	20 000 stable	low high
No. subpopulations	1	high
No. locations	1	high
Generation time	9.8 years	low
Global population share	100%	high

Garnett ST, Crowley GM (2000) *The Action Plan for Australian Birds 2000*. Environment Australia, Canberra.

James DJ, Retallick K (2007) 'Forest birds of Christmas Island: a baseline survey of abundance'. Parks Australia North Christmas Island Biodiversity Monitoring Programme. Report to the Department of Finance & Administration and the Department of the Environment & Water Resources, Canberra.

Stokes T (1988) 'A review of the birds of Christmas Island, Indian Ocean'. Australian National Parks and Wildlife Service Occasional Paper.

Stokes AS, Shiels W, Dunn K (1984) Birds of the Cocos (Keeling) Islands, Indian Ocean. *Emu* **84**, 23–28.

van Tets GF (1975) A report on the conservation of resident birds on Christmas Island. *Bulletin of the International Council for Bird Preservation* **12**, 238–242.

Comments received from

David James, Mike Smith, John Woinarski

Subspecies Conservation Summary

Silvereye (Lord Howe Island)

Zosterops lateralis tephropleurus Gould, 1855

Timaliidae

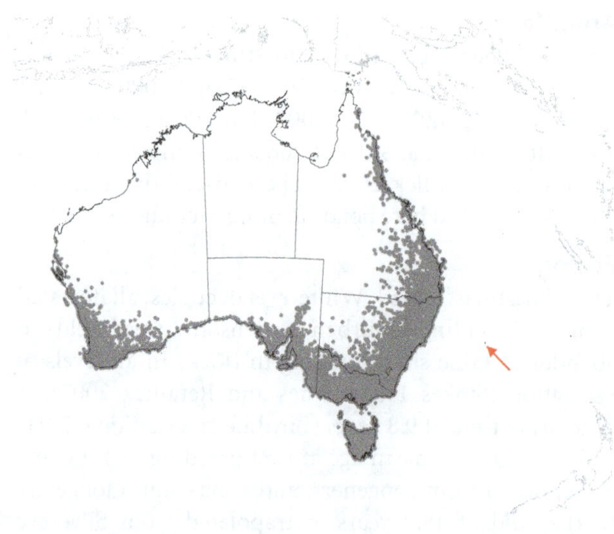

Conservation status

Near Threatened D2

Reasons for listing

Occurs at a single location where susceptible to introduction of predators, competitors or disease, although the risk is low

Status 2000

Near Threatened D2

Status 1990

Near Threatened D2

Taxonomy

Eight mainland Australian subspecies are Least Concern, as is the species. There are 8 extralimital subspecies

Taxonomic uniqueness: low (13 genera/family, 73 species/genus, 17 subspecies/species)

Range

Endemic to Lord Howe I. where it occurs throughout forested parts of island (Hutton 1991).

Abundance

There are no measures of abundance but, based on a density of about 10 birds/ha in Norfolk I. National Park (G. Dutson pers. obs.) and an occupied forest area on Lord Howe I. of about 1000 ha, there could be 10 000 mature individuals on the island. To be precautionary this estimate is halved as the many of forest types on Lord Howe I. are likely to be less productive than on Norfolk I. National Park and may not be as suitable.

Ecology

The Silvereye lives in forests throughout Lord Howe I., apparently nesting out of reach of most predators (Hutton

Current eligibility against IUCN Red List Criteria

IUCN category	Criteria eligibility
A	Not applicable: past, current or future population declines are thought unlikely to exceed 20% in any 3-generation period
B	Not applicable: EOO, AOO, habitat, locations or population not declining, population fluctuations not extreme
C	Not applicable: no continuing population decline or extreme fluctuations
D	Near Threatened: ≤5 locations with a plausible but low probability threat
E	Not applicable: no population viability analysis undertaken

IUCN Red List assessment data

	Estimate	Reliability
Extent of occurrence trend	18 km^2 stable	high high
Area of occupancy trend	12 km^2 stable	medium high
No. of mature individuals trend	5000 stable	low high
No. subpopulations	1	high
No. locations	1	high
Generation time	9.8 years	low
Global population share	100%	high

1991). A generation time of 9.8 years (BirdLife International 2011) is derived from a mean age at first breeding of 1.0 years, extrapolated from congeners, and a maximum longevity in the wild of 18.7 years (Australian Bird and Bat Banding Scheme).

Threats

Silvereyes on Lord Howe I. have survived the introduction of domestic and feral cats *Felis catus*, Black Rats *Rattus rattus*, feral pigs *Sus scrofa*, feral goats *Hircus capra* and Masked Owls *Tyto novaehollandiae*. The restricted area of occupancy, however, makes the subspecies susceptible to catastrophe, such as the introduction of another predator or disease, though the risk is low.

Conservation objectives

1. Maintain current breeding population

Information required

1. Population trends

Management actions required

1. Maintain quarantine for Lord Howe I.
2. Manage threats as required

Bibliography

BirdLife International (2011) 'Species factsheet: *Pachycephala pectoralis*'. Retrieved 17 January, 2011 from <http://www.birdlife.org/>.

Hutton I (1991) *Birds of Lord Howe Island, Past and Present*. The Author, Coffs Harbour.

McAllan IAW, Curtis BR, Hutton I, Cooper RM (2004) The birds of the Lord Howe Island Group: a review of records. *Australian Field Ornithology* **21**, 1–82.

Species Conservation Summary

Robust White-eye

Zosterops strenuus Gould, 1855

Timaliidae

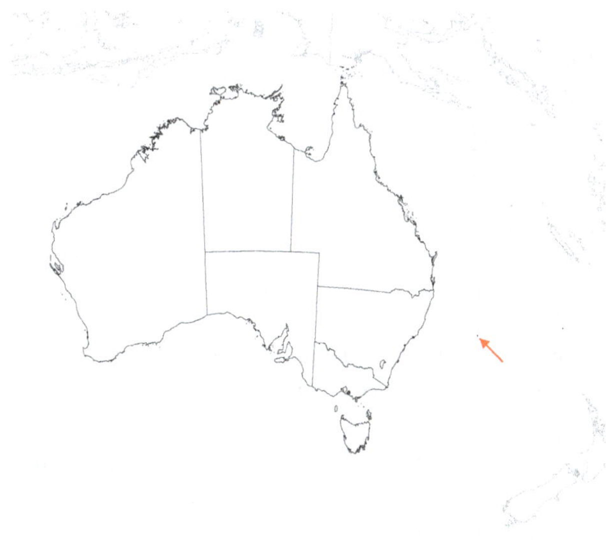

Conservation status

Extinct

Reasons for listing

Last recorded in 1918

Status 2000

Extinct

Status 1990

Extinct

Taxonomy

No infraspecific taxa described. Sometimes included with *Z. tenuirostris* of Norfolk I. (Schodde and Mason 1999).

Taxonomic uniqueness: medium (13 genera/family, 73 species/genus, 1 subspecies/species)

Former range

The Robust White-eye was endemic to Lord Howe I.

Former abundance

It was considered very common (McAllan *et al.* 2004). Its calls were heard all over the island in 1908 (Hull 1909) but it was extinct by 1928 (Sharland 1929).

Ecology

Occurred in lowland forest areas, palm glades and more open habitats including settled areas (Department of Sustainability, Environment, Water, Population and Communities 2011). It probably fed on fruit and insects, as well as eggs of other birds. It foraged on cultivated fruits and other crops, making it a pest to farmers (Hindwood 1940).

Reasons for extinction

Predation by Black Rats *Rattus rattus*, introduced accidentally in 1918 after a shipwreck, is the mostly likely cause of extinction (Hindwood 1940).

Bibliography

Department of Sustainability, Environment, Water, Population and Communities (2011) *Zosterops strenuous.* In 'Species Profile and Threats Database'. Department of Sustainability, Environment, Water, Population and Communities, Canberra. Retrieved 24 February, 2011 from <http://www.environment.gov.au/sprat>.

Hindwood KA (1940) The birds of Lord Howe Island. *Emu* **40**, 1–86.

Hull AFB (1909) The birds of Lord Howe and Norfolk Islands. *Proceedings of the Linnean Society of New South Wales* **34**, 636–693.

McAllan IAW, Curtis BR, Hutton I, Cooper RM (2004) The birds of the Lord Howe Island Group: a review of records. *Australian Field Ornithology* **21**, 1–82.

Schodde R, Mason IJ (1999) The Directory of Australian Birds: Passerines. CSIRO Publishing, Melbourne.

Sharland MSR (1929) Land birds of Lord Howe Island. *Emu* **29**, 5–11.

Species Conservation Summary

Slender-billed White-eye

Zosterops tenuirostris Gould, 1837
Timaliidae

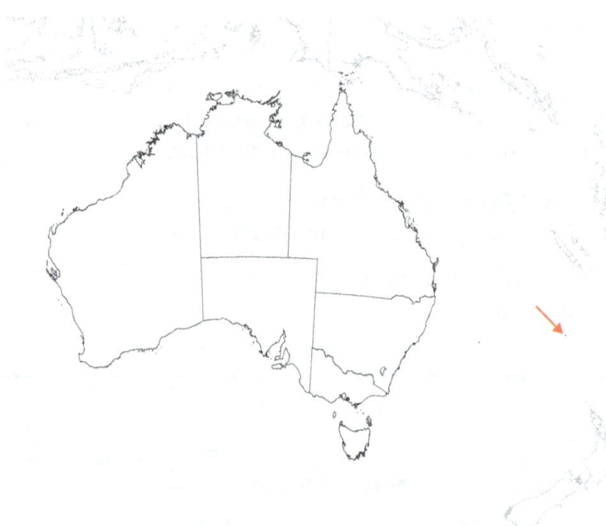

Conservation status

Near Threatened D2

Reasons for listing

Occurs at a single location where susceptible to introduction of predators, competitors or disease, although the risk is low

Status 2000

Near Threatened D2

Status 1990

Near Threatened D2

Taxonomy

No infraspecific taxa described. Sometimes included with *Z. t. strenuus* of Lord Howe I. (Schodde and Mason 1999).

Taxonomic uniqueness: medium (13 genera/family, 73 species/genus, 1 subspecies/species)

Range

Endemic to Norfolk I., where it is largely confined to the Norfolk Island National Park and nearby forested areas (Schodde *et al.* 1983).

Abundance

Has been declining since the 1960s, particularly outside the park, a decline which continued from 1987 to 1997 (Robinson 1997). However, numbers may now have stabilised. A survey in 2010 estimated a population of about 4000 birds in the park on the basis of 74 records in point counts at a density of about 900/km^2, with numbers detected along transects being similar to those detected by Robinson (1988, 1997; G. Dutson pers. obs.).

Ecology

They live in rainforest and tall secondary growth, using their long, down-curved bills to probe fissures in bark for insects, although they also take fruit, including exotic species (Mees 1969; Schodde *et al.* 1983; Robinson 1988). Forages in parties and appears to have a different ecological niche to that of the self-introduced Silvereye *Z. lateralis*, which forages in foliage (Robinson 1988). A generation time of 9.8 years (BirdLife International 2011) is derived from a mean age at first breeding of 1.0 years, extrapolated from congeners, and a maximum longevity

Current eligibility against IUCN Red List Criteria

IUCN category	Criteria eligibility
A	Not applicable: past, current or future population declines are thought unlikely to exceed 20% in any 3-generation period
B	Not applicable: EOO, AOO, habitat, locations or population not declining, population fluctuations not extreme
C	Not applicable: no continuing population decline or extreme fluctuations
D	Near Threatened: ≤5 locations with a plausible but low probability threat
E	Not applicable: no population viability analysis undertaken

IUCN Red List assessment data

	Estimate	Reliability
Extent of occurrence trend	10 km^2 stable	high high
Area of occupancy trend	5 km^2 stable	medium high
No. of mature individuals trend	4500 stable	medium medium
No. subpopulations	1	high
No. locations	1	high
Generation time	9.8 years	low
Global population share	100%	high

in the wild of 18.7 years, extrapolated from Silvereye (Australian Bird and Bat Banding Scheme).

Threats

Much suitable habitat has been cleared or fragmented, and the subspecies appears to be confined to the largest tract of remaining forest. The reason for the decline in the 1960s and 1970s is not known given that the island has been settled for 2 centuries. However, both Black Rats *Rattus rattus* (introduced in the mid 1940s; Robinson 1988) and feral cats *Felis catus* may take some birds (Bell 1990), and the white-eye may have been more vulnerable to Black Rats in disturbed environments. The addition of further predators or disease is also possible but all threats are deemed of low likelihood given the density of birds in the national park.

Conservation objectives

1. Maintain current population

Information required

1. Population trends

Management actions required

1. Maintain quarantine on Norfolk I.
2. Manage additional threats as they emerge
3. Consider reintroduction to Phillip I. when vegetation there is sufficiently mature

Bibliography

Bell BD (1990) 'The status and management of the White-breasted White-eye and other birds on Norfolk Island'. Report to Australian National Parks and Wildlife Service, Canberra.

BirdLife International (2011) 'Species factsheet: *Zosterops tenuirostris*'. Retrieved 17 January, 2011 from <http://www.birdlife.org/>.

Mees GF (1969) A systematic review of the Indo-Australian Zosteropidae (Part III). *Zoologische Verhandelingen* **102**, 1–390.

Robinson D (1988) 'Ecology and management of the Scarlet Robin, White-breasted White-eye and Long-billed White-eye on Norfolk Island'. Report to Australian National Parks and Wildlife Service, Canberra.

Robinson D (1997) 'An evaluation of the status of the Norfolk Island Robin following rat-control and weed-control works in the Norfolk Island National Park'. Report to Environment Australia, Canberra.

Schodde R, Fullagar P, Hermes N (1983) *A Review of Norfolk Island Birds: Past and Present.* Australian National Parks and Wildlife Service, Canberra.

Schodde R, Mason IJ (1999) *The Directory of Australian Birds: Passerines.* CSIRO Publishing, Melbourne.

Comments received from

Margaret Christian, Joe Forshaw, Penny Olsen

Species Conservation Summary

White-chested White-eye

Zosterops albogularis Gould, 1837

Timaliidae

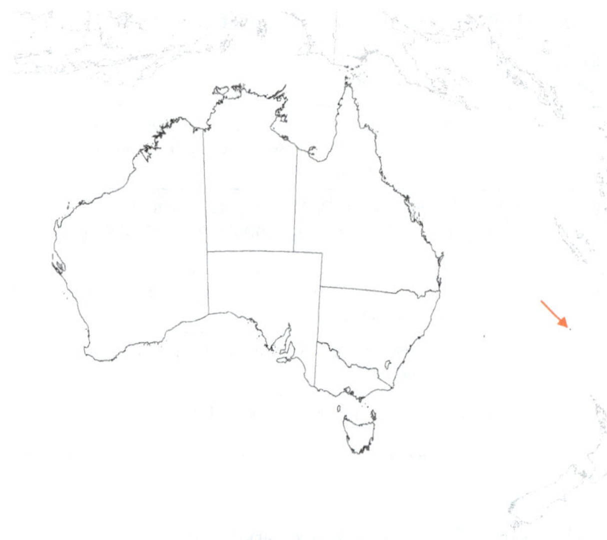

Conservation status

Critically Endangered (Possibly Extinct) D

Reasons for listing

Possibly Extinct since 1979, based on its documented decline, survey effort and presence of Black Rats, balanced against the continued existence of potentially suitable habitat and difficulties with identification

Status 2000

Critically Endangered (Possibly Extinct) D

Status 1990

Critically Endangered (Possibly Extinct) A2b, D

Taxonomy

No infraspecific taxa described

Taxonomic uniqueness: medium (13 genera/family, 73 species/genus, 1 subspecies/species)

Range

Endemic to Norfolk I. where the last records were in or near the Norfolk I. National Park.

Abundance

Reported as 'very plentiful' by Hull (1909) and 12 specimens taken in 1 week in 1926 by Correia (Mees 1969) but only 3 seen in 2 weeks in 1962, at which time the population was estimated at <50 birds (Mees 1969). Four records during an intensive survey in 1979 were the last multi-observer records (Schodde *et al.* 1983). Although previous formal searches in the 1980s (Hermes *et al.* 1986, Rooke 1986, Robinson 1988, Bell 1990) failed to find the species, there was a scatter of reports from the late 1970s to 2009. Some of these were published, but without supporting details (Moore 1981; Rooke 1986; Waugh 1988; Bell 1990; Anon. 1997, 1998; Moore 1999; Garnett and Crowley 2000; Director of National Parks 2009; M. Christian pers. comm.). However, the plumage of the White-chested White-eye is similar to that of the Silvereye *Z. lateralis* and many records are from observers new to Norfolk I., casting doubt on some of these records. A survey based on 353 point counts and 23 km of line-transects in 2009 failed to find the species, and it was calculated that the probability that a population of 10 birds was overlooked was less than 17% (G. Dutson pers. obs.). Given its documented decline, the likely threat from rat *Rattus* spp. and cat *Felis catus* predation, its predisposition to extinction from its tiny range, the extinction of congeners on Lord Howe Island (Hindwood 1940), and the number of negative surveys, it meets IUCN guidelines as Critically Endangered (Possibly Extinct; Butchart *et al.* 2006).

Ecology

Most records have been from rainforest and only a few from weedy forest (Higgins *et al.* 2006) which is mostly thickets of Red Guava *Pisidium cattleianum* or African Olive *Olea europaea cuspidata*, and there have been no reports from weedy forest in recent years (Schodde *et al.* 1983; R. Hill *in litt.*). However, they have been recorded nesting in orchards (Hull 1909), in Red Guava trees and feeding on olive fruits (Mees 1969). A generation time of 3.5 years (BirdLife International 2011) is derived from a mean age at first breeding of 1.0 years, a mean annual adult survival of 34.0% and maximum longevity of 7.8 years, all values extrapolated from congeners.

Threats

The principal reason for the species' decline has probably been predation by Black Rats *Rattus rattus* which is thought to have been introduced in the mid-1940s (Robinson 1988; Bell 1990). Competition from the self-introduced Silvereye, which was first recorded on the island in 1904 (North 1904), may also have contributed to the decline, as may predation by feral cats.

Conservation objectives

1. Maintain any extant population

Information required

1. Evidence it is extant
2. If extant, habitat requirements, threats, population size and trends

Management actions required

1. Continue rat baiting and cat trapping across Norfolk I. National Park
2. Establish population in captivity if any detected
3. Advocate for rat eradication across Norfolk I.
4. Continue to restore native habitat across Norfolk I.

Bibliography

Anon. (1997) Bird sightings. *Norfolk Nature Notes* **12**, 459.

Anon. (1998) Bird sightings. *Norfolk Nature Notes* **13**, 478–479.

Bell BD (1990) 'The status and management of the White-breasted White-eye and other birds on Norfolk Island'. Report to Australian National Parks and Wildlife Service, Canberra.

Current eligibility against IUCN Red List Criteria

IUCN category	Criteria eligibility
A	Not applicable: population declines are thought unlikely to have exceeded 20% in the last 3 generations (12 years)
B	Not applicable: EOO, AOO, habitat, locations or population not declining, population fluctuations not extreme
C	Not applicable: no continuing population decline or extreme fluctuations
D	Critically Endangered (Possibly Extinct): if extant, population <50 mature individuals
E	Not applicable: no population viability analysis undertaken

BirdLife International (2011) 'Species factsheet: *Zosterops albogularis*'. Retrieved 17 January, 2011 from <http://www.birdlife.org>.

Butchart SHM, Stattersfield AJ, Brooks TM (2006) Going or gone: defining 'Possibly Extinct' species to give a truer picture of recent extinctions. *Bulletin of the British Ornithological Club* **126**, 7–24.

Director of National Parks (2010) 'Norfolk Island region threatened species recovery plan'. Department of the Environment, Water, Heritage and the Arts, Canberra.

Gilmour P, Helman C (1989) *The Vegetation of Norfolk Island National Park*. Australian National Parks and Wildlife Service, Canberra.

Hermes N, Evans O, Evans B (1986) Norfolk Island birds: a review 1985. *Notornis* **33**, 141–149.

Higgins PJ, Peter JM, Cowling SJ (Eds) (2006) *Handbook of Australian, New Zealand and Antarctic Birds. Volume 7: Boatbill to Starlings*. Oxford University Press, Melbourne.

Hindwood KA (1940) The birds of Lord Howe Island. *Emu* **40**, 1–86.

Hull AFB (1909) The birds of Lord Howe and Norfolk Islands. *Proceedings of the Linnean Society of New South Wales* **34**, 636–693.

Mees GF (1969) A systematic review of the Indo-Australian Zosteropidae (Part III). *Zoologische Verhandelingen* **102**, 1–390.

Moore JL (1981) Norfolk Island notes 1971 to 1980. *Notornis* **28**, 50–56.

Moore JL (1999) Norfolk Island bird notes 1977 to 1997. *Notornis* **46**, 354–364.

North AJ (1904) Ornithological notes. *Records of the Australian Museum* **5**, 50–56.

Robinson D (1988) 'Ecology and management of the Scarlet Robin, White-breasted White-eye and Long-billed White-eye on Norfolk Island'. Report to Australian National Parks and Wildlife Service, Canberra.

Rooke I (1986) '*Survey of the White-breasted White-eye and the Norfolk Island Boobook Owl on Norfolk Island, October–November 1985*'. RAOU Report No. 20.

Schodde R, Fullagar P, Hermes N (1983) *A Review of Norfolk Island Birds: Past and Present*. Australian National Parks and Wildlife Service, Canberra.

Waugh P (1988) Sightings of the White-chested White-eye. *Norfolk Nature Notes* **4**, 252–253.

Comments received from

Margaret Christian, Penny Olsen, Coral Rowston, Ron Ward

Subspecies Conservation Summary

Bassian Thrush (South Australian)

Zoothera lunulata halmaturina (A. G. Campbell, 1906)
Turdidae

Conservation status
Vulnerable C2a(ii)

Reasons for listing
Population estimated to be <10 000 mature individuals and declining. Although population is divided between Kangaroo I. and mainland, mobility of *Zoothera* thrushes elsewhere suggests they are a single subpopulation; population declining on mainland and heard less commonly in the last decade on Kangaroo I.

Status 2000
Vulnerable C2a(ii)

Status 1990
Vulnerable C2a(ii)

Taxonomy
Z. l. lunulata (south-east Australia) and *Z. l. cuneata* (north-east Queensland) are both Least Concern, as is the species

Current eligibility against IUCN Red List Criteria

IUCN category	Criteria eligibility
A	Not applicable: past, current or future population declines are thought unlikely to exceed 20% in any 3-generation period
B	Not applicable: >10 locations, not severely fragmented, population fluctuations not extreme
C	Vulnerable: population 2500–10 000 mature individuals, suspected continuing decline, 100% in 1 subpopulation
D	Not applicable: population >1000 mature individuals, >5 locations
E	Not applicable: no population viability analysis undertaken

IUCN Red List assessment data

	Estimate	Reliability
Extent of occurrence trend	7900 km^2 stable	high medium
Area of occupancy trend	1000 km^2 decreasing	low medium
No. of mature individuals trend	4200 decreasing	low medium
No. subpopulations	1	medium
No. locations	>10	high
Generation time	5.2 years	medium
Global population share	100%	high

Taxonomic uniqueness: low (24 genera/family, 35 species/genus, 3 subspecies/species)

Range

Occurs in South Australia on Kangaroo I., the Mt Lofty Ranges as far north as Tarunda (Schodde and Mason 1999) and in southern Flinders Ranges from Wirrabara Forest reserve and Telowie Gorge Conservation Park to Mt Remarkable National Park (L. Pedler *in litt.*). On Kangaroo I. there are recent records from the Dudley Peninsula west to Flinders Chase and in forested habitats on both the north and south coast as well as in the catchment of the Cygnet River in the island's centre. Movements of other subspecies suggest that mainland and Kangaroo I. birds are likely to interchange enough to be considered a single subpopulation (Garnett 1992), although an individual collected at Wirrabra was markedly paler than more southerly specimens (L. Pedler *in litt.*). If only 100 mature individuals in the Mt Lofty Ranges (Possingham and Possingham 2008), AOO there unlikely to exceed 25 km^2 with a similar area in the southern Flinders Ranges. On Kangaroo I. potential habitat exists along streams over an area of about 1000 km^2 across the island. The total AOO from records in the Birds Australia database is 276 km^2, but many sites will not have been recorded.

Abundance

Based on a population density of 4 birds/km^2, and an AOO of 1000 km^2 the population on Kangaroo I. is probably about 4000 mature individuals. It is described as widespread there, being most common along the damper gullies of the western half of the island (Baxter 1995). While a substantial proportion of Flinders Chase, Western River Conservation Park and Cape Gantheaume Conservation Park were burnt in 2007, many of the damper areas occupied by the thrush outside Flinders Chase has remained intact. Anecdotal reports suggest a decline in calls heard on Kangaroo I. in the last decade (L. Pedler pers. comm.), but no other reports of decline. In the Mt Lofty Ranges it is now very rare, probably <100 birds (Possingham and Possingham 2008) and, while no decline was evident from 1974–1975 to 1984–1985 (Paton *et al.* 1994), it is now thought to have declined by 30% (Possingham and Possingham 2008). The population in the southern Flinders Ranges may also be about 100 birds based on the area of suitable habitat (L. Pedler *in litt.*), all of which is in conservation reserves.

Ecology

On the mainland, the Bassian Thrush favours damp eucalypt forest or woodland, and has also been recorded breeding in exotic Monterey Pine *Pinus radiata* plantations. On Kangaroo I., it occurs in the damp eucalypt forests, but also uses mature mallee eucalypt woodland. In much of its range, suitable habitat is confined to creeklines or dune swales, the birds foraging for worms among damp leaf litter. Damp habitats seem particularly important in summer. It lays 2–3 eggs in a bulky, bowl-shaped nest (Higgins *et al.* 2006). A generation time of 5.2 years (BirdLife International 2011) is derived from an age at first breeding of 1.0 years and an average annual survival of 46.0%, both values extrapolated from congeners.

Threats

Much of the best habitat has been cleared for agriculture in the past. On the mainland inappropriate fire regimes are currently the major threat: wildfire and prescribed burning in the spring can make habitat unsuitable (Willson and Bignall 2009). Remaining habitat is threatened by the damming of creeks, which results in the desiccation of the litter layer downstream, weed invasion, grazing by cattle, rabbits *Oryctolagus cuniculus* and kangaroos *Macropus* spp., competition with the introduced Common Blackbird *Turdus merula* and predation by fox *Vulpes vulpes* and feral and uncontrolled domestic cats *Felis catus* have all been identified as threats to the subspecies (Willson and Bignall 2009). On Kangaroo I. feral cats and inflated populations of Common Brushtail Possums *Trichosurus vulpecula* are likely to take some adults and nests and much habitat has been cleared but gullies are often undisturbed and rarely burnt. Much of the decline on Kangaroo I. in the last decade can be attributed to below-average rainfall; it is not yet known whether sites that dried out entirely will be re-occupied (L. Pedler *in litt.*)

Conservation objectives

1. Populations retained in the Mt Lofty and southern Flinders Ranges

Information required

1. Trends in numbers at key sites, including Kangaroo I.

2. Key sites for conservation action in the Mt Lofty Ranges
3. The major threats to the key sites and the subspecies, and techniques to manage them

Management actions required

1. Monitor numbers on the mainland
2. Survey Kangaroo I. to determine density and Abundance
3. Characterise favoured habitats and maintain these habitat values in all mainland reserves

Bibliography

Baxter C (1995) *An Annotated List of Birds of Kangaroo Island.* Department of Environment and Natural Resources, Kingscote.

BirdLife International (2011) 'Species factsheet: *Zoothera lunulata*'. Retrieved 17 January, 2011 from <http://www.birdlife.org/>.

Garnett ST (1992) *The Action Plan for Australian Birds.* Australian National Parks and Wildlife Service, Canberra.

Higgins PJ, Peter JM, Cowling SJ (Eds) (2006) *Handbook of Australian, New Zealand and Antarctic Birds. Volume 7: Boatbill to Starlings.* Oxford University Press, Melbourne.

Paton DC, Carpenter G, Sinclair RG (1994) A second bird atlas of the Adelaide region. Part 1: Changes in the distribution of birds: 1974–75 vs 1984–85. *South Australian Ornithologist* **31**, 151–193.

Possingham ML, Possingham HP (2008) Species richness and abundance of birds in Mt Lofty Ranges Gum woodland habitat: Year 2000 survey. *South Australian Ornithologist* **35**, 69–85.

Schodde R, Mason IJ (1999) *The Directory of Australian Birds: Passerines.* CSIRO Publishing, Melbourne.

Willson A,Bignall J (2009) 'Regional recovery plan for threatened species and ecological communities of Adelaide and the Mount Lofty Ranges, South Australia'. Department for Environment and Heritage, South Australia.

Comments received from

Lynn Pedler, Hugh Possingham, Adrian Stokes

Subspecies Conservation Summary

Island Thrush (Christmas Island)

Turdus poliocephalus erythropleurus Sharpe, 1887
Turdidae

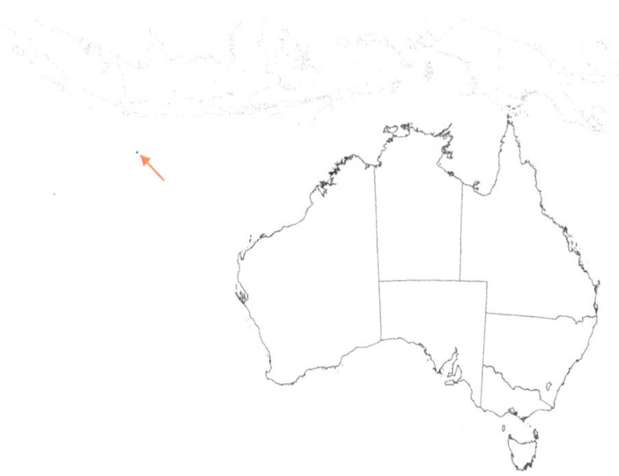

Conservation status

Near Threatened D2

Reasons for listing

Occurs at a single location where susceptible to introduction of predators, competitors or disease, although the risk is now considered low

Status 2000

Near Threatened A3ce, D2

Status 1990

Near Threatened D2

Taxonomy

T. p. poliocephalus (Norfolk I.) and *T. p. vinitinctus* (Lord Howe I.) are Extinct. The species is Least Concern. There are 49 extralimital subspecies.
Taxonomic uniqueness: low (24 genera/family, 67 species/genus, 52 subspecies/species)

Range

Endemic to Christmas I., Indian Ocean where it occurs throughout the island (Department of Sustainability, Environment, Water, Population and Communities 2011). Area of occupancy was reduced by clearance for phosphate mining from the 1890s to the 1980s, but the species remains relatively abundant in remaining forest and in regrowth sites.

Abundance

There are no good estimates of the population size. The species was recorded in 97% of 123 sites surveyed by James and Retallick (2007). The most recent estimates, of which the average is used here with low reliability, was 20 000–50 000 individuals (Beeton *et al.* 2010). There is no information on trends.

Ecology

Live in most habitats, including forests and settlements, but not in post-mining clearing or *Pandanus* thickets. Island Thrushes consume a wide range of food items, including invertebrates such as insects, spiders, snails, and earthworms, as well as carrion and even small reptiles. They also take fruit and seeds (Department of Sustainability, Environment, Water, Population and Communities 2011). A generation time of 6.1 years (BirdLife International 2011) is derived from an average age at first breeding of 1.3 years and a maximum longevity in the wild of 9.4 years, both values extrapolated from congeners.

Current eligibility against IUCN Red List Criteria

IUCN category	Criteria eligibility
A	Not applicable: past, current or future population declines are thought unlikely to exceed 20% in any 3-generation period
B	Not applicable: EOO, AOO, habitat, locations or population not declining, population fluctuations not extreme
C	Not applicable: no continuing population decline or extreme fluctuations
D	Near Threatened: ≤5 locations with a plausible but low probability threat
E	Not applicable: no population viability analysis undertaken

IUCN Red List assessment data

	Estimate	Reliability
Extent of occurrence trend	134 km^2 stable	high high
Area of occupancy trend	134 km^2 stable	high high
No. of mature individuals trend	35 000 stable	low low
No. subpopulations	1	high
No. locations	1	high
Generation time	6.1 years	low
Global population share	100%	high

Threats

Like other Christmas I. species, Island Thrushes are potentially threatened by the introduced Yellow Crazy Ant *Anoplolepis gracilipes*, which can alter the whole ecology of sites it occupies by killing the super-numerous Red Crab *Gecaroidea natalis*, as well as by tending scale insects that damage trees (O'Dowd *et al.* 2003; Abbott and Green 2007). Nest success and the number of juveniles encountered was lower in areas with ants than without (Davis *et al.* 2008). Faltering of ant control and infestation of all rainforest could reduce productivity below sustainable levels but the ants have been controlled, at least temporarily, and concerns held by Garnett and Crowley (2000), when the ant problem was first appreciated, have not been borne out. Feral cats *Felis catus* and Black Rats *Rattus rattus* are also likely to kill birds and reduce nesting success: rats in particular have been implicated in the extinction of Island Thrush subspecies on Norfolk and Lord Howe Is. The species may also be affected by the pesticide Fipronil used for ant control but this remains unproven (Beeton *et al.* 2010). Quarantine measures on the island are not currently sufficient to prevent invasion by additional unspecified predators.

Conservation objectives

1. Existing population maintained
2. Yellow Crazy Ant populations constrained to levels that do not threaten subspecies
3. Black Rats eradicated and feral cats controlled

Information required

1. Alternatives to Fipronil for controlling Yellow Crazy Ants
2. Population trends
3. Levels of Yellow Crazy Ant abundance that will threaten subspecies

Management actions required

1. Continue population monitoring
2. Upgrade quarantine standards for Christmas I.
3. Continue control of Yellow Crazy Ants
4. Eradicate Black Rats
5. Control or eradicate feral cats
6. Create a multi-stakeholder team with responsibility for ensuring threatened species are conserved on Christmas I.

Bibliography

Abbott KL, Green PT (2007) Collapse of an ant-scale mutualism in a rainforest on Christmas Island. *Oikos* **116**, 1238–1246.

Beeton B, Burbidge AA, Grigg G, Harrison P, How RA, Humphries B, McKenzie N, Woinarski J (2010) 'Final report, Christmas Island Expert Working Group to Minister for the Environment, Heritage and the Arts'. Retrieved 1 April, 2010 from <http://www.environment.gov.au/parks/publications/christmas/pubs/final-report.pdf>.

BirdLife International (2011) 'Species factsheet: *Turdus poliocephalus*'. Retrieved 17 January, 2011 from <http://www.birdlife.org/>.

Davis NE, O'Dowd DJ, Green PT, Mac Nally R (2008) Effects of an alien ant invasion on abundance, behavior, and reproductive success of endemic island birds. *Conservation Biology* **22**, 1165–1176.

Department of Sustainability, Environment, Water, Population and Communities (2011) *Turdus poliocephalus erythropleurus*. In 'Species Profile and Threats Database'. Department of Sustainability, Environment, Water, Population and Communities, Canberra. Retrieved 24 February, 2011 from <http://www.environment.gov.au/sprat>.

Garnett S, Crowley GM (2000) *The Action Plan for Australian Birds 2000*. Environment Australia, Canberra.

James DJ, Retallick K (2007) 'Forest birds of Christmas Island: a baseline survey of abundance'. Parks Australia North Christmas Island Biodiversity Monitoring Programme. Report to the Department of Finance & Administration and the Department of the Environment and Water Resources, Canberra.

O'Dowd DJ, Green PT, Lake PS (2003) Invasional 'meltdown' on an oceanic island. *Ecology Letters* **6**, 812–817.

Comments received from

Andrew Burbidge, David James, Mike Smith, John Woinarski

Subspecies Conservation Summary

Island Thrush (Norfolk Island; Grey-headed Blackbird)

Turdus poliocephalus poliocephalus Latham, 1802
Turdidae

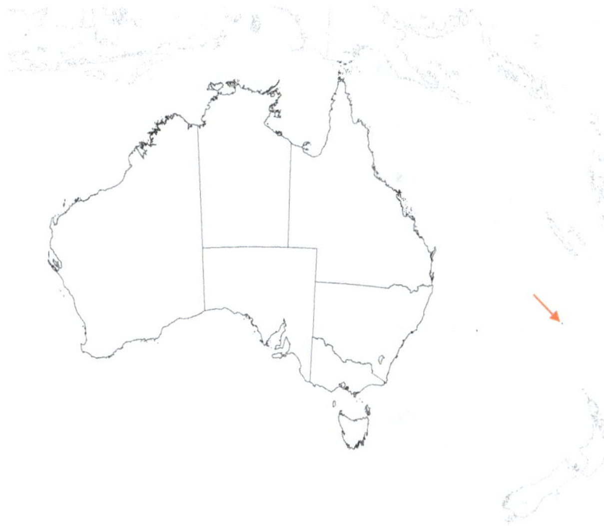

Conservation status
Extinct

Reasons for listing
There have been no records of this subspecies since 1975

Status 2000
Extinct

Status 1990
Extinct

Taxonomy
T. p. erythropleurus (Christmas I.) is Near Threatened, *T. p. vinitinctus* (Lord Howe I.) is Extinct. The species is Least Concern. There are 49 extralimital subspecies. Taxonomic uniqueness: low (24 genera/family, 67 species/genus, 52 subspecies/species)

Former range
Endemic to Norfolk I. Presumably occurred throughout the island, but became increasingly confined as habitat was cleared and fragmented. Persisted in Norfolk Island National Park until at least 1968, when 2 were mist-netted (Smithers and Disney 1969), and probably 1975, when 1 was seen (Schodde *et al.* 1983), but intensive searches by residents and visiting birdwatchers (McKean *et al.* 1976; Hermes *et al.* 1986; Bell 1990) have failed to detect any for over 20 years. Probably became extinct in the 1970s.

Former abundance
Formerly considered common and widespread, and was particularly common in gardens (Smithers and Disney 1969) until the 1940s (Schodde *et al.* 1983; Robinson 1988).

Ecology
A confiding forest bird that fed among leaf litter on the ground on insects. It was common in gardens near rainforest but, as it declined, was increasingly confined to the rainforest in the Norfolk I. National Park (Higgins *et al.* 2006).

Reasons for extinction
The principal reason for its disappearance was probably the arrival of Black Rats *Rattus rattus* in the 1940s (Robinson 1988). Though clearance would have reduced the area of suitable habitat, the subspecies remained common on the island until the 1940s by which time feral cats *Felis catus* had been present for more than a century and Common Blackbirds *Turdus merula* (introduced in 1918) and Song Thrushes *T. philomelos* (introduced in 1913; Smithers and Disney 1969; Schodde *et al.* 1983) had been established for 3 decades. Given that the Island Thrush is closely related to the Common Blackbird (Schodde and Mason 1999), hybridisation may have played a part in the subspecies' decline. Genes of the subspecies may persist in the population of Common Blackbirds. On Norfolk I., immature Common Blackbirds have a lighter head than adults and a substantial proportion of the population is pied (Schodde and Mason 1999).

Bibliography

Bell BD (1990) 'The status and management of the White-breasted White-eye and other birds on Norfolk Island'. Report to Australian National Parks and Wildlife Service, Canberra.

Hermes N, Evans O, Evans B (1986) Norfolk Island birds: a review 1985. *Notornis* **33**, 141–149.

Higgins PJ, Peter JM, Cowling SJ (Eds) (2006) *Handbook of Australian, New Zealand and Antarctic Birds. Volume 7: Boatbill to Starlings*. Oxford University Press, Melbourne.

McKean JL, Evans O, Lewis JH (1976) Notes on the birds of Norfolk Island. *Notornis* **23**, 299–301.

Robinson D (1988) 'Ecology and management of the Scarlet Robin, White-breasted White-eye and Long-billed White-eye on Norfolk Island'. Report to Australian National Parks and Wildlife Service, Canberra.

Schodde R, Fullagar P, Hermes N (1983) *A Review of Norfolk Island Birds: Past and Present*. Australian National Parks and Wildlife Service, Canberra.

Schodde R, Mason IJ (1999) *The Directory of Australian Birds: Passerines*. CSIRO Publishing, Melbourne.

Smithers CN, Disney HJdeS (1969) The distribution of terrestrial and freshwater birds on Norfolk Island. *Australian Journal of Zoology* **15**,127–140.

Subspecies Conservation Summary

Island Thrush (Lord Howe Island; Vinous-tinted)

Turdus poliocephalus vinitinctus (Gould, 1855)
Turdidae

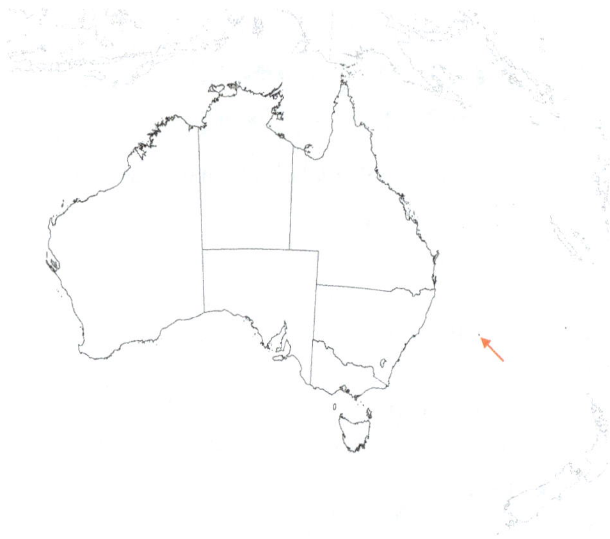

Conservation status
Extinct

Reasons for listing
Last recorded in 1913

Status 2000
Extinct

Status 1990
Extinct

Taxonomy
T. p. erythropleurus (Christmas I.) is Near Threatened, *T. p. poliocephalus* (Norfolk I.) is Extinct. The species is Least Concern. There are 49 extralimital subspecies. Taxonomic uniqueness: low (24 genera/family, 67 species/genus, 52 subspecies/species)

Former range
Endemic to Lord Howe I.

Former abundance
It was considered common (McAllan *et al.* 2004), partly because of its tameness (Hull 1909), but the population may never have been very large (Hindwood 1940).

Ecology
Appears to have occupied all forested habitats, nesting on or near the ground (Hindwood 1940, McAllan *et al.* 2004).

Reasons for extinction
The last reports in the literature were from 1913. It could not be found in 1928 (Sharland 1929). The subspecies presumably succumbed rapidly after Black Rats *Rattus rattus* colonised the island from a shipwreck in 1918 (Hindwood 1940).

Bibliography

Hindwood KA (1940) The birds of Lord Howe Island. *Emu* **40**, 1–86.

Hull AFB (1909) The birds of Lord Howe and Norfolk Islands. *Proceedings of the Linnean Society of New South Wales* **34**, 636–693.

McAllan IAW, Curtis BR, Hutton I, Cooper RM (2004) The birds of the Lord Howe Island Group: a review of records. *Australian Field Ornithology* **21**, 1–82.

Sharland MSR (1929) Land birds of Lord Howe Island. *Emu* **29**, 5–11.

Species Conservation Summary

Singing Starling

Aplonis cantoroides (G. R. Gray, 1862)
Sturnidae

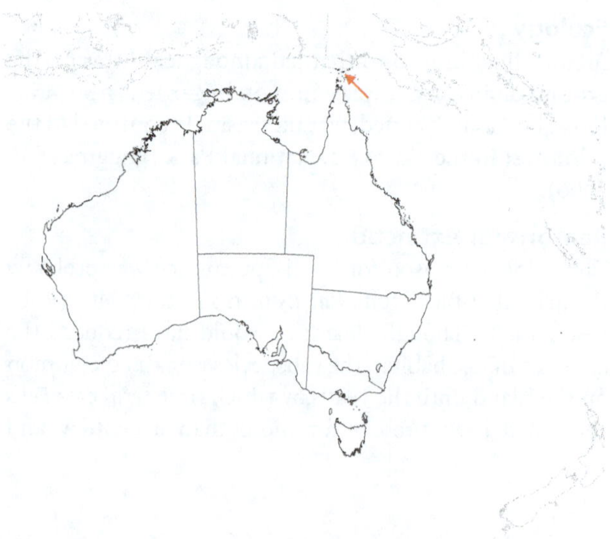

Conservation status
Australian breeding population: Near Threatened D1

Reasons for listing
Australian breeding population small enough for Endangered but not isolated from the main population in New Guinea that is Least Concern, so downgraded 2 categories to Near Threatened

Status 2000
Near Threatened D1

Current eligibility against IUCN Red List Criteria

IUCN category	Criteria eligibility
Australian breeding population	
A	Not applicable: past, current or future population declines are thought unlikely to exceed 20% in any 3-generation period
B	Not applicable: EOO, AOO, habitat, locations or population not declining, population fluctuations not extreme
C	Not applicable: no continuing population decline or extreme fluctuations
D	Endangered: population 50–250 mature individuals (but downgraded to Near Threatened as global status Least Concern)
E	Not applicable: no population viability analysis undertaken

IUCN Red List assessment data

	Estimate	Reliability
Australian breeding population		
Extent of occurrence trend	400 km^2 stable	high high
Area of occupancy trend	168 km^2 stable	high high
No. of mature individuals trend	100 stable	low medium
No. subpopulations	1	high
No. locations	2	high
Generation time	4.1 years	low
Global population share	<1 %	high
Level of genetic exchange	high	high

Status 1990

Near Threatened D1

Taxonomy

No infraspecific taxa described. The species is Least Concern.

Taxonomic uniqueness: medium (27 genera/family, 20 species/genus, 1 subspecies/species)

Range

Often recorded from Saibai and Boigu Is. in the north of Torres Strait adjacent to Papua New Guinea (Clarke 2004). Extralimitally it occurs across much of New Guinea and the Solomon Is. (Coates 1990).

Abundance

Regular records of up to 40 around the town on Boigu I., where recorded breeding (Clarke 2004) and 1 record of 3 individuals on Saibai I. (R. Clarke pers. comm.). The population density is likely to be similar to that in Papua New Guinea where it is locally common in open habitats, often flocking in towns, suggesting that the population in Australia may be around 100 individuals.

Ecology

Inhabits open lowland habitats, including forest edge, gardens, savanna and coasts, often around towns and buildings, and nests in tree hollows and on ledges or cavities in rocks and buildings (Coates 1990). Given the lack of subspecific variation between many relatively distant islands in its range, it is considered to be a nomadic or dispersive species (Coates 1990) and there is likely to be regular exchange between the Torres Strait islands and New Guinea. A generation time of 4.1 years (BirdLife International 2011) is derived from a mean age at first breeding of 1.2 years, extrapolated from mean values for *Acridotheres* and *Sturnus*, and a maximum longevity in the wild of 7.0 years, extrapolated from congeners.

Threats

The subspecies is listed as threatened because it has a very small population. There are no plausible threats within 3 generations; though Saibai and Boigu are low-lying, they are too far north for catastrophic cyclones.

Conservation objectives

1. Population maintained

Information required

2. Population size

Management actions required

1. Initiate community-based ecotourism business promoting the islands' unique avifauna. Monitoring of subpopulations and habitat could then be undertaken by visiting bird-watchers and local rangers.

Bibliography

BirdLife International (2011) 'Species factsheet: *Aplonis cantoroides*'. Retrieved 17 January, 2011 from <http://www.birdlife.org>.

Clarke RH (2004) The avifauna of northern Torres Strait: notes on a wet season visit. *Australian Field Ornithology* **21**, 49–66.

Coates BJ (1990) *The Birds of Papua New Guinea. Volume II.* Dove Publications, Alderley.

Comments received from

Rohan Clarke

Species Conservation Summary

Tasman Starling

Aplonis fusca Gould, 1836
Sturnidae

Conservation status
Extinct

Reasons for listing
There have been no records of this species since 1923

Status 2000
Extinct

Status 1990
Extinct

Taxonomy
A. f. fusca (Norfolk I.) and *A. f. hulliana* (Lord Howe I.) are both Extinct
Taxonomic uniqueness: low (27 genera/family, 24 species/genus. 2 subspecies/species)

Former range
The Tasman Starling was endemic to Norfolk and Lord Howe Is.

Former abundance
This was considered the most common bird on Lord Howe I. before Black Rats *Rattus rattus* arrived (Hindwood 1940). There is no information for Norfolk I. but it did occur around the settlement (Hull 1909).

Ecology
The Tasman Starling nested in tree hollows, often very low down, and presumably occupied all forested habitats on the 2 islands (Hindwood 1940). The diet was probably largely native fruit but later included at least bananas (Hull 1909) and a variety of other exotic fruits (Department of Sustainability Environment Water Population and Communities 2011).

Reasons for extinction
On both islands the most likely cause of extinction was the introduction of Black Rats *Rattus rattus* which were introduced to Lord Howe I. in 1918 and to Norfolk I. in the 1940s (Hindwood 1940; Robinson 1988).

Bibliography
Department of Sustainability, Environment, Water, Population and Communities (2011) *Aplonis fusca*. In 'Species Profile and Threats Database'. Department of Sustainability, Environment, Water, Population and Communities, Canberra. Available from: http://www.environment.gov.au/cgi-bin/sprat/public/publicspecies.pl?taxon_id=83432#sprat_profile. Accessed Thu, 24 Feb 2011.

Hindwood KA (1940) The birds of Lord Howe Island. *Emu* **40**, 1–86.

Hull AFB (1909) The birds of Lord Howe and Norfolk Islands. *Proceedings of the Linnean Society of New South Wales* **34**, 636–693.

Robinson D (1988) 'Ecology and management of the Scarlet Robin, White-breasted White-eye and Long-billed White-eye on Norfolk Island'. Report to Australian National Parks and Wildlife Service, Canberra.

Subspecies Conservation Summary

Tasman Starling (Norfolk Island)

Aplonis fusca fusca Gould, 1836
Sturnidae

Conservation status
Extinct

Reasons for listing
Last recorded in 1923

Status 2000
Extinct

Status 1990
Extinct

Taxonomy
A. f. hulliana (Lord Howe Island) is Extinct, as is the species
Taxonomic uniqueness: low (27 genera/family, 24 species/genus, 2 subspecies/species)

Former range
Endemic to Norfolk I.

Former abundance

Hull (1909) recorded them in gardens but there appear to have been no observations relating to the subspecies' abundance before extinction.

Ecology

The Tasman Starling nested in tree hollows, often very low down, and presumably occupied all forested habitats on Norfolk I. The diet was probably largely native fruit but later included at least bananas *Musa paradisica* (Hull 1909).

Reasons for extinction

There is a gap between when the species was last seen on Norfolk I. (1923) and when its absence was noted (Smithers and Disney 1969). Black Rats *Rattus rattus* were introduced in the 1940s (Robinson 1988) and probably caused the subspecies' extinction (Department of Sustainability, Environment, Water, Population and Communities 2011). Clearance for agriculture would have reduced the area of available native habitat and hence population size. In the 10 years before extinction, Common Blackbird *Turdus merula*, Song Thrush *T. philomelos* and Common Starling *Sturnus vulgaris* were introduced to the island, and may also have competed with the native starling (Schodde *et al.* 1983).

Bibliography

Department of Sustainability, Environment, Water, Population and Communities (2011) *Aplonis fusca*. In 'Species Profile and Threats Database'. Department of Sustainability, Environment, Water, Population and Communities, Canberra. Available from: http://www.environment.gov.au/cgi-bin/sprat/public/publicspecies.pl?taxon_id=83432#sprat_profile. Accessed Thu, 24 Feb 2011.

Hull AFB (1909) The birds of Lord Howe and Norfolk Islands. *Proceedings of the Linnean Society of New South Wales* **34**, 636–693.

Robinson D (1988) 'Ecology and management of the Scarlet Robin, White-breasted White-eye and Long-billed White-eye on Norfolk Island'. Report to Australian National Parks and Wildlife Service, Canberra.

Schodde R, Fullagar P, Hermes N (1983) *A Review of Norfolk Island Birds: Past and Present*. Australian National Parks and Wildlife Service, Canberra.

Smithers CN, Disney HJdeS (1969) The distribution of terrestrial and freshwater birds on Norfolk Island. *Australian Journal of Zoology* **15**,127–140.

Subspecies Conservation Summary

Tasman Starling (Lord Howe Island)

Aplonis fusca hulliana Mathews, 1912

Sturnidae

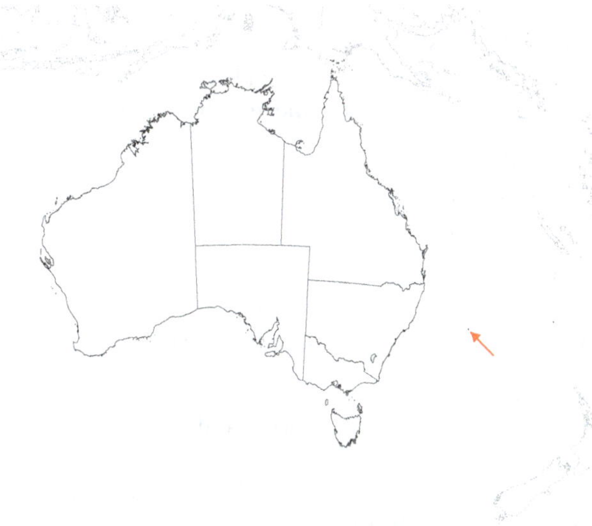

Conservation status

Extinct

Reasons for listing

Last recorded in 1918

Status 2000

Extinct

Status 1990

Extinct

Taxonomy

A. f. fusca (Norfolk Island) is Extinct, as is the species

Taxonomic uniqueness: low (27 genera/family, 24 species/genus, 2 subspecies/species)

Former range

This subspecies of Tasman Starling was endemic to Lord Howe Island.

Former abundance

In 1913–1915 it was described as occurring in 'thousands', but has not been seen since 1918 (Hindwood 1940).

Ecology

Nested in tree hollows and presumably occupied all forested habitats on Lord Howe I. (Hindwood 1940). The diet was probably largely native fruit but later included cultivated fruit, including bananas, peaches and strawberries (Department of Sustainability, Environment, Water, Population and Communities 2011). They also fed on snails and the eggs of other birds (Hindwood 1940; Hutton 1991; McAllan *et al.* 2004).

Reasons for extinction

The last reports in the literature are from 1915 (Hindwood 1940). It was not recorded in 1928 (Sharland

1929). The subspecies presumably succumbed rapidly after Black Rats *Rattus rattus* colonised the island from a shipwreck in 1918 (Hindwood 1940). They nested in hollows, often close to the ground, where they would have been particularly vulnerable to rat predation. The subspecies was extinct before the Common Starling *Sturnus vulgaris* arrived on the island (Department of Sustainability, Environment, Water, Population and Communities 2011).

Bibliography

Department of Sustainability Environment Water Population and Communities (2011). *Aplonis fusca*. In 'Species Profile and Threats Database'. Department of Sustainability, Environment, Water, Population and Communities, Canberra. Available from: http://www.environment.gov.au/cgi-bin/sprat/public/publicspecies.pl?taxon_id=83432#sprat_profile. Accessed Thu, 24 Feb 2011.

Hindwood KA (1940) The birds of Lord Howe Island. *Emu* **40**, 1–86.

Hutton I (1991) *Birds of Lord Howe Island, Past and Present*. Coffs Harbour, The Author.

McAllan IAW, Curtis BR, Hutton I, Cooper RM (2004) The birds of the Lord Howe Island Group: a review of records. *Australian Field Ornithology* **21**, 1–82.

Sharland MSR (1929) Land birds of Lord Howe Island. *Emu* **29**, 5–11.

Subspecies Conservation Summary

Black-throated Finch (southern)

Poephila cincta cincta (Gould, 1837)

Estrildidae

Conservation status

Vulnerable C2a(i)

Reasons for listing

Population <10 000 mature individuals and continuing to decline, <1000 in largest known subpopulation

Status 2000

Vulnerable A2c+4c

Status 1990

Near Threatened A2c+4c

Reasons for change in 2000: Increased rate of decline in EOO as distant subpopulations disappeared

Taxonomy

P. c. atropygialis (far north Queensland) is Least Concern, as is the species

Taxonomic uniqueness: medium (29 genera/family, 3 species/genus, 2 subspecies/species)

Range

Eastern Queensland from as far north as Townsville on the coast, and possibly Ingham, south to the Burdekin River and possibly Rockhampton, extending inland to Aramac. Since 1998, recorded in Townsville region (Giru, Serpentine Lagoon, Toonpan, near Ross River Dam and Bohle, Alice and Black River catchments), Ingham region (Ollera Creek near Mutarnee and Mt Fox) and scattered sites in central-eastern Qld (Great Basalt Wall, Yarrowmere Station, Moonoomoo Station, Doongmabulla Station, Fortuna Station and Aramac) (Black-throated Finch Recovery Team 2007), assessed here as 3 subpopulations. Historically from Inverell, north-eastern New South Wales, through eastern Queensland to the Burdekin-Lynd Divide, where it intergrades over a broad zone with *P. c. atropygialis* (Schodde and Mason 1999). Range has been contracting for several decades (Franklin 1999), including disappearance from Brisbane region by 1940s, most of NSW by 1970s (but last record in 1994), Rockhampton region in 1970s (but last record in 1989) and New England Tableland region in the 1990s (Black-throated Finch Recovery Team 2007). This resulted in a severe decline in EOO up to the 1990s, after which EOO has declined much more slowly. The AOO is estimated nominally at 1% of the EOO.

Abundance

Numbers counted at coordinated waterhole counts were stable 2004–2009 (450 sightings at 16 sites in 2005–2006 to 489 sightings at 25 sites in 2009) but the finches have disappeared from a number of sites around Townsville in the last decade (Black-throated Finch Recovery Team 2007). Given the mobility of the species across the landscape the average density is assessed as 1/km^2. The Townsville subpopulation is assessed as having no more than 600 mature individuals (T. Grice *in litt.*) and is assumed here to be the largest surviving subpopulation.

Current eligibility against IUCN Red List Criteria

IUCN category	Criteria eligibility
A	Not applicable: past, current or future population declines are thought unlikely to exceed 20% in any 3-generation period
B	Near Threatened: AOO near 2000 km^2, possibly severely fragmented, inferred decline in AOO, habitat condition, no. locations, no. mature individuals from urban expansion and mining
C	Vulnerable: population 500–2500 mature individuals, suspected continuing decline, <1000 mature individuals in largest subpopulation but <95% in largest subpopulation
D	Not applicable: population >1000 mature individuals, >5 locations
E	Not applicable: no population viability analysis undertaken

IUCN Red List assessment data

	Estimate	Reliability
Extent of occurrence trend	290 000 km^2 decreasing	medium low
Area of occupancy trend	2900 km^2 decreasing	low low
No. of mature individuals trend	1400 decreasing	low medium
No. subpopulations	3	low
No. locations	>10	medium
Largest subpopulation	600	low
Generation time	3.5 years	low
Global population share	100%	high

Poorly known subpopulations in central-eastern Qld are assumed to have no more than 400 mature individuals each.

Ecology

Inhabits open grassy woodland, usually dominated by eucalypts or ti-trees, where there is access to seeding grasses and water. They appear to feed preferentially in lightly grassed areas (Zann 1976; P. Buosi *in litt.*). It appears to be relatively sedentary, with birds present throughout the year near Townsville, but may move in response to drought (Ley and Cook 2001). In the dry season it feeds on seeds of annual grasses like *Schizachyrium* spp., switching in the wet season to half-ripe grass seeds and insects (Immelman 1982; Zann 1976). Probably needs a mosaic of different habitats in which to find seed during the wet season (Mitchell 1996). It builds domed nests in trees, sometimes in tree hollows, laying 5–9 eggs (Immelman 1982). A generation time of 3.5 years (BirdLife International 2011) is derived from an average age at first breeding of 1.0 years, an annual survival of adults of 50%, both extrapolated from mean values for Estrildidae, and a maximum longevity in the wild of 6.9 years, extrapolated from Masked Finch *P. personata* (Australian Bird and Bat Banding Scheme).

Threats

The synergistic effects of pastoralism, drought and fire are thought to have had the greatest impact on the species by increasing the period of resource shortage and decreasing the quality and availability of breeding habitat. The decline in eastern Qld began early in the 20th century under pastoralism (Franklin 1999), particularly in the southern part of the subspecies' range, where there has been extensive clearing for sheep grazing and rabbits *Oryctolagus cunciculus* were prevalent. Probable current threats are clearing and fragmentation of woodland, riverside habitats and shrubland; degradation of habitat by domestic stock and rabbits, including alterations to fuel load, vegetation structure and wet season food availability; alteration of habitat by changes in fire regime; invasion of habitat by exotic weeds and grasses; and perhaps illegal trapping of birds and predation by introduced mammals (Black-throated Finch Recovery Team 2007; Department of Sustainability Environment Water Population and Communities 2011). Intensification of settlement near Townsville is impinging on the largest subpopulation and the birds do not persist for long once pastoral land is subdivided into rural lifestyle blocks. Also some occupied areas in central Qld are likely to be mined in the near future.

Conservation objectives

1. Stable populations in monitored regions

Information required

1. Breeding requirements and threats to key breeding areas
2. Size of inland subpopulations
3. How populations use the landscape at different times of year
4. Effectiveness of management interventions
5. Suitability of birds currently in captivity for reintroduction in southern Queensland

Management actions required

1. Continue monitoring and analyse the results
2. Secure selected sites for conservation
3. Address threats on grazing lands
4. Ensure requirements of the finch are incorporated into planning around Townsville

Bibliography

BirdLife International (2011) 'Species factsheet: *Poephila cincta*'. Downloaded from http://www.birdlife.org on 10 April 2011.

Black-throated Finch Recovery Team, Department of Environment and Climate Change (NSW) and Queensland Parks and Wildlife Service (2007) 'National recovery plan for the black-throated finch southern subspecies *Poephila cincta cincta*'. Report to the Department of the Environment and Water Resources, Canberra. Department of Environment and Climate Change (NSW), Hurstville and Queensland Parks and Wildlife Service, Brisbane.

Department of Sustainability, Environment, Water, Population and Communities (2011) *Poephila cincta cincta*. In 'Species Profile and Threats Database'. Department of Sustainability, Environment, Water, Population and Communities, Canberra. Retrieved 24 February, 2011 from <http://www.environment.gov.au/sprat>.

Franklin DC (1999) Evidence of disarray amongst granivorous bird assemblages in the savannas of northern Australia, a region of sparse human settlement. *Biological Conservation* **90**, 53–68.

Immelman K (1982) *Australian Finches.* 2nd edn. Angus and Robertson, Sydney.

Ley AJ, Cook SM (2001) The Black-throated finch *Poephila cincta* in New South Wales. *Australian Bird Watcher* **19**, 115–120.

Schodde R, Mason IJ (1999) *The Directory of Australian Birds: Passerines.* CSIRO Publishing, Melbourne.

Zann R (1976) Distribution, status and breeding of Black-throated finches *Poephila cincta* in northern Queensland. *Emu* **76**, 201–206.

Comments received from

Peter Buosi, Tony Grice, Alex Kutt, Kim Maute, James Moloney

Subspecies Conservation Summary

Crimson Finch (white-bellied)

Neochmia phaeton evangelinae d'Albertis & Salvadori, 1879

Estrildidae

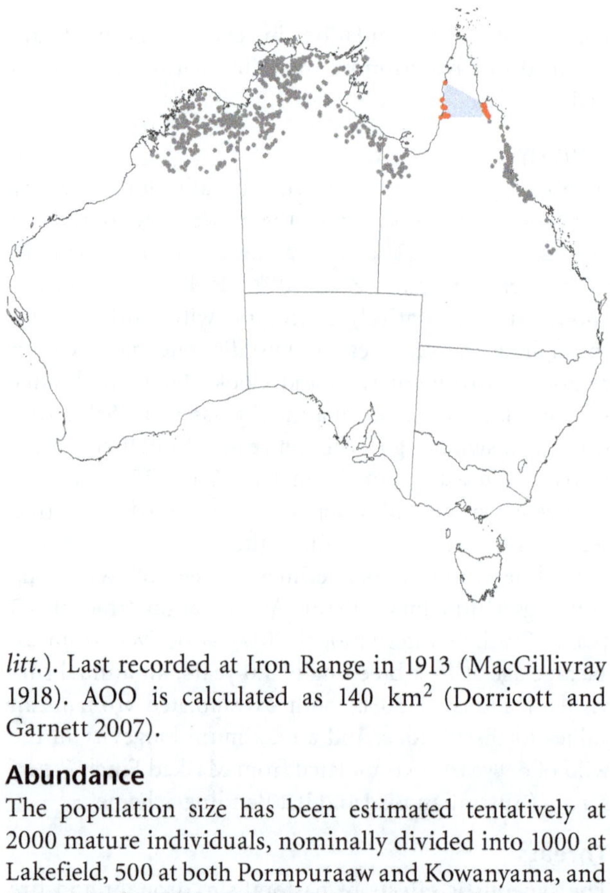

Conservation status

Australian breeding population: Near Threatened D1+2

Reasons for listing

Population approaching 1000 individuals at lowest point in year; and present at about 8 locations, all at risk from through fire

Status 2000

Near Threatened D1+2

Status 1990

Near Threatened D1+2

Taxonomy

N. p. phaeton (central-eastern and north-western Queensland, northern Northern Territory, northern Western Australia) is Least Concern, as is the species. No morphological differences have been detected between Australian and New Guinea populations.

Taxonomic uniqueness: medium (29 genera/family, 4 species/genus, 2 subspecies/species)

Range

Southern New Guinea and Cape York Peninsula, Queensland (Schodde and Mason 1999). On the west coast of Cape York Peninsula there are subpopulations at the mouth of the Archer River near Aurukun, along the coast between the mouths of the Edward and Munkun Rivers near Pormpuraaw, and along Magnificent Creek near Kowanyama. On east coast they occur along the Normanby, Kennedy and Laura Rivers in Lakefield National Park, with a few records just south of the National Park on Olivevale Station (about 5 locations; Dorricott and Garnett 2007; M. Ahmet pers. comm.; A. Freeman *in litt.*). Last recorded at Iron Range in 1913 (MacGillivray 1918). AOO is calculated as 140 km^2 (Dorricott and Garnett 2007).

Abundance

The population size has been estimated tentatively at 2000 mature individuals, nominally divided into 1000 at Lakefield, 500 at both Pormpuraaw and Kowanyama, and 50 at Aurukun (Dorricott and Garnett 2007). No recent declines evident in monitored populations (Freeman and Freeman 2007; K. Maute pers. comm.).

Ecology

The birds occupy 2 types of habitat: woodland dominated by *Pandanus* spp. at Pormpuraaw and Aurukun, and by the palms *Corypha* spp. and *Livistona* spp. over cane-grasses *Chionachne cyathopoda* type at Kowanyama and Lakefield (Dorricott and Garnett 2007). Although this habitat is regularly burnt, the birds persist in shrubs and unburnt remnants, particularly in and around rivers and, at Pormpuraaw, the crocodile breeding lagoon. Feeds primarily on immature and ripe seeds of grasses and herbaceous plants (Todd *et al.* 2003), including those of

Current eligibility against IUCN Red List Criteria

IUCN category	Criteria eligibility
Australian breeding population	
A	Not applicable: past, current or future population declines are thought unlikely to exceed 20% in any 3-generation period
B	Not applicable: EOO, AOO, habitat, locations or population not declining, population fluctuations not extreme
C	Not applicable: no continuing population decline or extreme fluctuations
D	Near Threatened: population near 1000 mature individuals, near 5 locations with plausible future threat
E	Not applicable: no population viability analysis undertaken

IUCN Red List assessment data

	Estimate	Reliability
Australian breeding population		
Extent of occurrence trend	44 000 km^2 stable	high high
Area of occupancy trend	140 km^2 stable	medium high
No. of mature individuals trend	2000 stable	medium medium
No. subpopulations	4	high
No. locations	8	medium
Largest subpopulation	1000	medium
Generation time	3.5 years	low
Global population share	1–10%	high
Level of genetic exchange	low	high

introduced species, and small invertebrates (Dorricott and Garnett 2007). Builds a domed grass nest in *Pandanus* or *Corypha* palms (Todd 2002) and lays up to 6 eggs (Immelman 1982). A generation time of 3.5 years (BirdLife International 2011) is derived from an average age at first breeding of 1.0 years, an annual survival of adults of 50%, both extrapolated from mean values for Estrildidae, and a maximum longevity in the wild of 6.9 years, extrapolated from Masked Finch *P. personata* (Australian Bird and Bat Banding Scheme).

Threats

The key threats are increased hot fire in canegrass habitats, invasion of riparian habitats by Rubber Vine *Cryptostegia grandis,* and probably overgrazing by feral pigs *Sus scrofa* and cattle (Dorricott and Garnett 2007). Fires that would be extensive enough to remove all habitat are infrequent in the riparian habitat along the Laura River and Magnificent Creek, but rare, extensive, hot, late dry season fires may have been responsible for restricting the finch's range to small segments of creek bank. However, fire is also an essential tool for controlling Rubber Vine and maintaining vegetation structure. Crimson Finches are no longer found at sites near Aurukun where tall grasses were removed by overgrazing, near Claudie River where grassland was replaced by rainforest, and along parts of Laura River now infested with rubber vine (Garnett and Crowley 2000). Large areas of apparently suitable habitat on both sides of Cape York Peninsula are unoccupied.

Conservation objectives

1. All 4 subpopulations stable

Information required

1. Status of birds near Aurukun, recorded in 1913, known to be present in 1999 but no records since
2. Population trends at the other 3 locations
3. Rate of loss of grasslands on Lakefield

Management actions required

1. Continue to monitor subpopulations at appropriate intervals
2. Manage Rubber Vine where threatening finches
3. Manage fire for this species to reduce incidence of intense fires affecting riparian habitat
4. Fence key sites to reduce grazing by cattle and pigs
5. Assess potential for reintroduction at Lockhart River

Bibliography

BirdLife International (2011) 'Species factsheet: *Neochmia phaeton*'. Retrieved 17 January, 2011 from <http://www.birdlife.org/>.

Dorricott KE, Garnett ST (2007) 'National recovery plan for the white-bellied subspecies of the crimson finch *Neochmia phaeton evangelinae* and the Northern subspecies of the star finch *Neochmia ruficauda clarescens*'. Report to the Australian Government Department of the Environment and Water Resources, Canberra. Queensland Parks and Wildlife Service, Brisbane.

Freeman A, Freeman A (2007) 'A proposed monitoring program for star and crimson finches in Lakefield National Park'. Report to Queensland Parks and Wildlife Service, Cairns.

Garnett S, Crowley GM (2000) *The Action Plan for Australian Birds 2000.* Environment Australia, Canberra.

Immelman K (1982) *Australian Finches.* 2nd edn. Angus and Robertson, Sydney.

MacGillivray W (1918) Ornithologists in north Queensland. *Emu* **17**, 180–212.

Schodde R, Mason IJ (1999) The Directory of Australian Birds: Passerines. CSIRO Publishing, Melbourne.

Todd MK (2002) Nest site and breeding season data for the Crimson Finch *Neochmia phaeton* in Australia. *Australian Bird Watcher* **19**, 161–171.

Todd MK, Felton A, Garnett ST (2003) Morphological and dietary differences between common and uncommon subspecies of Crimson Finch, *Neochmia phaeton*, and Star Finch, *Neochmia ruficauda*, in northern Australia. *Emu* **103**, 141–148.

Comments received from

Alastair Freeman, Sarah Legge, Kim Maute, Mick Todd

Subspecies Conservation Summary

Star Finch (Cape York Peninsula)

Neochmia ruficauda clarescens (Hartert, 1899)
Estrildidae

Conservation status
Near Threatened D1+2

Reasons for listing
Population approaching 1000 individuals at lowest point in year

Status 2000
Near Threatened D1+2

Status 1990
Near Threatened D1+2

Taxonomy
N. r. ruficauda (central eastern Queensland) is Critically Endangered (Possibly Extinct). *N. r. subclarescens* (north Western Australia and Northern Territory) is Least Concern, as is the species.
Taxonomic uniqueness: medium (29 genera/family, 4 species/genus, 3 subspecies/species)

Range
Cape York Peninsula, once intergrading with *N. r. ruficauda* across the Burdekin-Lynd Divide and with *N. r. subclarescens* across the Gulf of Carpentaria (Schodde and Mason 1999), though the former possibly no longer exists and is now well separated from the latter. Historical decline in range, with recent records from northern Lakefield National Park in the east and sites near Aurukun, Pormpuraaw, Kowanyama and Karumba in the west (Holmes 1998; Dorricott and Garnett 2007). AOO calculated as 540 km^2 by Dorricott and Garnett (2007).

Abundance
Population estimated with low reliability as c.3000 at Lakefield, c.500 at Pormpuraaw and small numbers at Karumba and Aurukun (Dorricott and Garnett 2007). There are no data to indicate that the subspecies has declined markedly in the last decade (Freeman and Freeman 2007, Department of Sustainability, Environment, Water, Population and Communities 2011).

Ecology
Mostly occurs in seasonally-flooded tussock grasslands, usually near the coast. Long rank grass without a significant overstorey is probably the key habitat factor but surface water is a common feature (Dorricott and Garnett 2007). In coastal situations, saltmarsh can be used in the wet season (Garnett *et al.* 2005). Feeds primarily on seeds of native grasses and agricultural weeds, and small quantities of small invertebrates (Immelman 1982; Holmes 1998; Todd *et al.* 2003). Dependent on small-seeded plants that are restricted in availability during the late dry season and early wet season e.g. *Xerochloa imberbis* at Lakefield (Todd *et al.* 2003; Garnett *et al.* 2005). At Pormpuraaw also feeds for extended periods on seeds of Beach She-oak *Casuarina equisetifolia* (Garnett *et al.* 2005). Builds domed nests in patches of rank grass, reeds or low in trees, usually near water (Holmes 1998). Breeds in colonies and forms large flocks during the dry season. A generation time of 3.5 years (BirdLife International 2011) is derived from an average age at first breeding of 1.0 years, an annual survival of adults of 50%, both extrapolated from mean values for Estrildidae, and a maximum longevity in the wild of 6.9 years, extrapolated from Masked Finch *Poephila personata* (Australian Bird and Bat Banding Scheme).

Threats
Altered fire regimes and cattle grazing are thought to be responsible for causing the invasion of coastal grassland by woody species such as Broad-leaved Ti-tree *Melaleuca viridiflora*, at a rate of about 1% per year (Neldner *et al.* 1997; Crowley and Garnett 1998). However, recent monitoring has indicated that in Star Finch habitat in northern Lakefield National Park invasion has largely ceased with little new recruitment of invasive ti-tree as a result of appropriate fire management (Ahmet 2006, 2007). Physical trampling can be a problem where cattle and feral pig *Sus scrofa* numbers are high, particularly in areas of long grass used for breeding in the wet season and for shelter in the dry season (Holmes 1998). While the population at Karumba is protected against grazing by fencing, and at Pormpuraaw rank grasses are additionally protected by being within an extensive crocodile breeding pen, grasslands on Lakefield are still grazed by straying cattle. Excessive fire may degrade habitat but saline herblands that support *Xerochloa imberbis* are protected from fire by saline flats, though these saline herblands may be threatened if sea levels rise due to global climate change (Garnett *et al.* 2005).

Conservation objectives
1. All 4 subpopulations persisting and stable

Current eligibility against IUCN Red List Criteria

IUCN category	Criteria eligibility
A	Not applicable: past, current or future population declines are thought unlikely to exceed 20% in any 3-generation period
B	Not applicable: EOO, AOO, habitat, locations or population not declining, population fluctuations not extreme
C	Not applicable: no continuing population decline or extreme fluctuations
D	Near Threatened: population near 1000 mature individuals, near 5 locations with plausible future threat
E	Not applicable: no population viability analysis undertaken

IUCN Red List assessment data

	Estimate	Reliability
Extent of occurrence trend	65 000 km^2 stable	high high
Area of occupancy trend	540 km^2 stable	low medium
No. of mature individuals trend	3500 stable	low medium
No. subpopulations	4	high
No. locations	4	low
Largest subpopulation	3000	low
Generation time	3.5 years	low
Global population share	1–10%	low

2. Former locations are re-colonised

Information required

1. Populations trends at all locations
2. Rate of loss of grasslands
3. Strategy for protection of habitat at Pormpuraaw and Karumba

Management actions required

1. Monitor subpopulations at appropriate intervals
2. Manage fire to ensure rank grass near water persist through the breeding season
3. In Lakefield National Park fence off dry season waterholes to reduce grazing by cattle and pigs

Bibliography

Ahmet M (2006) 'An analysis report on the vegetation dataplots and photo monitoring sites. Lakefield National Park. August 2006'. Report to Queensland Parks and Wildlife Service, Cairns.

Ahmet M (2007) 'A photo analysis report on the long-term photo monitoring sites, Lakefield National Park'. Report to Queensland Parks and Wildlife service, Cairns.

BirdLife International (2011) 'Species factsheet: *Neochmia ruficauda*'. Retrieved 17 January, 2011 from <http://www.birdlife.org>.

Crowley GM, Garnett ST (1998) Vegetation change in the grasslands and grassy woodlands of east-central Cape York Peninsula, Australia. *Pacific Conservation Biology* **4**, 132–148.

Department of Sustainability, Environment, Water, Population and Communities (2011) *Neochmia ruficauda clarescens*. In 'Species Profile and Threats Database'. Department of Sustainability, Environment, Water, Population and Communities, Canberra. Retrieved 24 February, 2011 from <http://www.environment.gov.au/sprat>.

Dorricott KE, Garnett ST (2007) 'National recovery plan for the white-bellied subspecies of the crimson finch *Neochmia phaeton evangelinae* and the Northern subspecies of the star finch *Neochmia ruficauda clarescens*'. Report to Australian Government Department of the Environment and Water Resources, Canberra and Queensland Parks and Wildlife Service, Brisbane.

Freeman A, Freeman A (2007) 'A proposed monitoring program for star and crimson finches in Lakefield National Park'. Report to Queensland Parks and Wildlife Service, Cairns.

Garnett ST, Clarkson JR, Felton A, Harrington GN, Freeman AND (2005) The feeding ecology of the Star Finch *Neochmia ruficauda clarescens* at Princess Charlotte Bay, Cape York Peninsula in the early wet season. *Emu* **105**, 81–85.

Holmes G (1998) A review of the distribution, status and ecology of the Star Finch *Neochmia ruficauda* in Queensland. *Australian Bird Watcher* **17**, 278–289.

Immelman K (1982) *Australian Finches*. 2nd edn. Angus and Robertson, Sydney.

Neldner VJ, Fensham RJ, Clarkson JR, Stanton JP (1997) The natural grasslands of Cape York Peninsula, Australia: description, distribution and conservation status. *Biological Conservation* **81**, 121–136.

Schodde R, Mason IJ (1999) *The Directory of Australian Birds: Passerines*. CSIRO Publishing, Melbourne.

Todd MK, Felton A, Garnett ST (2003) Morphological and dietary differences between common and uncommon subspecies of Crimson Finch, *Neochmia phaeton*, and Star Finch, *Neochmia ruficauda*, in northern Australia. *Emu* **103**, 141–148.

Comments received from

Alastair Freeman, Sarah Legge, Kim Maute, Mick Todd

Subspecies Conservation Summary

Star Finch (southern)

Neochmia ruficauda ruficauda (Gould, 1837)
Estrildidae

Conservation status

Critically Endangered (Possibly Extinct) D

Reasons for listing

Possibly Extinct since 1995, based on its documented decline, the decline across most of its range, survey effort, relatively conspicuous nature of the species and plausible threat of overgrazed drought refuges, balanced against the lack of known attributes predisposing it to extinction and continued existence of potentially suitable habitat

Status 2000

Critically Endangered (Possibly Extinct) A2b, D

Status 1990

Critically Endangered A3bc, B1ab(i,ii,iii,iv,v)+2ab(i,ii,iii, iv,v), C2a(i,ii), D

Taxonomy

N. r. clarescens (Cape York Peninsula) is Near Threatened, *N. r. subclarescens* (north Western Australia and Northern Territory) is Least Concern, as is the species Taxonomic uniqueness: medium (29 genera/family, 4 species/genus, 3 subspecies/species)

Former range

From 1835 to 1990, 25 records accepted unequivocally, 7 other records were possibly of hybrids with 1 of the northern subspecies or escapees. Most records were from the Rockhampton area, with 2 from the Connors R. (Holmes 1996, 1998; Department of Sustainability, Environment, Water, Population and Communities 2011).

Former abundance

The paucity of records suggests it was either never common or was highly sensitive to the first wave of European settlement. There have been 4 definite and 4 probably valid records since 1990, the last in 1995. They were scattered across central Queensland with no 2 records from the 1 place: 3 birds at Mt Surprise in October 1995, 12 birds at Bumeru Road near Wowan in August 1995, 15–20 birds in Aramac township in September or October 1994, 3 birds at 'Archervale' near Middleton in July 1993, 8 birds at Alton Downs in July 1993, 3 birds at Surprise Creek southwest of Winton in May 1992, 6 birds at Lions Lagoons near Rockhampton in November 1991, 3 birds in Bowen in January 1990 and 1 bird in Mt Isa in 1990 (Holmes 1996, 1998). None have been reported since then despite searching and a heightened awareness of their rarity. Formerly occurred from the Namoi Valley, New South Wales west to Blackall and north to the Burdekin-Lynd, Qld, including a record from Cardwell on the coast, where it intergraded with *N. r. clarescens* (Gould 1865; Holmes 1986; Schodde and Mason 1999).

Ecology

Largely sedentary inhabitants of grasslands or grassy woodland near water (Gould 1865; Holmes 1996, 1998), although the 3 most recent sightings have been in cleared, or even suburban, settings with very few remnant trees. The finches build domed nests in patches of rank grass, reeds or low in trees, usually near water (Holmes 1998). A generation time of 3.5 years (BirdLife International 2011) is derived from an average age at first breeding of 1.0 years, an annual survival of adults of 50%, both extrapolated from mean values for Estrildidae, and a maximum longevity in the wild of 6.9 years, extrapolated from Masked Finch, *Poephila personata* (Australian Bird and Bat Banding Scheme).

Reasons for possible extinction

The southern subspecies of Star Finch disappeared from much of its range before the process of decline could be described (Holmes 1996, 1998; Franklin 1999). However, the riparian habitat it occupied would have been highly vulnerable to overgrazing and trampling, particularly during drought. Remaining subpopulations may persist in fragments of suitable habitat that have chanced to escape damage at critical times (Garnett 1992). Despite recent sightings and the finch's largely sedentary behaviour, permanently occupied habitat has not been found. If birds are rediscovered, they may be threatened by trapping for aviculture. Large numbers were trapped in the past, although the subspecies is unlikely to have persisted in captivity (Holmes 1998). Invasion of habitat by weeds (Garnett 1992) and cyanide in mining tailings dams (Holmes 1998) may also be local threats.

Conservation objectives

1. A viable population

Current eligibility against IUCN Red List Criteria

IUCN category	Criteria eligibility
A	Not applicable: population declines are thought unlikely to have exceeded 20% in the last 3 generations (11 years)
B	Not applicable: EOO, AOO, habitat, locations or population not declining, population fluctuations not extreme
C	Not applicable: no continuing population decline or extreme fluctuations
D	Critically Endangered (Possibly Extinct): if extant, population <50 mature individuals
E	Not applicable: no population viability analysis undertaken

Information required

1. Evidence of a wild population
2. If extant, habitat requirements, threats, population size and trends

Management actions required

1. Undertake surveys of habitat that might still hold populations, especially during drier times when any surviving birds are likely to be confined to small areas
2. Publicise the bird and its fate in towns through western Queensland
3. Develop a contingency plan should the bird be rediscovered, including establishing a captive breeding population

Bibliography

BirdLife International (2011) 'Species factsheet: *Neochmia ruficauda*'. Retrieved 17 January, 2011 from <http://www.birdlife.org>.

Department of Sustainability Environment Water Population and Communities (2011) *Neochmia ruficauda clarescens.* In 'Species Profile and Threats Database'. Department of Sustainability, Environment, Water, Population and Communities, Canberra. Retrieved 24 February, 2011 from <http://www.environment.gov.au/sprat>

Franklin DC (1999) Evidence of disarray amongst granivorous bird assemblages in the savannas of northern Australia, a region of sparse human settlement. *Biological Conservation* **90**, 53–68.

Garnett ST (Ed.) (1992) Threatened and extinct birds of Australia. *RAOU Report* **82**.

Gould J (1865) *Handbook to the Birds of Australia.* The Author, London.

Holmes G (1996) Distribution and status of the southern Star Finch. *Sunbird* **26**, 49–59.

Holmes G (1998) A review of the distribution, status and ecology of the Star Finch *Neochmia ruficauda* in Queensland. *Australian Bird Watcher* **17**, 278–289.

Schodde R, Mason IJ (1999) *The Directory of Australian Birds: Passerines.* CSIRO Publishing, Melbourne.

Comments received from

Glenn Holmes, Mick Todd

Species Conservation Summary

Gouldian Finch

Erythrura gouldiae (Gould, 1844)
Estrildidae

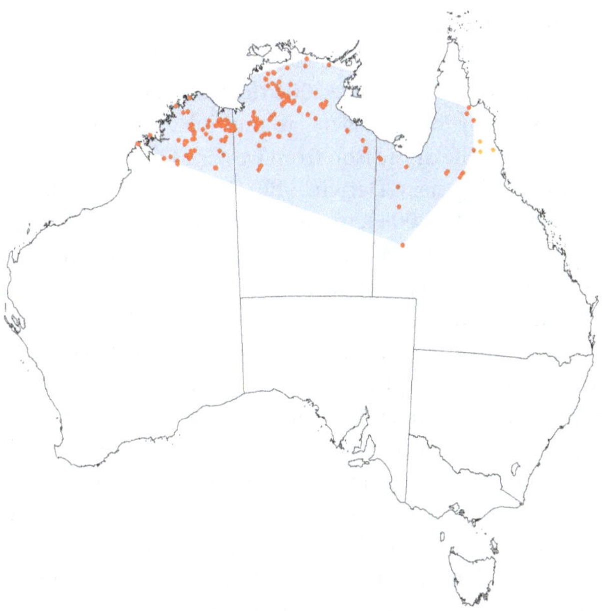

Conservation status

Near Threatened D1

Reasons for listing

Although well below historical levels, population appears to have stopped declining but may contract to near 1000 mature individuals at the end of the dry season

Status 2000

Endangered C2a(ii)
Reason for change in 2010: sustained evidence that declines have ceased

Status 1990

Endangered C2a(ii)

Taxonomy

No infraspecific taxa described
Taxonomic uniqueness: medium (29 genera/family, 12 species/genus, 1 subspecies/species)

Range

Northern Australia from Cape York Peninsula through north-west Queensland and the northern Northern Territory to the Kimberley Region of Western Australia (Higgins *et al.* 2006; O'Malley *et al.* 2006). Non-breeding birds disperse widely: a juveniles banded in Wyndham (WA) was

Current eligibility against IUCN Red List Criteria

IUCN category	Criteria eligibility
A	Not applicable: past, current or future population declines are thought unlikely to exceed 20% in any 10-year period
B	Not applicable: >10 locations, not severely fragmented, population fluctuations not extreme
C	Not applicable: no continuing population decline or extreme fluctuations
D	Near Threatened: population near 1000 mature individuals
E	Not applicable: no population viability analysis undertaken

IUCN Red List assessment data

	Estimate	Reliability
Extent of occurrence trend	1 370 000 km^2 stable	high high
Area of occupancy trend	2000 km^2 increasing	low medium
No. of mature individuals trend	2400 increasing	low medium
No. subpopulations	1	medium
No. locations	>10	medium
Generation time	2.7 years	low
Global population share	100%	high

recaptured a few weeks later 200 km away at Newry (NT) (S. Pryke, pers. comm.) and 4 birds were recorded on the Georgina River at Ethabouka Reserve in the Simpson Desert (M. Mathieson pers. comm.). Genetic analyses of mitochondrial markers indicate no population structuring across the species range, however analyses of nuclear markers indicate contemporary gene flow from Kimberley to the NT (Esparza-Salas *in litt.*). In Qld there are no recent breeding records but since 2005 birds have been sighted at at least 3 sites on Cape York Peninsula, on the Atherton Tablelands, and at several sites in and around Boodjamulla National Park. In the NT there are recent breeding records at well-known sites in the Yinberrie Hills and Newry, as well as at Wollogorang (D. Baker-Gabb *in litt.*) and near Maningrida (R. Noske pers. comm.). In the eastern Kimberley, 8–12 isolated breeding populations of 40 to 120 adults are known (S. Pryke *in litt.*) as well as populations at Mornington Sanctuary.

Abundance

The population size of the Gouldian Finch is extremely difficult to estimate. Since 2004 flocks of 100+ birds have been seen by birdwatchers at multiple, well-separated sites during the dry season from western Qld through the NT, including near Darwin (400+) and Maningrida in Arnhem Land (1000+), to the Kimberley (700+ at Mornington Sanctuary). However, >80% of these are juveniles (Woinarski and Tidemann 1992; Franklin *et al.* 1998; D. Franklin *in litt.*), there is a strongly skewed sex ratio towards males and a genetic incompatibility between different colour morphs (S. Pryke *in litt.*), all of which suggest that the effective population size is much smaller. Because of the uncertainty a structured elicitation process was employed using an expert panel (16 members) to assess all available information using a 2-stage Delphi technique to minimise biases (Burgman 2005). The panel concluded that the population was likely to be well above 1000 mature individuals (2400) but, to be precautionary, agreed that the total, across multiple locations, could be close to 1000 when finch numbers reach their annual minimum during the wet season. Trend data from annual counts in the Yinberrie Hills for 1996–2004 suggest numbers fluctuate but have not declined (O. Price *in litt.*). Data collected since have not been analysed. There was also no evidence of a decline at Mornington Wildlife Sanctuary from 2004 to 2007 (S. Legge *in litt.*). The expert committee concluded that the population is not currently declining.

Ecology

Gouldian Finches feed almost exclusively on grass seed and depend on a restricted variety of grasses which seed at different times through the year (Dostine and Franklin 2002; O'Malley *et al.* 2006). They nest in tree hollows, lay an average clutch of 5 eggs, and may raise several clutches in a season, but productivity averages 1.5 fledglings per pair (Tidemann *et al.* 1999). When breeding, they use small patches of suitable open woodland areas usually on ridges dominated by cavity bearing trees (such as White Northern Gum *Eucalyptus brevifolia* in the west and Territory Salmon Gum *E. tintinnans* in the east; Tidemann *et al.* 1992a) and seeding grasses (e.g. sorghum *Sarga* spp., spinifex *Trodia* spp.) within 2–4 km of perennial waterholes or springs (Dostine *et al.* 2001; O'Malley 2006). After breeding they move to areas with different seeding grasses (e.g. annual spear grasses or native sorghum *Sarga* spp.), flock and move across the broader landscape (O'Malley 2006). A generation time of 2.7 years (BirdLife International 2011) is derived from an age at first breeding of 1.0 years and a maximum longevity in the wild of 4.4 years, both values elicited from an expert committee.

Threats

Regular extensive fires are known to reduce body condition and other health indices, particularly during the wet season and early breeding season (S. Legge *in litt.*). Heavy grazing by cattle is known to reduce seed yields in grasses important to the finches (Crowley and Garnett 2001). In some areas, fire affects hollow availability, although nest hollow availability and productivity of juveniles appear not be limiting at most sites (Tidemann *et al.* 1999; Brazill-Boast *et al.* 2010, 2011; S. Pryke *in litt.*). However, although grazing is intensifying (Garnett *et al.* 2009) and c.50% of the finch's EOO is burnt annually (Russell-Smith *et al.* 1997; Edwards *et al.* 2001, 2003), both threats have been mitigated at some sites through active management for finch conservation (S. Legge *in litt.*). Historically an air-sac mite *Sternostoma tracheacolum* (Tidemann et al. 1992b; Bell 1996) also appeared to increase mortality in the wild and may have been a proximal cause of decline in a population already stressed by environmental change.

Trapping for aviculture was substantial in the past but has not occurred to any extent for 30 years. There may be ongoing localised threats to some breeding habitat from development such as mining. Given the level of ongoing threats, any indication from monitoring or the frequency of birdwatcher reports that the species is again declining would lead to a reappraisal of the status.

Conservation objectives

1. Ongoing increase in sightings
2. Key sites with improved fire and grazing regimes
3. Stable or improved trends in population and health indices at key sites across entire range

Information required

1. Numerical trends and health indices at key sites across entire range

Management actions required

1. Continue monitoring of numbers and health indices at key sites
2. Promote as an indicator of sustainable cattle and fire management

Bibliography

Bell PJ (1996) Survey of the nasal mite fauna (Rhinonyssidae and Kytoditidae) of the Gouldian Finch, *Erythrura gouldiae*, and some co-occurring birds in the Northern Territory. *Wildlife Research* **23**, 675–686.

Brazill-Boast J, Pryke SR, Griffith SC (2010) Nest-site utilisation and niche overlap in two sympatric, cavity-nesting finches. *Emu* **110**, 170–177.

Brazill-Boast J, van Rooij E, Pryke SR, Griffith SC (2011) Interference from long-tailed finches constrains reproduction in the endangered Gouldian finch. *Journal of Animal Ecology* **80**, 39–48.

Burgman MA (2005) *Risks and Decisions for Conservation and Environmental Management.* Cambridge University Press, Cambridge, New York.

Crowley GM, Garnett ST (2001) Growth, seed production and effect of defoliation in an early flowering perennial grass, *Alloteropsis semialata* (Poaceae), on Cape York Peninsula, Australia. *Australian Journal of Botany* **49**, 735–743.

Dostine PL, Franklin DC (2002) A comparison of the diet of three finch species in the Yinberrie Hills area, Northern Territory. *Emu* **102**,159–164.

Dostine PL, Johnson GC, Franklin DC, Zhang Y, Hempel C (2001) Seasonal use of savanna landscapes by the Gouldian finch, *Erythrura gouldiae*, in the Yinberrie Hills area, Northern Territory. *Wildlife Research* **28**, 445–458.

Edwards A, Hauser P, Anderson M, McCartney J, Armstrong M, Thackway R, Allan GE, Hempel C, Russell-Smith J (2001) A tale of two parks: contemporary fire regimes of Litchfield and Nitmiluk National Parks, monsoonal northern Australia. *International Journal of Wildland Fire* **10**, 79–89.

Edwards A, Kennett R, Price O, Russell-Smith J, Spiers G, Woinarski J (2003) Monitoring the impacts of fire on vegetation in northern Australia: an example from Kakadu National Park. *International Journal of Wildland Fire* **12**, 427–440.

Franklin DC, Dostine PL, Tidemann SC (1998) Post-juvenile moult strategies of co-existing Gouldian, Long-tailed and Masked Finches. *Corella* **22**, 73–79.

Garnett ST, Woinarski JCZ, Crowley GM, Kutt AS (2010) Biodiversity conservation in Australian tropical rangelands. In *Can Rangelands be Wildlands? Wildlife and Livestock in Semi-arid Ecosystems.* (Eds J du Toit, R Kock, J Deutsch) pp. 191–234. Blackwell Scientific, London.

Haynes CD (1985) The pattern of *munwag*: traditional Aboriginal fire regimes in north-central Arnhemland. *Proceedings of the Ecological Society of Australia* **13**, 203–214.

Higgins PJ, Peter JM, Cowling SJ (Eds) (2006) *Handbook of Australian, New Zealand and Antarctic Birds. Volume 7: Boatbill to Starlings.* Oxford University Press, Melbourne.

O'Malley C (2006) 'National recovery plan for the Gouldian Finch (*Erythrura gouldiae*)'. Report to WWF-Australia, Sydney, Parks and Wildlife NT, Department of Natural Resources, Environment and the Arts, Palmerston.

Rossiter NA, Setterfield SA, Douglas MM, Hutley LB (2003) Testing the grass-fire cycle: alien grass invasion in the tropical savannas of northern Australia. *Diversity and Distributions* **9**, 169–176.

Russell-Smith J, Ryan PG, Durieu R (1997) A LANDSAT MSS-derived fire history of Kakadu National Park, monsoonal northern Australia, 1980–94: seasonal extent, frequency and patchiness. A LANDSAT MSS-derived fire history of Kakadu National Park, monsoonal northern Australia, 1980–94: seasonal extent, frequency and patchiness. *Journal of Applied Ecology* **34**, 748–766.

Tidemann SC, Boyden J, Elvish R, Elvish J, O'Gorman (1992a) Comparison of the breeding sites and habitat of two hole-nesting estrildid finches, one endangered, in northern Australia. *Journal of Tropical Ecology* **8**, 373–388.

Tidemann SC, McOrist S, Woinarski JCZ, Freeland WJ (1992b) Parasitism of wild Gouldian Finches *Erythrura gouldiae* by the air sac mite *Sternostoma tracheacolum. Journal of Wildlife Diseases* **20**, 80–84.

Tidemann SC, Lawson C, Elvish R, Boyden J, Elvish J (1999) Breeding biology of the Gouldian Finch *Erythrura gouldiae*, an endangered finch of northern Australia. *Emu* **99**, 191–199.

Woinarski JCZ, Tidemann S (1992) Survivorship and some population parameters for the endangered Gouldian Finch *Erythrura gouldiae* and two other finch species at two sites in tropical northern Australia. *Emu* **92**, 33–38.

Woinarski JCZ, Williams RJ, Price O, Rankmore B (2005) Landscapes without boundaries: wildlife and their environments in northern Australia. *Wildlife Research* **25**, 377–388.

Comments received from

Barry Baker, David Baker-Gabb, Phil Bell, Allan Burbidge, Andrew Burbidge, Stuart Butchart, Les Christidis, Bruce Doran, Hugh Ford, Don Franklin, Simon Griffiths, Ian Gynther, Tim Holmes, Arthur Keates, Sheryl Keates, Alex Kutt, Sarah Legge, Richard Loyn, Peter Menkhorst, Charles Meredith, Tim Nerard, Richard Noske, Penny Olsen, Sarah Pryke, Del Richards, Andy Symes, Simon Ward, David Watson, John Westaway, John Woinarski

Appendix 1: Distribution of Threatened and Near Threatened birds in Australia

The Extent of Occurrence of all taxa are overlaid. Darker areas indicate greater index values at a given location from 1 (white) to a maximum of 54 (black). Shading is weighted by conservation status: Critically Endangered taxa have a weighting of 4, Endangered of 3, Vulnerable of 2, Near Threatened of 1.

A. Terrestrial, coastal and freshwater taxa

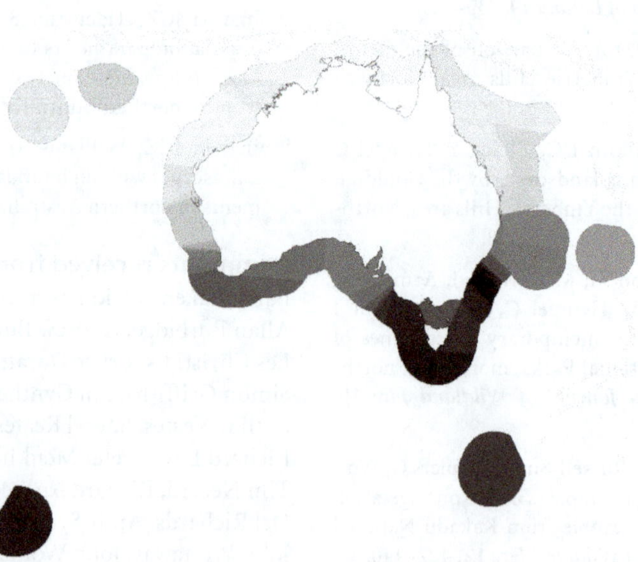

B. Marine taxa

Appendix 2: Taxa listed in the *Action Plan for Australian Birds 2000* and omitted from this volume

Most changes are self-explanatory. However the reasons for listing a taxon as Near Threatened in 2000 cannot be understood without reference to the following 4 criteria:

a. Reduced Area of Occupancy and/or Extent of Occurrence: Taxa that have disappeared from over 50% of their former Area of Occupancy and/or Extent of Occurrence and are at risk of further decline.
c. Declined in abundance: Taxa that have experienced a significant and continuing decline in abundance in over 50% of their former Area of Occupancy and/or Extent of Occurrence.
d. Small populations: Taxa with populations smaller than 3000 mature individuals, or of unknown population size but suspected to be small.
e. Regionally secure populations: Taxa with small or restricted populations in Australia (hence could qualify as Vulnerable D1 or D2) that are genetically continuous with secure regional populations. Such taxa were downgraded.

The latest version of the IUCN Red List Guidelines (v.8.1) is much more specific about the criteria for Near Threatened.

Taxon name	Reason provided for listing in 2000	Reason not listed in 2010
Cotton Pygmy-Goose (Australian) *Nettapus coromandelianus albipennis*	The population of this subspecies seems small, and appears to have declined in density over at least the southern half of its historical range (Near Threatened c)	Revised criteria for Near Threatened: small population but historical decline has stopped well before the 3-generation limit (21 years) and now appears stable
Flock Bronzewing *Phaps histrionica*	This species experiences large seasonal fluctuations in both range and population size, but analysis suggests its density has decreased substantially (Near Threatened c). Although the range may have contracted, not by as much as 50% (so not Near Threatened a)	New knowledge: recent research demonstrates fluctuations but no decline
Spinifex Pigeon (eastern) *Geophaps plumifera leucogaster*	The subspecies has disappeared from the southern part of its former range and occurs at a lower density in north (Near Threatened c), but is still too numerous and widespread to be considered Vulnerable	Revised criteria for Near Threatened: although historical contraction in SA range, Qld and NT populations show no recent declines
Squatter Pigeon (southern) *Geophaps scripta scripta*	This subspecies has disappeared from the southern half of its historical range (Near Threatened a), a trend that does not appear to be continuing	Revised criteria for Near Threatened: no recent declines and persists at numerous sites across a broad distribution
Peaceful Dove (Papuan) *Geopelia striata papua*	The Australian population of this subspecies is confined to 3 small islands with a combined area of about 100 km^2 (Vulnerable D2). The global status, however, is Least Concern and genetic interchange is likely to be substantial. The Australian status has therefore been downgraded to Near Threatened e (genetic interchange beyond borders)	Correct application of regional guidelines: could be Near Threatened D2 as ≤5 locations and low probability future threat but downgraded 2 levels because genetic interchange likely
Banded Fruit-Dove (Australian) *Ptilinopus cinctus alligator*	This subspecies has a small population size (Near Threatened d), but there is no suggestion of a decline	Revised criteria for Near Threatened: population probably much greater than 1000 individuals and no evidence of decline. Population scattered through Arnhem Land stone country in many fire-proof pockets
Marbled Frogmouth (Plumed) *Podargus ocellatus plumiferus*	In the past, clearance has reduced the area occupied by this subspecies by at least 50% (Near Threatened a). This decline had probably stopped, as the birds are now confined to forests where clearing is no longer likely	New knowledge: much larger population than realised in south-east Qld, all habitat protected

Taxon name	Reason provided for listing in 2000	Reason not listed in 2010
Australian Owlet-nightjar (Tasmanian) *Aegotheles cristatus tasmanicus*	There are probably only 5000 mature individuals in the single subpopulation of this subspecies, which is declining (Vulnerable C2b)	New knowledge: an increase in sightings suggests that, even if there was a decline, it is not continuing
Southern Giant-Petrel *Macronectes giganteus*	The Australian population of the species has decreased in number by 50% (Endangered A1) and the global population by 20–50% (Vulnerable A1) in the last 3 generations, as determined from direct observation (a), an appropriate index of abundance (b), probably as a result of longline fishing (d) and possibly introduced predators (e). There is a high probability that the global decrease will continue at the same rate (b) as a result of fishing bycatch (2d) and predators (e). Site fidelity is high, so the immigration rate is assumed to be low. The national status of the breeding population is therefore independent of the global status	New knowledge: Australian population occupies 4 locations and threats, though unlikely, are plausible; could therefore be Near Threatened D2, but downgraded 1 level because there is potential for genetic exchange with Least Concern global population, for which fears of a decline have not been borne out by monitoring
Northern Giant-Petrel *Macronectes halli*	The species breeds at a single location within Australian territory (Vulnerable D2). Globally, its population size is increasing, but has the potential to decrease to Vulnerable levels (Near Threatened as per Birdlife International, 2000)	New knowledge: Australian population breeds at 2 locations and threats, though unlikely, are plausible; could therefore be Near Threatened D2, but downgraded 1 level because there is potential for genetic exchange with Least Concern global population, for which fears of a decline have not been borne out by monitoring
Cape Petrel (southern) *Daption capense capense*	The Australian population breeds at a single location (Vulnerable D2)	Revised criteria for Near Threatened: Australian population breeds at 1 location and threats, though unlikely, are plausible; could therefore be Near Threatened D2, but downgraded 1 level because there is potential for genetic exchange with Least Concern global population
King Penguin *Aptenodytes patagonicus*	Although the Australian population is large and increasing, it is confined to 2 breeding locations (Vulnerable D2). Heard I. has been recolonised after localised extinction, implying substantial genetic exchange. Since the global status is Least Concern, the Australian status has been downgraded to Near Threatened: e	Revised criteria for Near Threatened: Australian population breeds at 4 locations and threats, though highly unlikely, are plausible; could therefore be Near Threatened D2, but downgraded 1 level because there is genetic exchange with Least Concern global population and population is increasing rapidly
Masked Booby (Indo-Pacific) *Sula dactylatra personata*	The Australian population breeds at fewer than 5 locations (Vulnerable D2. Global status of the subspecies is probably also Vulnerable C2a, as extralimital breeding colonies are probably hunted for food. Site fidelity is probably high, especially given the high rate of subspeciation in Masked Booby, so the immigration rate is assumed to be low	Not applicable: *S. d. bedouti* no longer considered valid; both eastern and western populations of *S. d. personata* are Least Concern. Although those in the west only at 5 sites (Rowley Shoals, Bedout, Ashmore, Adele, N. Keeling), numbers are increasing and there is currently no plausible future threat
Little Bittern (Australo-Papuan) *Ixobrychus minutus dubius*	Although still widespread, the area of suitable habitat within the subspecies' range is possibly less than half its pre-European extent (Near Threatened c). The population probably contains only 5000 mature individuals, but is possibly divided between east and south-west with 1 subpopulation of more than 1000 individuals (so not Vulnerable C2a), and there is no real evidence of a recent decline (so not C2b). Given that most of the population occurs within Australia, the subspecies' global status is probably the same, regardless of the high probability of genetic interchange	Revised criteria for Near Threatened: still recorded across former range with no evidence of a decline
Lewin's Rail (eastern Australian) *Lewinia pectoralis pectoralis*	A decline in density over at least half the subspecies' range (Near Threatened c) is suggested by the subspecies' scarcity in comparison to the Tasmanian subspecies, and the contraction of mainland wetlands	New knowledge: range too large and any decline too slow to meet criteria, anecdotal information on changes in recording frequency inconsistent
Australian Bustard *Ardeotis australis*	Though the species' population size is still substantial, there has been a massive historical decline in abundance in southern Australia and a patchy mix of declines in northern Australia that are only partly offset by localised increases (Near Threatened c). Because such a large proportion of the population is in Australia, global status is also Near Threatened	Revised criteria for Near Threatened: population loss in southern Australia occurred before the 3 generation limit (47 years); population in north now relatively large and stable
Bush Stone-curlew *Burhinus grallarius*	Although the species is still relatively common in the north and on islands, the population occurs at a fraction of its former density in southern mainland Australia (Near Threatened c). Because most of the population occurs in Australia, global status is also Near Threatened	Revised criteria for Near Threatened: most of the ongoing population loss in southern Australia occurred before the 3-generation limit (31 years); population in northern Australia is now large with no suggestion of decline

Taxon name	Reason provided for listing in 2000	Reason not listed in 2010
Chestnut-backed Button-quail *Turnix castanotus*	Population density has probably been reduced over half of this species' range (Near Threatened c). There has also been some contraction in range. However, the species cannot be listed as Vulnerable, as the rate of decline has not been sufficient, and both the population size and range are too great.	Revised criteria for Near Threatened: no recent declines even though population sparse and scattered; Population declines are unlikely in the last 3 generations (10 years) or in the next decade
Brown Skua *Stercorarius lonnbergi*	In Australian territory, this subspecies breeds in only 2 locations (Vulnerable D2). Global status is Least Concern, but there is no evidence of genetic interchange between populations, so the Australian status is assessed independently	Revised criteria for Near Threatened: Australian population breeds at few locations and threats, though unlikely, are plausible; could therefore be Near Threatened D2 but downgraded 1 level because there is potential for genetic exchange with Least Concern global population
Grey Ternlet (western Pacific) *Procelsterna cerulea albivitta*	Within Australian territory, this subspecies is confined to a small area of occupancy at 2 breeding localities (Endangered B1) and a continuing decline is projected in the number of mature individuals. Levels of genetic interchange are assumed to be low so the status of the Australian population has been assessed independently from the global status	Revised criteria for Near Threatened: Australian population at few locations and threats, though unlikely, are plausible; could therefore be Near Threatened D2 but downgraded 1 level because there is potential for genetic exchange with Least Concern global population
Western Corella (southern, Muir's) *Cacatua pastinator pastinator*	Thanks to enforcement of conservation legislation, the population of this subspecies has been increasing since at least the 1970s and now numbers several thousand mature birds	Genuine recovery: thanks to enforcement of conservation legislation, the population of this subspecies has been increasing since at least the 1970s and now numbers several thousand mature individuals
Eclectus Parrot (Papuan) *Eclectus roratus polychloros*	The Australian population of this subspecies is small (Vulnerable D1) and confined to 3 small islands with an area of about 100 km^2 (D2). The global status, however, is Least Concern and genetic interchange is likely to be substantial. The Australian status has therefore been downgraded to Near Threatened: e	New knowledge: birds visit Torres Strait islands but do not breed there; taxon not otherwise threatened in New Guinea
Red-cheeked Parrot (Cape York Peninsula) *Geoffroyus geoffroyi maclennani*	This subspecies has a small population (Near Threatened d), but is facing no threats	Revised criteria for Near Threatened: Australian population likely to exceed that of Eclectus Parrot *Eclectus roratus macgillivrayi*, and is probably stable
Superb Parrot *Polytelis swainsonii*	The single subpopulation of this species is thought to contain about 6500 mature individuals and to be declining (Vulnerable C2b)	New Knowledge: recent evidence that population is well over 10 000 mature individuals and no evidence of continuing decline; projected declines nest hollows of 20–29% for 1 part of the population over the next 3 generations (19 years), but not yet a link between hollow availability and parrot abundance
Eastern Rosella (Tasmanian) *Platycercus eximius diemenensis*	Though this subspecies still has a substantial population, observations suggest that this may have at least halved over the last century (Near Threatened c)	Revised criteria for Near Threatened: population too large and decline too slow to be close to threatened
Western Rosella (inland) *Platycercus icterotis xanthogenys*	There has been a decline in density over at least half the area of occupancy of this subspecies that is likely to continue (Near Threatened c)	Revised criteria for Near Threatened; population too large, particularly east of wheatbelt, and decline too slow to be close to threatened
Turquoise Parrot *Neophema pulchella*	Although currently expanding, the area of occupancy of this species is still probably less than half of the size that it was a century ago (Near Threatened a)	Revised criteria for Near Threatened, population too large and no evidence of recent decline
Rufous Owl (Cape York Peninsula) *Ninox rufa meesi*	Although the population probably contains around 1500 mature individuals (Near Threatened d), this subspecies is in no immediate danger.	New knowledge. Population was thought to be restricted to Iron Range area and likely to be near 1000 individuals given home range size. However records also along west coast and in riparian strips along rivers so now thought to be well in excess of 1000 individuals and not declining
Rufous Owl (eastern) *Ninox rufa queenslandica*	This subspecies probably has a small population (Near Threatened d), but is in no imminent danger	Revised criteria for Near Threatened: declines from clearing historical; population probably stable across range in the last 3 generations (26 years)
Masked Owl (southern) *Tyto novaehollandiae novaehollandiae*	The area occupied by this subspecies is thought to have declined by at least half, particularly in the semi-arid zone (Near Threatened a)	Revised criteria for Near Threatened: no evidence of recent or ongoing decline and population probably well over 1000 individuals
Little Kingfisher (Papuan) *Ceyx pusilla pusilla*	The Australian population of this subspecies probably contains fewer than 1000 mature individuals (Vulnerable D1) and is confined to 2 small islands with a combined area of less than 100 km^2 (D2). The global status, however, is Least Concern and genetic interchange is likely to be substantial. The Australian status has therefore been downgraded to Near Threatened e (genetic interchange beyond borders)	Correct application of regional guidelines: could be Vulnerable D1 as, 1000 mature individuals or Near Threatened D2 as ≤5 locations and low probability future threat, but downgraded 2 levels because genetic interchange likely

Taxon name	Reason provided for listing in 2000	Reason not listed in 2010
Sacred Kingfisher (Tasman Sea) *Todiramphus sanctus vagans*	Within Australia, this subspecies has a small population (Vulnerable: D1), which is restricted to a single location (D2). The global status, however, is Least Concern, and self introduction implies genetic interchange. The Australian status has therefore been downgraded to Near Threatened e (genetic interchange beyond borders)	Correct application of regional guidelines: could be Vulnerable D1 given 1000 mature individuals or Near Threatened D2 given ≤5 locations and low probability future threat, but downgraded 2 levels because genetic interchange likely
White-browed Treecreeper (eastern) *Climacteris affinis superciliosa*	Density has decreased over more than 50% of this subspecies' historical range (Near Threatened c)	Revised criteria for Near Threatened: although there have undoubtedly been declines in east of range, the population and range remain above thresholds in the west
Brown Treecreeper (Cape York Peninsula) *Climacteris picumnus melanotus*	This subspecies' relatively large and widely distributed population appears to have declined in density over more than half its historical range (Near Threatened c)	Revised criteria for Near Threatened: although possibly continuing to decline on central Cape York Peninsula it is still widespread and numerous across the southern Peninsula
Western Bowerbird (North-west Cape) *Ptilonorhynchus guttatus carteri*	The population of this subspecies probably contains no more than 2000 mature individuals (Near Threatened d)	New knowledge: population larger than had been appreciated and no immediate threat
Western Grasswren (Shark Bay) *Amytornis textilis textilis*	This subspecies suffered a massive decline early in 20th century (Near Threatened a), which inexplicably stopped before extinction was reached. Remnant subpopulations are stable	New knowledge: population now known to be well above 1000 individuals and no recent declines, even though very large declines in the past
Rufous Bristlebird (Otways) *Dasyornis broadbenti caryochrous*	The population of this subspecies contains about 4000 mature individuals and is probably decreasing in size (Vulnerable C2). Arguably, all individuals belong to a single subpopulation (b). Although the Extent of Occurrence is about 1300 km^2 and Area of Occupancy about 200 km^2, the subspecies is not considered Endangered: B, because the distribution is neither fragmented nor at fewer than 6 locations	Revised criteria for Near Threatened: no evidence of recent population decline and occurs in too many locations to meet criteria
Shy Heathwren (Riverina) *Hylacola cauta macrorhyncha*	This is an isolated subspecies that probably occupies less than 50% of its former Area of Occupancy (Near Threatened a), as its distribution has been fragmented by clearing	Revised criteria for Near Threatened: population too large and decline too small in the last 3 generations (15 years) for listing
Shy Heathwren (western) *Hylacola cauta whitlocki*	More than half of the habitat of this subspecies has been cleared (Near Threatened a), but there has probably not been a population decline of 20% over in last 3 generations (so not Vulnerable A1)	Revised criteria for Near Threatened: population too large and decline too small in the last 3 generations (15 years) for listing
Rufous Fieldwren (western wheatbelt) *Calamanthus campestris montanellus*	More than half of the habitat of this subspecies has been cleared (Near Threatened a), but there has probably not been a population decline of 20% over the last 3 generations (so not Vulnerable A1)	Revised criteria for Near Threatened: population too large and decline too small in the last 3 generations (15 years) for listing
Speckled Warbler *Chthonicola sagittata*	This species has declined over most of its range (Near Threatened c), though not at a rate or scale that warrants its listing as Vulnerable	New knowledge: population too large and decline too small in last 3 generations (17 years) for listing; has proved more resilient in patches than originally anticipated
Large-billed Gerygone (Papuan) *Gerygone magnirostris brunneipectus*	The Australian population of this subspecies is confined to 2 small islands with a combined area of less than 100 km^2 (D2). The global status, however, is Least Concern and genetic interchange is likely to be substantial. The Australian status has therefore been downgraded to Near Threatened e (genetic interchange beyond borders)	Correct application of regional guidelines: could be Near Threatened D2 as ≤5 locations and low probability future threat, but downgraded 2 levels because genetic interchange likely
Slender-billed Thornbill (eastern) *Acanthiza iredalei hedleyi*	Habitat suitable for this subspecies has been reduced by at least half (Near Threatened a)	Revised criteria for Near Threatened: although affected by fire, recovers quickly and also occurs in many sites not recently burnt in Vic and SA
Dusky Honeyeater (Papuan) *Myzomela obscura fumata*	The Australian population of this subspecies is confined to 2 small islands with a combined area of about 100 km^2 (Vulnerable D2). The global status, however, is Least Concern and genetic interchange is likely to be substantial. The Australian status has therefore been downgraded to Near Threatened e (genetic interchange beyond borders)	Correct application of regional guidelines: could be Near Threatened D2 as ≤5 locations and low probability future threat but downgraded 2 levels because genetic interchange likely
Red-headed Honeyeater (Papuan) *Myzomela erythrocephala infuscata*	The Australian population of this subspecies is confined to 3 small islands with a combined area of about 100 km^2 (Vulnerable D2). The global status, however, is Least Concern and genetic interchange is likely to be substantial. The Australian status has therefore been downgraded to Near Threatened e (genetic interchange beyond borders)	Correct application of regional guidelines: could be Near Threatened D2 as ≤5 locations and low probability future threat but downgraded 2 levels because genetic interchange likely

Taxon name	Reason provided for listing in 2000	Reason not listed in 2010
Tawny-breasted Honeyeater (Papuan) *Xanthotis flaviventer saturatior*	The Australian population of this subspecies is probably stable, but limited to 2 islands with a combined area of less than 100 km^2 (Vulnerable D2). The global status, however, is Least Concern and genetic interchange is likely to be substantial. The Australian status has therefore been downgraded to Near Threatened e (genetic interchange beyond borders)	Correct application of regional guidelines: could be Near Threatened D2 as ≤5 locations and low probability future threat but downgraded 2 levels because genetic interchange likely
Grey-crowned Babbler (eastern) *Pomatostomus temporalis temporalis*	Although the subspecies is abundant and widespread in northern Qld, there has been a considerable and continuing decline in southern part of its range (Near Threatened c), though not at a rate or scale that warrants listing as Vulnerable. The status is assessed separately for Australian and New Guinea populations, which are genetically isolated. However, since at least 90% of the population occurs in Australia, the global status of the subspecies is probably also Near Threatened	Revised criteria for Near Threatened: population too large and decline too small in the last 3 generations (25 years) for listing
White-browed Babbler (south-western) *Pomatostomus superciliosus ashbyi*	Over half the subspecies' habitat has been cleared (Near Threatened a). Continuing declines are inevitable, even though the subspecies is still widespread and is more persistent in fragments than other wheatbelt taxa	Revised criteria for Near Threatened: population too large and decline too small in the last 3 generations (25 years) for listing
Chestnut Quail-thrush (eastern) *Cinclosoma castanotus castanotus*	More than half of the habitat used by this subspecies has been cleared and remaining subpopulations are fragmented (Near Threatened a). Although the subspecies has proved resilient to both fire and grazing in the remaining habitat, the bulk of which is protected from clearing, losses from fragments are inevitable	Revised criteria for Near Threatened: population too large and decline too small in the last 3 generations (15 years) for listing
Western Whipbird (western wheatbelt) *Psophodes nigrogularis oberon*	At least half the habitat of this subspecies has been cleared for agriculture (Near Threatened a). The subspecies exists at 12 locations, and although the Area of Occupancy appears to have been reduced to about 800 km^2 and may still be declining, the population is not severely fragmented (so not Vulnerable B1). Occasional fires do not result in extreme fluctuations in population size (B3). Similarly, although the population probably contains only 5000 mature individuals and could be decreasing, security is ensured by the high number of subpopulations (so not C2b), including some with more than 1000 individuals (so not C2a)	Revised criteria for Near Threatened: population too large and decline too small in the last 3 generations (19 years) for listing
Crested Shrike-tit (western) *Falcunculus frontatus leucogaster*	Substantial areas of habitat have been lost to clearing and this species is thought to have declined in density over at least half its Area of Occupancy (Near Threatened c)	Revised criteria for Near Threatened: population too large and decline too small in the last 3 generations (29 years) for listing
Crested Shrike-tit (northern) *Falcunculus frontatus whitei*	The scarce records available suggest that this species' population contains about 2500 mature individuals, is severely fragmented, has no subpopulation more than 250, and is probably declining (Endangered C2a)	New knowledge: new search techniques have found birds at many more sites; there is no evidence of decline
Olive Whistler (Glenelg) *Pachycephala olivacea hesperus*	Clearance has probably halved the subspecies' distribution (Near Threatened a). Three subpopulations contain about 5000 mature birds (so not Vulnerable C2b, though possibly decreasing), 1 containing over 1000 mature birds (so not C2a). Despite a small Area of Occupancy, the population is not severely fragmented (so neither Endangered nor Vulnerable B)	New knowledge: population now known to be much larger than Near Threatened thresholds
Crested Bellbird (southern) *Oreoica gutturalis gutturalis*	This subspecies has disappeared from well in excess of 50% of its historical range, particularly along the periphery (Near Threatened a)	Revised criteria for Near Threatened: population too large and decline too small in the last 3 generations (24 years) for listing
Black-faced Woodswallow (Cape York Peninsula) *Artamus cinereus normani*	Scant data suggests that the density of this subspecies has decreased by at least 50% (Near Threatened c)	Revised criteria for Near Threatened: although declines on Cape York Peninsula may be continuing, still widespread and abundant on the southern edge of its range
Black Butcherbird (Papuan) *Cracticus quoyi alecto*	The Australian population of this subspecies is confined to 3 small islands with a combined area of less than 100 km^2 (Vulnerable D2). The global status, however, is Least Concern and genetic interchange is likely to be substantial. The Australian status has therefore been downgraded to Near Threatened e (genetic interchange beyond borders)	Correct application of regional guidelines: could be Vulnerable D1 as, 1000 mature individuals or Near Threatened D2 as ≤5 locations and low probability future threat but downgraded 2 levels because genetic interchange likely

Taxon name	Reason provided for listing in 2000	Reason not listed in 2010
Spangled Drongo (Papuan) *Dicrurus bracteatus carbonarius*	The Australian population of this subspecies is confined to 3 small islands with a combined area of less than 100 km^2 (Vulnerable D2). The global status, however, is Least Concern and genetic interchange is likely to be substantial. The Australian status has therefore been downgraded to Near Threatened e (genetic interchange beyond borders)	Correct application of regional guidelines: could be Vulnerable D1 as, 1000 mature individuals or Near Threatened D2 as ≤5 locations and low probability future threat but downgraded 2 levels because genetic interchange likely
Northern Fantail (Papuan) *Rhipidura rufiventris gularis*	The Australian population of this subspecies is confined to 3 small islands with a combined area of less than 100 km^2 (Vulnerable D2). The global status, however, is Least Concern and genetic interchange is likely to be substantial. The Australian status has therefore been downgraded to Near Threatened e (genetic interchange beyond borders)	Correct application of regional guidelines: could be Near Threatened D2 as ≤5 locations and low probability future threat but downgraded 2 levels because genetic interchange likely
Willie Wagtail (Papuan) *Rhipidura leucophrys melaleuca*	The Australian population of this subspecies is confined to 3 small islands with a combined area of less than 100 km^2 (Vulnerable D2). The global status, however, is Least Concern and genetic interchange is likely to be substantial. The Australian status has therefore been downgraded to Near Threatened e (genetic interchange beyond borders)	Correct application of regional guidelines: could be Near Threatened D2 as ≤5 locations and low probability future threat but downgraded 2 levels because genetic interchange likely
Forest Raven (New England) *Corvus tasmanicus boreus*	Although there is little information available regarding either population size or trends, the subspecies is listed as Near Threatened on the basis that habitat loss within the historical range is likely to have caused a reduction in population density of over 50% (c)	Revised criteria for Near Threatened: no evidence of recent decline and population probably much greater than 1000 individuals
Torresian Crow (Papuan) *Corvus orru orru*	The Australian population of this subspecies is confined to 3 small islands with a combined area of less than 100 km^2 (Vulnerable D2). The global status, however, is Least Concern and genetic interchange is likely to be substantial. The Australian status has therefore been downgraded to Near Threatened e (genetic interchange beyond borders)	Correct application of regional guidelines: could be Vulnerable D1 as, 1000 mature individuals or Near Threatened D2 as ≤5 locations and low probability future threat but downgraded 2 levels because genetic interchange likely
Star Finch (western) *Neochmia ruficauda subclarescens*	A decrease in density across the historical range is indicated by frequency of recording (Near Threatened c)	New knowledge: population has been large and stable over the last 10 years with no immediate threat
Diamond Firetail *Stagonopleura guttata*	This species has declined over most of its historical range (Near Threatened c), though not at a rate or scale that warrants its listing as Vulnerable	Revised criteria for Near Threatened: population too large and widespread and decline not approaching 20% in 10 years
Pictorella Mannikin *Heteromunia pectoralis*	The species' distribution is patchy but, its density has arguably declined by at least half (Near Threatened c)	New knowledge: although numbers at individual sites vary irregularly, no declines apparent in recent decades

Index of birds and threats

www.ingramcontent.com/pod-product-compliance
Lightning Source LLC
LaVergne TN
LVHW060210210926
848481LV00008B/2682